Preface to the Fifth Edition

When preparing the first edition of this book, more than 10 years ago, we tried to accomplish two objectives: it should be useful as an advanced graduate textbook but also as a reference work for research. With each new edition, we have to decide how the book can be improved further. Of course, it is less and less possible to describe the growing area comprehensively.

If we included everything that we like, the book would grow beyond a single volume. Since the book is used for many courses, now even sometimes at undergraduate level, we thought that adding some classical material might be more useful than including a selection of the latest results.

In this edition, we added a proof of Cayley's formula, more details on blocking flows, the new faster b-matching separation algorithm, an approximation scheme for multidimensional knapsack, and results concerning the multicommodity max-flow min-cut ratio and the sparsest cut problem. There are further small improvements in numerous places and more than 60 new exercises. Of course, we also updated the references to point to the most recent results and corrected some minor errors that were discovered.

We would like to thank Takao Asano, Maxim Babenko, Ulrich Brenner, Benjamin Bolten, Christoph Buchheim, Jean Fonlupt, András Frank, Michael Gester, Stephan Held, Stefan Hougardy, Hiroshi Iida, Klaus Jansen, Alexander Karzanov, Levin Keller, Alexander Kleff, Niko Klewinghaus, Stefan Knauf, Barbara Langfeld, Jens Maßberg, Marc Pfetsch, Klaus Radke, Rabe von Randow, Tomás Salles, Jan Schneider, Christian Schulte, András Sebő, Martin Skutella, Jácint Szabó, and Simon Wedeking for valuable feedback on the previous edition.

We are pleased that this book has been received so well, and further translations are on their way. Editions in Japanese, French, Italian, German, Russian, and Chinese have appeared since 2009 or are scheduled to appear soon. We hope that our book will continue to serve its purpose in teaching and research in combinatorial optimization.

Bonn, Germany
September 2011

Bernhard Korte
Jens Vygen

Algorithms and Combinatorics

Volume 21

More information about this series at http://www.springer.com/series/13

Bernhard Korte
Research Institute for Discrete
Mathematics
University of Bonn
Bonn, Germany

Jens Vygen
Research Institute for Discrete
Mathematics
University of Bonn
Bonn, Germany

ISSN 0937-5511   ISSN 2197-6783 (electronic)
Algorithms and Combinatorics
ISBN 978-3-662-58566-5   ISBN 978-3-662-56039-6 (eBook)
https://doi.org/10.1007/978-3-662-56039-6

Mathematics Subject Classification (2010): 90C27, 68R10, 05C85, 68Q25

Printed on acid-free paper

This Springer imprint is published by Springer Nature
The registered company is Springer-Verlag GmbH, DE
The registered company address is: Heidelberger Platz 3, 14197 Berlin, Germany

Bernhard Korte • Jens Vygen

Combinatorial Optimization

Theory and Algorithms

Sixth Edition

 Springer

Preface to the Sixth Edition

After six years, it was again time for a new edition. Besides updates, new exercises, and a few corrections, it contains the following new material.

Section 7.4 is devoted to shallow-light trees. Section 14.6 contains the recent two-factor approximation algorithm for submodular function maximization. Section 17.5 discusses the Nemhauser-Ullmann algorithm and smoothed analysis. In Section 20.3, we present the $(\ln 4 + \epsilon)$-factor approximation algorithm for the Steiner tree problem. Finally, Section 20.7 contains the VPN theorem. There are also small additions, e.g. on the integrality ratio in Section 5.1 and kernelization in Section 15.7.

We would like to thank Maxim Babenko, Steffen Böhmer, Ulrich Brenner, György Dósa, Michael Etscheid, Jean Fonlupt, Michel Goemans, Stephan Held, Stefan Hougardy, Jochen Könemann, Solomon Lo, Jens Maßberg, Neil Olver, Dieter Rautenbach, Heiko Röglin, Jan Schneider, Sophie Spirkl, and Uri Zwick for feedback on the previous edition or proofreading new parts.

We hope that, with this new edition, the book remains useful for research and teaching for many years to come.

Bonn, Germany
September 2017

Bernhard Korte
Jens Vygen

Preface to the Fourth Edition

With four English editions, and translations into four other languages forthcoming, we are very happy with the development of our book. Again, we have revised, updated, and significantly extended it for this fourth edition. We have added some classical material that may have been missed so far, in particular on linear programming, the network simplex algorithm, and the max-cut problem. We have also added a number of new exercises and up-to-date references. We hope that these changes serve to make our book an even better basis for teaching and research.

We gratefully acknowledge the continuous support of the Union of the German Academies of Sciences and Humanities and the North Rhine-Westphalian (NRW) Academy of Sciences via the long-term research project "Discrete Mathematics and Its Applications." We also thank those who gave us feedback on the third edition, in particular Takao Asano, Christoph Bartoschek, Bert Besser, Ulrich Brenner, Jean Fonlupt, Satoru Fujishige, Marek Karpinski, Jens Maßberg, Denis Naddef, Sven Peyer, Klaus Radke, Rabe von Randow, Dieter Rautenbach, Martin Skutella, Markus Struzyna, Jürgen Werber, Minyi Yue, and Guochuan Zhang, for their valuable comments. At `http://www.or.uni-bonn.de/~vygen/co.html`, we will continue to maintain updated information about this book.

Bonn, Germany
August 2007

Bernhard Korte
Jens Vygen

Preface to the Third Edition

After 5 years, it was time for a thoroughly revised and substantially extended edition. The most significant feature is a completely new chapter on facility location. No constant-factor approximation algorithms were known for this important class of *NP*-hard problems until 8 years ago. Today there are several interesting and very different techniques that lead to good approximation guarantees, which makes this area particularly appealing also for teaching. In fact, the chapter has arisen from a special course on facility location.

Many of the other chapters have also been extended significantly. The new material includes Fibonacci heaps, Fujishige's new maximum flow algorithm, flows over time, Schrijver's algorithm for submodular function minimization, and the Robins-Zelikovsky Steiner tree approximation algorithm. Several proofs have been streamlined, and many new exercises and references have been added.

We thank those who gave us feedback on the second edition, in particular Takao Asano, Yasuhito Asano, Ulrich Brenner, Stephan Held, Tomio Hirata, Dirk Müller, Kazuo Murota, Dieter Rautenbach, Martin Skutella, Markus Struzyna, and Jürgen Werber, for their valuable comments. Eminently, Takao Asano's notes and Jürgen Werber's proofreading of Chapter 22 helped to improve the presentation at various places.

Again we would like to mention the Union of the German Academies of Sciences and Humanities and the North Rhine-Westphalian Academy of Sciences. Their continuous support via the long-term project "Discrete Mathematics and Its Applications" funded by the German Federal Ministry of Education and Research and the state of North Rhine-Westphalia is gratefully acknowledged.

Bonn, Germany
May 2005

Bernhard Korte
Jens Vygen

Preface to the Second Edition

It was more than a surprise to us that the first edition of this book already went out of print about a year after its first appearance. We were flattered by the many positive and even enthusiastic comments and letters from colleagues and the general readership. Several of our colleagues helped us in finding typographical and other errors. In particular, we thank Ulrich Brenner, András Frank, Bernd Gärtner, and Rolf Möhring. Of course, all errors detected so far have been corrected in this second edition, and references have been updated.

Moreover, the first preface had a flaw. We listed all individuals who helped us in preparing this book. But we forgot to mention the institutional support, for which we make amends here.

It is evident that a book project which took 7 years benefited from many different grants. We would like to mention explicitly the bilateral Hungarian-German Research Project, sponsored by the Hungarian Academy of Sciences and the Deutsche Forschungsgemeinschaft, two Sonderforschungsbereiche (special research units) of the Deutsche Forschungsgemeinschaft, the Ministère Français de la Recherche et de la Technologie and the Alexander von Humboldt Foundation for support via the Prix Alexandre de Humboldt, and the Commission of the European Communities for participation in two projects of DONET. Our most sincere thanks go to the Union of the German Academies of Sciences and Humanities and to the North Rhine-Westphalian Academy of Sciences. Their long-term project "Discrete Mathematics and Its Applications" supported by the German Federal Ministry of Education and Research (BMBF) and the state of North Rhine-Westphalia was of decisive importance for this book.

Bonn, Germany
October 2001

Bernhard Korte
Jens Vygen

Preface to the First Edition

Combinatorial optimization is one of the youngest and most active areas of discrete mathematics and is probably its driving force today. It became a subject in its own right about 50 years ago.

This book describes the most important ideas, theoretical results, and algorithms in combinatorial optimization. We have conceived it as an advanced graduate text which can also be used as an up-to-date reference work for current research. The book includes the essential fundamentals of graph theory, linear and integer programming, and complexity theory. It covers classical topics in combinatorial optimization as well as very recent ones. The emphasis is on theoretical results and algorithms with provably good performance. Applications and heuristics are mentioned only occasionally.

Combinatorial optimization has its roots in combinatorics, operations research, and theoretical computer science. A main motivation is that thousands of real-life problems can be formulated as abstract combinatorial optimization problems. We focus on the detailed study of classical problems which occur in many different contexts, together with the underlying theory.

Most combinatorial optimization problems can be formulated naturally in terms of graphs and as (integer) linear programs. Therefore this book starts, after an introduction, by reviewing basic graph theory and proving those results in linear and integer programming which are most relevant for combinatorial optimization.

Next, the classical topics in combinatorial optimization are studied: minimum spanning trees, shortest paths, network flows, matchings, and matroids. Most of the problems discussed in Chapters 6–14 have polynomial-time ("efficient") algorithms, while most of the problems studied in Chapters 15–21 are *NP*-hard, i.e., a polynomial-time algorithm is unlikely to exist. In many cases, one can at least find approximation algorithms that have a certain performance guarantee. We also mention some other strategies for coping with such "hard" problems.

This book goes beyond the scope of a normal textbook on combinatorial optimization in various aspects. For example, we cover the equivalence of optimization and separation (for full-dimensional polytopes), $O(n^3)$-implementations of matching algorithms based on ear decompositions, Turing machines, the perfect graph theorem, *MAXSNP*-hardness, the Karmarkar-Karp algorithm for bin packing, recent approximation algorithms for multicommodity flows, survivable network design, and the Euclidean traveling salesman problem. All results are accompanied by detailed proofs.

Of course, no book on combinatorial optimization can be absolutely comprehensive. Examples of topics which we mention only briefly or do not cover at all are tree decompositions, separators, submodular flows, path matchings, delta-matroids, the matroid parity problem, location and scheduling problems, nonlinear problems, semidefinite programming, the average-case analysis of algorithms, advanced data structures, parallel and randomized algorithms, and the theory of probabilistically checkable proofs (we cite the *PCP* theorem without proof).

At the end of each chapter, there are a number of exercises containing additional results and applications of the material in that chapter. Some exercises which might be more difficult are marked with an asterisk. Each chapter ends with a list of references, including texts recommended for further reading.

This book arose from several courses on combinatorial optimization and from special classes on topics like polyhedral combinatorics or approximation algorithms. Thus, material for basic and advanced courses can be selected from this book.

We have benefited from discussions and suggestions of many colleagues and friends and – of course – from other texts on this subject. Especially we owe sincere thanks to András Frank, László Lovász, András Recski, Alexander Schrijver, and Zoltán Szigeti. Our colleagues and students in Bonn, Christoph Albrecht, Ursula Bünnagel, Thomas Emden-Weinert, Mathias Hauptmann, Sven Peyer, Rabe von Randow, André Rohe, Martin Thimm, and Jürgen Werber, have carefully read several versions of the manuscript and helped to improve it. Last but not least, we thank Springer-Verlag for the most efficient cooperation.

Bonn, Germany   Bernhard Korte
January 2000   Jens Vygen

Table of Contents

1 Introduction

Let us start with two examples.

A company has a machine which drills holes into printed circuit boards. Since it produces many of these boards it wants the machine to complete one board as fast as possible. We cannot optimize the drilling time but we can try to minimize the time the machine needs to move from one point to another. Usually drilling machines can move in two directions: the table moves horizontally while the drilling arm moves vertically. Since both movements can be done simultaneously, the time needed to adjust the machine from one position to another is proportional to the maximum of the horizontal and the vertical distance. This is often called the ℓ_∞-distance. (Older machines can only move either horizontally or vertically at a time; in this case the adjusting time is proportional to the ℓ_1-distance, the sum of the horizontal and the vertical distance.)

An optimum drilling path is given by an ordering of the hole positions $p_1, \ldots, p_n$ such that $\sum_{i=1}^{n-1} d(p_i, p_{i+1})$ is minimum, where d is the ℓ_∞-distance: for two points $p = (x, y)$ and $p' = (x', y')$ in the plane we write $d(p, p') := \max\{|x - x'|, |y - y'|\}$. An order of the holes can be represented by a permutation, i.e. a bijection $\pi : \{1, \ldots, n\} \to \{1, \ldots, n\}$.

Which permutation is best of course depends on the hole positions; for each list of hole positions we have a different problem instance. We say that one instance of our problem is a list of points in the plane, i.e. the coordinates of the holes to be drilled. Then the problem can be stated formally as follows:

DRILLING PROBLEM

Instance: A set of points $p_1, \ldots, p_n \in \mathbb{R}^2$.

Task: Find a permutation $\pi : \{1, \ldots, n\} \to \{1, \ldots, n\}$ such that $\sum_{i=1}^{n-1} d(p_{\pi(i)}, p_{\pi(i+1)})$ is minimum.

We now explain our second example. We have a set of jobs to be done, each having a specified processing time. Each job can be done by a subset of the employees, and we assume that all employees who can do a job are equally efficient. Several employees can contribute to the same job at the same time, and one employee can contribute to several jobs (but not at the same time). The objective is to get all jobs done as early as possible.

© Springer-Verlag GmbH Germany 2018
Bernhard Korte and Jens Vygen, *Combinatorial Optimization*, Algorithms
and Combinatorics 21, https://doi.org/10.1007/978-3-662-56039-6_1

In this model it suffices to prescribe for each employee how long he or she should work on which job. The order in which the employees carry out their jobs is not important, since the time when all jobs are done obviously depends only on the maximum total working time we have assigned to one employee. Hence we have to solve the following problem:

JOB ASSIGNMENT PROBLEM

Instance: A set of numbers $t_1, \ldots, t_n \in \mathbb{R}_+$ (the processing times for n jobs), a number $m \in \mathbb{N}$ of employees, and a nonempty subset $S_i \subseteq \{1, \ldots, m\}$ of employees for each job $i \in \{1, \ldots, n\}$.

Task: Find numbers $x_{ij} \in \mathbb{R}_+$ for all $i = 1, \ldots, n$ and $j \in S_i$ such that $\sum_{j \in S_i} x_{ij} = t_i$ for $i = 1, \ldots, n$ and $\max_{j \in \{1, \ldots, m\}} \sum_{i: j \in S_i} x_{ij}$ is minimum.

These are two typical problems arising in combinatorial optimization. How to model a practical problem as an abstract combinatorial optimization problem is not described in this book; indeed there is no general recipe for this task. Besides giving a precise formulation of the input and the desired output it is often important to ignore irrelevant components (e.g. the drilling time which cannot be optimized or the order in which the employees carry out their jobs).

Of course we are not interested in a solution to a particular drilling problem or job assignment problem in some company, but rather we are looking for a way how to solve all problems of these types. We first consider the DRILLING PROBLEM.

1.1 Enumeration

How can a solution to the DRILLING PROBLEM look like? There are infinitely many instances (finite sets of points in the plane), so we cannot list an optimum permutation for each instance. Instead, what we look for is an algorithm which, given an instance, computes an optimum solution. Such an algorithm exists: Given a set of n points, just try all possible $n!$ orders, and for each compute the ℓ_∞-length of the corresponding path.

There are different ways of formulating an algorithm, differing mostly in the level of detail and the formal language they use. We certainly would not accept the following as an algorithm: "Given a set of n points, find an optimum path and output it." It is not specified at all how to find the optimum solution. The above suggestion to enumerate all possible $n!$ orders is more useful, but still it is not clear how to enumerate all the orders. Here is one possible way:

We enumerate all n-tuples of numbers $1, \ldots, n$, i.e. all n^n vectors of $\{1, \ldots, n\}^n$. This can be done similarly to counting: we start with $(1, \ldots, 1, 1)$, $(1, \ldots, 1, 2)$ up to $(1, \ldots, 1, n)$ then switch to $(1, \ldots, 1, 2, 1)$, and so on. At each step we increment the last entry unless it is already n, in which case we go back to the last entry that is smaller than n, increment it and set all subsequent entries to 1.

This technique is sometimes called backtracking. The order in which the vectors of $\{1, \ldots, n\}^n$ are enumerated is called the lexicographical order:

Definition 1.1. *Let $x, y \in \mathbb{R}^n$ be two vectors. We say that a vector x is **lexicographically smaller** than y if there exists an index $j \in \{1, \ldots, n\}$ such that $x_i = y_i$ for $i = 1, \ldots, j-1$ and $x_j < y_j$.*

Knowing how to enumerate all vectors of $\{1, \ldots, n\}^n$ we can simply check for each vector whether its entries are pairwise distinct and, if so, whether the path represented by this vector is shorter than the best path encountered so far.

Since this algorithm enumerates n^n vectors it will take at least n^n steps (in fact, even more). This is not best possible. There are only $n!$ permutations of $\{1, \ldots, n\}$, and $n!$ is significantly smaller than n^n. (By Stirling's formula $n! \approx \sqrt{2\pi n}\,\frac{n^n}{e^n}$ (Stirling [1730]); see Exercise 1.) We shall show how to enumerate all paths in approximately $n^2 \cdot n!$ steps. Consider the following algorithm which enumerates all permutations in lexicographical order:

PATH ENUMERATION ALGORITHM

Input: A natural number $n \geq 3$. A set $\{p_1, \ldots, p_n\}$ of points in the plane.

Output: A permutation $\pi^* : \{1, \ldots, n\} \to \{1, \ldots, n\}$ with $cost(\pi^*) := \sum_{i=1}^{n-1} d(p_{\pi^*(i)}, p_{\pi^*(i+1)})$ minimum.

①   Set $\pi(i) := i$ and $\pi^*(i) := i$ for $i = 1, \ldots, n$. Set $i := n - 1$.

②   Let $k := \min(\{\pi(i) + 1, \ldots, n + 1\} \setminus \{\pi(1), \ldots, \pi(i - 1)\})$.

③   **If** $k \leq n$ **then:**
     Set $\pi(i) := k$.
     If $i = n$ and $cost(\pi) < cost(\pi^*)$ **then** set $\pi^* := \pi$.
     If $i < n$ **then** set $\pi(i + 1) := 0$ and $i := i + 1$.
   If $k = n + 1$ **then** set $i := i - 1$.
   If $i \geq 1$ **then** go to ②.

Starting with $(\pi(i))_{i=1,\ldots,n} = (1, 2, 3, \ldots, n-1, n)$ and $i = n-1$, the algorithm finds at each step the next possible value of $\pi(i)$ (not using $\pi(1), \ldots, \pi(i-1)$). If there is no more possibility for $\pi(i)$ (i.e. $k = n + 1$), then the algorithm decrements i (backtracking). Otherwise it sets $\pi(i)$ to the new value. If $i = n$, the new permutation is evaluated, otherwise the algorithm will try all possible values for $\pi(i + 1), \ldots, \pi(n)$ and starts by setting $\pi(i + 1) := 0$ and incrementing i.

So all permutation vectors $(\pi(1), \ldots, \pi(n))$ are generated in lexicographical order. For example, the first iterations in the case $n = 6$ are shown below:

$$\pi := (1, 2, 3, 4, 5, 6), \quad i := 5$$

$$k := 6, \quad \pi := (1, 2, 3, 4, 6, 0), \quad i := 6$$

$$k := 5, \quad \pi := (1, 2, 3, 4, 6, 5), \qquad\qquad cost(\pi) < cost(\pi^*)?$$

$$k := 7, \qquad\qquad\qquad\qquad\qquad i := 5$$

$$k := 7, \qquad\qquad\qquad\qquad\qquad i := 4$$

$$k := 5, \quad \pi := (1, 2, 3, 5, 0, 5), \quad i := 5$$

$$k := 4, \quad \pi := (1, 2, 3, 5, 4, 0), \quad i := 6$$

$$k := 6, \quad \pi := (1, 2, 3, 5, 4, 6), \qquad\qquad cost(\pi) < cost(\pi^*)?$$

Since the algorithm compares the cost of each path to π^*, the best path encountered so far, it indeed outputs the optimum path. But how many steps will this algorithm perform? Of course, the answer depends on what we call a single step. Since we do not want the number of steps to depend on the actual implementation we ignore constant factors. On any reasonable computer, ① will take at least $2n+1$ steps (this many variable assignments are done) and at most cn steps for some constant c. The following common notation is useful for ignoring constant factors:

Definition 1.2. *Let* $f, g : D \to \mathbb{R}_+$ *be two functions. We say that* f *is* $O(g)$ *(and sometimes write* $f = O(g)$, *and also* $g = \Omega(f)$) *if there exist constants* $\alpha, \beta > 0$ *such that* $f(x) \le \alpha g(x) + \beta$ *for all* $x \in D$. *If* $f = O(g)$ *and* $g = O(f)$ *we also say that* $f = \Theta(g)$ *(and of course* $g = \Theta(f)$). *In this case,* f *and* g *have the same* **rate of growth**.

Note that the use of the equation sign in the O-notation is not symmetric. To illustrate this definition, let $D = \mathbb{N}$, and let $f(n)$ be the number of elementary steps in ① and $g(n) = n$ ($n \in \mathbb{N}$). Clearly we have $f = O(g)$ (in fact $f = \Theta(g)$) in this case; we say that ① takes $O(n)$ time (or linear time). A single execution of ③ takes a constant number of steps (we speak of $O(1)$ time or constant time) except in the case $k \le n$ and $i = n$; in this case the cost of two paths have to be compared, which takes $O(n)$ time.

What about ②? A naive implementation, checking for each $j \in \{\pi(i) + 1, \ldots, n\}$ and each $h \in \{1, \ldots, i - 1\}$ whether $j = \pi(h)$, takes $O((n - \pi(i))i)$ steps, which can be as big as $\Theta(n^2)$. A better implementation of ② uses an auxiliary array indexed by $1, \ldots, n$:

②   **For** $j := 1$ **to** n **do** $aux(j) := 0$.
    For $j := 1$ **to** $i - 1$ **do** $aux(\pi(j)) := 1$.
    Set $k := \pi(i) + 1$.
    While $k \le n$ **and** $aux(k) = 1$ **do** $k := k + 1$.

Obviously with this implementation a single execution of ② takes only $O(n)$ time. Simple techniques like this are usually not elaborated in this book; we assume that the reader can find such implementations himself or herself.

Having computed the running time for each single step we now estimate the total amount of work. Since the number of permutations is $n!$ we only have to estimate the amount of work which is done between two permutations. The counter i might move back from n to some index i' where a new value $\pi(i') \le n$ is found. Then it moves forward again up to $i = n$. While the counter i is constant each of ② and ③ is performed once, except in the case $k \leq n$ and $i = n$; in this case ② and ③ are performed twice. So the total amount of work between two permutations consists of at most $4n$ times ② and ③, i.e. $O(n^2)$. So the overall running time of the PATH ENUMERATION ALGORITHM is $O(n^2 n!)$.

One can do slightly better; a more careful analysis shows that the running time is only $O(n \cdot n!)$ (Exercise 4).

Still the algorithm is too time-consuming if n is large. The problem with the enumeration of all paths is that the number of paths grows exponentially with the number of points; already for 20 points there are $20! = 2432902008176640000 \approx 2.4 \cdot 10^{18}$ different paths and even the fastest computer needs several years to evaluate all of them. So complete enumeration is impossible even for instances of moderate size.

The main subject of combinatorial optimization is to find better algorithms for problems like this. Often one has to find the best element of some finite set of feasible solutions (in our example: drilling paths or permutations). This set is not listed explicitly but implicitly depends on the structure of the problem. Therefore an algorithm must exploit this structure.

In the case of the DRILLING PROBLEM all information of an instance with n points is given by $2n$ coordinates. While the naive algorithm enumerates all $n!$ paths it might be possible that there is an algorithm which finds the optimum path much faster, say in n^2 computation steps. It is not known whether such an algorithm exists (though results of Chapter 15 suggest that it is unlikely). Nevertheless there are much better algorithms than the naive one.

1.2 Running Time of Algorithms

One can give a formal definition of an algorithm, and we shall in fact give one in Section 15.1. However, such formal models lead to very long and tedious descriptions as soon as algorithms are a bit more complicated. This is quite similar to mathematical proofs: Although the concept of a proof can be formalized nobody uses such a formalism for writing down proofs since they would become very long and almost unreadable.

Therefore all algorithms in this book are written in an informal language. Still the level of detail should allow a reader with a little experience to implement the algorithms on any computer without too much additional effort.

Since we are not interested in constant factors when measuring running times we do not have to fix a concrete computing model. We count elementary steps, but we are not really interested in how elementary steps look like. Examples of elementary steps are variable assignments, random access to a variable whose index is stored in another variable, conditional jumps (if – then – go to), and simple arithmetic operations like addition, subtraction, multiplication, division and comparison of numbers.

An algorithm consists of a set of valid inputs and a sequence of instructions each of which can be composed of elementary steps, such that for each valid input the computation of the algorithm is a uniquely defined finite series of elementary steps which produces a certain output. Usually we are not satisfied with finite computation but rather want a good upper bound on the number of elementary steps performed, depending on the input size.

The input to an algorithm usually consists of a list of numbers. If all these numbers are integers, we can code them in binary representation, using $O(\log(|a|+2))$ bits for storing an integer a. Rational numbers can be stored by coding the numerator and the denominator separately. The **input size** $\mathrm{size}(x)$ of an instance x with rational data is the total number of bits needed for the binary representation.

Definition 1.3. *Let A be an algorithm which accepts inputs from a set X, and let $f : \mathbb{N} \to \mathbb{R}_+$. If there exist constants $\alpha, \beta > 0$ such that A terminates its computation after at most $\alpha f(\mathrm{size}(x)) + \beta$ elementary steps (including arithmetic operations) for each input $x \in X$, then we say that A **runs in** $O(f)$ **time**. We also say that the **running time** (or the **time complexity**) of A is $O(f)$.*

Definition 1.4. *An algorithm with rational input is said to run in **polynomial time** if there is an integer k such that it runs in $O(n^k)$ time, where n is the input size, and all numbers in intermediate computations can be stored with $O(n^k)$ bits.*

*An algorithm with arbitrary input is said to run in **strongly polynomial time** if there is an integer k such that it runs in $O(n^k)$ time for any input consisting of n numbers and it runs in polynomial time for rational input. In the case $k = 1$ we have a **linear-time algorithm**.*

*An algorithm which runs in polynomial but not strongly polynomial time is called **weakly polynomial**.*

Note that the running time might be different for several instances of the same size (this was not the case with the PATH ENUMERATION ALGORITHM). We consider the worst-case running time, i.e. the function $f : \mathbb{N} \to \mathbb{N}$ where $f(n)$ is the maximum running time of an instance with input size n. For some algorithms we do not know the rate of growth of f but only have an upper bound.

The worst-case running time might be a pessimistic measure if the worst case occurs rarely. In some cases an average-case running time with some probabilistic model might be appropriate, but we shall not consider this.

If A is an algorithm which for each input $x \in X$ computes the output $f(x) \in Y$, then we say that A **computes** $f : X \to Y$. If a function is computed by some polynomial-time algorithm, it is said to be **computable in polynomial time**.

Polynomial-time algorithms are sometimes called "good" or "efficient". This concept was introduced by Cobham [1964] and Edmonds [1965]. Table 1.1 motivates this by showing hypothetical running times of algorithms with various time complexities. For various input sizes n we show the running time of algorithms that take $100n \log n$, $10n^2$, $n^{3.5}$, $n^{\log n}$, 2^n, and $n!$ elementary steps; we assume that one elementary step takes one nanosecond. As always in this book, log denotes the logarithm with basis 2.

Table 1.1.

n	$100n \log n$	$10n^2$	$n^{3.5}$	$n^{\log n}$	2^n	$n!$
10	3 μs	1 μs	3 μs	2 μs	1 μs	4 ms
20	9 μs	4 μs	36 μs	420 μs	1 ms	76 years
30	15 μs	9 μs	148 μs	20 ms	1 s	$8 \cdot 10^{15}$ y.
40	21 μs	16 μs	404 μs	340 ms	1100 s	
50	28 μs	25 μs	884 μs	4 s	13 days	
60	35 μs	36 μs	2 ms	32 s	37 years	
80	50 μs	64 μs	5 ms	1075 s	$4 \cdot 10^7$ y.	
100	66 μs	100 μs	10 ms	5 hours	$4 \cdot 10^{13}$ y.	
200	153 μs	400 μs	113 ms	12 years		
500	448 μs	2.5 ms	3 s	$5 \cdot 10^5$ y.		
1000	1 ms	10 ms	32 s	$3 \cdot 10^{13}$ y.		
10^4	13 ms	1 s	28 hours			
10^5	166 ms	100 s	10 years			
10^6	2 s	3 hours	3169 y.			
10^7	23 s	12 days	10^7 y.			
10^8	266 s	3 years	$3 \cdot 10^{10}$ y.			
10^{10}	9 hours	$3 \cdot 10^4$ y.				
10^{12}	46 days	$3 \cdot 10^8$ y.				

As Table 1.1 shows, polynomial-time algorithms are faster for large enough instances. The table also illustrates that constant factors of moderate size are not very important when considering the asymptotic growth of the running time.

Table 1.2 shows the maximum input sizes solvable within one hour with the above six hypothetical algorithms. In (a) we again assume that one elementary step takes one nanosecond, (b) shows the corresponding figures for a ten times faster machine. Polynomial-time algorithms can handle larger instances in reasonable time. Moreover, even a speedup by a factor of 10 of the computers does not increase the size of solvable instances significantly for exponential-time algorithms, but it does for polynomial-time algorithms.

Table 1.2.

	$100n \log n$	$10n^2$	$n^{3.5}$	$n^{\log n}$	2^n	$n!$
(a)	$1.19 \cdot 10^9$	60000	3868	87	41	15
(b)	$10.8 \cdot 10^9$	189737	7468	104	45	16

(Strongly) polynomial-time algorithms, if possible linear-time algorithms, are what we look for. There are some problems where it is known that no polynomial-time algorithm exists, and there are problems for which no algorithm exists at all. (For example, a problem which can be solved in finite time but not in polynomial time is to decide whether a so-called regular expression defines the empty set; see Aho, Hopcroft and Ullman [1974]. A problem for which there exists no algorithm at all, the HALTING PROBLEM, is discussed in Exercise 1 of Chapter 15.)

However, almost all problems considered in this book belong to the following two classes. For the problems of the first class we have a polynomial-time algorithm. For each problem of the second class it is an open question whether a polynomial-time algorithm exists. However, we know that if one of these problems has a polynomial-time algorithm, then all problems of this class do. A precise formulation and a proof of this statement will be given in Chapter 15.

The JOB ASSIGNMENT PROBLEM belongs to the first class, the DRILLING PROBLEM belongs to the second class.

These two classes of problems divide this book roughly into two parts. We first deal with tractable problems for which polynomial-time algorithms are known. Then, starting with Chapter 15, we discuss hard problems. Although no polynomial-time algorithms are known, there are often much better methods than complete enumeration. Moreover, for many problems (including the DRILLING PROBLEM), one can find approximate solutions within a certain percentage of the optimum in polynomial time.

1.3 Linear Optimization Problems

We now consider our second example given initially, the JOB ASSIGNMENT PROBLEM, and briefly address some central topics which will be discussed in later chapters.

The JOB ASSIGNMENT PROBLEM is quite different to the DRILLING PROBLEM since there are infinitely many feasible solutions for each instance (except for trivial cases). We can reformulate the problem by introducing a variable T for the time when all jobs are done:

$$\min \quad T$$

$$
\begin{aligned}
\text{s.t.} \quad \sum_{j \in S_i} x_{ij} &= t_i & (i \in \{1, \ldots, n\}) \\
x_{ij} &\geq 0 & (i \in \{1, \ldots, n\}, \; j \in S_i) \\
\sum_{i : j \in S_i} x_{ij} &\leq T & (j \in \{1, \ldots, m\})
\end{aligned}
\tag{1.1}
$$

The numbers t_i and the sets S_i $(i = 1, \ldots, n)$ are given, the variables x_{ij} and T are what we look for. Such an optimization problem with a linear objective function and linear constraints is called a **linear program**. The set of feasible solutions of (1.1), a so-called **polyhedron**, is easily seen to be convex, and one can prove that there always exists an optimum solution which is one of the finitely many extreme points of this set. Therefore a linear program can, theoretically, also be solved by complete enumeration. But there are much better ways as we shall see later.

Although there are several algorithms for solving linear programs in general, such general techniques are usually less efficient than special algorithms exploiting the structure of the problem. In our case it is convenient to model the sets S_i, $i = 1, \ldots, n$, by a **graph**. For each job i and for each employee j we have a point (called vertex), and we connect employee j with job i by an edge if he or she can contribute to this job (i.e. if $j \in S_i$). Graphs are a fundamental combinatorial structure; many combinatorial optimization problems are described most naturally in terms of graph theory.

Suppose for a moment that the processing time of each job is one hour, and we ask whether we can finish all jobs within one hour. So we look for numbers x_{ij} ($i \in \{1, \ldots, n\}$, $j \in S_i$) such that $0 \le x_{ij} \le 1$ for all i and j, $\sum_{j \in S_i} x_{ij} = 1$ for $i = 1, \ldots, n$, and $\sum_{i:j \in S_i} x_{ij} \le 1$ for $j = 1, \ldots, n$. One can show that if such a solution exists, then in fact an integral solution exists, i.e. all x_{ij} are either 0 or 1. This is equivalent to assigning each job to one employee, such that no employee has to do more than one job. In the language of graph theory we then look for a **matching** covering all jobs. The problem of finding optimal matchings is one of the best-known combinatorial optimization problems.

We review the basics of graph theory and linear programming in Chapters 2 and 3. In Chapter 4 we prove that linear programs can be solved in polynomial time, and in Chapter 5 we discuss integral polyhedra. In the subsequent chapters we discuss some classical combinatorial optimization problems in detail.

1.4 Sorting

Let us conclude this chapter by considering a special case of the DRILLING PROBLEM where all holes to be drilled are on one horizontal line. So we are given just one coordinate for each point p_i, $i = 1, \ldots, n$. Then a solution to the drilling problem is easy, all we have to do is sort the points by their coordinates: the drill will just move from left to right. Although there are still $n!$ permutations, it is clear that we do not have to consider all of them to find the optimum drilling path, i.e. the sorted list. It is very easy to sort n numbers in nondecreasing order in $O(n^2)$ time.

To sort n numbers in $O(n \log n)$ time requires a little more skill. There are several algorithms accomplishing this; we present the well-known MERGE-SORT ALGORITHM. It proceeds as follows. First the list is divided into two sublists of approximately equal size. Then each sublist is sorted (this is done recursively by the same algorithm). Finally the two sorted sublists are merged together. This general strategy, often called "divide and conquer", can be used quite often. See e.g. Section 17.1 for another example.

We did not discuss recursive algorithms so far. In fact, it is not necessary to discuss them, since any recursive algorithm can be transformed into a sequential algorithm without increasing the running time. But some algorithms are easier to formulate (and implement) using recursion, so we shall use recursion when it is convenient.

MERGE-SORT ALGORITHM

Input: A list $a_1, \ldots, a_n$ of real numbers.

Output: A permutation $\pi : \{1, \ldots, n\} \to \{1, \ldots, n\}$ such that $a_{\pi(i)} \leq a_{\pi(i+1)}$ for all $i = 1, \ldots, n-1$.

①   **If** $n = 1$ **then** set $\pi(1) := 1$ and **stop** (**return** π).

②   Set $m := \lfloor \frac{n}{2} \rfloor$.
     Let $\rho :=$ MERGE-SORT$(a_1, \ldots, a_m)$.
     Let $\sigma :=$ MERGE-SORT$(a_{m+1}, \ldots, a_n)$.

③   Set $k := 1, l := 1$.
     While $k \leq m$ and $l \leq n - m$ **do**:
       If $a_{\rho(k)} \leq a_{m+\sigma(l)}$ **then** set $\pi(k+l-1) := \rho(k)$ and $k := k+1$
                             else set $\pi(k+l-1) := m + \sigma(l)$ and $l := l+1$.
     While $k \leq m$ **do**: Set $\pi(k+l-1) := \rho(k)$ and $k := k+1$.
     While $l \leq n - m$ **do**: Set $\pi(k+l-1) := m + \sigma(l)$ and $l := l+1$.

As an example, consider the list "69,32,56,75,43,99,28". The algorithm first splits this list into two, "69,32,56" and "75,43,99,28" and recursively sorts each of the two sublists. We get the permutations $\rho = (2, 3, 1)$ and $\sigma = (4, 2, 1, 3)$ corresponding to the sorted lists "32,56,69" and "28,43,75,99". Now these lists are merged as shown below:

					$k := 1, \quad l := 1$
$\rho(1) = 2,$	$\sigma(1) = 4,$	$a_{\rho(1)} = 32,$	$a_{\sigma(1)} = 28,$	$\pi(1) := 7,$	$l := 2$
$\rho(1) = 2,$	$\sigma(2) = 2,$	$a_{\rho(1)} = 32,$	$a_{\sigma(2)} = 43,$	$\pi(2) := 2,$	$k := 2$
$\rho(2) = 3,$	$\sigma(2) = 2,$	$a_{\rho(2)} = 56,$	$a_{\sigma(2)} = 43,$	$\pi(3) := 5,$	$l := 3$
$\rho(2) = 3,$	$\sigma(3) = 1,$	$a_{\rho(2)} = 56,$	$a_{\sigma(3)} = 75,$	$\pi(4) := 3,$	$k := 3$
$\rho(3) = 1,$	$\sigma(3) = 1,$	$a_{\rho(3)} = 69,$	$a_{\sigma(3)} = 75,$	$\pi(5) := 1,$	$k := 4$
	$\sigma(3) = 1,$		$a_{\sigma(3)} = 75,$	$\pi(6) := 4,$	$l := 4$
	$\sigma(4) = 3,$		$a_{\sigma(4)} = 99,$	$\pi(7) := 6,$	$l := 5$

Theorem 1.5. *The* MERGE-SORT ALGORITHM *works correctly and runs in* $O(n \log n)$ *time.*

Proof: The correctness is obvious. We denote by $T(n)$ the running time (number of steps) needed for instances consisting of n numbers and observe that $T(1) = 1$ and $T(n) = T(\lfloor \frac{n}{2} \rfloor) + T(\lceil \frac{n}{2} \rceil) + 3n + 6$. (The constants in the term $3n + 6$ depend on how exactly a computation step is defined; but they do not really matter.)

We claim that this yields $T(n) \leq 12n \log n + 1$. Since this is trivial for $n = 1$ we proceed by induction. For $n \geq 2$, assuming that the inequality is true for $1, \ldots, n - 1$, we get

$$
\begin{aligned}
T(n) &\leq 12 \left\lfloor \frac{n}{2} \right\rfloor \log \left(\frac{2}{3}n \right) + 1 + 12 \left\lceil \frac{n}{2} \right\rceil \log \left(\frac{2}{3}n \right) + 1 + 3n + 6 \\
&= 12n (\log n + 1 - \log 3) + 3n + 8 \\
&\leq 12n \log n - \frac{13}{2}n + 3n + 8 \leq 12n \log n + 1,
\end{aligned}
$$

because $\log 3 \geq \frac{37}{24}$. $\square$

Of course the algorithm works for sorting the elements of any totally ordered set, assuming that we can compare any two elements in constant time. Can there be a faster, a linear-time algorithm? Suppose that the only way we can get information on the unknown order is to compare two elements. Then we can show that any algorithm needs at least $\Theta(n \log n)$ comparisons in the worst case. The outcome of a comparison can be regarded as a zero or one; the outcome of all comparisons an algorithm does is a 0-1-string (a sequence of zeros and ones). Note that two different orders in the input of the algorithm must lead to two different 0-1-strings (otherwise the algorithm could not distinguish between the two orders). For an input of n elements there are $n!$ possible orders, so there must be $n!$ different 0-1-strings corresponding to the computation. Since the number of 0-1-strings with length less than $\left\lfloor \frac{n}{2} \log \frac{n}{2} \right\rfloor$ is $2^{\left\lfloor \frac{n}{2} \log \frac{n}{2} \right\rfloor} - 1 < 2^{\frac{n}{2} \log \frac{n}{2}} = (\frac{n}{2})^{\frac{n}{2}} \leq n!$ we conclude that the maximum length of the 0-1-strings, and hence of the computation, must be at least $\frac{n}{2} \log \frac{n}{2} = \Theta(n \log n)$.

In the above sense, the running time of the MERGE-SORT ALGORITHM is optimal up to a constant factor. However, there is an algorithm for sorting integers (or sorting strings lexicographically) whose running time is linear in the input size; see Exercise 8. An algorithm to sort n integers in $O(n \log \log n)$ time was proposed by Han [2004].

Lower bounds like the one above are known only for very few problems (except trivial linear bounds). Often a restriction on the set of operations is necessary to derive a superlinear lower bound.

Exercises

1. Prove that for all $n \in \mathbb{N}$:

$$
e \left(\frac{n}{e} \right)^n \leq n! \leq en \left(\frac{n}{e} \right)^n .
$$

   Hint: Use $1 + x \leq e^x$ for all $x \in \mathbb{R}$.
2. Prove that $\log(n!) = \Theta(n \log n)$.
3. Prove that $n \log n = O(n^{1+\epsilon})$ for any $\epsilon > 0$.

4. Show that the running time of the PATH ENUMERATION ALGORITHM is $O(n \cdot n!)$.

5. Show that there is a polynomial-time algorithm for the DRILLING PROBLEM where d is the ℓ_1-distance if and only if there is one for ℓ_∞-distance.
   Note: Both is unlikely as the problems were proved to be *NP*-hard (this will be explained in Chapter 15) by Garey, Graham and Johnson [1976].

6. Suppose we have an algorithm whose running time is $\Theta(n(t + n^{1/t}))$, where n is the input length and t is a positive parameter we can choose arbitrarily. How should t be chosen (depending on n) such that the running time (as a function of n) has a minimum rate of growth?

7. Let s, t be binary strings, both of length m. We say that s is lexicographically smaller than t if there exists an index $j \in \{1, \ldots, m\}$ such that $s_i = t_i$ for $i = 1, \ldots, j - 1$ and $s_j < t_j$. Now given n strings of length m, we want to sort them lexicographically. Prove that there is a linear-time algorithm for this problem (i.e. one with running time $O(nm)$).
   Hint: Group the strings according to the first bit and sort each group.

8. Describe an algorithm which sorts a list of natural numbers $a_1, \ldots, a_n$ in linear time; i.e. which finds a permutation π with $a_{\pi(i)} \leq a_{\pi(i+1)}$ ($i = 1, \ldots, n - 1$) and runs in $O(\log(a_1 + 1) + \cdots + \log(a_n + 1))$ time.
   Hint: First sort the strings encoding the numbers according to their length. Then apply the algorithm of Exercise 7.
   Note: The algorithm discussed in this and the previous exercise is often called radix sorting.

References

General Literature

Cormen, T.H., Leiserson, C.E., Rivest, R.L., and Stein, C. [2009]: Introduction to Algorithms. Third Edition. MIT Press, Cambridge 2009

Hougardy, S., and Vygen, J. [2016]: Algorithmic Mathematics. Springer, Cham 2016

Knuth, D.E. [1968]: The Art of Computer Programming; Vol. 1. Fundamental Algorithms. Addison-Wesley, Reading 1968 (third edition: 1997)

Mehlhorn, K., and Sanders, P. [2008]: Algorithms and Data Structures: The Basic Toolbox. Springer, Berlin 2008

Cited References

Aho, A.V., Hopcroft, J.E., and Ullman, J.D. [1974]: The Design and Analysis of Computer Algorithms. Addison-Wesley, Reading 1974

Cobham, A. [1964]: The intrinsic computational difficulty of functions. Proceedings of the 1964 Congress for Logic Methodology and Philosophy of Science (Y. Bar-Hillel, ed.), North-Holland, Amsterdam 1964, pp. 24–30

Edmonds, J. [1965]: Paths, trees, and flowers. Canadian Journal of Mathematics 17 (1965), 449–467

Garey, M.R., Graham, R.L., and Johnson, D.S. [1976]: Some NP-complete geometric problems. Proceedings of the 8th Annual ACM Symposium on the Theory of Computing (1976), 10–22

Han, Y. [2004]: Deterministic sorting in $O(n \log \log n)$ time and linear space. Journal of Algorithms 50 (2004), 96–105

Stirling, J. [1730]: Methodus Differentialis. London 1730

2 Graphs

Graphs are a fundamental combinatorial structure used throughout this book. In this chapter we not only review the standard definitions and notation, but also prove some basic theorems and mention some fundamental algorithms.

After some basic definitions in Section 2.1 we consider fundamental objects occurring very often in this book: trees, circuits, and cuts. We prove some important properties and relations, and we also consider tree-like set systems in Section 2.2. The first graph algorithms, determining connected and strongly connected components, appear in Section 2.3. In Section 2.4 we prove Euler's Theorem on closed walks using every edge exactly once. Finally, in Sections 2.5 and 2.6 we consider graphs that can be drawn in the plane without crossings.

2.1 Basic Definitions

An **undirected graph** is a triple (V, E, Ψ), where V is a nonempty finite set, E is a finite set, and $\Psi : E \to \{X \subseteq V : |X| = 2\}$. A **directed graph** or **digraph** is a triple (V, E, Ψ), where V is a nonempty finite set, E is a finite set, and $\Psi : E \to \{(v, w) \in V \times V : v \neq w\}$. By a **graph** we mean either an undirected graph or a digraph. The elements of V are called **vertices**, the elements of E are the **edges**.

Edges e, e' with $e \neq e'$ and $\Psi(e) = \Psi(e')$ are called **parallel**. Graphs without parallel edges are called **simple**. For simple graphs we usually identify an edge e with its image $\Psi(e)$ and write $G = (V(G), E(G))$, where $E(G) \subseteq \binom{V}{2} := \{X \subseteq V(G) : |X| = 2\}$ or $E(G) \subseteq V(G) \times V(G)$. We often use this simpler notation even in the presence of parallel edges, then the "set" $E(G)$ may contain several "identical" elements. $|E(G)|$ denotes the number of edges, and for two edge sets E and F we always have $|E \,\dot\cup\, F| = |E| + |F|$ even if parallel edges arise. We write $e = \{v, w\}$ or $e = (v, w)$ for each edge e with $\Psi(e) = \{v, w\}$ or $\Psi(e) = (v, w)$, respectively.

We say that an edge $e = \{v, w\}$ or $e = (v, w)$ **joins** v and w. In this case, v and w are **adjacent**. v is a **neighbour** of w (and vice versa). v and w are the **endpoints** of e. If v is an endpoint of an edge e, we say that v is **incident** with e. In the directed case we say that $e = (v, w)$ **leaves** v (the **tail** of e) and **enters** w (the **head** of e). Two edges which share at least one endpoint are called **adjacent**.

© Springer-Verlag GmbH Germany 2018
Bernhard Korte and Jens Vygen, *Combinatorial Optimization*, Algorithms and Combinatorics 21, https://doi.org/10.1007/978-3-662-56039-6_2

This terminology for graphs is not the only one. Sometimes vertices are called nodes or points, other names for edges are arcs (especially in the directed case) or lines. In some texts, a graph is what we call a simple undirected graph, in the presence of parallel edges they speak of multigraphs. Sometimes edges whose endpoints coincide, so-called loops, are considered. However, unless otherwise stated, we do not use them.

For a digraph G we sometimes consider the **underlying undirected graph**, i.e. the undirected graph G' on the same vertex set which contains an edge $\{v, w\}$ for each edge (v, w) of G (so $|E(G')| = |E(G)|$). We also say that G is an **orientation** of G'.

A **subgraph** of a graph $G = (V(G), E(G))$ is a graph $H = (V(H), E(H))$ with $V(H) \subseteq V(G)$ and $E(H) \subseteq E(G)$. We also say that G **contains** H. H is an **induced subgraph** of G if it is a subgraph of G and $E(H) = \{\{x, y\} \in E(G) : x, y \in V(H)\}$ or $E(H) = \{(x, y) \in E(G) : x, y \in V(H)\}$. Here H is the subgraph of G **induced by** $V(H)$. We also write $H = G[V(H)]$. A subgraph H of G is called **spanning** if $V(H) = V(G)$.

If $v \in V(G)$, we write $G - v$ for the subgraph of G induced by $V(G) \setminus \{v\}$. If $e \in E(G)$, we define $G - e := (V(G), E(G) \setminus \{e\})$. We also use this notation for deleting a set X of vertices or edges and write $G - X$. Furthermore, the addition of a new edge e is abbreviated by $G + e := (V(G), E(G) \,\dot\cup\, \{e\})$. If G and H are two graphs, we denote by $G + H$ the graph with $V(G + H) = V(G) \cup V(H)$ and $E(G + H)$ being the disjoint union of $E(G)$ and $E(H)$ (parallel edges may arise). A family of graphs is called **vertex-disjoint** or **edge-disjoint** if their vertex sets or edge sets are pairwise disjoint, respectively.

Two graphs G and H are called **isomorphic** if there are bijections $\Phi_V : V(G) \to V(H)$ and $\Phi_E : E(G) \to E(H)$ such that $\Phi_E((v, w)) = (\Phi_V(v), \Phi_V(w))$ for all $(v, w) \in E(G)$, or $\Phi_E(\{v, w\}) = \{\Phi_V(v), \Phi_V(w)\}$ for all $\{v, w\} \in E(G)$ in the undirected case. We normally do not distinguish between isomorphic graphs; for example we say that G contains H if G has a subgraph isomorphic to H.

Suppose we have an undirected graph G and $X \subseteq V(G)$. By **contracting** (or **shrinking**) X we mean deleting the vertices in X and the edges in $G[X]$, adding a new vertex x and replacing each edge $\{v, w\}$ with $v \in X$, $w \notin X$ by an edge $\{x, w\}$ (parallel edges may arise). Similarly for digraphs. We often call the result G/X.

For a graph G and $X, Y \subseteq V(G)$ we define $E(X, Y) := \{\{x, y\} \in E(G) : x \in X \setminus Y, y \in Y \setminus X\}$ if G is undirected and $E^+(X, Y) := \{(x, y) \in E(G) : x \in X \setminus Y, y \in Y \setminus X\}$ if G is directed. For undirected graphs G and $X \subseteq V(G)$ we define $\delta(X) := E(X, V(G) \setminus X)$. The **set of neighbours** of X is defined by $\Gamma(X) := \{v \in V(G) \setminus X : E(X, \{v\}) \neq \emptyset\}$. For digraphs G and $X \subseteq V(G)$ we define $\delta^+(X) := E^+(X, V(G) \setminus X)$, $\delta^-(X) := \delta^+(V(G) \setminus X)$ and $\delta(X) := \delta^+(X) \cup \delta^-(X)$. We use subscripts (e.g. $\delta_G(X)$) to specify the graph G if necessary.

For **singletons**, i.e. one-element vertex sets $\{v\}$ ($v \in V(G)$) we write $\delta(v) := \delta(\{v\})$, $\Gamma(v) := \Gamma(\{v\})$, $\delta^+(v) := \delta^+(\{v\})$ and $\delta^-(v) := \delta^-(\{v\})$. The **degree** of a vertex v is $|\delta(v)|$, the number of edges incident to v. In the directed case, the **in-degree** is $|\delta^-(v)|$, the **out-degree** is $|\delta^+(v)|$, and the degree is $|\delta^+(v)|+|\delta^-(v)|$. A vertex with degree zero is called **isolated**. A graph where all vertices have degree k is called **k-regular**.

For any graph, $\sum_{v \in V(G)} |\delta(v)| = 2|E(G)|$. In particular, the number of vertices with odd degree is even. In a digraph, $\sum_{v \in V(G)} |\delta^+(v)| = \sum_{v \in V(G)} |\delta^-(v)|$. To prove these statements, please observe that each edge is counted twice on each side of the first equation and once on each side of the second equation. With just a little more effort we get the following useful statements:

Lemma 2.1. *For a digraph G and any two sets $X, Y \subseteq V(G)$:*

(a) $|\delta^+(X)|+|\delta^+(Y)| = |\delta^+(X \cap Y)|+|\delta^+(X \cup Y)|+|E^+(X,Y)|+|E^+(Y,X)|;$
(b) $|\delta^-(X)|+|\delta^-(Y)| = |\delta^-(X \cap Y)|+|\delta^-(X \cup Y)|+|E^+(X,Y)|+|E^+(Y,X)|.$

For an undirected graph G and any two sets $X, Y \subseteq V(G)$:

(c) $|\delta(X)|+|\delta(Y)| = |\delta(X \cap Y)|+|\delta(X \cup Y)|+2|E(X,Y)|;$
(d) $|\delta(X)|+|\delta(Y)| = |\delta(X \setminus Y)|+|\delta(Y \setminus X)|+2|E(X \cap Y, V(G) \setminus (X \cup Y))|;$
(e) $|\Gamma(X)|+|\Gamma(Y)| \geq |\Gamma(X \cap Y)|+|\Gamma(X \cup Y)|.$

Proof: All parts can be proved by simple counting arguments. Let $Z := V(G) \setminus (X \cup Y)$.

To prove (a), observe that $|\delta^+(X)|+|\delta^+(Y)| = |E^+(X,Z)|+|E^+(X,Y \setminus X)|+|E^+(Y,Z)|+|E^+(Y,X \setminus Y)| = |E^+(X \cup Y, Z)|+|E^+(X \cap Y, Z)|+|E^+(X, Y \setminus X)|+|E^+(Y, X \setminus Y)| = |\delta^+(X \cup Y)|+|\delta^+(X \cap Y)|+|E^+(X,Y)|+|E^+(Y,X)|.$

(b) follows from (a) by reversing each edge (replace (v,w) by (w,v)). (c) follows from (a) by replacing each edge $\{v,w\}$ by a pair of oppositely directed edges (v,w) and (w,v). Substituting Y by $V(G) \setminus Y$ in (c) yields (d).

To show (e), observe that $|\Gamma(X)|+|\Gamma(Y)| = |\Gamma(X \cup Y)|+|\Gamma(X) \cap \Gamma(Y)|+|\Gamma(X) \cap Y|+|\Gamma(Y) \cap X| \geq |\Gamma(X \cup Y)|+|\Gamma(X \cap Y)|.$ $\square$

A function $f : 2^U \to \mathbb{R}$ (where U is some finite set and 2^U denotes its power set) is called

- **submodular** if $f(X \cap Y) + f(X \cup Y) \leq f(X) + f(Y)$ for all $X, Y \subseteq U$;
- **supermodular** if $f(X \cap Y) + f(X \cup Y) \geq f(X) + f(Y)$ for all $X, Y \subseteq U$;
- **modular** if $f(X \cap Y) + f(X \cup Y) = f(X) + f(Y)$ for all $X, Y \subseteq U$.

So Lemma 2.1 implies that $|\delta^+|, |\delta^-|, |\delta|$ and $|\Gamma|$ are submodular. This will be useful later.

A **complete graph** is a simple undirected graph where each pair of vertices is adjacent. We denote the complete graph on n vertices by K_n. The **complement** of a simple undirected graph G is the graph H for which $V(G) = V(H)$ and $G + H$ is a complete graph.

A **matching** in an undirected graph G is a set of pairwise disjoint edges (i.e. the endpoints are all different). A **vertex cover** in G is a set $S \subseteq V(G)$ of vertices such that every edge of G is incident to at least one vertex in S. An **edge cover** in

G is a set $F \subseteq E(G)$ of edges such that every vertex of G is incident to at least one edge in F. A **stable set** in G is a set of pairwise non-adjacent vertices. A graph containing no edges is called **empty**. A **clique** is a set of pairwise adjacent vertices.

Proposition 2.2. *Let G be a graph and $X \subseteq V(G)$. Then the following three statements are equivalent:*

(a) *X is a vertex cover in G,*
(b) *$V(G) \setminus X$ is a stable set in G,*
(c) *$V(G) \setminus X$ is a clique in the complement of G.* $\square$

If $\mathcal{F}$ is a family of sets or graphs, we say that F is a **minimal** element of $\mathcal{F}$ if $\mathcal{F}$ contains F but no proper subset/subgraph of F. Similarly, F is **maximal** in $\mathcal{F}$ if $F \in \mathcal{F}$ and F is not a proper subset/subgraph of any element of $\mathcal{F}$. When we speak of a **minimum** or **maximum** element, we mean one of minimum/maximum cardinality.

For example, a minimal vertex cover is not necessarily a minimum vertex cover (see e.g. the graph in Figure 13.1), and a maximal matching is in general not maximum. The problems of finding a maximum matching, stable set or clique, or a minimum vertex cover or edge cover in an undirected graph will play important roles in later chapters.

The **line graph** of a simple undirected graph G is the graph $(E(G), F)$, where $F = \{\{e_1, e_2\} : e_1, e_2 \in E(G), |e_1 \cap e_2| = 1\}$. Obviously, matchings in a graph G correspond to stable sets in the line graph of G.

For the following notation, let G be a graph, directed or undirected. An **edge progression** W in G (**from** v_1 **to** v_{k+1}) is a sequence $v_1, e_1, v_2, \ldots, v_k, e_k, v_{k+1}$ such that $k \geq 0$, and $e_i = (v_i, v_{i+1}) \in E(G)$ or $e_i = \{v_i, v_{i+1}\} \in E(G)$ for $i = 1, \ldots, k$. If in addition $e_i \neq e_j$ for all $1 \leq i < j \leq k$, W is called a **walk** in G. W is **closed** if $v_1 = v_{k+1}$.

A **path** is a graph $P = (\{v_1, \ldots, v_{k+1}\}, \{e_1, \ldots, e_k\})$ such that $v_i \neq v_j$ for $1 \leq i < j \leq k+1$ and the sequence $v_1, e_1, v_2, \ldots, v_k, e_k, v_{k+1}$ is a walk. P is also called a path **from** v_1 **to** v_{k+1} or a v_1-v_{k+1}-path. v_1 and v_{k+1} are the **endpoints** of P, $v_2, \ldots, v_k$ are its **internal vertices**. By $P_{[x,y]}$ with $x, y \in V(P)$ we mean the (unique) subgraph of P which is an x-y-path.

For any graph G and vertices $v, w \in V(G)$, there is an edge progression from v to w in G if and only if there is a v-w-path in G (meaning: contained in G as a subgraph). For undirected graphs this defines an equivalence relation.

A **circuit** or a **cycle** is a graph $(\{v_1, \ldots, v_k\}, \{e_1, \ldots, e_k\})$ such that the sequence $v_1, e_1, v_2, \ldots, v_k, e_k, v_1$ is a (closed) walk with $k \geq 2$ and $v_i \neq v_j$ for $1 \leq i < j \leq k$. An easy induction argument shows that the edge set of a closed walk can be partitioned into edge sets of circuits.

By an **undirected path** or an **undirected circuit** in a digraph, we mean a subgraph corresponding to a path or circuit, respectively, in the underlying undirected graph.

The **length** of a path or circuit is the number of its edges. If it is a subgraph of G, we speak of a path or circuit in G. A spanning path in G is called a **Hamiltonian path** while a spanning circuit in G is called a **Hamiltonian circuit** or a **tour**. A graph containing a Hamiltonian circuit is a **Hamiltonian graph**.

For two vertices v and w we write $\text{dist}(v, w)$ or $\text{dist}_G(v, w)$ for the length of a shortest v-w-path (the **distance** from v to w) in G. If there is no v-w-path at all, i.e. w is not **reachable** from v, we set $\text{dist}(v, w) := \infty$. In the undirected case, $\text{dist}(v, w) = \text{dist}(w, v)$ for all $v, w \in V(G)$.

We shall often have a weight (or cost) function $c : E(G) \to \mathbb{R}$. Then for $F \subseteq E(G)$ we write $c(F) := \sum_{e \in F} c(e)$ (and $c(\emptyset) = 0$). This extends c to a modular function $c : 2^{E(G)} \to \mathbb{R}$. Moreover, $\text{dist}_{(G,c)}(v, w)$ denotes the minimum $c(E(P))$ over all v-w-paths P in G.

2.2 Trees, Circuits, and Cuts

An undirected graph G is called **connected** if there is a v-w-path for all $v, w \in V(G)$; otherwise G is **disconnected**. A digraph is called connected if the underlying undirected graph is connected. The maximal connected subgraphs of a graph are its **connected components**. Sometimes we identify the connected components with the vertex sets inducing them. A set of vertices X is called connected if the subgraph induced by X is connected. A vertex v with the property that $G - v$ has more connected components than G is called an **articulation vertex**. An edge e is called a **bridge** if $G - e$ has more connected components than G.

An undirected graph without a circuit (as a subgraph) is called a **forest**. A connected forest is a **tree**. A vertex of degree at most 1 in a tree is called a **leaf**. A **star** is a tree where at most one vertex is not a leaf.

In the following we shall give some equivalent characterizations of trees and their directed counterparts, arborescences. We need the following connectivity criterion:

Proposition 2.3.

(a) *An undirected graph G is connected if and only if $\delta(X) \neq \emptyset$ for all $\emptyset \neq X \subset V(G)$.*

(b) *Let G be a digraph and $r \in V(G)$. Then there exists an r-v-path for every $v \in V(G)$ if and only if $\delta^+(X) \neq \emptyset$ for all $X \subset V(G)$ with $r \in X$.*

Proof: (a): If there is a set $X \subset V(G)$ with $r \in X$, $v \in V(G) \setminus X$, and $\delta(X) = \emptyset$, there can be no r-v-path, so G is not connected. On the other hand, if G is not connected, there is no r-v-path for some r and v. Let R be the set of vertices reachable from r. We have $r \in R$, $v \notin R$ and $\delta(R) = \emptyset$.

(b) is proved analogously. $\square$

Theorem 2.4. *Let G be an undirected graph on n vertices. Then the following statements are equivalent:*

(a) *G is a tree (i.e. is connected and has no circuits).*
(b) *G has n − 1 edges and no circuits.*
(c) *G has n − 1 edges and is connected.*
(d) *G is connected and every edge is a bridge.*
(e) *G satisfies $\delta(X) \neq \emptyset$ for all $\emptyset \neq X \subset V(G)$, but deleting any edge would destroy this property.*
(f) *G is a forest, but the addition of an arbitrary edge would create a circuit.*
(g) *G contains a unique path between any pair of vertices.*

Proof: (a)⇒(g) follows from the fact that the union of two distinct paths with the same endpoints contains a circuit.

(g)⇒(e)⇒(d) follows from Proposition 2.3(a).

(d)⇒(f) is trivial.

(f)⇒(b)⇒(c): This follows from the fact that for forests with n vertices, m edges and p connected components $n = m + p$ holds. (The proof is a trivial induction on m.)

(c)⇒(a): Let G be connected with $n-1$ edges. As long as there are any circuits in G, we destroy them by deleting an edge of the circuit. Suppose we have deleted k edges. The resulting graph G' is still connected and has no circuits. G' has $m = n - 1 - k$ edges. So $n = m + p = n - 1 - k + 1$, implying $k = 0$.   □

In particular, (d)⇒(a) implies that a graph is connected if and only if it contains a **spanning tree** (a spanning subgraph which is a tree).

A digraph is a **branching** if the underlying undirected graph is a forest and each vertex v has at most one entering edge. A connected branching is an **arborescence**. By Theorem 2.4 an arborescence with n vertices has $n - 1$ edges, hence it has exactly one vertex r with $\delta^-(r) = \emptyset$. This vertex is called its **root**; we also speak of an arborescence **rooted at** r. For a vertex v in a branching, the vertices w for which (v, w) is an edge are called the **children** of v. For a child w of v, v is called the **parent** or **predecessor** of w. Vertices without children are called **leaves**.

Theorem 2.5. *Let G be a digraph on n vertices. Then the following statements are equivalent:*

(a) *G is an arborescence rooted at r (i.e. a connected branching with $\delta^-(r) = \emptyset$).*
(b) *G is a branching with n − 1 edges and $\delta^-(r) = \emptyset$.*
(c) *G has n − 1 edges and every vertex is reachable from r.*
(d) *Every vertex is reachable from r, but deleting any edge would destroy this property.*
(e) *G satisfies $\delta^+(X) \neq \emptyset$ for all $X \subset V(G)$ with $r \in X$, but deleting any edge would destroy this property.*
(f) *$\delta^-(r) = \emptyset$, and there is a unique walk from r to v for each $v \in V(G) \setminus \{r\}$.*
(g) *$\delta^-(r) = \emptyset$, $|\delta^-(v)| = 1$ for all $v \in V(G) \setminus \{r\}$, and G contains no circuit.*

Proof: (a)$\Rightarrow$(b) and (c)$\Rightarrow$(d) follow from Theorem 2.4.

(b)$\Rightarrow$(c): We have that $|\delta^-(v)| = 1$ for all $v \in V(G) \setminus \{r\}$. So for any v we have an r-v-path (start at v and always follow the entering edge until r is reached).

(d)$\Leftrightarrow$(e) is implied by Proposition 2.3(b).

(d)$\Rightarrow$(f): Any edge in $\delta^-(r)$ could be deleted without destroying reachability from r. Suppose that, for some $v \in V(G)$, there are two r-v-walks P and Q. Let e be the last edge of P that does not belong to Q (if such an edge does not exist, exchange P and Q). Then after deleting e, every vertex is still reachable from r.

(f)$\Rightarrow$(g): If every vertex is reachable from r and $|\delta^-(v)| > 1$ for some vertex $v \in V(G) \setminus \{r\}$, then we have two walks from r to v. If G contains a circuit C, let $v \in V(C)$, consider the r-v-path P, and let x be the first vertex on P belonging to C. Then there are two walks from r to x: $P_{[r,x]}$, and $P_{[r,x]}$ plus C.

(g)$\Rightarrow$(a): If $|\delta^-(v)| \leq 1$ for all $v \in V(G)$, then every undirected circuit is a (directed) circuit. $\qquad\square$

A **cut** in an undirected graph G is an edge set of type $\delta(X)$ for some $\emptyset \neq X \subset V(G)$. In a digraph G, $\delta^+(X)$ is a **directed cut** if $\emptyset \neq X \subset V(G)$ and $\delta^-(X) = \emptyset$, i.e. no edge enters the set X.

We say that an edge set $F \subseteq E(G)$ **separates** two vertices s and t if t is reachable from s in G but not in $(V(G), E(G) \setminus F)$. An **s-t-cut** in an undirected graph is a cut $\delta(X)$ for some $X \subset V(G)$ with $s \in X$ and $t \notin X$. In a digraph, an s-t-cut is an edge set $\delta^+(X)$ with $s \in X$ and $t \notin X$. An **r-cut** in a digraph is an edge set $\delta^+(X)$ for some $X \subset V(G)$ with $r \in X$.

An **undirected cut** in a digraph is an edge set corresponding to a cut in the underlying undirected graph, i.e., $\delta(X)$ for some $\emptyset \neq X \subset V(G)$.

Lemma 2.6.   (Minty [1960])   *Let G be a digraph and $e \in E(G)$. Suppose e is coloured black, while all other edges are coloured red, black or green. Then exactly one of the following statements holds:*

(a) *There is an undirected circuit containing e and only red and black edges such that all black edges have the same orientation.*

(b) *There is an undirected cut containing e and only green and black edges such that all black edges have the same orientation.*

Proof: Let $e = (x, y)$. We label the vertices of G by the following procedure. First label y. In case v is already labelled and w is not, we label w if there is a black edge (v, w), a red edge (v, w) or a red edge (w, v). In this case, we write $pred(w) := v$.

When the labelling procedure stops, there are two possibilities:

Case 1: x has been labelled. Then the vertices x, $pred(x)$, $pred(pred(x))$, ..., y form an undirected circuit with the property (a).

Case 2: x has not been labelled. Then let R consist of all labelled vertices. Obviously, the undirected cut $\delta^+(R) \cup \delta^-(R)$ has the property (b).

Suppose that an undirected circuit C as in (a) and an undirected cut $\delta^+(X) \cup \delta^-(X)$ as in (b) both exist. All edges in their (nonempty) intersection are black, they all have the same orientation with respect to C, and they all leave X or all enter X. This is a contradiction.   □

A digraph is called **strongly connected** if there is a path from s to t and a path from t to s for all $s, t \in V(G)$. The **strongly connected components** of a digraph are the maximal strongly connected subgraphs.

Corollary 2.7. *In a digraph G, each edge belongs either to a (directed) circuit or to a directed cut. Moreover the following statements are equivalent:*

(a) *G is strongly connected.*
(b) *G contains no directed cut.*
(c) *G is connected and each edge of G belongs to a circuit.*

Proof: The first statement follows directly from Minty's Lemma 2.6 by colouring all edges black. This also proves (b)$\Rightarrow$(c).

(a)$\Rightarrow$(b) follows from Proposition 2.3(b).

(c)$\Rightarrow$(a): Let $r \in V(G)$ be an arbitrary vertex. We prove that there is an r-v-path for each $v \in V(G)$. Suppose this is not true, then by Proposition 2.3(b) there is some $X \subset V(G)$ with $r \in X$ and $\delta^+(X) = \emptyset$. Since G is connected, we have $\delta^+(X) \cup \delta^-(X) \neq \emptyset$ (by Proposition 2.3(a)), so let $e \in \delta^-(X)$. But then e cannot belong to a circuit since no edge leaves X.   □

Corollary 2.7 and Theorem 2.5 imply that a digraph is strongly connected if and only if it contains for each vertex v a spanning arborescence rooted at v.

A digraph is called **acyclic** if it contains no (directed) circuit. So by Corollary 2.7 a digraph is acyclic if and only if each edge belongs to a directed cut. Moreover, a digraph is acyclic if and only if its strongly connected components are the singletons. The vertices of an acyclic digraph can be ordered in a nice way:

Definition 2.8. *Let G be a digraph. A **topological order** of G is an order of the vertices $V(G) = \{v_1, \ldots, v_n\}$ such that for each edge $(v_i, v_j) \in E(G)$ we have $i < j$.*

Proposition 2.9. *A digraph has a topological order if and only if it is acyclic.*

Proof: If a digraph has a circuit, it clearly cannot have a topological order. We show the converse by induction on the number of edges. If there are no edges, every order is topological. Otherwise let $e \in E(G)$; by Corollary 2.7 e belongs to a directed cut $\delta^+(X)$. Then a topological order of $G[X]$ followed by a topological order of $G - X$ (both exist by the induction hypothesis) is a topological order of G.   □

Circuits and cuts also play an important role in algebraic graph theory. For a graph G we associate a vector space $\mathbb{R}^{E(G)}$ whose elements are vectors $(x_e)_{e \in E(G)}$ with $|E(G)|$ real components. Following Berge [1985] we shall now briefly discuss two linear subspaces which are particularly important.

Let G be a digraph. We associate a vector $\zeta(C) \in \{-1, 0, 1\}^{E(G)}$ with each undirected circuit C in G by setting $\zeta(C)_e = 0$ for $e \notin E(C)$, and setting $\zeta(C)_e \in \{-1, 1\}$ for $e \in E(C)$ such that reorienting all edges e with $\zeta(C)_e = -1$ results in a directed circuit. Similarly, we associate a vector $\zeta(D) \in \{-1, 0, 1\}^{E(G)}$ with each undirected cut $D = \delta(X)$ in G by setting $\zeta(D)_e = 0$ for $e \notin D$, $\zeta(D)_e = -1$ for $e \in \delta^-(X)$ and $\zeta(D)_e = 1$ for $e \in \delta^+(X)$. Note that these vectors are properly defined only up to multiplication by -1. However, the subspaces of the vector space $\mathbb{R}^{E(G)}$ generated by the set of vectors associated with the undirected circuits and by the set of vectors associated with the undirected cuts in G are properly defined; they are called the **cycle space** and the **cocycle space** of G, respectively.

Proposition 2.10. *The cycle space and the cocycle space are orthogonal to each other.*

Proof: Let C be any undirected circuit and $D = \delta(X)$ be any undirected cut. We claim that the scalar product of $\zeta(C)$ and $\zeta(D)$ is zero. Since reorienting any edge does not change the scalar product we may assume that D is a directed cut. But then the result follows from observing that any circuit enters a set X the same number of times as it leaves X. $\square$

We shall now show that the sum of the dimensions of the cycle space and the cocycle space is $|E(G)|$, the dimension of the whole space. A set of undirected circuits (undirected cuts) is called a **cycle basis** (a **cocycle basis**) if the associated vectors form a basis of the cycle space (the cocycle space, respectively). Let G be a graph (directed or undirected) and T a maximal subgraph without an undirected circuit. For each $e \in E(G) \setminus E(T)$ we call the unique undirected circuit in $T + e$ the **fundamental circuit** of e with respect to T. Moreover, for each $e \in E(T)$ there is a set $X \subseteq V(G)$ with $\delta_G(X) \cap E(T) = \{e\}$ (consider a component of $T - e$); we call $\delta_G(X)$ the **fundamental cut** of e with respect to T.

Theorem 2.11. *Let G be a digraph and T a maximal subgraph without an undirected circuit. The $|E(G) \setminus E(T)|$ fundamental circuits with respect to T form a cycle basis of G, and the $|E(T)|$ fundamental cuts with respect to T form a cocycle basis of G.*

Proof: The vectors associated with the fundamental circuits are linearly independent since each fundamental circuit contains an element not belonging to any other. The same holds for the fundamental cuts. Since the vector spaces are orthogonal to each other by Proposition 2.10, the sum of their dimensions cannot exceed $|E(G)| = |E(G) \setminus E(T)| + |E(T)|$. $\square$

The fundamental cuts have a nice property which we shall exploit quite often and which we shall discuss now. Let T be a digraph whose underlying undirected graph is a tree. Consider the family $\mathcal{F} := \{C_e : e \in E(T)\}$, where for $e = (x, y) \in E(T)$ we denote by C_e the connected component of $T - e$ containing y (so $\delta(C_e)$ is the fundamental cut of e with respect to T). If T is an arborescence, then any two elements of $\mathcal{F}$ are either disjoint or one is a subset of the other. In general $\mathcal{F}$ is at least cross-free:

(a)

(b)

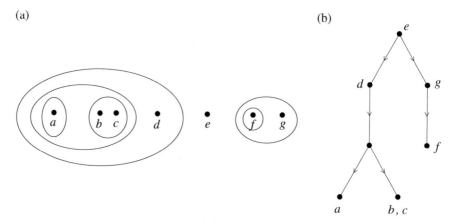

Fig. 2.1.

Definition 2.12. *A* **set system** *is a pair* $(U, \mathcal{F})$, *where* U *is a nonempty finite set and* $\mathcal{F}$ *a family of subsets of* U. $(U, \mathcal{F})$ *is* **cross-free** *if for any two sets* $X, Y \in \mathcal{F}$, *at least one of the four sets* $X \setminus Y$, $Y \setminus X$, $X \cap Y$, $U \setminus (X \cup Y)$ *is empty.* $(U, \mathcal{F})$ *is* **laminar** *if for any two sets* $X, Y \in \mathcal{F}$, *at least one of the three sets* $X \setminus Y$, $Y \setminus X$, $X \cap Y$ *is empty.*

In the literature set systems are also known as **hypergraphs**. See Figure 2.1(a) for an illustration of the laminar family $\{\{a\}, \{b, c\}, \{a, b, c\}, \{a, b, c, d\}, \{f\}, \{f, g\}\}$. Another word used for laminar is **nested**.

Whether a set system $(U, \mathcal{F})$ is laminar does not depend on U, so we sometimes simply say that $\mathcal{F}$ is a laminar family. However, whether a set system is cross-free can depend on the ground set U. If U contains an element that does not belong to any set of $\mathcal{F}$, then $\mathcal{F}$ is cross-free if and only if it is laminar. Let $r \in U$ be arbitrary. It follows directly from the definition that a set system $(U, \mathcal{F})$ is cross-free if and only if

$$\mathcal{F}' := \{X \in \mathcal{F} : r \notin X\} \cup \{U \setminus X : X \in \mathcal{F}, r \in X\}$$

is laminar. Hence cross-free families are sometimes depicted similarly to laminar families: for example, Figure 2.2(a) shows the cross-free family $\{\{b, c, d, e, f\}, \{c\}, \{a, b, c\}, \{e\}, \{a, b, c, d, f\}, \{e, f\}\}$; a square corresponds to the set containing all elements outside.

While oriented trees lead to cross-free families the converse is also true: every cross-free family can be represented by a tree in the following sense:

Definition 2.13. *Let* T *be a digraph such that the underlying undirected graph is a tree. Let* U *be a finite set and* $\varphi : U \to V(T)$. *Let* $\mathcal{F} := \{S_e : e \in E(T)\}$, *where for* $e = (x, y)$ *we define*

$$S_e := \{s \in U : \varphi(s) \text{ is in the same connected component of } T - e \text{ as } y\}.$$

Then (T, φ) *is called a* **tree-representation** *of* $(U, \mathcal{F})$.

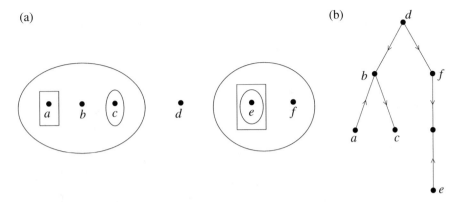

Fig. 2.2.

See Figures 2.1(b) and 2.2(b) for examples.

Proposition 2.14. *Let $(U, \mathcal{F})$ be a set system with a tree-representation (T, φ). Then $(U, \mathcal{F})$ is cross-free. If T is an arborescence, then $(U, \mathcal{F})$ is laminar. Moreover, every cross-free family has a tree-representation, and for laminar families, an arborescence can be chosen as T.*

Proof: If (T, φ) is a tree-representation of $(U, \mathcal{F})$ and $e = (v, w)$, $f = (x, y) \in E(T)$, we have an undirected v-x-path P in T (ignoring the orientations). There are four cases: If $w, y \notin V(P)$ then $S_e \cap S_f = \emptyset$ (since T contains no circuit). If $w \notin V(P)$ and $y \in V(P)$ then $S_e \subseteq S_f$. If $y \notin V(P)$ and $w \in V(P)$ then $S_f \subseteq S_e$. If $w, y \in V(P)$ then $S_e \cup S_f = U$. Hence $(U, \mathcal{F})$ is cross-free. If T is an arborescence, the last case cannot occur (otherwise at least one vertex of P would have two entering edges), so $\mathcal{F}$ is laminar.

To prove the converse, let $\mathcal{F}$ first be a laminar family. We define $V(T) := \mathcal{F} \cup \{r\}$ and $E(T) :=$

$$\{(X, Y) \in \mathcal{F} \times \mathcal{F} : X \supset Y \neq \emptyset \text{ and there is no } Z \in \mathcal{F} \text{ with } X \supset Z \supset Y\}$$
$$\cup \ \{(r, X) : X = \emptyset \in \mathcal{F} \text{ or } X \text{ is a maximal element of } \mathcal{F}\}.$$

We set $\varphi(x) := X$, where X is the minimal set in $\mathcal{F}$ containing x, and $\varphi(x) := r$ if no set in $\mathcal{F}$ contains x. Obviously, T is an arborescence rooted at r, and (T, φ) is a tree-representation of $\mathcal{F}$.

Now let $\mathcal{F}$ be a cross-free family of subsets of U. Let $r \in U$. As noted above,

$$\mathcal{F}' := \{X \in \mathcal{F} : r \notin X\} \cup \{U \setminus X : X \in \mathcal{F}, r \in X\}$$

is laminar, so let (T, φ) be a tree-representation of $(U, \mathcal{F}')$. Now for an edge $e \in E(T)$ there are three cases: If $S_e \in \mathcal{F}$ and $U \setminus S_e \in \mathcal{F}$, we replace the edge $e = (x, y)$ by two edges (x, z) and (y, z), where z is a new vertex. If $S_e \notin \mathcal{F}$ and $U \setminus S_e \in \mathcal{F}$, we replace the edge $e = (x, y)$ by (y, x). If $S_e \in \mathcal{F}$ and $U \setminus S_e \notin \mathcal{F}$, we do nothing. Let T' be the resulting graph. Then (T', φ) is a tree-representation of $(U, \mathcal{F})$. □

The above result is mentioned by Edmonds and Giles [1977] but was probably known earlier.

Corollary 2.15. *A laminar family of distinct subsets of U has at most $2|U|$ elements. A cross-free family of distinct subsets of U has at most $4|U| - 2$ elements.*

Proof: We first consider a laminar family $\mathcal{F}$ of distinct nonempty proper subsets of U. We prove that $|\mathcal{F}| \leq 2|U| - 2$. Let (T, φ) be a tree-representation, where T is an arborescence whose number of vertices is as small as possible. For every $w \in V(T)$ we have either $|\delta^+(w)| \geq 2$ or there exists an $x \in U$ with $\varphi(x) = w$ or both. (For the root this follows from $U \notin \mathcal{F}$, for the leaves from $\emptyset \notin \mathcal{F}$, for all other vertices from the minimality of T.)

There can be at most $|U|$ vertices w with $\varphi(x) = w$ for some $x \in U$ and at most $\left\lfloor \frac{|E(T)|}{2} \right\rfloor$ vertices w with $|\delta^+(w)| \geq 2$. So $|E(T)| + 1 = |V(T)| \leq |U| + \frac{|E(T)|}{2}$ and thus $|\mathcal{F}| = |E(T)| \leq 2|U| - 2$.

Now let $(U, \mathcal{F})$ be a cross-free family with $\emptyset, U \notin \mathcal{F}$, and let $r \in U$. Since

$$\mathcal{F}' := \{X \in \mathcal{F} : r \notin X\} \cup \{U \setminus X : X \in \mathcal{F}, r \in X\}$$

is laminar, we have $|\mathcal{F}'| \leq 2|U| - 2$. Hence $|\mathcal{F}| \leq 2|\mathcal{F}'| \leq 4|U| - 4$. The proof is concluded by taking $\emptyset$ and U as possible members of $\mathcal{F}$ into account. □

2.3 Connectivity

Connectivity is a very important concept in graph theory. For many problems it suffices to consider connected graphs, since otherwise we can solve the problem for each connected component separately. So it is a fundamental task to detect the connected components of a graph. The following simple algorithm finds a path from a specified vertex s to all other vertices that are reachable from s. It works for both directed and undirected graphs. In the undirected case it builds a maximal tree containing s; in the directed case it constructs a maximal arborescence rooted at s.

GRAPH SCANNING ALGORITHM

Input: A graph G (directed or undirected) and some vertex s.

Output: The set R of vertices reachable from s, and a set $T \subseteq E(G)$ such that (R, T) is an arborescence rooted at s, or a tree.

---

①   Set $R := \{s\}$, $Q := \{s\}$ and $T := \emptyset$.

②   **If** $Q = \emptyset$ **then stop,**
            else choose a $v \in Q$.

③ Choose a $w \in V(G) \setminus R$ with $e = (v, w) \in E(G)$ or $e = \{v, w\} \in E(G)$. **If** there is no such w **then** set $Q := Q \setminus \{v\}$ and **go to** ②.

④ Set $R := R \cup \{w\}$, $Q := Q \cup \{w\}$ and $T := T \cup \{e\}$. **Go to** ②.

Proposition 2.16. *The* GRAPH SCANNING ALGORITHM *works correctly.*

Proof: At any time, (R, T) is a tree or an arborescence rooted at s. Suppose at the end there is a vertex $w \in V(G) \setminus R$ that is reachable from s. Let P be an s-w-path, and let $\{x, y\}$ or (x, y) be an edge of P with $x \in R$ and $y \notin R$. Since x has been added to R, it also has been added to Q at some time during the execution of the algorithm. The algorithm does not stop before removing x from Q. But this is done in ③ only if there is no edge $\{x, y\}$ or (x, y) with $y \notin R$. □

Since this is the first graph algorithm in this book we discuss some implementation issues. The first question is how the graph is given. There are several natural ways. For example, one can think of a matrix with a row for each vertex and a column for each edge. The **incidence matrix** of an undirected graph G is the matrix $A = (a_{v,e})_{v \in V(G), \, e \in E(G)}$ where

$$a_{v,e} = \begin{cases} 1 & \text{if } v \in e \\ 0 & \text{if } v \notin e \end{cases}.$$

The **incidence matrix** of a digraph G is the matrix $A = (a_{v,e})_{v \in V(G), \, e \in E(G)}$ where

$$a_{v,(x,y)} = \begin{cases} -1 & \text{if } v = x \\ 1 & \text{if } v = y \\ 0 & \text{if } v \notin \{x, y\} \end{cases}.$$

Of course this is not very efficient since each column contains only two nonzero entries. The space needed for storing an incidence matrix is obviously $O(nm)$, where $n := |V(G)|$ and $m := |E(G)|$.

A better way is a matrix whose rows and columns are indexed by the vertex set. The **adjacency matrix** of a simple graph G is the 0-1-matrix $A = (a_{v,w})_{v,w \in V(G)}$ with $a_{v,w} = 1$ iff $\{v, w\} \in E(G)$ or $(v, w) \in E(G)$. For graphs with parallel edges we can define $a_{v,w}$ to be the number of edges from v to w. An adjacency matrix requires $O(n^2)$ space for simple graphs.

The adjacency matrix is appropriate if the graph is **dense**, i.e. has $\Theta(n^2)$ edges (or more). For **sparse** graphs, say with $O(n)$ edges only, one can do much better. Besides storing the number of vertices we can simply store a list of the edges, for each edge noting its endpoints. If we address each vertex by a number from 1 to n, the space needed for each edge is $O(\log n)$. Hence we need $O(m \log n)$ space altogether.

Just storing the edges in an arbitrary order is not very convenient. Almost all graph algorithms require finding the edges incident to a given vertex. Thus one should have a list of incident edges for each vertex. In case of directed graphs, two lists, one for entering edges and one for leaving edges, are appropriate. This data structure is called **adjacency list**; it is the most customary one for graphs. For direct access to the list(s) of each vertex we have pointers to the heads of all lists; these can be stored with $O(n \log m)$ additional bits. Hence the total number of bits required for an adjacency list is $O(n \log m + m \log n)$.

Whenever a graph is part of the input of an algorithm in this book, we assume that the graph is given by an adjacency list.

As for elementary operations on numbers (see Section 1.2), we assume that elementary operations on vertices and edges take constant time only. This includes scanning an edge, identifying its ends and accessing the head of the adjacency list for a vertex. The running time will be measured by the parameters n and m, and an algorithm running in $O(m + n)$ time is called linear.

We shall always use the letters n and m for the number of vertices and the number of edges. For many graph algorithms it causes no loss of generality to assume that the graph at hand is simple and connected; hence $n - 1 \le m < n^2$. Among parallel edges we often have to consider only one, and different connected components can often be analyzed separately. The preprocessing can be done in linear time in advance; see Exercise 17 and the following.

We can now analyze the running time of the GRAPH SCANNING ALGORITHM:

Proposition 2.17. *The* GRAPH SCANNING ALGORITHM *can be implemented to run in $O(m)$ time. The connected components of an undirected graph can be determined in linear time.*

Proof: We assume that G is given by an adjacency list. Implement Q by a simple list, such that ② takes constant time. For each vertex x that we insert into Q we introduce a pointer *current*(x), indicating the current edge in the list containing all edges in $\delta(x)$ or $\delta^+(x)$ (this list is part of the input). Initially *current*(x) is set to the first element of the list. In ③, the pointer moves forward. When the end of the list is reached, x is removed from Q and will never be inserted again. So the overall running time is proportional to the number of vertices reachable from s plus the number of edges, i.e. $O(m)$.

To identify the connected components of a graph, we apply the algorithm once and check if $R = V(G)$. If so, the graph is connected. Otherwise R is a connected component, and we apply the algorithm to (G, s') for an arbitrary vertex $s' \in V(G) \setminus R$ (and iterate until all vertices have been scanned, i.e. added to R). Again, no edge is scanned more than twice, so the overall running time remains linear.   □

An interesting question is in which order the vertices are chosen in ③. Obviously we cannot say much about this order if we do not specify how to choose a $v \in Q$ in ②. Two methods are frequently used; they are called DEPTH-FIRST SEARCH (DFS) and BREADTH-FIRST SEARCH (BFS). In DFS we choose the $v \in Q$ that was the last to enter Q. In other words, Q is implemented as a LIFO-stack (last-in-first-out). In BFS we choose the $v \in Q$ that was the first to enter Q. Here Q is implemented by a FIFO-queue (first-in-first-out).

An algorithm similar to DFS has been described already before 1900 by Trémaux and Tarry; see König [1936]. BFS seems to have been mentioned first by Moore [1959]. Trees (in the directed case: arborescences) (R, T) computed by DFS and BFS are called **DFS-tree** and **BFS-tree**, respectively. For BFS-trees we note the following important property:

Proposition 2.18. *A* BFS-*tree contains a shortest path from s to each vertex reachable from s. The values* $\text{dist}_G(s, v)$ *for all* $v \in V(G)$ *can be determined in linear time.*

Proof: We apply BFS to (G, s) and add two statements: initially (in ① of the GRAPH SCANNING ALGORITHM) we set $l(s) := 0$, and in ④ we set $l(w) := l(v) + 1$. We obviously have that $l(v) = \text{dist}_{(R,T)}(s, v)$ for all $v \in R$, at any stage of the algorithm. Moreover, if v is the currently scanned vertex (chosen in ②), at this time there is no vertex $w \in R$ with $l(w) > l(v) + 1$ (because the vertices are scanned in an order with nondecreasing l-values).

Suppose that when the algorithm terminates there is a vertex $w \in V(G)$ with $\text{dist}_G(s, w) < \text{dist}_{(R,T)}(s, w)$; let w have minimum distance from s in G with this property. Let P be a shortest s-w-path in G, and let $e = (v, w)$ or $e = \{v, w\}$ be the last edge in P. We have $\text{dist}_G(s, v) = \text{dist}_{(R,T)}(s, v)$, but e does not belong to T. Moreover, $l(w) = \text{dist}_{(R,T)}(s, w) > \text{dist}_G(s, w) = \text{dist}_G(s, v) + 1 = \text{dist}_{(R,T)}(s, v) + 1 = l(v) + 1$. This inequality combined with the above observation proves that w did not belong to R when v was removed from Q. But this contradicts ③ because of edge e. $\square$

This result will also follow from the correctness of DIJKSTRA'S ALGORITHM for the SHORTEST PATH PROBLEM, which can be thought of as a generalization of BFS to the case where we have nonnegative weights on the edges (see Section 7.1).

We now show how to identify the strongly connected components of a digraph. Of course, this can easily be done by using n times DFS (or BFS). However, it is possible to find the strongly connected components by visiting every edge only twice:

STRONGLY CONNECTED COMPONENT ALGORITHM

Input: A digraph G.

Output: A function $comp : V(G) \rightarrow \mathbb{N}$ indicating the membership of the strongly connected components.

---

①   Set $R := \emptyset$. Set $N := 0$.

②   **For all** $v \in V(G)$ **do: If** $v \notin R$ **then** VISIT1(v).

③   Set $R := \emptyset$. Set $K := 0$.

④   **For** $i := |V(G)|$ **down to** 1 **do:**
        If $\psi^{-1}(i) \notin R$ **then** set $K := K + 1$ and VISIT2$(\psi^{-1}(i))$.

VISIT1(v)

①   Set $R := R \cup \{v\}$.

②   **For** all w with $(v, w) \in E(G)$ **do**:
         If $w \notin R$ **then** VISIT1(w).

③   Set $N := N + 1$, $\psi(v) := N$ and $\psi^{-1}(N) := v$.

VISIT2(v)

①   Set $R := R \cup \{v\}$.

②   **For** all w with $(w, v) \in E(G)$ **do**:
         If $w \notin R$ **then** VISIT2(w).

③   Set $comp(v) := K$.

Figure 2.3 shows an example: The first DFS scans the vertices in the order a, g, b, d, e, f and produces the arborescence shown in the middle; the numbers are the ψ-labels. Vertex c is the only one that is not reachable from a; it gets the highest label $\psi(c) = 7$. The second DFS starts with c but cannot reach any other vertex via a reverse edge. So it proceeds with vertex a because $\psi(a) = 6$. Now b, g and f can be reached. Finally e is reached from d. The strongly connected components are $\{c\}$, $\{a, b, f, g\}$ and $\{d, e\}$.

In summary, one DFS is needed to find an appropriate numbering, while in the second DFS the reverse graph is considered and the vertices are processed in decreasing order with respect to this numbering. Each connected component of the second DFS-forest is an **anti-arborescence**, a graph arising from an arborescence by reversing every edge. We show that these anti-arborescences identify the strongly connected components.

Theorem 2.19. *The* STRONGLY CONNECTED COMPONENT ALGORITHM *identifies the strongly connected components correctly in linear time.*

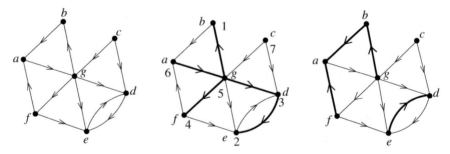

Fig. 2.3.

Proof: The running time is obviously $O(n + m)$. Of course, vertices of the same strongly connected component are always in the same component of any DFS-forest, so they get the same *comp*-value. We have to prove that two vertices u and v with $comp(u) = comp(v)$ indeed lie in the same strongly connected component. Let $r(u)$ and $r(v)$ be the vertex reachable from u and v with the highest ψ-label, respectively. Since $comp(u) = comp(v)$, i.e. u and v lie in the same anti-arborescence of the second DFS-forest, $r := r(u) = r(v)$ is the root of this anti-arborescence. So r is reachable from both u and v.

Since r is reachable from u and $\psi(r) \geq \psi(u)$, r has not been added to R after u in the first DFS, and the first DFS-forest contains an r-u-path. In other words, u is reachable from r. Analogously, v is reachable from r. Altogether, u is reachable from v and vice versa, proving that indeed u and v belong to the same strongly connected component. $\square$

It is interesting that this algorithm also solves another problem: finding a topological order of an acyclic digraph. Observe that contracting the strongly connected components of any digraph yields an acyclic digraph. By Proposition 2.9 this acyclic digraph has a topological order. In fact, such an order is given by the numbers $comp(v)$ computed by the STRONGLY CONNECTED COMPONENT ALGORITHM:

Theorem 2.20.   *The* STRONGLY CONNECTED COMPONENT ALGORITHM *determines a topological order of the digraph resulting from contracting each strongly connected component of G. In particular, we can for any given digraph either find a topological order or decide that none exists in linear time.*

Proof: Let X and Y be two strongly connected components of a digraph G, and suppose the STRONGLY CONNECTED COMPONENT ALGORITHM computes $comp(x) = k_1$ for $x \in X$ and $comp(y) = k_2$ for $y \in Y$ with $k_1 < k_2$. We claim that $E_G^+(Y, X) = \emptyset$.

Suppose that there is an edge $(y, x) \in E(G)$ with $y \in Y$ and $x \in X$. All vertices in X are added to R in the second DFS before the first vertex of Y is added. In particular we have $x \in R$ and $y \notin R$ when the edge (y, x) is scanned in the second DFS. But this means that y is added to R before K is incremented, contradicting $comp(y) \neq comp(x)$.

Hence the *comp*-values computed by the STRONGLY CONNECTED COMPONENT ALGORITHM determine a topological order of the digraph resulting from contracting the strongly connected components. The second statement of the theorem now follows from Proposition 2.9 and the observation that a digraph is acyclic if and only if its strongly connected components are the singletons. $\square$

A linear-time algorithm that identifies the strongly connected components was first given by Karzanov [1970] and Tarjan [1972]. The problem of finding a topological order (or deciding that none exists) was solved earlier (Kahn [1962], Knuth [1968]). Both BFS and DFS occur as subroutines in many other combinatorial algorithms. Some examples will appear in later chapters.

Sometimes one is interested in higher connectivity. Let $k \geq 2$. An undirected graph with more than k vertices and the property that it remains connected even if we delete any $k - 1$ vertices, is called **k-connected**. A graph with at least two vertices is **k-edge-connected** if it remains connected after deleting any $k - 1$ edges. So a connected graph with at least three vertices is 2-connected (2-edge-connected) if and only if it has no articulation vertex (no bridge, respectively).

The largest k and l such that a graph G is k-connected and l-edge-connected are called the **vertex-connectivity** and **edge-connectivity** of G. Here we say that a graph is 1-connected (and 1-edge-connected) if it is connected. A disconnected graph has vertex-connectivity and edge-connectivity zero.

The **blocks** of an undirected graph are its maximal connected subgraphs without articulation vertex. Thus each block is either a maximal 2-connected subgraph, or consists of a bridge or an isolated vertex. Two blocks have at most one vertex in common, and a vertex belonging to more than one block is an articulation vertex. The blocks of an undirected graph can be determined in linear time quite similarly to the STRONGLY CONNECTED COMPONENT ALGORITHM; see Exercise 21. Here we prove a nice structure theorem for 2-connected graphs. We construct graphs from a single vertex by sequentially adding ears:

Definition 2.21. *Let G be a graph (directed or undirected). An* **ear-decomposition** *of G is a sequence $r, P_1, \ldots, P_k$ with $G = (\{r\}, \emptyset) + P_1 + \cdots + P_k$, such that each P_i is either a path where exactly the endpoints belong to $\{r\} \cup V(P_1) \cup \cdots \cup V(P_{i-1})$, or a circuit where exactly one of its vertices belongs to $\{r\} \cup V(P_1) \cup \cdots \cup V(P_{i-1})$ $(i \in \{1, \ldots, k\})$.*

$P_1, \ldots, P_k$ are called **ears**. *If $k \geq 1$, P_1 is a circuit of length at least three, and $P_2, \ldots, P_k$ are paths, then the ear-decomposition is called* **proper**.

Theorem 2.22. (Whitney [1932]) *An undirected graph is 2-connected if and only if it has a proper ear-decomposition.*

Proof: Evidently a circuit of length at least three is 2-connected. Moreover, if G is 2-connected, then so is $G + P$, where P is an x-y-path, $x, y \in V(G)$ and $x \neq y$: deleting any vertex does not destroy connectivity. We conclude that a graph with a proper ear-decomposition is 2-connected.

To show the converse, let G be a 2-connected graph. Let G' be the maximal simple subgraph of G; evidently G' is also 2-connected. Hence G' cannot be a tree; i.e. it contains a circuit. Since it is simple, G', and thus G, contains a circuit of length at least three. So let H be a maximal subgraph of G that has a proper ear-decomposition; H exists by the above consideration.

Suppose H is not spanning. Since G is connected, we then know that there exists an edge $e = \{x, y\} \in E(G)$ with $x \in V(H)$ and $y \notin V(H)$. Let z be a vertex in $V(H) \setminus \{x\}$. Since $G - x$ is connected, there exists a path P from y to z in $G - x$. Let z' be the first vertex on this path, when traversed from y, that belongs to $V(H)$. Then $P_{[y,z']} + e$ can be added as an ear, contradicting the maximality of H.

Thus H is spanning. Since each edge of $E(G) \setminus E(H)$ can be added as an ear, we conclude that $H = G$.   □

See Exercise 22 for similar characterizations of 2-edge-connected graphs and strongly connected digraphs.

2.4 Eulerian and Bipartite Graphs

Euler's work on the problem of traversing each of the seven bridges of Königsberg exactly once was the origin of graph theory. He showed that the problem had no solution by defining a graph, asking for a walk containing all edges, and observing that more than two vertices had odd degree.

Definition 2.23. *An **Eulerian walk** in a graph G is a closed walk containing every edge. An undirected graph G is called **Eulerian** if the degree of each vertex is even. A digraph G is Eulerian if $|\delta^-(v)| = |\delta^+(v)|$ for each $v \in V(G)$.*

Although Euler neither proved sufficiency nor considered the case explicitly in which we ask for a closed walk, the following famous result is usually attributed to him:

Theorem 2.24. (Euler [1736], Hierholzer [1873])   *A connected (directed or undirected) graph has an Eulerian walk if and only if it is Eulerian.*

Proof: The necessity of the degree conditions is obvious, as a vertex appearing k times in an Eulerian walk (or $k + 1$ times if it is the first and the last vertex) must have in-degree k and out-degree k, or degree $2k$ in the undirected case.

For the sufficiency, let $W = v_1, e_1, v_2, \ldots, v_k, e_k, v_{k+1}$ be a longest walk in G, i.e. one with maximum number of edges. In particular, W must contain all edges leaving v_{k+1}, which implies $v_{k+1} = v_1$ by the degree conditions. So W is closed. Suppose that W does not contain all edges. As G is connected, we then conclude that there is an edge $e \in E(G)$ for which e does not appear in W, but at least one of its endpoints appears in W, say v_i. Then e can be combined with $v_i, e_i, v_{i+1}, \ldots, e_k, v_{k+1} = v_1, e_1, v_2, \ldots, e_{i-1}, v_i$ to a walk which is longer than W.   □

The following algorithm accepts as input only connected Eulerian graphs. Note that one can check in linear time whether a given graph is connected (Theorem 2.17) and Eulerian (trivial). The algorithm first chooses an initial vertex, then calls a recursive procedure. We first describe it for undirected graphs:

EULER'S ALGORITHM

Input: An undirected connected Eulerian graph G.

Output: An Eulerian walk W in G.

①   Choose $v_1 \in V(G)$ arbitrarily. **Return** $W := \text{EULER}(v_1)$.

EULER(v_1)

①   Set $W := v_1$ and $x := v_1$.

②   **If** $\delta(x) = \emptyset$ **then go to** ④.
               Else let $e \in \delta(x)$, say $e = \{x, y\}$.

③   Set $W := W, e, y$ and $x := y$. Set $E(G) := E(G) \setminus \{e\}$ and **go to** ②.

④   Let $v_1, e_1, v_2, e_2, \ldots, v_k, e_k, v_{k+1}$ be the sequence W.
          For $i := 2$ **to** k **do:** Set $W_i := \text{EULER}(v_i)$.

⑤   **Return** $W := v_1, e_1, W_2, e_2, \ldots, W_k, e_k, v_{k+1}$.

For digraphs, ② has to be replaced by:

②   **If** $\delta^+(x) = \emptyset$ **then go to** ④.
               Else let $e \in \delta^+(x)$, say $e = (x, y)$.

We can analyze both versions (undirected and directed) simultaneously:

Theorem 2.25. EULER'S ALGORITHM *works correctly. Its running time is* $O(m)$,
where $m = |E(G)|$.

Proof: We show that EULER(v_1), if called for an Eulerian graph G and $v_1 \in V(G)$,
returns an Eulerian walk in the connected component G_1 of G that contains v_1.
We use induction on $|E(G)|$, the case $E(G) = \emptyset$ being trivial.

Because of the degree conditions, $v_{k+1} = x = v_1$ when ④ is reached. So at
this stage W is a closed walk. Let G' be the graph G at this stage. G' is also
Eulerian.

For each edge $e \in E(G_1) \cap E(G')$ there exists a minimum $i \in \{2, \ldots, k\}$ such
that e is in the same connected component of G' as v_i (note that $v_1 = v_{k+1}$ is
isolated in G'). Then by the induction hypothesis e belongs to W_i. So the closed
walk W composed in ⑤ is indeed an Eulerian walk in G_1.

The running time is linear because each edge is deleted immediately after being
examined. $\square$

EULER'S ALGORITHM will be used several times as a subroutine in later
chapters.

Sometimes one is interested in making a given graph Eulerian by adding or
contracting edges. Let G be an undirected graph and F a family of unordered pairs
of $V(G)$ (edges or not). F is called an **odd join** if $(V(G), E(G) \stackrel{.}{\cup} F)$ is Eulerian.
F is called an **odd cover** if the graph which results from G by contracting the
vertex set of each connected component of $(V(G), F)$ is Eulerian. Both concepts
are equivalent in the following sense.

Theorem 2.26. (Aoshima and Iri [1977]) *For any undirected graph we have:*

(a) *Every odd join is an odd cover.*
(b) *Every minimal odd cover is an odd join.*

Proof: Let G be an undirected graph.

To prove (a), let F be an odd join. We build a graph G' by contracting the connected components of $(V(G), F)$ in G. Each of these connected components contains an even number of odd-degree vertices (with respect to F and thus with respect to G, because F is an odd join). So the resulting graph has even degrees only. Thus F is an odd cover.

To prove (b), let F be a minimal odd cover. Because of the minimality, $(V(G), F)$ is a forest. We have to show that $|\delta_F(v)| \equiv |\delta_G(v)|$ (mod 2) for each $v \in V(G)$. So let $v \in V(G)$. Let $C_1, \ldots, C_k$ be the connected components of $(V(G), F) - v$ that contain a vertex w with $\{v, w\} \in F$. Since F is a forest, $k = |\delta_F(v)|$.

As F is an odd cover, contracting $X := V(C_1) \cup \cdots \cup V(C_k) \cup \{v\}$ in G yields a vertex of even degree, i.e. $|\delta_G(X)|$ is even. On the other hand, because of the minimality of F, $F \setminus \{\{v, w\}\}$ is not an odd cover (for any w with $\{v, w\} \in F$), so $|\delta_G(V(C_i))|$ is odd for $i = 1, \ldots, k$. Since

$$\sum_{i=1}^{k} |\delta_G(V(C_i))| \;=\; |\delta_G(X)| + |\delta_G(v)|$$

$$- 2|E_G(\{v\}, V(G) \setminus X)| + 2 \sum_{1 \leq i < j \leq k} |E_G(C_i, C_j)|,$$

we conclude that k has the same parity as $|\delta_G(v)|$. $\square$

We shall return to the problem of making a graph Eulerian in Section 12.2.

A **bipartition** of an undirected graph G consists of disjoint sets A and B whose union is $V(G)$ such that every edge has an endpoint in A and an endpoint in B. A graph is called **bipartite** if it has a bipartition. The simple bipartite graph G with $V(G) = A \cup B$, $|A| = n$, $|B| = m$ and $E(G) = \{\{a, b\} : a \in A, b \in B\}$ is denoted by $K_{n,m}$ (the complete bipartite graph). When we write $G = (A \cup B, E(G))$, we mean that $A \cup B$ is a bipartition of G.

Proposition 2.27. (König [1916]) *An undirected graph is bipartite if and only if it contains no odd circuit (circuit of odd length). There is a linear-time algorithm which, given an undirected graph G, either finds a bipartition or an odd circuit.*

Proof: Suppose G is bipartite with bipartition $V(G) = A \cup B$, and the closed walk $v_1, e_1, v_2, \ldots, v_k, e_k, v_{k+1}$ defines some circuit in G. W.l.o.g. $v_1 \in A$. But then $v_2 \in B$, $v_3 \in A$, and so on. We conclude that $v_i \in A$ if and only if i is odd. But $v_{k+1} = v_1 \in A$, so k must be even.

To prove the sufficiency, we may assume that G is connected, since a graph is bipartite iff each connected component is (and the connected components can be determined in linear time; Proposition 2.17). We choose an arbitrary vertex $s \in V(G)$ and apply BFS to (G, s) in order to obtain the distances from s to v for all $v \in V(G)$ (see Proposition 2.18). Let T be the resulting BFS-tree. Define $A := \{v \in V(G) : \text{dist}_G(s, v) \text{ is even}\}$ and $B := V(G) \setminus A$.

If there is an edge $e = \{x, y\}$ in $G[A]$ or $G[B]$, the x-y-path in T together with e forms an odd circuit in G. If there is no such edge, we have a bipartition.

$\square$

2.5 Planarity

We often draw graphs in the plane. A graph is called planar if it can be drawn such that no pair of edges intersect. To formalize this concept we need the following topological terms:

Definition 2.28. *A* **simple Jordan curve** *is the image of a continuous injective function* $\varphi : [0, 1] \to \mathbb{R}^2$; *its* **endpoints** *are* $\varphi(0)$ *and* $\varphi(1)$. *A* **closed Jordan curve** *is the image of a continuous function* $\varphi : [0, 1] \to \mathbb{R}^2$ *with* $\varphi(0) = \varphi(1)$ *and* $\varphi(\tau) \neq \varphi(\tau')$ *for* $0 \leq \tau < \tau' < 1$. *A* **polygonal arc** *is a simple Jordan curve which is the union of finitely many intervals (straight line segments). A* **polygon** *is a closed Jordan curve which is the union of finitely many intervals.*

Let $R = \mathbb{R}^2 \setminus J$, *where* J *is the union of finitely many intervals. We define the* **connected regions** *of* R *as equivalence classes where two points in* R *are equivalent if they can be joined by a polygonal arc within* R.

Definition 2.29. *A* **planar embedding** *of a graph* G *consists of an injective mapping* $\psi : V(G) \to \mathbb{R}^2$ *and for each* $e = \{x, y\} \in E(G)$ *a polygonal arc* J_e *with endpoints* $\psi(x)$ *and* $\psi(y)$, *such that for each* $e = \{x, y\} \in E(G)$:

$$(J_e \setminus \{\psi(x), \psi(y)\}) \cap \left(\{\psi(v) : v \in V(G)\} \cup \bigcup_{e' \in E(G) \setminus \{e\}} J_{e'} \right) = \emptyset.$$

A graph is called **planar** *if it has a planar embedding.*

Let G *be a (planar) graph with some fixed planar embedding* $\Phi = (\psi, (J_e)_{e \in E(G)})$. *After removing the points and polygonal arcs from the plane, the remainder,*

$$R := \mathbb{R}^2 \setminus \left(\{\psi(v) : v \in V(G)\} \cup \bigcup_{e \in E(G)} J_e \right),$$

splits into open connected regions, called **faces** *of* Φ.

For example, K_4 is obviously planar but it will turn out that K_5 is not planar. Exercise 29 shows that restricting ourselves to polygonal arcs instead of arbitrary Jordan curves makes no substantial difference. We will show later that for simple graphs it is indeed sufficient to consider straight line segments only.

Our aim is to characterize planar graphs. Following Thomassen [1981], we first prove the following topological fact, a version of the Jordan curve theorem:

Theorem 2.30. *If J is a polygon, then $\mathbb{R}^2 \setminus J$ splits into exactly two connected regions, each of which has J as its boundary. If J is a polygonal arc, then $\mathbb{R}^2 \setminus J$ has only one connected region.*

Proof: Let J be a polygon, $p \in \mathbb{R}^2 \setminus J$ and $q \in J$. Then there exists a polygonal arc in $(\mathbb{R}^2 \setminus J) \cup \{q\}$ joining p and q: starting from p, one follows the straight line towards q until one gets close to J, then one proceeds within the vicinity of J. (We use the elementary topological fact that disjoint compact sets, in particular non-adjacent intervals of J, have a positive distance from each other.) We conclude that p is in the same connected region of $\mathbb{R}^2 \setminus J$ as some points arbitrarily close to q.

J is the union of finitely many intervals; one or two of these intervals contain q. Let $\epsilon > 0$ such that the ball with center q and radius ϵ intersects no other interval of J; then clearly this ball intersects at most two connected regions. Since $p \in \mathbb{R}^2 \setminus J$ and $q \in J$ were chosen arbitrarily, we conclude that there are at most two regions and each region has J as its boundary.

Since the above also holds if J is a polygonal arc and q is an endpoint of J, $\mathbb{R}^2 \setminus J$ has only one connected region in this case.

Returning to the case when J is a polygon, it remains to prove that $\mathbb{R}^2 \setminus J$ has more than one region. For any $p \in \mathbb{R}^2 \setminus J$ and any angle α we consider the ray l_α starting at p with angle α. $J \cap l_\alpha$ is a set of points or closed intervals. Let $cr(p, l_\alpha)$ be the number of these points or intervals that J enters from a different side of l_α than to which it leaves (the number of times J "crosses" l_α; e.g. in Figure 2.4 we have $cr(p, l_\alpha) = 2$).

Note that for any angle α,

$$\left| \lim_{\epsilon \to 0, \, \epsilon > 0} cr(p, l_{\alpha - \epsilon}) - cr(p, l_\alpha) \right| \quad \text{and} \quad \left| \lim_{\epsilon \to 0, \, \epsilon > 0} cr(p, l_{\alpha + \epsilon}) - cr(p, l_\alpha) \right|$$

are even integers: twice the number of points and intervals of $J \cap l_\alpha$ that J enters from the same side to which it leaves (left side and right side, respectively). Therefore $g(p, \alpha) := (cr(p, l_\alpha) \bmod 2)$ is a continuous function in α, so it is constant and we denote it by $g(p)$. Clearly $g(p)$ is constant for points p on each straight line

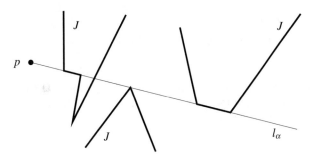

Fig. 2.4.

not intersecting J, so it is constant within each region. However, $g(p) \neq g(q)$ for points p, q such that the straight line segment joining p and q intersects J exactly once. Hence there are indeed two regions.   □

Exactly one of the faces, the **outer face**, is unbounded.

Proposition 2.31. *Let G be a 2-connected graph with a planar embedding Φ. Then every face is bounded by a circuit, and every edge is on the boundary of exactly two faces. Moreover, the number of faces is $|E(G)| - |V(G)| + 2$.*

Proof: By Theorem 2.30 both assertions are true if G is a circuit. For general 2-connected graphs we use induction on the number of edges, using Theorem 2.22. Consider a proper ear-decomposition of G, and let P be the last ear, a path with endpoints x and y, say. Let G' be the graph before adding the last ear, and let Φ' be the restriction of Φ to G'.

Let $\Phi = (\psi, (J_e)_{e \in E(G)})$. Let F' be the face of Φ' containing $\bigcup_{e \in E(P)} J_e \setminus \{\psi(x), \psi(y)\}$. By induction, F' is bounded by a circuit C. C contains x and y, so C is the union of two x-y-paths Q_1, Q_2 in G'. Now we apply Theorem 2.30 to each of the circuits $Q_1 + P$ and $Q_2 + P$. We conclude that

$$F' \cup \{\psi(x), \psi(y)\} = F_1 \dot{\cup} F_2 \dot{\cup} \bigcup_{e \in E(P)} J_e$$

and F_1 and F_2 are two faces of G bounded by the circuits $Q_1 + P$ and $Q_2 + P$, respectively. Hence G has one more face than G'. Using $|E(G) \setminus E(G')| = |V(G) \setminus V(G')| + 1$, this completes the induction step.   □

This proof is due to Tutte. It also implies easily that the circuits bounding the finite faces constitute a cycle basis (Exercise 30). The last statement of Proposition 2.31 is known as Euler's formula; it holds for general connected graphs:

Theorem 2.32. (Euler [1958], Legendre [1794]) *For any planar connected graph G with any embedding, the number of faces is $|E(G)| - |V(G)| + 2$.*

Proof: We have already proved the statement for 2-connected graphs (Proposition 2.31). Moreover, the assertion is trivial if $|V(G)| = 1$ and follows from Theorem 2.30 if $|E(G)| = 1$. If $|V(G)| = 2$ and $|E(G)| \geq 2$, then we can subdivide one edge e, thereby increasing the number of vertices and the number of edges by one and making the graph 2-connected, and apply Proposition 2.31.

So we may now assume that G has an articulation vertex x; we proceed by induction on the number of vertices. Let Φ be an embedding of G. Let $C_1, \ldots, C_k$ be the connected components of $G - x$; and let Φ_i be the restriction of Φ to $G_i := G[V(C_i) \cup \{x\}]$ for $i = 1, \ldots, k$.

The set of inner (bounded) faces of Φ is the disjoint union of the sets of inner faces of Φ_i, $i = 1, \ldots, k$. By applying the induction hypothesis to (G_i, Φ_i), $i = 1, \ldots, k$, we get that the total number of inner faces of (G, Φ) is

$$\sum_{i=1}^{k}(|E(G_i)|-|V(G_i)|+1) \;=\; |E(G)|-\sum_{i=1}^{k}|V(G_i)\setminus\{x\}| \;=\; |E(G)|-|V(G)|+1.$$

Taking the outer face into account concludes the proof. $\square$

In particular, the number of faces is independent of the embedding. The average degree of a simple planar graph is less than 6:

Corollary 2.33. *Let G be a 2-connected simple planar graph whose minimum circuit has length k (we also say that G has **girth** k). Then G has at most $(n-2)\frac{k}{k-2}$ edges. Any simple planar graph with $n \geq 3$ vertices has at most $3n - 6$ edges.*

Proof: First assume that G is 2-connected. Let some embedding Φ of G be given, and let r be the number of faces. By Euler's formula (Theorem 2.32), $r = |E(G)| - |V(G)| + 2$. By Proposition 2.31, each face is bounded by a circuit, i.e. by at least k edges, and each edge is on the boundary of exactly two faces. Hence $kr \leq 2|E(G)|$. Combining the two results we get $|E(G)|-|V(G)|+2 \leq \frac{2}{k}|E(G)|$, implying $|E(G)| \leq (n - 2)\frac{k}{k-2}$.

If G is not 2-connected we add edges between non-adjacent vertices to make it 2-connected while preserving planarity. By the first part we have at most $(n-2)\frac{3}{3-2}$ edges, including the new ones. $\square$

Now we show that certain graphs are non-planar:

Corollary 2.34.   *Neither K_5 nor $K_{3,3}$ is planar.*

Proof: This follows directly from Corollary 2.33: K_5 has five vertices but $10 > 3 \cdot 5 - 6$ edges; $K_{3,3}$ is 2-connected, has girth 4 (as it is bipartite) and $9 > (6-2)\frac{4}{4-2}$ edges. $\square$

Figure 2.5 shows these two graphs, which are the smallest non-planar graphs. We shall prove that every non-planar graph contains, in a certain sense, K_5 or $K_{3,3}$. To make this precise we need the following notion:

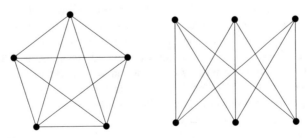

Fig. 2.5.

Definition 2.35. *Let G and H be two undirected graphs. G is a* **minor** *of H if there exists a subgraph H' of H and a partition $V(H') = V_1 \,\dot\cup\, \cdots \,\dot\cup\, V_k$ of its vertex set into connected subsets such that contracting each of $V_1, \ldots, V_k$ yields a graph which is isomorphic to G.*

In other words, G is a minor of H if it can be obtained from H by a series of operations of the following type: delete a vertex, delete an edge or contract an edge. Since neither of these operations destroys planarity, any minor of a planar graph is planar. Hence a graph which contains K_5 or $K_{3,3}$ as a minor cannot be planar. Kuratowski's Theorem says that the converse is also true. We first consider 3-connected graphs and start with the following lemma (which is the heart of Tutte's so-called wheel theorem):

Lemma 2.36. (Tutte [1961], Thomassen [1980]) *Let G be a 3-connected graph with at least five vertices. Then there exists an edge e such that G/e is also 3-connected.*

Proof: Suppose there is no such edge. Then for each edge $e = \{v, w\}$ there exists a vertex x such that $G - \{v, w, x\}$ is disconnected, i.e. has a connected component C with $|V(C)| < |V(G)| - 3$. Choose e, x and C such that $|V(C)|$ is minimum.

x has a neighbour y in C, because otherwise C is a connected component of $G - \{v, w\}$ (but G is 3-connected). By our assumption, $G/\{x, y\}$ is not 3-connected, i.e. there exists a vertex z such that $G - \{x, y, z\}$ is disconnected. Since $\{v, w\} \in E(G)$, there exists a connected component D of $G - \{x, y, z\}$ which contains neither v nor w.

But D contains a neighbour d of y, since otherwise D is a connected component of $G - \{x, z\}$ (again contradicting the fact that G is 3-connected). So $d \in V(D) \cap V(C)$, and thus D is a subgraph of C. Since $y \in V(C) \setminus V(D)$, we have a contradiction to the minimality of $|V(C)|$. $\square$

Theorem 2.37. (Kuratowski [1930], Wagner [1937]) *A 3-connected graph is planar if and only if it contains neither K_5 nor $K_{3,3}$ as a minor.*

Proof: As the necessity is evident (see above), we prove the sufficiency. Since K_4 is obviously planar, we proceed by induction on the number of vertices: let G be a 3-connected graph with more than four vertices but no K_5 or $K_{3,3}$ minor.

By Lemma 2.36, there exists an edge $e = \{v, w\}$ such that G/e is 3-connected. Let $\Phi = \left(\psi, (J_{e'})_{e' \in E(G/e)}\right)$ be a planar embedding of G/e, which exists by induction. Let x be the vertex in G/e which arises by contracting e. Consider $(G/e) - x$ with the restriction of Φ as a planar embedding. Since $(G/e) - x$ is 2-connected, every face is bounded by a circuit (Proposition 2.31). In particular, the face containing the point $\psi(x)$ is bounded by a circuit C.

Let $y_1, \ldots, y_k \in V(C)$ be the neighbours of v that are distinct from w, numbered in cyclic order, and partition C into edge-disjoint paths P_i, $i = 1, \ldots, k$, such that P_i is a y_i-y_{i+1}-path ($y_{k+1} := y_1$).

Suppose there exists an index $i \in \{1, \ldots, k\}$ such that $\Gamma(w) \subseteq \{v\} \cup V(P_i)$. Then a planar embedding of G can be constructed easily by modifying Φ.

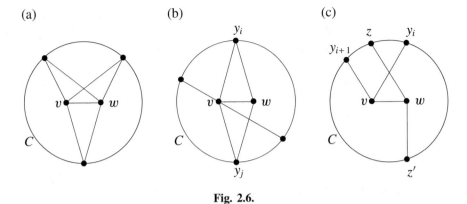

Fig. 2.6.

We shall prove that all other cases are impossible. First, if w has three neighbours among $y_1, \ldots, y_k$, we have a K_5 minor (Figure 2.6(a)).

Next, if $\Gamma(w) = \{v, y_i, y_j\}$ for some $i < j$, then we must have $i + 1 < j$ and $(i, j) \neq (1, k)$ (otherwise y_i and y_j would both lie on P_i or P_j); see Figure 2.6(b). Otherwise there is a neighbour z of w in $V(P_i) \setminus \{y_i, y_{i+1}\}$ for some i and another neighbour $z' \notin V(P_i)$ (Figure 2.6(c)). In both cases, there are four vertices y, z, y', z' on C, in this cyclic order, with $y, y' \in \Gamma(v)$ and $z, z' \in \Gamma(w)$. This implies that we have a $K_{3,3}$ minor. □

The proof implies quite directly that every 3-connected simple planar graph has a planar embedding where each edge is embedded by a straight line and each face, except the outer face, is convex (Exercise 33(a)). The general case of Kuratowski's Theorem can be reduced to the 3-connected case by gluing together planar embeddings of the maximal 3-connected subgraphs, or by the following lemma:

Lemma 2.38. (Thomassen [1980]) *Let G be a graph with at least five vertices which is not 3-connected and which contains neither K_5 nor $K_{3,3}$ as a minor. Then there exist two non-adjacent vertices $v, w \in V(G)$ such that $G + e$, where $e = \{v, w\}$ is a new edge, does not contain a K_5 or $K_{3,3}$ minor either.*

Proof: We use induction on $|V(G)|$. Let G be as above. Without loss of generality, G is simple. If G is disconnected, we can simply add an edge e joining two different connected components. So henceforth we assume that G is connected. Since G is not 3-connected, there exists a set $X = \{x, y\}$ of two vertices such that $G - X$ is disconnected. (If G is not even 2-connected we may choose x to be an articulation vertex and y a neighbour of x.) Let C be a connected component of $G - X$, $G_1 := G[V(C) \cup X]$ and $G_2 := G - V(C)$. We first prove the following:

Claim: Let $v, w \in V(G_1)$ be two vertices such that adding an edge $e = \{v, w\}$ to G creates a $K_{3,3}$ or K_5 minor. Then at least one of $G_1 + e + f$ and $G_2 + f$ contains a K_5 or $K_{3,3}$ minor, where f is a new edge joining x and y.

To prove this claim, let $v, w \in V(G_1)$, $e = \{v, w\}$ and suppose that there are pairwise disjoint connected vertex sets $Z_1, \ldots, Z_t$ of $G + e$ such that after contracting each of them we have a K_5 ($t = 5$) or $K_{3,3}$ ($t = 6$) subgraph.

Note that it is impossible that $Z_i \subseteq V(G_1) \setminus X$ and $Z_j \subseteq V(G_2) \setminus X$ for some $i, j \in \{1, \ldots, t\}$: in this case the set of those Z_k with $Z_k \cap X \neq \emptyset$ (there are at most two of these) separate Z_i and Z_j, contradicting the fact that both K_5 and $K_{3,3}$ are 3-connected.

Hence there are two cases: If none of $Z_1, \ldots, Z_t$ is a subset of $V(G_2) \setminus X$, then $G_1 + e + f$ also contains a K_5 or $K_{3,3}$ minor: just consider $Z_i \cap V(G_1)$ ($i = 1, \ldots, t$).

Analogously, if none of $Z_1, \ldots, Z_t$ is a subset of $V(G_1) \setminus X$, then $G_2 + f$ contains a K_5 or $K_{3,3}$ minor (consider $Z_i \cap V(G_2)$ ($i = 1, \ldots, t$)).

The claim is proved. Now we first consider the case when G contains an articulation vertex x, and y is a neighbour of x. We choose a second neighbour z of x such that y and z are in different connected components of $G - x$. W.l.o.g. say that $z \in V(G_1)$. Suppose that the addition of $e = \{y, z\}$ creates a K_5 or $K_{3,3}$ minor. By the claim, at least one of $G_1 + e$ and G_2 contains a K_5 or $K_{3,3}$ minor (an edge $\{x, y\}$ is already present). But then G_1 or G_2, and thus G, contains a K_5 or $K_{3,3}$ minor, contradicting our assumption.

Hence we may assume that G is 2-connected. Recall that $x, y \in V(G)$ were chosen such that $G - \{x, y\}$ is disconnected. If $\{x, y\} \notin E(G)$ we simply add an edge $f = \{x, y\}$. If this creates a K_5 or $K_{3,3}$ minor, the claim implies that $G_1 + f$ or $G_2 + f$ contains such a minor. Since there is an x-y-path in each of G_1, G_2 (otherwise we would have an articulation vertex of G), this implies that there is a K_5 or $K_{3,3}$ minor in G which is again a contradiction.

Thus we can assume that $f = \{x, y\} \in E(G)$. Suppose now that at least one of the graphs G_i ($i \in \{1, 2\}$) is not planar. Then this G_i has at least five vertices. Since it does not contain a K_5 or $K_{3,3}$ minor (this would also be a minor of G), we conclude from Theorem 2.37 that G_i is not 3-connected. So we can apply the induction hypothesis to G_i. By the claim, if adding an edge within G_i does not introduce a K_5 or $K_{3,3}$ minor in G_i, it cannot introduce such a minor in G either.

So we may assume that both G_1 and G_2 are planar; let Φ_1 and Φ_2 be planar embeddings. Let F_i be a face of Φ_i with f on its boundary, and let z_i be another vertex on the boundary of F_i, $z_i \notin \{x, y\}$ ($i = 1, 2$). We claim that adding an edge $\{z_1, z_2\}$ (cf. Figure 2.7) does not introduce a K_5 or $K_{3,3}$ minor.

Suppose, on the contrary, that adding $\{z_1, z_2\}$ and contracting some pairwise disjoint connected vertex sets $Z_1, \ldots, Z_t$ would create a K_5 ($t = 5$) or $K_{3,3}$ ($t = 6$) subgraph.

First suppose that at most one of the sets Z_i is a subset of $V(G_1) \setminus \{x, y\}$. Then the graph G_2', arising from G_2 by adding one vertex w and edges from w to x, y and z_2, also contains a K_5 or $K_{3,3}$ minor. (Here w corresponds to the contracted set $Z_i \subseteq V(G_1) \setminus \{x, y\}$.) This is a contradiction since there is a planar embedding of G_2': just supplement Φ_2 by placing w within F_2.

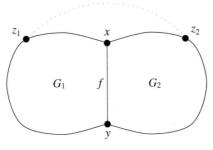

Fig. 2.7.

So we may assume that $Z_1, Z_2 \subseteq V(G_1) \setminus \{x, y\}$. Analogously, we may assume that $Z_3, Z_4 \subseteq V(G_2) \setminus \{x, y\}$. W.l.o.g. we have $z_1 \notin Z_1$ and $z_2 \notin Z_3$. Then we cannot have a K_5, because Z_1 and Z_3 are not adjacent. Moreover, the only possible common neighbours of Z_1 and Z_3 are Z_5 and Z_6. Since in $K_{3,3}$ two vertices are either adjacent or have three common neighbours, a $K_{3,3}$ minor is also impossible. $\qquad \square$

Theorem 2.37 and Lemma 2.38 yield Kuratowski's Theorem:

Theorem 2.39. (Kuratowski [1930], Wagner [1937]) *An undirected graph is planar if and only if it contains neither K_5 nor $K_{3,3}$ as a minor.* $\qquad \square$

Indeed, Kuratowski proved a stronger version (Exercise 34). The proof can be turned into a polynomial-time algorithm quite easily (Exercise 33(b)). In fact, a linear-time algorithm exists:

Theorem 2.40. (Hopcroft and Tarjan [1974]) *There is a linear-time algorithm for finding a planar embedding of a given graph or deciding that it is not planar.*

2.6 Planar Duality

We shall now introduce an important duality concept. In this section, graphs may contain loops, i.e. edges whose endpoints coincide. In a planar embedding loops are of course represented by polygons instead of polygonal arcs.

Note that Euler's formula (Theorem 2.32) also holds for graphs with loops: this follows from the observation that subdividing a loop e (i.e. replacing $e = \{v, v\}$ by two parallel edges $\{v, w\}$, $\{w, v\}$ where w is a new vertex) and adjusting the embedding (replacing the polygon J_e by two polygonal arcs whose union is J_e) increases the number of edges and vertices each by one but does not change the number of faces.

Definition 2.41. *Let G be a directed or undirected graph, possibly with loops, and let $\Phi = (\psi, (J_e)_{e \in E(G)})$ be a planar embedding of G. We define the **planar***

dual G^* *whose vertices are the faces of* Φ *and whose edge set is* $\{e^* : e \in E(G)\}$, *where* e^* *connects the faces that are adjacent to* J_e *(if* J_e *is adjacent to only one face, then* e^* *is a loop). In the directed case, say for* $e = (v, w)$, *we orient* $e^* = (F_1, F_2)$ *in such a way that* F_1 *is the face "to the right" when traversing* J_e *from* $\psi(v)$ *to* $\psi(w)$.

G^* is again planar. In fact, there obviously exists a planar embedding $(\psi^*, (J_{e^*})_{e^* \in E(G^*)})$ of G^* such that $\psi^*(F) \in F$ for all faces F of Φ and, for each $e \in E(G)$,

$$
J_{e^*} \cap \left(\{\psi(v) : v \in V(G)\} \cup \bigcup_{f \in E(G) \setminus \{e\}} J_f \right) = \emptyset,
$$

$|J_{e^*} \cap J_e| = 1$, and if e^* is a loop then the face bounded by J_{e^*} contains exactly one endpoint of e. Such an embedding is called a **standard embedding** of G^*.

The planar dual of a graph really depends on the embedding: consider the two embeddings of the same graph shown in Figure 2.8. The resulting planar duals are not isomorphic, since the second one has a vertex of degree four (corresponding to the outer face) while the first one is 3-regular.

Proposition 2.42. *Let G be an undirected connected planar graph with a fixed embedding. Let* G^* *be its planar dual with a standard embedding. Then* $(G^*)^* = G$.

Proof: Let $(\psi, (J_e)_{e \in E(G)})$ be a fixed embedding of G and $(\psi^*, (J_{e^*})_{e^* \in E(G^*)})$ a standard embedding of G^*. Let F be a face of G^*. The boundary of F contains J_{e^*} for at least one edge e^*, so F must contain $\psi(v)$ for one endpoint v of e. So every face of G^* contains at least one vertex of G.

By applying Euler's formula (Theorem 2.32) to G^* and to G, we get that the number of faces of G^* is $|E(G^*)| - |V(G^*)| + 2 = |E(G)| - (|E(G)| - |V(G)| + 2) + 2 = |V(G)|$. Hence each face of G^* contains exactly one vertex of G. From this we conclude that the planar dual of G^* is isomorphic to G.   □

The requirement that G is connected is essential here: note that G^* is always connected, even if G is disconnected.

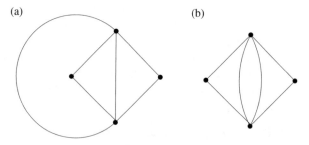

(a)   (b)

Fig. 2.8.

Theorem 2.43. *Let G be a connected planar undirected graph with arbitrary embedding. The edge set of any circuit in G corresponds to a minimal cut in G^*, and any minimal cut in G corresponds to the edge set of a circuit in G^*.*

Proof: Let $\Phi = (\psi, (J_e)_{e \in E(G)})$ be a fixed planar embedding of G. Let C be a circuit in G. By Theorem 2.30, $\mathbb{R}^2 \setminus \bigcup_{e \in E(C)} J_e$ splits into exactly two connected regions. Let A and B be the set of faces of Φ in the inner and outer region, respectively. We have $V(G^*) = A \dot\cup B$ and $E_{G^*}(A, B) = \{e^* : e \in E(C)\}$. Since A and B form connected sets in G^*, this is indeed a minimal cut.

Conversely, let $\delta_G(A)$ be a minimal cut in G. Let $\Phi^* = (\psi^*, (J_e)_{e \in E(G^*)})$ be a standard embedding of G^*. Let $a \in A$ and $b \in V(G) \setminus A$. Observe that there is no polygonal arc in

$$R := \mathbb{R}^2 \setminus \left(\{\psi^*(v) : v \in V(G^*)\} \cup \bigcup_{e \in \delta_G(A)} J_{e^*} \right)$$

which connects $\psi(a)$ and $\psi(b)$: the sequence of faces of G^* passed by such a polygonal arc would define an edge progression from a to b in G not using any edge of $\delta_G(A)$.

So R consists of at least two connected regions. Then, obviously, the boundary of each region must contain a circuit. Hence $F := \{e^* : e \in \delta_G(A)\}$ contains the edge set of a circuit C in G^*. We have $\{e^* : e \in E(C)\} \subseteq \{e^* : e \in F\} = \delta_G(A)$, and, by the first part, $\{e^* : e \in E(C)\}$ is a minimal cut in $(G^*)^* = G$ (cf. Proposition 2.42). We conclude that $\{e^* : e \in E(C)\} = \delta_G(A)$. $\square$

In particular, e^* is a loop if and only if e is a bridge, and vice versa. For digraphs the above proof yields:

Corollary 2.44. *Let G be a connected planar digraph with some fixed planar embedding. The edge set of any circuit in G corresponds to a minimal directed cut in G^*, and vice versa.* $\square$

Another interesting consequence of Theorem 2.43 is:

Corollary 2.45. *Let G be a connected undirected graph with arbitrary planar embedding. Then G is bipartite if and only if G^* is Eulerian, and G is Eulerian if and only if G^* is bipartite.*

Proof: Observe that a connected graph is Eulerian if and only if every minimal cut has even cardinality. By Theorem 2.43, G is bipartite if G^* is Eulerian, and G is Eulerian if G^* is bipartite. By Proposition 2.42, the converse is also true. $\square$

An **abstract dual** of G is a graph G' for which there is a bijection $\chi : E(G) \to E(G')$ such that F is the edge set of a circuit iff $\chi(F)$ is a minimal cut in G' and vice versa. Theorem 2.43 shows that any planar dual is also an abstract dual. The converse is not true. However, Whitney [1933] proved that a graph has an abstract dual if and only if it is planar (Exercise 40). We shall return to this duality relation when dealing with matroids in Section 13.3.

Exercises

1. Let G be a simple undirected graph on n vertices which is isomorphic to its complement. Show that $n \bmod 4 \in \{0, 1\}$.

2. Let G be an undirected graph. For $X \subseteq V(G)$ let $f(x) := |E(G[X])|$. Show that $f : 2^{V(G)} \to \mathbb{Z}$ is supermodular.

3. Prove that every simple undirected graph G with $|\delta(v)| \geq \frac{1}{2}|V(G)|$ for all $v \in V(G)$ is Hamiltonian.
   Hint: Consider a longest path in G and the neighbours of its endpoints.
   (Dirac [1952])

4. Prove that any simple undirected graph G with $|E(G)| > \binom{|V(G)|-1}{2}$ is connected.

5. Let G be a simple undirected graph. Show that G or its complement is connected.

6. Prove that every simple undirected graph with more than one vertex contains two vertices that have the same degree. Prove that every tree (except a single vertex) contains at least two leaves.

7. Let T be a tree with k leaves. Show that T contains at most $k - 2$ vertices of degree at least 3.

8. Prove that every tree T contains a vertex v such that no connected component of $T - v$ contains more than $\frac{|V(T)|}{2}$ vertices. Can you find such a vertex in linear time?

9. Let G be a connected undirected graph, and let $(V(G), F)$ be a forest in G. Prove that there is a spanning tree $(V(G), T)$ with $F \subseteq T \subseteq E(G)$.

10. Let (V, F_1) and (V, F_2) be two forests with $|F_1| < |F_2|$. Prove that there exists an edge $e \in F_2 \setminus F_1$ such that $(V, F_1 \cup \{e\})$ is a forest.

11. Let (V, F_1) and (V, F_2) be two branchings with $2|F_1| < |F_2|$. Prove that there exists an edge $e \in F_2 \setminus F_1$ such that $(V, F_1 \cup \{e\})$ is a branching.

12. Prove that any cut in an undirected graph is the disjoint union of minimal cuts.

13. Let G be an undirected graph, C a circuit and D a cut. Show that $|E(C) \cap D|$ is even.

14. Show that any undirected graph has a cut containing at least half of the edges.

15. Let $(U, \mathcal{F})$ be a cross-free set system with $|U| \geq 2$. Prove that $\mathcal{F}$ contains at most $4|U| - 4$ distinct elements.

16. Let G be a connected undirected graph. Show that there exists an orientation G' of G and a spanning arborescence T of G' such that the set of fundamental circuits with respect to T is precisely the set of directed circuits in G'.
    Hint: Consider a DFS-tree.
    (Camion [1968])

17. Describe a linear-time algorithm for the following problem: Given an adjacency list of a graph G, compute an adjacency list of the maximal simple subgraph of G. Do not assume that parallel edges appear consecutively in the input.

18. Given a graph G (directed or undirected), show that there is a linear-time algorithm to find a circuit or decide that none exists.

19. Describe a simple linear-time algorithm that finds a topological order in a given acyclic digraph. (Do not use the STRONGLY CONNECTED COMPONENT ALGORITHM).

20. Let G be a connected undirected graph, $s \in V(G)$ and T a DFS-tree resulting from running DFS on (G, s). s is called the root of T. x is an ancestor of y in T if x lies on the (unique) s-y-path in T. x is the parent of y if the edge $\{x, y\}$ lies on the s-y-path in T. y is a child (successor) of x if x is the parent (an ancestor) of y. Note that with this definition each vertex is an ancestor and a successor of itself. Every vertex except s has exactly one parent. Prove:
   (a) For any edge $\{v, w\} \in E(G)$, v is an ancestor or a successor of w in T.
   (b) A vertex v is an articulation vertex of G if and only if
   • either $v = s$ and $|\delta_T(v)| > 1$
   • or $v \neq s$ and there is a child w of v such that no edge in G connects a proper ancestor of v (that is, excluding v) with a successor of w.

∗ 21. Use Exercise 20 to design a linear-time algorithm which finds the blocks of an undirected graph. It will be useful to compute numbers

$$\alpha(x) := \min\{f(w) : w = x \text{ or } \{w, y\} \in E(G) \backslash T \text{ for some successor } y \text{ of } x\}$$

recursively during the DFS. Here (R, T) is the DFS-tree (with root s), and the f-values represent the order in which the vertices are added to R (see the GRAPH SCANNING ALGORITHM). If for some vertex $x \in R \setminus \{s\}$ we have $\alpha(x) \geq f(w)$, where w is the parent of x, then w must be either the root or an articulation vertex.

22. Prove:
   (a) An undirected graph is 2-edge-connected if and only if it has at least two vertices and an ear-decomposition.
   (b) A digraph is strongly connected if and only if it has an ear-decomposition.
   (c) The edges of an undirected graph G with at least two vertices can be oriented such that the resulting digraph is strongly connected if and only if G is 2-edge-connected.
   (Robbins [1939])

23. A tournament is a digraph such that the underlying undirected graph is a (simple) complete graph. Prove that every tournament contains a Hamiltonian path (Rédei [1934]). Prove that every strongly connected tournament is Hamiltonian (Camion [1959]).

24. Let G be an undirected graph. Prove that there exists an orientation G' of G such that $||\delta_{G'}^+(v)| - |\delta_{G'}^-(v)|| \leq 1$ for all $v \in V(G')$.

25. Prove that if a connected undirected simple graph is Eulerian then its line graph is Hamiltonian. What about the converse?

26. Prove that any connected bipartite graph has a unique bipartition. Prove that any non-bipartite undirected graph contains an odd circuit as an induced subgraph. Prove that an undirected graph G is bipartite if and only if $E(G)$ can be partitioned into cuts.

27. Prove that a strongly connected digraph whose underlying undirected graph is non-bipartite contains a (directed) circuit of odd length.

* 28. Let G be an undirected graph. A **tree-decomposition** of G is a pair (T, φ), where T is a tree and $\varphi : V(T) \to 2^{V(G)}$ satisfies the following conditions:
    - for each $e \in E(G)$ there exists a $t \in V(T)$ with $e \subseteq \varphi(t)$;
    - for each $v \in V(G)$ the set $\{t \in V(T) : v \in \varphi(t)\}$ is connected in T.

    We say that the width of (T, φ) is $\max_{t \in V(T)} |\varphi(t)| - 1$. The **tree-width** of a graph G is the minimum width of a tree-decomposition of G. This notion is due to Robertson and Seymour [1986].

    Show that the simple graphs of tree-width at most 1 are the forests. Moreover, prove that the following statements are equivalent for an undirected graph G:
    (a) G has tree-width at most 2;
    (b) G does not contain K_4 as a minor;
    (c) G can be obtained from an empty graph by successively adding bridges and doubling and subdividing edges. (Doubling an edge $e = \{v, w\} \in E(G)$ means adding another edge with endpoints v and w; subdividing an edge $e = \{v, w\} \in E(G)$ means adding a vertex x and replacing e by two edges $\{v, x\}, \{x, w\}$.)

    Note: Because of the construction in (c) such graphs are called series-parallel.
29. Show that if a graph G has a planar embedding where the edges are embedded by arbitrary Jordan curves, then it also has a planar embedding with polygonal arcs only.
30. Let G be a 2-connected graph with a planar embedding. Show that the set of circuits bounding the finite faces constitutes a cycle basis of G.
31. Can you generalize Euler's formula (Theorem 2.32) to disconnected graphs?
32. Show that there are exactly five Platonic graphs (corresponding to the Platonic solids; cf. Exercise 11 of Chapter 4), i.e. 3-connected planar regular graphs whose faces are all bounded by the same number of edges.
    Hint: Use Euler's formula (Theorem 2.32).
33. Deduce from the proof of Kuratowski's Theorem 2.39:
    (a) Every 3-connected simple planar graph has a planar embedding where each edge is embedded by a straight line and each face, except the outer face, is convex.
    (b) There is a polynomial-time algorithm for checking whether a given graph is planar.
* 34. Given a graph G and an edge $e = \{v, w\} \in E(G)$, we say that H results from G by subdividing e if $V(H) = V(G) \cup \{x\}$ and $E(H) = (E(G) \setminus \{e\}) \cup \{\{v, x\}, \{x, w\}\}$. A graph resulting from G by successively subdividing edges is called a **subdivision** of G.
    (a) Trivially, if H contains a subdivision of G then G is a minor of H. Show that the converse is not true.
    (b) Prove that a graph containing a $K_{3,3}$ or K_5 minor also contains a subdivision of $K_{3,3}$ or K_5.
        Hint: Consider what happens when contracting one edge.
    (c) Conclude that a graph is planar if and only if no subgraph is a subdivision of $K_{3,3}$ or K_5.
    (Kuratowski [1930])

35. Prove that each of the following statements implies the other:
    (a) For every infinite sequence of graphs $G_1, G_2, \ldots$ there are two indices $i < j$ such that G_i is a minor of G_j.
    (b) Let $\mathcal{G}$ be a class of graphs such that for each $G \in \mathcal{G}$ and each minor H of G we have $H \in \mathcal{G}$ (i.e. membership in $\mathcal{G}$ is a hereditary graph property). Then there exists a finite set $\mathcal{X}$ of graphs such that $\mathcal{G}$ consists of all graphs that do not contain any element of $\mathcal{X}$ as a minor.
    Note: The statements have been proved by Robertson and Seymour [2004]; they are a main result of their series of papers on graph minors. Theorem 2.39 and Exercise 28 give examples of forbidden minor characterizations as in (b).
36. Let G be a planar graph with an embedding Φ, and let C be a circuit of G bounding some face of Φ. Prove that then there is an embedding Φ' of G such that C bounds the outer face.
37. (a) Let G be disconnected with an arbitrary planar embedding, and let G^* be the planar dual with a standard embedding. Prove that $(G^*)^*$ arises from G by successively applying the following operation, until the graph is connected: Choose two vertices x and y which belong to different connected components and which are adjacent to the same face; contract $\{x, y\}$.
    (b) Generalize Corollary 2.45 to arbitrary planar graphs.
    Hint: Use (a) and Theorem 2.26.
38. Let G be a connected digraph with a fixed planar embedding, and let G^* be the planar dual with a standard embedding. How are G and $(G^*)^*$ related?
39. Prove that if a planar digraph is acyclic (strongly connected), then its planar dual is strongly connected (acyclic). What about the converse?
40. (a) Show that if G has an abstract dual and H is a minor of G then H also has an abstract dual.
  * (b) Show that neither K_5 nor $K_{3,3}$ has an abstract dual.
    (c) Conclude that a graph is planar if and only if it has an abstract dual. (Whitney [1933])

References

General Literature

Berge, C. [1985]: Graphs. Second Edition. Elsevier, Amsterdam 1985
Bollobás, B. [1998]: Modern Graph Theory. Springer, New York 1998
Bondy, J.A. [1995]: Basic graph theory: paths and circuits. In: Handbook of Combinatorics; Vol. 1 (R.L. Graham, M. Grötschel, L. Lovász, eds.), Elsevier, Amsterdam 1995
Bondy, J.A., and Murty, U.S.R. [2008]: Graph Theory. Springer, New York 2008
Diestel, R. [2010]: Graph Theory. Fourth Edition. Springer, New York 2010
Wilson, R.J. [2010]: Introduction to Graph Theory. Fifth Edition. Addison-Wesley, Reading 2010

Cited References

Aoshima, K., and Iri, M. [1977]: Comments on F. Hadlock's paper: finding a maximum cut of a planar graph in polynomial time. SIAM Journal on Computing 6 (1977), 86–87

Camion, P. [1959]: Chemins et circuits hamiltoniens des graphes complets. Comptes Rendus Hebdomadaires des Séances de l'Académie des Sciences (Paris) 249 (1959), 2151–2152

Camion, P. [1968]: Modulaires unimodulaires. Journal of Combinatorial Theory A 4 (1968), 301–362

Dirac, G.A. [1952]: Some theorems on abstract graphs. Proceedings of the London Mathematical Society 2 (1952), 69–81

Edmonds, J., and Giles, R. [1977]: A min-max relation for submodular functions on graphs. In: Studies in Integer Programming; Annals of Discrete Mathematics 1 (P.L. Hammer, E.L. Johnson, B.H. Korte, G.L. Nemhauser, eds.), North-Holland, Amsterdam 1977, pp. 185–204

Euler, L. [1736]: Solutio problematis ad geometriam situs pertinentis. Commentarii Academiae Petropolitanae 8 (1736), 128–140

Euler, L. [1758]: Demonstratio nonnullarum insignium proprietatum quibus solida hedris planis inclusa sunt praedita. Novi Commentarii Academiae Petropolitanae 4 (1758), 140–160

Hierholzer, C. [1873]: Über die Möglichkeit, einen Linienzug ohne Wiederholung und ohne Unterbrechung zu umfahren. Mathematische Annalen 6 (1873), 30–32

Hopcroft, J.E., and Tarjan, R.E. [1974]: Efficient planarity testing. Journal of the ACM 21 (1974), 549–568

Kahn, A.B. [1962]: Topological sorting of large networks. Communications of the ACM 5 (1962), 558–562

Karzanov, A.V. [1970]: An efficient algorithm for finding all the bi-components of a graph. In: Trudy 3-ĭ Zimneĭ Shkoly po Matematicheskomu Programmirovaniyu i Smezhnym Voprosam (Drogobych, 1970), Issue 2, Moscow Institute for Construction Engineering (MISI) Press, Moscow, 1970, pp. 343–347 [in Russian]

Knuth, D.E. [1968]: The Art of Computer Programming; Vol. 1. Fundamental Algorithms. Addison-Wesley, Reading 1968 (third edition: 1997)

König, D. [1916]: Über Graphen und ihre Anwendung auf Determinantentheorie und Mengenlehre. Mathematische Annalen 77 (1916), 453–465

König, D. [1936]: Theorie der endlichen und unendlichen Graphen. Teubner, Leipzig 1936; reprint: Chelsea Publishing Co., New York 1950

Kuratowski, K. [1930]: Sur le problème des courbes gauches en topologie. Fundamenta Mathematicae 15 (1930), 271–283

Legendre, A.M. [1794]: Éléments de Géométrie. Firmin Didot, Paris 1794

Minty, G.J. [1960]: Monotone networks. Proceedings of the Royal Society of London A 257 (1960), 194–212

Moore, E.F. [1959]: The shortest path through a maze. Proceedings of the International Symposium on the Theory of Switching; Part II. Harvard University Press 1959, pp. 285–292

Rédei, L. [1934]: Ein kombinatorischer Satz. Acta Litt. Szeged 7 (1934), 39–43

Robbins, H.E. [1939]: A theorem on graphs with an application to a problem of traffic control. American Mathematical Monthly 46 (1939), 281–283

Robertson, N., and Seymour, P.D. [1986]: Graph minors II: algorithmic aspects of tree-width. Journal of Algorithms 7 (1986), 309–322

Robertson, N., and Seymour, P.D. [2004]: Graph minors XX: Wagner's conjecture. Journal of Combinatorial Theory B 92 (2004), 325–357

Tarjan, R.E. [1972]: Depth first search and linear graph algorithms. SIAM Journal on Computing 1 (1972), 146–160

Thomassen, C. [1980]: Planarity and duality of finite and infinite graphs. Journal of Combinatorial Theory B 29 (1980), 244–271

Thomassen, C. [1981]: Kuratowski's theorem. Journal of Graph Theory 5 (1981), 225–241

Tutte, W.T. [1961]: A theory of 3-connected graphs. Proceedings of the Koninklijke Nederlandse Akademie van Wetenschappen A 64 (1961), 441–455

Wagner, K. [1937]: Über eine Eigenschaft der ebenen Komplexe. Mathematische Annalen 114 (1937), 570–590

Whitney, H. [1932]: Non-separable and planar graphs. Transactions of the American Mathematical Society 34 (1932), 339–362

Whitney, H. [1933]: Planar graphs. Fundamenta Mathematicae 21 (1933), 73–84

3 Linear Programming

In this chapter we review the most important facts about Linear Programming. Although this chapter is self-contained, it cannot be considered to be a comprehensive treatment of the field. The reader unfamiliar with Linear Programming is referred to the textbooks mentioned at the end of this chapter.

The general problem reads as follows:

LINEAR PROGRAMMING

Instance: A matrix $A \in \mathbb{R}^{m \times n}$ and column vectors $b \in \mathbb{R}^m, c \in \mathbb{R}^n$.

Task: Find a column vector $x \in \mathbb{R}^n$ such that $Ax \leq b$ and $c^\top x$ is maximum, decide that $\{x \in \mathbb{R}^n : Ax \leq b\}$ is empty, or decide that for all $\alpha \in \mathbb{R}$ there is an $x \in \mathbb{R}^n$ with $Ax \leq b$ and $c^\top x > \alpha$.

Here $c^\top x$ denotes the scalar product of the vectors. The notion $x \leq y$ for vectors x and y (of equal size) means that the inequality holds in each component. If no sizes are specified, the matrices and vectors are always assumed to be compatible in size. We often omit indicating the transposition of column vectors and write e.g. cx for the scalar product. By 0 we denote the number zero as well as all-zero vectors and all-zero matrices (the order will always be clear from the context).

A **linear program (LP)** is an instance of the above problem. We often write a linear program as $\max\{cx : Ax \leq b\}$. A **feasible solution** of an LP $\max\{cx : Ax \leq b\}$ is a vector x with $Ax \leq b$. A feasible solution attaining the maximum is called an **optimum solution**.

As the problem formulation indicates, there are two possibilities when an LP has no solution: The problem can be **infeasible** (i.e. $P := \{x \in \mathbb{R}^n : Ax \leq b\} = \emptyset$) or **unbounded** (i.e. for all $\alpha \in \mathbb{R}$ there is an $x \in P$ with $cx > \alpha$). If an LP is neither infeasible nor unbounded it has an optimum solution:

Proposition 3.1. *Let* $P = \{x \in \mathbb{R}^n : Ax \leq b\} \neq \emptyset$ *and* $c \in \mathbb{R}^n$ *with* $\delta := \sup\{c^\top x : x \in P\} < \infty$. *Then there exists a vector* $z \in P$ *with* $c^\top z = \delta$.

Proof: Let U be a matrix whose columns are an orthonormal basis of the kernel of A, i.e. $U^\top U = I$, $AU = 0$, and $\mathrm{rank}(A') = n$ where $A' := \begin{pmatrix} A \\ U^\top \end{pmatrix}$. Let $b' := \begin{pmatrix} b \\ 0 \end{pmatrix}$.

We show that for every $y \in P$ there exists a subsystem $A''x \leq b''$ of $A'x \leq b'$ such that A'' is nonsingular, $y' := (A'')^{-1}b'' \in P$, and $c^\top y' \geq c^\top y$. As there

© Springer-Verlag GmbH Germany 2018
Bernhard Korte and Jens Vygen, *Combinatorial Optimization*, Algorithms and Combinatorics 21, https://doi.org/10.1007/978-3-662-56039-6_3

are only finitely many such subsystems, one of these y' attains the maximum $(c^\top y' = \delta)$, and the assertion follows.

So let $y \in P$, and denote by $k(y)$ the rank of A'' for the maximal subsystem $A''x \le b''$ of $A'x \le b'$ with $A''y = b''$. Suppose that $k(y) < n$. We show how to find a $y' \in P$ with $c^\top y' \ge c^\top y$ and $k(y') > k(y)$. After at most n steps we have a vector y' with $k(y') = n$ as required.

If $U^\top y \ne 0$, we set $y' := y - UU^\top y$. Since $y + \lambda UU^\top c \in P$ for all $\lambda \in \mathbb{R}$ we have $\sup\{c^\top (y + \lambda UU^\top c) : \lambda \in \mathbb{R}\} \le \delta < \infty$ and hence $c^\top U = 0$ and $c^\top y' = c^\top y$. Moreover, $Ay' = Ay - AUU^\top y = Ay$ and $U^\top y' = U^\top y - U^\top UU^\top y = 0$.

Now suppose that $U^\top y = 0$. Let $v \ne 0$ with $A''v = 0$. Denote by $a_i x \le \beta_i$ the i-th row of $Ax \le b$. Let $\mu := \min\left\{\frac{\beta_i - a_i y}{a_i v} : a_i v > 0\right\}$ and $\kappa := \max\left\{\frac{\beta_i - a_i y}{a_i v} : a_i v < 0\right\}$, where $\min \emptyset = \infty$ and $\max \emptyset = -\infty$. We have $\kappa \le 0 \le \mu$, and at least one of κ and μ is finite (because $A'v \ne 0$ but $U^\top v = 0$).

For $\lambda \in \mathbb{R}$ with $\kappa \le \lambda \le \mu$ we have $A''(y + \lambda v) = A''y + \lambda A''v = A''y = b''$ and $A(y + \lambda v) = Ay + \lambda Av \le b$, i.e. $y + \lambda v \in P$. Thus, as $\sup\{c^\top x : x \in P\} < \infty$, we have $\mu < \infty$ if $c^\top v > 0$ and $\kappa > -\infty$ if $c^\top v < 0$.

Moreover, if $c^\top v \ge 0$ and $\mu < \infty$, we have $a_i(y + \mu v) = \beta_i$ for some i. Analogously, if $c^\top v \le 0$ and $\kappa > -\infty$, we have $a_i(y + \kappa v) = \beta_i$ for some i.

Thus in each case we have found a vector $y' \in P$ with $c^\top y' \ge c^\top y$ and $k(y') \ge k(y) + 1$. □

This justifies the notation $\max\{c^\top x : Ax \le b\}$ instead of $\sup\{c^\top x : Ax \le b\}$.

Many combinatorial optimization problems can be formulated as LPs. To do this, we encode the feasible solutions as vectors in $\mathbb{R}^n$ for some n. In Section 3.5 we show that one can optimize a linear objective function over a finite set S of vectors by solving a linear program. Although the feasible set of this LP contains not only the vectors in S but also all their convex combinations, one can show that among the optimum solutions there is always an element of S.

In Section 3.1 we compile some terminology and basic facts about polyhedra, the sets $P = \{x \in \mathbb{R}^n : Ax \le b\}$ of feasible solutions of LPs. In Sections 3.2 and 3.3 we present the SIMPLEX ALGORITHM, which we also use to derive the Duality Theorem and related results (Section 3.4). LP duality is a most important concept which explicitly or implicitly appears in almost all areas of combinatorial optimization; we shall often refer to the results in Sections 3.4 and 3.5.

3.1 Polyhedra

Linear Programming deals with maximizing or minimizing a linear objective function of finitely many variables subject to finitely many linear inequalities. So the set of feasible solutions is the intersection of finitely many halfspaces. Such a set is called a polyhedron:

Definition 3.2. *A **polyhedron** in $\mathbb{R}^n$ is a set of type $P = \{x \in \mathbb{R}^n : Ax \leq b\}$ for some matrix $A \in \mathbb{R}^{m \times n}$ and some vector $b \in \mathbb{R}^m$. If A and b are rational, then P is a **rational** polyhedron. A bounded polyhedron is also called a **polytope**.*

*We denote by $\operatorname{rank}(A)$ the rank of a matrix A. The **dimension** $\dim X$ of a nonempty set $X \subseteq \mathbb{R}^n$ is defined to be*

$$n - \max\{\operatorname{rank}(A) : A \text{ is an } n \times n\text{-matrix with } Ax = Ay \text{ for all } x, y \in X\}.$$

*A polyhedron $P \subseteq \mathbb{R}^n$ is called **full-dimensional** if $\dim P = n$.*

Equivalently, a polyhedron is full-dimensional if and only if there is a point in its interior. For most of this chapter it makes no difference whether we are in the rational or real space. We need the following standard terminology:

Definition 3.3. *Let $P := \{x : Ax \leq b\}$ be a nonempty polyhedron. If c is a nonzero vector for which $\delta := \max\{cx : x \in P\}$ is finite, then $\{x : cx = \delta\}$ is called a **supporting hyperplane** of P. A **face** of P is P itself or the intersection of P with a supporting hyperplane of P. A point x for which $\{x\}$ is a face is called a **vertex** of P, and also a **basic solution** of the system $Ax \leq b$.*

Proposition 3.4. *Let $P = \{x : Ax \leq b\}$ be a polyhedron and $F \subseteq P$. Then the following statements are equivalent:*

(a) *F is a face of P.*
(b) *There exists a vector c such that $\delta := \max\{cx : x \in P\}$ is finite and $F = \{x \in P : cx = \delta\}$.*
(c) *$F = \{x \in P : A'x = b'\} \neq \emptyset$ for some subsystem $A'x \leq b'$ of $Ax \leq b$.*

Proof: (a) and (b) are obviously equivalent.

(c)$\Rightarrow$(b): If $F = \{x \in P : A'x = b'\}$ is nonempty, let c be the sum of the rows of A', and let δ be the sum of the components of b'. Then obviously $cx \leq \delta$ for all $x \in P$ and $F = \{x \in P : cx = \delta\}$.

(b)$\Rightarrow$(c): Assume that c is a vector, $\delta := \max\{cx : x \in P\}$ is finite and $F = \{x \in P : cx = \delta\}$. Let $A'x \leq b'$ be the maximal subsystem of $Ax \leq b$ such that $A'x = b'$ for all $x \in F$. Let $A''x \leq b''$ be the rest of the system $Ax \leq b$.

We first observe that for each inequality $a_i''x \leq \beta_i''$ of $A''x \leq b''$ ($i = 1, \ldots, k$) there is a point $x_i \in F$ such that $a_i''x_i < \beta_i''$. Let $x^* := \frac{1}{k}\sum_{i=1}^{k} x_i$ be the center of gravity of these points (if $k = 0$, we can choose an arbitrary $x^* \in F$); we have $x^* \in F$ and $a_i''x^* < \beta_i''$ for all i.

We have to prove that $A'y = b'$ cannot hold for any $y \in P \setminus F$. So let $y \in P \setminus F$. We have $cy < \delta$. Now consider $z := x^* + \epsilon(x^* - y)$ for some small $\epsilon > 0$; in particular let ϵ be smaller than $\frac{\beta_i'' - a_i''x^*}{a_i''(x^* - y)}$ for all $i \in \{1, \ldots, k\}$ with $a_i''x^* > a_i''y$.

We have $cz > \delta$ and thus $z \notin P$. So there is an inequality $ax \leq \beta$ of $Ax \leq b$ such that $az > \beta$. Thus $ax^* > ay$. The inequality $ax \leq \beta$ cannot belong to $A''x \leq b''$, since otherwise we have $az = ax^* + \epsilon a(x^* - y) < ax^* + \frac{\beta - ax^*}{a(x^* - y)}a(x^* - y) = \beta$

(by the choice of ϵ). Hence the inequality $ax \leq \beta$ belongs to $A'x \leq b'$. Since $ay = a(x^* + \frac{1}{\epsilon}(x^* - z)) < \beta$, this completes the proof. $\qquad \square$

As a trivial but important corollary we remark:

Corollary 3.5. *If* $\max\{cx : x \in P\}$ *is bounded for a nonempty polyhedron P and a vector c, then the set of points where the maximum is attained is a face of P.* $\qquad \square$

The relation "is a face of" is transitive:

Corollary 3.6. *Let P be a polyhedron and F a face of P. Then F is again a polyhedron. Furthermore, a set* $F' \subseteq F$ *is a face of P if and only if it is a face of F.* $\qquad \square$

The maximal faces distinct from P are particularly important:

Definition 3.7. *Let P be a polyhedron. A* **facet** *of P is a maximal face distinct from P. An inequality* $cx \leq \delta$ *is* **facet-defining** *for P if* $cx \leq \delta$ *for all* $x \in P$ *and* $\{x \in P : cx = \delta\}$ *is a facet of P.*

Proposition 3.8. *Let* $P \subseteq \{x \in \mathbb{R}^n : Ax = b\}$ *be a nonempty polyhedron of dimension* $n - \mathrm{rank}(A)$. *Let* $A'x \leq b'$ *be a minimal inequality system such that* $P = \{x : Ax = b, A'x \leq b'\}$. *Then each inequality of* $A'x \leq b'$ *is facet-defining for P, and each facet of P is defined by an inequality of* $A'x \leq b'$.

Proof: If $P = \{x \in \mathbb{R}^n : Ax = b\}$, then there are no facets and the statement is trivial. So let $A'x \leq b'$ be a minimal inequality system with $P = \{x : Ax = b, A'x \leq b'\}$, let $a'x \leq \beta'$ be one of its inequalities and $A''x \leq b''$ be the rest of the system $A'x \leq b'$. Let y be a vector with $Ay = b$, $A''y \leq b''$ and $a'y > \beta'$ (such a vector y exists as the inequality $a'x \leq \beta'$ is not redundant). Let $x \in P$ such that $A'x < b'$ (such a vector must exist because $\dim P = n - \mathrm{rank}(A)$).

Consider $z := x + \frac{\beta' - a'x}{a'y - a'x}(y - x)$. We have $a'z = \beta'$, $A''z < b''$, and, since $0 < \frac{\beta' - a'x}{a'y - a'x} < 1$, $z \in P$. Therefore $F := \{x \in P : a'x = \beta'\} \neq 0$ and $F \neq P$ (as $x \in P \setminus F$). We conclude that F is a facet of P.

By Proposition 3.4 each facet is defined by an inequality of $A'x \leq b'$. $\qquad \square$

The other important class of faces (beside facets) are minimal faces (i.e. faces not containing any other face). Here we have:

Proposition 3.9. (Hoffman and Kruskal [1956]) *Let* $P = \{x : Ax \leq b\}$ *be a polyhedron. A nonempty subset* $F \subseteq P$ *is a minimal face of P if and only if* $F = \{x : A'x = b'\}$ *for some subsystem* $A'x \leq b'$ *of* $Ax \leq b$.

Proof: If F is a minimal face of P, by Proposition 3.4 there is a subsystem $A'x \leq b'$ of $Ax \leq b$ such that $F = \{x \in P : A'x = b'\}$. We choose $A'x \leq b'$ maximal. Let $A''x \leq b''$ be a minimal subsystem of $Ax \leq b$ such that $F = \{x : A'x = b', A''x \leq b''\}$. We claim that $A''x \leq b''$ does not contain any inequality.

Suppose, on the contrary, that $a''x \leq \beta''$ is an inequality of $A''x \leq b''$. Since it is not redundant for the description of F, Proposition 3.8 implies that $F' := \{x : A'x = b', A''x \leq b'', a''x = \beta''\}$ is a facet of F. By Corollary 3.6 F' is also a face of P, contradicting the assumption that F is a minimal face of P.

Now let $\emptyset \neq F = \{x : A'x = b'\} \subseteq P$ for some subsystem $A'x \leq b'$ of $Ax \leq b$. Obviously F has no faces except itself. By Proposition 3.4, F is a face of P. It follows by Corollary 3.6 that F is a minimal face of P. □

Corollary 3.5 and Proposition 3.9 imply that LINEAR PROGRAMMING can be solved in finite time by solving the linear equation system $A'x = b'$ for each subsystem $A'x \leq b'$ of $Ax \leq b$. A more intelligent way is the SIMPLEX ALGORITHM which is described in the next section.

Another consequence of Proposition 3.9 is:

Corollary 3.10. *Let $P = \{x \in \mathbb{R}^n : Ax \leq b\}$ be a polyhedron. Then all minimal faces of P have dimension $n - \text{rank}(A)$. The minimal faces of polytopes are vertices.* □

This is why polyhedra $\{x \in \mathbb{R}^n : Ax \leq b\}$ with $\text{rank}(A) = n$ are called **pointed**: their minimal faces are points.

Let us close this section with some remarks on polyhedral cones.

Definition 3.11. *A (convex) **cone** is a set $C \subseteq \mathbb{R}^n$ for which $x, y \in C$ and $\lambda, \mu \geq 0$ implies $\lambda x + \mu y \in C$. A **cone** C is said to be **generated** by $x_1, \ldots, x_k$ if $x_1, \ldots, x_k \in C$ and for any $x \in C$ there are numbers $\lambda_1, \ldots, \lambda_k \geq 0$ with $x = \sum_{i=1}^k \lambda_i x_i$. A cone is called **finitely generated** if some finite set of vectors generates it. A **polyhedral cone** is a polyhedron of type $\{x : Ax \leq 0\}$.*

It is immediately clear that polyhedral cones are indeed cones. We shall now show that polyhedral cones are finitely generated. I always denotes an identity matrix.

Lemma 3.12. *(Minkowski [1896]) Let $C = \{x \in \mathbb{R}^n : Ax \leq 0\}$ be a polyhedral cone. Then C is generated by a subset of the set of solutions to the systems $My = b'$, where M consists of n linearly independent rows of $\left(\begin{smallmatrix} A \\ I \end{smallmatrix} \right)$ and $b' = \pm e_j$ for some unit vector e_j.*

Proof: Let A be an $m \times n$-matrix. Consider the systems $My = b'$ where M consists of n linearly independent rows of $\left(\begin{smallmatrix} A \\ I \end{smallmatrix} \right)$ and $b' = \pm e_j$ for some unit vector e_j. Let $y_1, \ldots, y_t$ be those solutions of these equality systems that belong to C. We claim that C is generated by $y_1, \ldots, y_t$.

First suppose $C = \{x : Ax = 0\}$, i.e. C is a linear subspace. Write $C = \{x : A'x = 0\}$ where A' consists of a maximal set of linearly independent rows of A. Let I' consist of some rows of I such that $\left(\begin{smallmatrix} A' \\ I' \end{smallmatrix} \right)$ is a nonsingular square matrix.

Then C is generated by the solutions of

$$\begin{pmatrix} A' \\ I' \end{pmatrix} x = \begin{pmatrix} 0 \\ b \end{pmatrix}, \quad \text{for } b = \pm e_j, \ j = 1, \dots, \dim C.$$

For the general case we use induction on the dimension of C. If C is not a linear subspace, then there is a vector $z \in C$ such that $-z \notin C$. Then there is a row a of A such that $az < 0$.

Let A' consist of any maximal set of rows of A such that (i) the rows of $\begin{pmatrix} A' \\ a \end{pmatrix}$ are linearly independent, and (ii) there exists a vector $z \in C$ with $A'z = 0$ and $az < 0$.

Let y be any vector with $A'y = 0$ and $ay = -1$. We claim that $y \in C$.

Let z satisfy (ii), i.e., $z \in C$, $A'z = 0$ and $az < 0$. Let B be the set of rows b of A with $by > 0$. Each $b \in B$ must be linearly independent of a and A': otherwise $b = cA' + \delta a$ for a vector c and a number δ, but then $0 \ge bz = cA'z + \delta az = \delta az$, hence $\delta \ge 0$, contradicting $0 < by = cA'y + \delta ay = -\delta$.

Suppose that B is nonempty. Let $\mu := \min\{\frac{bz}{by} : b \in B\}$. We have $\mu \le 0$. Then $z' := z - \mu y \in C$, $A'z' = A'z - \mu A'y = 0$, $az' = az - \mu ay < 0$, and there is a $b' \in B$ with $b'z' = 0$. This contradicts the maximality of A'. So $B = \emptyset$, i.e., $y \in C$.

Hence by construction there is an index $s \in \{1, \dots, t\}$ such that $A'y_s = 0$ and $ay_s = -1$.

Now let an arbitrary $z \in C$ be given. Let $a_1, \dots, a_m$ be the rows of A and $\mu := \min\left\{\frac{a_i z}{a_i y_s} : i = 1, \dots, m, \ a_i y_s < 0\right\}$. We have $\mu \ge 0$. Let k be an index where the minimum is attained. Consider $z' := z - \mu y_s$. By the definition of μ we have $a_j z' = a_j z - \frac{a_k z}{a_k y_s} a_j y_s$ for $j = 1, \dots, m$, and hence $z' \in C' := \{x \in C : a_k x = 0\}$. C' is a cone whose dimension is one less than that of C (because $a_k y_s < 0$ and $y_s \in C$). By induction, C' is generated by a subset of $y_1, \dots, y_t$, so $z' = \sum_{i=1}^{t} \lambda_i y_i$ for some $\lambda_1, \dots, \lambda_t \ge 0$. By setting $\lambda'_s := \lambda_s + \mu$ (observe that $\mu \ge 0$) and $\lambda'_i := \lambda_i$ $(i \ne s)$, we obtain $z = z' + \mu y_s = \sum_{i=1}^{t} \lambda'_i y_i$. $\qquad\square$

Thus any polyhedral cone is finitely generated. We shall show the converse at the end of Section 3.4.

3.2 The Simplex Algorithm

The oldest and best-known algorithm for LINEAR PROGRAMMING is Dantzig's [1951] simplex method. We first assume that the polyhedron has a vertex, and that some vertex is given as input. Later we shall show how general LPs can be solved with this method.

For a set J of row indices we write A_J for the submatrix of A consisting of the rows in J only, and b_J for the subvector of b consisting of the components with indices in J. We abbreviate $a_i := A_{\{i\}}$ and $\beta_i := b_{\{i\}}$.

SIMPLEX ALGORITHM

Input: A matrix $A \in \mathbb{R}^{m \times n}$ and column vectors $b \in \mathbb{R}^m, c \in \mathbb{R}^n$.
             A vertex x of $P := \{x \in \mathbb{R}^n : Ax \leq b\}$.

Output: A vertex x of P attaining $\max\{cx : x \in P\}$ or a vector $w \in \mathbb{R}^n$ with
             $Aw \leq 0$ and $cw > 0$ (i.e. the LP is unbounded).

①   Choose a set of n row indices J such that A_J is nonsingular and $A_J x = b_J$.

②   Compute $c (A_J)^{-1}$ and add zeros in order to obtain a vector y with $c = yA$
     such that all entries of y outside J are zero.
     If $y \geq 0$ then stop. Return x and y.

③   Choose the minimum index i with $y_i < 0$.
     Let w be the column of $-(A_J)^{-1}$ with index i, so $A_{J \setminus \{i\}} w = 0$ and
     $a_i w = -1$.
     If $Aw \leq 0$ then stop.
     Return w.

④   Let $\lambda := \min \left\{ \dfrac{\beta_j - a_j x}{a_j w} : j \in \{1, \dots, m\}, a_j w > 0 \right\}$,
     and let j be the smallest row index attaining this minimum.

⑤   Set $J := (J \setminus \{i\}) \cup \{j\}$ and $x := x + \lambda w$.
     Go to ②.

Step ① relies on Proposition 3.9 and can be implemented with GAUSSIAN ELIMINATION (Section 4.3). The selection rules for i and j in ③ and ④ (often called pivot rule) are due to Bland [1977]. If one just chose an arbitrary i with $y_i < 0$ and an arbitrary j attaining the minimum in ④ the algorithm would run into cyclic repetitions for some instances. Bland's pivot rule is not the only one that avoids cycling; another one (the so-called lexicographic rule) was proved to avoid cycling already by Dantzig, Orden and Wolfe [1955]. Before proving the correctness of the SIMPLEX ALGORITHM, let us make the following observation (sometimes known as "weak duality"):

Proposition 3.13. *Let x and y be feasible solutions of the LPs*

$$\max\{cx : Ax \leq b\} \qquad and \qquad (3.1)$$
$$\min\{yb : y^\top A = c^\top, y \geq 0\}, \qquad (3.2)$$

respectively. Then $cx \leq yb$.

Proof: $cx = (yA)x = y(Ax) \leq yb$. □

Theorem 3.14. (Dantzig [1951], Dantzig, Orden and Wolfe [1955], Bland [1977]) *The SIMPLEX ALGORITHM terminates after at most $\binom{m}{n}$ iterations. If it returns x and y in ②, these vectors are optimum solutions of the LPs (3.1) and (3.2), respectively, with $cx = yb$. If the algorithm returns w in ③ then $cw > 0$ and the LP (3.1) is unbounded.*

Proof: We first prove that the following conditions hold at any stage of the algorithm:

(a) $x \in P$;
(b) $A_J x = b_J$;
(c) A_J is nonsingular;
(d) $cw > 0$;
(e) $\lambda \geq 0$.

(a) and (b) hold initially. ② and ③ guarantee $cw = yAw = -y_i > 0$. By ④, $x \in P$ implies $\lambda \geq 0$. (c) follows from the fact that $A_{J \setminus \{i\}} w = 0$ and $a_j w > 0$. It remains to show that ⑤ preserves (a) and (b).

We show that if $x \in P$, then also $x + \lambda w \in P$. For a row index k we have two cases: If $a_k w \leq 0$ then (using $\lambda \geq 0$) $a_k(x + \lambda w) \leq a_k x \leq \beta_k$. Otherwise $\lambda \leq \frac{\beta_k - a_k x}{a_k w}$ and hence $a_k(x + \lambda w) \leq a_k x + a_k w \frac{\beta_k - a_k x}{a_k w} = \beta_k$. (Indeed, λ is chosen in ④ to be the largest number such that $x + \lambda w \in P$.)

To show (b), note that after ④ we have $A_{J \setminus \{i\}} w = 0$ and $\lambda = \frac{\beta_j - a_j x}{a_j w}$, so $A_{J \setminus \{i\}}(x + \lambda w) = A_{J \setminus \{i\}} x = b_{J \setminus \{i\}}$ and $a_j(x + \lambda w) = a_j x + a_j w \frac{\beta_j - a_j x}{a_j w} = \beta_j$. Therefore after ⑤, $A_J x = b_J$ holds again.

So we indeed have (a)–(e) at any stage. If the algorithm returns x and y in ②, x and y are feasible solutions of (3.1) and (3.2), respectively. x is a vertex of P by (a), (b) and (c). Moreover, $cx = yAx = yb$ since the components of y are zero outside J. This proves the optimality of x and y by Proposition 3.13.

If the algorithm stops in ③, the LP (3.1) is indeed unbounded because in this case $x + \mu w \in P$ for all $\mu \geq 0$, and $cw > 0$ by (d).

We finally show that the algorithm terminates. Let $J^{(k)}$ and $x^{(k)}$ be the set J and the vector x in iteration k of the SIMPLEX ALGORITHM, respectively. If the algorithm did not terminate after $\binom{m}{n}$ iterations, there are iterations $k < l$ with $J^{(k)} = J^{(l)}$. By (b) and (c), $x^{(k)} = x^{(l)}$. By (d) and (e), cx never decreases, and it strictly increases if $\lambda > 0$. Hence λ is zero in all the iterations $k, k+1, \ldots, l-1$, and $x^{(k)} = x^{(k+1)} = \cdots = x^{(l)}$.

Let h be the highest index leaving J in one of the iterations $k, \ldots, l-1$, say in iteration p. Index h must also have been added to J in some iteration $q \in \{k, \ldots, l-1\}$. Now let y' be the vector y at iteration p, and let w' be the vector w at iteration q. We have $y'Aw' = cw' > 0$. So let r be an index for which $y'_r a_r w' > 0$. Since $y'_r \neq 0$, index r belongs to $J^{(p)}$. If $r > h$, index r would also belong to $J^{(q)}$ and $J^{(q+1)}$, implying $a_r w' = 0$. So $r \leq h$. But by the choice of i in iteration p we have $y'_r < 0$ iff $r = h$, and by the choice of j in iteration q we have $a_r w' > 0$ iff $r = h$ (recall that $\lambda = 0$ and $a_r x^{(q)} = a_r x^{(p)} = \beta_r$ as $r \in J^{(p)}$). This is a contradiction.   □

Klee and Minty [1972] and Avis and Chvátal [1978] found examples where the SIMPLEX ALGORITHM (with Bland's rule) needs 2^n iterations on LPs with n variables and $2n$ constraints, proving that it is not a polynomial-time algorithm. It is not known whether there is a pivot rule that leads to a polynomial-time algorithm. However, Borgwardt [1982] showed that the average running time (for random instances in a certain natural probabilistic model) can be bounded by a polynomial. Spielman and Teng [2004] introduced a so-called smoothed analysis (cf. Section 17.5): for each input they consider the expected running time with respect to small random perturbations of the input. The maximum of all these expectations is polynomially bounded. Kelner and Spielman [2006] proposed a randomized polynomial-time algorithm for LINEAR PROGRAMMING that is similar to the SIMPLEX ALGORITHM. The SIMPLEX ALGORITHM is also quite fast in practice if implemented skilfully; see Section 3.3.

We now show how to solve general linear programs with the SIMPLEX AL-GORITHM. More precisely, we show how to find an initial vertex. Since there are polyhedra that do not have vertices at all, we put a given LP into a different form first.

Let $\max\{cx : Ax \leq b\}$ be an LP. We substitute x by $y - z$ and write it equivalently in the form

$$\max\left\{ (c \quad -c)\begin{pmatrix} y \\ z \end{pmatrix} : (A \quad -A)\begin{pmatrix} y \\ z \end{pmatrix} \leq b, \ y, z \geq 0 \right\}.$$

So w.l.o.g. we assume that our LP has the form

$$\max\{cx : A'x \leq b', A''x \leq b'', x \geq 0\} \tag{3.3}$$

with $b' \geq 0$ and $b'' < 0$. We first run the SIMPLEX ALGORITHM on the instance

$$\min\{(\mathbb{1}A'')x + \mathbb{1}y : A'x \leq b', A''x + y \geq b'', x, y \geq 0\}, \tag{3.4}$$

where $\mathbb{1}$ denotes a vector whose entries are all 1. Since $\begin{pmatrix} x \\ y \end{pmatrix} = 0$ defines a vertex, this is possible. The LP is obviously not unbounded since the minimum must be at least $\mathbb{1}b''$. For any feasible solution x of (3.3), $\begin{pmatrix} x \\ b''-A''x \end{pmatrix}$ is an optimum solution of (3.4) of value $\mathbb{1}b''$. Hence if the minimum of (3.4) is greater than $\mathbb{1}b''$, then (3.3) is infeasible.

In the contrary case, let $\begin{pmatrix} x \\ y \end{pmatrix}$ be an optimum vertex of (3.4) of value $\mathbb{1}b''$. We claim that x is a vertex of the polyhedron defined by (3.3). To see this, first observe that $A''x + y = b''$. Let n and m be the dimensions of x and y, respectively; then by Proposition 3.9 there is a set S of $n + m$ inequalities of (3.4) satisfied with equality, such that the submatrix corresponding to these $n + m$ inequalities is nonsingular.

Let S' be the inequalities of $A'x \leq b'$ and of $x \geq 0$ that belong to S. Let S'' consist of those inequalities of $A''x \leq b''$ for which the corresponding inequalities of $A''x + y \geq b''$ and $y \geq 0$ both belong to S. Obviously $|S' \cup S''| \geq |S| - m = n$, and the inequalities of $S' \cup S''$ are linearly independent and satisfied by x with equality. Hence x satisfies n linearly independent inequalities of (3.3) with equality; thus x is indeed a vertex. Therefore we can start the SIMPLEX ALGORITHM with (3.3) and x.

3.3 Implementation of the Simplex Algorithm

The previous description of the SIMPLEX ALGORITHM is simple but not suitable for an efficient implementation. As we will see, it is not necessary to solve a linear equation system in each iteration. To motivate the main idea, we start with a proposition (which is actually not needed later): for LPs of the form $\max\{cx : Ax = b, x \geq 0\}$, vertices can be represented not only by subsets of rows but also by subsets of columns.

For a matrix A and a set J of column indices we denote by A^J the submatrix consisting of the columns in J only. Consequently, A_I^J denotes the submatrix of A with rows in I and columns in J. Sometimes the order of the rows and columns is important: if $J = (j_1, \ldots, j_k)$ is a vector of row (column) indices, we denote by A_J (A^J) the matrix whose i-th row (column) is the j_i-th row (column) of A ($i = 1, \ldots, k$).

Proposition 3.15. *Let $P := \{x : Ax = b, x \geq 0\}$, where A is a matrix and b is a vector. Then x is a vertex of P if and only if $x \in P$ and the columns of A corresponding to positive entries of x are linearly independent.*

Proof: Let A be an $m \times n$-matrix. Let $X := \left(\begin{smallmatrix} -I & 0 \\ A & I \end{smallmatrix}\right)$ and $b' := \left(\begin{smallmatrix} 0 \\ b \end{smallmatrix}\right)$. Let $N := \{1, \ldots, n\}$ and $M := \{n+1, \ldots, n+m\}$. For an index set $J \subseteq N \cup M$ with $|J| = n$ let $\bar{J} := (N \cup M) \setminus J$. Then X_J^N is nonsingular iff $X_{M \cap J}^{N \cap \bar{J}}$ is nonsingular iff $X_M^{\bar{J}}$ is nonsingular.

If x is a vertex of P, then – by Proposition 3.9 – there exists a set $J \subseteq N \cup M$ such that $|J| = n$, X_J^N is nonsingular, and $X_J^N x = b_J'$. Then the components of x corresponding to $N \cap J$ are zero. Moreover, $X_M^{\bar{J}}$ is nonsingular, and hence the columns of $A^{N \cap \bar{J}}$ are linearly independent.

Conversely, let $x \in P$, and let the set of columns of A corresponding to positive entries of x be linearly independent. By adding suitable unit column vectors to these columns we obtain a nonsingular submatrix X_M^B with $x_i = 0$ for $i \in N \setminus B$. Then $X_{\bar{B}}^N$ is nonsingular and $X_{\bar{B}}^N x = b_{\bar{B}}'$. Hence, by Proposition 3.9, x is a vertex of P. □

Corollary 3.16. *Let $\left(\begin{smallmatrix} x \\ y \end{smallmatrix}\right) \in P := \left\{\left(\begin{smallmatrix} x \\ y \end{smallmatrix}\right) : Ax + y = b, x \geq 0, y \geq 0\right\}$. Then $\left(\begin{smallmatrix} x \\ y \end{smallmatrix}\right)$ is a vertex of P if and only if the columns of $(A\ I)$ corresponding to positive components of $\left(\begin{smallmatrix} x \\ y \end{smallmatrix}\right)$ are linearly independent. Moreover, x is a vertex of $\{x : Ax \leq b, x \geq 0\}$ if and only if $\left(\begin{smallmatrix} x \\ b-Ax \end{smallmatrix}\right)$ is a vertex of P.* □

We will now analyze the behaviour of the SIMPLEX ALGORITHM when applied to an LP of the form $\max\{cx : Ax \leq b, x \geq 0\}$.

Theorem 3.17. *Let $A \in \mathbb{R}^{m \times n}$, $b \in \mathbb{R}^m$, and $c \in \mathbb{R}^n$. Let $A' := \left(\begin{smallmatrix} -I \\ A \end{smallmatrix}\right)$, $b' := \left(\begin{smallmatrix} 0 \\ b \end{smallmatrix}\right)$ and $\bar{c} := (c^\top, 0)$. Let $B \in \{1, \ldots, n+m\}^m$ such that $(A\ I)^B$ is nonsingular. Let $J \subseteq \{1, \ldots, n+m\}$ be the set of the remaining n indices. Let $Q_B := ((A\ I)^B)^{-1}$.*

Then:

(a) A'_J *is nonsingular.*

(b) $(b' - A'x)_J = 0$ *and* $(b' - A'x)_B = Q_B b$ *and* $c^\top x = \bar{c}^B Q_B b$, *where* $x :=$ $(A'_J)^{-1} b'_J$.

(c) *Let* y *be the vector with* $y_B = 0$ *and* $y^\top A' = c^\top$. *Then* $y^\top = \bar{c}^B Q_B (A\ I) - \bar{c}$.

(d) *Let* $i \in J$. *Let* w *be the vector with* $A'_i w = -1$ *and* $A'_{J \setminus \{i\}} w = 0$. *Then* $A'_B w = Q_B (A\ I)^i$.

(e) *Define*

$$T_B := \left(\begin{array}{c|c} Q_B (A\ I) & Q_B b \\ \hline \bar{c}^B Q_B (A\ I) - \bar{c} & c^\top x \end{array} \right).$$

Given B *and* T_B, *we can compute* B' *and* $T_{B'}$ *in* $O(m(n+m))$ *time, where* B' *arises from* B *by replacing* j *by* i, *and* i *and* j *are given as in* ②–④ *of the* SIMPLEX ALGORITHM *(applied to* A', b', c, *and index set* J).

T_B *is called the* **simplex tableau** *with respect to the* **basis** B.

Proof: (a): Let $N := \{1, \ldots, n\}$. As $(A\ I)^B$ is nonsingular, also $(A')^{N \setminus J}_{J \setminus N}$ is nonsingular, and thus A'_J is nonsingular.

(b): The first statement follows directly from $A'_J x = b'_J$. Then $b = Ax + I(b - Ax) = (A\ I)(b' - A'x) = (A\ I)^B (b' - A'x)_B$ and $c^\top x = \bar{c}(b' - A'x) = \bar{c}^B (b' - A'x)_B = \bar{c}^B Q_B b$.

(c): This follows from $(\bar{c}^B Q_B (A\ I) - \bar{c})^B = \bar{c}^B Q_B (A\ I)^B - \bar{c}^B = 0$ and $(\bar{c}^B Q_B (A\ I) - \bar{c}) A' = \bar{c}^B Q_B (A\ I) A' - c^\top (-I) = c^\top$.

(d): This follows from $0 = (A\ I) A'w = (A\ I)^B (A'_B w) + (A\ I)^{J \setminus \{i\}} (A'_{J \setminus \{i\}} w) + (A\ I)^i (A'_i w) = (A\ I)^B (A'_B w) - (A\ I)^i$.

(e): By (c), y as in ② of the SIMPLEX ALGORITHM is given by the last row of T_B. If $y \geq 0$, we stop (x and y are optimal). Otherwise i is the first index with $y_i < 0$, found in $O(n + m)$ time. If the i-th column of T_B has no positive entry, we stop (the LP is unbounded, and w is given by (d)). Otherwise, by (b) and (d), we have that λ in ④ of the SIMPLEX ALGORITHM is given by

$$\lambda = \min \left\{ \frac{(Q_B b)_j}{(Q_B (A\ I)^i)_j} : j \in \{1, \ldots, m\}, (Q_B (A\ I)^i)_j > 0 \right\},$$

and among the indices attaining this minimum, j is the one for which the j-th component of B is minimum. So we can compute j in $O(m)$ time by considering the i-th and the last column of T_B. This yields B'.

We can compute the updated tableau $T_{B'}$ as follows: Divide the j-th row by the entry in row j and column i. Then add a suitable multiple of the j-th row to all other rows, such that the i-th column has zeros only outside row j.

Note that these row operations do not destroy the property that the tableau has the form

$$\left(\begin{array}{c|c} Q(A\ I) & Qb \\ \hline v(A\ I) - \bar{c} & vb \end{array} \right)$$

for some nonsingular matrix Q and some vector v, and in addition we have $Q(A\ I)^{B'} = I$ and $(v(A\ I) - \bar{c})^{B'} = 0$. Since there is only one choice for Q and v, namely $Q = Q_{B'}$ and $v = \bar{c}^{B'} Q_{B'}$, the updated tableau $T_{B'}$ is computed correctly by the above operations in $O(m(n+m))$ time.   □

To start the SIMPLEX ALGORITHM we consider an LP of the form

$$\max\{cx : A'x \le b',\ A''x \le b'',\ x \ge 0\}$$

with $A' \in \mathbb{R}^{m' \times n}$, $A'' \in \mathbb{R}^{m'' \times n}$, $b' \ge 0$ and $b'' < 0$. We first run the SIMPLEX ALGORITHM on the instance

$$\min\{(\mathbb{1}A'')x + \mathbb{1}y : A'x \le b',\ A''x + y \ge b'',\ x, y \ge 0\},$$

starting with the tableau

$$\begin{pmatrix} A' & 0 & I & 0 & b' \\ -A'' & -I & 0 & I & -b'' \\ \hline \mathbb{1}A'' & \mathbb{1} & 0 & 0 & 0 \end{pmatrix}, \tag{3.5}$$

corresponding to the basic solution $x = 0$, $y = 0$. Then we run the iterations of the SIMPLEX ALGORITHM as in Theorem 3.17(e).

If the algorithm terminates with optimum value $\mathbb{1}b$, we modify the final simplex tableau as follows. Multiply some rows by -1 such that none of the columns $n + m'' + m' + 1, \ldots, n + m'' + m' + m''$ (the fourth section in (3.5)) is a unit vector, delete the fourth section of the tableau (i.e. columns $n + m'' + m' + 1, \ldots, n + m'' + m' + m''$), and replace the last row by $(-c, 0, 0, 0)$. Then add suitable multiples of the other rows to the last row in order to get zeros at $m' + m''$ places corresponding to columns with distinct unit vectors; these will form our basis. The result is the simplex tableau with respect to the original LP and this basis. Therefore we can continue running the iterations of the SIMPLEX ALGORITHM as in Theorem 3.17(e).

In fact, one can often do even more efficiently. Suppose we want to solve an LP $\min\{cx : Ax \ge b,\ x \ge 0\}$ with a very large number of inequalities which are implicitly given in a way that allows us to solve the following problem efficiently: Given a vector $x \ge 0$, decide if $Ax \ge b$ and find a violated inequality otherwise. We apply the SIMPLEX ALGORITHM to the dual LP $\max\{yb : yA \le c,\ y \ge 0\} = \max\{by : A^\top y \le c,\ y \ge 0\}$. Let $\bar{b} := (b^\top, 0)$. For a basis B we set $Q_B := ((A^\top\ I)^B)^{-1}$ and store only the right-hand part of the simplex tableau

$$\begin{pmatrix} Q_B & Q_B c \\ \hline \bar{b}^B Q_B & b^\top x \end{pmatrix}.$$

The last row of the full simplex tableau is $\bar{b}^B Q_B (A^\top\ I) - \bar{b}$. To perform an iteration, we must check if $\bar{b}^B Q_B \ge 0$ and $\bar{b}^B Q_B A^\top - b \ge 0$, and find a negative component if one exists. This reduces to solving the above problem for $x = (\bar{b}^B Q_B)^\top$. Then we generate the corresponding column of the full simplex tableau, but only for the current iteration. After updating the reduced tableau we can delete it again. This technique is known under the names *revised simplex* and *column generation*. We will see applications later.

3.4 Duality

Theorem 3.14 shows that the LPs (3.1) and (3.2) are related. This motivates the following definition:

Definition 3.18. *Given a linear program* $\max\{cx : Ax \leq b\}$, *we define the* **dual LP** *to be the linear program* $\min\{yb : yA = c, y \geq 0\}$.

In this case, the original LP $\max\{cx : Ax \leq b\}$ is often called the **primal LP**.

Proposition 3.19. *The dual of the dual of an LP is (equivalent to) the original LP.*

Proof: Let the primal LP $\max\{cx : Ax \leq b\}$ be given. Its dual is $\min\{yb : yA = c, y \geq 0\}$, or equivalently

$$- \max \left\{ -by : \begin{pmatrix} A^\top \\ -A^\top \\ -I \end{pmatrix} y \leq \begin{pmatrix} c \\ -c \\ 0 \end{pmatrix} \right\}.$$

(Each equality constraint has been split up into two inequality constraints.) So the dual of the dual is

$$- \min \left\{ zc - z'c : \begin{pmatrix} A & -A & -I \end{pmatrix} \begin{pmatrix} z \\ z' \\ w \end{pmatrix} = -b, \ z, z', w \geq 0 \right\}$$

which is equivalent to $-\min\{-cx : -Ax - w = -b, w \geq 0\}$ (where we have substituted x for $z' - z$). By eliminating the slack variables w we see that this is equivalent to the primal LP. □

We now obtain the most important theorem in LP theory, the Duality Theorem:

Theorem 3.20. (von Neumann [1947], Gale, Kuhn and Tucker [1951]) *If the polyhedra* $P := \{x : Ax \leq b\}$ *and* $D := \{y : yA = c, y \geq 0\}$ *are both nonempty, then* $\max\{cx : x \in P\} = \min\{yb : y \in D\}$.

Proof: If D is nonempty, it has a vertex y. We run the SIMPLEX ALGORITHM for $\min\{yb : y \in D\}$ and y. By Proposition 3.13, the existence of some $x \in P$ guarantees that $\min\{yb : y \in D\}$ is not unbounded. Thus by Theorem 3.14, the SIMPLEX ALGORITHM returns optimum solutions y and z of the LP $\min\{yb : y \in D\}$ and its dual. However, the dual is $\max\{cx : x \in P\}$ by Proposition 3.19. We have $yb = cz$, as required. □

We can say even more about the relation between the optimum solutions of the primal and dual LP:

Corollary 3.21. *Let* $\max\{cx : Ax \leq b\}$ *and* $\min\{yb : yA = c, y \geq 0\}$ *be a primal-dual pair of LPs. Let* x *and* y *be feasible solutions, i.e.* $Ax \leq b$, $yA = c$ *and* $y \geq 0$. *Then the following statements are equivalent:*

(a) x *and* y *are both optimum solutions.*
(b) $cx = yb$.
(c) $y(b - Ax) = 0$.

Proof: The Duality Theorem 3.20 immediately implies the equivalence of (a) and (b). The equivalence of (b) and (c) follows from $y(b - Ax) = yb - yAx = yb - cx$. $\quad\square$

The property (c) of optimum solutions is often called **complementary slackness**. It can also be formulated as follows: a point $x^* \in P = \{x : Ax \leq b\}$ is an optimum solution of $\max\{cx : x \in P\}$ if and only if c is a nonnegative combination of those rows of A which correspond to inequalities of $Ax \leq b$ that are satisfied by x^* with equality. It also implies:

Corollary 3.22. *Let* $P = \{x : Ax \leq b\}$ *be a polyhedron and* $\emptyset \neq Z \subseteq P$. *Then the set of vectors* c *for which each* $z \in Z$ *is an optimum solution of* $\max\{cx : x \in P\}$ *is the cone generated by the rows of* A', *where* $A'x \leq b'$ *is the maximal subsystem of* $Ax \leq b$ *with* $A'z = b'$ *for all* $z \in Z$.

Proof: There is a $z \in \text{conv}(Z)$ that satisfies all other inequalities of $Ax \leq b$ strictly. Let c be a vector for which each element of Z, and hence also z, is an optimum solution of $\max\{cx : x \in P\}$. Then by Corollary 3.21 there exists an $y \geq 0$ with $c = yA'$, i.e. c is a nonnegative linear combination of the rows of A'.
    Conversely, for a row $a'x \leq \beta'$ of $A'x \leq b'$ and $z \in Z$ we have $a'z = \beta' = \max\{a'x : x \in P\}$. $\quad\square$

Let us write Corollary 3.21 in another form:

Corollary 3.23. *Let* $\min\{cx : Ax \geq b, x \geq 0\}$ *and* $\max\{yb : yA \leq c, y \geq 0\}$ *be a primal-dual pair of LPs. Let* x *and* y *be feasible solutions, i.e.* $Ax \geq b$, $yA \leq c$ *and* $x, y \geq 0$. *Then the following statements are equivalent:*

(a) x *and* y *are both optimum solutions.*
(b) $cx = yb$.
(c) $(c - yA)x = 0$ *and* $y(b - Ax) = 0$.

Proof: The equivalence of (a) and (b) is obtained by applying the Duality Theorem 3.20 to $\max\left\{(-c)x : \binom{-A}{-I}x \leq \binom{-b}{0}\right\}$.
    To prove that (b) and (c) are equivalent, observe that we have $y(b - Ax) \leq 0 \leq (c - yA)x$ for any feasible solutions x and y, and that $y(b - Ax) = (c - yA)x$ iff $yb = cx$. $\quad\square$

The two conditions in (c) are sometimes called **primal** and **dual complementary slackness conditions**.

The Duality Theorem has many applications in combinatorial optimization. One reason for its importance is that the optimality of a solution can be proved by giving a feasible solution of the dual LP with the same objective value. We shall show now how to prove that an LP is unbounded or infeasible:

Theorem 3.24.   *There exists a vector x with $Ax \leq b$ if and only if $yb \geq 0$ for each vector $y \geq 0$ for which $yA = 0$.*

Proof: If there is a vector x with $Ax \leq b$, then $yb \geq yAx = 0$ for each $y \geq 0$ with $yA = 0$.

Consider the LP

$$- \min\{ \mathbb{1}w : Ax - w \leq b, \ w \geq 0\}. \tag{3.6}$$

Writing it in standard form we have

$$\max \left\{ (0 \ \ -\mathbb{1}) \begin{pmatrix} x \\ w \end{pmatrix} : \begin{pmatrix} A & -I \\ 0 & -I \end{pmatrix} \begin{pmatrix} x \\ w \end{pmatrix} \leq \begin{pmatrix} b \\ 0 \end{pmatrix} \right\}.$$

The dual of this LP is

$$\min \left\{ (b \ \ 0) \begin{pmatrix} y \\ z \end{pmatrix} : \begin{pmatrix} A^\top & 0 \\ -I & -I \end{pmatrix} \begin{pmatrix} y \\ z \end{pmatrix} = \begin{pmatrix} 0 \\ -\mathbb{1} \end{pmatrix}, \ y, z \geq 0 \right\},$$

or, equivalently,

$$\min\{yb : yA = 0, \ 0 \leq y \leq \mathbb{1}\}. \tag{3.7}$$

Since both (3.6) and (3.7) have a solution ($x = 0$, $w = |b|$, $y = 0$), we can apply Theorem 3.20. So the optimum values of (3.6) and (3.7) are the same. Since the system $Ax \leq b$ has a solution iff the optimum value of (3.6) is zero, the proof is complete. $\square$

So the fact that a linear inequality system $Ax \leq b$ has no solution can be proved by giving a vector $y \geq 0$ with $yA = 0$ and $yb < 0$. We mention two equivalent formulations of Theorem 3.24:

Corollary 3.25.   *There is a vector $x \geq 0$ with $Ax \leq b$ if and only if $yb \geq 0$ for each vector $y \geq 0$ with $yA \geq 0$.*

Proof: Apply Theorem 3.24 to the system $\begin{pmatrix} A \\ -I \end{pmatrix} x \leq \begin{pmatrix} b \\ 0 \end{pmatrix}$. $\square$

Corollary 3.26.   (Farkas [1894]) *There is a vector $x \geq 0$ with $Ax = b$ if and only if $yb \geq 0$ for each vector y with $yA \geq 0$.*

Proof: Apply Corollary 3.25 to the system $\begin{pmatrix} A \\ -A \end{pmatrix} x \leq \begin{pmatrix} b \\ -b \end{pmatrix}$, $x \geq 0$. $\square$

Corollary 3.26 is usually known as Farkas' Lemma. The above results in turn imply the Duality Theorem 3.20 which is interesting since they have quite easy direct proofs (in fact they were known before the SIMPLEX ALGORITHM); see Exercises 11 and 12.

We have seen how to prove that an LP is infeasible. How can we prove that an LP is unbounded? The next theorem answers this question.

Theorem 3.27. *If an LP is unbounded, then its dual LP is infeasible. If an LP has an optimum solution, then its dual also has an optimum solution.*

Proof: The first statement follows immediately from Proposition 3.13.

To prove the second statement, suppose that the (primal) LP $\max\{cx : Ax \leq b\}$ has an optimum solution x^*, but the dual $\min\{yb : yA = c, y \geq 0\}$ is infeasible (it cannot be unbounded due to the first statement). In other words, there is no $y \geq 0$ with $A^\top y = c$, and we apply Farkas' Lemma (Corollary 3.26) to get a vector z with $zA^\top \geq 0$ and $zc < 0$. But then $x^* - z$ is feasible for the primal, because $A(x^* - z) = Ax^* - Az \leq b$. The observation $c(x^* - z) > cx^*$ therefore contradicts the optimality of x^*. $\square$

So there are four cases for a primal-dual pair of LPs: either both have an optimum solution (in which case the optimum values are the same), or one is infeasible and the other one is unbounded, or both are infeasible. We also note:

Corollary 3.28. *A feasible LP $\max\{cx : Ax \leq b\}$ is bounded if and only if c belongs to the cone generated by the rows of A.*

Proof: The LP is bounded iff its dual is feasible, i.e. there is a $y \geq 0$ with $yA = c$. $\square$

Farkas' Lemma also enables us to prove that each finitely generated cone is polyhedral:

Theorem 3.29. (Minkowski [1896], Weyl [1935]) *A cone is polyhedral if and only if it is finitely generated.*

Proof: The only-if direction is given by Lemma 3.12. So consider the cone C generated by $a_1, \ldots, a_t$. We have to show that C is polyhedral. Let A be the matrix whose rows are $a_1, \ldots, a_t$.

By Lemma 3.12, the cone $D := \{x : Ax \leq 0\}$ is generated by some vectors $b_1, \ldots, b_s$. Let B be the matrix whose rows are $b_1, \ldots, b_s$. We prove that $C = \{x : Bx \leq 0\}$.

As $b_j a_i = a_i b_j \leq 0$ for all i and j, we have $C \subseteq \{x : Bx \leq 0\}$. Now suppose there is a vector $w \notin C$ with $Bw \leq 0$. $w \notin C$ means that there is no $v \geq 0$ such that $A^\top v = w$. By Farkas' Lemma (Corollary 3.26) this means that there is a vector y with $yw < 0$ and $Ay \geq 0$. So $-y \in D$. Since D is generated by $b_1, \ldots, b_s$ we have $-y = zB$ for some $z \geq 0$. But then $0 < -yw = zBw \leq 0$, a contradiction. $\square$

3.5 Convex Hulls and Polytopes

In this section we collect some more facts on polytopes. In particular, we show that polytopes are precisely those sets that are the convex hull of a finite number of points. We start by recalling some basic definitions:

Definition 3.30. *Given vectors $x_1, \ldots, x_k \in \mathbb{R}^n$ and $\lambda_1, \ldots, \lambda_k \geq 0$ with $\sum_{i=1}^{k} \lambda_i = 1$, we call $x = \sum_{i=1}^{k} \lambda_i x_i$ a* **convex combination** *of $x_1, \ldots, x_k$. A set $X \subseteq \mathbb{R}^n$ is* **convex** *if $\lambda x + (1 - \lambda)y \in X$ for all $x, y \in X$ and $\lambda \in [0, 1]$. The* **convex hull** *$\mathrm{conv}(X)$ of a set X is defined as the set of all convex combinations of points in X. An* **extreme point** *of a set X is an element $x \in X$ with $x \notin \mathrm{conv}(X \setminus \{x\})$.*

So a set X is convex if and only if all convex combinations of points in X are again in X. The convex hull of a set X is the smallest convex set containing X. Moreover, the intersection of convex sets is convex. Hence polyhedra are convex. Now we prove the "finite basis theorem for polytopes", a fundamental result which seems to be obvious but is not trivial to prove directly:

Theorem 3.31. (Minkowski [1896], Steinitz [1916], Weyl [1935]) *A set P is a polytope if and only if it is the convex hull of a finite set of points.*

Proof: (Schrijver [1986]) Let $P = \{x \in \mathbb{R}^n : Ax \leq b\}$ be a nonempty polytope. Obviously,

$$P = \left\{x : \begin{pmatrix} x \\ 1 \end{pmatrix} \in C\right\}, \quad \text{where} \quad C = \left\{\begin{pmatrix} x \\ \lambda \end{pmatrix} \in \mathbb{R}^{n+1} : \lambda \geq 0,\ Ax - \lambda b \leq 0\right\}.$$

C is a polyhedral cone, so by Theorem 3.29 it is generated by finitely many nonzero vectors, say by $\begin{pmatrix} x_1 \\ \lambda_1 \end{pmatrix}, \ldots, \begin{pmatrix} x_k \\ \lambda_k \end{pmatrix}$. Since P is bounded, all λ_i are nonzero; w.l.o.g. all λ_i are 1. So $x \in P$ if and only if

$$\begin{pmatrix} x \\ 1 \end{pmatrix} = \mu_1 \begin{pmatrix} x_1 \\ 1 \end{pmatrix} + \cdots + \mu_k \begin{pmatrix} x_k \\ 1 \end{pmatrix}$$

for some $\mu_1, \ldots, \mu_k \geq 0$. In other words, P is the convex hull of $x_1, \ldots, x_k$.

Now let P be the convex hull of $x_1, \ldots, x_k \in \mathbb{R}^n$. Then $x \in P$ if and only if $\begin{pmatrix} x \\ 1 \end{pmatrix} \in C$, where C is the cone generated by $\begin{pmatrix} x_1 \\ 1 \end{pmatrix}, \ldots, \begin{pmatrix} x_k \\ 1 \end{pmatrix}$. By Theorem 3.29, C is polyhedral, so

$$C = \left\{\begin{pmatrix} x \\ \lambda \end{pmatrix} : Ax + b\lambda \leq 0\right\}.$$

We conclude that $P = \{x \in \mathbb{R}^n : Ax + b \leq 0\}$. $\square$

Corollary 3.32. *A polytope is the convex hull of its vertices.*

Proof: Let P be a polytope. By Theorem 3.31, the convex hull of its vertices is a polytope Q. Obviously $Q \subseteq P$. Suppose there is a point $z \in P \setminus Q$. Then there is a vector c with $cz > \max\{cx : x \in Q\}$. The supporting hyperplane

$\{x : cx = \max\{cy : y \in P\}\}$ of P defines a face of P containing no vertex. This is impossible by Corollary 3.10. □

The previous two and the following result are the starting point of polyhedral combinatorics; they will be used very often in this book. For a given ground set E and a subset $X \subseteq E$, the **incidence vector** of X (with respect to E) is defined as the vector $x \in \{0, 1\}^E$ with $x_e = 1$ for $e \in X$ and $x_e = 0$ for $e \in E \setminus X$.

Corollary 3.33. *Let $(E, \mathcal{F})$ be a set system, P the convex hull of the incidence vectors of the elements of $\mathcal{F}$, and $c : E \to \mathbb{R}$. Then $\max\{cx : x \in P\} = \max\{c(X) : X \in \mathcal{F}\}$.*

Proof: Since $\max\{cx : x \in P\} \geq \max\{c(X) : X \in \mathcal{F}\}$ is trivial, let x be an optimum solution of $\max\{cx : x \in P\}$ (note that P is a polytope by Theorem 3.31). By definition of P, x is a convex combination of incidence vectors $y_1, \ldots, y_k$ of elements of $\mathcal{F}$: $x = \sum_{i=1}^{k} \lambda_i y_i$ for some $\lambda_1, \ldots, \lambda_k \geq 0$ with $\sum_{i=1}^{k} \lambda_i = 1$. Since $cx = \sum_{i=1}^{k} \lambda_i c y_i$, we have $c y_i \geq cx$ for at least one $i \in \{1, \ldots, k\}$. This y_i is the incidence vector of a set $Y \in \mathcal{F}$ with $c(Y) = c y_i \geq cx$. □

Exercises

1. Let H be a hypergraph, $F \subseteq V(H)$, and $x, y : F \to \mathbb{R}$. The task is to find $x, y : V(H) \setminus F \to \mathbb{R}$ such that $\sum_{e \in E(H)} (\max_{v \in e} x(v) - \min_{v \in e} x(v) + \max_{v \in e} y(v) - \min_{v \in e} y(v))$ is minimum. Show that this can be formulated as an LP.

   Note: This is a relaxation of a placement problem in VLSI design. Here H is called the netlist, and its vertices correspond to modules that need to placed on the chip. Some (those in F) are pre-placed in advance. The main difficulty (ignored in this relaxation) is that modules must not overlap.

2. A set of vectors $x_1, \ldots, x_k$ is called affinely independent if there is no $\lambda \in \mathbb{R}^k \setminus \{0\}$ with $\lambda^\top \mathbb{1} = 0$ and $\sum_{i=1}^{k} \lambda_i x_i = 0$. Let $\emptyset \neq X \subseteq \mathbb{R}^n$. Show that the maximum cardinality of an affinely independent set of elements of X equals $\dim X + 1$.

3. Let $P, Q \in \mathbb{R}^n$ be polyhedra. Prove that the closure of $\mathrm{conv}(P \cup Q)$ is a polyhedron. Show polyhedra P and Q for which $\mathrm{conv}(P \cup Q)$ is not a polyhedron.

4. Show that the problem to compute the largest ball that is a subset of a given polyhedron can be formulated as a linear program.

5. Let P be a polyhedron. Prove that the dimension of any facet of P is one less than the dimension of P.

6. Let F be a minimal face of a polyhedron $\{x : Ax \leq b\}$. Prove that then $Ax = Ay$ for all $x, y \in F$.

7. Let $A \in \mathbb{R}^{m \times n}$, $b \in \mathbb{R}^m$, $c \in \mathbb{R}^n$, and $u \in \mathbb{Z}^n$. Consider the LP $\max\{cx : Ax \leq b, 0 \leq x \leq u\}$. Prove: if this LP has an optimum solution, then it has an optimum solution with at most m components that are not integers.

8. Formulate the dual of the LP formulation (1.1) of the JOB ASSIGNMENT PROBLEM. Show how to solve the primal and the dual LP in the case when there are only two jobs (by a simple algorithm).

9. Let G be a digraph, $c : E(G) \to \mathbb{R}_+$, $E_1, E_2 \subseteq E(G)$, and $s, t \in V(G)$. Consider the following linear program

$$
\begin{array}{rll}
\min & \displaystyle\sum_{e \in E(G)} c(e) y_e & \\[2ex]
\text{s.t.} & y_e \geq z_w - z_v & (e = (v, w) \in E(G)) \\
& z_t - z_s = 1 & \\
& y_e \geq 0 & (e \in E_1) \\
& y_e \leq 0 & (e \in E_2).
\end{array}
$$

Prove that there is an optimum solution (y, z) and $s \in X \subseteq V(G) \setminus \{t\}$ with $y_e = 1$ for $e \in \delta^+(X)$, $y_e = -1$ for $e \in \delta^-(X) \setminus E_1$, and $y_e = 0$ for all other edges e.
Hint: Consider the complementary slackness conditions for the edges entering or leaving $\{v \in V(G) : z_v \leq z_s\}$.

10. Let $Ax \leq b$ be a linear inequality system in n variables. By multiplying each row by a positive constant we may assume that the first column of A is a vector with entries $0, -1$ and 1 only. So we can write $Ax \leq b$ equivalently as

$$
\begin{array}{rll}
a_i'x' & \leq & b_i \qquad (i = 1, \ldots, m_1), \\
-x_1 + a_j'x' & \leq & b_j \qquad (j = m_1 + 1, \ldots, m_2), \\
x_1 + a_k'x' & \leq & b_k \qquad (k = m_2 + 1, \ldots, m),
\end{array}
$$

where $x' = (x_2, \ldots, x_n)$ and $a_1', \ldots, a_m'$ are the rows of A without the first entry. Then one can eliminate x_1: Prove that $Ax \leq b$ has a solution if and only if the system

$$
\begin{array}{rll}
a_i'x' & \leq & b_i \qquad\qquad (i = 1, \ldots, m_1), \\
a_j'x' - b_j & \leq & b_k - a_k'x' \qquad (j = m_1 + 1, \ldots, m_2, \; k = m_2 + 1, \ldots, m)
\end{array}
$$

has a solution. Show that this technique, when iterated, leads to an algorithm for solving a linear inequality system $Ax \leq b$ (or proving infeasibility).
Note: This method is known as Fourier-Motzkin elimination because it was proposed by Fourier and studied by Motzkin [1936]. One can prove that it is not a polynomial-time algorithm.

11. Use Fourier-Motzkin elimination (Exercise 10) to prove Theorem 3.24 directly. (Kuhn [1956])

12. Show that Theorem 3.24 implies the Duality Theorem 3.20.

13. Prove the decomposition theorem for polyhedra: Any polyhedron P can be written as $P = \{x + c : x \in X, c \in C\}$, where X is a polytope and C is a polyhedral cone.
(Motzkin [1936])

* 14. Let P be a rational polyhedron and F a face of P. Show that

$$\{c : cz = \max\{cx : x \in P\} \text{ for all } z \in F\}$$

is a rational polyhedral cone.

15. Prove Carathéodory's theorem:
    If $X \subseteq \mathbb{R}^n$ and $y \in \text{conv}(X)$, then there are $x_1, \ldots, x_{n+1} \in X$ such that $y \in \text{conv}(\{x_1, \ldots, x_{n+1}\})$.
    (Carathéodory [1911])

16. Prove the following extension of Carathéodory's theorem (Exercise 15):
    If $X \subseteq \mathbb{R}^n$ and $y, z \in \text{conv}(X)$, then there are $x_1, \ldots, x_n \in X$ such that $y \in \text{conv}(\{z, x_1, \ldots, x_n\})$.

17. Prove that the extreme points of a polyhedron are precisely its vertices.

18. Let P be a nonempty polytope. Consider the graph $G(P)$ whose vertices are the vertices of P and whose edges correspond to the 1-dimensional faces of P. Let x be any vertex of P, and c a vector with $c^\top x < \max\{c^\top z : z \in P\}$. Prove that then there is a neighbour y of x in $G(P)$ with $c^\top x < c^\top y$.

* 19. Use Exercise 18 to prove that $G(P)$ is n-connected for any n-dimensional polytope P ($n \geq 1$).

20. Let $P \subseteq \mathbb{R}^n$ be a polytope (not necessarily rational) and $y \notin P$. Prove that there exists a rational vector c with $\max\{cx : x \in P\} < cy$. Show that the statement does not hold for general polyhedra.

21. Let $X \subset \mathbb{R}^n$ be a nonempty convex set, $\bar{X}$ the closure of X, and $y \notin X$. Prove:
    (a) There is a unique point in $\bar{X}$ that has minimum distance to y.
    (b) There exists a vector $a \in \mathbb{R}^n \setminus \{0\}$ with $a^\top x \leq a^\top y$ for all $x \in X$.
    (c) If $y \notin \bar{X}$, then there exists a vector $a \in \mathbb{R}^n$ with $a^\top x < a^\top y$ for all $x \in X$.
    (d) If X is bounded and $y \notin \bar{X}$, then there exists a vector $a \in \mathbb{Q}^n$ with $a^\top x < a^\top y$ for all $x \in X$.
    (e) A closed convex set is the intersection of all closed half-spaces containing it.

References

General Literature

Bertsimas, D., and Tsitsiklis, J.N. [1997]: Introduction to Linear Optimization. Athena Scientific, Belmont 1997

Chvátal, V. [1983]: Linear Programming. Freeman, New York 1983

Matoušek, J., and Gärtner, B. [2007]: Understanding and Using Linear Programming. Springer, Berlin 2007

Padberg, M. [1999]: Linear Optimization and Extensions. Second Edition. Springer, Berlin 1999

Schrijver, A. [1986]: Theory of Linear and Integer Programming. Wiley, Chichester 1986

Cited References

Avis, D., and Chvátal, V. [1978]: Notes on Bland's pivoting rule. Mathematical Programming Study 8 (1978), 24–34

Bland, R.G. [1977]: New finite pivoting rules for the simplex method. Mathematics of Operations Research 2 (1977), 103–107

Borgwardt, K.-H. [1982]: The average number of pivot steps required by the simplex method is polynomial. Zeitschrift für Operations Research 26 (1982), 157–177

Carathéodory, C. [1911]: Über den Variabilitätsbereich der Fourierschen Konstanten von positiven harmonischen Funktionen. Rendiconto del Circolo Matematico di Palermo 32 (1911), 193–217

Dantzig, G.B. [1951]: Maximization of a linear function of variables subject to linear inequalities. In: Activity Analysis of Production and Allocation (T.C. Koopmans, ed.), Wiley, New York 1951, pp. 359–373

Dantzig, G.B., Orden, A., and Wolfe, P. [1955]: The generalized simplex method for minimizing a linear form under linear inequality restraints. Pacific Journal of Mathematics 5 (1955), 183–195

Farkas, G. [1894]: A Fourier-féle mechanikai elv alkalmazásai. Mathematikai és Természettudományi Értesitö 12 (1894), 457–472

Gale, D., Kuhn, H.W., and Tucker, A.W. [1951]: Linear programming and the theory of games. In: Activity Analysis of Production and Allocation (T.C. Koopmans, ed.), Wiley, New York 1951, pp. 317–329

Hoffman, A.J., and Kruskal, J.B. [1956]: Integral boundary points of convex polyhedra. In: Linear Inequalities and Related Systems; Annals of Mathematical Study 38 (H.W. Kuhn, A.W. Tucker, eds.), Princeton University Press, Princeton 1956, pp. 223–246

Kelner, J.A., and Spielman, D.A. [2006]: A randomized polynomial-time simplex algorithm for linear programming. Proceedings of the 38th Annual ACM Symposium on Theory of Computing (2006), 51–60

Klee, V., and Minty, G.J. [1972]: How good is the simplex algorithm? In: Inequalities III (O. Shisha, ed.), Academic Press, New York 1972, pp. 159–175

Kuhn, H.W. [1956]: Solvability and consistency for linear equations and inequalities. The American Mathematical Monthly 63 (1956), 217–232

Minkowski, H. [1896]: Geometrie der Zahlen. Teubner, Leipzig 1896

Motzkin, T.S. [1936]: Beiträge zur Theorie der linearen Ungleichungen (Dissertation). Azriel, Jerusalem 1936

von Neumann, J. [1947]: Discussion of a maximum problem. Working paper. Published in: John von Neumann, Collected Works; Vol. VI (A.H. Taub, ed.), Pergamon Press, Oxford 1963, pp. 27–28

Spielman, D.A., and Teng, S.-H. [2004]: Smoothed analysis of algorithms: why the simplex algorithm usually takes polynomial time. Journal of the ACM 51 (2004), 385–463

Steinitz, E. [1916]: Bedingt konvergente Reihen und konvexe Systeme. Journal für die reine und angewandte Mathematik 146 (1916), 1–52

Weyl, H. [1935]: Elementare Theorie der konvexen Polyeder. Commentarii Mathematici Helvetici 7 (1935), 290–306

4 Linear Programming Algorithms

Three types of algorithms for LINEAR PROGRAMMING had the most impact: the SIMPLEX ALGORITHM (see Section 3.2), interior point algorithms, and the ELLIPSOID METHOD.

Each of these has a disadvantage: In contrast to the other two, so far no variant of the SIMPLEX ALGORITHM has been shown to have a polynomial running time. In Sections 4.4 and 4.5 we present the ELLIPSOID METHOD and prove that it leads to a polynomial-time algorithm for LINEAR PROGRAMMING. However, the ELLIPSOID METHOD is too inefficient to be used in practice. Interior point algorithms and, despite its exponential worst-case running time, the SIMPLEX ALGORITHM are far more efficient, and they are both used in practice to solve LPs. In fact, both the ELLIPSOID METHOD and interior point algorithms can be used for more general convex optimization problems, e.g. for so-called semidefinite programs.

An advantage of the SIMPLEX ALGORITHM and the ELLIPSOID METHOD is that they do not require the LP to be given explicitly. It suffices to have an oracle (a subroutine) which decides whether a given vector is feasible and, if not, returns a violated constraint. We shall discuss this in detail with respect to the ELLIPSOID METHOD in Section 4.6, because it implies that many combinatorial optimization problems can be solved in polynomial time; for some problems this is in fact the only known way to show polynomial solvability. This is the reason why we discuss the ELLIPSOID METHOD but not interior point algorithms in this book.

A prerequisite for polynomial-time algorithms is that there exists an optimum solution that has a binary representation whose length is bounded by a polynomial in the input size. We prove in Section 4.1 that this condition holds for LINEAR PROGRAMMING. In Sections 4.2 and 4.3 we review some basic algorithms needed later, including the well-known Gaussian elimination method for solving systems of equations.

4.1 Size of Vertices and Faces

Instances of LINEAR PROGRAMMING are vectors and matrices. Since no strongly polynomial-time algorithm for LINEAR PROGRAMMING is known we have to restrict attention to rational instances when analyzing the running time of algorithms.

© Springer-Verlag GmbH Germany 2018

Bernhard Korte and Jens Vygen, *Combinatorial Optimization*, Algorithms and Combinatorics 21, https://doi.org/10.1007/978-3-662-56039-6_4

We assume that all numbers are coded in binary. To estimate the size (number of bits) of this representation we define $\text{size}(n) := 1 + \lceil \log(|n| + 1) \rceil$ for integers $n \in \mathbb{Z}$ and $\text{size}(r) := \text{size}(p) + \text{size}(q)$ for rational numbers $r = \frac{p}{q}$, where p, q are relatively prime integers (i.e. their greatest common divisor is 1). For vectors $x = (x_1, \ldots, x_n) \in \mathbb{Q}^n$ we store the components and have $\text{size}(x) := n + \text{size}(x_1) + \ldots + \text{size}(x_n)$. For a matrix $A \in \mathbb{Q}^{m \times n}$ with entries a_{ij} we have $\text{size}(A) := mn + \sum_{i,j} \text{size}(a_{ij})$.

Of course these precise values are a somewhat random choice, but remember that we are not really interested in constant factors. For polynomial-time algorithms it is important that the sizes of numbers do not increase too much by elementary arithmetic operations. We note:

Proposition 4.1. *If $r_1, \ldots, r_n$ are rational numbers, then*

$$\text{size}(r_1 \cdots r_n) \leq \text{size}(r_1) + \cdots + \text{size}(r_n);$$
$$\text{size}(r_1 + \cdots + r_n) \leq 2(\text{size}(r_1) + \cdots + \text{size}(r_n)).$$

Proof: For integers $s_1, \ldots, s_n$ we obviously have $\text{size}(s_1 \cdots s_n) \leq \text{size}(s_1) + \cdots + \text{size}(s_n)$ and $\text{size}(s_1 + \cdots + s_n) \leq \text{size}(s_1) + \cdots + \text{size}(s_n)$.

Let now $r_i = \frac{p_i}{q_i}$, where p_i and q_i are nonzero integers $(i = 1, \ldots, n)$. Then $\text{size}(r_1 \cdots r_n) \leq \text{size}(p_1 \cdots p_n) + \text{size}(q_1 \cdots q_n) \leq \text{size}(r_1) + \cdots + \text{size}(r_n)$.

For the second statement, observe that the denominator $q_1 \cdots q_n$ has size at most $\text{size}(q_1) + \cdots + \text{size}(q_n)$. The numerator is the sum of the numbers $q_1 \cdots q_{i-1} p_i q_{i+1} \cdots q_n$ $(i = 1, \ldots, n)$, so its absolute value is at most $(|p_1| + \cdots + |p_n|)|q_1 \cdots q_n|$. Therefore the size of the numerator is at most $\text{size}(r_1) + \cdots + \text{size}(r_n)$. $\square$

The first part of this proposition also implies that we can often assume w.l.o.g. that all numbers in a problem instance are integers, since otherwise we can multiply each of them with the product of all denominators. For addition and inner product of vectors we have:

Proposition 4.2. *If $x, y \in \mathbb{Q}^n$ are rational vectors, then*

$$\text{size}(x + y) \leq 2(\text{size}(x) + \text{size}(y));$$
$$\text{size}(x^\top y) \leq 2(\text{size}(x) + \text{size}(y)).$$

Proof: Using Proposition 4.1 we have $\text{size}(x + y) = n + \sum_{i=1}^n \text{size}(x_i + y_i) \leq n + 2\sum_{i=1}^n \text{size}(x_i) + 2\sum_{i=1}^n \text{size}(y_i) = 2(\text{size}(x) + \text{size}(y)) - 3n$ and $\text{size}(x^\top y) = \text{size}\left(\sum_{i=1}^n x_i y_i\right) \leq 2\sum_{i=1}^n \text{size}(x_i y_i) \leq 2\sum_{i=1}^n \text{size}(x_i) + 2\sum_{i=1}^n \text{size}(y_i) = 2(\text{size}(x) + \text{size}(y)) - 4n$. $\square$

Even under more complicated operations the numbers involved do not grow fast. Recall that the determinant of a matrix $A = (a_{ij})_{1 \leq i, j \leq n}$ is defined by

$$\det A := \sum_{\pi \in S_n} \text{sgn}(\pi) \prod_{i=1}^n a_{i, \pi(i)}, \tag{4.1}$$

where S_n is the set of all permutations of $\{1, \ldots, n\}$ and $\text{sgn}(\pi)$ is the sign of the permutation π (defined to be 1 if π can be obtained from the identity map by an even number of transpositions, and -1 otherwise).

Proposition 4.3. *For any rational square matrix A we have* $\text{size}(\det A) \leq 2\,\text{size}(A)$.

Proof: We write $a_{ij} = \frac{p_{ij}}{q_{ij}}$ with relatively prime integers p_{ij}, q_{ij}. Now let $\det A = \frac{p}{q}$ where p and q are relatively prime integers. Then $|\det A| \leq \prod_{i,j}(|p_{ij}| + 1)$ and $|q| \leq \prod_{i,j} |q_{ij}|$. We obtain $\text{size}(q) \leq \text{size}(A)$ and, using $|p| = |\det A||q| \leq \prod_{i,j}(|p_{ij}| + 1)|q_{ij}|$,

$$\text{size}(p) \leq \sum_{i,j}(\text{size}(p_{ij}) + 1 + \text{size}(q_{ij})) = \text{size}(A).$$

$\square$

With this observation we can prove:

Theorem 4.4. *Suppose the rational LP $\max\{cx : Ax \leq b\}$ has an optimum solution. Then it also has an optimum solution x with $\text{size}(x) \leq 4n(\text{size}(A) + \text{size}(b))$, with components of size at most $4(\text{size}(A) + \text{size}(b))$. If $b = e_i$ or $b = -e_i$ for some unit vector e_i, then there is a nonsingular submatrix A' of A and an optimum solution x with $\text{size}(x) \leq 4n\,\text{size}(A')$.*

Proof: By Corollary 3.5, the maximum is attained in a face F of $\{x : Ax \leq b\}$. Let $F' \subseteq F$ be a minimal face. By Proposition 3.9, $F' = \{x : A'x = b'\}$ for some subsystem $A'x \leq b'$ of $Ax \leq b$. W.l.o.g., we may assume that the rows of A' are linearly independent. We then take a maximal set of linear independent columns (call this matrix A'') and set all other components to zero. Then $x = (A'')^{-1}b'$, filled up with zeros, is an optimum solution to our LP. By Cramer's rule the entries of x are given by $x_j = \frac{\det A'''}{\det A''}$, where A''' arises from A'' by replacing the j-th column by b'. By Proposition 4.3 we obtain $\text{size}(x) \leq n + 2n(\text{size}(A''') + \text{size}(A'')) \leq 4n(\text{size}(A'') + \text{size}(b'))$. If $b = \pm e_i$ then $|\det(A''')|$ is the absolute value of a subdeterminant of A''. $\square$

The encoding length of the faces of a polytope given by its vertices can be estimated as follows:

Lemma 4.5. *Let $P \subseteq \mathbb{R}^n$ be a rational polytope and $T \in \mathbb{N}$ such that $\text{size}(x) \leq T$ for each vertex x. Then $P = \{x : Ax \leq b\}$ for some inequality system $Ax \leq b$, each of whose inequalities $ax \leq \beta$ satisfies $\text{size}(a) + \text{size}(\beta) \leq 75n^2T$.*

Proof: First assume that P is full-dimensional. Let $F = \{x \in P : ax = \beta\}$ be a facet of P, where $P \subseteq \{x : ax \leq \beta\}$.

Let $y_1, \ldots, y_t$ be the vertices of F (by Proposition 3.6 they are also vertices of P). Let c be the solution of $Mc = e_1$, where M is a $t \times n$-matrix whose i-th row is $y_i - y_1$ $(i = 2, \ldots, t)$ and whose first row is some unit vector that is linearly independent of the other rows. Observe that $\mathrm{rank}(M) = n$ (because $\dim F = n-1$). So we have $c^\top = \kappa a$ for some $\kappa \in \mathbb{R} \setminus \{0\}$.

By Theorem 4.4 $\mathrm{size}(c) \leq 4n\,\mathrm{size}(M')$, where M' is a nonsingular $n \times n$-submatrix of M. By Proposition 4.2 we have $\mathrm{size}(M') \leq 4nT$ and $\mathrm{size}(c^\top y_1) \leq 2(\mathrm{size}(c) + \mathrm{size}(y_1))$. So the inequality $c^\top x \leq \delta$ (or $c^\top x \geq \delta$ if $\kappa < 0$), where $\delta := c^\top y_1 = \kappa \beta$, satisfies $\mathrm{size}(c) + \mathrm{size}(\delta) \leq 3\,\mathrm{size}(c) + 2T \leq 48n^2 T + 2T \leq 50n^2 T$. Collecting these inequalities for all facets F yields a description of P.

If $P = \emptyset$, the assertion is trivial, so we now assume that P is neither full-dimensional nor empty. Let V be the set of vertices of P. For $s = (s_1, \ldots, s_n) \in \{-1, 1\}^n$ let P_s be the convex hull of $V \cup \{x + s_i e_i : x \in V, i = 1, \ldots, n\}$. Each P_s is a full-dimensional polytope (Theorem 3.31), and the size of any of its vertices is at most $T + n$ (cf. Corollary 3.32). By the above, P_s can be described by inequalities of size at most $50n^2(T + n) \leq 75n^2 T$ (note that $T \geq 2n$). Since $P = \bigcap_{s \in \{-1,1\}^n} P_s$, this completes the proof.   □

4.2 Continued Fractions

When we say that the numbers occurring in a certain algorithm do not grow too fast, we often assume that for each rational $\frac{p}{q}$ the numerator p and the denominator q are relatively prime. This assumption causes no problem if we can easily find the greatest common divisor of two natural numbers. This is accomplished by one of the oldest algorithms:

EUCLIDEAN ALGORITHM

Input: Two natural numbers p and q.

Output: The greatest common divisor d of p and q, i.e. $\frac{p}{d}$ and $\frac{q}{d}$ are relatively prime integers.

---

① **While** $p > 0$ and $q > 0$ **do**:
   If $p < q$ **then** set $q := q - \lfloor \frac{q}{p} \rfloor p$ **else** set $p := p - \lfloor \frac{p}{q} \rfloor q$.

② **Return** $d := \max\{p, q\}$.

Theorem 4.6. *The* EUCLIDEAN ALGORITHM *works correctly. The number of iterations is at most* $\mathrm{size}(p) + \mathrm{size}(q)$.

Proof: The correctness follows from the fact that the set of common divisors of p and q does not change throughout the algorithm, until one of the numbers becomes zero. One of p or q is reduced by at least a factor of two in each iteration, hence there are at most $\log p + \log q + 1$ iterations.   □

Since no number occurring in an intermediate step is greater than p and q, we have a polynomial-time algorithm.

A similar algorithm is the so-called CONTINUED FRACTION EXPANSION. This can be used to approximate any number by a rational number whose denominator is not too large. For any positive real number x we define $x_0 := x$ and $x_{i+1} := \frac{1}{x_i - \lfloor x_i \rfloor}$ for $i = 1, 2, \ldots$, until $x_k \in \mathbb{N}$ for some k. Then we have

$$x = x_0 = \lfloor x_0 \rfloor + \frac{1}{x_1} = \lfloor x_0 \rfloor + \frac{1}{\lfloor x_1 \rfloor + \frac{1}{x_2}} = \lfloor x_0 \rfloor + \frac{1}{\lfloor x_1 \rfloor + \frac{1}{\lfloor x_2 \rfloor + \frac{1}{x_3}}} = \cdots$$

We claim that this sequence is finite if and only if x is rational. One direction follows immediately from the observation that x_{i+1} is rational if and only if x_i is rational. The other direction is also easy: If $x = \frac{p}{q}$, the above procedure is equivalent to the EUCLIDEAN ALGORITHM applied to p and q. This also shows that for a given rational number $\frac{p}{q}$ with $p, q > 0$ the (finite) sequence $x_1, x_2, \ldots, x_k$ as above can be computed in polynomial time. The following algorithm is almost identical to the EUCLIDEAN ALGORITHM except for the computation of the numbers g_i and h_i; we shall prove that the sequence $\left(\frac{g_i}{h_i}\right)_{i \in \mathbb{N}}$ converges to x.

CONTINUED FRACTION EXPANSION

Input: Natural numbers p and q (let $x := \frac{p}{q}$).

Output: The sequence $\left(x_i = \frac{p_i}{q_i}\right)_{i=0,1,\ldots}$ with $x_0 = \frac{p}{q}$ and $x_{i+1} := \frac{1}{x_i - \lfloor x_i \rfloor}$.

①   Set $i := 0$, $p_0 := p$ and $q_0 := q$.
    Set $g_{-2} := 0$, $g_{-1} := 1$, $h_{-2} := 1$, and $h_{-1} := 0$.

②   **While** $q_i \neq 0$ **do:**
        Set $a_i := \lfloor \frac{p_i}{q_i} \rfloor$.
        Set $g_i := a_i g_{i-1} + g_{i-2}$.
        Set $h_i := a_i h_{i-1} + h_{i-2}$.
        Set $q_{i+1} := p_i - a_i q_i$.
        Set $p_{i+1} := q_i$.
        Set $i := i + 1$.

We claim that the sequence $\frac{g_i}{h_i}$ yields good approximations of x. Before we can prove this, we need some preliminary observations:

Proposition 4.7.   *The following statements hold for all iterations i in the above algorithm:*

(a) $a_i \geq 1$ *(except possibly for $i = 0$) and $h_i \geq h_{i-1}$.*
(b) $g_{i-1} h_i - g_i h_{i-1} = (-1)^i$.
(c) $\dfrac{p_i g_{i-1} + q_i g_{i-2}}{p_i h_{i-1} + q_i h_{i-2}} = x$.
(d) $\frac{g_i}{h_i} \leq x$ *if i is even and $\frac{g_i}{h_i} \geq x$ if i is odd.*

Proof: (a) is obvious. (b) is easily shown by induction: For $i = 0$ we have $g_{i-1}h_i - g_i h_{i-1} = g_{-1}h_0 = 1$, and for $i \geq 1$ we have

$$
\begin{aligned}
g_{i-1}h_i - g_i h_{i-1} &= g_{i-1}(a_i h_{i-1} + h_{i-2}) - h_{i-1}(a_i g_{i-1} + g_{i-2}) \\
&= g_{i-1}h_{i-2} - h_{i-1}g_{i-2}.
\end{aligned}
$$

(c) is also proved by induction: For $i = 0$ we have

$$
\frac{p_i g_{i-1} + q_i g_{i-2}}{p_i h_{i-1} + q_i h_{i-2}} = \frac{p_i \cdot 1 + 0}{0 + q_i \cdot 1} = x.
$$

For $i \geq 1$ we have

$$
\begin{aligned}
\frac{p_i g_{i-1} + q_i g_{i-2}}{p_i h_{i-1} + q_i h_{i-2}} &= \frac{q_{i-1}(a_{i-1}g_{i-2} + g_{i-3}) + (p_{i-1} - a_{i-1}q_{i-1})g_{i-2}}{q_{i-1}(a_{i-1}h_{i-2} + h_{i-3}) + (p_{i-1} - a_{i-1}q_{i-1})h_{i-2}} \\
&= \frac{q_{i-1}g_{i-3} + p_{i-1}g_{i-2}}{q_{i-1}h_{i-3} + p_{i-1}h_{i-2}}.
\end{aligned}
$$

We finally prove (d). We note $\frac{g_{-2}}{h_{-2}} = 0 < x < \infty = \frac{g_{-1}}{h_{-1}}$ and proceed by induction. The induction step follows easily from the fact that the function $f(\alpha) := \frac{\alpha g_{i-1} + g_{i-2}}{\alpha h_{i-1} + h_{i-2}}$ is monotone for $\alpha > 0$, and $f(\frac{p_i}{q_i}) = x$ by (c). $\qquad\square$

Theorem 4.8. (Khintchine [1956]) *Given a rational number α and a natural number n, a rational number β with denominator at most n such that $|\alpha - \beta|$ is minimum can be found in polynomial time (polynomial in size(n) + size(α)).*

Proof: We run the CONTINUED FRACTION EXPANSION with $x := \alpha$. If the algorithm stops with $q_i = 0$ and $h_{i-1} \leq n$, we can set $\beta = \frac{g_{i-1}}{h_{i-1}} = \alpha$ by Proposition 4.7(c). Otherwise let i be the last index with $h_i \leq n$, and let t be the maximum integer such that $th_i + h_{i-1} \leq n$ (cf. Proposition 4.7(a)). Since $a_{i+1}h_i + h_{i-1} = h_{i+1} > n$, we have $t < a_{i+1}$. We claim that

$$
y := \frac{g_i}{h_i} \quad \text{or} \quad z := \frac{tg_i + g_{i-1}}{th_i + h_{i-1}}
$$

is an optimum solution. Both numbers have denominators at most n.

If i is even, then $y \leq x < z$ by Proposition 4.7(d). Similarly, if i is odd, we have $y \geq x > z$. We show that any rational number $\frac{p}{q}$ between y and z has denominator greater than n.

Observe that

$$
|z - y| = \frac{|h_i g_{i-1} - h_{i-1}g_i|}{h_i(th_i + h_{i-1})} = \frac{1}{h_i(th_i + h_{i-1})}
$$

(using Proposition 4.7(b)). On the other hand,

$$
|z - y| = \left| z - \frac{p}{q} \right| + \left| \frac{p}{q} - y \right| \geq \frac{1}{(th_i + h_{i-1})q} + \frac{1}{h_i q} = \frac{h_{i-1} + (t+1)h_i}{qh_i(th_i + h_{i-1})},
$$

so $q \geq h_{i-1} + (t+1)h_i > n$. $\qquad\square$

The above proof is from the book of Grötschel, Lovász and Schrijver [1988], which also contains important generalizations.

4.3 Gaussian Elimination

The most important algorithm in linear algebra is the so-called Gaussian elimination. It has been applied by Gauss but was known much earlier (see Schrijver [1986] for historical notes). Gaussian elimination is used to determine the rank of a matrix, to compute the determinant and to solve a system of linear equations. It occurs very often as a subroutine in linear programming algorithms; e.g. in ① of the SIMPLEX ALGORITHM.

Given a matrix $A \in \mathbb{Q}^{m \times n}$, our algorithm for Gaussian Elimination works with an extended matrix $Z = (B \quad C) \in \mathbb{Q}^{m \times (n+m)}$; initially $B = A$ and $C = I$. The algorithm transforms B to the form $\left(\begin{smallmatrix} I & R \\ 0 & 0 \end{smallmatrix}\right)$ by the following elementary operations: permuting rows and columns, adding a multiple of one row to another row, and (in the final step) multiplying rows by nonzero constants. At each iteration C is modified accordingly, such that the property $C\tilde{A} = B$ is maintained throughout where $\tilde{A}$ results from A by permuting rows and columns.

The first part of the algorithm, consisting of ② and ③, transforms B to an upper triangular matrix. Consider for example the matrix Z after two iterations; it has the form

$$
\left(
\begin{array}{cccccc|cccccc}
z_{11} \neq 0 & z_{12} & z_{13} & \cdot & \cdot & z_{1n} & 1 & 0 & 0 & \cdot & \cdot & 0 \\
0 & z_{22} \neq 0 & z_{23} & \cdot & \cdot & z_{2n} & z_{2,n+1} & 1 & 0 & \cdot & \cdot & 0 \\
0 & 0 & z_{33} & \cdot & \cdot & z_{3n} & z_{3,n+1} & z_{3,n+2} & 1 & 0 & \cdot & 0 \\
\cdot & \cdot & \cdot & & & \cdot & \cdot & & & 0 & & \cdot \\
\cdot & \cdot & \cdot & & & \cdot & \cdot & & & & I & \cdot \\
\cdot & \cdot & \cdot & & & \cdot & \cdot & & & & & 0 \\
0 & 0 & z_{m3} & \cdot & \cdot & z_{mn} & z_{m,n+1} & z_{m,n+2} & 0 & \cdot & \cdot & 0 & 1 \\
\end{array}
\right).
$$

If $z_{33} \neq 0$, then the next step just consists of subtracting $\frac{z_{i3}}{z_{33}}$ times the third row from the i-th row, for $i = 4, \ldots, m$. If $z_{33} = 0$ we first exchange the third row and/or the third column with another one. Note that if we exchange two rows, we have to exchange also the two corresponding columns of C in order to maintain the property $C\tilde{A} = B$. To have $\tilde{A}$ available at each point we store the permutations of the rows and columns in variables $row(i)$, $i = 1, \ldots, m$ and $col(j)$, $j = 1, \ldots, n$. Then $\tilde{A} = (A_{row(i),col(j)})_{i \in \{1,\ldots,m\}, j \in \{1,\ldots,n\}}$.

The second part of the algorithm, consisting of ④ and ⑤, is simpler since no rows or columns are exchanged anymore.

GAUSSIAN ELIMINATION

Input: A matrix $A = (a_{ij}) \in \mathbb{Q}^{m \times n}$.

Output: Its rank r,
            a maximal nonsingular submatrix $A' = (a_{row(i),col(j)})_{i,j \in \{1,\ldots,r\}}$ of A,
            its determinant $d = \det A'$, and its inverse $(A')^{-1} = (z_{i,n+j})_{i,j \in \{1,\ldots,r\}}$.

①   Set $r := 0$ and $d := 1$.
      Set $z_{ij} := a_{ij}$, $row(i) := i$ and $col(j) := j$ $(i = 1, \ldots, m, \ j = 1, \ldots, n)$.
      Set $z_{i,n+j} := 0$ and $z_{i,n+i} := 1$ for $1 \leq i, j \leq m, \ i \neq j$.

②   Let $p \in \{r+1, \ldots, m\}$ and $q \in \{r+1, \ldots, n\}$ with $z_{pq} \neq 0$. **If** no such p
      and q exist, **then go to** ④.
      Set $r := r + 1$.
      If $p \neq r$ **then** exchange z_{pj} and z_{rj} $(j = 1, \ldots, n+m)$, exchange $z_{i,n+p}$
      and $z_{i,n+r}$ $(i = 1, \ldots, m)$, and exchange $row(p)$ and $row(r)$.
      If $q \neq r$ **then** exchange z_{iq} and z_{ir} $(i = 1, \ldots, m)$, and exchange $col(q)$
      and $col(r)$.

③   Set $d := d \cdot z_{rr}$.
      For $i := r + 1$ **to** m **do:**
         Set $\alpha := \frac{z_{ir}}{z_{rr}}$.
         For $j := r$ **to** $n+r$ **do:** $z_{ij} := z_{ij} - \alpha z_{rj}$.
      Go to ②.

④   **For** $k := r$ **down to** 2 **do:**
         For $i := 1$ **to** $k - 1$ **do:**
             Set $\alpha := \frac{z_{ik}}{z_{kk}}$.
             For $j := k$ **to** $n+r$ **do** $z_{ij} := z_{ij} - \alpha z_{kj}$.

⑤   **For** $k := 1$ **to** r **do:**
         Set $\alpha := \frac{1}{z_{kk}}$.
         For $j := 1$ **to** $n+r$ **do** $z_{kj} := \alpha z_{kj}$.

Theorem 4.9. GAUSSIAN ELIMINATION *works correctly and terminates after*
$O(mnr)$ *steps.*

Proof: First observe that each time before ② we have $z_{ii} \neq 0$ for $i \in \{1, \ldots, r\}$
and $z_{ij} = 0$ for all $j \in \{1, \ldots, r\}$ and $i \in \{j+1, \ldots, m\}$. Hence

$$\det\left((z_{ij})_{i,j \in \{1,2,\ldots,r\}}\right) \ = \ z_{11} z_{22} \cdots z_{rr} \ = \ d \ \neq \ 0.$$

Since adding a multiple of one row to another row of a square matrix does not
change the value of the determinant (this well-known fact follows directly from
the definition (4.1)) we have

$$\det\left((z_{ij})_{i,j \in \{1,2,\ldots,r\}}\right) \ = \ \det\left((a_{row(i),col(j)})_{i,j \in \{1,2,\ldots,r\}}\right)$$

at any stage before ⑤, and hence the determinant d is computed correctly. A' is
a nonsingular $r \times r$-submatrix of A. Since $(z_{ij})_{i \in \{1,\ldots,m\}, j \in \{1,\ldots,n\}}$ has rank r at
termination and the operations did not change the rank, A has also rank r.

Moreover, $\sum_{j=1}^{m} z_{i,n+j} a_{row(j),col(k)} = z_{ik}$ for all $i \in \{1, \ldots, m\}$ and $k \in$
$\{1, \ldots, n\}$ (i.e. $C\tilde{A} = B$ in our above notation) holds throughout. (Note that for
$j = r+1, \ldots, m$ we have at any stage $z_{j,n+j} = 1$ and $z_{i,n+j} = 0$ for $i \neq j$.)
Since $(z_{ij})_{i,j \in \{1,2,\ldots,r\}}$ is the unit matrix at termination this implies that $(A')^{-1}$ is

also computed correctly. The number of steps is obviously $O(rmn + r^2(n+r)) = O(mnr)$.   □

In order to prove that GAUSSIAN ELIMINATION is a polynomial-time algorithm we have to guarantee that all numbers that occur are polynomially bounded by the input size. This is not trivial but can be shown:

Theorem 4.10.   (Edmonds [1967]) GAUSSIAN ELIMINATION *is a polynomial-time algorithm. Each number occurring in the course of the algorithm can be stored with* $O(m(m+n)\,\mathrm{size}(A))$ *bits.*

Proof: We first show that in ② and ③ all numbers are 0, 1, or quotients of subdeterminants of A. First observe that entries z_{ij} with $i \leq r$ or $j \leq r$ are not modified anymore. Entries z_{ij} with $j > n+r$ are 0 (if $j \neq n+i$) or 1 (if $j = n+i$). Furthermore, we have for all $s \in \{r+1, \ldots, m\}$ and $t \in \{r+1, \ldots, n+m\}$

$$z_{st} = \frac{\det\left((z_{ij})_{i \in \{1,2,\ldots,r,s\}, j \in \{1,2,\ldots,r,t\}}\right)}{\det\left((z_{ij})_{i,j \in \{1,2,\ldots,r\}}\right)}.$$

(This follows from evaluating the determinant $\det\left((z_{ij})_{i \in \{1,2,\ldots,r,s\}, j \in \{1,2,\ldots,r,t\}}\right)$ along the last row because $z_{sj} = 0$ for all $s \in \{r+1, \ldots, m\}$ and all $j \in \{1, \ldots, r\}$.)

We have already observed in the proof of Theorem 4.9 that

$$\det\left((z_{ij})_{i,j \in \{1,2,\ldots,r\}}\right) = \det\left((a_{row(i),col(j)})_{i,j \in \{1,2,\ldots,r\}}\right),$$

because adding a multiple of one row to another row of a square matrix does not change the value of the determinant. By the same argument we have

$$\det\left((z_{ij})_{i \in \{1,2,\ldots,r,s\}, j \in \{1,2,\ldots,r,t\}}\right) = \det\left((a_{row(i),col(j)})_{i \in \{1,2,\ldots,r,s\}, j \in \{1,2,\ldots,r,t\}}\right)$$

for $s \in \{r+1, \ldots, m\}$ and $t \in \{r+1, \ldots, n\}$. Furthermore,

$$\det\left((z_{ij})_{i \in \{1,2,\ldots,r,s\}, j \in \{1,2,\ldots,r,n+t\}}\right) = \det\left((a_{row(i),col(j)})_{i \in \{1,2,\ldots,r,s\} \setminus \{t\}, j \in \{1,2,\ldots,r\}}\right)$$

for all $s \in \{r+1, \ldots, m\}$ and $t \in \{1, \ldots, r\}$, which is checked by evaluating the left-hand side determinant (after ①) along column $n+t$.

We conclude that at any stage in ② and ③ all numbers z_{ij} are 0, 1, or quotients of subdeterminants of A. Hence, by Proposition 4.3, each number occurring in ② and ③ can be stored with $O(\mathrm{size}(A))$ bits.

Finally observe that ④ is equivalent to applying ② and ③ again, choosing p and q appropriately (reversing the order of the first r rows and columns). Hence each number occurring in ④ can be stored with $O\left(\mathrm{size}\left((z_{ij})_{i \in \{1,\ldots,m\}, j \in \{1,\ldots,m+n\}}\right)\right)$ bits, which is $O(m(m+n)\,\mathrm{size}(A))$.

The easiest way to keep the representations of the numbers z_{ij} small enough is to guarantee that the numerator and denominator of each of these numbers are relatively prime at any stage. This can be accomplished by applying the EUCLIDEAN ALGORITHM after each computation. This gives an overall polynomial running time.   □

In fact, we can easily implement GAUSSIAN ELIMINATION to be a strongly polynomial-time algorithm (Exercise 4).

So we can check in polynomial time whether a set of vectors is linearly independent, and we can compute the determinant and the inverse of a nonsingular matrix in polynomial time (exchanging two rows or columns changes just the sign of the determinant). Moreover we get:

Corollary 4.11. *Given a matrix $A \in \mathbb{Q}^{m \times n}$ and a vector $b \in \mathbb{Q}^m$ we can in polynomial time find a vector $x \in \mathbb{Q}^n$ with $Ax = b$ or decide that no such vector exists.*

Proof: We compute a maximal nonsingular submatrix $A' = (a_{row(i),col(j)})_{i,j \in \{1,...,r\}}$ of A and its inverse $(A')^{-1} = (z_{i,n+j})_{i,j \in \{1,...,r\}}$ by GAUSSIAN ELIMINATION. Then we set $x_{col(j)} := \sum_{k=1}^{r} z_{j,n+k} b_{row(k)}$ for $j = 1, \ldots, r$ and $x_k := 0$ for $k \notin \{col(1), \ldots, col(r)\}$. We obtain for $i = 1, \ldots r$:

$$
\begin{aligned}
\sum_{j=1}^{n} a_{row(i),j} x_j &= \sum_{j=1}^{r} a_{row(i),col(j)} x_{col(j)} \\
&= \sum_{j=1}^{r} a_{row(i),col(j)} \sum_{k=1}^{r} z_{j,n+k} b_{row(k)} \\
&= \sum_{k=1}^{r} b_{row(k)} \sum_{j=1}^{r} a_{row(i),col(j)} z_{j,n+k} \\
&= b_{row(i)}.
\end{aligned}
$$

Since the other rows of A with indices not in $\{row(1), \ldots, row(r)\}$ are linear combinations of these, either x satisfies $Ax = b$ or no vector satisfies this system of equations. $\qquad\square$

4.4 The Ellipsoid Method

In this section we describe the so-called ellipsoid method, developed by Iudin and Nemirovskii [1976] and Shor [1977] for nonlinear optimization. Khachiyan [1979] observed that it can be modified in order to solve LPs in polynomial time. Most of our presentation is based on (Grötschel, Lovász and Schrijver [1981]), (Bland, Goldfarb and Todd [1981]) and the book of Grötschel, Lovász and Schrijver [1988], which is also recommended for further study.

The idea of the ellipsoid method is very roughly the following. In Proposition 4.16 we will show that it is sufficient to find a feasible solution to an LP or decide that none exists. We start with an ellipsoid that we know a priori to contain the basic solutions (e.g. a large ball). At each iteration k, we check if the center x_k of the current ellipsoid is a feasible solution. Otherwise, we take a hyperplane containing x_k such that all the solutions lie on one side of this hyperplane. Now we

have a half-ellipsoid which contains all solutions. We take the smallest ellipsoid
completely containing this half-ellipsoid and continue.

Definition 4.12.   *An* **ellipsoid** *is a set* $E(A, x) = \{z \in \mathbb{R}^n : (z-x)^\top A^{-1}(z-x) \leq 1\}$ *for some symmetric positive definite $n \times n$-matrix A.*

Note that $B(x, r) := E(r^2 I, x)$ (with I being the $n \times n$ unit matrix) is the
n-dimensional Euclidean ball with center x and radius r.

The volume of an ellipsoid $E(A, x)$ is known to be

$$\text{volume}\,(E(A, x)) \;=\; \sqrt{\det A}\;\,\text{volume}\,(B(0, 1))$$

(see Exercise 7). Given an ellipsoid $E(A, x)$ and a hyperplane $\{z : az = ax\}$,
the smallest ellipsoid $E(A', x')$ containing the half-ellipsoid $E' = \{z \in E(A, x) : az \geq ax\}$ is called the Löwner-John ellipsoid of E' (see Figure 4.1). It can be
computed by the following formulas:

$$A' \;=\; \frac{n^2}{n^2 - 1}\left(A - \frac{2}{n+1}bb^\top\right),$$

$$x' \;=\; x + \frac{1}{n+1}b,$$

$$b \;=\; \frac{1}{\sqrt{a^\top A a}}Aa.$$

One difficulty of the ellipsoid method is caused by the square root in the
computation of b. Because we have to tolerate rounding errors, it is necessary to
increase the radius of the next ellipsoid a little bit. Here is an algorithmic scheme
that takes care of this problem:

ELLIPSOID METHOD

Input: A number $n \in \mathbb{N}$, $n \geq 2$. A number $N \in \mathbb{N}$. $x_0 \in \mathbb{Q}^n$ and $R \in \mathbb{Q}_+$,
           $R \geq 2$.

Output: An ellipsoid $E(A_N, x_N)$.

---

① Set $p := \lceil 6N + \log(9n^3)\rceil$.
   Set $A_0 := R^2 I$, where I is the $n \times n$ unit matrix.
   Set $k := 0$.

② Choose any $a_k \in \mathbb{Q}^n \setminus \{0\}$.

③ Set $b_k := \dfrac{1}{\sqrt{a_k^\top A_k a_k}}A_k a_k$.

   Set $x_{k+1} :\approx x_{k+1}^* := x_k + \dfrac{1}{n+1}b_k$.

   Set $A_{k+1} :\approx A_{k+1}^* := \dfrac{2n^2 + 3}{2n^2}\left(A_k - \dfrac{2}{n+1}b_k b_k^\top\right)$.

   (Here $:\approx$ means computing the entries up to p binary digits behind
   the point, taking care that A_{k+1} is symmetric).

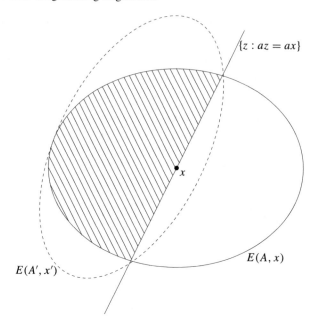

$\{z : az = ax\}$

$E(A, x)$

$E(A', x')$

Fig. 4.1.

④   Set $k := k + 1$.
    If $k < N$ then go to ② **else stop**.

So in each of the N iterations an approximation $E(A_{k+1}, x_{k+1})$ of the smallest ellipsoid containing $E(A_k, x_k) \cap \{z : a_k z \geq a_k x_k\}$ is computed. Two main issues, how to obtain the a_k and how to choose N, will be addressed in the next section. But let us first prove some lemmas.

Let $||x||$ denote the Euclidean norm of vector x, while $||A|| := \max\{||Ax|| : ||x|| = 1\}$ shall denote the norm of the matrix A. For symmetric matrices, $||A||$ is the maximum absolute value of an eigenvalue and $||A|| = \max\{x^\top Ax : ||x|| = 1\}$.

The first lemma says that each $E_k := E(A_k, x_k)$ is indeed an ellipsoid. Furthermore, the absolute values of the numbers involved remain smaller than $R^2 2^N + 2^{\text{size}(x_0)}$. Therefore each iteration of the ELLIPSOID METHOD consists of $O(n^2)$ computational steps, each involving numbers with $O(p + \text{size}(a_k) + \text{size}(R) + \text{size}(x_0))$ bits.

Lemma 4.13. (Grötschel, Lovász and Schrijver [1981]) *Let $k \in \{0, 1, \dots, N\}$. Then A_k is positive definite, and we have*

$$||x_k|| \leq ||x_0|| + R2^k, \qquad ||A_k|| \leq R^2 2^k, \qquad and \qquad ||A_k^{-1}|| \leq R^{-2} 4^k.$$

Proof: We use induction on k. For $k = 0$ all the statements are obvious. Assume that they are true for some $k \geq 0$. By a straightforward computation one verifies that

$$(A_{k+1}^*)^{-1} = \frac{2n^2}{2n^2 + 3} \left(A_k^{-1} + \frac{2}{n-1} \frac{a_k a_k^\top}{a_k^\top A_k a_k} \right). \tag{4.2}$$

So $(A_{k+1}^*)^{-1}$ is the sum of a positive definite and a positive semidefinite matrix; thus it is positive definite. Hence A_{k+1}^* is also positive definite.

Note that for positive semidefinite matrices A and B we have $||A|| \leq ||A + B||$. Therefore

$$||A_{k+1}^*|| = \frac{2n^2 + 3}{2n^2} \left\| A_k - \frac{2}{n+1} b_k b_k^\top \right\| \leq \frac{2n^2 + 3}{2n^2} ||A_k|| \leq \frac{11}{8} R^2 2^k.$$

Since the $n \times n$ all-one matrix has norm n, the matrix $A_{k+1} - A_{k+1}^*$, each of whose entries has absolute value at most 2^{-p}, has norm at most $n2^{-p}$. We conclude

$$||A_{k+1}|| \leq ||A_{k+1}^*|| + ||A_{k+1} - A_{k+1}^*|| \leq \frac{11}{8} R^2 2^k + n2^{-p} \leq R^2 2^{k+1}$$

(here we used the very rough estimate $2^{-p} \leq \frac{1}{n}$).

It is well-known from linear algebra that for any symmetric positive definite $n \times n$-matrix A there exists a symmetric positive definite matrix B with $A = BB$. Writing $A_k = BB$ with $B = B^\top$ we obtain

$$||b_k|| = \frac{||A_k a_k||}{\sqrt{a_k^\top A_k a_k}} = \sqrt{\frac{a_k^\top A_k^2 a_k}{a_k^\top A_k a_k}} = \sqrt{\frac{(Ba_k)^\top A_k (Ba_k)}{(Ba_k)^\top (Ba_k)}} \leq \sqrt{||A_k||} \leq R2^{k-1}.$$

Using this (and again the induction hypothesis) we get

$$||x_{k+1}|| \leq ||x_k|| + \frac{1}{n+1} ||b_k|| + ||x_{k+1} - x_{k+1}^*||$$

$$\leq ||x_0|| + R2^k + \frac{1}{n+1} R2^{k-1} + \sqrt{n}2^{-p} \leq ||x_0|| + R2^{k+1}.$$

Using (4.2) and $||a_k a_k^\top|| = a_k^\top a_k$ we compute

$$\left\| (A_{k+1}^*)^{-1} \right\| \leq \frac{2n^2}{2n^2 + 3} \left(\left\| A_k^{-1} \right\| + \frac{2}{n-1} \frac{a_k^\top a_k}{a_k^\top A_k a_k} \right) \tag{4.3}$$

$$= \frac{2n^2}{2n^2 + 3} \left(\left\| A_k^{-1} \right\| + \frac{2}{n-1} \frac{a_k^\top B A_k^{-1} B a_k}{a_k^\top B B a_k} \right)$$

$$\leq \frac{2n^2}{2n^2 + 3} \left(\left\| A_k^{-1} \right\| + \frac{2}{n-1} \left\| A_k^{-1} \right\| \right) < \frac{n+1}{n-1} \left\| A_k^{-1} \right\|$$

$$\leq 3R^{-2} 4^k.$$

Let λ be the smallest eigenvalue of A_{k+1}, and let v be a corresponding eigenvector with $||v|| = 1$. Then – writing $A_{k+1}^* = CC$ for a symmetric matrix C – we have

$$
\begin{aligned}
\lambda \;=\; & v^\top A_{k+1} v \;=\; v^\top A_{k+1}^* v + v^\top (A_{k+1} - A_{k+1}^*) v \\
=\; & \frac{v^\top C C v}{v^\top C \left(A_{k+1}^*\right)^{-1} C v} + v^\top (A_{k+1} - A_{k+1}^*) v \\
\geq\; & \left\| \left(A_{k+1}^*\right)^{-1} \right\|^{-1} - \|A_{k+1} - A_{k+1}^*\| \;>\; \frac{1}{3} R^2 4^{-k} - n 2^{-p} \;\geq\; R^2 4^{-(k+1)},
\end{aligned}
$$

where we used $2^{-p} \leq \frac{1}{3n} 4^{-k}$. Since $\lambda > 0$, A_{k+1} is positive definite. Furthermore,

$$
\left\| \left(A_{k+1}\right)^{-1} \right\| \;=\; \frac{1}{\lambda} \;\leq\; R^{-2} 4^{k+1}.
$$

$\square$

Next we show that in each iteration the ellipsoid contains the intersection of E_0 and the previous half-ellipsoid:

Lemma 4.14. *For $k = 0, \ldots, N-1$ we have $E_{k+1} \supseteq \{x \in E_k \cap E_0 : a_k x \geq a_k x_k\}$.*

Proof: Let $x \in E_k \cap E_0$ with $a_k x \geq a_k x_k$. We first compute (using (4.2))

$$
\begin{aligned}
& (x - x_{k+1}^*)^\top (A_{k+1}^*)^{-1} (x - x_{k+1}^*) \\
=\; & \frac{2n^2}{2n^2+3} \left(x - x_k - \frac{1}{n+1} b_k\right)^\top \left(A_k^{-1} + \frac{2}{n-1} \frac{a_k a_k^\top}{a_k^\top A_k a_k}\right) \left(x - x_k - \frac{1}{n+1} b_k\right) \\
=\; & \frac{2n^2}{2n^2+3} \Bigg((x - x_k)^\top A_k^{-1} (x - x_k) + \frac{2}{n-1} (x - x_k)^\top \frac{a_k a_k^\top}{a_k^\top A_k a_k} (x - x_k) \\
& + \frac{1}{(n+1)^2} \left(b_k^\top A_k^{-1} b_k + \frac{2}{n-1} \frac{b_k^\top a_k a_k^\top b_k}{a_k^\top A_k a_k}\right) \\
& - \frac{2(x - x_k)^\top}{n+1} \left(A_k^{-1} b_k + \frac{2}{n-1} \frac{a_k a_k^\top b_k}{a_k^\top A_k a_k}\right) \Bigg) \\
=\; & \frac{2n^2}{2n^2+3} \Bigg((x - x_k)^\top A_k^{-1} (x - x_k) + \frac{2}{n-1} (x - x_k)^\top \frac{a_k a_k^\top}{a_k^\top A_k a_k} (x - x_k) + \\
& \frac{1}{(n+1)^2} \left(1 + \frac{2}{n-1}\right) - \frac{2}{n+1} \frac{(x - x_k)^\top a_k}{\sqrt{a_k^\top A_k a_k}} \left(1 + \frac{2}{n-1}\right) \Bigg).
\end{aligned}
$$

Since $x \in E_k$, we have $(x - x_k)^\top A_k^{-1} (x - x_k) \leq 1$. By abbreviating $t := \dfrac{a_k^\top (x - x_k)}{\sqrt{a_k^\top A_k a_k}}$ we obtain

$$
(x - x_{k+1}^*)^\top (A_{k+1}^*)^{-1} (x - x_{k+1}^*) \;\leq\; \frac{2n^2}{2n^2+3} \left(1 + \frac{2}{n-1} t^2 + \frac{1}{n^2-1} - \frac{2}{n-1} t\right).
$$

Since $b_k^\top A_k^{-1} b_k = 1$ and $b_k^\top A_k^{-1}(x - x_k) = t$, we have

$$
\begin{aligned}
1 &\geq (x - x_k)^\top A_k^{-1}(x - x_k) \\
&= (x - x_k - tb_k)^\top A_k^{-1}(x - x_k - tb_k) + t^2 \\
&\geq t^2,
\end{aligned}
$$

because A_k^{-1} is positive definite. So (using $a_k x \geq a_k x_k$) we have $0 \leq t \leq 1$ and obtain

$$
(x - x_{k+1}^*)^\top (A_{k+1}^*)^{-1}(x - x_{k+1}^*) \leq \frac{2n^4}{2n^4 + n^2 - 3}.
$$

It remains to estimate the rounding error

$$
\begin{aligned}
Z :=&\ \left| (x - x_{k+1})^\top (A_{k+1})^{-1}(x - x_{k+1}) - (x - x_{k+1}^*)^\top (A_{k+1}^*)^{-1}(x - x_{k+1}^*) \right| \\
\leq&\ \left| (x - x_{k+1})^\top (A_{k+1})^{-1}(x_{k+1}^* - x_{k+1}) \right| \\
&+ \left| (x_{k+1}^* - x_{k+1})^\top (A_{k+1})^{-1}(x - x_{k+1}^*) \right| \\
&+ \left| (x - x_{k+1}^*)^\top \left((A_{k+1})^{-1} - (A_{k+1}^*)^{-1} \right)(x - x_{k+1}^*) \right| \\
\leq&\ \|x - x_{k+1}\|\, \|(A_{k+1})^{-1}\|\, \|x_{k+1}^* - x_{k+1}\| \\
&+ \|x_{k+1}^* - x_{k+1}\|\, \|(A_{k+1})^{-1}\|\, \|x - x_{k+1}^*\| \\
&+ \|x - x_{k+1}^*\|^2\, \|(A_{k+1})^{-1}\|\, \|(A_{k+1}^*)^{-1}\|\, \|A_{k+1}^* - A_{k+1}\|.
\end{aligned}
$$

Using Lemma 4.13 and $x \in E_0$ we get $\|x - x_{k+1}\| \leq \|x - x_0\| + \|x_{k+1} - x_0\| \leq R + R2^N$ and $\|x - x_{k+1}^*\| \leq \|x - x_{k+1}\| + \sqrt{n}2^{-P} \leq R2^{N+1}$. We also use (4.3) and obtain

$$
\begin{aligned}
Z &\leq 2(R2^{N+1})(R^{-2}4^N)(\sqrt{n}2^{-P}) + (R^2 4^{N+1})(R^{-2}4^N)(3R^{-2}4^{N-1})(n2^{-P}) \\
&= 4R^{-1}2^{3N}\sqrt{n}2^{-P} + 3R^{-2}2^{6N}n2^{-P} \\
&\leq 2^{6N}n2^{-P} \\
&\leq \frac{1}{9n^2},
\end{aligned}
$$

by definition of p. Altogether we have

$$
(x - x_{k+1})^\top (A_{k+1})^{-1}(x - x_{k+1}) \leq \frac{2n^4}{2n^4 + n^2 - 3} + \frac{1}{9n^2} \leq 1.
$$

$\square$

The volumes of the ellipsoids decrease by a constant factor in each iteration:

Lemma 4.15. *For $k = 0, \ldots, N - 1$ we have* $\dfrac{\text{volume}(E_{k+1})}{\text{volume}(E_k)} < e^{-\frac{1}{5n}}.$

Proof: (Grötschel, Lovász and Schrijver [1988]) We write

$$\frac{\text{volume}(E_{k+1})}{\text{volume}(E_k)} = \sqrt{\frac{\det A_{k+1}}{\det A_k}} = \sqrt{\frac{\det A_{k+1}^*}{\det A_k}} \sqrt{\frac{\det A_{k+1}}{\det A_{k+1}^*}}$$

and estimate the two factors independently. First observe that

$$\frac{\det A_{k+1}^*}{\det A_k} = \left(\frac{2n^2+3}{2n^2}\right)^n \det\left(I - \frac{2}{n+1}\frac{a_k a_k^\top A_k}{a_k^\top A_k a_k}\right).$$

The matrix $\frac{a_k a_k^\top A_k}{a_k^\top A_k a_k}$ has rank one and 1 as its only nonzero eigenvalue (eigenvector a_k). Since the determinant is the product of the eigenvalues, we conclude that

$$\frac{\det A_{k+1}^*}{\det A_k} = \left(\frac{2n^2+3}{2n^2}\right)^n \left(1 - \frac{2}{n+1}\right) < e^{\frac{3}{2n}} e^{-\frac{2}{n}} = e^{-\frac{1}{2n}},$$

where we used $1 + x \le e^x$ for all x and $\left(\frac{n-1}{n+1}\right)^n < e^{-2}$ for $n \ge 2$.

For the second estimation we use (4.3) and the well-known fact that $\det B \le \|B\|^n$ for any matrix B:

$$\begin{aligned}
\frac{\det A_{k+1}}{\det A_{k+1}^*} &= \det\left(I + (A_{k+1}^*)^{-1}(A_{k+1} - A_{k+1}^*)\right) \\
&\le \left\|I + (A_{k+1}^*)^{-1}(A_{k+1} - A_{k+1}^*)\right\|^n \\
&\le \left(\|I\| + \|(A_{k+1}^*)^{-1}\| \, \|A_{k+1} - A_{k+1}^*\|\right)^n \\
&\le \left(1 + (R^{-2}4^{k+1})(n2^{-p})\right)^n \\
&\le \left(1 + \frac{1}{10n^2}\right)^n \\
&\le e^{\frac{1}{10n}}
\end{aligned}$$

(we used $2^{-p} \le \frac{4}{10n^3 4^N} \le \frac{R^2}{10n^3 4^{k+1}}$).

We conclude that

$$\frac{\text{volume}(E_{k+1})}{\text{volume}(E_k)} = \sqrt{\frac{\det A_{k+1}^*}{\det A_k}} \sqrt{\frac{\det A_{k+1}}{\det A_{k+1}^*}} \le e^{-\frac{1}{4n}} e^{\frac{1}{20n}} = e^{-\frac{1}{5n}}.$$

$\square$

4.5 Khachiyan's Theorem

In this section we shall prove Khachiyan's theorem: the ELLIPSOID METHOD can be applied to LINEAR PROGRAMMING in order to obtain a polynomial-time algorithm. Let us first prove that it suffices to have an algorithm for checking feasibility of linear inequality systems:

Proposition 4.16. *Suppose there is a polynomial-time algorithm for the following problem: "Given a matrix $A \in \mathbb{Q}^{m \times n}$ and a vector $b \in \mathbb{Q}^m$, decide if $\{x : Ax \leq b\}$ is empty." Then there is a polynomial-time algorithm for* LINEAR PROGRAMMING *which finds an optimum basic solution if there exists one.*

Proof: Let an LP max$\{cx : Ax \leq b\}$ be given. We first check if the primal and dual LPs are both feasible. If at least one of them is infeasible, we are done by Theorem 3.27. Otherwise, by Corollary 3.21, it is sufficient to find an element of $\{(x, y) : Ax \leq b, yA = c, y \geq 0, cx = yb\}$.

We show (by induction on k) that a solution of a feasible system of k inequalities and l equalities can be found by k calls to the subroutine checking emptiness of polyhedra plus additional polynomial-time work. For $k = 0$ a solution can be found easily by GAUSSIAN ELIMINATION (Corollary 4.11).

Now let $k > 0$. Let $ax \leq \beta$ be an inequality of the system. By a call to the subroutine we check whether the system becomes infeasible by replacing $ax \leq \beta$ by $ax = \beta$. If so, the inequality is redundant and can be removed (cf. Proposition 3.8). If not, we replace it by the equality. In both cases we reduced the number of inequalities by one, so we are done by induction.

If there exists an optimum basic solution, the above procedure generates one, because the final equality system contains a maximal feasible subsystem of $Ax = b$. □

Before we can apply the ELLIPSOID METHOD, we have to take care that the polyhedron is bounded and full-dimensional:

Proposition 4.17.   (Khachiyan [1979], Gács and Lovász [1981]) *Let $A \in \mathbb{Q}^{m \times n}$ and $b \in \mathbb{Q}^m$. The system $Ax \leq b$ has a solution if and only if the system*

$$Ax \leq b + \epsilon \mathbb{1}, \quad -R\mathbb{1} \leq x \leq R\mathbb{1}$$

has a solution, where $\mathbb{1}$ is the all-one vector, $\frac{1}{\epsilon} = 2n2^{4(\text{size}(A)+\text{size}(b))}$ and $R = 1 + 2^{4(\text{size}(A)+\text{size}(b))}$.

If $Ax \leq b$ has a solution, then volume $(\{x \in \mathbb{R}^n : Ax \leq b + \epsilon \mathbb{1}, -R\mathbb{1} \leq x \leq R\mathbb{1}\}) \geq \left(\frac{2\epsilon}{n2^{\text{size}(A)}}\right)^n$.

Proof: The box constraints $-R\mathbb{1} \leq x \leq R\mathbb{1}$ do not change the solvability by Theorem 4.4. Now suppose that $Ax \leq b$ has no solution. By Theorem 3.24 (a version of Farkas' Lemma), there is a vector $y \geq 0$ with $yA = 0$ and $yb = -1$. By applying Theorem 4.4 to min$\{\mathbb{1}y : y \geq 0, A^\top y = 0, b^\top y = -1\}$ we conclude that y can be chosen such that its components are of absolute value less than $2^{4(\text{size}(A)+\text{size}(b))}$. Therefore $y(b + \epsilon \mathbb{1}) < -1 + (n + 1)2^{4(\text{size}(A)+\text{size}(b))}\epsilon \leq 0$. Again by Theorem 3.24, this proves that $Ax \leq b + \epsilon \mathbb{1}$ has no solution.

For the second statement, if $x \in \mathbb{R}^n$ with $Ax \leq b$ has components of absolute value at most $R-1$ (cf. Theorem 4.4), then $\{x \in \mathbb{R}^n : Ax \leq b+\epsilon \mathbb{1}, -R\mathbb{1} \leq x \leq R\mathbb{1}\}$ contains all points z with $||z - x||_\infty \leq \frac{\epsilon}{n2^{\text{size}(A)}}$. □

Note that the construction of this proposition increases the size of the system of inequalities by at most a factor of $O(m + n)$.

Theorem 4.18. (Khachiyan [1979]) *There exists a polynomial-time algorithm for* LINEAR PROGRAMMING *(with rational input), and this algorithm finds an optimum basic solution if there exists one.*

Proof: By Proposition 4.16 it suffices to check feasibility of a system $Ax \leq b$. We transform the system as in Proposition 4.17 in order to obtain a polytope P which is either empty or has volume at least $\left(\frac{2\epsilon}{n2^{\text{size}(A)}}\right)^n$.

We run the ELLIPSOID METHOD with $x_0 = 0$, $R = n\left(1 + 2^{4(\text{size}(A)+\text{size}(b))}\right)$, $N = \lceil 10n^2(2\log n + 5(\text{size}(A) + \text{size}(b)))\rceil$. Each time in ② we check whether $x_k \in P$. If yes, we are done. Otherwise we take a violated inequality $ax \leq \beta$ of the system $Ax \leq b$ and set $a_k := -a$.

We claim that if the algorithm does not find an $x_k \in P$ before iteration N, then P must be empty. To see this, we first observe that $P \subseteq E_k$ for all k: for $k = 0$ this is clear by the construction of P and R; the induction step is Lemma 4.14. So we have $P \subseteq E_N$.

By Lemma 4.15, we have, abbreviating $s := \text{size}(A) + \text{size}(b)$,

$$
\begin{aligned}
\text{volume}(E_N) &\leq \text{volume}(E_0)e^{-\frac{N}{5n}} \leq (2R)^n e^{-\frac{N}{5n}} \\
&< \left(2n\left(1 + 2^{4s}\right)\right)^n n^{-4n} e^{-10ns} < n^{-2n}2^{-5ns}.
\end{aligned}
$$

On the other hand, $P \neq \emptyset$ implies

$$
\text{volume}(P) \geq \left(\frac{2\epsilon}{n2^s}\right)^n = \left(\frac{1}{n^2 2^{5s}}\right)^n = n^{-2n}2^{-5ns},
$$

which is a contradiction. $\square$

If we estimate the running time for solving an LP $\max\{cx : Ax \leq b\}$ with the above method, we get the bound $O((n + m)^9(\text{size}(A) + \text{size}(b) + \text{size}(c))^2)$ (Exercise 9), which is polynomial but completely useless for practical purposes. In practice, either the SIMPLEX ALGORITHM or interior point algorithms are used. Karmarkar [1984] was the first to describe a polynomial-time interior point algorithm for LINEAR PROGRAMMING. The currently fastest version is due to Lee and Sidford [2014]. We shall not go into the details here.

A strongly polynomial-time algorithm for LINEAR PROGRAMMING is not known. However, Tardos [1986] showed that there is an algorithm for solving $\max\{cx : Ax \leq b\}$ with a running time that polynomially depends on $\text{size}(A)$ only. For many combinatorial optimization problems, where A is a 0-1-matrix, this gives a strongly polynomial-time algorithm. Tardos' result was extended by Frank and Tardos [1987].

4.6 Separation and Optimization

The above method (in particular Proposition 4.16) requires that the polyhedron be given explicitly by a list of inequalities. However, a closer look shows that this is not really necessary. It is sufficient to have a subroutine which – given a vector x – decides if $x \in P$ or otherwise returns a separating hyperplane, i.e. a vector a such that $ax > \max\{ay : y \in P\}$. We shall prove this for full-dimensional polytopes; for the general (more complicated) case we refer to Grötschel, Lovász and Schrijver [1988] (or Padberg [1999]). The results in this section are due to Grötschel, Lovász and Schrijver [1981] and independently to Karp and Papadimitriou [1982] and Padberg and Rao [1981].

With the results of this section one can solve certain linear programs in polynomial time although the polytope has an exponential number of facets. Many examples will be discussed later in this book; see e.g. Corollary 12.22 or Theorem 20.40. By considering the dual LP one can also deal with linear programs with a huge number of variables.

Let $P \subseteq \mathbb{R}^n$ be a full-dimensional polytope, or more generally, a full-dimensional bounded convex set. We assume that we know the dimension n and two balls $B(x_0, r)$ and $B(x_0, R)$ such that $B(x_0, r) \subseteq P \subseteq B(x_0, R)$. But we do not assume that we know a linear inequality system defining P. In fact, this would not make sense if we want to solve linear programs with an exponential number of constraints in polynomial time, or even optimize linear objective functions over nonlinearly constrained convex sets.

Below we shall prove that, under some reasonable assumptions, we can optimize a linear function over a polyhedron P in polynomial time (independent of the number of constraints) if we have a so-called **separation oracle**: a subroutine for the following problem:

SEPARATION PROBLEM

Instance: A convex set $P \subseteq \mathbb{R}^n$. A vector $y \in \mathbb{Q}^n$.

Task: Either decide that $y \in P$
   or find a vector $d \in \mathbb{Q}^n$ such that $dx < dy$ for all $x \in P$.

Note that such a vector d exists if P is a rational polyhedron or a compact convex set (cf. Exercise 21 of Chapter 3). Given a convex set P by such a separation oracle, we look for an **oracle algorithm** using this as a black box. In an oracle algorithm we may ask the oracle at any time and we get a correct answer in one step. We can regard this concept as a subroutine whose running time we do not take into account. (In Chapter 15 we shall give a formal definition.)

Indeed, it often suffices to have an oracle which solves the SEPARATION PROBLEM approximately. More precisely we assume an oracle for the following problem:

WEAK SEPARATION PROBLEM

Instance: A convex set $P \subseteq \mathbb{R}^n$, a vector $c \in \mathbb{Q}^n$ and a number $\epsilon > 0$. A vector $y \in \mathbb{Q}^n$.

Task: Either find a vector $y' \in P$ with $cy \leq cy' + \epsilon$
or find a vector $d \in \mathbb{Q}^n$ such that $dx < dy$ for all $x \in P$.

Using a weak separation oracle we first solve linear programs approximately:

WEAK OPTIMIZATION PROBLEM

Instance: A number $n \in \mathbb{N}$. A vector $c \in \mathbb{Q}^n$. A number $\epsilon > 0$.
A convex set $P \subseteq \mathbb{R}^n$ given by an oracle for the WEAK SEPARATION PROBLEM for P, c and $\frac{\epsilon}{2}$.

Task: Find a vector $y \in P$ with $cy \geq \sup\{cx : x \in P\} - \epsilon$.

Note that the above two definitions differ from the ones given e.g. in Grötschel, Lovász and Schrijver [1981]. However, they are basically equivalent, and we shall need the above form again in Section 18.3.

The following variant of the ELLIPSOID METHOD solves the WEAK OPTIMIZATION PROBLEM for bounded full-dimensional convex sets:

GRÖTSCHEL-LOVÁSZ-SCHRIJVER ALGORITHM

Input: A number $n \in \mathbb{N}$, $n \geq 2$. A vector $c \in \mathbb{Q}^n$. A number $0 < \epsilon \leq 1$.
A convex set $P \subseteq \mathbb{R}^n$ given by an oracle for the WEAK SEPARATION PROBLEM for P, c and $\frac{\epsilon}{2}$.
$x_0 \in \mathbb{Q}^n$ and r, $R \in \mathbb{Q}_+$ such that $B(x_0, r) \subseteq P \subseteq B(x_0, R)$.

Output: A vector $y^* \in P$ with $cy^* \geq \sup\{cx : x \in P\} - \epsilon$.

①   Set $R := \max\{R, 2\}$, $r := \min\{r, 1\}$ and $\gamma := \max\{\|c\|, 1\}$.
Set $N := 5n^2 \left\lceil \ln \frac{6R^2\gamma}{r\epsilon} \right\rceil$. Set $y^* := x_0$.

②   Run the ELLIPSOID METHOD, with a_k in ② being computed as follows:
Run the oracle for the WEAK SEPARATION PROBLEM with $y = x_k$.
If it returns a $y' \in P$ with $cy \leq cy' + \frac{\epsilon}{2}$ **then:**
  If $cy' > cy^*$ **then** set $y^* := y'$.
  Set $a_k := c$.
If it returns a $d \in \mathbb{Q}^n$ with $dx < dy$ for all $x \in P$ **then:**
  Set $a_k := -d$.

Theorem 4.19. *The* GRÖTSCHEL-LOVÁSZ-SCHRIJVER ALGORITHM *correctly solves the* WEAK OPTIMIZATION PROBLEM *for bounded full-dimensional convex sets. Its running time is bounded by*

$$O\left(n^6\alpha^2 + n^4\alpha f(\text{size}(c), \text{size}(\epsilon), n\,\text{size}(x_0) + n^3\alpha)\right),$$

where $\alpha = \log \frac{R^2\gamma}{r\epsilon}$ and $f(\text{size}(c), \text{size}(\epsilon), \text{size}(y))$ *is an upper bound of the running time of the oracle for the* WEAK SEPARATION PROBLEM *for P with input* c, ϵ, y.

Proof: (Grötschel, Lovász and Schrijver [1981]) The running time in each of the $N = O(n^2\alpha)$ iterations of the ELLIPSOID METHOD is $O(n^2(n^2\alpha + \text{size}(R) + \text{size}(x_0) + q))$ plus one oracle call, where q is the size of the output of the oracle. As $\text{size}(y) \leq n(\text{size}(x_0) + \text{size}(R) + N)$ by Lemma 4.13, the overall running time is $O(n^4\alpha(n^2\alpha + \text{size}(x_0) + f(\text{size}(c), \text{size}(\epsilon), n\,\text{size}(x_0) + n^3\alpha)))$, as stated.

By Lemma 4.14, we have

$$\left\{ x \in P : cx \geq cy^* + \frac{\epsilon}{2} \right\} \subseteq E_N.$$

Let $z \in P$ with $cz \geq \sup\{cx : x \in P\} - \frac{\epsilon}{6}$. We may assume that $cz > cy^* + \frac{\epsilon}{2}$; otherwise we are done.

Consider the convex hull U of z and the $(n-1)$-dimensional ball $B(x_0, r) \cap \{x : cx = cx_0\}$ (see Figure 4.2). We have $U \subseteq P$ and hence $U' := \{x \in U : cx \geq cy^* + \frac{\epsilon}{2}\}$ is contained in E_N. The volume of U' is

$$\begin{aligned}
\text{volume}(U') &= \text{volume}(U) \left(\frac{cz - cy^* - \frac{\epsilon}{2}}{cz - cx_0} \right)^n \\
&= V_{n-1} r^{n-1} \frac{cz - cx_0}{n\|c\|} \left(\frac{cz - cy^* - \frac{\epsilon}{2}}{cz - cx_0} \right)^n,
\end{aligned}$$

where V_n denotes the volume of the n-dimensional unit ball. Since $\text{volume}(U') \leq \text{volume}(E_N)$, and Lemma 4.15 yields

$$\text{volume}(E_N) \leq e^{-\frac{N}{5n}} \text{volume}(E_0) = e^{-\frac{N}{5n}} V_n R^n,$$

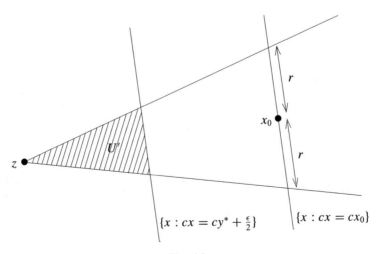

$\{x : cx = cy^* + \frac{\epsilon}{2}\}$ $\{x : cx = cx_0\}$

Fig. 4.2.

we have

$$cz - cy^* - \frac{\epsilon}{2} \leq e^{-\frac{N}{5n^2}} R \left(\frac{V_n(cz - cx_0)^{n-1} n ||c||}{V_{n-1} r^{n-1}} \right)^{\frac{1}{n}}.$$

Since $cz - cx_0 \leq ||c|| \cdot ||z - x_0|| \leq ||c|| R$ we obtain

$$cz - cy^* - \frac{\epsilon}{2} \leq ||c|| e^{-\frac{N}{5n^2}} R \left(\frac{n V_n R^{n-1}}{V_{n-1} r^{n-1}} \right)^{\frac{1}{n}} < 2||c|| e^{-\frac{N}{5n^2}} \frac{R^2}{r} \leq \frac{\epsilon}{3}$$

and hence $cy^* \geq cz - \frac{5}{6}\epsilon \geq \sup\{cx : x \in P\} - \epsilon$. □

Of course we are usually interested in the exact optimum. For this we restrict ourselves to rational full-dimensional polytopes. We need some assumption on the size of the vertices of the polytope.

Lemma 4.20. *Let $n \in \mathbb{N}$, let $P \subseteq \mathbb{R}^n$ be a rational polytope, and let $x_0 \in \mathbb{Q}^n$ be a point in the interior of P. Let $T \in \mathbb{N}$ such that $\mathrm{size}(x_0) \leq \log T$ and $\mathrm{size}(x) \leq \log T$ for all vertices x of P. Then $B(x_0, r) \subseteq P \subseteq B(x_0, R)$, where $r := \frac{1}{n} T^{-379n^2}$ and $R := 2nT$.*

Moreover, let $K := 4T^{2n+1}$. Let $c \in \mathbb{Z}^n$, and define $c' := K^n c + (1, K, \ldots, K^{n-1})$. Then $\max\{c'x : x \in P\}$ is attained by a unique vector x^*, for all other vertices y of P we have $c'(x^* - y) > T^{-2n}$, and x^* is also an optimum solution of $\max\{cx : x \in P\}$.

Proof: For any vertex x of P we have $||x|| \leq nT$ and $||x_0|| \leq nT$, so $||x - x_0|| \leq 2nT$ and $x \in B(x_0, R)$.

To show that $B(x_0, r) \subseteq P$, let $F = \{x \in P : ax = \beta\}$ be a facet of P, where by Lemma 4.5 we may assume that $\mathrm{size}(a) + \mathrm{size}(\beta) < 75n^2 \log T$. Suppose there is a point $y \in F$ with $||y - x_0|| < r$. Then

$$|ax_0 - \beta| = |ax_0 - ay| \leq ||a|| \cdot ||y - x_0|| < n 2^{\mathrm{size}(a)} r \leq T^{-304n^2}$$

But on the other hand the size of $ax_0 - \beta$ can by estimated by

$$\begin{aligned} \mathrm{size}(ax_0 - \beta) &\leq 4(\mathrm{size}(a) + \mathrm{size}(x_0) + \mathrm{size}(\beta)) \\ &\leq 300n^2 \log T + 4\log T \leq 304n^2 \log T. \end{aligned}$$

Since $ax_0 \neq \beta$ (x_0 is in the interior of P), this implies $|ax_0 - \beta| \geq T^{-304n^2}$, a contradiction.

To prove the last statements, let x^* be a vertex of P maximizing $c'x$, and let y be another vertex of P. By the assumption on the size of the vertices of P we may write $x^* - y = \frac{1}{\alpha} z$, where $\alpha \in \{1, 2, \ldots, T^{2n} - 1\}$ and z is an integral vector whose components have absolute value at most $\frac{K}{2}$. Then

$$0 \leq c'(x^* - y) = \frac{1}{\alpha} \left(K^n cz + \sum_{i=1}^{n} K^{i-1} z_i \right).$$

Since $K^n > \sum_{i=1}^n K^{i-1}|z_i|$, we must have $cz \geq 0$ and hence $cx^* \geq cy$. So x^* indeed maximizes cx over P. Moreover, since $z \neq 0$, we obtain

$$c'(x^* - y) \geq \frac{1}{\alpha} > T^{-2n},$$

as required. $\square$

Theorem 4.21. *Let $n \in \mathbb{N}$ and $c \in \mathbb{Q}^n$. Let $P \subseteq \mathbb{R}^n$ be a rational polytope, and let $x_0 \in \mathbb{Q}^n$ be a point in the interior of P. Let $T \in \mathbb{N}$ such that $\mathrm{size}(x_0) \leq \log T$ and $\mathrm{size}(x) \leq \log T$ for all vertices x of P.*

Given n, c, x_0, T and a polynomial-time oracle for the SEPARATION PROBLEM for P, a vertex x^* of P attaining $\max\{c^\top x : x \in P\}$ can be found in time polynomial in n, $\log T$ and $\mathrm{size}(c)$.

Proof: (Grötschel, Lovász and Schrijver [1981]) We first use the GRÖTSCHEL-LOVÁSZ-SCHRIJVER ALGORITHM to solve the WEAK OPTIMIZATION PROBLEM; we set c', r and R according to Lemma 4.20 and $\epsilon := \frac{1}{8nT^{2n+3}}$. (We first have to make c integral by multiplying with the product of its denominators; this increases its size by at most a factor $2n$.)

The GRÖTSCHEL-LOVÁSZ-SCHRIJVER ALGORITHM returns a vector $y \in P$ with $c'y \geq c'x^* - \epsilon$, where x^* is an optimum solution of $\max\{c'x : x \in P\}$. By Theorem 4.19 the running time is $O\left(n^6\alpha^2 + n^4\alpha f(\mathrm{size}(c'), \mathrm{size}(\epsilon), n\,\mathrm{size}(x_0) + n^3\alpha)\right) = O\left(n^6\alpha^2 + n^4\alpha f(\mathrm{size}(c'), 6n\log T, n\log T + n^3\alpha)\right)$, where $\alpha = \log\frac{R^2\max\{\|c'\|,1\}}{r\epsilon} \leq \log(16n^5T^{400n^2}2^{\mathrm{size}(c')}) = O(n^2\log T + \mathrm{size}(c'))$ and f is a polynomial upper bound of the running time of the oracle for the SEPARATION PROBLEM for P. Since $\mathrm{size}(c') \leq 6n^2\log T + 2\,\mathrm{size}(c)$, we have an overall running time that is polynomial in n, $\log T$ and $\mathrm{size}(c)$.

We claim that $\|x^* - y\| \leq \frac{1}{2T^2}$. To see this, write y as a convex combination of the vertices $x^*, x_1, \ldots, x_k$ of P:

$$y = \lambda_0 x^* + \sum_{i=1}^k \lambda_i x_i, \quad \lambda_i \geq 0, \quad \sum_{i=0}^k \lambda_i = 1.$$

Now – using Lemma 4.20 –

$$\epsilon \geq c'(x^* - y) = \sum_{i=1}^k \lambda_i c'\left(x^* - x_i\right) > \sum_{i=1}^k \lambda_i T^{-2n} = (1 - \lambda_0)T^{-2n},$$

so $1 - \lambda_0 < \epsilon T^{2n}$. We conclude that

$$\|y - x^*\| \leq \sum_{i=1}^k \lambda_i \|x_i - x^*\| \leq (1 - \lambda_0)2R < 4nT^{2n+1}\epsilon \leq \frac{1}{2T^2}.$$

So when rounding each entry of y to the next rational number with denominator at most T, we obtain x^*. The rounding can be done in polynomial time by Theorem 4.8. $\square$

We have proved that, under certain assumptions, optimizing over a polytope can be done whenever there is a separation oracle. We close this chapter by noting that the converse is also true. We need the concept of polarity: If $X \subseteq \mathbb{R}^n$, we define the **polar** of X to be the set

$$X^\circ := \{y \in \mathbb{R}^n : y^\top x \leq 1 \text{ for all } x \in X\}.$$

When applied to full-dimensional polytopes, this operation has some nice properties:

Theorem 4.22. *Let P be a polytope in $\mathbb{R}^n$ with 0 in the interior. Then:*

(a) P° *is a polytope with 0 in the interior;*
(b) $(P^\circ)^\circ = P$;
(c) x *is a vertex of P if and only if $x^\top y \leq 1$ is a facet-defining inequality of P°.*

Proof: (a): Let P be the convex hull of $x_1, \ldots, x_k$ (cf. Theorem 3.31). By definition, $P^\circ = \{y \in \mathbb{R}^n : y^\top x_i \leq 1 \text{ for all } i \in \{1, \ldots, k\}\}$, i.e. P° is a polyhedron and the facet-defining inequalities of P° are given by vertices of P. Moreover, 0 is in the interior of P° because 0 satisfies all of the finitely many inequalities strictly. Suppose P° is unbounded, i.e. there exists a $w \in \mathbb{R}^n \setminus \{0\}$ with $\alpha w \in P^\circ$ for all $\alpha > 0$. Then $\alpha w x \leq 1$ for all $\alpha > 0$ and all $x \in P$, so $wx \leq 0$ for all $x \in P$. But then 0 cannot be in the interior of P.

(b): Trivially, $P \subseteq (P^\circ)^\circ$. To show the converse, suppose that $z \in (P^\circ)^\circ \setminus P$. Then there is an inequality $c^\top x \leq \delta$ satisfied by all $x \in P$ but not by z. We have $\delta > 0$ since 0 is in the interior of P. Then $\frac{1}{\delta} c \in P^\circ$ but $\frac{1}{\delta} c^\top z > 1$, contradicting the assumption that $z \in (P^\circ)^\circ$.

(c): We have already seen in (a) that the facet-defining inequalities of P° are given by vertices of P. Conversely, if $x_1, \ldots, x_k$ are the vertices of P, then $\bar{P} := \mathrm{conv}(\{\frac{1}{2} x_1, x_2, \ldots, x_k\}) \neq P$, and 0 is in the interior of $\bar{P}$. Now (b) implies $\bar{P}^\circ \neq P^\circ$. Hence $\{y \in \mathbb{R}^n : y^\top x_1 \leq 2, y^\top x_i \leq 1 (i = 2, \ldots, k)\} = \bar{P}^\circ \neq P^\circ = \{y \in \mathbb{R}^n : y^\top x_i \leq 1 (i = 1, \ldots, k)\}$. We conclude that $x_1^\top y \leq 1$ is a facet-defining inequality of P°. $\qquad\square$

Now we can prove:

Theorem 4.23. *Let $n \in \mathbb{N}$ and $y \in \mathbb{Q}^n$. Let $P \subseteq \mathbb{R}^n$ be a rational polytope, and let $x_0 \in \mathbb{Q}^n$ be a point in the interior of P. Let $T \in \mathbb{N}$ such that $\mathrm{size}(x_0) \leq \log T$ and $\mathrm{size}(x) \leq \log T$ for all vertices x of P.*

Given n, y, x_0, T and an oracle which for any given $c \in \mathbb{Q}^n$ returns a vertex x^* of P attaining $\max\{c^\top x : x \in P\}$, we can solve the SEPARATION PROBLEM *for P and y in time polynomial in n, $\log T$ and $\mathrm{size}(y)$. Indeed, in the case $y \notin P$ we can find a facet-defining inequality of P that is violated by y.*

Proof: Consider $Q := \{x - x_0 : x \in P\}$ and its polar Q°. If $x_1, \ldots, x_k$ are the vertices of P, we have

$$Q^\circ = \{z \in \mathbb{R}^n : z^\top (x_i - x_0) \leq 1 \text{ for all } i \in \{1, \ldots, k\}\}.$$

By Theorem 4.4 we have size$(z) \leq 4n(4n \log T + 3n) \leq 28n^2 \log T$ for all vertices z of Q°.

Observe that the SEPARATION PROBLEM for P and y is equivalent to the SEPARATION PROBLEM for Q and $y - x_0$. Since by Theorem 4.22

$$Q = (Q^\circ)^\circ = \{x : zx \leq 1 \text{ for all } z \in Q^\circ\},$$

the SEPARATION PROBLEM for Q and $y - x_0$ is equivalent to solving $\max\{(y - x_0)^\top x : x \in Q^\circ\}$. Since each vertex of Q° corresponds to a facet-defining inequality of Q (and thus of P), it remains to show how to find a vertex attaining $\max\{(y - x_0)^\top x : x \in Q^\circ\}$.

To do this, we apply Theorem 4.21 to Q°. By Theorem 4.22, Q° is full-dimensional with 0 in the interior. We have shown above that the size of the vertices of Q° is at most $28n^2 \log T$. So it remains to show that we can solve the SEPARATION PROBLEM for Q° in polynomial time. However, this reduces to the optimization problem for Q which can be solved using the oracle for optimizing over P.   □

We finally mention that a new algorithm which is faster than the ELLIPSOID METHOD and also implies the equivalence of optimization and separation has been proposed by Vaidya [1996]. However, this algorithm does not seem to be of practical use either.

Exercises

1. Let A be a nonsingular rational $n \times n$-matrix. Prove that size$(A^{-1}) \leq 4n^2$ size(A).

* 2. Let $n \geq 2$, $c \in \mathbb{R}^n$ and $y_1, \ldots, y_k \in \{-1, 0, 1\}^n$ such that $0 < c^\top y_{i+1} \leq \frac{1}{2} c^\top y_i$ for $i = 1, \ldots, k - 1$. Prove that then $k \leq 3n \log n$.
   Hint: Consider the linear program $\max\{y_1^\top x : y_k^\top x = 1, (y_i - 2y_{i+1})^\top x \geq 0$ $(i = 1, \ldots, k - 1)\}$ and recall the proof of Theorem 4.4.
   (M. Goemans)

3. Consider the numbers h_i in the CONTINUED FRACTION EXPANSION. Prove that $h_i \geq F_{i+1}$ for all i, where F_i is the i-th Fibonacci number ($F_1 = F_2 = 1$ and $F_n = F_{n-1} + F_{n-2}$ for $n \geq 3$). Observe that

$$F_n = \frac{1}{\sqrt{5}} \left(\left(\frac{1 + \sqrt{5}}{2} \right)^n - \left(\frac{1 - \sqrt{5}}{2} \right)^n \right).$$

Conclude that the number of iterations of the CONTINUED FRACTION EXPANSION is $O(\log q)$.
   (Grötschel, Lovász and Schrijver [1988])

4. Show that GAUSSIAN ELIMINATION can be made a strongly polynomial-time algorithm.

Hint: First assume that A is integral. Recall the proof of Theorem 4.10 and observe that we can choose d as the common denominator of the entries. (Edmonds [1967])

* 5. Let $x_1, \ldots, x_k \in \mathbb{R}^l$, $d := 1 + \dim\{x_1, \ldots, x_k\}$, $\lambda_1, \ldots, \lambda_k \in \mathbb{R}_+$ with $\sum_{i=1}^k \lambda_i = 1$, and $x := \sum_{i=1}^k \lambda_i x_i$. Show how to compute numbers $\mu_1, \ldots, \mu_k \in \mathbb{R}_+$, at most d of which are nonzero, such that $\sum_{i=1}^k \mu_i = 1$ and $x = \sum_{i=1}^k \mu_i x_i$ (cf. Exercise 15 of Chapter 3). Show that all computations can be performed in $O((k+l)^3)$ time.

     Hint: Run GAUSSIAN ELIMINATION with the matrix $A \in \mathbb{R}^{(l+1)\times k}$ whose i-th column is $\binom{1}{x_i}$. If $d < k$, let $w \in \mathbb{R}^k$ be the vector with $w_{col(i)} := z_{i,d+1}$ $(i = 1, \ldots, d)$, $w_{col(d+1)} := -1$ and $w_{col(i)} := 0$ $(i = d+2, \ldots, k)$; observe that $Aw = 0$. Add a multiple of w to λ, eliminate at least one vector and iterate.

  6. Let $A = \begin{pmatrix} \alpha & b^\top \\ b & C \end{pmatrix} \in \mathbb{R}^{n\times n}$ be a symmetric positive semidefinite matrix with $\alpha > 0$ and $b \in \mathbb{R}^{n-1}$. Let $A' := \begin{pmatrix} \alpha & 0 \\ 0 & C - \frac{1}{\alpha}bb^\top \end{pmatrix}$ and $U := \begin{pmatrix} 1 & \frac{1}{\alpha}b \\ 0 & I \end{pmatrix}$. Prove that $A = U^\top A' U$ and $C - \frac{1}{\alpha}bb^\top$ is positive semidefinite. Iterate and conclude that for any positive semidefinite matrix A there is a matrix U with $A = U^\top U$, and such a matrix can be computed with arbitrary precision in $O(n^3)$ steps (some of which consist of computing approximate square roots).

     Note: This is called Cholesky factorization. It cannot be computed exactly as U can be irrational.

* 7. Let A be a symmetric positive definite $n \times n$-matrix. Let $v_1, \ldots, v_n$ be n orthogonal eigenvectors of A, with corresponding eigenvalues $\lambda_1, \ldots, \lambda_n$. W.l.o.g. $\|v_i\| = 1$ for $i = 1, \ldots, n$. Prove that then

$$E(A, 0) = \left\{ \mu_1 \sqrt{\lambda_1} v_1 + \cdots + \mu_n \sqrt{\lambda_n} v_n : \mu \in \mathbb{R}^n, \|\mu\| \le 1 \right\}.$$

     (The eigenvectors correspond to the axes of symmetry of the ellipsoid.) Conclude that volume$(E(A, 0)) = \sqrt{\det A}$ volume$(B(0, 1))$.

  8. Let $E(A, x) \subseteq \mathbb{R}^n$ be an ellipsoid and $a \in \mathbb{R}^n$, and let $E(A', x'))$ be as defined on page 85. Prove that $\{z \in E(A, x) : az \ge ax\} \subseteq E(A', x')$.

  9. Prove that the algorithm of Theorem 4.18 solves a linear program $\max\{cx : Ax \le b\}$ in $O((n+m)^9(\text{size}(A) + \text{size}(b) + \text{size}(c))^2)$ time.

  10. Show that the assumption that P is bounded can be omitted in Theorem 4.21. One can detect if the LP is unbounded and otherwise find an optimum solution.

* 11. Let $P \subseteq \mathbb{R}^3$ be a 3-dimensional polytope with 0 in its interior. Consider again the graph $G(P)$ whose vertices are the vertices of P and whose edges correspond to the 1-dimensional faces of P (cf. Exercises 18 and 19 of Chapter 3). Show that $G(P^\circ)$ is the planar dual of $G(P)$.

     Note: Steinitz [1922] proved that for every simple 3-connected planar graph G there is a 3-dimensional polytope P with $G = G(P)$.

* 12. Let G be a simple connected undirected graph. Show that the LP

$$\min \quad \sum_{e=\{v,w\}\in E(G)} x_{vw}$$

$$\text{s.t.} \quad \sum_{w\in S} x_{vw} \;\geq\; \left\lceil \tfrac{1}{4}|S|^2 + \tfrac{1}{2}|S| \right\rceil \quad (v\in V(G),\; S\subseteq V(G)\setminus\{v\})$$

$$x_{uw} \;\leq\; x_{uv} + x_{vw} \qquad (u,v,w\in V(G))$$

$$x_{vw} \;\geq\; 0 \qquad\qquad\quad (v,w\in V(G))$$

$$x_{vv} \;=\; 0 \qquad\qquad\quad (v\in V(G))$$

can be solved in time polynomial in $|V(G)|$.

Note: This can be regarded as a relaxation of the OPTIMAL LINEAR AR-RANGEMENT PROBLEM; cf. Exercise 8 of Chapter 19.

13. Prove that the polar of a polyhedron is always a polyhedron. For which polyhedra P is $(P^\circ)^\circ = P$?

References

General Literature

Grötschel, M., Lovász, L., and Schrijver, A. [1988]: Geometric Algorithms and Combinatorial Optimization. Springer, Berlin 1988

Padberg, M. [1999]: Linear Optimization and Extensions. Second edition. Springer, Berlin 1999

Schrijver, A. [1986]: Theory of Linear and Integer Programming. Wiley, Chichester 1986

Cited References

Bland, R.G., Goldfarb, D., and Todd, M.J. [1981]: The ellipsoid method: a survey. Operations Research 29 (1981), 1039–1091

Edmonds, J. [1967]: Systems of distinct representatives and linear algebra. Journal of Research of the National Bureau of Standards B 71 (1967), 241–245

Frank, A., and Tardos, É. [1987]: An application of simultaneous Diophantine approximation in combinatorial optimization. Combinatorica 7 (1987), 49–65

Gács, P., and Lovász, L. [1981]: Khachiyan's algorithm for linear programming. Mathematical Programming Study 14 (1981), 61–68

Grötschel, M., Lovász, L., and Schrijver, A. [1981]: The ellipsoid method and its consequences in combinatorial optimization. Combinatorica 1 (1981), 169–197

Iudin, D.B., and Nemirovskii, A.S. [1976]: Informational complexity and effective methods of solution for convex extremal problems. Ekonomika i Matematicheskie Metody 12 (1976), 357–369 [in Russian]

Karmarkar, N. [1984]: A new polynomial-time algorithm for linear programming. Combinatorica 4 (1984), 373–395

Karp, R.M., and Papadimitriou, C.H. [1982]: On linear characterizations of combinatorial optimization problems. SIAM Journal on Computing 11 (1982), 620–632

Khachiyan, L.G. [1979]: A polynomial algorithm in linear programming [in Russian]. Doklady Akademii Nauk SSSR 244 (1979) 1093–1096. English translation: Soviet Mathematics Doklady 20 (1979), 191–194

Khintchine, A. [1956]: Kettenbrüche. Teubner, Leipzig 1956

Lee, Y.T., and Sidford, A. [2014]: Path finding methods for linear programming. Proceedings of the 55th Annual IEEE Symposium on Foundations of Computer Science (2014), 424–433

Padberg, M.W., and Rao, M.R. [1981]: The Russian method for linear programming III: Bounded integer programming. Research Report 81-39, New York University 1981

Shor, N.Z. [1977]: Cut-off method with space extension in convex programming problems. Cybernetics 13 (1977), 94–96

Steinitz, E. [1922]: Polyeder und Raumeinteilungen. Enzyklopädie der Mathematischen Wissenschaften, Band 3 (1922), 1–139

Tardos, É. [1986]: A strongly polynomial algorithm to solve combinatorial linear programs. Operations Research 34 (1986), 250–256

Vaidya, P.M. [1996]: A new algorithm for minimizing convex functions over convex sets. Mathematical Programming 73 (1996), 291–341

5 Integer Programming

In this chapter, we consider linear programs with integrality constraints:

INTEGER PROGRAMMING

Instance: A matrix $A \in \mathbb{Z}^{m \times n}$ and vectors $b \in \mathbb{Z}^m, c \in \mathbb{Z}^n$.

Task: Find a vector $x \in \mathbb{Z}^n$ such that $Ax \leq b$ and cx is maximum, decide that $\{x \in \mathbb{Z}^n : Ax \leq b\} = \emptyset$, or decide that $\sup\{cx : x \in \mathbb{Z}^n, Ax \leq b\} = \infty$.

We do not consider mixed integer programs, i.e. linear programs with integrality constraints for only a subset of the variables. Most of the theory of linear and integer programming can be extended to mixed integer programming in a natural way.

Virtually all combinatorial optimization problems can be formulated as integer programs. The set of feasible solutions can be written as $\{x : Ax \leq b, x \in \mathbb{Z}^n\}$ for some matrix A and some vector b. The set $P := \{x \in \mathbb{R}^n : Ax \leq b\}$ is a polyhedron, so let us define by $P_I = \{x : Ax \leq b\}_I$ the convex hull of the integral vectors in P. We call P_I the **integer hull** of P. Obviously $P_I \subseteq P$.

If P is bounded, then P_I is also a polytope by Theorem 3.31 (see Figure 5.1). Meyer [1974] proved:

Theorem 5.1. *For any rational polyhedron P, its integer hull P_I is a rational polyhedron.*

Proof: Let $P = \{x : Ax \leq b\}$. By Lemma 3.12 the rational polyhedral cone $C := \{(x, \xi) : x \in \mathbb{R}^n, \xi \geq 0, Ax - \xi b \leq 0\}$ is generated by a finite set of rational vectors. We may assume that $(x_1, 1), \ldots, (x_k, 1), (y_1, 0), \ldots, (y_l, 0)$ generate C, where $x_1, \ldots, x_k$ are rational and $y_1, \ldots, y_l$ are integral (by multiplying the elements of a finite set of generators by suitable positive scalars).

Consider the polytope

$$Q := \left\{ \sum_{i=1}^{k} \kappa_i x_i + \sum_{i=1}^{l} \lambda_i y_i \ : \ \kappa_i \geq 0 \ (i = 1, \ldots, k), \ \sum_{i=1}^{k} \kappa_i = 1, \right.$$
$$\left. 0 \leq \lambda_i \leq 1 \ (i = 1, \ldots, l) \right\}.$$

© Springer-Verlag GmbH Germany 2018
Bernhard Korte and Jens Vygen, *Combinatorial Optimization*, Algorithms and Combinatorics 21, https://doi.org/10.1007/978-3-662-56039-6_5

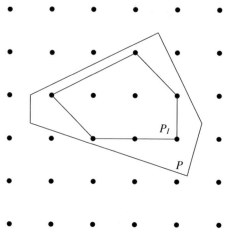

Fig. 5.1.

Note that $Q \subseteq P$. Let $z_1, \ldots, z_m$ be the integral points in Q. By Theorem 3.29 the cone C' generated by $(y_1, 0), \ldots, (y_l, 0), (z_1, 1), \ldots, (z_m, 1)$ is polyhedral, i.e. can be written as $\{(x, \xi) : Mx + \xi d \leq 0\}$ for a rational matrix M and a rational vector d.

We claim that $P_I = \{x : Mx \leq -d\}$.

To show "$\subseteq$", let $x \in P \cap \mathbb{Z}^n$. We have $(x, 1) \in C$, i.e. $x = \sum_{i=1}^{k} \kappa_i x_i + \sum_{i=1}^{l} \lambda_i y_i$ for some $\kappa_1, \ldots, \kappa_k \geq 0$ with $\sum_{i=1}^{k} \kappa_i = 1$ and $\lambda_1, \ldots, \lambda_l \geq 0$. Then $c := \sum_{i=1}^{l} \lfloor \lambda_i \rfloor y_i$ is integral, and hence $x - c$ is integral. Moreover, $x - c = \sum_{i=1}^{k} \kappa_i x_i + \sum_{i=1}^{l} (\lambda_i - \lfloor \lambda_i \rfloor) y_i \in Q$, and hence $x - c = z_i$ for some i. Thus $(x, 1) = (c, 0) + (x - c, 1) \in C'$ and hence $Mx + d \leq 0$.

To show "$\supseteq$", let x be a vector satisfying $Mx \leq -d$, i.e. $(x, 1) \in C'$. Then $x = \sum_{i=1}^{l} \lambda_i y_i + \sum_{i=1}^{m} \mu_i z_i$ for some $\lambda_1, \ldots, \lambda_l, \mu_1, \ldots, \mu_m \geq 0$ with $\sum_{i=1}^{m} \mu_i = 1$. W.l.o.g. assume $l \geq 1$ and $\mu_1 > 0$. Let $\lambda_i' := \frac{\lambda_i}{\mu_1}$ for $i = 1, \ldots, l$. Then $(z_1 + \nu y_i, 1) \in C$ for all $i = 1, \ldots, l$ and $\nu > 0$ and hence

$$x = \sum_{i=1}^{l} \frac{\mu_1}{l} \left(\lfloor \lambda_i' \rfloor + 1 - \lambda_i' \right) \left(z_1 + \lfloor \lambda_i' \rfloor y_i \right)$$

$$+ \sum_{i=1}^{l} \frac{\mu_1}{l} \left(\lambda_i' - \lfloor \lambda_i' \rfloor \right) \left(z_1 + (\lfloor \lambda_i' \rfloor + 1) y_i \right) + \sum_{i=2}^{m} \mu_i z_i$$

is a convex combination of integral points in P. $\qquad\square$

This does in general not hold for irrational polyhedra; see Exercise 1. Due to Theorem 5.1 we can write an instance of INTEGER PROGRAMMING as $\max\{c^\top x : x \in P_I\}$ where $P = \{x : Ax \leq b\}$.

We prove a generalization of Meyer's Theorem 5.1 in Section 5.1 (Theorem 5.9). After some preparation in Section 5.2 we study conditions under which polyhedra are integral (i.e. $P = P_I$) in Sections 5.3 and 5.4. Note that in this case the integer linear program is equivalent to its LP relaxation (arising by omitting the integrality constraints), and can hence be solved in polynomial time. We shall encounter this situation for several combinatorial optimization problems in later chapters.

In general, however, INTEGER PROGRAMMING is much harder than LINEAR PROGRAMMING, and polynomial-time algorithms are not known. This is indeed not surprising since we can formulate many apparently hard problems as integer programs. Nevertheless we discuss a general method for finding the integer hull by successively cutting off parts of $P \setminus P_I$ in Section 5.5. Although it does not yield a polynomial-time algorithm it is a useful technique in some cases. Finally Section 5.6 contains an efficient way of approximating the optimal value of an integer linear program.

5.1 The Integer Hull of a Polyhedron

As linear programs, integer programs can be infeasible or unbounded. It is not easy to decide whether $P_I = \emptyset$ for a polyhedron P. But if an integer program is feasible we can decide whether it is bounded by simply considering the LP relaxation.

Proposition 5.2. *Let $P = \{x : Ax \leq b\}$ be some rational polyhedron whose integer hull is nonempty, and let c be some vector (not necessarily rational). Then $\max \{cx : x \in P\}$ is bounded if and only if $\max \{cx : x \in P_I\}$ is bounded.*

Proof: Suppose $\max \{cx : x \in P\}$ is unbounded. Then Corollary 3.28 says that the system $yA = c$, $y \geq 0$ has no solution. By Corollary 3.26 there is a vector z with $cz < 0$ and $Az \geq 0$. Then the LP $\min\{cz : Az \geq 0, -\mathbb{1} \leq z \leq \mathbb{1}\}$ is feasible. Let z^* be an optimum basic solution of this LP. z^* is rational as it is a vertex of a rational polytope. Multiply z^* by a suitable natural number to obtain an integral vector w with $Aw \geq 0$ and $cw < 0$. Let $v \in P_I$ be some integral vector. Then $v - kw \in P_I$ for all $k \in \mathbb{N}$, and thus $\max \{cx : x \in P_I\}$ is unbounded.

The other direction is trivial.   □

There are various ways of measuring how different P and P_I are. For approximation algorithms (to be discussed later in this book), the following notion is the most useful. For $P \subseteq \mathbb{R}^n_+$ we call $\sup\left\{\frac{\min\{cx : x \in P_I\}}{\min\{cx : x \in P\}} : c \in \mathbb{R}^n_+\right\}$ the **integrality ratio** of P. This number can be characterized in a different way. The following is often attributed to Carr and Vempala [2004], but was observed earlier by Goemans [1995].

Theorem 5.3. *Let $P \subseteq \mathbb{R}^n_+$ be a rational polyhedron and $r \geq 0$. Then the integrality ratio of P is at most r if and only if for every $x \in P$ there is a $y \in P_I$ with $rx \geq y$.*

Proof: "⇐": Suppose that for every $x \in P$ there is a $y \in P_I$ with $rx \geq y$. To show that the integrality ratio is at most r, let $c \in \mathbb{R}^n_+$ and x^* an optimum solution to the LP min $\{cx : x \in P\}$. There is a $y \in P_I$ with $y \leq rx^*$ and hence $cy \leq r(cx^*)$.

"⇒": Now suppose that there exists an $x \in P$ such that for all $y \in P_I$ we have $rx \not\geq y$. To show that the integrality ratio is greater than r, let $Q := \{y + z : y \in P_I, z \geq 0\}$. We have $rx \notin Q$. As Q is a polyhedron by Theorem 5.1, there exists a vector w with $w(rx) < \min\{wq : q \in Q\}$. The vector w is nonnegative (otherwise $\min\{wq : q \in Q\} = -\infty$); so $r(wx) = w(rx) < \min\{wq : q \in Q\} = \min\{wy : y \in P_I\}$, and the integrality ratio is greater than r.   □

Next we would like to bound the distance between an optimum fractional and an optimum integral solution. We need some information about rational polyhedral cones.

Definition 5.4.   *Let A be an integral matrix. A **subdeterminant** of A is det B for some square submatrix B of A (defined by arbitrary row and column indices). We write $\Xi(A)$ for the maximum absolute value of the subdeterminants of A.*

Lemma 5.5.   *Let $C = \{x : Ax \leq 0\}$ be a polyhedral cone, where A is an integral matrix. Then C is generated by a finite set of integral vectors, each having components with absolute value at most $\Xi(A)$.*

Proof: By Lemma 3.12, C is generated by some of the vectors $y_1, \ldots, y_t$, such that for each i, y_i is the solution to a system $My = b'$ where M consists of n linearly independent rows of $\binom{A}{I}$ and $b' = \pm e_j$ for some unit vector e_j. Set $z_i := |\det M| y_i$. By Cramer's rule, z_i is integral with $\|z_i\|_\infty \leq \Xi(A)$. Since this holds for each i, the set $\{z_1, \ldots, z_t\}$ has the required properties.   □

A similar lemma will be used in the next section:

Lemma 5.6.   *Each rational polyhedral cone C is generated by a finite set of integral vectors $\{a_1, \ldots, a_t\}$ such that each integral vector in C is a nonnegative integral combination of $a_1, \ldots, a_t$. (Such a set is called a **Hilbert basis** for C.)*

Proof: Let C be generated by the integral vectors $b_1, \ldots, b_k$. Let $a_1, \ldots, a_t$ be all integral vectors in the polytope

$$\{\lambda_1 b_1 + \ldots + \lambda_k b_k : 0 \leq \lambda_i \leq 1 \ \ (i = 1, \ldots, k)\}$$

We show that $\{a_1, \ldots, a_t\}$ is a Hilbert basis for C. They indeed generate C, because $b_1, \ldots, b_k$ occur among the $a_1, \ldots, a_t$.

For any integral vector $x \in C$ there are $\mu_1, \ldots, \mu_k \geq 0$ with

$$
\begin{aligned}
x = \mu_1 b_1 + \ldots + \mu_k b_k &= \lfloor \mu_1 \rfloor b_1 + \ldots + \lfloor \mu_k \rfloor b_k + \\
&\quad (\mu_1 - \lfloor \mu_1 \rfloor) b_1 + \ldots + (\mu_k - \lfloor \mu_k \rfloor) b_k,
\end{aligned}
$$

so x is a nonnegative integral combination of $a_1, \ldots, a_t$.   □

An important basic fact in integer programming is that optimum integral and fractional solutions are not too far away from each other:

Theorem 5.7. (Cook et al. [1986]) *Let A be an integral $m \times n$-matrix and $b \in \mathbb{R}^m$, $c \in \mathbb{R}^n$ arbitrary vectors. Let $P := \{x : Ax \le b\}$ and suppose that $P_I \ne \emptyset$.*

(a) *Suppose y is an optimum solution of $\max \{cx : x \in P\}$. Then there exists an optimum integral solution z of $\max \{cx : x \in P_I\}$ with $||z - y||_\infty \le n \, \Xi(A)$.*

(b) *Suppose y is a feasible integral solution of $\max \{cx : x \in P_I\}$, but not an optimal one. Then there exists a feasible integral solution $z \in P_I$ with $cz > cy$ and $||z - y||_\infty \le n \, \Xi(A)$.*

Proof: The proof is almost the same for both parts. Let first $y \in P$ arbitrary. Let $z^* \in P \cap \mathbb{Z}^n$ be (a) an optimum solution of $\max \{cx : x \in P_I\}$ (note that $P_I = \{x : Ax \le \lfloor b \rfloor\}_I$ is a polyhedron by Theorem 5.1, and thus the maximum is attained) or (b) a vector with $cz^* > cy$.

We split $Ax \le b$ into two subsystems $A_1 x \le b_1$, $A_2 x \le b_2$ such that $A_1 z^* \ge A_1 y$ and $A_2 z^* < A_2 y$. Then $z^* - y$ belongs to the polyhedral cone $C := \{x : A_1 x \ge 0, A_2 x \le 0\}$.

C is generated by some vectors x_i ($i = 1, \ldots, s$). By Lemma 5.5, we may assume that x_i is integral and $||x_i||_\infty \le \Xi(A)$ for all i.

Since $z^* - y \in C$, there are nonnegative numbers $\lambda_1, \ldots, \lambda_s$ with $z^* - y = \sum_{i=1}^s \lambda_i x_i$. We may assume that at most n of the λ_i are nonzero.

For $\mu = (\mu_1, \ldots, \mu_s)$ with $0 \le \mu_i \le \lambda_i$ ($i = 1, \ldots, s$) we define

$$z_\mu := z^* - \sum_{i=1}^s \mu_i x_i = y + \sum_{i=1}^s (\lambda_i - \mu_i) x_i$$

and observe that $z_\mu \in P$: the first representation of z_μ implies $A_1 z_\mu \le A_1 z^* \le b_1$; the second one implies $A_2 z_\mu \le A_2 y \le b_2$.

Case 1: There is some $i \in \{1, \ldots, s\}$ with $\lambda_i \ge 1$ and $cx_i > 0$. Let $z := y + x_i$. We have $cz > cy$, showing that this case cannot occur in case (a). In case (b), when y is integral, z is an integral solution of $Ax \le b$ such that $cz > cy$ and $||z - y||_\infty = ||x_i||_\infty \le \Xi(A)$.

Case 2: For all $i \in \{1, \ldots, s\}$, $\lambda_i \ge 1$ implies $cx_i \le 0$. Let

$$z := z_{\lfloor \lambda \rfloor} = z^* - \sum_{i=1}^s \lfloor \lambda_i \rfloor x_i.$$

z is an integral vector of P with $cz \ge cz^*$ and

$$||z - y||_\infty \le \sum_{i=1}^s (\lambda_i - \lfloor \lambda_i \rfloor) ||x_i||_\infty \le n \, \Xi(A).$$

Hence in both (a) and (b) this vector z does the job. $\qquad\square$

As a corollary we can bound the size of optimum solutions of integer programming problems:

Corollary 5.8. *If $P = \{x \in \mathbb{Q}^n : Ax \leq b\}$ is a rational polyhedron and $\max\{cx : x \in P_I\}$ has an optimum solution, then it also has an optimum integral solution x with $\text{size}(x) \leq 13n(\text{size}(A) + \text{size}(b))$.*

Proof: By Proposition 5.2 and Theorem 4.4, $\max\{cx : x \in P\}$ has an optimum solution y with $\text{size}(y) \leq 4n(\text{size}(A) + \text{size}(b))$. By Theorem 5.7(a) there is an optimum solution x of $\max\{cx : x \in P_I\}$ with $||x - y||_\infty \leq n\,\Xi(A)$. By Propositions 4.1 and 4.3 we have

$$
\begin{aligned}
\text{size}(x) &\leq\; 2\,\text{size}(y) + 2n\,\text{size}(n\,\Xi(A)) \\
&\leq\; 8n(\text{size}(A) + \text{size}(b)) + 2n\log n + 4n\,\text{size}(A) \\
&\leq\; 13n(\text{size}(A) + \text{size}(b)).
\end{aligned}
$$

$\square$

Theorem 5.7(b) implies the following: given any feasible solution of an integer program, optimality of a vector x can be checked simply by testing $x + y$ for a finite set of vectors y that depend on the matrix A only. Such a finite test set (whose existence has been proved first by Graver [1975]) enables us to prove a fundamental theorem on integer programming:

Theorem 5.9. (Wolsey [1981], Cook et al. [1986]) *For each integral $m \times n$-matrix A there exists an integral matrix M whose entries have absolute value at most $n^{2n}\,\Xi(A)^n$, such that for each vector $b \in \mathbb{Q}^m$ there exists a rational vector d with*

$$
\{x : Ax \leq b\}_I \;=\; \{x : Mx \leq d\}.
$$

Proof: We may assume $A \neq 0$. Let C be the cone generated by the rows of A. Let

$$
L \;:=\; \{z \in \mathbb{Z}^n : ||z||_\infty \leq n\Xi(A)\}.
$$

For each $K \subseteq L$, consider the cone

$$
C_K \;:=\; C \cap \{y : zy \leq 0 \;\text{ for all }\; z \in K\}.
$$

By the proof of Theorem 3.29 and Lemma 5.5, $C_K = \{y : Uy \leq 0\}$ for some integral matrix U (whose rows are generators of $\{x : Ax \leq 0\}$ and elements of K) whose entries have absolute value at most $n\Xi(A)$. Hence, again by Lemma 5.5, there is a finite set $G(K)$ of integral vectors generating C_K, each having components with absolute value at most $\Xi(U) \leq n!(n\Xi(A))^n \leq n^{2n}\,\Xi(A)^n$.

Let M be the matrix with rows $\bigcup_{K \subseteq L} G(K)$. Since $C_\emptyset = C$, we may assume that the rows of A are also rows of M.

Now let b be some fixed vector. If $Ax \leq b$ has no solution, we can complete b to a vector d arbitrarily and have $\{x : Mx \leq d\} \subseteq \{x : Ax \leq b\} = \emptyset$.

If $Ax \leq b$ contains a solution, but no integral solution, we set $b' := b - A'1\!\!1$, where A' arises from A by taking the absolute value of each entry. Then $Ax \leq b'$ has no solution, since any such solution yields an integral solution of $Ax \leq b$ by rounding. Again, we complete b' to d arbitrarily.

Now we may assume that $Ax \leq b$ has an integral solution. For $y \in C$ we define

$$\delta_y := \max\{yx : Ax \leq b, \ x \text{ integral}\}$$

(by Corollary 3.28 this maximum is bounded for $y \in C$). It suffices to show that

$$\{x : Ax \leq b\}_I = \left\{x : yx \leq \delta_y \text{ for each } y \in \bigcup_{K \subseteq L} G(K)\right\}. \qquad (5.1)$$

Here "$\subseteq$" is trivial. To show the converse, let c be any vector for which

$$\max\{cx : Ax \leq b, \ x \text{ integral}\}$$

is bounded, and let x^* be a vector attaining this maximum. We show that $cx \leq cx^*$ for all x satisfying the inequalities on the right-hand side of (5.1).

By Proposition 5.2 the LP $\max\{cx : Ax \leq b\}$ is bounded, so by Corollary 3.28 we have $c \in C$.

Let $\bar{K} := \{z \in L : A(x^* + z) \leq b\}$. By definition $cz \leq 0$ for all $z \in \bar{K}$, so $c \in C_{\bar{K}}$. Thus there are nonnegative numbers λ_y ($y \in G(\bar{K})$) such that

$$c = \sum_{y \in G(\bar{K})} \lambda_y y.$$

Next we claim that x^* is an optimum solution for

$$\max\{yx : Ax \leq b, \ x \text{ integral}\}$$

for each $y \in G(\bar{K})$: the contrary assumption would, by Theorem 5.7(b), yield a vector $z \in \bar{K}$ with $yz > 0$, which is impossible since $y \in C_{\bar{K}}$. We conclude that

$$\sum_{y \in G(\bar{K})} \lambda_y \delta_y = \sum_{y \in G(\bar{K})} \lambda_y y x^* = \left(\sum_{y \in G(\bar{K})} \lambda_y y\right) x^* = cx^*.$$

Thus the inequality $cx \leq cx^*$ is a nonnegative linear combination of the inequalities $yx \leq \delta_y$ for $y \in G(\bar{K})$. Hence (5.1) is proved. $\qquad\qquad\square$

See Lasserre [2004] for a similar result.

5.2 Unimodular Transformations

In this section we shall prove two lemmas for later use. A square matrix is called **unimodular** if it is integral and has determinant 1 or -1. Three types of unimodular matrices will be of particular interest: For $n \in \mathbb{N}$, $p \in \{1, \ldots, n\}$ and

$q \in \{1, \ldots, n\} \setminus \{p\}$ consider the matrices $(a_{ij})_{i,j \in \{1,\ldots,n\}}$ defined in one of the following ways:

$$a_{ij} = \begin{cases} 1 & \text{if } i = j \neq p \\ -1 & \text{if } i = j = p \\ 0 & \text{otherwise} \end{cases} \qquad a_{ij} = \begin{cases} 1 & \text{if } i = j \notin \{p, q\} \\ 1 & \text{if } \{i, j\} = \{p, q\} \\ 0 & \text{otherwise} \end{cases}$$

$$a_{ij} = \begin{cases} 1 & \text{if } i = j \\ -1 & \text{if } (i, j) = (p, q) \\ 0 & \text{otherwise} \end{cases}$$

These matrices are evidently unimodular. If U is one of the above matrices, then replacing an arbitrary matrix A (with n columns) by AU is equivalent to applying one of the following elementary column operations to A:

- multiply a column by -1;
- exchange two columns;
- subtract one column from another column.

A series of the above operations is called a **unimodular transformation**. Obviously the product of unimodular matrices is unimodular. It can be shown that a matrix is unimodular if and only if it arises from an identity matrix by a unimodular transformation (equivalently, it is the product of matrices of the above three types); see Exercise 6. Here we do not need this fact.

Proposition 5.10. *The inverse of a unimodular matrix is also unimodular. For each unimodular matrix U the mappings $x \mapsto Ux$ and $x \mapsto xU$ are bijections on $\mathbb{Z}^n$.*

Proof: Let U be a unimodular matrix. By Cramer's rule the inverse of a unimodular matrix is integral. Since $(\det U)(\det U^{-1}) = \det(UU^{-1}) = \det I = 1$, U^{-1} is also unimodular. The second statement follows directly from this. $\qquad\square$

Lemma 5.11. *For each rational matrix A whose rows are linearly independent there exists a unimodular matrix U such that AU has the form $(B \quad 0)$, where B is a nonsingular square matrix.*

Proof: Suppose we have found a unimodular matrix U such that

$$AU = \begin{pmatrix} B & 0 \\ C & D \end{pmatrix}$$

for some nonsingular square matrix B. (Initially $U = I$, $D = A$, and the parts B, C and 0 have no entries.)

Let $(\delta_1, \ldots, \delta_k)$ be the first row of D. Apply unimodular transformations such that all δ_i are nonnegative and $\sum_{i=1}^{k} \delta_i$ is minimum. W.l.o.g. $\delta_1 \geq \delta_2 \geq \cdots \geq \delta_k$. Then $\delta_1 > 0$ since the rows of A (and hence those of AU) are linearly independent.

If $\delta_2 > 0$, then subtracting the second column of D from the first one would decrease $\sum_{i=1}^{k} \delta_i$. So $\delta_2 = \delta_3 = \ldots = \delta_k = 0$. We can expand B by one row and one column and continue.   □

The matrix B that we get is in fact a lower diagonal matrix. With a little more effort one can obtain the so-called Hermite normal form of A. The following lemma gives a criterion for integral solvability of equation systems, similar to Farkas' Lemma.

Lemma 5.12. *Let A be a rational matrix and b a rational column vector. Then $Ax = b$ has an integral solution if and only if yb is an integer for each rational vector y for which yA is integral.*

Proof: Necessity is obvious: if x and yA are integral vectors and $Ax = b$, then $yb = yAx$ is an integer.

To prove sufficiency, suppose yb is an integer whenever yA is integral. We may assume that $Ax = b$ contains no redundant equalities, i.e. $yA = 0$ implies $yb \neq 0$ for all $y \neq 0$. Let m be the number of rows of A. If $\mathrm{rank}(A) < m$ then $\{y : yA = 0\}$ contains a nonzero vector y' and $y'' := \frac{1}{2y'b} y'$ satisfies $y''A = 0$ and $y''b = \frac{1}{2} \notin \mathbb{Z}$. So the rows of A are linearly independent.

By Lemma 5.11 there exists a unimodular matrix U with $AU = (B \quad 0)$, where B is a nonsingular $m \times m$-matrix. Since $B^{-1}AU = (I \quad 0)$ is an integral matrix, we have for each row y of B^{-1} that yAU is integral and thus by Proposition 5.10 yA is integral. Hence yb is an integer for each row y of B^{-1}, implying that $B^{-1}b$ is an integral vector. So $U \begin{pmatrix} B^{-1}b \\ 0 \end{pmatrix}$ is an integral solution of $Ax = b$.   □

5.3 Total Dual Integrality

In this and the next section we focus on integral polyhedra:

Definition 5.13. *A polyhedron P is **integral** if $P = P_I$.*

Theorem 5.14. (Hoffman [1974], Edmonds and Giles [1977]) *Let P be a rational polyhedron. Then the following statements are equivalent:*

(a) *P is integral.*
(b) *Each face of P contains integral vectors.*
(c) *Each minimal face of P contains integral vectors.*
(d) *Each supporting hyperplane of P contains integral vectors.*
(e) *Each rational supporting hyperplane of P contains integral vectors.*
(f) *max $\{cx : x \in P\}$ is attained by an integral vector for each c for which the maximum is finite.*
(g) *max $\{cx : x \in P\}$ is an integer for each integral c for which the maximum is finite.*

Proof: We first prove (a)$\Rightarrow$(b)$\Rightarrow$(f)$\Rightarrow$(a), then (b)$\Rightarrow$(d)$\Rightarrow$(e)$\Rightarrow$(c)$\Rightarrow$(b), and finally (f)$\Rightarrow$(g)$\Rightarrow$(e).

(a)$\Rightarrow$(b): Let F be a face, say $F = P \cap H$, where H is a supporting hyperplane, and let $x \in F$. If $P = P_I$, then x is a convex combination of integral points in P, and these must belong to H and thus to F.

(b)$\Rightarrow$(f) follows directly from Proposition 3.4, because $\{y \in P : cy = \max\{cx : x \in P\}\}$ is a face of P for each c for which the maximum is finite.

(f)$\Rightarrow$(a): Suppose there is a vector $y \in P \setminus P_I$. Then (since P_I is a polyhedron by Theorem 5.1) there is an inequality $ax \leq \beta$ valid for P_I for which $ay > \beta$. Then clearly (f) is violated, since $\max\{ax : x \in P\}$ (which is finite by Proposition 5.2) is not attained by any integral vector.

(b)$\Rightarrow$(d) is also trivial since the intersection of a supporting hyperplane with P is a face of P. (d)$\Rightarrow$(e) and (c)$\Rightarrow$(b) are trivial.

(e)$\Rightarrow$(c): Let $P = \{x : Ax \leq b\}$. We may assume that A and b are integral. Let $F = \{x : A'x = b'\}$ be a minimal face of P, where $A'x \leq b'$ is a subsystem of $Ax \leq b$ (we use Proposition 3.9). If $A'x = b'$ has no integral solution, then – by Lemma 5.12 – there exists a rational vector y such that $c := yA'$ is integral but $\delta := yb'$ is not an integer. Adding integers to components of y does not destroy this property (A' and b' are integral), so we may assume that all components of y are positive. Observe that $H := \{x : cx = \delta\}$ is a rational hyperplane which contains no integral vectors.

We finally show that H is a supporting hyperplane by proving that $H \cap P = F$. Since $F \subseteq H$ is trivial, it remains to show that $H \cap P \subseteq F$. But for $x \in H \cap P$ we have $yA'x = cx = \delta = yb'$, so $y(A'x - b') = 0$. Since $y > 0$ and $A'x \leq b'$, this implies $A'x = b'$, so $x \in F$.

(f)$\Rightarrow$(g) is trivial, so we finally show (g)$\Rightarrow$(e). Let $H = \{x : cx = \delta\}$ be a rational supporting hyperplane of P, so $\max\{cx : x \in P\} = \delta$. Suppose H contains no integral vectors. Then – by Lemma 5.12 – there exists a number γ such that γc is integral but $\gamma \delta \notin \mathbb{Z}$. Then

$$\max\{(\lfloor \gamma \rfloor c)x : x \in P\} = |\gamma| \max\{cx : x \in P\} = |\gamma| \delta \notin \mathbb{Z},$$

contradicting our assumption.   □

See also Gomory [1963], Fulkerson [1971] and Chvátal [1973] for earlier partial results. By (a)$\Leftrightarrow$(b) and Corollary 3.6 every face of an integral polyhedron is integral. The equivalence of (f) and (g) of Theorem 5.14 motivated Edmonds and Giles to define TDI-systems:

Definition 5.15. (Edmonds and Giles [1977]) *A system $Ax \leq b$ of linear inequalities is called* **totally dual integral (TDI)** *if the minimum in the LP duality equation*

$$\max\{cx : Ax \leq b\} = \min\{yb : yA = c, \ y \geq 0\}$$

has an integral optimum solution y for each integral vector c for which the minimum is finite.

With this definition we get an easy corollary of (g)⇒(a) of Theorem 5.14:

Corollary 5.16.   *Let $Ax \leq b$ be a TDI-system where A is rational and b is integral. Then the polyhedron $\{x : Ax \leq b\}$ is integral.*   □

But total dual integrality is not a property of polyhedra (cf. Exercise 8). In general, a TDI-system contains more inequalities than necessary for describing the polyhedron. Adding valid inequalities does not destroy total dual integrality:

Proposition 5.17.   *If $Ax \leq b$ is TDI and $ax \leq \beta$ is a valid inequality for $\{x : Ax \leq b\}$, then the system $Ax \leq b$, $ax \leq \beta$ is also TDI.*

Proof: Let c be an integral vector such that $\min\{yb + \gamma\beta : yA + \gamma a = c,\ y \geq 0,\ \gamma \geq 0\}$ is finite. Since $ax \leq \beta$ is valid for $\{x : Ax \leq b\}$,

$$
\begin{aligned}
\min\{yb : yA = c,\ y \geq 0\} &= \max\{cx : Ax \leq b\} \\
&= \max\{cx : Ax \leq b,\ ax \leq \beta\} \\
&= \min\{yb + \gamma\beta : yA + \gamma a = c,\ y \geq 0,\ \gamma \geq 0\}.
\end{aligned}
$$

The first minimum is attained by some integral vector y^*, so $y = y^*$, $\gamma = 0$ is an integral optimum solution for the second minimum.   □

Theorem 5.18.   (Giles and Pulleyblank [1979]) *For each rational polyhedron P there exists a rational TDI-system $Ax \leq b$ with A integral and $P = \{x : Ax \leq b\}$. Here b can be chosen to be integral if and only if P is integral.*

Proof: Let $P = \{x : Cx \leq d\}$ with C and d integral. W.l.o.g., $P \neq \emptyset$. For each minimal face F of P let

$$K_F := \{c : cz = \max\{cx : x \in P\} \text{ for all } z \in F\}.$$

By Corollary 3.22 and Theorem 3.29, K_F is a rational polyhedral cone. By Lemma 5.6 there exists an integral Hilbert basis $a_1, \ldots, a_t$ generating K_F. Let $\mathcal{S}_F$ be the system of inequalities

$$a_1 x \leq \max\{a_1 x : x \in P\},\ \ldots,\ a_t x \leq \max\{a_t x : x \in P\}.$$

Let $Ax \leq b$ be the collection of all these systems $\mathcal{S}_F$ (for all minimal faces F). Note that if P is integral then b is integral. Moreover, $P \subseteq \{x : Ax \leq b\}$.

Let c be an integral vector for which $\max\{cx : x \in P\}$ is finite. The set of vectors attaining this maximum is a face of P, so let F be a minimal face such that $cz = \max\{cx : x \in P\}$ for all $z \in F$. Let $\mathcal{S}_F$ be the system $a_1 x \leq \beta_1, \ldots, a_t x \leq \beta_t$. Then $c = \lambda_1 a_1 + \cdots + \lambda_t a_t$ for some nonnegative integers $\lambda_1, \ldots, \lambda_t$. We add zero components to $\lambda_1, \ldots, \lambda_t$ in order to get an integral vector $\bar{\lambda} \geq 0$ with $\bar{\lambda}A = c$ and thus $cx = (\bar{\lambda}A)x = \bar{\lambda}(Ax) \leq \bar{\lambda}b = \bar{\lambda}(Az) = (\bar{\lambda}A)z = cz$ for all x with $Ax \leq b$ and all $z \in F$.

Applying this for each row c of C yields $Cx \leq d$ for all x with $Ax \leq b$; and thus $P = \{x : Ax \leq b\}$. Moreover, for general c we conclude that $\bar{\lambda}$ is an optimum solution of the dual LP $\min\{yb : y \geq 0, yA = c\}$. Hence $Ax \leq b$ is TDI.

If P is integral, we have chosen b to be integral. Conversely, if b can be chosen integral, by Corollary 5.16 P must be integral.   □

Indeed, for full-dimensional rational polyhedra there is a unique minimal TDI-system describing it (Schrijver [1981]). For later use, we prove that each "face" of a TDI-system is again TDI:

Theorem 5.19. (Cook [1983]) *Let $Ax \leq b$, $ax \leq \beta$ be a TDI-system, where a is integral. Then the system $Ax \leq b$, $ax = \beta$ is also TDI.*

Proof: (Schrijver [1986]) Let c be an integral vector such that

$$\begin{aligned} &\max\{cx : Ax \leq b, ax = \beta\} \\ =\ &\min\{yb + (\lambda - \mu)\beta : y, \lambda, \mu \geq 0, yA + (\lambda - \mu)a = c\} \end{aligned} \tag{5.2}$$

is finite. Let $x^*, y^*, \lambda^*, \mu^*$ attain these optima. We set $c' := c + \lceil \mu^* \rceil a$ and observe that

$$\max\{c'x : Ax \leq b, ax \leq \beta\} \ =\ \min\{yb + \lambda\beta : y, \lambda \geq 0, yA + \lambda a = c'\} \tag{5.3}$$

is finite, because $x := x^*$ is feasible for the maximum and $y := y^*$, $\lambda := \lambda^* + \lceil \mu^* \rceil - \mu^*$ is feasible for the minimum.

Since $Ax \leq b$, $ax \leq \beta$ is TDI, the minimum in (5.3) has an integral optimum solution $\tilde{y}, \tilde{\lambda}$. We finally set $y := \tilde{y}$, $\lambda := \tilde{\lambda}$ and $\mu := \lceil \mu^* \rceil$ and claim that (y, λ, μ) is an integral optimum solution for the minimum in (5.2).

Obviously (y, λ, μ) is feasible for the minimum in (5.2). Furthermore,

$$\begin{aligned} yb + (\lambda - \mu)\beta \ &=\ \tilde{y}b + \tilde{\lambda}\beta - \lceil \mu^* \rceil \beta \\ &\leq\ y^*b + (\lambda^* + \lceil \mu^* \rceil - \mu^*)\beta - \lceil \mu^* \rceil \beta \end{aligned}$$

since $(y^*, \lambda^* + \lceil \mu^* \rceil - \mu^*)$ is feasible for the minimum in (5.3), and $(\tilde{y}, \tilde{\lambda})$ is an optimum solution. We conclude that

$$yb + (\lambda - \mu)\beta \ \leq\ y^*b + (\lambda^* - \mu^*)\beta,$$

proving that (y, λ, μ) is an integral optimum solution for the minimum in (5.2).
     □

The following statements are straightforward consequences of the definition of TDI-systems: A system $Ax = b$, $x \geq 0$ is TDI if $\min\{yb : yA \geq c\}$ has an integral optimum solution y for each integral vector c for which the minimum is finite. A system $Ax \leq b$, $x \geq 0$ is TDI if $\min\{yb : yA \geq c, y \geq 0\}$ has an integral optimum solution y for each integral vector c for which the minimum is finite. One may ask whether there are matrices A such that $Ax \leq b$, $x \geq 0$ is TDI for each integral vector b. It will turn out that these matrices are exactly the totally unimodular matrices.

5.4 Totally Unimodular Matrices

Definition 5.20. *A matrix A is **totally unimodular** if each subdeterminant of A is 0, +1, or −1.*

In particular, each entry of a totally unimodular matrix must be 0, +1, or −1. The main result of this section is:

Theorem 5.21. (Hoffman and Kruskal [1956]) *An integral matrix A is totally unimodular if and only if the polyhedron $\{x : Ax \leq b, \ x \geq 0\}$ is integral for each integral vector b.*

Proof: Let A be an $m \times n$-matrix and $P := \{x : Ax \leq b, \ x \geq 0\}$. Observe that the minimal faces of P are vertices.

To prove necessity, suppose that A is totally unimodular. Let b be some integral vector and x a vertex of P. x is the solution of $A'x = b'$ for some subsystem $A'x \leq b'$ of $\left(\begin{smallmatrix} A \\ -I \end{smallmatrix} \right) x \leq \left(\begin{smallmatrix} b \\ 0 \end{smallmatrix} \right)$, with A' being a nonsingular $n \times n$-matrix. Since A is totally unimodular, $|\det A'| = 1$, so by Cramer's rule $x = (A')^{-1}b'$ is integral.

We now prove sufficiency. Suppose that the vertices of P are integral for each integral vector b. Let A' be some nonsingular $k \times k$-submatrix of A. We have to show $|\det A'| = 1$. W.l.o.g., A' contains the elements of the first k rows and columns of A.

Consider the integral $m \times m$-matrix B consisting of the first k and the last $m - k$ columns of $(A \ \ I)$ (see Figure 5.2). Obviously, $|\det B| = |\det A'|$.

To prove $|\det B| = 1$, we shall prove that B^{-1} is integral. Since $\det B \det B^{-1} = 1$, this implies that $|\det B| = 1$, and we are done.

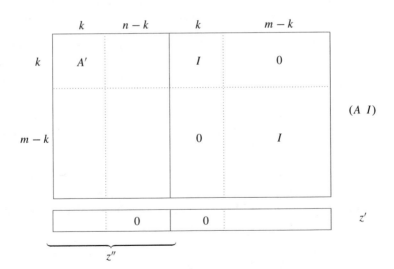

Fig. 5.2.

Let $i \in \{1, \ldots, m\}$; we prove that $B^{-1}e_i$ is integral. Choose an integral vector y such that $z := y + B^{-1}e_i \geq 0$. Then $b := Bz = By + e_i$ is integral. We add zero components to z in order to obtain z' with

$$\begin{pmatrix} A & I \end{pmatrix} z' = Bz = b.$$

Now z'', consisting of the first n components of z', belongs to P. Furthermore, n linearly independent constraints are satisfied with equality, namely the first k and the last $n - k$ inequalities of

$$\begin{pmatrix} A \\ -I \end{pmatrix} z'' \leq \begin{pmatrix} b \\ 0 \end{pmatrix}.$$

Hence z'' is a vertex of P. By our assumption z'' is integral. But then z' must also be integral: its first n components are the components of z'', and the last m components are the slack variables $b - Az''$ (and A and b are integral). So z is also integral, and hence $B^{-1}e_i = z - y$ is integral.   □

The above proof is due to Veinott and Dantzig [1968].

Corollary 5.22. *An integral matrix A is totally unimodular if and only if for all integral vectors b and c both optima in the LP duality equation*

$$\max\{cx : Ax \leq b,\, x \geq 0\} = \min\{yb : y \geq 0,\, yA \geq c\}$$

are attained by integral vectors (if they are finite).

Proof: This follows from the Hoffman-Kruskal Theorem 5.21 by using the fact that the transpose of a totally unimodular matrix is also totally unimodular.   □

Let us reformulate these statements in terms of total dual integrality:

Corollary 5.23. *An integral matrix A is totally unimodular if and only if the system $Ax \leq b$, $x \geq 0$ is TDI for each vector b.*

Proof: If A (and thus $A^\top$) is totally unimodular, then by the Hoffman-Kruskal Theorem $\min\{yb : yA \geq c,\, y \geq 0\}$ is attained by an integral vector for each vector b and each integral vector c for which the minimum is finite. In other words, the system $Ax \leq b$, $x \geq 0$ is TDI for each vector b.

To show the converse, suppose $Ax \leq b$, $x \geq 0$ is TDI for each integral vector b. Then by Corollary 5.16, the polyhedron $\{x : Ax \leq b,\, x \geq 0\}$ is integral for each integral vector b. By Theorem 5.21 this means that A is totally unimodular.   □

This is not the only way how total unimodularity can be used to prove that a certain system is TDI. The following lemma contains another proof technique; this will be used several times later (Theorems 6.14, 19.17 and 14.12).

Lemma 5.24. *Let $Ax \leq b$, $x \geq 0$ be an inequality system, where $A \in \mathbb{R}^{m \times n}$ and $b \in \mathbb{R}^m$. Suppose that for each $c \in \mathbb{Z}^n$ for which $\min\{yb : yA \geq c, \, y \geq 0\}$ has an optimum solution, it has one y^* such that the rows of A corresponding to nonzero components of y^* form a totally unimodular matrix. Then $Ax \leq b$, $x \geq 0$ is TDI.*

Proof: Let $c \in \mathbb{Z}^n$, and let y^* be an optimum solution of $\min\{yb : yA \geq c, \, y \geq 0\}$ such that the rows of A corresponding to nonzero components of y^* form a totally unimodular matrix A'. We claim that

$$\min\{yb : yA \geq c, y \geq 0\} = \min\{yb' : yA' \geq c, y \geq 0\}, \tag{5.4}$$

where b' consists of the components of b corresponding to the rows of A'. To see the inequality "$\leq$" of (5.4), observe that the LP on the right-hand side arises from the LP on the left-hand side by setting some variables to zero. The inequality "$\geq$" follows from the fact that y^* without zero components is a feasible solution for the LP on the right-hand side.

Since A' is totally unimodular, the second minimum in (5.4) has an integral optimum solution (by the Hoffman-Kruskal Theorem 5.21). By filling this solution with zeros we obtain an integral optimum solution to the first minimum in (5.4), completing the proof. □

A very useful criterion for total unimodularity is the following:

Theorem 5.25. (Ghouila-Houri [1962]) *A matrix $A = (a_{ij}) \in \mathbb{Z}^{m \times n}$ is totally unimodular if and only if for every $R \subseteq \{1, \ldots, m\}$ there are R_1 and R_2 with $R = R_1 \,\dot\cup\, R_2$ such that*

$$\sum_{i \in R_1} a_{ij} - \sum_{i \in R_2} a_{ij} \in \{-1, 0, 1\}$$

for all $j = 1, \ldots, n$.

Proof: Let A be totally unimodular, and let $R \subseteq \{1, \ldots, m\}$. Let $d_r := 1$ for $r \in R$ and $d_r := 0$ for $r \in \{1, \ldots, m\} \setminus R$. The matrix $\begin{pmatrix} A^\top \\ -A^\top \\ I \end{pmatrix}$ is also totally unimodular, so by Theorem 5.21 the polytope

$$\left\{ x : xA \leq \left\lceil \frac{1}{2} dA \right\rceil, \, xA \geq \left\lfloor \frac{1}{2} dA \right\rfloor, \, x \leq d, \, x \geq 0 \right\}$$

is integral. Moreover it is nonempty because it contains $\frac{1}{2} d$. So it has an integral vertex, say z. Setting $R_1 := \{r \in R : z_r = 0\}$ and $R_2 := \{r \in R : z_r = 1\}$ we obtain

$$\left(\sum_{i \in R_1} a_{ij} - \sum_{i \in R_2} a_{ij} \right)_{1 \leq j \leq n} = (d - 2z)A \in \{-1, 0, 1\}^n,$$

as required.

We now prove the converse. By induction on k we prove that every $k \times k$-submatrix has determinant 0, 1 or -1. For $k = 1$ this is directly implied by the criterion for $|R| = 1$.

Now let $k > 1$, and let $B = (b_{ij})_{i,j \in \{1,\ldots,k\}}$ be a nonsingular $k \times k$-submatrix of A. By Cramer's rule, each entry of B^{-1} is $\frac{\det B'}{\det B}$, where B' arises from B by replacing a column by a unit vector. By the induction hypothesis, $\det B' \in \{-1, 0, 1\}$. So $B^* := (\det B)B^{-1}$ is a matrix with entries $-1, 0, 1$ only.

Let b_1^* be the first row of B^*. We have $b_1^* B = (\det B)e_1$, where e_1 is the first unit vector. Let $R := \{i : b_{1i}^* \neq 0\}$. Then for $j = 2, \ldots, k$ we have $0 = (b_1^* B)_j = \sum_{i \in R} b_{1i}^* b_{ij}$, so $|\{i \in R : b_{ij} \neq 0\}|$ is even.

By the hypothesis there is a partition $R = R_1 \dot{\cup} R_2$ with $\sum_{i \in R_1} b_{ij} - \sum_{i \in R_2} b_{ij} \in \{-1, 0, 1\}$ for all j. So for $j = 2, \ldots, k$ we have $\sum_{i \in R_1} b_{ij} - \sum_{i \in R_2} b_{ij} = 0$. If also $\sum_{i \in R_1} b_{i1} - \sum_{i \in R_2} b_{i1} = 0$, then the sum of the rows in R_1 equals the sum of the rows in R_2, contradicting the assumption that B is nonsingular (because $R \neq \emptyset$).

So $\sum_{i \in R_1} b_{i1} - \sum_{i \in R_2} b_{i1} \in \{-1, 1\}$ and we have $yB \in \{e_1, -e_1\}$, where

$$
y_i := \begin{cases} 1 & \text{if } i \in R_1 \\ -1 & \text{if } i \in R_2 \, . \\ 0 & \text{if } i \notin R \end{cases}
$$

Since $b_1^* B = (\det B)e_1$ and B is nonsingular, we have $b_1^* \in \{(\det B)y, -(\det B)y\}$. Since both y and b_1^* are nonzero vectors with entries $-1, 0, 1$ only, this implies that $|\det B| = 1$. □

We apply this criterion to the incidence matrices of graphs:

Theorem 5.26. *The incidence matrix of an undirected graph G is totally uni-modular if and only if G is bipartite.*

Proof: By Theorem 5.25 the incidence matrix M of G is totally unimodular if and only if for any $X \subseteq V(G)$ there is a partition $X = A \dot{\cup} B$ such that $E(G[A]) = E(G[B]) = \emptyset$. By definition, such a partition exists iff $G[X]$ is bipartite. □

Theorem 5.27. *The incidence matrix of any digraph is totally unimodular.*

Proof: Using Theorem 5.25, it suffices to set $R_1 := R$ and $R_2 := \emptyset$ for any $R \subseteq V(G)$. □

Applications of Theorems 5.26 and 5.27 will be discussed in later chapters. Theorem 5.27 has an interesting generalization to cross-free families:

Definition 5.28. *Let G be a digraph and $\mathcal{F}$ a family of subsets of $V(G)$. The* **one-way cut-incidence matrix** *of $\mathcal{F}$ is the matrix $M = (m_{X,e})_{X \in \mathcal{F}, e \in E(G)}$*

where

$$m_{X,e} = \begin{cases} 1 & \text{if } e \in \delta^+(X) \\ 0 & \text{if } e \notin \delta^+(X) \end{cases}.$$

The **two-way cut-incidence matrix** *of $\mathcal{F}$ is the matrix $M = (m_{X,e})_{X \in \mathcal{F}, e \in E(G)}$*
where

$$m_{X,e} = \begin{cases} -1 & \text{if } e \in \delta^-(X) \\ 1 & \text{if } e \in \delta^+(X) \\ 0 & \text{otherwise} \end{cases}.$$

Theorem 5.29. *Let G be a digraph and $(V(G), \mathcal{F})$ a cross-free set system. Then the two-way cut-incidence matrix of $\mathcal{F}$ is totally unimodular. If $\mathcal{F}$ is laminar, then also the one-way cut-incidence matrix of $\mathcal{F}$ is totally unimodular.*

Proof: Let $\mathcal{F}$ be some cross-free family of subsets of $V(G)$. We first consider the case when $\mathcal{F}$ is laminar.

We use Theorem 5.25. To see that the criterion is satisfied, let $\mathcal{R} \subseteq \mathcal{F}$, and consider the tree-representation (T, φ) of $\mathcal{R}$, where T is an arborescence rooted at r (Proposition 2.14). With the notation of Definition 2.13, $\mathcal{R} = \{S_e : e \in E(T)\}$. Set $\mathcal{R}_1 := \{S_{(v,w)} \in \mathcal{R} : \text{dist}_T(r, w) \text{ even}\}$ and $\mathcal{R}_2 := \mathcal{R} \setminus \mathcal{R}_1$. Now for any edge $f \in E(G)$, the edges $e \in E(T)$ with $f \in \delta^+(S_e)$ form a path P_f in T (possibly of zero length). So

$$|\{X \in \mathcal{R}_1 : f \in \delta^+(X)\}| - |\{X \in \mathcal{R}_2 : f \in \delta^+(X)\}| \in \{-1, 0, 1\},$$

as required for the one-way cut-incidence matrix.

Moreover, for any edge f the edges $e \in E(T)$ with $f \in \delta^-(S_e)$ form a path Q_f in T. Since P_f and Q_f have a common endpoint, we have

$$|\{X \in \mathcal{R}_1 : f \in \delta^+(X)\}| - |\{X \in \mathcal{R}_2 : f \in \delta^+(X)\}|$$
$$-|\{X \in \mathcal{R}_1 : f \in \delta^-(X)\}| + |\{X \in \mathcal{R}_2 : f \in \delta^-(X)\}| \in \{-1, 0, 1\},$$

as required for the two-way cut-incidence matrix.

Now if $(V(G), \mathcal{F})$ is a general cross-free set system, consider

$$\mathcal{F}' := \{X \in \mathcal{F} : r \notin X\} \cup \{V(G) \setminus X : X \in \mathcal{F}, r \in X\}$$

for some fixed $r \in V(G)$. $\mathcal{F}'$ is laminar. Since the two-way cut-incidence matrix of $\mathcal{F}$ is a submatrix of $\begin{pmatrix} M \\ -M \end{pmatrix}$, where M is the two-way cut-incidence matrix of $\mathcal{F}'$, it is totally unimodular, too. $\qquad\square$

For general cross-free families the one-way cut-incidence matrix is not totally unimodular; see Exercise 13. For a necessary and sufficient condition, see Schrijver [1983]. The two-way cut-incidence matrices of cross-free families are also known as network matrices (Exercise 14).

Seymour [1980] showed that all totally unimodular matrices can be constructed in a certain way from these network matrices and two other totally unimodular matrices. This deep result implies a polynomial-time algorithm which decides whether a given matrix is totally unimodular (see Schrijver [1986]).

5.5 Cutting Planes

In the previous sections we considered integral polyhedra. For general polyhedra P we have $P \supset P_I$. If we want to solve an integer linear program max $\{cx : x \in P_I\}$, it is a natural idea to cut off certain parts of P such that the resulting set is again a polyhedron P' and we have $P \supset P' \supset P_I$. Hopefully max $\{cx : x \in P'\}$ is attained by an integral vector; otherwise we can repeat this cutting-off procedure for P' in order to obtain P'' and so on. This is the basic idea behind the cutting plane method, first proposed for a special problem (the TSP) by Dantzig, Fulkerson and Johnson [1954].

Gomory [1958, 1963] found an algorithm which solves general integer programs with the cutting plane method. In this section we restrict ourselves to the theoretical background. Gomory's algorithm does not run in polynomial time and has little practical relevance in its original form. However, the general idea of cutting plane methods is used very often and successfully in practice. We shall discuss this in Section 21.6. The following presentation is mainly based on Schrijver [1986].

Definition 5.30. *Let $P \subseteq \mathbb{R}^n$ be a convex set. Then we define*

$$P' := \bigcap_{P \subseteq H} H_I,$$

where the intersection ranges over all rational affine half-spaces $H = \{x : cx \le \delta\}$ containing P. We set $P^{(0)} := P$ and $P^{(i+1)} := \left(P^{(i)}\right)'$. $P^{(i)}$ is called the i-th **Gomory-Chvátal-truncation** *of P.*

If P is a rational polyhedron, then P' is also a rational polyhedron, as we will show below. Therefore $P \supseteq P' \supseteq P^{(2)} \supseteq \cdots \supseteq P_I$ and $P_I = (P')_I$. We remark that Dadush, Dey and Vielma [2014] proved that if P is any compact convex set, then P' is a rational polytope.

Proposition 5.31. *For any rational polyhedron $P = \{x : Ax \le b\}$,*

$$P' = \{x : uAx \le \lfloor ub \rfloor \text{ for all } u \ge 0 \text{ with } uA \text{ integral}\}.$$

Proof: We first make two observations. For any rational affine half-space $H = \{x : cx \le \delta\}$ with c integral we obviously have

$$H' = H_I \subseteq \{x : cx \le \lfloor \delta \rfloor\}. \tag{5.5}$$

If in addition the components of c are relatively prime (i.e., their greatest common divisor is 1), we claim that

$$H' = H_I = \{x : cx \le \lfloor \delta \rfloor\}. \tag{5.6}$$

To prove (5.6), let c be an integral vector whose components are relatively prime. By Lemma 5.12 the hyperplane $\{x : cx = \lfloor \delta \rfloor\}$ contains an integral vector y. For any rational vector $x \in \{x : cx \leq \lfloor \delta \rfloor\}$ let $\alpha \in \mathbb{N}$ such that αx is integral. Then we can write

$$x = \frac{1}{\alpha}(\alpha x - (\alpha - 1)y) + \frac{\alpha - 1}{\alpha}y,$$

i.e. x is a convex combination of integral points in H. Hence $x \in H_I$, implying (5.6).

We now turn to the main proof. To see "$\subseteq$", observe that for any $u \geq 0$, $\{x : uAx \leq ub\}$ is a half-space containing P, so by (5.5) $P' \subseteq \{x : uAx \leq \lfloor ub \rfloor\}$ if uA is integral.

We now prove "$\supseteq$". For $P = \emptyset$ this is easy, so we assume $P \neq \emptyset$. Let $H = \{x : cx \leq \delta\}$ be some rational affine half-space containing P. W.l.o.g. c is integral and the components of c are relatively prime. We observe that

$$\delta \geq \max\{cx : Ax \leq b\} = \min\{ub : uA = c, u \geq 0\}.$$

Now let u^* be any optimum solution for the minimum. Then for any

$$z \in \{x : uAx \leq \lfloor ub \rfloor \text{ for all } u \geq 0 \text{ with } uA \text{ integral}\} \subseteq \{x : u^*Ax \leq \lfloor u^*b \rfloor\}$$

we have:

$$cz = u^*Az \leq \lfloor u^*b \rfloor \leq \lfloor \delta \rfloor$$

which, using (5.6), implies $z \in H_I$. $\square$

Below we shall prove that for any rational polyhedron P there is a number t with $P_I = P^{(t)}$. So Gomory's cutting plane method successively solves the linear programs over P, P', P'', and so on, until the optimum is integral. At each step only a finite number of new inequalities have to be added, namely those corresponding to a TDI-system defining the current polyhedron (recall Theorem 5.18):

Theorem 5.32.   (Schrijver [1980]) *Let $P = \{x : Ax \leq b\}$ be a polyhedron with $Ax \leq b$ TDI, A integral and b rational. Then $P' = \{x : Ax \leq \lfloor b \rfloor\}$. In particular, for any rational polyhedron P, P' is also a rational polyhedron.*

Proof: The statement is trivial if P is empty, so let $P \neq \emptyset$. Obviously $P' \subseteq \{x : Ax \leq \lfloor b \rfloor\}$. To show the other inclusion, let $u \geq 0$ be a vector with uA integral. By Proposition 5.31 it suffices to show that $uAx \leq \lfloor ub \rfloor$ for all x with $Ax \leq \lfloor b \rfloor$.

We know that

$$ub \geq \max\{uAx : Ax \leq b\} = \min\{yb : y \geq 0, yA = uA\}.$$

Since $Ax \leq b$ is TDI, the minimum is attained by some integral vector y^*. Now $Ax \leq \lfloor b \rfloor$ implies

$$uAx = y^*Ax \leq y^*\lfloor b \rfloor \leq \lfloor y^*b \rfloor \leq \lfloor ub \rfloor.$$

The second statement follows from Theorem 5.18. $\square$

To prove the main theorem of this section, we need two more lemmas:

Lemma 5.33. *If F is a face of a rational polyhedron P, then $F' = P' \cap F$. More generally, $F^{(i)} = P^{(i)} \cap F$ for all $i \in \mathbb{N}$.*

Proof: Let $P = \{x : Ax \leq b\}$ with A integral, b rational, and $Ax \leq b$ TDI (recall Theorem 5.18).

Now let $F = \{x : Ax \leq b, ax = \beta\}$ be a face of P, where $ax \leq \beta$ is a valid inequality for P with a and β integral.

By Proposition 5.17, $Ax \leq b$, $ax \leq \beta$ is TDI, so by Theorem 5.19, $Ax \leq b$, $ax = \beta$ is also TDI. As β is an integer,

$$
\begin{aligned}
P' \cap F &= \{x : Ax \leq \lfloor b \rfloor, ax = \beta\} \\
&= \{x : Ax \leq \lfloor b \rfloor, ax \leq \lfloor \beta \rfloor, ax \geq \lceil \beta \rceil\} \\
&= F'.
\end{aligned}
$$

Here we have used Theorem 5.32 twice.

Note that F' is either empty or a face of P'. Now the statement follows by induction on i: For each i, $F^{(i)}$ is either empty or a face of $P^{(i)}$, and $F^{(i)} = P^{(i)} \cap F^{(i-1)} = P^{(i)} \cap (P^{(i-1)} \cap F) = P^{(i)} \cap F$. $\qquad\square$

Lemma 5.34. *Let P be a polyhedron in $\mathbb{R}^n$ and U a unimodular $n \times n$-matrix. Write $f(P) := \{Ux : x \in P\}$. Then $f(P)$ is again a polyhedron. Moreover, if P is a rational polyhedron, then $(f(P))' = f(P')$ and $(f(P))_I = f(P_I)$.*

Proof: Since $f : \mathbb{R}^n \to \mathbb{R}^n$, $x \mapsto Ux$ is a bijective linear function, the first statement is obviously true. Since also the restrictions of f and f^{-1} to $\mathbb{Z}^n$ are bijections (by Proposition 5.10) we have

$$
\begin{aligned}
(f(P))_I &= \operatorname{conv}(\{y \in \mathbb{Z}^n : y = Ux, x \in P\}) \\
&= \operatorname{conv}(\{y \in \mathbb{R}^n : y = Ux, x \in P, x \in \mathbb{Z}^n\}) \\
&= \operatorname{conv}(\{y \in \mathbb{R}^n : y = Ux, x \in P_I\}) \\
&= f(P_I).
\end{aligned}
$$

Let $P = \{x : Ax \leq b\}$ with $Ax \leq b$ TDI, A integral, b rational (cf. Theorem 5.18). Then by definition $AU^{-1}x \leq b$ is also TDI. Using Theorem 5.32 twice we get

$$
(f(P))' = \{x : AU^{-1}x \leq b\}' = \{x : AU^{-1}x \leq \lfloor b \rfloor\} = f(P').
$$

$\qquad\square$

Theorem 5.35. (Schrijver [1980]) *For each rational polyhedron P there exists a number t such that $P^{(t)} = P_I$.*

Proof: Let P be a rational polyhedron in $\mathbb{R}^n$. We prove the theorem by induction on $n + \dim P$. The case $P = \emptyset$ is trivial, the case $\dim P = 0$ is easy.

First suppose that P is not full-dimensional. Then $P \subseteq K$ for some rational hyperplane K.

If K contains no integral vectors, $K = \{x : ax = \beta\}$ for some integral vector a and some non-integer β (by Lemma 5.12). But then $P' \subseteq \{x : ax \leq \lfloor \beta \rfloor, ax \geq \lceil \beta \rceil\} = \emptyset = P_I$.

If K contains integral vectors, say $K = \{x : ax = \beta\}$ with a integral, β an integer, we may assume $\beta = 0$, because the theorem is invariant under translations by integral vectors. By Lemma 5.11 there exists a unimodular matrix U with $aU = \alpha e_1$. Since the theorem is also invariant under the transformation $x \mapsto U^{-1}x$ (by Lemma 5.34), we may assume $a = \alpha e_1$. Then the first component of each vector in P is zero, and thus we can reduce the dimension of the space by one and apply the induction hypothesis (observe that $(\{0\} \times Q)_I = \{0\} \times Q_I$ and $(\{0\} \times Q)^{(t)} = \{0\} \times Q^{(t)}$ for any polyhedron Q in $\mathbb{R}^{n-1}$ and any $t \in \mathbb{N}$).

Let now $P = \{x : Ax \leq b\}$ be full-dimensional, and w.l.o.g. A integral. By Theorem 5.1 there is some integral matrix C and some vector d with $P_I = \{x : Cx \leq d\}$. In the case $P_I = \emptyset$ we set $C := A$ and $d := b - A' \mathbb{1}$, where A' arises from A by taking the absolute value of each entry. (Note that $\{x : Ax \leq b - A' \mathbb{1}\} = \emptyset$.)

Let $cx \leq \delta$ be an inequality of $Cx \leq d$. We claim that $P^{(s)} \subseteq H := \{x : cx \leq \delta\}$ for some $s \in \mathbb{N}$. This claim obviously implies the theorem.

First observe that there is some $\beta \geq \delta$ such that $P \subseteq \{x : cx \leq \beta\}$: in the case $P_I = \emptyset$ this follows from the choice of C and d; in the case $P_I \neq \emptyset$ this follows from Proposition 5.2.

Suppose our claim is false, i.e. there is an integer γ with $\delta < \gamma \leq \beta$ for which there exists an $s_0 \in \mathbb{N}$ with $P^{(s_0)} \subseteq \{x : cx \leq \gamma\}$, but there is no $s \in \mathbb{N}$ with $P^{(s)} \subseteq \{x : cx \leq \gamma - 1\}$.

Observe that $\max\{cx : x \in P^{(s)}\} = \gamma$ for all $s \geq s_0$, because if $\max\{cx : x \in P^{(s)}\} < \gamma$ for some s, then $P^{(s+1)} \subseteq \{x : cx \leq \gamma - 1\}$.

Let $F := P^{(s_0)} \cap \{x : cx = \gamma\}$. F is a face of $P^{(s_0)}$, and $\dim F < n = \dim P$. By the induction hypothesis, there is a number s_1 such that

$$F^{(s_1)} = F_I \subseteq P_I \cap \{x : cx = \gamma\} = \emptyset.$$

By applying Lemma 5.33 to F and $P^{(s_0)}$ we obtain

$$\emptyset = F^{(s_1)} = P^{(s_0+s_1)} \cap F = P^{(s_0+s_1)} \cap \{x : cx = \gamma\}.$$

Hence $\max\{cx : x \in P^{(s_0+s_1)}\} < \gamma$, a contradiction. $\qquad\square$

This theorem also implies the following:

Theorem 5.36. (Chvátal [1973]) *For each polytope P there is a number t such that $P^{(t)} = P_I$.*

Proof: As P is bounded, there exists some rational polytope $Q \supseteq P$ with $Q_I = P_I$ (take a hypercube containing P and intersect it with a rational half-space containing P but not z for each integral point z which belongs to the hypercube but not to P; cf. Exercise 20 of Chapter 3). By Theorem 5.35, $Q^{(t)} = Q_I$ for some t. Hence $P_I \subseteq P^{(t)} \subseteq Q^{(t)} = Q_I = P_I$, implying $P^{(t)} = P_I$. $\square$

This number t is called the Chvátal rank of P. If P is neither bounded nor rational, one cannot have an analogous theorem: see Exercises 1 and 17.

A more efficient algorithm which computes the integer hull of a two-dimensional polyhedron has been found by Harvey [1999]. A version of the cutting plane method which, in polynomial time, approximates a linear objective function over an integral polytope given by a separation oracle was described by Boyd [1997]. Cook, Kannan and Schrijver [1990], generalized the Gomory-Chvátal procedure to mixed integer programming. Eisenbrand [1999] showed that it is *coNP*-complete to decide whether a given rational vector is in P' for a given rational polyhedron P. Cornuéjols and Li [2016] proved that deciding whether P' is empty is *NP*-complete, even if P is rational and $P_I = \emptyset$.

5.6 Lagrangean Relaxation

Suppose we have an integer linear program $\max\{cx : Ax \leq b, A'x \leq b', x \text{ integral}\}$ that becomes substantially easier to solve when the constraints $A'x \leq b'$ are omitted. We write $Q := \{x \in \mathbb{Z}^n : Ax \leq b\}$ and assume that we can optimize linear objective functions over Q (for example if $\text{conv}(Q) = \{x : Ax \leq b\}$). Lagrangean relaxation is a technique to get rid of some troublesome constraints (in our case $A'x \leq b'$). Instead of explicitly enforcing the constraints we modify the objective function in order to punish infeasible solutions. More precisely, instead of optimizing

$$\max\{c^\top x : A'x \leq b', x \in Q\} \tag{5.7}$$

we consider, for any vector $\lambda \geq 0$,

$$LR(\lambda) := \max\{c^\top x + \lambda^\top(b' - A'x) : x \in Q\}. \tag{5.8}$$

For each $\lambda \geq 0$, $LR(\lambda)$ is an upper bound for (5.7) which is relatively easy to compute. (5.8) is called the **Lagrangean relaxation** of (5.7), and the components of λ are called **Lagrange multipliers**.

Lagrangean relaxation is a useful technique in nonlinear programming; but here we restrict ourselves to (integer) linear programming.

Of course one is interested in as good an upper bound as possible. Observe that $\lambda \mapsto LR(\lambda)$ is a convex function. The following procedure (called subgradient optimization) can be used to minimize $LR(\lambda)$:

Start with an arbitrary vector $\lambda^{(0)} \geq 0$. In iteration i, given $\lambda^{(i)}$, find a vector $x^{(i)}$ maximizing $c^\top x + (\lambda^{(i)})^\top(b' - A'x)$ over Q (i.e. compute $LR(\lambda^{(i)})$). Note that $LR(\lambda) - LR(\lambda^{(i)}) \geq (\lambda - \lambda^{(i)})^\top(b' - A'x^{(i)})$ for all λ, i.e. $b' - A'x^{(i)}$ is a subgradient of LR at $\lambda^{(i)}$. Set $\lambda^{(i+1)} := \max\{0, \lambda^{(i)} - t_i(b' - A'x^{(i)})\}$ for some $t_i > 0$. Polyak [1967] showed that if $\lim_{i\to\infty} t_i = 0$ and $\sum_{i=0}^{\infty} t_i = \infty$, then

$\lim_{i \to \infty} LR(\lambda^{(i)}) = \min\{LR(\lambda) : \lambda \geq 0\}$. For more results on the convergence of subgradient optimization, see Goffin [1977] and Anstreicher and Wolsey [2009].

The problem to find the best such upper bound, i.e.

$$\min\{LR(\lambda) : \lambda \geq 0\},$$

is sometimes called the **Lagrangean dual** of (5.7). We will show that the minimum is always attained unless $\{x : Ax \leq b, A'x \leq b'\} = \emptyset$. The second question that we address is how good this upper bound is. Of course this depends on the structure of the original problem. In Section 21.5 we shall meet an application to the TSP, where Lagrangean relaxation is very effective. The following theorem helps to estimate the quality of the upper bound:

Theorem 5.37. (Geoffrion [1974]) *Let $c \in \mathbb{R}^n$, $A' \in \mathbb{R}^{m \times n}$ and $b' \in \mathbb{R}^m$. Let $Q \subseteq \mathbb{R}^n$ such that $\mathrm{conv}(Q)$ is a polyhedron. Suppose that $\max\{c^\top x : A'x \leq b', x \in \mathrm{conv}(Q)\}$ has an optimum solution. Let $LR(\lambda) := \max\{c^\top x + \lambda^\top(b' - A'x) : x \in Q\}$. Then $\inf\{LR(\lambda) : \lambda \geq 0\}$ (the optimum value of the Lagrangean dual of $\max\{c^\top x : A'x \leq b', x \in Q\}$) is attained by some λ, and the minimum is equal to $\max\{c^\top x : A'x \leq b', x \in \mathrm{conv}(Q)\}$.*

Proof: Let $\mathrm{conv}(Q) = \{x : Ax \leq b\}$. By using the LP Duality Theorem 3.20 twice and reformulating we get

$$\max\{c^\top x : x \in \mathrm{conv}(Q), A'x \leq b'\}$$
$$= \max\{c^\top x : Ax \leq b, A'x \leq b'\}$$
$$= \min\{\lambda^\top b' + y^\top b : y^\top A + \lambda^\top A' = c^\top, y \geq 0, \lambda \geq 0\}$$
$$= \min\{\lambda^\top b' + \min\{y^\top b : y^\top A = c^\top - \lambda^\top A', y \geq 0\} : \lambda \geq 0\}$$
$$= \min\{\lambda^\top b' + \max\{(c^\top - \lambda^\top A')x : Ax \leq b\} : \lambda \geq 0\}$$
$$= \min\{\max\{c^\top x + \lambda^\top(b' - A'x) : x \in \mathrm{conv}(Q)\} : \lambda \geq 0\}$$
$$= \min\{\max\{c^\top x + \lambda^\top(b' - A'x) : x \in Q\} : \lambda \geq 0\}$$
$$= \min\{LR(\lambda) : \lambda \geq 0\}.$$

The third line, an LP, shows that there is a λ attaining the minimum. $\square$

In particular, if we have an integer linear program $\max\{cx : A'x \leq b', Ax \leq b, x \text{ integral}\}$ where $\{x : Ax \leq b\}$ is integral, then the Lagrangean dual (when relaxing $A'x \leq b'$ as above) yields the same upper bound as the standard LP relaxation $\max\{cx : A'x \leq b', Ax \leq b\}$. If $\{x : Ax \leq b\}$ is not integral, the upper bound is in general stronger (but can be difficult to compute). See Exercise 21 for an example.

Lagrangean relaxation can also be used to approximate linear programs. For example, consider the JOB ASSIGNMENT PROBLEM (see (1.1) in Section 1.3). The problem can be rewritten equivalently as

$$\min\left\{T : \sum_{j \in S_i} x_{ij} \geq t_i \ (i = 1, \ldots, n), (x, T) \in P\right\} \tag{5.9}$$

where P is the polytope

$$\left\{ (x, T) \ : \ 0 \le x_{ij} \le t_i \ (i = 1, \ldots, n, \ j \in S_i), \right.$$

$$\sum_{i:j\in S_i} x_{ij} \le T \ (j = 1, \ldots, m),$$

$$\left. T \le \sum_{i=1}^{n} t_i \right\}.$$

Now we apply Lagrangean relaxation and consider

$$LR(\lambda) \ := \ \min \left\{ T + \sum_{i=1}^{n} \lambda_i \left(t_i - \sum_{j\in S_i} x_{ij} \right) : (x, T) \in P \right\}. \tag{5.10}$$

Because of its special structure this LP can be solved by a simple combinatorial algorithm (Exercise 23), for arbitrary λ. If we let Q be the set of vertices of P (cf. Corollary 3.32), then we can apply Theorem 5.37 and conclude that the optimum value of the Lagrangean dual $\max\{LR(\lambda) : \lambda \ge 0\}$ equals the optimum of (5.9).

Exercises

1. Let $P := \left\{ (x, y) \in \mathbb{R}^2 : y \le \sqrt{2}x \right\}$. Prove that P_I is not a polyhedron. Show a polyhedron P where even the closure of P_I is not a polyhedron.

2. Let $P = \{x \in \mathbb{R}^{k+l} : Ax \le b\}$ be a rational polyhedron. Show that conv($P \cap (\mathbb{Z}^k \times \mathbb{R}^l)$) is a polyhedron.
   Hint: Generalize the proof of Theorem 5.1.
   Note: This is the basis of mixed integer programming. See Schrijver [1986].

* 3. Prove the following integer analogue of Carathéodory's theorem (Exercise 15 of Chapter 3): For each pointed polyhedral cone C, each Hilbert basis $\{a_1, \ldots, a_t\}$ of C, and each integral point $x \in C$ there are $2n - 1$ vectors among $a_1, \ldots, a_t$ such that x is a nonnegative integer combination of those.
   Hint: Consider an optimum basic solution of the LP $\max\{y\mathbb{1} : yA = x, y \ge 0\}$ (where the rows of A are $a_1, \ldots, a_t$), and round the components down.
   Note: The number $2n - 1$ was improved by Sebő [1990] to $2n - 2$. It cannot be improved to less than $\lfloor \frac{7}{6}n \rfloor$ (Bruns et al. [1999]).
   (Cook, Fonlupt and Schrijver [1986])

4. Let $C = \{x : Ax \ge 0\}$ be a rational polyhedral cone and b some vector with $bx > 0$ for all $x \in C \setminus \{0\}$. Show that there exists a unique minimal integral Hilbert basis generating C.
   (Schrijver [1981])

5. Let A be an integral $m \times n$-matrix, and let b and c be vectors, and y an optimum solution of $\max \{cx : Ax \le b, x \text{ integral}\}$. Prove that there exists an

optimum solution z of $\max\{cx : Ax \leq b\}$ with $||y - z||_\infty \leq n\Xi(A)$.
(Cook et al. [1986])

6. Prove that each unimodular matrix arises from an identity matrix by a unimodular transformation.

    Hint: Recall the proof of Lemma 5.11.

* 7. Prove that there is a polynomial-time algorithm which, given an integral matrix A and an integral vector b, finds an integral vector x with $Ax = b$ or decides that none exists.

    Hint: See the proofs of Lemma 5.11 and Lemma 5.12.

8. Consider the two systems

$$\begin{pmatrix} 1 & 1 \\ 1 & 0 \\ 1 & -1 \end{pmatrix} \begin{pmatrix} x_1 \\ x_2 \end{pmatrix} \leq \begin{pmatrix} 0 \\ 0 \\ 0 \end{pmatrix} \quad \text{and} \quad \begin{pmatrix} 1 & 1 \\ 1 & -1 \end{pmatrix} \begin{pmatrix} x_1 \\ x_2 \end{pmatrix} \leq \begin{pmatrix} 0 \\ 0 \end{pmatrix}.$$

    They clearly define the same polyhedron. Prove that the first one is TDI but the second one is not.

9. Let $a \neq 0$ be an integral vector and β a rational number. Prove that the inequality $ax \leq \beta$ is TDI if and only if the components of a are relatively prime.

10. Let $Ax \leq b$ be TDI, $k \in \mathbb{N}$ and $\alpha > 0$ rational. Show that $\frac{1}{k}Ax \leq \alpha b$ is again TDI. Moreover, prove that $\alpha Ax \leq \alpha b$ is not necessarily TDI.

11. Use Theorem 5.26 in order to prove König's Theorem 10.2 (cf. Exercise 2 of Chapter 11):

    The maximum cardinality of a matching in a bipartite graph equals the minimum cardinality of a vertex cover.

12. Show that $A = \begin{pmatrix} 1 & 1 & 1 \\ -1 & 1 & 0 \\ 1 & 0 & 0 \end{pmatrix}$ is not totally unimodular, but $\{x : Ax = b\}$ is integral for all integral vectors b.

    (Nemhauser and Wolsey [1988])

13. Let G be the digraph $(\{1, 2, 3, 4\}, \{(1, 3), (2, 4), (2, 1), (4, 1), (4, 3)\})$, and let $\mathcal{F} := \{\{1, 2, 4\}, \{1, 2\}, \{2\}, \{2, 3, 4\}, \{4\}\}$. Prove that $(V(G), \mathcal{F})$ is cross-free but the one-way cut-incidence matrix of $\mathcal{F}$ is not totally unimodular.

* 14. Let G and T be digraphs such that $V(G) = V(T)$ and the undirected graph underlying T is a tree. For $v, w \in V(G)$ let $P(v, w)$ be the unique undirected path from v to w in T. Let $M = (m_{f,e})_{f \in E(T), e \in E(G)}$ be the matrix defined by

$$m_{(x,y),(v,w)} := \begin{cases} 1 & \text{if } (x, y) \in E(P(v, w)) \text{ and } (x, y) \in E(P(v, y)) \\ -1 & \text{if } (x, y) \in E(P(v, w)) \text{ and } (x, y) \in E(P(v, x)) . \\ 0 & \text{if } (x, y) \notin E(P(v, w)) \end{cases}$$

    Matrices arising this way are called network matrices. Show that the network matrices are precisely the two-way cut-incidence matrices of cross-free set systems.

15. An interval matrix is a 0-1-matrix such that in each row the 1-entries are consecutive. Prove that interval matrices are totally unimodular.
    Note: Hochbaum and Levin [2006] showed how to solve optimization problems with such matrices very efficiently.

16. Consider the following interval packing problem: Given a list of intervals $[a_i, b_i]$, $i = 1, \ldots, n$, with weights $c_1, \ldots, c_n$ and a number $k \in \mathbb{N}$, find a maximum weight subset of the intervals such that no point is contained in more than k of them.

    (a) Find an LP formulation (without integrality constraints) of this problem.

    (b) Consider the case $k = 1$. What combinatorial meaning has the dual LP? Show how to solve the dual LP by a simple combinatorial algorithm.

    (c) Use (b) to obtain an algorithm for the interval packing problem in the case $k = 1$ which runs in $O(n \log n)$ time.

    (d) Find a simple $O(n \log n)$-algorithm for general k and unit weights.
    Note: See also Exercise 11 of Chapter 9.

17. Let $P := \{(x, y) \in \mathbb{R}^2 : y = \sqrt{2}x, x \geq 0\}$ and $Q := \{(x, y) \in \mathbb{R}^2 : y = \sqrt{2}x\}$. Prove that $P^{(t)} = P \neq P_I$ for all $t \in \mathbb{N}$ and $Q' = \mathbb{R}^2$.

18. Let P be the convex hull of the three points $(0, 0)$, $(0, 1)$ and $(k, \frac{1}{2})$ in $\mathbb{R}^2$, where $k \in \mathbb{N}$. Show that $P^{(2k-1)} \neq P_I$ but $P^{(2k)} = P_I$.

* 19. Let $P \subseteq [0, 1]^n$ be a polytope in the unit hypercube with $P_I = \emptyset$. Prove that then $P^{(n)} = \emptyset$.
    Note: Eisenbrand and Schulz [2003] proved that $P^{(\lceil n^2(1+\log n)\rceil)} = P_I$ for any polytope $P \subseteq [0, 1]^n$. See also Pokutta and Schulz [2010], and Rothvoß and Sanità [2017].

20. In this exercise we apply Lagrangean relaxation to linear equation systems. Let Q be a finite set of vectors in $\mathbb{R}^n$, $c \in \mathbb{R}^n$ and $A' \in \mathbb{R}^{m \times n}$ and $b' \in \mathbb{R}^m$. Prove that

$$\min \{\max\{c^\top x + \lambda^\top (b' - A'x) : x \in Q\} : \lambda \in \mathbb{R}^m\}$$
$$= \max\{c^\top y : y \in \mathrm{conv}(Q), A'y = b'\}.$$

21. Consider the following facility location problem: Given a set of n customers with demands $d_1, \ldots, d_n$, and m optional facilities each of which can be opened or not. For each facility $i = 1, \ldots, m$ we have a cost f_i for opening it, a capacity u_i and a distance c_{ij} to each customer $j = 1, \ldots, n$. The task is to decide which facilities should be opened and to assign each customer to an open facility. The total demand of the customers assigned to one facility must not exceed its capacity. The objective is to minimize the facility opening costs plus the sum of the distances of each customer to its facility. In terms of INTEGER PROGRAMMING the problem can be formulated as

$$\min \left\{ \sum_{i,j} c_{ij} x_{ij} + \sum_i f_i y_i : \sum_j d_j x_{ij} \leq u_i y_i, \sum_i x_{ij} = 1, x_{ij}, y_i \in \{0, 1\} \right\}.$$

Apply Lagrangean relaxation in two ways, once relaxing $\sum_j d_j x_{ij} \leq u_i y_i$ for all i, then relaxing $\sum_i x_{ij} = 1$ for all j. Which Lagrangean dual yields a tighter bound?

Note: Both Lagrangean relaxations can be dealt with: see Exercise 11 of Chapter 17.

* 22. Consider the UNCAPACITATED FACILITY LOCATION PROBLEM: given numbers n, m, f_i and c_{ij} ($i = 1, \ldots, m$, $j = 1, \ldots, n$), the problem can be formulated as

$$\min \left\{ \sum_{i,j} c_{ij} x_{ij} + \sum_i f_i y_i : \sum_i x_{ij} = 1, \ x_{ij} \leq y_i, \ x_{ij}, y_i \in \{0, 1\} \right\}.$$

For $S \subseteq \{1, \ldots, n\}$ we denote by $c(S)$ the cost of supplying facilities for the customers in S, i.e.

$$\min \left\{ \sum_{i,j} c_{ij} x_{ij} + \sum_i f_i y_i : \sum_i x_{ij} = 1 \text{ for } j \in S, \ x_{ij} \leq y_i, \ x_{ij}, y_i \in \{0, 1\} \right\}.$$

The cost allocation problem asks whether the total cost $c(\{1, \ldots, n\})$ can be distributed among the customers such that no subset S pays more than $c(S)$. In other words: are there numbers $p_1, \ldots, p_n$ such that $\sum_{j=1}^n p_j = c(\{1, \ldots, n\})$ and $\sum_{j \in S} p_j \leq c(S)$ for all $S \subseteq \{1, \ldots, n\}$? Show that this is the case if and only if $c(\{1, \ldots, n\})$ equals

$$\min \left\{ \sum_{i,j} c_{ij} x_{ij} + \sum_i f_i y_i : \sum_i x_{ij} = 1, \ x_{ij} \leq y_i, \ x_{ij}, y_i \geq 0 \right\},$$

i.e. if the integrality conditions can be left out.

Hint: Apply Lagrangean relaxation to the above LP. For each set of Lagrange multipliers decompose the resulting minimization problem to minimization problems over polyhedral cones. What are the vectors generating these cones? (Goemans and Skutella [2004])

23. Describe a combinatorial algorithm (without using LINEAR PROGRAMMING) to solve (5.10) for arbitrary (but fixed) Lagrange multipliers λ. What running time can you achieve?

References

General Literature

Bertsimas, D., and Weismantel, R. [2005]: Optimization Over Integers. Dynamic Ideas, Belmont 2005

Cook, W.J., Cunningham, W.H., Pulleyblank, W.R., and Schrijver, A. [1998]: Combinatorial Optimization. Wiley, New York 1998, Chapter 6

Jünger, M., Liebling, T., Naddef, D., Nemhauser, G., Pulleyblank, W., Reinelt, G., Rinaldi, G. and Wolsey, L. (Eds.) [2010]: 50 Years of Integer Programming 1958–2008. Springer, Berlin 2010

Nemhauser, G.L., and Wolsey, L.A. [1988]: Integer and Combinatorial Optimization. Wiley, New York 1988

Schrijver, A. [1986]: Theory of Linear and Integer Programming. Wiley, Chichester 1986

Wolsey, L.A. [1998]: Integer Programming. Wiley, New York 1998

Cited References

Anstreicher, K.M., and Wolsey, L.A. [2009]: Two "well-known" properties of subgradient optimization. Mathematical Programming B 120 (2009), 213–220

Boyd, E.A. [1997]: A fully polynomial epsilon approximation cutting plane algorithm for solving combinatorial linear programs containing a sufficiently large ball. Operations Research Letters 20 (1997), 59–63

Bruns, W., Gubeladze, J., Henk, M., Martin, A., and Weismantel, R. [1999]: A counterexample to an integral analogue of Carathéodory's theorem. Journal für die Reine und Angewandte Mathematik 510 (1999), 179–185

Carr, R., and Vempala, S. [2004]: On the Held-Karp relaxation for the asymmetric and symmetric traveling salesman problems. Mathematical Programming A 100 (2004), 569–587

Chvátal, V. [1973]: Edmonds' polytopes and a hierarchy of combinatorial problems. Discrete Mathematics 4 (1973), 305–337

Cook, W. [1983]: Operations that preserve total dual integrality. Operations Research Letters 2 (1983), 31–35

Cook, W., Fonlupt, J., and Schrijver, A. [1986]: An integer analogue of Carathéodory's theorem. Journal of Combinatorial Theory B 40 (1986), 63–70

Cook, W., Gerards, A.M.H., Schrijver, A., and Tardos, É. [1986]: Sensitivity theorems in integer linear programming. Mathematical Programming 34 (1986), 251–264

Cook, W., Kannan, R., and Schrijver, A. [1990]: Chvátal closures for mixed integer programming problems. Mathematical Programming 47 (1990), 155–174

Cornuéjols, G. and Li, Y. [2016]: Deciding emptiness of the Gomory-Chvátal closure is NP-complete, even for a rational polyhedron containing no integer point. In: Integer Programming and Combinatorial Optimization; Proceedings of the 18th International IPCO Conference; LNCS 9682 (Q. Louveaux, M. Skutella, eds.), Springer, Switzerland 2016, pp. 387–397

Dadush, D., Dey, S.S., and Vielma, J.P. [2014]: On the Chvátal-Gomory closure of a compact convex set. Mathematical Programming A 145 (2014), 327–348

Dantzig, G., Fulkerson, R., and Johnson, S. [1954]: Solution of a large-scale traveling-salesman problem. Operations Research 2 (1954), 393–410

Edmonds, J., and Giles, R. [1977]: A min-max relation for submodular functions on graphs. In: Studies in Integer Programming; Annals of Discrete Mathematics 1

(P.L. Hammer, E.L. Johnson, B.H. Korte, G.L. Nemhauser, eds.), North-Holland, Amsterdam 1977, pp. 185–204

Eisenbrand, F. [1999]: On the membership problem for the elementary closure of a polyhedron. Combinatorica 19 (1999), 297–300

Eisenbrand, F., and Schulz, A.S. [2003]: Bounds on the Chvátal rank of polytopes in the 0/1-cube. Combinatorica 23 (2003), 245–261

Fulkerson, D.R. [1971]: Blocking and anti-blocking pairs of polyhedra. Mathematical Programming 1 (1971), 168–194

Geoffrion, A.M. [1974]: Lagrangean relaxation for integer programming. Mathematical Programming Study 2 (1974), 82–114

Giles, F.R., and Pulleyblank, W.R. [1979]: Total dual integrality and integer polyhedra. Linear Algebra and Its Applications 25 (1979), 191–196

Ghouila-Houri, A. [1962]: Caractérisation des matrices totalement unimodulaires. Comptes Rendus Hebdomadaires des Séances de l'Académie des Sciences (Paris) 254 (1962), 1192–1194

Goemans, M.X. [1995]: Worst-case comparison of valid inequalities for the TSP. Mathematical Programming 69 (1995), 335–349

Goemans, M.X., and Skutella, M. [2004]: Cooperative facility location games. Journal of Algorithms 50 (2004), 194–214

Goffin, J.L. [1977]: On convergence rates of subgradient optimization methods. Mathematical Programming 13 (1977), 329–347

Gomory, R.E. [1958]: Outline of an algorithm for integer solutions to linear programs. Bulletin of the American Mathematical Society 64 (1958), 275–278

Gomory, R.E. [1963]: An algorithm for integer solutions of linear programs. In: Recent Advances in Mathematical Programming (R.L. Graves, P. Wolfe, eds.), McGraw-Hill, New York, 1963, pp. 269–302

Graver, J.E. [1975]: On the foundations of linear and integer programming I. Mathematical Programming 9 (1975), 207–226

Harvey, W. [1999]: Computing two-dimensional integer hulls. SIAM Journal on Computing 28 (1999), 2285–2299

Hochbaum, D.S., and Levin, A. [2006]: Optimizing over consecutive 1's and circular 1's constraints. SIAM Journal on Optimization 17 (2006), 311–330

Hoffman, A.J. [1974]: A generalization of max flow-min cut. Mathematical Programming 6 (1974), 352–359

Hoffman, A.J., and Kruskal, J.B. [1956]: Integral boundary points of convex polyhedra. In: Linear Inequalities and Related Systems; Annals of Mathematical Study 38 (H.W. Kuhn, A.W. Tucker, eds.), Princeton University Press, Princeton 1956, 223–246

Lasserre, J.B. [2004]: The integer hull of a convex rational polytope. Discrete & Computational Geometry 32 (2004), 129–139

Meyer, R.R. [1974]: On the existence of optimal solutions to integer and mixed-integer programming problems. Mathematical Programming 7 (1974), 223–235

Pokutta, S., and Schulz, A.S. [2010]: On the rank of cutting-plane proof systems. In: Integer Programming and Combinatorial Optimization; Proceedings of the 14th International IPCO Conference; LNCS 6080 (F. Eisenbrand, F.B. Shepherd, eds.), Springer, Berlin 2010, pp. 450–463

Polyak, B.T. [1967]: A general method for solving extremal problems. Doklady Akademii Nauk SSSR 174 (1967), 33–36 [in Russian]. English translation: Soviet Mathematics Doklady 8 (1967), 593–597

Rothvoß, T., and Sanità, L. [2017]: 0/1 polytopes with quadratic Chvátal rank. Operations Research 65 (2017), 212–220

Schrijver, A. [1980]: On cutting planes. In: Combinatorics 79; Part II; Annals of Discrete Mathematics 9 (M. Deza, I.G. Rosenberg, eds.), North-Holland, Amsterdam 1980, pp. 291–296

Schrijver, A. [1981]: On total dual integrality. Linear Algebra and its Applications 38 (1981), 27–32

Schrijver, A. [1983]: Packing and covering of crossing families of cuts. Journal of Combinatorial Theory B 35 (1983), 104–128

Sebő, A. [1990]: Hilbert bases, Carathéodory's theorem and combinatorial optimization. In: Integer Programming and Combinatorial Optimization (R. Kannan and W.R. Pulleyblank, eds.), University of Waterloo Press, 1990

Seymour, P.D. [1980]: Decomposition of regular matroids. Journal of Combinatorial Theory B 28 (1980), 305–359

Veinott, A.F., Jr., and Dantzig, G.B. [1968]: Integral extreme points. SIAM Review 10 (1968), 371–372

Wolsey, L.A. [1981]: The b-hull of an integer program. Discrete Applied Mathematics 3 (1981), 193–201

6 Spanning Trees and Arborescences

Consider a telephone company that wants to rent a subset from an existing set of cables, each of which connects two cities. The rented cables should suffice to connect all cities and they should be as cheap as possible. It is natural to model the network by a graph: the vertices are the cities and the edges correspond to the cables. By Theorem 2.4 the minimal connected spanning subgraphs of a given graph are its spanning trees. So we look for a spanning tree of minimum weight, where we say that a subgraph T of a graph G with weights $c : E(G) \to \mathbb{R}$ has weight $c(E(T)) = \sum_{e \in E(T)} c(e)$. We shall refer to $c(e)$ also as the cost of e.

This is a simple but very important combinatorial optimization problem. It is also among the combinatorial optimization problems with the longest history; the first algorithm was given by Borůvka [1926a,b]; see Nešetřil, Milková, and Nešetřilová [2001] .

Compared to the DRILLING PROBLEM which asks for a shortest path containing all vertices of a complete graph, we now look for a shortest spanning tree. Although the number of spanning trees is even bigger than the number of paths (K_n contains $\frac{n!}{2}$ Hamiltonian paths, but as many as n^{n-2} different spanning trees; cf. Theorem 6.2), the problem turns out to be much easier. In fact, a simple greedy strategy works as we shall see in Section 6.1.

Arborescences can be considered as the directed counterparts of trees; by Theorem 2.5 they are the minimal spanning subgraphs of a digraph such that all vertices are reachable from a root. The directed version of the MINIMUM SPANNING TREE PROBLEM, the MINIMUM WEIGHT ARBORESCENCE PROBLEM, is more difficult since a greedy strategy no longer works. In Section 6.2 we show how to solve this problem.

Since there are very efficient combinatorial algorithms it is not recommended to solve these problems with LINEAR PROGRAMMING. Nevertheless it is interesting that the corresponding polytopes (the convex hull of the incidence vectors of spanning trees or arborescences; cf. Corollary 3.33) can be described in a nice way, which we shall show in Section 6.3. In Section 6.4 we prove some classical results concerning the packing of spanning trees and arborescences.

© Springer-Verlag GmbH Germany 2018
Bernhard Korte and Jens Vygen, *Combinatorial Optimization*, Algorithms and Combinatorics 21, https://doi.org/10.1007/978-3-662-56039-6_6

6.1 Minimum Spanning Trees

In this section, we consider the following two problems:

MAXIMUM WEIGHT FOREST PROBLEM

Instance: An undirected graph G, weights $c : E(G) \to \mathbb{R}$.

Task: Find a forest in G of maximum weight.

MINIMUM SPANNING TREE PROBLEM

Instance: An undirected graph G, weights $c : E(G) \to \mathbb{R}$.

Task: Find a spanning tree in G of minimum weight or decide that G is not connected.

We claim that both problems are equivalent. To make this precise, we say that a problem $\mathcal{P}$ **linearly reduces** to a problem $\mathcal{Q}$ if there are functions f and g, each computable in linear time, such that f transforms an instance x of $\mathcal{P}$ to an instance $f(x)$ of $\mathcal{Q}$, and g transforms a solution of $f(x)$ to a solution of x. If $\mathcal{P}$ linearly reduces to $\mathcal{Q}$ and $\mathcal{Q}$ linearly reduces to $\mathcal{P}$, then both problems are called **equivalent**.

Proposition 6.1. *The* MAXIMUM WEIGHT FOREST PROBLEM *and the* MINIMUM SPANNING TREE PROBLEM *are equivalent.*

Proof: Given an instance (G, c) of the MAXIMUM WEIGHT FOREST PROBLEM, delete all edges of negative weight, let $c'(e) := -c(e)$ for all $e \in E(G)$, and add a minimum set F of edges (with arbitrary weight) to make the graph connected; let us call the resulting graph G'. Then instance (G', c') of the MINIMUM SPANNING TREE PROBLEM is equivalent in the following sense: Deleting the edges of F from a minimum weight spanning tree in (G', c') yields a maximum weight forest in (G, c).

Conversely, given an instance (G, c) of the MINIMUM SPANNING TREE PROBLEM, let $c'(e) := K - c(e)$ for all $e \in E(G)$, where $K = 1 + \max_{e \in E(G)} c(e)$. Then the instance (G, c') of the MAXIMUM WEIGHT FOREST PROBLEM is equivalent, since all spanning trees have the same number of edges (Theorem 2.4).   □

We shall return to different reductions of one problem to another in Chapter 15. In the rest of this section we consider the MINIMUM SPANNING TREE PROBLEM only. Let us first count the number of feasible solutions. The following is known as Cayley's Theorem:

Theorem 6.2.   (Sylvester [1857], Cayley [1889]) *For $n \in \mathbb{N}$ the number of spanning trees in K_n is n^{n-2}.*

Proof: Let t_n be the number of spanning trees in K_n. Let

$$\mathcal{B}_{n,k} := \{(B, f) : B \text{ branching}, V(B) = \{1, \ldots, n\}, |E(B)| = k,$$
$$f : E(B) \to \{1, \ldots, k\} \text{ bijective}\}.$$

Since each spanning tree has n orientations as an arborescence and $(n-1)!$ orderings of the edge set, we have $|\mathcal{B}_{n,n-1}| = (n-1)! \cdot n \cdot t_n$. Moreover, $|\mathcal{B}_{n,0}| = 1$. We show $|\mathcal{B}_{n,i+1}| = n(n-i-1)|\mathcal{B}_{n,i}|$, which implies $|\mathcal{B}_{n,n-1}| = n^{n-1}(n-1)!|\mathcal{B}_{n,0}|$, and hence $t_n = n^{n-2}$.

For each element $(B, f) \in \mathcal{B}_{n,i+1}$ we define $g(B, f) \in \mathcal{B}_{n,i}$ by deleting the edge $f^{-1}(i+1)$. Now each $(B', f) \in \mathcal{B}_{n,i}$ is the image of exactly $n(n-i-1)$ elements of $\mathcal{B}_{n,i+1}$: we add an edge $e = (v, w)$ and set $f(e) := i+1$, and there are n choices for v (all vertices), and then $n - i - 1$ choices for w (the roots of the connected components of B' that do not contain v). □

Next we prove optimality conditions:

Theorem 6.3. *Let (G, c) be an instance of the* MINIMUM SPANNING TREE PROB-LEM, *and let T be a spanning tree in G. Then the following statements are equivalent:*

(a) *T is optimum.*
(b) *For every $e = \{x, y\} \in E(G) \setminus E(T)$, no edge on the x-y-path in T has higher cost than e.*
(c) *For every $e \in E(T)$, e is a minimum cost edge of $\delta(V(C))$, where C is a connected component of $T - e$.*
(d) *We can order $E(T) = \{e_1, \ldots, e_{n-1}\}$ such that for each $i \in \{1, \ldots, n-1\}$ there exists a set $X \subseteq V(G)$ such that e_i is a minimum cost edge of $\delta(X)$ and $e_j \notin \delta(X)$ for all $j \in \{1, \ldots, i-1\}$.*

Proof: (a)$\Rightarrow$(b): Suppose (b) is violated: Let $e = \{x, y\} \in E(G) \setminus E(T)$ and let f be an edge on the x-y-path in T with $c(f) > c(e)$. Then $(T - f) + e$ is a spanning tree with lower cost.

(b)$\Rightarrow$(c): Suppose (c) is violated: let $e \in E(T)$, C a connected component of $T - e$ and $f = \{x, y\} \in \delta(V(C))$ with $c(f) < c(e)$. Observe that the x-y-path in T must contain an edge of $\delta(V(C))$, but the only such edge is e. So (b) is violated.

(c)$\Rightarrow$(d): Take an arbitrary order and $X := V(C)$.

(d)$\Rightarrow$(a): Suppose $E(T) = \{e_1, \ldots, e_{n-1}\}$ satisfies (d), and let T^* be an optimum spanning tree such that $i := \min\{h \in \{1, \ldots, n-1\} : e_h \notin E(T^*)\}$ is maximum. We show that $i = \infty$, i.e. $T = T^*$. Suppose not, then let $X \subseteq V(G)$ such that e_i is a minimum cost edge of $\delta(X)$ and $e_j \notin \delta(X)$ for all $j \in \{1, \ldots, i-1\}$. $T^* + e_i$ contains a circuit C. Since $e_i \in E(C) \cap \delta(X)$, at least one more edge f $(f \neq e_i)$ of C must belong to $\delta(X)$ (see Exercise 13 of Chapter 2). Observe that $(T^* + e_i) - f$ is a spanning tree. Since T^* is optimum, $c(e_i) \geq c(f)$. But since $f \in \delta(X)$, we also have $c(f) \geq c(e_i)$. Moreover, if $f = e_j \in E(T)$, then $j > i$. So $c(f) = c(e_i)$, and $(T^* + e_i) - f$ is another optimum spanning tree which contradicts the maximality of i. □

The following "greedy" algorithm for the MINIMUM SPANNING TREE PROB-
LEM was proposed by Kruskal [1956]. It can be regarded as a special case of a
quite general greedy algorithm which will be discussed in Section 13.4. In the
following let $n := |V(G)|$ and $m := |E(G)|$.

KRUSKAL'S ALGORITHM

Input: A connected undirected graph G, weights $c : E(G) \to \mathbb{R}$.

Output: A spanning tree T of minimum weight.

---

①   Sort the edges such that $c(e_1) \le c(e_2) \le \ldots \le c(e_m)$.

②   Set $T := (V(G), \emptyset)$.

③   **For** $i := 1$ **to** m **do**:
        If $T + e_i$ contains no circuit **then** set $T := T + e_i$.

Theorem 6.4. KRUSKAL'S ALGORITHM *works correctly.*

Proof: It is clear that the algorithm constructs a spanning tree T. It also guarantees
condition (b) of Theorem 6.3, so T is optimum.   □

The running time of KRUSKAL'S ALGORITHM is $O(mn)$: the edges can be
sorted in $O(m \log m)$ time (Theorem 1.5), and testing for a circuit in a graph with
at most n edges can be implemented in $O(n)$ time (just apply DFS (or BFS) and
check if there is any edge not belonging to the DFS-tree). Since this is repeated
m times, we get a total running time of $O(m \log m + mn) = O(mn)$. However, a
more efficient implementation is possible:

Theorem 6.5. KRUSKAL'S ALGORITHM *can be implemented to run in*
$O(m \log n)$ *time.*

Proof: Parallel edges can be eliminated first: all but the cheapest edges are
redundant. So we may assume that $m = O(n^2)$. Since the running time of ①
is obviously $O(m \log m) = O(m \log n)$ we concentrate on ③. We study a data
structure maintaining the connected components of T. In ③ we have to test whether
the addition of an edge $e_i = \{v, w\}$ to T results in a circuit. This is equivalent to
testing if v and w are in the same connected component.

Our implementation maintains a branching B with $V(B) = V(G)$. At any time
the connected components of B will be induced by the same vertex sets as the
connected components of T. (Note however that B is in general not an orientation
of T.)

When checking an edge $e_i = \{v, w\}$ in ③, we find the root r_v of the arborescence
in B containing v and the root r_w of the arborescence in B containing w. The
time needed for this is proportional to the length of the r_v-v-path plus the length
of the r_w-w-path in B. We shall show later that this length is always at most $\log n$.

Next we check if $r_v = r_w$. If $r_v \neq r_w$, we insert e_i into T and we have
to add an edge to B. Let $h(r)$ be the maximum length of a path from r in B.

If $h(r_v) \geq h(r_w)$, then we add an edge (r_v, r_w) to B, otherwise we add (r_w, r_v) to B. If $h(r_v) = h(r_w)$, this operation increases $h(r_v)$ by one, otherwise the new root has the same h-value as before. So the h-values of the roots can be maintained easily. Of course initially $B := (V(G), \emptyset)$ and $h(v) := 0$ for all $v \in V(G)$.

We claim that an arborescence of B with root r contains at least $2^{h(r)}$ vertices. This implies that $h(r) \leq \log n$, concluding the proof. At the beginning, the claim is clearly true. We have to show that this property is maintained when adding an edge (x, y) to B. This is trivial if $h(x)$ does not change. Otherwise we have $h(x) = h(y)$ before the operation, implying that each of the two arborescences contains at least $2^{h(x)}$ vertices. So the new arborescence rooted at x contains at least $2 \cdot 2^{h(x)} = 2^{h(x)+1}$ vertices, as required. $\qquad\square$

The above implementation can be improved by another trick: whenever the root r_v of the arborescence in B containing v has been determined, all the edges on the r_v-v-path P are deleted and an edge (r_x, x) is inserted for each $x \in V(P) \setminus \{r_v\}$. A complicated analysis shows that this so-called path compression heuristic makes the running time of ③ almost linear: it is $O(m\alpha(m, n))$, where $\alpha(m, n)$ is the functional inverse of Ackermann's function (see Tarjan [1975, 1983]).

We now mention another well-known algorithm for the MINIMUM SPANNING TREE PROBLEM, due to Jarník [1930] (see Korte and Nešetřil [2001]), Dijkstra [1959] and Prim [1957]:

PRIM'S ALGORITHM

Input: A connected undirected graph G, weights $c : E(G) \rightarrow \mathbb{R}$.

Output: A spanning tree T of minimum weight.

---

①   Choose $v \in V(G)$. Set $T := (\{v\}, \emptyset)$.

②   **While** $V(T) \neq V(G)$ **do**:
          Choose an edge $e \in \delta_G(V(T))$ of minimum weight. Set $T := T + e$.

Theorem 6.6. PRIM'S ALGORITHM *works correctly. Its running time is* $O(m + n^2)$.

Proof: The correctness follows from the fact that condition (d) of Theorem 6.3 is guaranteed (order the edges of T as chosen by the algorithm).

We again eliminate parallel edges first. To obtain the $O(n^2)$ running time, we maintain for each vertex $v \in V(G) \setminus V(T)$ a cheapest edge $e \in E(V(T), \{v\})$. Let us call these edges the candidates. The initialization of the candidates after ① takes $O(n)$ time. Each selection of a cheapest edge among the candidates takes $O(n)$ time. The update of the candidates can be done by scanning the edges incident to the vertex which is added to $V(T)$ and thus also takes $O(n)$ time. Since the while-loop of ② has $n - 1$ iterations, the $O(n^2)$ bound is proved. $\qquad\square$

The running time can be improved by efficient data structures. Denote $l_{T,v} := \min\{c(e) : e \in E(V(T), \{v\})\}$. We maintain the set $\{(v, l_{T,v}) : v \in V(G) \setminus$

$V(T)$, $l_{T,v} < \infty\}$ in a data structure, called priority queue or heap, that allows inserting an element, finding and deleting an element (v, l) with minimum l, and decreasing the so-called key l of an element (v, l). Then PRIM'S ALGORITHM can be written as follows:

①   Choose $v \in V(G)$. Set $T := (\{v\}, \emptyset)$.
     Let $l_w := \infty$ for $w \in V(G) \setminus \{v\}$.
②   **While** $V(T) \neq V(G)$ **do:**
         For $e = \{v, w\} \in E(\{v\}, V(G) \setminus V(T))$ **do:**
             If $c(e) < l_w < \infty$ **then** set $l_w := c(e)$ and DECREASEKEY(w, l_w).
             If $l_w = \infty$ **then** set $l_w := c(e)$ and INSERT(w, l_w).
         $(v, l_v) :=$ DELETEMIN.
         Let $e \in E(V(T), \{v\})$ with $c(e) = l_v$. Set $T := T + e$.

There are several possible ways to implement a heap. A very efficient way, the so-called Fibonacci heap, has been proposed by Fredman and Tarjan [1987]. Our presentation is based on Schrijver [2003]:

Theorem 6.7. *It is possible to maintain a data structure for a finite set (initially empty), where each element u is associated with a real number $d(u)$, called its key, and perform any sequence of*

- *p INSERT-operations (adding an element u with key $d(u)$);*
- *n DELETEMIN-operations (finding and deleting an element u with $d(u)$ minimum);*
- *m DECREASEKEY-operations (decreasing $d(u)$ to a specified value for an element u)*

in $O(m + p + n \log p)$ time.

Proof: The set, denoted by U, is stored in a Fibonacci heap, i.e. a branching (U, E) with a function $\varphi : U \rightarrow \{0, 1\}$ with the following properties:

(i) If $(u, v) \in E$ then $d(u) \leq d(v)$. (This is called the heap order.)
(ii) For each $u \in U$ the children of u can be numbered $1, \ldots, |\delta^+(u)|$ such that the i-th child v satisfies $|\delta^+(v)| + \varphi(v) \geq i - 1$.
(iii) If u and v are distinct roots $(\delta^-(u) = \delta^-(v) = \emptyset)$, then $|\delta^+(u)| \neq |\delta^+(v)|$.

Condition (ii) implies:

(iv) If a vertex u has out-degree at least k, then at least $\sqrt{2}^k$ vertices are reachable from u.

We prove (iv) by induction on k, the cases $k = 0$ and $k = 1$ being trivial. So let u be a vertex with $|\delta^+(u)| \geq k \geq 1$, and let v be a child of u with $|\delta^+(v)| \geq k - 2$ (v exists due to (ii)). We apply the induction hypothesis to v in (U, E) and to u in $(U, E \setminus \{(u, v)\})$ and conclude that at least $\sqrt{2}^{k-2}$ and $\sqrt{2}^{k-1}$ vertices are reachable. (iv) follows from observing that $\sqrt{2}^k \leq \sqrt{2}^{k-2} + \sqrt{2}^{k-1}$.

In particular, (iv) implies that $|\delta^+(u)| \le 2 \log |U|$ for all $u \in U$. Thus, using (iii), we can store the roots of (U, E) by a function $b : \{0, 1, \ldots, \lfloor 2 \log |U| \rfloor\} \to U$ with $b(|\delta^+(u)|) = u$ for each root u. Note that $b(i) = u$ neither implies that $|\delta^+(u)| = i$ nor that u is a root.

In addition to this, we keep track of a doubly-linked list of children (in arbitrary order), a pointer to the parent (if existent), and the out-degree of each vertex. We now show how the INSERT-, DELETEMIN- and DECREASEKEY-operations are implemented.

INSERT$(v, d(v))$ is implemented by setting $\varphi(v) := 0$ and applying

PLANT(v):

①   Set $r := b(|\delta^+(v)|)$.
        if r is a root with $r \ne v$ and $|\delta^+(r)| = |\delta^+(v)|$ **then**:
            if $d(r) \le d(v)$ **then** add (r, v) to E and PLANT(r).
            if $d(v) < d(r)$ **then** add (v, r) to E and PLANT(v).
        else set $b(|\delta^+(v)|) := v$.

As (U, E) is always a branching, the recursion terminates. Note also that (i), (ii) and (iii) are maintained.

DELETEMIN is implemented by scanning $b(i)$ for $i = 0, \ldots, \lfloor 2 \log |U| \rfloor$ in order to find an element u with $\delta^-(u) = \emptyset$ and $d(u)$ minimum, deleting u and its incident edges, and successively applying PLANT(v) for each (former) child v of u.

DECREASEKEY$(v, d(v))$ is a bit more complicated. Let P be the longest path in (U, E) ending in v such that each internal vertex u satisfies $\varphi(u) = 1$. We set $\varphi(u) := 1 - \varphi(u)$ for all $u \in V(P) \setminus \{v\}$, delete all edges of P from E, and apply PLANT(z) for each $z \in V(P)$ which is a root of the new forest.

To see that this maintains (ii) we only have to consider the parent of the start vertex x of P, if existent. But then x is not a root, and thus $\varphi(x)$ changes from 0 to 1, making up for the lost child.

We finally estimate the running time. As φ increases at most m times (at most once in each DECREASEKEY), φ decreases at most m times. Thus the sum of the length of the paths P in all DECREASEKEY-operations is at most $m + m$. So at most $2m + 2n \log p$ edges are deleted overall (as each DELETEMIN-operation may delete up to $2 \log p$ edges). Thus at most $2m + 2n \log p + p - 1$ edges are inserted in total. This proves the overall $O(m + p + n \log p)$ running time. □

Corollary 6.8. PRIM'S ALGORITHM *implemented with Fibonacci heap solves the* MINIMUM SPANNING TREE PROBLEM *in* $O(m + n \log n)$ *time.*

Proof: We have at most $n - 1$ INSERT-, $n - 1$ DELETEMIN-, and m DECREASEKEY-operations. □

With a more sophisticated implementation, the running time can be improved to $O(m \log \beta(n, m))$, where $\beta(n, m) = \min \{i : \log^{(i)} n \le \frac{m}{n}\}$; see Fredman and Tarjan [1987], Gabow, Galil and Spencer [1989], and Gabow et al. [1986]. The fastest deterministic algorithm is due to Pettie and Ramachandran [2002] . It is at least as fast as the algorithm by Chazelle [2000], whose running time is $O(m\alpha(m, n))$, where α is the functional inverse of Ackermann's function.

On a different computational model Fredman and Willard [1994] achieved linear running time. Moreover, there is a randomized algorithm which finds a minimum weight spanning tree and has linear expected running time (Karger, Klein and Tarjan [1995]; such an algorithm which always finds an optimum solution is called a Las Vegas algorithm). This algorithm uses a (deterministic) procedure for testing whether a given spanning tree is optimum; a linear-time algorithm for this problem has been found by Dixon, Rauch and Tarjan [1992] ; see also King [1997].

The MINIMUM SPANNING TREE PROBLEM for planar graphs can be solved (deterministically) in linear time (Cheriton and Tarjan [1976]). The problem of finding a minimum spanning tree for a set of n points in the plane can be solved in $O(n \log n)$ time (Exercise 14). PRIM'S ALGORITHM can be quite efficient for such instances since one can use suitable data structures for finding nearest neighbours in the plane effectively.

6.2 Minimum Weight Arborescences

Natural directed generalizations of the MAXIMUM WEIGHT FOREST PROBLEM and the MINIMUM SPANNING TREE PROBLEM read as follows:

MAXIMUM WEIGHT BRANCHING PROBLEM

Instance: A digraph G, weights $c : E(G) \to \mathbb{R}$.

Task: Find a maximum weight branching in G.

MINIMUM WEIGHT ARBORESCENCE PROBLEM

Instance: A digraph G, weights $c : E(G) \to \mathbb{R}$.

Task: Find a minimum weight spanning arborescence in G or decide that none exists.

Sometimes we want to specify the root in advance:

MINIMUM WEIGHT ROOTED ARBORESCENCE PROBLEM

Instance: A digraph G, a vertex $r \in V(G)$, weights $c : E(G) \to \mathbb{R}$.

Task: Find a minimum weight spanning arborescence rooted at r in G or decide that none exists.

As for the undirected case, these three problems are equivalent:

Proposition 6.9. *The* MAXIMUM WEIGHT BRANCHING PROBLEM, *the* MINIMUM WEIGHT ARBORESCENCE PROBLEM *and the* MINIMUM WEIGHT ROOTED ARBORESCENCE PROBLEM *are all equivalent.*

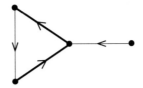

Fig. 6.1.

Proof: Given an instance (G, c) of the MINIMUM WEIGHT ARBORESCENCE PROBLEM, let $c'(e) := K - c(e)$ for all $e \in E(G)$, where $K = 1 + \sum_{e \in E(G)} |c(e)|$. Then the instance (G, c') of the MAXIMUM WEIGHT BRANCHING PROBLEM is equivalent, because for any two branchings B, B' with $|E(B)| > |E(B')|$ we have $c'(E(B)) > c'(E(B'))$ (and branchings with $n - 1$ edges are exactly the spanning arborescences).

Given an instance (G, c) of the MAXIMUM WEIGHT BRANCHING PROBLEM, let $G' := (V(G) \,\dot\cup\, \{r\}, E(G) \cup \{(r, v) : v \in V(G)\})$. Let $c'(e) := -c(e)$ for $e \in E(G)$ and $c(e) := 0$ for $e \in E(G') \setminus E(G)$. Then the instance (G', r, c') of the MINIMUM WEIGHT ROOTED ARBORESCENCE PROBLEM is equivalent.

Finally, given an instance (G, r, c) of the MINIMUM WEIGHT ROOTED ARBORESCENCE PROBLEM, let $G' := (V(G) \,\dot\cup\, \{s\}, E(G) \cup \{(s, r)\})$ and $c((s, r)) := 0$. Then the instance (G', c) of the MINIMUM WEIGHT ARBORESCENCE PROBLEM is equivalent. $\square$

In the rest of this section we shall deal with the MAXIMUM WEIGHT BRANCHING PROBLEM only. This problem is not as easy as its undirected version, the MAXIMUM WEIGHT FOREST PROBLEM. For example any maximal forest is maximum, but the bold edges in Figure 6.1 form a maximal branching which is not maximum.

Recall that a branching is a graph B with $|\delta_B^-(x)| \leq 1$ for all $x \in V(B)$, such that the underlying undirected graph is a forest. Equivalently, a branching is an acyclic digraph B with $|\delta_B^-(x)| \leq 1$ for all $x \in V(B)$; see Theorem 2.5(g):

Proposition 6.10.   *Let B be a digraph with $|\delta_B^-(x)| \leq 1$ for all $x \in V(B)$. Then B contains a circuit if and only if the underlying undirected graph contains a circuit.* $\square$

Now let G be a digraph and $c : E(G) \rightarrow \mathbb{R}_+$. We can ignore negative weights since such edges will never appear in an optimum branching. A first idea towards an algorithm could be to take the best entering edge for each vertex. Of course the resulting graph may contain circuits. Since a branching cannot contain circuits, we must delete at least one edge of each circuit. The following lemma says that one is enough.

Lemma 6.11.   (Karp [1972]) *Let B_0 be a maximum weight subgraph of G with $|\delta_{B_0}^-(v)| \leq 1$ for all $v \in V(B_0)$. Then there exists an optimum branching B of G such that for each circuit C in B_0, $|E(C) \setminus E(B)| = 1$.*

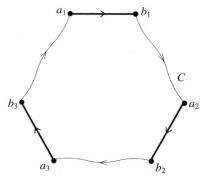

Fig. 6.2.

Proof: Let B be an optimum branching of G containing as many edges of B_0 as possible. Let C be some circuit in B_0. Let $E(C) \setminus E(B) = \{(a_1, b_1), \ldots, (a_k, b_k)\}$; suppose that $k \geq 2$ and $a_1, b_1, a_2, b_2, a_3, \ldots, b_k$ lie in this order on C (see Figure 6.2).

We claim that B contains a b_i-b_{i-1}-path for each $i = 1, \ldots, k$ ($b_0 := b_k$). This, however, is a contradiction because these paths form a closed edge progression in B, and a branching cannot have a closed edge progression.

Let $i \in \{1, \ldots, k\}$. It remains to show that B contains a b_i-b_{i-1}-path. Consider B' with $V(B') = V(G)$ and $E(B') := \{(x, y) \in E(B) : y \neq b_i\} \cup \{(a_i, b_i)\}$.

B' cannot be a branching since it would be optimum and contain more edges of B_0 than B. So (by Proposition 6.10) B' contains a circuit, i.e. B contains a b_i-a_i-path P. Since $k \geq 2$, P is not completely on C, so let e be the last edge of P not belonging to C. Obviously $e = (x, b_{i-1})$ for some x, so P (and thus B) contains a b_i-b_{i-1}-path. □

The main idea of Edmonds'[1967] algorithm is to find first B_0 as above, and then contract every circuit of B_0 in G. If we choose the weights of the resulting graph G_1 correctly, any optimum branching in G_1 will correspond to an optimum branching in G.

EDMONDS' BRANCHING ALGORITHM

Input: A digraph G, weights $c : E(G) \rightarrow \mathbb{R}_+$.

Output: A maximum weight branching B of G.

①   Set $i := 0$, $G_0 := G$, and $c_0 := c$.

②   Let B_i be a maximum weight subgraph of G_i with $|\delta^-_{B_i}(v)| \leq 1$ for all $v \in V(B_i)$.

③   **If** B_i contains no circuit **then** set $B := B_i$ and **go to** ⑤.

④   Let $\mathcal{C}$ be the set of circuits in B_i. Contract these circuits:
Let $V(G_{i+1}) := \mathcal{C} \cup (V(G_i) \setminus \bigcup_{C \in \mathcal{C}} V(C))$.
For $e = (v, w) \in E(G_i)$ let $e' = (v', w')$ and $\Phi_{i+1}(e') := e$, where
    $v' = C$ if $v \in V(C)$ for $C \in \mathcal{C}$, and $v' = v$ if $v \notin \bigcup_{C \in \mathcal{C}} V(C)$, and
    $w' = C$ if $w \in V(C)$ for $C \in \mathcal{C}$, and $w' = w$ if $w \notin \bigcup_{C \in \mathcal{C}} V(C)$.
Let $E(G_{i+1}) := \{e' = (v', w') : e \in E(G_i), v' \neq w'\}$
    (parallel edges may arise).
For $e = (v, w) \in E(G_i)$ with $e' = (v', w') \in E(G_{i+1})$ set
    $c_{i+1}(e') := c_i(e)$ if $w' \notin C$, and
    $c_{i+1}(e') := c_i(e) - c_i(\alpha(e, C)) + c_i(e_C)$ if $w' = C \in \mathcal{C}$, where
    $\alpha(e, C) \in \delta_C^-(w)$ and e_C is some cheapest edge of C.
Set $i := i + 1$ and **go to** ②.

⑤   **While** $i > 0$ **do:**
Set $B' := (V(G_{i-1}), \{\Phi_i(e) : e \in E(B)\})$.
For each circuit C of B_{i-1} **do:**
    If there is an edge $e \in \delta_{B'}^-(V(C))$
        then set $E(B') := E(B') \cup (E(C) \setminus \{\alpha(e, C)\})$
        else set $E(B') := E(B') \cup (E(C) \setminus \{e_C\})$.
Set $B := B'$ and $i := i - 1$.

This algorithm was also discovered independently by Chu and Liu [1965] and Bock [1971].

Theorem 6.12. (Edmonds [1967]) EDMONDS' BRANCHING ALGORITHM *works correctly.*

Proof: We show that in ⑤, B is always an optimum branching of G_i. This is trivial for the first time we reach ⑤. So we have to show that one iteration of ⑤ transforms an optimum branching B of G_i into an optimum branching B' of G_{i-1}.

Let B_{i-1}^* be any branching of G_{i-1} such that $|E(C) \setminus E(B_{i-1}^*)| = 1$ for each circuit C of B_{i-1}. Let B_i^* result from B_{i-1}^* by contracting the circuits of B_{i-1}. B_i^* is a branching of G_i. Furthermore we have

$$c_{i-1}(B_{i-1}^*) \leq c_i(B_i^*) + \sum_{C: \text{ circuit of } B_{i-1}} (c_{i-1}(E(C)) - c_{i-1}(e_C)).$$

By the induction hypothesis, B is an optimum branching of G_i, so we have $c_i(B) \geq c_i(B_i^*)$. We conclude that

$$c_{i-1}(B_{i-1}^*) \leq c_i(B) + \sum_{C: \text{ circuit of } B_{i-1}} (c_{i-1}(E(C)) - c_{i-1}(e_C))$$

$$= c_{i-1}(B').$$

This, together with Lemma 6.11, implies that B' is an optimum branching of G_{i-1}.

□

This proof is due to Karp [1972]. Edmonds' original proof was based on a linear programming formulation (see Corollary 6.15). The running time of EDMONDS' BRANCHING ALGORITHM is easily seen to be $O(mn)$, where $m = |E(G)|$ and $n = |V(G)|$: there are at most n iterations (i.e. $i \leq n$ at any stage of the algorithm), and each iteration can be implemented in $O(m)$ time.

The best known bound has been obtained by Gabow et al. [1986] using a Fibonacci heap: their branching algorithm runs in $O(m + n \log n)$ time.

6.3 Polyhedral Descriptions

A polyhedral description of the MINIMUM SPANNING TREE PROBLEM is as follows:

Theorem 6.13. (Edmonds [1970]) *Given a connected undirected graph G, $n := |V(G)|$, the polytope $P :=$*

$$\left\{ x \in [0, 1]^{E(G)} : \sum_{e \in E(G)} x_e = n - 1, \quad \sum_{e \in E(G[X])} x_e \leq |X| - 1 \text{ for } \emptyset \neq X \subset V(G) \right\}$$

*is integral. Its vertices are exactly the incidence vectors of spanning trees of G. (P is called the **spanning tree polytope** of G.)*

Proof: Let T be a spanning tree of G, and let x be the incidence vector of $E(T)$. Obviously (by Theorem 2.4), $x \in P$. Furthermore, since $x \in \{0, 1\}^{E(G)}$, it must be a vertex of P.

On the other hand let x be an integral vertex of P. Then x is the incidence vector of the edge set of some subgraph H with $n - 1$ edges and no circuit. Again by Theorem 2.4 this implies that H is a spanning tree.

So it suffices to show that P is integral (recall Theorem 5.14). Let $c : E(G) \to \mathbb{R}$, and let T be the tree produced by KRUSKAL'S ALGORITHM when applied to (G, c) (ties are broken arbitrarily when sorting the edges). Denote $E(T) = \{f_1, \ldots, f_{n-1}\}$, where the f_i were taken in this order by the algorithm. In particular, $c(f_1) \leq \cdots \leq c(f_{n-1})$. Let $X_k \subseteq V(G)$ be the connected component of $(V(G), \{f_1, \ldots, f_k\})$ containing f_k $(k = 1, \ldots, n-1)$.

Let x^* be the incidence vector of $E(T)$. We show that x^* is an optimum solution to the LP

$$\min \quad \sum_{e \in E(G)} c(e) x_e$$

$$\text{s.t.} \quad \sum_{e \in E(G)} x_e \; = \; n - 1$$

$$\sum_{e \in E(G[X])} x_e \; \leq \; |X| - 1 \qquad (\emptyset \neq X \subset V(G))$$

$$x_e \; \geq \; 0 \qquad (e \in E(G)).$$

We introduce a dual variable z_X for each $\emptyset \neq X \subset V(G)$ and one additional dual variable $z_{V(G)}$ for the equality constraint. Then the dual LP is

$$\max \quad - \sum_{\emptyset \neq X \subseteq V(G)} (|X| - 1)z_X$$

$$\text{s.t.} \quad - \sum_{e \subseteq X \subseteq V(G)} z_X \;\leq\; c(e) \qquad (e \in E(G))$$

$$z_X \;\geq\; 0 \qquad (\emptyset \neq X \subset V(G)).$$

Note that the dual variable $z_{V(G)}$ is not forced to be nonnegative. For $k = 1, \ldots, n-2$ let $z^*_{X_k} := c(f_l) - c(f_k)$, where l is the first index greater than k for which $f_l \cap X_k \neq \emptyset$. Let $z^*_{V(G)} := -c(f_{n-1})$, and let $z^*_X := 0$ for all $X \notin \{X_1, \ldots, X_{n-1}\}$.

For each $e = \{v, w\}$ we have that

$$- \sum_{e \subseteq X \subseteq V(G)} z^*_X \;=\; c(f_i),$$

where i is the smallest index such that $v, w \in X_i$. Moreover $c(f_i) \leq c(e)$ since v and w are in different connected components of $(V(G), \{f_1, \ldots, f_{i-1}\})$. Hence z^* is a feasible dual solution.

For each $e \in E(G)$ with $x^*_e > 0$ we have $e \in E(T)$ and therefore

$$- \sum_{e \subseteq X \subseteq V(G)} z^*_X \;=\; c(e),$$

i.e. the corresponding dual constraint is satisfied with equality. Finally, $z^*_X > 0$ implies that $T[X]$ is connected, so the corresponding primal constraint is satisfied with equality. In other words, the primal and dual complementary slackness conditions are satisfied, thus (by Corollary 3.23) x^* and z^* are optimum solutions for the primal and dual LP, respectively. $\square$

Indeed, we have proved that the inequality system in Theorem 6.13 is TDI. We remark that the above is also an alternative proof of the correctness of KRUSKAL'S ALGORITHM (Theorem 6.4). Another description of the spanning tree polytope is the subject of Exercise 19. There is also a polynomial-size LP formulation; see Exercise 20.

If we replace the constraint $\sum_{e \in E(G)} x_e = n - 1$ by $\sum_{e \in E(G)} x_e \leq n - 1$, we obtain the convex hull of the incidence vectors of all forests in G (Exercise 21). A generalization of these results is Edmonds' characterization of the matroid polytope (Theorem 13.21).

We now turn to a polyhedral description of the MINIMUM WEIGHT ROOTED ARBORESCENCE PROBLEM. First we prove a classical result of Fulkerson. Recall that an r-cut is a set of edges $\delta^+(S)$ for some $S \subset V(G)$ with $r \in S$.

Theorem 6.14.   (Fulkerson [1974]) *Let G be a digraph with weights $c : E(G) \to \mathbb{Z}_+$, and $r \in V(G)$ such that G contains a spanning arborescence rooted at r. Then*

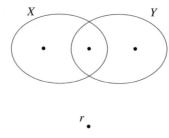

Fig. 6.3.

*the minimum weight of a spanning arborescence rooted at r equals the maximum
number t of r-cuts $C_1, \ldots, C_t$ (repetitions allowed) such that no edge e is contained
in more than c(e) of these cuts.*

Proof: Let A be the matrix whose columns are indexed by the edges and whose
rows are all incidence vectors of r-cuts. Consider the LP

$$\min\{cx : Ax \geq \mathbb{1}, x \geq 0\},$$

and its dual

$$\max\{\mathbb{1}y : yA \leq c, y \geq 0\}.$$

Then (by part (e) of Theorem 2.5) we have to show that for any nonnegative integral
c, both the primal and dual LP have integral optimum solutions. By Corollary 5.16
it suffices to show that the system $Ax \geq \mathbb{1}, x \geq 0$ is TDI. We use Lemma 5.24.

Since the dual LP is feasible if and only if c is nonnegative, let $c : E(G) \to \mathbb{Z}_+$.
Let y be an optimum solution of $\max\{\mathbb{1}y : yA \leq c, y \geq 0\}$ for which

$$\sum_{\emptyset \neq X \subseteq V(G)\setminus\{r\}} y_{\delta^-(X)}|X|^2 \tag{6.1}$$

is as large as possible. We claim that $\mathcal{F} := \{X : y_{\delta^-(X)} > 0\}$ is laminar. To see
this, suppose $X, Y \in \mathcal{F}$ with $X \cap Y \neq \emptyset$, $X \setminus Y \neq \emptyset$ and $Y \setminus X \neq \emptyset$ (Figure 6.3).
Let $\epsilon := \min\{y_{\delta^-(X)}, y_{\delta^-(Y)}\}$. Set $y'_{\delta^-(X)} := y_{\delta^-(X)} - \epsilon$, $y'_{\delta^-(Y)} := y_{\delta^-(Y)} - \epsilon$,
$y'_{\delta^-(X \cap Y)} := y_{\delta^-(X \cap Y)} + \epsilon$, $y'_{\delta^-(X \cup Y)} := y_{\delta^-(X \cup Y)} + \epsilon$, and $y'(S) := y(S)$ for all
other r-cuts S. Observe that $y'A \leq yA$, so y' is a feasible dual solution. Since
$\mathbb{1}y = \mathbb{1}y'$, it is also optimum and contradicts the choice of y, because (6.1) is
larger for y'. (For any numbers $a > b \geq c > d > 0$ with $a + d = b + c$ we have
$a^2 + d^2 > b^2 + c^2$.)

Now let A' be the submatrix of A consisting of the rows corresponding to the
elements of $\mathcal{F}$. A' is the one-way cut-incidence matrix of a laminar family (to be
precise, we must consider the graph resulting from G by reversing each edge). So
by Theorem 5.29 A' is totally unimodular, as required. □

The above proof also yields the promised polyhedral description:

Corollary 6.15. (Edmonds [1967]) *Let G be a digraph with weights $c : E(G) \to \mathbb{R}_+$, and $r \in V(G)$ such that G contains a spanning arborescence rooted at r. Then the LP*

$$\min \left\{ cx : x \geq 0, \sum_{e \in \delta^+(X)} x_e \geq 1 \text{ for all } X \subset V(G) \text{ with } r \in X \right\}$$

has an integral optimum solution (which is the incidence vector of a minimum weight spanning arborescence rooted at r, plus possibly some edges of zero weight). □

For a description of the convex hull of the incidence vectors of all branchings or spanning arborescences rooted at r, see Exercises 22 and 23.

6.4 Packing Spanning Trees and Arborescences

If we are looking for more than one spanning tree or arborescence, classical theorems of Tutte, Nash-Williams and Edmonds are of help. We first give a proof of Tutte's Theorem on packing spanning trees which is essentially due to Mader (see Diestel [1997]) and which uses the following lemma:

Lemma 6.16. *Let G be an undirected graph, and let $F = (F_1, \ldots, F_k)$ be a k-tuple of edge-disjoint forests in G such that $|E(F)|$ is maximum, where $E(F) := \bigcup_{i=1}^{k} E(F_i)$. Let $e \in E(G) \setminus E(F)$. Then there exists a set $X \subseteq V(G)$ with $e \subseteq X$ such that $F_i[X]$ is connected for each $i \in \{1, \ldots, k\}$.*

Proof: For two k-tuples $F' = (F'_1, \ldots, F'_k)$ and $F'' = (F''_1, \ldots, F''_k)$ of edge-disjoint forests we say that F'' arises from F' by exchanging e' for e'' if $F''_j = (F'_j \setminus e') \mathbin{\dot{\cup}} e''$ for some j and $F''_i = F'_i$ for all $i \neq j$. Let $\mathcal{F}$ be the set of all k-tuples of edge-disjoint forests arising from F by a sequence of such exchanges. Let $\overline{E} := E(G) \setminus \left(\bigcap_{F' \in \mathcal{F}} E(F') \right)$ and $\overline{G} := (V(G), \overline{E})$. We have $F \in \mathcal{F}$ and thus $e \in \overline{E}$. Let X be the vertex set of the connected component of $\overline{G}$ containing e. We shall prove that $F_i[X]$ is connected for each i.

Claim: For any $F' = (F'_1, \ldots, F'_k) \in \mathcal{F}$ and any $\bar{e} = \{v, w\} \in E(\overline{G}[X]) \setminus E(F')$ there exists a v-w-path in $F'_i[X]$ for all $i \in \{1, \ldots, k\}$.

To prove this, let $i \in \{1, \ldots, k\}$ be fixed. Since $F' \in \mathcal{F}$ and $|E(F')| = |E(F)|$ is maximum, $F'_i + \bar{e}$ contains a circuit C. Now for all $e' \in E(C) \setminus \{\bar{e}\}$ we have $F'_{e'} \in \mathcal{F}$, where $F'_{e'}$ arises from F' by exchanging e' for $\bar{e}$. This shows that $E(C) \subseteq \overline{E}$, and so $C - \bar{e}$ is a v-w-path in $F'_i[X]$. This proves the claim.

Since $\overline{G}[X]$ is connected, it suffices to prove that for each $\bar{e} = \{v, w\} \in E(\overline{G}[X])$ and each i there is a v-w-path in $F_i[X]$.

So let $\bar{e} = \{v, w\} \in E(\overline{G}[X])$. Since $\bar{e} \in \overline{E}$, there is some $F' = (F'_1, \ldots, F'_k) \in \mathcal{F}$ with $\bar{e} \notin E(F')$. By the claim there is a v-w-path in $F'_i[X]$ for each i.

Now there is a sequence $F = F^{(0)}, F^{(1)} \ldots, F^{(s)} = F'$ of elements of $\mathcal{F}$ such that $F^{(r+1)}$ arises from $F^{(r)}$ by exchanging one edge ($r = 0, \ldots, s-1$). It suffices to show that the existence of a v-w-path in $F_i^{(r+1)}[X]$ implies the existence of a v-w-path in $F_i^{(r)}[X]$ ($r = 0, \ldots, s-1$).

To see this, suppose that $F_i^{(r+1)}$ arises from $F_i^{(r)}$ by exchanging e_r for e_{r+1}, and let P be the v-w-path in $F_i^{(r+1)}[X]$. If P does not contain $e_{r+1} = \{x, y\}$, it is also a path in $F_i^{(r)}[X]$. Otherwise we can replace e_{r+1}: consider the x-y-path Q in $F_i^{(r)}[X]$, which exists by the claim because $e_{r+1} \in E(\overline{G}[X])$. Since $(E(P) \setminus \{e_{r+1}\}) \cup E(Q)$ contains a v-w-path in $F_i^{(r)}[X]$, the proof is complete. $\square$

Now we can prove Tutte's theorem on disjoint spanning trees. A **multicut** in an undirected graph G is a set of edges $\delta(X_1, \ldots, X_p) := \delta(X_1) \cup \cdots \cup \delta(X_p)$ for some partition $V(G) = X_1 \,\dot\cup\, X_2 \,\dot\cup\, \cdots \,\dot\cup\, X_p$ of the vertex set into nonempty subsets. For $p = 3$ we also speak of 3-cuts. Observe that cuts are multicuts with $p = 2$.

Theorem 6.17. (Tutte [1961], Nash-Williams [1961]) *An undirected graph G contains k edge-disjoint spanning trees if and only if*

$$|\delta(X_1, \ldots, X_p)| \geq k(p-1)$$

for every multicut $\delta(X_1, \ldots, X_p)$.

Proof: To prove necessity, let $T_1, \ldots, T_k$ be edge-disjoint spanning trees in G, and let $\delta(X_1, \ldots, X_p)$ be a multicut. Contracting each of the vertex subsets $X_1, \ldots, X_p$ yields a graph G' whose vertices are $X_1, \ldots, X_p$ and whose edges correspond to the edges of the multicut. $T_1, \ldots, T_k$ correspond to edge-disjoint connected subgraphs $T_1', \ldots, T_k'$ in G'. Each of the $T_1', \ldots, T_k'$ has at least $p-1$ edges, so G' (and thus the multicut) has at least $k(p-1)$ edges.

To prove sufficiency we use induction on $|V(G)|$. For $n := |V(G)| \leq 2$ the statement is true. Now assume $n > 2$, and suppose that $|\delta(X_1, \ldots, X_p)| \geq k(p-1)$ for every multicut $\delta(X_1, \ldots, X_p)$. In particular (consider the partition into singletons) G has at least $k(n-1)$ edges. Moreover, the condition is preserved when contracting vertex sets, so by the induction hypothesis G/X contains k edge-disjoint spanning trees for each $X \subset V(G)$ with $|X| \geq 2$.

Let $F = (F_1, \ldots, F_k)$ be a k-tuple of edge-disjoint forests in G such that $|E(F)|$ is maximum, where again $E(F) := \bigcup_{i=1}^{k} E(F_i)$. We claim that each F_i is a spanning tree. Otherwise $|E(F)| < k(n-1)$, so there is an edge $e \in E(G) \setminus E(F)$. By Lemma 6.16 there is an $X \subseteq V(G)$ with $e \subseteq X$ such that $F_i[X]$ is connected for each i. Since $|X| \geq 2$, G/X contains k edge-disjoint spanning trees $F_1', \ldots, F_k'$. Now F_i' together with $F_i[X]$ forms a spanning tree in G for each i, and all these k spanning trees are edge-disjoint. $\square$

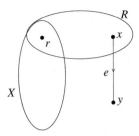

Fig. 6.4.

We now turn to the corresponding problem in digraphs, packing spanning arborescences:

Theorem 6.18.   (Edmonds [1973]) *Let G be a digraph and $r \in V(G)$. Then the maximum number of edge-disjoint spanning arborescences rooted at r equals the minimum cardinality of an r-cut.*

Proof: Let k be the minimum cardinality of an r-cut. Obviously there are at most k edge-disjoint spanning arborescences rooted at r. We prove the existence of k edge-disjoint spanning arborescences rooted at r by induction on k. The case $k = 0$ is trivial.

If we can find one spanning arborescence A rooted at r such that

$$\min_{r \in S \subset V(G)} \left| \delta_G^+(S) \setminus E(A) \right| \geq k - 1, \tag{6.2}$$

then we are done by induction. Suppose we have already found some arborescence A rooted at r (but not necessarily spanning) such that (6.2) holds. Let $R \subseteq V(G)$ be the set of vertices covered by A. Initially, $R = \{r\}$; if $R = V(G)$, we are done.

If $R \neq V(G)$, we call a set $X \subseteq V(G)$ critical if

(a) $r \in X$;
(b) $X \cup R \neq V(G)$;
(c) $|\delta_G^+(X) \setminus E(A)| = k - 1$.

If there is no critical vertex set, we can augment A by any edge leaving R. Otherwise let X be a maximal critical set, and let $e = (x, y)$ be an edge such that $x \in R \setminus X$ and $y \in V(G) \setminus (R \cup X)$ (see Figure 6.4). Such an edge must exist because

$$|\delta_{G-E(A)}^+(R \cup X)| = |\delta_G^+(R \cup X)| \geq k > k - 1 = |\delta_{G-E(A)}^+(X)|.$$

We now add e to A. Obviously $A + e$ is an arborescence rooted at r. We have to show that (6.2) continues to hold.

Suppose there is some Y such that $r \in Y \subset V(G)$ and $|\delta_G^+(Y) \setminus E(A+e)| < k-1$. Then $x \in Y$, $y \notin Y$, and $|\delta_G^+(Y) \setminus E(A)| = k - 1$. Now Lemma 2.1(a) implies

$$
\begin{aligned}
k - 1 + k - 1 \;&=\; |\delta_{G-E(A)}^+(X)| + |\delta_{G-E(A)}^+(Y)| \\
&\geq\; |\delta_{G-E(A)}^+(X \cup Y)| + |\delta_{G-E(A)}^+(X \cap Y)| \\
&\geq\; k - 1 + k - 1 \,,
\end{aligned}
$$

because $r \in X \cap Y$ and $y \in V(G) \setminus (X \cup Y)$. So equality must hold throughout, in particular $|\delta_{G-E(A)}^+(X \cup Y)| = k - 1$. Since $y \in V(G) \setminus (X \cup Y \cup R)$ we conclude that $X \cup Y$ is critical. But since $x \in Y \setminus X$, this contradicts the maximality of X. $\square$

This proof is due to Lovász [1976]. Fujishige [2010] generalized this result to packing arborescences with given (but not necessarily identical) roots, spanning given subsets of vertices. A common generalization of Theorems 6.17 and 6.18 was found by Frank [1981]. A good characterization (see Chapter 15 for an explanation of this term) of the problem of packing spanning arborescences with arbitrary roots is given by the following theorem, which we cite without proof:

Theorem 6.19. (Frank [1979]) *A digraph G contains k edge-disjoint spanning arborescences if and only if*

$$
\sum_{i=1}^{p} |\delta^-(X_i)| \;\geq\; k(p - 1)
$$

for every collection of pairwise disjoint nonempty subsets $X_1, \ldots, X_p \subseteq V(G)$.

Another question is how many forests are needed to cover a graph. This is answered by the following theorem:

Theorem 6.20. (Nash-Williams [1964]) *The edge set of an undirected graph G is the union of k forests if and only if $|E(G[X])| \leq k(|X| - 1)$ for all $\emptyset \neq X \subseteq V(G)$.*

Proof: The necessity is clear since no forest can contain more than $|X| - 1$ edges within a vertex set X. To prove the sufficiency, assume that $|E(G[X])| \leq k(|X| - 1)$ for all $\emptyset \neq X \subseteq V(G)$, and let $F = (F_1, \ldots, F_k)$ be a k-tuple of edge-disjoint forests in G such that $|E(F)| = \left| \bigcup_{i=1}^{k} E(F_i) \right|$ is maximum. We claim that $E(F) = E(G)$. To see this, suppose there is an edge $e \in E(G) \setminus E(F)$. By Lemma 6.16 there exists a set $X \subseteq V(G)$ with $e \subseteq X$ such that $F_i[X]$ is connected for each i. In particular,

$$
|E(G[X])| \;\geq\; \left| \{e\} \,\dot\cup\, \bigcup_{i=1}^{k} E(F_i[X]) \right| \;\geq\; 1 + k(|X| - 1),
$$

contradicting the assumption. $\square$

Exercise 29 gives a directed version. A generalization of Theorems 6.17 and 6.20 to matroids can be found in Exercise 20 of Chapter 13.

Exercises

1. Prove Cayley's Theorem 6.2 by showing that the following defines a one-to-one correspondence between the spanning trees in K_n and the vectors in $\{1, \ldots, n\}^{n-2}$: For a tree T with $V(T) = \{1, \ldots, n\}$, $n \geq 3$, let v be the leaf with the smallest index and let a_1 be the neighbour of v. We recursively define $a(T) := (a_1, \ldots, a_{n-2})$, where $(a_2, \ldots, a_{n-2}) = a(T - v)$.
   (Prüfer [1918])

2. Prove that there are exactly $(n+1)^{n-1}$ branchings B with $V(B) = \{1, \ldots, n\}$.

3. Let p_n be the probability that vertex 1 is a leaf in T if T is chosen randomly among all trees with vertex set $\{1, \ldots, n\}$ (with uniform distribution). What is $\lim_{n \to \infty} p_n$?

4. Let (V, T_1) and (V, T_2) be two trees on the same vertex set V. Prove that for any edge $e \in T_1$ there is an edge $f \in T_2$ such that both $(V, (T_1 \setminus \{e\}) \cup \{f\})$ and $(V, (T_2 \setminus \{f\}) \cup \{e\})$ are trees.

5. Let (G, c) be an instance of the MINIMUM SPANNING TREE PROBLEM where G is connected and $c(e) \neq c(e')$ for any two distinct edges e and e'. Prove that then there is exactly one optimum solution.

6. Given an undirected graph G with weights $c : E(G) \to \mathbb{R}$ and a vertex $v \in V(G)$, we ask for a minimum weight spanning tree in G where v is not a leaf. Can you solve this problem in polynomial time?

7. We want to determine the set of edges e in an undirected graph G with weights $c : E(G) \to \mathbb{R}$ for which there exists a minimum weight spanning tree in G containing e (in other words, we are looking for the union of all minimum weight spanning trees in G). Show how this problem can be solved in $O(mn)$ time.

8. Given an undirected graph G with arbitrary weights $c : E(G) \to \mathbb{R}$, we ask for a minimum weight connected spanning subgraph. Can you solve this problem efficiently?

9. Consider the following algorithm (sometimes called WORST-OUT-GREEDY ALGORITHM, see Section 13.4). Examine the edges in order of non-increasing weights. Delete an edge unless it is a bridge. Does this algorithm solve the MINIMUM SPANNING TREE PROBLEM?

10. Consider the following "colouring" algorithm. Initially all edges are uncoloured. Then apply the following rules in arbitrary order until all edges are coloured:
    Blue rule: Select a cut containing no blue edge. Among the uncoloured edges in the cut, select one of minimum cost and colour it blue.
    Red rule: Select a circuit containing no red edge. Among the uncoloured edges in the circuit, select one of maximum cost and colour it red.
    Show that one of the rules is always applicable as long as there are uncoloured edges left (cf. Lemma 2.6). Moreover, show that the algorithm maintains the "colour invariant": there always exists an optimum spanning tree containing

all blue edges but no red edge. (So the algorithm solves the MINIMUM SPAN-NING TREE PROBLEM optimally.) Observe that KRUSKAL'S ALGORITHM and PRIM'S ALGORITHM are special cases.
(Tarjan [1983])

11. Suppose we wish to find a spanning tree T in an undirected graph such that the maximum weight of an edge in T is as small as possible. How can this be done?

12. Is it true that the maximum length of a path in a branching implementing a Fibonacci heap is $O(\log n)$, where n is the number of elements?

13. Show that two Fibonacci heaps with n_1 and n_2 elements can be merged in $O(\log(n_1 + n_2))$ time. The resulting Fibonacci heap shall contain all $n_1 + n_2$ elements.

14. For a finite set $V \subset \mathbb{R}^2$, the Voronoï diagram consists of the regions

$$P_v := \left\{ x \in \mathbb{R}^2 : ||x - v||_2 = \min_{w \in V} ||x - w||_2 \right\}$$

for $v \in V$. The Delaunay triangulation of V is the graph

$$(V, \{\{v, w\} \subseteq V, v \neq w, |P_v \cap P_w| > 1\}).$$

A minimum spanning tree for V is a tree T with $V(T) = V$ whose length $\sum_{\{v,w\} \in E(T)} ||v - w||_2$ is minimum. Prove that every minimum spanning tree is a subgraph of the Delaunay triangulation.
Note: Using the fact that the Delaunay triangulation can be computed in $O(n \log n)$ time (where $n = |V|$; see e.g. Fortune [1987], Knuth [1992]), this implies an $O(n \log n)$ algorithm for the MINIMUM SPANNING TREE PROB-LEM for point sets in the plane.
(Shamos and Hoey [1975]; see also Zhou, Shenoy and Nicholls [2002])

15. Can you decide in linear time whether a digraph contains a spanning arbores-cence?
Hint: To find a possible root, start at an arbitrary vertex and traverse edges backwards as long as possible. When encountering a circuit, contract it.

16. Can you find a maximum cardinality branching in a given digraph in linear time?
Hint: First find the strongly connected components.

17. The MINIMUM WEIGHT ROOTED ARBORESCENCE PROBLEM can be re-duced to the MAXIMUM WEIGHT BRANCHING PROBLEM by Proposition 6.9. However, it can also be solved directly by a modified version of EDMONDS' BRANCHING ALGORITHM. Show how.

18. Prove that the spanning tree polytope of an undirected graph G (see Theo-rem 6.13) with $n := |V(G)|$ is in general a proper subset of the polytope

$$\left\{ x \in [0, 1]^{E(G)} : \sum_{e \in E(G)} x_e = n - 1, \sum_{e \in \delta(X)} x_e \geq 1 \text{ for } \emptyset \subset X \subset V(G) \right\}.$$

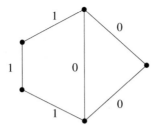

Fig. 6.5.

Hint: To prove that this polytope is not integral, consider the graph shown in Figure 6.5 (the numbers are edge weights).

(Magnanti and Wolsey [1995])

19. In Exercise 18 we saw that cut constraints do not suffice to describe the spanning tree polytope. However, if we consider multicuts instead, we obtain a complete description: Prove that the spanning tree polytope of an undirected graph G with $n := |V(G)|$ consists of all vectors $x \in [0, 1]^{E(G)}$ with

$$\sum_{e \in E(G)} x_e = n - 1 \quad \text{and} \quad \sum_{e \in C} x_e \geq k - 1 \quad \text{for all multicuts } C = \delta(X_1, \ldots, X_k).$$

(Magnanti and Wolsey [1995])

* 20. Let G be an undirected graph and $n := |V(G)|$. Prove that the following linear inequality system with $O(n^3)$ variables and constraints describes a polytope whose orthogonal projection onto the x-variables yields the spanning tree polytope of G: $x_e \geq 0$ ($e \in E(G)$), $z_{u,v,w} \geq 0$ ($\{u, v\} \in E(G)$, $w \in V(G) \setminus \{u, v\}$), $\sum_{e \in E(G)} x_e = n - 1$, $x_e = z_{u,v,w} + z_{v,u,w}$ ($e = \{u, v\} \in E(G)$, $w \in V(G) \setminus \{u, v\}$), and $x_e + \sum_{\{u,v\} \in \delta(v) \setminus \{e\}} z_{u,v,w} = 1$ ($v \in V(G)$, $e = \{v, w\} \in \delta(v)$).

Note: See Conforti, Cornuéjols and Zambelli [2010] for a survey of such extended formulations of combinatorial LPs.

21. Prove that the convex hull of the incidence vectors of all forests in an undirected graph G is the polytope

$$P := \left\{ x \in [0, 1]^{E(G)} : \sum_{e \in E(G[X])} x_e \leq |X| - 1 \text{ for } \emptyset \neq X \subseteq V(G) \right\}.$$

Note: This statement implies Theorem 6.13 since $\sum_{e \in E(G[X])} x_e = |V(G)| - 1$ is a supporting hyperplane. Moreover, it is a special case of Theorem 13.21.

* 22. Prove that the convex hull of the incidence vectors of all branchings in a digraph G is the set of all vectors $x \in [0, 1]^{E(G)}$ with

$$\sum_{e \in E(G[X])} x_e \leq |X| - 1 \text{ for } \emptyset \neq X \subseteq V(G) \quad \text{and} \quad \sum_{e \in \delta^-(v)} x_e \leq 1 \text{ for } v \in V(G).$$

Note: This is a special case of Theorem 14.13.

* 23. Let G be a digraph and $r \in V(G)$. Prove that the polytopes

$$\left\{ x \in [0, 1]^{E(G)} \; : \; x_e = 0 \; (e \in \delta^-(r)), \; \sum_{e \in \delta^-(v)} x_e = 1 \; (v \in V(G) \setminus \{r\}), \right.$$

$$\left. \sum_{e \in E(G[X])} x_e \leq |X| - 1 \text{ for } \emptyset \neq X \subseteq V(G) \right\}$$

and

$$\left\{ x \in [0, 1]^{E(G)} \; : \; x_e = 0 \; (e \in \delta^-(r)), \; \sum_{e \in \delta^-(v)} x_e = 1 \; (v \in V(G) \setminus \{r\}), \right.$$

$$\left. \sum_{e \in \delta^+(X)} x_e \geq 1 \text{ for } r \in X \subset V(G) \right\}$$

are both equal to the convex hull of the incidence vectors of all spanning arborescences rooted at r.

24. Prove that every $2k$-edge-connected graph contains k pairwise edge-disjoint spanning trees.

25. Let G be a digraph and $r \in V(G)$. Prove that G is the disjoint union of k spanning arborescences rooted at r if and only if the underlying undirected graph is the disjoint union of k spanning trees and $|\delta^-(x)| = k$ for all $x \in V(G) \setminus \{r\}$.
(Edmonds)

26. Let G be a digraph and $r \in V(G)$. Suppose that G contains k edge-disjoint paths from r to every other vertex, but removing any edge destroys this property. Prove that every vertex of G except r has exactly k entering edges.
Hint: Use Theorem 6.18.

* 27. Prove the statement of Exercise 26 without using Theorem 6.18. Formulate and prove a vertex-disjoint version.
Hint: If a vertex v has more than k entering edges, take k edge-disjoint r-v-paths. Show that an edge entering v that is not used by these paths can be deleted.

28. Give a polynomial-time algorithm for finding a maximum set of edge-disjoint spanning arborescences (rooted at r) in a digraph G.
Note: The most efficient algorithm is due to Gabow [1995]; see also (Gabow and Manu [1998]).

29. Prove that the edges of a digraph G can be covered by k branchings if and only if the following two conditions hold:
(a) $|\delta^-(v)| \leq k$ for all $v \in V(G)$;
(b) $|E(G[X])| \leq k(|X| - 1)$ for all $X \subseteq V(G)$.
Hint: Use Theorem 6.18.
(Frank [1979])

References

General Literature

Ahuja, R.K., Magnanti, T.L., and Orlin, J.B. [1993]: Network Flows. Prentice-Hall, Englewood Cliffs 1993, Chapter 13

Balakrishnan, V.K. [1995]: Network Optimization. Chapman and Hall, London 1995, Chapter 1

Cormen, T.H., Leiserson, C.E., Rivest, R.L., and Stein, C. [2009]: Introduction to Algorithms. Third Edition. MIT Press, Cambridge 2009, Chapter 23

Gondran, M., and Minoux, M. [1984]: Graphs and Algorithms. Wiley, Chichester 1984, Chapter 4

Magnanti, T.L., and Wolsey, L.A. [1995]: Optimal trees. In: Handbooks in Operations Research and Management Science; Volume 7: Network Models (M.O. Ball, T.L. Magnanti, C.L. Monma, G.L. Nemhauser, eds.), Elsevier, Amsterdam 1995, pp. 503–616

Schrijver, A. [2003]: Combinatorial Optimization: Polyhedra and Efficiency. Springer, Berlin 2003, Chapters 50–53

Tarjan, R.E. [1983]: Data Structures and Network Algorithms. SIAM, Philadelphia 1983, Chapter 6

Wu, B.Y., and Chao, K.-M. [2004]: Spanning Trees and Optimization Problems. Chapman & Hall/CRC, Boca Raton 2004

Cited References

Bock, F.C. [1971]: An algorithm to construct a minimum directed spanning tree in a directed network. In: Avi-Itzhak, B. (Ed.): Developments in Operations Research, Volume I. Gordon and Breach, New York 1971, pp. 29–44

Borůvka, O. [1926a]: O jistém problému minimálním. Práca Moravské Přírodo-vědecké Spolnečnosti 3 (1926), 37–58 [in Czech]

Borůvka, O. [1926b]: Příspevěk k řešení otázky ekonomické stavby. Elektrovodních sítí. Elektrotechnicky Obzor 15 (1926), 153–154 [in Czech]

Cayley, A. [1889]: A theorem on trees. Quarterly Journal on Mathematics 23 (1889), 376–378

Chazelle, B. [2000]: A minimum spanning tree algorithm with inverse-Ackermann type complexity. Journal of the ACM 47 (2000), 1028–1047

Cheriton, D., and Tarjan, R.E. [1976]: Finding minimum spanning trees. SIAM Journal on Computing 5 (1976), 724–742

Chu, Y., and Liu, T. [1965]: On the shortest arborescence of a directed graph. Scientia Sinica 4 (1965), 1396–1400; Mathematical Review 33, # 1245

Conforti, M., Cornuéjols, G., and Zambelli, G. [2010]: Extended formulations in combinatorial optimization. 4OR 8 (2010), 1–48

Diestel, R. [1997]: Graph Theory. Springer, New York 1997

Dijkstra, E.W. [1959]: A note on two problems in connexion with graphs. Numerische Mathematik 1 (1959), 269–271

Dixon, B., Rauch, M., and Tarjan, R.E. [1992]: Verification and sensitivity analysis of minimum spanning trees in linear time. SIAM Journal on Computing 21 (1992), 1184–1192

Edmonds, J. [1967]: Optimum branchings. Journal of Research of the National Bureau of Standards B 71 (1967), 233–240

Edmonds, J. [1970]: Submodular functions, matroids and certain polyhedra. In: Combinatorial Structures and Their Applications; Proceedings of the Calgary International Conference on Combinatorial Structures and Their Applications 1969 (R. Guy, H. Hanani, N. Sauer, J. Schönheim, eds.), Gordon and Breach, New York 1970, pp. 69–87

Edmonds, J. [1973]: Edge-disjoint branchings. In: Combinatorial Algorithms (R. Rustin, ed.), Algorithmic Press, New York 1973, pp. 91–96

Fortune, S. [1987]: A sweepline algorithm for Voronoi diagrams. Algorithmica 2 (1987), 153–174

Frank, A. [1981]: On disjoint trees and arborescences. In: Algebraic Methods in Graph Theory; Colloquia Mathematica Societatis János Bolyai 25 (L. Lovász, V.T. Sós, eds.), North-Holland, Amsterdam 1981, pp. 159–169

Frank, A. [1979]: Covering branchings. Acta Scientiarum Mathematicarum (Szeged) 41 (1979), 77–82

Fredman, M.L., and Tarjan, R.E. [1987]: Fibonacci heaps and their uses in improved network optimization problems. Journal of the ACM 34 (1987), 596–615

Fredman, M.L., and Willard, D.E. [1994]: Trans-dichotomous algorithms for minimum spanning trees and shortest paths. Journal of Computer and System Sciences 48 (1994), 533–551

Fujishige, S. [2010]: A note on disjoint arborescences. Combinatorica 30 (2010), 247–252

Fulkerson, D.R. [1974]: Packing rooted directed cuts in a weighted directed graph. Mathematical Programming 6 (1974), 1–13

Gabow, H.N. [1995]: A matroid approach to finding edge connectivity and packing arborescences. Journal of Computer and System Sciences 50 (1995), 259–273

Gabow, H.N., Galil, Z., and Spencer, T. [1989]: Efficient implementation of graph algorithms using contraction. Journal of the ACM 36 (1989), 540–572

Gabow, H.N., Galil, Z., Spencer, T., and Tarjan, R.E. [1986]: Efficient algorithms for finding minimum spanning trees in undirected and directed graphs. Combinatorica 6 (1986), 109–122

Gabow, H.N., and Manu, K.S. [1998]: Packing algorithms for arborescences (and spanning trees) in capacitated graphs. Mathematical Programming B 82 (1998), 83–109

Jarník, V. [1930]: O jistém problému minimálním. Práca Moravské Přírodovědecké Společnosti 6 (1930), 57–63

Karger, D., Klein, P.N., and Tarjan, R.E. [1995]: A randomized linear-time algorithm to find minimum spanning trees. Journal of the ACM 42 (1995), 321–328

Karp, R.M. [1972]: A simple derivation of Edmonds' algorithm for optimum branchings. Networks 1 (1972), 265–272

King, V. [1997]: A simpler minimum spanning tree verification algorithm. Algorithmica 18 (1997), 263–270

Knuth, D.E. [1992]: Axioms and hulls; LNCS 606. Springer, Berlin 1992

Korte, B., and Nešetřil, J. [2001]: Vojtěch Jarník's work in combinatorial optimization. Discrete Mathematics 235 (2001), 1–17

Kruskal, J.B. [1956]: On the shortest spanning subtree of a graph and the travelling salesman problem. Proceedings of the AMS 7 (1956), 48–50

Lovász, L. [1976]: On two minimax theorems in graph. Journal of Combinatorial Theory B 21 (1976), 96–103

Nash-Williams, C.S.J.A. [1961]: Edge-disjoint spanning trees of finite graphs. Journal of the London Mathematical Society 36 (1961), 445–450

Nash-Williams, C.S.J.A. [1964]: Decompositions of finite graphs into forests. Journal of the London Mathematical Society 39 (1964), 12

Nešetřil, J., Milková, E., and Nešetřilová, H. [2001]: Otakar Borůvka on minimum spanning tree problem. Translation of both the 1926 papers, comments, history. Discrete Mathematics 233 (2001), 3–36

Pettie, S., and Ramachandran, V. [2002]: An optimal minimum spanning tree algorithm. Journal of the ACM 49 (2002), 16–34

Prim, R.C. [1957]: Shortest connection networks and some generalizations. Bell System Technical Journal 36 (1957), 1389–1401

Prüfer, H. [1918]: Neuer Beweis eines Satzes über Permutationen. Arch. Math. Phys. 27 (1918), 742–744

Shamos, M.I., and Hoey, D. [1975]: Closest-point problems. Proceedings of the 16th Annual IEEE Symposium on Foundations of Computer Science (1975), 151–162

Sylvester, J.J. [1857]: On the change of systems of independent variables. Quarterly Journal of Mathematics 1 (1857), 42–56

Tarjan, R.E. [1975]: Efficiency of a good but not linear set union algorithm. Journal of the ACM 22 (1975), 215–225

Tutte, W.T. [1961]: On the problem of decomposing a graph into n connected factor. Journal of the London Mathematical Society 36 (1961), 221–230

Zhou, H., Shenoy, N., and Nicholls, W. [2002]: Efficient minimum spanning tree construction without Delaunay triangulation. Information Processing Letters 81 (2002), 271–276

7 Shortest Paths

One of the best-known combinatorial optimization problems is to find a shortest path between two specified vertices of a directed or undirected graph:

SHORTEST PATH PROBLEM

Instance: A graph G (directed or undirected), weights $c : E(G) \to \mathbb{R}$ and two vertices $s, t \in V(G)$.

Task: Find a shortest s-t-path P, i.e. one of minimum weight $c(E(P))$, or decide that t is not reachable from s.

Obviously this problem has many practical applications. Like the MINIMUM SPANNING TREE PROBLEM it also often appears as a subproblem when one deals with more difficult combinatorial optimization problems.

In fact, the problem is not easy to solve if we allow arbitrary weights. For example, if all weights are -1 then the s-t-paths of weight $1 - |V(G)|$ are precisely the Hamiltonian s-t-paths. Deciding whether such a path exists is a difficult problem (see Exercise 17(b) of Chapter 15).

The problem becomes much easier if we restrict ourselves to nonnegative weights or at least exclude negative circuits:

Definition 7.1. *Let G be a (directed or undirected) graph with weights $c : E(G) \to \mathbb{R}$. c is called **conservative** if there is no circuit of negative total weight.*

We shall present algorithms for the SHORTEST PATH PROBLEM in digraphs in Section 7.1. The first one allows nonnegative weights only while the second algorithm can deal with arbitrary conservative weights.

The algorithms of Section 7.1 in fact compute a shortest s-v-path for all $v \in V(G)$ without using significantly more running time. Sometimes one is interested in the distance for every pair of vertices; Section 7.2 shows how to deal with this problem.

Since negative circuits cause problems we also show how to detect them. If none exists, a circuit of minimum total weight can be computed quite easily. Another interesting problem asks for a circuit whose mean weight is minimum. As we shall see in Section 7.3, the directed version of this problem can also be solved efficiently by similar techniques.

© Springer-Verlag GmbH Germany 2018
Bernhard Korte and Jens Vygen, *Combinatorial Optimization*, Algorithms and Combinatorics 21, https://doi.org/10.1007/978-3-662-56039-6_7

In Section 7.4 we show that undirected graphs with nonnegative weights always contain trees that are not much more expensive than minimum spanning trees but also contain short paths from a source s to all other vertices.

Finding shortest paths in undirected graphs is more difficult unless the edge weights are nonnegative. Undirected edges of nonnegative weights can be replaced equivalently by a pair of oppositely directed edges of the same weight; this reduces the undirected problem to a directed one. However, this construction does not work for edges of negative weight since it would introduce negative circuits. We shall return to the problem of finding shortest paths in undirected graphs with conservative weights in Section 12.2 (Corollary 12.13).

Henceforth we work with a digraph G. Without loss of generality we may assume that G is connected and simple; among parallel edges we have to consider only the one with least weight.

7.1 Shortest Paths From One Source

All shortest path algorithms we present are based on the following observation, sometimes called Bellman's principle of optimality, which is indeed the core of dynamic programming:

Proposition 7.2. *Let G be a digraph with conservative weights $c : E(G) \to \mathbb{R}$, let $k \in \mathbb{N}$, and let s and w be two vertices. Let P be a shortest one among all s-w-paths with at most k edges, and let $e = (v, w)$ be its final edge. Then $P_{[s,v]}$ (i.e. P without the edge e) is a shortest one among all s-v-paths with at most $k - 1$ edges.*

Proof: Suppose Q is a shorter s-v-path than $P_{[s,v]}$, and $|E(Q)| \le k - 1$. Then $c(E(Q)) + c(e) < c(E(P))$. If Q does not contain w, then $Q + e$ is a shorter s-w-path than P, otherwise $Q_{[s,w]}$ has length $c(E(Q_{[s,w]})) = c(E(Q)) + c(e) - c(E(Q_{[w,v]} + e)) < c(E(P)) - c(E(Q_{[w,v]} + e)) \le c(E(P))$, because $Q_{[w,v]} + e$ is a circuit and c is conservative. In both cases we have a contradiction to the assumption that P is a shortest s-w-path with at most k edges. $\square$

The same result holds for undirected graphs with nonnegative weights and also for acyclic digraphs with arbitrary weights. It yields the recursion formulas $\text{dist}(s, s) = 0$ and $\text{dist}(s, w) = \min\{\text{dist}(s, v) + c((v, w)) : (v, w) \in E(G)\}$ for $w \in V(G) \setminus \{s\}$ which immediately solve the SHORTEST PATH PROBLEM for acyclic digraphs (Exercise 7).

Proposition 7.2 is also the reason why most algorithms compute the shortest paths from s to all other vertices. If one computes a shortest s-t-path P, one has already computed a shortest s-v-path for each vertex v on P. Since we cannot know in advance which vertices belong to P, it is only natural to compute shortest s-v-paths for all v. We can store these s-v-paths very efficiently by just storing the final edge of each path.

We first consider nonnegative edge weights, i.e. $c : E(G) \rightarrow \mathbb{R}_+$. The SHORT-EST PATH PROBLEM can be solved by BFS if all weights are 1 (Proposition 2.18). For weights $c : E(G) \rightarrow \mathbb{N}$ one could replace an edge e by a path of length $c(e)$ and again use BFS. However, this might introduce an exponential number of edges; recall that the input size is $\Theta \left(n \log m + m \log n + \sum_{e \in E(G)} \log c(e) \right)$, where $n = |V(G)|$ and $m = |E(G)|$.

A much better idea is to use the following algorithm, due to Dijkstra [1959]. It is quite similar to PRIM'S ALGORITHM for the MINIMUM SPANNING TREE PROBLEM (Section 6.1).

DIJKSTRA'S ALGORITHM

Input: A digraph G, weights $c : E(G) \rightarrow \mathbb{R}_+$ and a vertex $s \in V(G)$.

Output: Shortest paths from s to all $v \in V(G)$ and their lengths. More precisely, we get the outputs $l(v)$ and $p(v)$ for all $v \in V(G) \setminus \{s\}$. $l(v)$ is the length of a shortest s-v-path, which consists of a shortest s-$p(v)$-path together with the edge $(p(v), v)$. If v is not reachable from s, then $l(v) = \infty$ and $p(v)$ is undefined.

---

①   Set $l(s) := 0$. Set $l(v) := \infty$ for all $v \in V(G) \setminus \{s\}$.
      Set $R := \emptyset$.

②   Find a vertex $v \in V(G) \setminus R$ such that $l(v) = \min_{w \in V(G) \setminus R} l(w)$.

③   Set $R := R \cup \{v\}$.

④   **For all** $w \in V(G) \setminus R$ such that $(v, w) \in E(G)$ **do:**
      If $l(w) > l(v) + c((v, w))$ **then**
          set $l(w) := l(v) + c((v, w))$ and $p(w) := v$.

⑤   **If** $R \neq V(G)$ **then go to** ②.

Theorem 7.3. (Dijkstra [1959]) DIJKSTRA'S ALGORITHM *works correctly.*

Proof: We prove that the following statements hold at any stage of the algorithm:

(a) For each $v \in V(G) \setminus \{s\}$ with $l(v) < \infty$ we have $p(v) \in R$, $l(p(v)) + c((p(v), v)) = l(v)$, and the sequence $v, p(v), p(p(v)), \ldots$ contains s.
(b) For all $v \in R$: $l(v) = \text{dist}_{(G,c)}(s, v)$.

The statements trivially hold after ①. $l(w)$ is decreased to $l(v) + c((v, w))$ and $p(w)$ is set to v in ④ only if $v \in R$ and $w \notin R$. As the sequence $v, p(v), p(p(v)), \ldots$ contains s but no vertex outside R, in particular not w, (a) is preserved by ④.

(b) is trivial for $v = s$. Suppose that $v \in V(G) \setminus \{s\}$ is added to R in ③, and there is an s-v-path P in G that is shorter than $l(v)$. Let y be the first vertex on

P that belongs to $(V(G) \setminus R) \cup \{v\}$, and let x be the predecessor of y on P. Since $x \in R$, we have by ④ and the induction hypothesis:

$$l(y) \leq l(x) + c((x, y)) = \text{dist}_{(G,c)}(s, x) + c((x, y))$$
$$\leq c(E(P_{[s,y]})) \leq c(E(P)) < l(v),$$

contradicting the choice of v in ②.   □

The running time is obviously $O(n^2)$. Using a Fibonacci heap we can do better:

Theorem 7.4. (Fredman and Tarjan [1987]) DIJKSTRA'S ALGORITHM *implemented with a Fibonacci heap runs in* $O(m + n \log n)$ *time, where* $n = |V(G)|$ *and* $m = |E(G)|$.

Proof: We apply Theorem 6.7 to maintain the set $\{(v, l(v)) : v \in V(G) \setminus R, l(v) < \infty\}$. Then ② and ③ are one DELETEMIN-operation, while the update of $l(w)$ in ④ is an INSERT-operation if $l(w)$ was infinite and a DECREASEKEY-operation otherwise.   □

This is the best known strongly polynomial running time for the SHORTEST PATH PROBLEM with nonnegative weights. (On different computational models, Fredman and Willard [1994], Thorup [2000] and Raman [1997] achieved slightly better running times.)

If the weights are integers within a fixed range there is a simple linear-time algorithm (Exercise 3). In general, running times of $O(m \log \log c_{\max})$ (Johnson [1982]) and $O\left(m + n\sqrt{\log c_{\max}}\right)$ (Ahuja et al. [1990]) are possible for weights $c : E(G) \to \{0, \ldots, c_{\max}\}$. This has been improved by Thorup [2004] to $O(m + n \log \log c_{\max})$ and $O(m + n \log \log n)$, but even the latter bound applies to integral edge weights only, and the algorithm is not strongly polynomial. For general nonnegative weights, Orlin et al. [2010] described an $O(m \log(2 + nk/m))$-time algorithm, where k is the number of distinct edge weights.

For planar digraphs there is a linear-time algorithm due to Henzinger et al. [1997]. Finally we mention that Thorup [1999] found a linear-time algorithm for finding a shortest path in an undirected graph with nonnegative integral weights. See also Pettie and Ramachandran [2005]; this paper also contains more references.

We now turn to an algorithm for general conservative weights:

MOORE-BELLMAN-FORD ALGORITHM

Input: A digraph G, weights $c : E(G) \to \mathbb{R}$, and a vertex $s \in V(G)$.

Output: A negative circuit C in G, or shortest paths from s to all $v \in V(G)$ and their lengths.
           More precisely, in the second case we get the outputs $l(v)$ and $p(v)$ for all $v \in V(G) \setminus \{s\}$. $l(v)$ is the length of a shortest s-v-path, which consists of a shortest s-$p(v)$-path together with the edge $(p(v), v)$. If v is not reachable from s, then $l(v) = \infty$ and $p(v)$ is undefined.

---

①   Set $l(s) := 0$ and $l(v) := \infty$ for all $v \in V(G) \setminus \{s\}$. Let $n := |V(G)|$.

②   **For** $i := 1$ **to** $n - 1$ **do**:
    For each edge $(v, w) \in E(G)$ **do**:
      If $l(w) > l(v) + c((v, w))$ **then**
        set $l(w) := l(v) + c((v, w))$ and $p(w) := v$.

③   **If** there is an edge $(v, w) \in E(G)$ with $l(w) > l(v) + c((v, w))$ **then** set
    $x_n := w$, $x_{n-1} := v$, and $x_{n-i-1} := p(x_{n-i})$ for $i = 1, \ldots, n - 1$, and
    output any circuit C in $(V(G), \{(x_{i-1}, x_i) : i = 1, \ldots, n\})$.

Theorem 7.5.   (Moore [1959], Bellman [1958], Ford [1956]) *The* MOORE-BELL-
MAN-FORD ALGORITHM *works correctly. Its running time is* $O(nm)$.

Proof: The $O(nm)$ running time is obvious. At any stage of the algorithm, for
$v \in V(G)$, let $k(v)$ be the iteration in which $l(v)$ was decreased to its present
value, and $k(v) := 0$ if $l(v)$ has not been changed after ①. Let $F := \{(p(y), y) :
y \in V(G), k(y) > 0\}$, and $F' := \{(v, w) \in E(G) : l(w) > l(v) + c((v, w))\}$. We
claim that the following statements always hold:

(a) $l(y) \geq l(x) + c((x, y))$ and $k(x) \geq k(y) - 1$ for all $(x, y) \in F$;
(b) If $(V(G), F \cup F')$ contains a circuit C, then C has negative total weight.

To prove (a), observe that $l(y) = l(x) + c((x, y))$ and $k(x) \geq k(y) - 1$ when
$p(y)$ is set to x, and $l(x)$ is never increased, and $k(x)$ is never decreased.

To prove (b), consider a circuit C in $(V(G), F \cup F')$. By (a) we have
$\sum_{(v,w) \in E(C)} c((v, w)) = \sum_{(v,w) \in E(C)} (c((v, w)) + l(v) - l(w)) \leq 0$. This proves
(b) except in the case that $E(C) \subseteq F$. If at some stage a circuit C in $(V(G), F)$
is created by setting $p(y) := x$, then we had $(x, y) \in F'$ just before this step, and
hence C is a negative circuit.

If the algorithm finds an edge $(v, w) \in F'$ in ③, then $k(v) = n - 1$, and thus
$k(x_i) \geq i$ for $i = n - 2, \ldots, 1$. Hence the sequence $x_0, \ldots, x_n$ is well-defined, and
it must contain a repetition. Hence ③ indeed finds a circuit C, which has negative
total weight due to (b).

If the algorithm ends with $l(w) \leq l(v) + c((v, w))$ for all $(v, w) \in E(G)$,
then we have $\sum_{(v,w) \in E(C)} c((v, w)) = \sum_{(v,w) \in E(C)} (c((v, w)) + l(v) - l(w)) \geq 0$
for every circuit C in $G[R]$, where $R := \{v \in V(G) : l(v) < \infty\}$. Hence $G[R]$
contains no negative circuit. Then (b) implies that (R, F) is acyclic. Moreover,
$x \in R \setminus \{s\}$ implies $p(x) \in R$, so (R, F) is an arborescence rooted at s.

By (a), $l(x)$ is at least the length of the s-x-path in (R, F) for any $x \in R$ (at
any stage of the algorithm).

We claim that after k iterations of the algorithm, $l(x)$ is at most the length
of a shortest s-x-path with at most k edges. This statement is easily proved by
induction: Let P be a shortest s-x-path with at most k edges and let (w, x) be
the last edge of P. Then, by applying Proposition 7.2 to $G[R]$, $P_{[s,w]}$ must be
a shortest s-w-path with at most $k - 1$ edges, and by the induction hypothesis

we have $l(w) \leq c(E(P_{[s,w]}))$ after $k - 1$ iterations. But in the k-th iteration edge (w, x) is also examined, after which $l(x) \leq l(w) + c((w, x)) \leq c(E(P))$.

Since no path has more than $n-1$ edges, the above claim implies the correctness of the algorithm. $\qquad\square$

Note that if c is conservative, (R, F) is an arborescence containing a shortest path from s to each vertex that is reachable from s. This is sometimes called a *shortest paths tree*.

This algorithm is still the fastest known strongly polynomial-time algorithm for the SHORTEST PATH PROBLEM in digraphs with conservative weights. A scaling algorithm due to Goldberg [1995] has a running time of $O\left(\sqrt{n}m \log(C + 2)\right)$ if the edge weights are integral and have absolute value at most C. For planar digraphs, Mozes and Wulff-Nilsen [2010] described an $O(n \log^2 n/\log\log n)$-time algorithm.

If G contains negative circuits, no polynomial-time algorithm for the SHORTEST PATH PROBLEM is known (the problem becomes *NP*-hard; see Exercise 17(b) of Chapter 15). The main difficulty is that Proposition 7.2 does not hold for general weights. It is not clear how to construct a path instead of an arbitrary edge progression. If there are no negative circuits, any shortest edge progression is a path, plus possibly some circuits of zero weight that can be deleted. In view of this it is also an important question how to prove that there are no negative circuits. The following concept due to Edmonds and Karp [1972] is useful:

Definition 7.6. *Let G be a digraph with weights $c : E(G) \to \mathbb{R}$, and let $\pi : V(G) \to \mathbb{R}$. Then for any $(x, y) \in E(G)$ we define the **reduced cost** of (x, y) with respect to π by $c_\pi((x, y)) := c((x, y)) + \pi(x) - \pi(y)$. If $c_\pi(e) \geq 0$ for all $e \in E(G)$, π is called a **feasible potential**.*

Theorem 7.7. *Let G be a digraph with weights $c : E(G) \to \mathbb{R}$. There exists a feasible potential of (G, c) if and only if c is conservative. Given a digraph G with weights $c : E(G) \to \mathbb{R}$, we can find in $O(nm)$ time either a feasible potential or a negative circuit.*

Proof: If π is a feasible potential, we have for each circuit C:

$$0 \leq \sum_{e \in E(C)} c_\pi(e) = \sum_{e=(x,y)\in E(C)} (c(e) + \pi(x) - \pi(y)) = \sum_{e \in E(C)} c(e)$$

(the potentials cancel). So c is conservative.

To find either a negative circuit or a feasible potential, we add a new vertex s and edges (s, v) of zero cost for all $v \in V(G)$. Then we run the MOORE-BELLMAN-FORD ALGORITHM. The output is either a negative circuit (which of course cannot contain s) or numbers $l(v) < \infty$ for all $v \in V(G)$ constituting a feasible potential. $\qquad\square$

This can be regarded as a special form of LP duality; see Exercise 9. In practice there are more efficient methods to detect negative circuits; see Cherkassky and Goldberg [1999].

7.2 Shortest Paths Between All Pairs of Vertices

Suppose we now want to find a shortest s-t-path for all ordered pairs of vertices (s, t) in a digraph:

ALL PAIRS SHORTEST PATHS PROBLEM

Instance: A digraph G and conservative weights $c : E(G) \to \mathbb{R}$.

Task: Find numbers l_{st} and vertices p_{st} for all $s, t \in V(G)$ with $s \neq t$, such that l_{st} is the length of a shortest s-t-path, and (p_{st}, t) is the final edge of such a path (if it exists).

Of course we could run the MOORE-BELLMAN-FORD ALGORITHM n times, once for each choice of s. This immediately gives us an $O(n^2 m)$-algorithm. However, one can do better, as was observed by Bazaraa and Langley [1974] and Johnson [1977]:

Theorem 7.8. *The* ALL PAIRS SHORTEST PATHS PROBLEM *can be solved in* $O(mn + n^2 \log n)$ *time, where* $n = |V(G)|$ *and* $m = |E(G)|$.

Proof: Let (G, c) be an instance. First we compute a feasible potential π, which is possible in $O(nm)$ time by Corollary 7.7. Then for each $s \in V(G)$ we do a single-source shortest path computation from s using the reduced costs c_π instead of c. For any vertex t the resulting s-t-path is also a shortest path with respect to c, because the length of any s-t-path changes by $\pi(s) - \pi(t)$, a constant. Since the reduced costs are nonnegative, we can use DIJKSTRA'S ALGORITHM each time. So, by Theorem 7.4, the total running time is $O(mn + n(m + n \log n))$. $\qquad\square$

The same idea will be used again in Chapter 9 (in the proof of Theorem 9.14).

Pettie [2004] showed how to improve the running time to $O(mn + n^2 \log \log n)$; this is the best known time bound. For dense graphs with nonnegative weights, the $O(n^3 \log \log n / \log^2 n)$ bound by Han and Takaoka [2016] is slightly better. If all edge weights are small positive integers, this can be improved using fast matrix multiplication; see e.g. Zwick [2002].

The solution of the ALL PAIRS SHORTEST PATHS PROBLEM also enables us to compute the metric closure:

Definition 7.9. *Given a graph G (directed or undirected) with conservative weights $c : E(G) \to \mathbb{R}$. The* **metric closure** *of (G, c) is the pair $(\bar{G}, \bar{c})$, where $\bar{G}$ is the simple graph on $V(G)$ that, for $x, y \in V(G)$ with $x \neq y$, contains an edge $e = \{x, y\}$ (or $e = (x, y)$ if G is directed) with weight $\bar{c}(e) = \mathrm{dist}_{(G,c)}(x, y)$ if and only if y is reachable from x in G.*

Corollary 7.10. *Let G be a digraph with conservative weights $c : E(G) \to \mathbb{R}$, or an undirected graph with nonnegative weights $c : E(G) \to \mathbb{R}_+$. Then the metric closure of (G, c) can be computed in $O(mn + n^2 \log n)$ time.*

Proof: If G is undirected, we replace each edge by a pair of oppositely directed edges. Then we solve the resulting instance of the ALL PAIRS SHORTEST PATHS PROBLEM.   □

The rest of the section is devoted to the FLOYD-WARSHALL ALGORITHM, another $O(n^3)$-algorithm for the ALL PAIRS SHORTEST PATHS PROBLEM. The main advantage of the FLOYD-WARSHALL ALGORITHM is its simplicity. We assume w.l.o.g. that the vertices are numbered $1, \ldots, n$.

FLOYD-WARSHALL ALGORITHM

Input: A digraph G with $V(G) = \{1, \ldots, n\}$ and conservative weights $c :$ $E(G) \to \mathbb{R}$.

Output: Matrices $(l_{ij})_{1 \le i,j \le n}$ and $(p_{ij})_{1 \le i,j \le n}$ where l_{ij} is the length of a shortest path from i to j, and (p_{ij}, j) is the final edge of such a path (if it exists).

---

①   Set $l_{ij} := c((i, j))$ for all $(i, j) \in E(G)$.
       Set $l_{ij} := \infty$ for all $(i, j) \in (V(G) \times V(G)) \setminus E(G)$ with $i \ne j$.
       Set $l_{ii} := 0$ for all i.
       Set $p_{ij} := i$ for all $i, j \in V(G)$.

②   **For** $j := 1$ **to** n **do:**
       For $i := 1$ **to** n **do: If** $i \ne j$ **then:**
          For $k := 1$ **to** n **do: If** $k \ne j$ **then:**
             If $l_{ik} > l_{ij} + l_{jk}$ **then** set $l_{ik} := l_{ij} + l_{jk}$ and $p_{ik} := p_{jk}$.

Theorem 7.11.   (Floyd [1962], Warshall [1962]) *The* FLOYD-WARSHALL ALGORITHM *works correctly. Its running time is* $O(n^3)$.

Proof: The running time is obvious.
Claim: After the algorithm has run through the outer loop for $j = 1, 2, \ldots, j_0$, the variable l_{ik} contains the length of a shortest i-k-path with intermediate vertices $v \in \{1, \ldots, j_0\}$ only (for all i and k), and (p_{ik}, k) is the final edge of such a path.

This statement will be shown by induction for $j_0 = 0, \ldots, n$. For $j_0 = 0$ it is true by ①, and for $j_0 = n$ it implies the correctness of the algorithm.

Suppose the claim holds for some $j_0 \in \{0, \ldots, n - 1\}$. We have to show that it still holds for $j_0 + 1$. For any i and k, during processing the outer loop for $j = j_0 + 1$, l_{ik} (containing by the induction hypothesis the length of a shortest i-k-path with intermediate vertices $v \in \{1, \ldots, j_0\}$ only) is replaced by $l_{i,j_0+1} + l_{j_0+1,k}$ if this value is smaller. It remains to show that the corresponding i-$(j_0 + 1)$-path P and the $(j_0 + 1)$-k-path Q have no inner vertex in common.

Suppose that there is an inner vertex belonging to both P and Q. By shortcutting a maximal closed walk in $P + Q$ (which by our assumption has nonnegative weight because it is the union of circuits) we get an i-k-path R with intermediate vertices $v \in \{1, \ldots, j_0\}$ only. R is no longer than $l_{i,j_0+1} + l_{j_0+1,k}$ (and in particular shorter than the l_{ik} before processing the outer loop for $j = j_0 + 1$).

This contradicts the induction hypothesis since R has intermediate vertices $v \in \{1, \ldots, j_0\}$ only. ☐

Like the MOORE-BELLMAN-FORD ALGORITHM, the FLOYD-WARSHALL AL-GORITHM can also be used to detect the existence of negative circuits (see Exercise 13 and Hougardy [2010]).

The ALL PAIRS SHORTEST PATHS PROBLEM in undirected graphs with arbitrary conservative weights is more difficult; see Theorem 12.14.

7.3 Minimum Mean Cycles

We can easily find a circuit of minimum total weight in a digraph with conservative weights, using the above shortest path algorithms (see Exercise 14). Another problem asks for a circuit whose mean weight is minimum:

DIRECTED MINIMUM MEAN CYCLE PROBLEM

Instance: A digraph G, weights $c : E(G) \to \mathbb{R}$.

Task: Find a circuit C whose mean weight $\frac{c(E(C))}{|E(C)|}$ is minimum, or decide that G is acyclic.

In this section we show how to solve this problem with dynamic programming, quite similar to the shortest path algorithms. We could assume that G is strongly connected, since otherwise we can identify the strongly connected components in linear time (Theorem 2.19) and solve the problem for each strongly connected component separately. But for the following min-max theorem it suffices to assume that there is a vertex s from which all vertices are reachable. We consider not only paths, but arbitrary edge progressions (where vertices and edges may be repeated).

Theorem 7.12. (Karp [1978]) *Let G be a digraph with weights $c : E(G) \to \mathbb{R}$. Let $s \in V(G)$ such that each vertex is reachable from s. For $x \in V(G)$ and $k \in \mathbb{Z}_+$ let*

$$F_k(x) := \min \left\{ \sum_{i=1}^{k} c((v_{i-1}, v_i)) : v_0 = s, \ v_k = x, \ (v_{i-1}, v_i) \in E(G) \text{ for all } i \right\}$$

be the minimum weight of an edge progression of length k from s to x (and ∞ if there is none). Let $\mu(G, c)$ be the minimum mean weight of a circuit in G (and $\mu(G, c) = \infty$ if G is acyclic). Then

$$\mu(G, c) = \min_{\substack{x \in V(G)}} \max_{\substack{0 \le k \le n-1 \\ F_k(x) < \infty}} \frac{F_n(x) - F_k(x)}{n - k}.$$

Proof: If G is acyclic, then $F_n(x) = \infty$ for all $x \in V(G)$, so the theorem holds. We now assume that $\mu(G, c) < \infty$.

First we prove that if $\mu(G, c) = 0$ then also

$$\min_{x \in V(G)} \max_{\substack{0 \le k \le n-1 \\ F_k(x) < \infty}} \frac{F_n(x) - F_k(x)}{n - k} = 0.$$

Let G be a digraph with $\mu(G, c) = 0$. G contains no negative circuit. Since c is conservative, $F_n(x) \ge \operatorname{dist}_{(G,c)}(s, x) = \min_{0 \le k \le n-1} F_k(x)$, so

$$\max_{\substack{0 \le k \le n-1 \\ F_k(x) < \infty}} \frac{F_n(x) - F_k(x)}{n - k} \ge 0$$

for all $x \in V(G)$. We show that there is a vertex x for which equality holds, i.e. $F_n(x) = \operatorname{dist}_{(G,c)}(s, x)$. Let C be any zero-weight circuit in G, and let $w \in V(C)$. Let P be a shortest s-w-path followed by n repetitions of C. Let P' consist of the first n edges of P, and let x be the end-vertex of P'. Since P is a minimum-weight edge progression from s to w, any initial segment, in particular P', must be a minimum-weight edge progression. So $F_n(x) = c(E(P')) = \operatorname{dist}_{(G,c)}(s, x)$.

Having proved the theorem for the case $\mu(G, c) = 0$, we now turn to the general case. Note that adding a constant to all edge weights changes both $\mu(G, c)$ and

$$\min_{x \in V(G)} \max_{\substack{0 \le k \le n-1 \\ F_k(x) < \infty}} \frac{F_n(x) - F_k(x)}{n - k}$$

by the same amount, namely this constant. By choosing this constant to be $-\mu(G, c)$ we are back to the case $\mu(G, c) = 0$. □

This theorem suggests the following algorithm:

MINIMUM MEAN CYCLE ALGORITHM

Input: A digraph G, weights $c : E(G) \to \mathbb{R}$.

Output: A circuit C with minimum mean weight or the information that G is acyclic.

---

① Add to G a vertex s and edges (s, x) with $c((s, x)) := 0$ for all $x \in V(G)$.

② Set $n := |V(G)|$, $F_0(s) := 0$, and $F_0(x) := \infty$ for all $x \in V(G) \setminus \{s\}$.

③ **For** $k := 1$ **to** n **do:**
   For all $x \in V(G)$ **do:**
      Set $F_k(x) := \infty$.
      For all $(w, x) \in \delta^-(x)$ **do:**
         If $F_{k-1}(w) + c((w, x)) < F_k(x)$ **then:**
            Set $F_k(x) := F_{k-1}(w) + c((w, x))$ and $p_k(x) := w$.

④ **If** $F_n(x) = \infty$ for all $x \in V(G)$ **then stop** (G is acyclic).

⑤ Let x be a vertex for which $\max_{\substack{0 \leq k \leq n-1 \\ F_k(x) < \infty}} \dfrac{F_n(x) - F_k(x)}{n-k}$ is minimum.

⑥ Let C be any circuit in the edge progression given by
$$s = p_1(p_2(\cdots (p_n(x)) \cdots)), \ldots, p_{n-1}(p_n(x)), p_n(x), x.$$

Corollary 7.13. (Karp [1978]) *The* MINIMUM MEAN CYCLE ALGORITHM *works correctly. Its running time is* $O(nm)$.

Proof: ① does not create any new circuit in G but makes Theorem 7.12 applicable. It is obvious that ② and ③ compute the numbers $F_k(x)$ correctly. So if the algorithm stops in ④, G is indeed acyclic.

Consider the instance (G, c'), where $c'(e) := c(e) - \mu(G, c)$ for all $e \in E(G)$. On this instance the algorithm runs exactly the same way as with (G, c), the only difference being the change of the F-values to $F_k'(x) = F_k(x) - k\mu(G, c)$. By the choice of x in ⑤, Theorem 7.12 and $\mu(G, c') = 0$ we have $F_n'(x) = \min_{0 \leq k \leq n-1} F_k'(x)$. Hence any edge progression from s to x with n edges and length $F_n'(x)$ in (G, c') consists of a shortest s-x-path plus one or more circuits of zero weight. These circuits have mean weight $\mu(G, c)$ in (G, c).

Hence each circuit on a minimum weight edge progression of length n from s to x (for the vertex x chosen in ⑤) is a circuit of minimum mean weight. In ⑥ such a circuit is chosen.

The running time is dominated by ③ which obviously takes $O(nm)$ time. Note that ⑤ takes only $O(n^2)$ time. $\square$

This algorithm cannot be used for finding a circuit of minimum mean weight in an undirected graph with edge weights. See Exercises 10 and 18 of Chapter 12.

Algorithms for more general minimum ratio problems have been proposed by Megiddo [1979, 1983] and Radzik [1993]. See also Exercise 10 of Chapter 12.

7.4 Shallow-Light Trees

We saw efficient algorithms for computing minimum spanning trees and for shortest paths trees. These trees can look very different. In undirected graphs, however, one can make a reasonable tradeoff, computing a cheap tree that contains short paths from a source s to all vertices:

Theorem 7.14. (Khuller, Raghavachari and Young [1995]) *Let G be an undirected graph with edge weights $c : E(G) \to \mathbb{R}_+$ and a source $s \in V(G)$. Let $\epsilon > 0$. Then there is a spanning tree S in G such that* $\text{dist}_{(S,c)}(s, v) \leq (1 + \epsilon) \, \text{dist}_{(G,c)}(s, v)$ *for all $v \in V(G)$ and $\sum_{e \in E(S)} c(e)$ is at most $(1 + \frac{2}{\epsilon})$ times the cost of a minimum spanning tree in (G, c). Such an arborescence can be computed in $O(m + n \log n)$ time, where $n = |V(G)|$ and $m = |E(G)|$.*

Proof: Let S_0 be a minimum cost spanning tree in (G, c) and P a shortest paths tree, i.e., a tree containing a shortest s-v-path in (G, c) for all $v \in V(G)$. S_0 and P can be computed by PRIM'S ALGORITHM and DIJKSTRA'S ALGORITHM, respectively, both in $O(m + n \log n)$ time (Corollary 6.8 and Theorem 7.4).

Start with $S = S_0$. Double all edges in S_0 and find an Eulerian walk W (cf. Theorem 2.25). Traverse this walk from s. Whenever we visit a vertex v for the first time and $\text{dist}_{(S,c)}(s, v) > (1 + \epsilon)\text{dist}_{(G,c)}(s, v)$, we add $E(P_{[s,v]})$ to $E(S)$, where $P_{[s,v]}$ is the s-v-path in P.

The procedure clearly outputs a graph S with $\text{dist}_{(S,c)}(s, v) \le (1 + \epsilon)$ $\text{dist}_{(G,c)}(s, v)$ for all $v \in V(G)$. We now bound the total edge cost.

Let $v_1, \ldots, v_k$ be the vertices v for which we add a shortest s-v-path to S, and $v_0 := s$. Let $c_W(v_{i-1}, v_i)$ be the total length of the edges in the subwalk of W from the first occurrence of v_{i-1} to the first occurrence of v_i. When v_i is examined, the graph S contains a shortest s-v_{i-1}-path, so

$$\text{dist}_{(G,c)}(s, v_{i-1}) + c_W(v_{i-1}, v_i) \ge \text{dist}_{(S,c)}(s, v_i) > (1 + \epsilon)\text{dist}_{(G,c)}(s, v_i).$$

Summing up these inequalities yields

$$\sum_{i=1}^{k} \text{dist}_{(G,c)}(s, v_{i-1}) + 2 \sum_{e \in E(S_0)} c(e) > (1 + \epsilon) \sum_{i=1}^{k} \text{dist}_{(G,c)}(s, v_i).$$

We get

$$\epsilon \sum_{i=1}^{k} \text{dist}_{(G,c)}(s, v_i) \le \epsilon \sum_{i=1}^{k-1} \text{dist}_{(G,c)}(s, v_i) + (1 + \epsilon)\text{dist}_{(G,c)}(s, v_k) < 2 \sum_{e \in E(S_0)} c(e)$$

and hence

$$\sum_{e \in E(S)} c(e) = \sum_{e \in E(S_0)} c(e) + \sum_{i=1}^{k} \text{dist}_{(G,c)}(s, v_i) \le \left(1 + \frac{2}{\epsilon}\right) \sum_{e \in E(S_0)} c(e)$$

as required. Now S is normally not yet a tree, but computing a shortest path tree from s in S finishes the job.

To get the stated runtime, observe that we do not need to add all edges of $P_{[s,v]}$ but only those that haven't been added earlier. By storing for each node v whether S already contains $P_{[s,v]}$ we can avoid considering edges of P more than once.

The remaining difficulty is how to decide fast enough whether $\text{dist}_{(S,c)}(s, v) > (1 + \epsilon) \text{dist}_{(G,c)}(s, v)$ when visiting a vertex v. However, the above proof shows that we can replace this condition by $\text{dist}_{(G,c)}(s, v_{i-1}) + c_W(v_{i-1}, v_i) > (1 + \epsilon)$ $\text{dist}_{(G,c)}(s, v)$, where v_{i-1} is the last vertex u for which we added a shortest s-u-path to S (or s is that has never happened yet). $\square$

Such trees are called shallow-light trees. The tradeoff cannot be improved; see Exercise 19.

Exercises

1. Let G be a graph (directed or undirected) with weights $c : E(G) \to \mathbb{Z}_+$, and let $s, t \in V(G)$ such that t is reachable from s. Show that the minimum length of an s-t-path equals the maximum number of cuts separating s and t such that each edge e is contained in at most $c(e)$ of them.

2. Consider DIJKSTRA'S ALGORITHM, and suppose that we are only interested in a shortest s-t-path for a given vertex t. Show that we can stop as soon as $v = t$ or $l(v) = \infty$ in ②.

3. Suppose the weights are integers between 0 and C for some constant C. Can one implement DIJKSTRA'S ALGORITHM for this special case with linear running time?
(Dial [1969])

4. Given a digraph G, weights $c : E(G) \to \mathbb{R}_+$, and two vertices $s, t \in V(G)$. Suppose there is only one shortest s-t-path P. Can one then find the shortest s-t-path different from P in polynomial time?

5. Modify DIJKSTRA'S ALGORITHM in order to solve the bottleneck path problem: Given a digraph G, $c : E(G) \to \mathbb{R}$, and $s, t \in V(G)$, find an s-t-path whose longest edge is shortest possible.

6. Let G be a digraph with $s, t \in V(G)$. To each edge $e \in E(G)$ we assign a number $r(e)$ (its reliability), with $0 \le r(e) \le 1$. The reliability of a path P is defined to be the product of the reliabilities of its edges. The problem is to find an s-t-path of maximum reliability.
   (a) Show that by taking logarithms one can reduce this problem to a SHORTEST PATH PROBLEM.
   (b) Show how to solve this problem (in polynomial time) without taking logarithms.

7. Given an acyclic digraph G, $c : E(G) \to \mathbb{R}$ and $s, t, \in V(G)$. Show how to find a shortest s-t-path in G in linear time.

8. Given an acyclic digraph G, $c : E(G) \to \mathbb{R}$ and $s, t, \in V(G)$. Show how to find the union of all longest s-t-paths in G in linear time.

9. Prove Theorem 7.7 using LP duality, in particular Theorem 3.24.

10. Let G be a digraph with conservative weights $c : E(G) \to \mathbb{R}$. Let $s, t \in V(G)$ such that t is reachable from s. Prove that the minimum length of an s-t-path in G equals the maximum of $\pi(t) - \pi(s)$, where π is a feasible potential of (G, c).

11. Let G be a strongly connected digraph with weights $c : E(G) \to \mathbb{R}_+$. Let $\emptyset \ne L \subseteq V(G)$ such that every element of L is reachable from all vertices. Let $s, t \in V(G)$ and $\pi(v) := \min\{0, \min_{l \in L}(\operatorname{dist}_{(G,c)}(t, l) - \operatorname{dist}_{(G,c)}(v, l))\}$ for $v \in V(G)$. Prove:
   (a) π is a feasible potential.
   (b) Any shortest s-t-path in (G, c_π) is a shortest s-t-path in (G, c).
   (c) $\{v \in V(G) : \operatorname{dist}_{(G,c_\pi)}(s, v) < \operatorname{dist}_{(G,c_\pi)}(s, t)\} \subseteq \{v \in V(G) : \operatorname{dist}_{(G,c)}(s, v) < \operatorname{dist}_{(G,c)}(s, t)\}$.

Note: If a large number of shortest paths needs to be computed in a large graph, it may be worthwhile to precompute distances to a well-chosen set $L \subseteq V(G)$ (of so-called landmarks). By (a), (b), (c), and Exercise 2, one can then use the distances to the landmarks to speed up the search for shortest s-t-paths for many pairs s and t by applying DIJKSTRA'S ALGORITHM to (G, c_π). (Goldberg and Harrelson [2005])

12. Let G be a digraph, $V(G) = A \cup B$ and $E(G[B]) = \emptyset$. Moreover, suppose that $|\delta(v)| \leq k$ for all $v \in B$. Let $s, t \in V(G)$ and $c : E(G) \rightarrow \mathbb{R}$ conservative. Prove that then a shortest s-t-path can be found in $O(|A|k|E(G)|)$ time, and if c is nonnegative in $O(|A|^2)$ time. (Orlin [1993])

13. Suppose that we run the FLOYD-WARSHALL ALGORITHM on an instance (G, c) with arbitrary weights $c : E(G) \rightarrow \mathbb{R}$. Prove that all l_{ii} $(i = 1, \ldots, n)$ remain nonnegative if and only if c is conservative.

14. Given a digraph with conservative weights, show how to find a circuit of minimum total weight in polynomial time. Can you achieve an $O(n^3)$ running time?

    Hint: Modify the FLOYD-WARSHALL ALGORITHM slightly.

    Note: Orlin and Sedeño-Noda [2017] found an $O(nm)$-time algorithm. For general weights the problem includes the decision whether a given digraph is Hamiltonian (and is thus *NP*-hard; see Chapter 15). How to find the minimum weight circuit in an undirected graph (with conservative weights) is described in Section 12.2.

15. Let G be a complete (undirected) graph and $c : E(G) \rightarrow \mathbb{R}_+$. Show that (G, c) is its own metric closure if and only if the triangle inequality holds: $c(\{x, y\}) + c(\{y, z\}) \geq c(\{x, z\})$ for any three distinct vertices $x, y, z \in V(G)$.

16. The timing constraints of a logic chip can be modelled by a digraph G with edge weights $c : E(G) \rightarrow \mathbb{R}_+$. The vertices represent the storage elements, the edges represent paths through combinational logic, and the weights are worst-case estimations of the propagation time of a signal. An important task in the design of very large scale integrated (VLSI) circuits is to find an optimum clock schedule, i.e. a mapping $a : V(G) \rightarrow \mathbb{R}$ such that $a(v) + c((v, w)) \leq a(w) + T$ for all $(v, w) \in E(G)$ and a number T which is as small as possible. (T is the cycle time of the chip, and $a(v)$ and $a(v) + T$ are the "departure time" and latest feasible "arrival time" of a signal at v, respectively.)

    (a) Reduce the problem of finding the optimum T to the DIRECTED MINIMUM MEAN CYCLE PROBLEM.

    (b) Show how the numbers $a(v)$ of an optimum solution can be determined efficiently.

    (c) Typically, some of the numbers $a(v)$ are fixed in advance. Show how to solve the problem in this case.

    (Albrecht et al. [2002])

17. Show that the following variant of the MINIMUM MEAN CYCLE ALGORITHM also works correctly. No extra vertex s is added. Instead of ① and ② we only set $F_0(x) := 0$ for all $x \in V(G)$.
18. Find a polynomial-time algorithm that, given an undirected graph G, two vertices s and t, and an integer k, decides whether there exists an edge progression of length k from s to t in G.
    Hint: k could be much larger than n. Consider the case $k > 2n$ separately.
19. Show that for every $\epsilon > 0$ and $\delta < 1 + \frac{2}{\epsilon}$ there is a weighted graph (G, c) in which every spanning tree S with $\mathrm{dist}_{(S,c)}(s, v) \leq (1 + \epsilon)\, \mathrm{dist}_{(G,c)}(s, v)$ for all $v \in V(G)$ has total cost more than δ times the cost of a minimum spanning tree.
    (Khuller, Raghavachari and Young [1995])

References

General Literature

Ahuja, R.K., Magnanti, T.L., and Orlin, J.B. [1993]: Network Flows. Prentice-Hall, Englewood Cliffs 1993, Chapters 4 and 5

Cormen, T.H., Leiserson, C.E., Rivest, R.L., and Stein, C. [2009]: Introduction to Algorithms. Third Edition. MIT Press, Cambridge 2009, Chapters 24 and 25

Dreyfus, S.E. [1969]: An appraisal of some shortest path algorithms. Operations Research 17 (1969), 395–412

Gallo, G., and Pallottino, S. [1988]: Shortest paths algorithms. Annals of Operations Research 13 (1988), 3–79

Gondran, M., and Minoux, M. [1984]: Graphs and Algorithms. Wiley, Chichester 1984, Chapter 2

Lawler, E.L. [1976]: Combinatorial Optimization: Networks and Matroids. Holt, Rinehart and Winston, New York 1976, Chapter 3

Schrijver, A. [2003]: Combinatorial Optimization: Polyhedra and Efficiency. Springer, Berlin 2003, Chapters 6–8

Tarjan, R.E. [1983]: Data Structures and Network Algorithms. SIAM, Philadelphia 1983, Chapter 7

Cited References

Ahuja, R.K., Mehlhorn, K., Orlin, J.B., and Tarjan, R.E. [1990]: Faster algorithms for the shortest path problem. Journal of the ACM 37 (1990), 213–223

Albrecht, C., Korte, B., Schietke, J., and Vygen, J. [2002]: Maximum mean weight cycle in a digraph and minimizing cycle time of a logic chip. Discrete Applied Mathematics 123 (2002), 103–127

Bazaraa, M.S., and Langley, R.W. [1974]: A dual shortest path algorithm. SIAM Journal on Applied Mathematics 26 (1974), 496–501

Bellman, R.E. [1958]: On a routing problem. Quarterly of Applied Mathematics 16 (1958), 87–90

Cherkassky, B.V., and Goldberg, A.V. [1999]: Negative-cycle detection algorithms. Mathematical Programming A 85 (1999), 277–311

Dial, R.B. [1969]: Algorithm 360: shortest path forest with topological order. Communications of the ACM 12 (1969), 632–633

Dijkstra, E.W. [1959]: A note on two problems in connexion with graphs. Numerische Mathematik 1 (1959), 269–271

Edmonds, J., and Karp, R.M. [1972]: Theoretical improvements in algorithmic efficiency for network flow problems. Journal of the ACM 19 (1972), 248–264

Floyd, R.W. [1962]: Algorithm 97 – shortest path. Communications of the ACM 5 (1962), 345

Ford, L.R. [1956]: Network flow theory. Paper P-923, The Rand Corporation, Santa Monica 1956

Fredman, M.L., and Tarjan, R.E. [1987]: Fibonacci heaps and their uses in improved network optimization problems. Journal of the ACM 34 (1987), 596–615

Fredman, M.L., and Willard, D.E. [1994]: Trans-dichotomous algorithms for minimum spanning trees and shortest paths. Journal of Computer and System Sciences 48 (1994), 533–551

Goldberg, A.V. [1995]: Scaling algorithms for the shortest paths problem. SIAM Journal on Computing 24 (1995), 494–504

Goldberg, A.V., and Harrelson, C. [2005]: Computing the shortest path: A^* search meets graph theory. Proceedings of the 16th Annual ACM-SIAM Symposium on Discrete Algorithms (2005), 156–165

Han, Y., and Takaoka, T. [2016]: An $O(n^3 \log \log n / \log^2 n)$ time algorithm for all pairs shortest paths. Journal of Discrete Algorithms 38–41 (2016), 9–19

Henzinger, M.R., Klein, P., Rao, S., and Subramanian, S. [1997]: Faster shortest-path algorithms for planar graphs. Journal of Computer and System Sciences 55 (1997), 3–23

Hougardy, S. [2010]: The Floyd-Warshall algorithm on graphs with negative cycles. Information Processing Letters 110 (2010), 279–281

Johnson, D.B. [1977]: Efficient algorithms for shortest paths in sparse networks. Journal of the ACM 24 (1977), 1–13

Johnson, D.B. [1982]: A priority queue in which initialization and queue operations take $O(\log \log D)$ time. Mathematical Systems Theory 15 (1982), 295–309

Karp, R.M. [1978]: A characterization of the minimum cycle mean in a digraph. Discrete Mathematics 23 (1978), 309–311

Khuller, S., Raghavachari, B., and Young, N. [1995]: Balancing minimum spanning trees and shortest-path trees. Algorithmica 14 (1995), 305–321

Megiddo, N. [1979]: Combinatorial optimization with rational objective functions. Mathematics of Operations Research 4 (1979), 414–424

Megiddo, N. [1983]: Applying parallel computation algorithms in the design of serial algorithms. Journal of the ACM 30 (1983), 852–865

Moore, E.F. [1959]: The shortest path through a maze. Proceedings of the International Symposium on the Theory of Switching, Part II, Harvard University Press, 1959, 285–292

Mozes, S., and Wulff-Nilsen, C. [2010]: Shortest paths in planar graphs with real lengths in $O(n \log^2 n / \log \log n)$ time. In: Algorithms – Proceedings of the 18th European Symposium on Algorithms (ESA), Part II; LNCS 6347 (M. de Berg, U. Meyer, eds.), Springer, Berlin 2010, pp. 206–217

Orlin, J.B. [1993]: A faster strongly polynomial minimum cost flow algorithm. Operations Research 41 (1993), 338–350

Orlin, J.B., Madduri, K., Subramani, K., and Williamson, M. [2010]: A faster algorithm for the single source shortest path problem with few distinct positive lengths. Journal of Discrete Algorithms 8 (2010), 189–198

Orlin, J.B., and Sedeño-Noda, A. [2017]: An $O(nm)$ time algorithm for finding the min length directed cycle in a graph. Proceedings of the 28th Annual ACM-SIAM Symposium on Discrete Algorithms (2017), 1866–1879

Pettie, S. [2004]: A new approach to all-pairs shortest paths on real-weighted graphs. Theoretical Computer Science 312 (2004), 47–74

Pettie, S., and Ramachandran, V. [2005]: Computing shortest paths with comparisons and additions. SIAM Journal on Computing 34 (2005), 1398–1431

Radzik, T. [1993]: Parametric flows, weighted means of cuts, and fractional combinatorial optimization. In: Complexity in Numerical Optimization (P.M. Pardalos, ed.), World Scientific, Singapore 1993

Raman, R. [1997]: Recent results on the single-source shortest paths problem. ACM SIGACT News 28 (1997), 81–87

Thorup, M. [1999]: Undirected single-source shortest paths with positive integer weights in linear time. Journal of the ACM 46 (1999), 362–394

Thorup, M. [2000]: On RAM priority queues. SIAM Journal on Computing 30 (2000), 86–109

Thorup, M. [2004]: Integer priority queues with decrease key in constant time and the single source shortest paths problem. Journal of Computer and System Sciences 69 (2004), 330–353

Warshall, S. [1962]: A theorem on boolean matrices. Journal of the ACM 9 (1962), 11–12

Zwick, U. [2002]: All pairs shortest paths using bridging sets and rectangular matrix multiplication. Journal of the ACM 49 (2002), 289–317

8 Network Flows

In this and the next chapter we consider flows in networks. We have a digraph G with edge capacities $u : E(G) \rightarrow \mathbb{R}_+$ and two specified vertices s (the **source**) and t (the **sink**). The quadruple (G, u, s, t) is sometimes called a **network**.

Our main motivation is to transport as many units as possible simultaneously from s to t. A solution to this problem will be called a maximum flow. Formally we define:

Definition 8.1. *Given a digraph G with capacities $u : E(G) \rightarrow \mathbb{R}_+$, a **flow** is a function $f : E(G) \rightarrow \mathbb{R}_+$ with $f(e) \leq u(e)$ for all $e \in E(G)$. The **excess** of a flow f at $v \in V(G)$ is*

$$\mathrm{ex}_f(v) := \sum_{e \in \delta^-(v)} f(e) - \sum_{e \in \delta^+(v)} f(e).$$

*We say that f satisfies the **flow conservation rule** at vertex v if $\mathrm{ex}_f(v) = 0$. A flow satisfying the flow conservation rule at every vertex is called a **circulation**.*

*Now given a network (G, u, s, t), an **s-t-flow** is a flow f satisfying $\mathrm{ex}_f(s) \leq 0$ and $\mathrm{ex}_f(v) = 0$ for all $v \in V(G) \setminus \{s, t\}$. We define the **value** of an s-t-flow f by* value $(f) := -\mathrm{ex}_f(s)$.

Now we can formulate the basic problem of this chapter:

MAXIMUM FLOW PROBLEM

Instance: A network (G, u, s, t).

Task: Find an s-t-flow of maximum value.

It causes no loss of generality to assume that G is simple as parallel edges can be united beforehand.

This problem has numerous applications. For example, consider the JOB AS-SIGNMENT PROBLEM: given n jobs, their processing times $t_1, \ldots, t_n \in \mathbb{R}_+$ and a nonempty subset $S_i \subseteq \{1, \ldots, m\}$ of employees that can contribute to each job $i \in \{1, \ldots, n\}$, we ask for numbers $x_{ij} \in \mathbb{R}_+$ for all $i = 1, \ldots, n$ and $j \in S_i$ (meaning how long employee j works on job i) such that all jobs are finished, i.e. $\sum_{j \in S_i} x_{ij} = t_i$ for $i = 1, \ldots, n$. Our goal was to minimize the amount of time in

© Springer-Verlag GmbH Germany 2018
Bernhard Korte and Jens Vygen, *Combinatorial Optimization*, Algorithms and Combinatorics 21, https://doi.org/10.1007/978-3-662-56039-6_8

which all jobs are done, i.e. $T(x) := \max_{j=1}^{m} \sum_{i:j \in S_i} x_{ij}$. Instead of solving this problem with LINEAR PROGRAMMING we look for a combinatorial algorithm.

We apply binary search for the optimum $T(x)$. Then for one specific value T we have to find numbers $x_{ij} \in \mathbb{R}_+$ with $\sum_{j \in S_i} x_{ij} = t_i$ for all i and $\sum_{i:j \in S_i} x_{ij} \leq T$ for all j. We model the sets S_i by a (bipartite) digraph with a vertex v_i for each job i, a vertex w_j for each employee j and an edge (v_i, w_j) whenever $j \in S_i$. We introduce two additional vertices s and t and edges (s, v_i) for all i and (w_j, t) for all j. Let this graph be G. We define capacities $u : E(G) \to \mathbb{R}_+$ by $u((s, v_i)) := t_i$ and $u(e) := T$ for all other edges. Then the feasible solutions x with $T(x) \leq T$ evidently correspond to the s-t-flows of value $\sum_{i=1}^{n} t_i$ in (G, u). Indeed, these are maximum flows.

In Section 8.1 we describe a basic algorithm for the MAXIMUM FLOW PROBLEM and use it to prove the Max-Flow-Min-Cut Theorem, one of the best-known results in combinatorial optimization, which shows the relation to the problem of finding a minimum capacity s-t-cut. Moreover we show that, for integral capacities, there always exists an optimum flow which is integral. The combination of these two results also implies Menger's Theorem on disjoint paths as we discuss in Section 8.2.

Sections 8.3, 8.4 and 8.5 contain efficient algorithms for the MAXIMUM FLOW PROBLEM. Then we shift attention to the problem of finding minimum cuts. Section 8.6 describes an elegant way to store the minimum capacity of an s-t-cut (which equals the maximum value of an s-t-flow) for all pairs of vertices s and t. Section 8.7 shows how the edge-connectivity, or a minimum capacity cut in an undirected graph, can be determined more efficiently than by applying several network flow computations.

8.1 Max-Flow-Min-Cut Theorem

The definition of the MAXIMUM FLOW PROBLEM suggests the following LP formulation:

$$\max \quad \sum_{e \in \delta^+(s)} x_e - \sum_{e \in \delta^-(s)} x_e$$

$$\text{s.t.} \quad \sum_{e \in \delta^-(v)} x_e = \sum_{e \in \delta^+(v)} x_e \qquad (v \in V(G) \setminus \{s, t\})$$

$$x_e \leq u(e) \qquad (e \in E(G))$$

$$x_e \geq 0 \qquad (e \in E(G))$$

Since this LP is obviously bounded and the zero flow $f \equiv 0$ is always feasible, we have the following :

Proposition 8.2. *The* MAXIMUM FLOW PROBLEM *always has an optimum solution.* □

Furthermore, by Theorem 4.18 there exists a polynomial-time algorithm. However, we are not satisfied with this, but will rather look for a combinatorial algorithm (not using Linear Programming).

Recall that an s-t-cut in G is an edge set $\delta^+(X)$ with $s \in X$ and $t \in V(G) \setminus X$. The **capacity** of an s-t-cut is the sum of the capacities of its edges. By a minimum s-t-cut in (G, u) we mean an s-t-cut of minimum capacity (with respect to u) in G.

Lemma 8.3.   *For any $A \subseteq V(G)$ such that $s \in A, t \notin A$, and any s-t-flow f,*

(a) value $(f) = \sum_{e \in \delta^+(A)} f(e) - \sum_{e \in \delta^-(A)} f(e)$.
(b) value $(f) \leq \sum_{e \in \delta^+(A)} u(e)$.

Proof: (a): Since the flow conservation rule holds for $v \in A \setminus \{s\}$,

$$
\begin{aligned}
\text{value}\,(f) &= \sum_{e \in \delta^+(s)} f(e) - \sum_{e \in \delta^-(s)} f(e) \\
&= \sum_{v \in A} \left(\sum_{e \in \delta^+(v)} f(e) - \sum_{e \in \delta^-(v)} f(e) \right) \\
&= \sum_{e \in \delta^+(A)} f(e) - \sum_{e \in \delta^-(A)} f(e).
\end{aligned}
$$

(b): This follows from (a) by using $0 \leq f(e) \leq u(e)$ for $e \in E(G)$. $\square$

In other words, the value of a maximum flow cannot exceed the capacity of a minimum s-t-cut. In fact, we have equality here. To see this, we need the concept of augmenting paths which will reappear in several other chapters.

Definition 8.4.   *For a digraph G we define $\overleftrightarrow{G} := \left(V(G), E(G) \,\dot\cup\, \{ \overleftarrow{e} : e \in E(G) \} \right)$, where for $e = (v, w) \in E(G)$ we define $\overleftarrow{e}$ to be a new edge from w to v. We call $\overleftarrow{e}$ the **reverse edge** of e and vice versa. Note that if $e = (v, w), e' = (w, v) \in E(G)$, then $\overleftarrow{e}$ and e' are two distinct parallel edges in $\overleftrightarrow{G}$.*

Given a digraph G with capacities $u : E(G) \to \mathbb{R}_+$ and a flow f, we define **residual capacities** *$u_f : E(\overleftrightarrow{G}) \to \mathbb{R}_+$ by $u_f(e) := u(e) - f(e)$ and $u_f(\overleftarrow{e}) := f(e)$ for all $e \in E(G)$. The* **residual graph** *G_f is the graph $\left(V(G), \{ e \in E(\overleftrightarrow{G}) : u_f(e) > 0 \} \right)$.*

Given a flow f and a path (or circuit) P in G_f, to **augment** *f along P by γ means to do the following for each $e \in E(P)$: if $e \in E(G)$ then increase $f(e)$ by γ, otherwise – if $e = \overleftarrow{e_0}$ for $e_0 \in E(G)$ – decrease $f(e_0)$ by γ.*

Given a network (G, u, s, t) and an s-t-flow f, an **f-augmenting path** *is an s-t-path in the residual graph G_f.*

Using this concept, the following algorithm for the MAXIMUM FLOW PROBLEM, due to Ford and Fulkerson [1957], is natural. We first restrict ourselves to integral capacities.

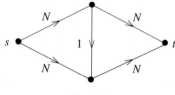

Fig. 8.1.

FORD-FULKERSON ALGORITHM

Input: A network (G, u, s, t) with $u : E(G) \rightarrow \mathbb{Z}_+$.

Output: An s-t-flow f of maximum value.

①   Set $f(e) := 0$ for all $e \in E(G)$.

②   Find an f-augmenting path P. **If** none exists **then stop**.

③   Compute $\gamma := \min\limits_{e \in E(P)} u_f(e)$. Augment f along P by γ and **go to** ②.

Edges where the minimum in ③ is attained are sometimes called bottleneck edges. The choice of γ guarantees that f continues to be a flow. Since P is an s-t-path, the flow conservation rule is preserved at all vertices except s and t.

To find an augmenting path is easy (we just have to find any s-t-path in G_f). However, we should be careful how to do this. In fact, if we allow irrational capacities (and have bad luck when choosing the augmenting paths), the algorithm might not terminate at all (Exercise 2).

Even in the case of integer capacities, we may have an exponential number of augmentations. This is illustrated by the simple network shown in Figure 8.1, where the numbers are the edge capacities ($N \in \mathbb{N}$). If we choose an augmenting path of length 3 in each iteration, we can augment the flow by just one unit each time, so we need $2N$ iterations. Observe that the input length is $O(\log N)$, since capacities are of course encoded in binary form. We shall overcome these problems in Section 8.3.

We now claim that when the algorithm stops, then f is indeed a maximum flow:

Theorem 8.5. *An s-t-flow f is maximum if and only if there is no f-augmenting path.*

Proof: If there is an augmenting path P, then ③ of the FORD-FULKERSON ALGORITHM computes a flow of greater value, so f is not maximum. If there is no augmenting path, this means that t is not reachable from s in G_f. Let R be the set of vertices reachable from s in G_f. By the definition of G_f, we have $f(e) = u(e)$ for all $e \in \delta_G^+(R)$ and $f(e) = 0$ for all $e \in \delta_G^-(R)$.

Now Lemma 8.3 (a) says that

$$\text{value}\,(f) \;=\; \sum_{e \in \delta_G^+(R)} u(e)$$

which by Lemma 8.3 (b) implies that f is a maximum flow.   □

In particular, for any maximum s-t-flow we have an s-t-cut whose capacity equals the value of the flow. Together with Lemma 8.3 (b) this yields the central result of network flow theory, the Max-Flow-Min-Cut Theorem:

Theorem 8.6. (Ford and Fulkerson [1956], Dantzig and Fulkerson [1956]) *In a network the maximum value of an s-t-flow equals the minimum capacity of an s-t-cut.*   □

An alternative proof was proposed by Elia, Feinstein and Shannon [1956]. The Max-Flow-Min-Cut Theorem also follows quite easily from LP duality; see Exercise 9 of Chapter 3.

If all capacities are integers, γ in ③ of the FORD-FULKERSON ALGORITHM is always integral. Since there is a maximum flow of finite value (Proposition 8.2), the algorithm terminates after a finite number of steps. Therefore we have the following important consequence:

Corollary 8.7. (Dantzig and Fulkerson [1956]) *If the capacities of a network are integers, then there exists an integral maximum flow.*   □

This corollary – sometimes called the Integral Flow Theorem – can also be proved easily by using the total unimodularity of the incidence matrix of a digraph (Exercise 3).

We close this section with another easy but useful observation, the Flow Decomposition Theorem:

Theorem 8.8. (Gallai [1958], Ford and Fulkerson [1962]) *Let (G, u, s, t) be a network and let f be an s-t-flow in G. Then there exists a family $\mathcal{P}$ of s-t-paths and a family $\mathcal{C}$ of circuits in G along with weights $w : \mathcal{P} \cup \mathcal{C} \to \mathbb{R}_+$ such that $f(e) = \sum_{P \in \mathcal{P} \cup \mathcal{C}:\, e \in E(P)} w(P)$ for all $e \in E(G)$, $\sum_{P \in \mathcal{P}} w(P) = \text{value}\,(f)$, and $|\mathcal{P}| + |\mathcal{C}| \le |E(G)|$.*
Moreover, if f is integral then w can be chosen to be integral.

Proof: We construct $\mathcal{P}$, $\mathcal{C}$ and w by induction on the number of edges with nonzero flow. Assume that there is an edge e^+ with $f(e^+) > 0$; otherwise the statement is trivial. Consider a maximal walk W containing e^+ in which every edge carries positive flow and no vertex appears twice except possibly if it is one of the endpoints. W contains at most n edges. Moreover, W contains a circuit P, or W is a path P. In the latter case, P begins at s, ends at t, and $f(\delta^-(s)) = f(\delta^+(t)) = 0$ (due to the maximality of W and the flow conservation rule).

Let $w(P) := \min_{e \in E(P)} f(e)$. Set $f'(e) := f(e) - w(P)$ for $e \in E(P)$ and $f'(e) := f(e)$ for $e \notin E(P)$. An application of the induction hypothesis to f' completes the proof. □

The proof also leads to an $O(mn)$-time algorithm for computing such a flow decomposition.

8.2 Menger's Theorem

Consider Corollary 8.7 and Theorem 8.8 in the special case where all capacities are 1. Here integral s-t-flows can be regarded as collections of edge-disjoint s-t-paths and circuits. We obtain the following important theorem:

Theorem 8.9. (Menger [1927]) *Let G be a graph (directed or undirected), let s and t be two vertices, and $k \in \mathbb{N}$. Then there are k edge-disjoint s-t-paths if and only if after deleting any $k - 1$ edges t is still reachable from s.*

Proof: Necessity is obvious. To prove sufficiency in the directed case, let (G, u, s, t) be a network with unit capacities $u \equiv 1$ such that t is reachable from s even after deleting any $k - 1$ edges. This implies that the minimum capacity of an s-t-cut is at least k. By the Max-Flow-Min-Cut Theorem 8.6 and Corollary 8.7 there is an integral s-t-flow of value at least k. By Theorem 8.8 this flow can be decomposed into integral flows on s-t-paths (and possibly some circuits). Since all capacities are 1 we must have at least k edge-disjoint s-t-paths.

To prove sufficiency in the undirected case, let G be an undirected graph with two vertices s and t such that t is reachable from s even after deleting any $k - 1$ edges. This property obviously remains true if we replace each undirected edge $e = \{v, w\}$ by five directed edges $(v, x_e), (w, x_e), (x_e, y_e), (y_e, v), (y_e, w)$ where x_e and y_e are new vertices (see Figure 8.2). Now we have a digraph G' and, by the first part, k edge-disjoint s-t-paths in G'. These can be easily transformed to k edge-disjoint s-t-paths in G. □

In turn it is easy to derive the Max-Flow-Min-Cut Theorem (at least for rational capacities) from Menger's Theorem. We now consider the vertex-disjoint version of Menger's Theorem. We call a set of paths **internally disjoint** if no two of

Fig. 8.2.

them have an edge or an internal vertex in common. Although they may share endpoints, internally disjoint paths are sometimes also called vertex-disjoint (if the set of endpoints is given).

Theorem 8.10. (Menger [1927]) *Let G be a graph (directed or undirected), let s and t be two non-adjacent vertices, and $k \in \mathbb{N}$. Then there are k internally disjoint s-t-paths if and only if after deleting any $k - 1$ vertices (distinct from s and t) t is still reachable from s.*

Proof: Necessity is again trivial. Sufficiency in the directed case follows from the directed part of Theorem 8.9 by the following elementary construction: we replace each vertex v of G by two vertices v' and v'' and an edge (v', v''). Each edge (v, w) of G is replaced by (v'', w'). Any set of $k - 1$ edges in the new graph G' whose deletion makes t' unreachable from s'' implies a set of at most $k - 1$ vertices in G, containing neither s nor t, whose deletion makes t unreachable from s. Moreover, edge-disjoint s''-t'-paths in the new graph correspond to internally disjoint s-t-paths in the old one.

The undirected version follows from the directed one by the same construction as in the proof of Theorem 8.9 (Figure 8.2).   □

The following corollary is an important consequence of Menger's Theorem:

Corollary 8.11. (Whitney [1932]) *An undirected graph G with at least two vertices is k-edge-connected if and only if for each pair $s, t \in V(G)$ with $s \neq t$ there are k edge-disjoint s-t-paths.*

An undirected graph G with more than k vertices is k-connected if and only if for each pair $s, t \in V(G)$ with $s \neq t$ there are k internally disjoint s-t-paths.

Proof: The first statement follows directly from Theorem 8.9.

To prove the second statement let G be an undirected graph with more than k vertices. If G has $k - 1$ vertices whose deletion makes the graph disconnected, then it cannot have k internally disjoint s-t-paths for each pair $s, t \in V(G)$.

Conversely, if G does not have k internally disjoint s-t-paths for some $s, t \in V(G)$, then we consider two cases. If s and t are non-adjacent, then by Theorem 8.10 G has $k - 1$ vertices whose deletion separates s and t.

If s and t are joined by a set F of parallel edges, $|F| \geq 1$, then $G - F$ has no $k - |F|$ internally disjoint s-t-paths, so by Theorem 8.10 it has a set X of $k - |F| - 1$ vertices whose deletion separates s and t. Let $v \in V(G) \setminus (X \cup \{s, t\})$. Then v cannot be reachable from s and from t in $(G - F) - X$, say v is not reachable from s. Then v and s are in different connected components of $G - (X \cup \{t\})$.   □

In many applications one looks for edge-disjoint or vertex-disjoint (or internally disjoint) paths between several pairs of vertices. The four versions of Menger's

Theorem (directed and undirected, vertex- and edge-disjoint) correspond to four versions of the DISJOINT PATHS PROBLEM:

DIRECTED/UNDIRECTED EDGE-/VERTEX-DISJOINT PATHS PROBLEM

Instance: Two directed/undirected graphs (G, H) on the same vertices.

Task: Find a family $(P_f)_{f \in E(H)}$ of edge-disjoint/internally disjoint paths in G such that for each $f = (t, s)$ or $f = \{t, s\}$ in H, P_f is an s-t-path.

Such a family is called a **solution** of (G, H). We say that P_f **realizes** f. The edges of G are called **supply edges**, the edges of H **demand edges**. A vertex incident to some demand edge is called a **terminal**.

Above we considered the special case when $E(H)$ is just a set of k parallel edges. The general DISJOINT PATHS PROBLEM will be discussed in Chapter 19. Here we only note the following useful special case of Menger's Theorem:

Proposition 8.12. *Let (G, H) be an instance of the* DIRECTED EDGE-DISJOINT PATHS PROBLEM *where $E(H)$ is just a set of parallel edges and $G + H$ is Eulerian. Then (G, H) has a solution.*

Proof: Since $G + H$ is Eulerian, every edge, in particular any $f \in E(H)$, belongs to some circuit C. We take $C - f$ as the first path of our solution, delete C, and apply induction.   □

8.3 The Edmonds-Karp Algorithm

In Exercise 2 it is shown that it is necessary to make ② of the FORD-FULKERSON ALGORITHM more precise. Instead of choosing an arbitrary augmenting path it is a good idea to look for a shortest one, i.e. an augmenting path with a minimum number of edges. With this simple idea Edmonds and Karp [1972] obtained the first polynomial-time algorithm for the MAXIMUM FLOW PROBLEM.

EDMONDS-KARP ALGORITHM

Input: A network (G, u, s, t).

Output: An s-t-flow f of maximum value.

①   Set $f(e) := 0$ for all $e \in E(G)$.

②   Find a shortest f-augmenting path P. **If there is none then stop.**

③   Compute $\gamma := \min_{e \in E(P)} u_f(e)$. Augment f along P by γ and **go to** ②.

This means that ② of the FORD-FULKERSON ALGORITHM should be implemented by BFS (see Section 2.3).

Lemma 8.13. *Let $f_1, f_2, \ldots$ be a sequence of flows such that f_{i+1} results from f_i by augmenting along P_i, where P_i is a shortest f_i-augmenting path. Then*

(a) $|E(P_k)| \leq |E(P_{k+1})|$ *for all k.*
(b) $|E(P_k)| + 2 \leq |E(P_l)|$ *for all $k < l$ such that $P_k \cup P_l$ contains a pair of reverse edges.*

Proof: (a): Consider the graph G_1 which results from $P_k \cup P_{k+1}$ by deleting pairs of reverse edges. (Edges appearing both in P_k and P_{k+1} appear twice in G_1.) Every simple subgraph of G_1 is a subgraph of G_{f_k}, since any edge in $E(G_{f_{k+1}}) \setminus E(G_{f_k})$ must be the reverse of an edge in P_k.

Let H_1 simply consist of two copies of (t, s). Obviously $G_1 + H_1$ is Eulerian. Thus by Proposition 8.12 there are two edge-disjoint s-t-paths Q_1 and Q_2. Since $E(G_1) \subseteq E(G_{f_k})$, both Q_1 and Q_2 are f_k-augmenting paths. Since P_k was a shortest f_k-augmenting path, $|E(P_k)| \leq |E(Q_1)|$ and $|E(P_k)| \leq |E(Q_2)|$. Thus,

$$2|E(P_k)| \leq |E(Q_1)| + |E(Q_2)| \leq |E(G_1)| \leq |E(P_k)| + |E(P_{k+1})|,$$

implying $|E(P_k)| \leq |E(P_{k+1})|$.

(b): By part (a) it is enough to prove the statement for those k, l such that for $k < i < l$, $P_i \cup P_l$ contains no pair of reverse edges.

As above, consider the graph G_1 which results from $P_k \cup P_l$ by deleting pairs of reverse edges. Again, we claim that every simple subgraph of G_1 is a subgraph of G_{f_k}. To see this, observe that $E(P_k) \subseteq E(G_{f_k})$, $E(P_l) \subseteq E(G_{f_l})$, and any edge of $E(G_{f_l}) \setminus E(G_{f_k})$ must be the reverse of an edge in one of $P_k, P_{k+1}, \ldots, P_{l-1}$. But – due to the choice of k and l – among these paths only P_k contains the reverse of an edge in P_l.

Let H_1 again consist of two copies of (t, s). Since $G_1 + H_1$ is Eulerian, Proposition 8.12 guarantees that there are two edge-disjoint s-t-paths Q_1 and Q_2. Again Q_1 and Q_2 are both f_k-augmenting. Since P_k was a a shortest f_k-augmenting path, $|E(P_k)| \leq |E(Q_1)|$ and $|E(P_k)| \leq |E(Q_2)|$. We conclude that

$$2|E(P_k)| \leq |E(Q_1)| + |E(Q_2)| \leq |E(P_k)| + |E(P_l)| - 2$$

(since we have deleted at least two edges). This completes the proof. $\square$

Theorem 8.14. (Edmonds and Karp [1972]) *Regardless of the edge capacities, the* EDMONDS-KARP ALGORITHM *stops after at most $\frac{mn}{2}$ augmentations, where m and n denote the number of edges and vertices, respectively.*

Proof: Let $P_1, P_2, \ldots$ be the augmenting paths chosen during the EDMONDS-KARP ALGORITHM. By the choice of γ in ③ of the algorithm, each augmenting path contains at least one bottleneck edge.

For any edge e, let $P_{i_1}, P_{i_2}, \ldots$ be the subsequence of augmenting paths containing e as a bottleneck edge. Obviously, between P_{i_j} and $P_{i_{j+1}}$ there must be an augmenting path P_k $(i_j < k < i_{j+1})$ containing $\overleftarrow{e}$. By Lemma 8.13 (b),

$|E(P_{i_j})| + 4 \le |E(P_k)| + 2 \le |E(P_{i_{j+1}})|$ for all j. If e has neither s nor t as endpoint, we have $3 \le |E(P_{i_j})| \le n - 1$ for all j, and there can be at most $\frac{n}{4}$ augmenting paths containing e as a bottleneck edge. Otherwise at most one of the augmenting paths contains e or $\overleftarrow{e}$ as bottleneck edge.

Since any augmenting path must contain at least one edge of $\overleftrightarrow{G}$ as a bottleneck edge, there can be at most $|E(\overleftrightarrow{G})|\frac{n}{4} = \frac{mn}{2}$ augmenting paths. $\qquad\square$

Corollary 8.15. *The* EDMONDS-KARP ALGORITHM *solves the* MAXIMUM FLOW PROBLEM *in* $O(m^2n)$ *time.*

Proof: By Theorem 8.14 there are at most $\frac{mn}{2}$ augmentations. Each augmentation uses BFS and thus takes $O(m)$ time. $\qquad\square$

8.4 Dinic's, Karzanov's, and Fujishige's Algorithm

Around the time when Edmonds and Karp observed how to obtain a polynomial-time algorithm for the MAXIMUM FLOW PROBLEM, Dinic [1970] independently found an even better algorithm. It is based on the following definition:

Definition 8.16. *Given a network* (G, u, s, t) *and an* s-t-*flow* f. *The* **level graph** G_f^L *of* G_f *is the digraph*

$$\left(V(G), \left\{e = (x, y) \in E(G_f) : \text{dist}_{G_f}(s, x) + 1 = \text{dist}_{G_f}(s, y)\right\}\right).$$

Note that the level graph is acyclic. The level graph can be constructed easily by BFS in $O(m)$ time. The s-t-paths in G_f^L are precisely the shortest s-t-paths in G_f.

Lemma 8.13(a) says that the length of the shortest augmenting paths in the EDMONDS-KARP ALGORITHM is non-decreasing. Let us call a sequence of augmenting paths of the same length a **phase** of the algorithm. Let f be the flow at the beginning of a phase. The proof of Lemma 8.13 (b) yields that all augmenting paths of this phase must already be augmenting paths in G_f. Therefore all these paths must be s-t-paths in the level graph of G_f. The total of all augmentations in a phase can be regarded as a blocking flow in G_f^L:

Definition 8.17. *Given a network* (G, u, s, t), *an* s-t-*flow* f *is called* **blocking** *if* $(V(G), \{e \in E(G) : f(e) < u(e)\})$ *contains no* s-t-*path.*

Note that a blocking flow is not necessarily maximum. The above considerations suggest the following algorithmic scheme:

DINIC'S ALGORITHM

Input: A network (G, u, s, t).

Output: An s-t-flow f of maximum value.

---

①   Set $f(e) := 0$ for all $e \in E(G)$.

②   Construct the level graph G_f^L of G_f.

③   Find a blocking s-t-flow f' in (G_f^L, u_f). **If $f' = 0$ then stop.**

④   Augment f by f' and **go to** ②.

Augmenting f by f' of course means increasing $f(e)$ by $f'(e)$ for each $e \in E(G_f^L) \cap E(G)$ and decreasing $f(e)$ by $f'(\overleftarrow{e})$ for each $e \in E(G)$ with $\overleftarrow{e} \in E(G_f^L)$.

Theorem 8.18. (Dinic [1970]) DINIC'S ALGORITHM *works correctly and stops after at most n iterations.*

Proof: If the algorithm stops, there is no s-t-path in G_f^L, and hence in G_f. It remains to prove that the length of a shortest augmenting path (which must be in $\{1, 2, \ldots, n - 1, \infty\}$) increases in each iteration. Consider the flow f at the beginning of some iteration, and let $\bar{f}$ be the augmented flow f after ④. Note that $\text{dist}_{G_f}(s, y) \leq \text{dist}_{G_f}(s, x)$ for all $(x, y) \in E(G_f) \setminus E(G_f^L)$ and also for reverse edges (x, y) of edges in G_f^L, while $\text{dist}_{G_f}(s, y) = \text{dist}_{G_f}(s, x) + 1$ for all $e = (x, y) \in E(G_f^L)$. Since each augmenting path with respect to $\bar{f}$ contains only edges from $E(G_f) \cup \{\overleftarrow{e} : e \in E(G_f^L)\}$, and at least one edge that is not in G_f^L, it must contain more than $\text{dist}_{G_f}(s, t)$ edges.   □

Note that ② (using BFS) and ④ can be implemented in linear time. So it remains to show how a blocking flow in an acyclic digraph can be found efficiently. Dinic obtained an $O(nm)$ bound for each phase, which is not very difficult to show (Exercise 20). We shall now describe Karzanov's faster algorithm. It is based on the following important definition:

Definition 8.19. (Karzanov [1974]) *Given a network (G, u, s, t), an s-t-**preflow** is a function $f : E(G) \to \mathbb{R}_+$ satisfying $f(e) \leq u(e)$ for all $e \in E(G)$ and $\text{ex}_f(v) \geq 0$ for all $v \in V(G) \setminus \{s\}$. We call a vertex $v \in V(G) \setminus \{s, t\}$ **active** if $\text{ex}_f(v) > 0$.*

Obviously, an s-t-preflow is an s-t-flow if and only if there are no active vertices. This concept will be used again in the next section.

Theorem 8.20. (Karzanov [1974]) *A blocking flow in a network (G, u, s, t) with an acyclic digraph G can be found in $O(n^2)$ time, where $n = |V(G)|$.*

Proof: First compute a topological order $V(G) = \{v_1, v_2, \ldots, v_n\}$ (cf. Theorem 2.20). For each vertex v except s and t, the algorithm maintains $\text{ex}_f(v)$, the list of its leaving edges, and a stack, initially empty, whose elements are pairs from $\delta^-(v) \times \mathbb{R}_+$. All vertices are marked non-frozen initially.

The algorithm maintains an s-t-preflow f throughout. We begin by setting $f(e) := u(e)$ for each edge $e = (s, v) \in \delta^+(s)$ and putting $(e, u(e))$ on the stack of v. For all other edges e we set $f(e) := 0$ initially. Then the following two steps alternate until there is no active vertex anymore:

The push step scans the vertices in topological order. While a vertex v is active and there is an edge $e = (v, w)$ with $f(e) < u(e)$ and w is not frozen, increase $f(e)$ by $\delta := \min\{u(e) - f(e), \mathrm{ex}_f(v)\}$ and put (e, δ) on the stack of w.

The balancing step deals with the active vertex v_i for which i is maximum. We remove the topmost pair (e, δ) from the stack of v_i (i.e., the one inserted last) and decrease $f(e)$ by $\delta' := \min\{\delta, \mathrm{ex}_f(v_i)\}$. If v_i is still active, we continue with the next pair from the stack. Finally, v_i is inactive and will be marked frozen.

If a vertex v_i is subject to balancing, then at that time all vertices v_j with $j > i$ are inactive, and so their entering edges will never carry less flow anymore than now. This is because flow is reduced during rebalancing according to the stack: first on edges where the flow was increased last. Hence v_i will never become active anymore. Thus each vertex is subject to balancing at most once, and the number of iterations is less than n. Each push step runs in $O(n + p)$ time, where p is the number of edges that receive a saturating push. Note that an edge that is saturated will not be considered anymore by a push step. Hence the total time for all push steps is $O(n^2 + m)$, and so is the total time for rebalancing.

At any stage, t is not reachable in $(V(G), \{e \in E(G) : f(e) < u(e)\})$ from s or from any frozen vertex. Hence f is a blocking flow at termination. $\square$

We give another proof, due to Malhotra, Kumar and Maheshwari [1978] :

Second Proof of Theorem 8.20:   First compute a topological order of G (cf. Theorem 2.20). For $v \in V(G)$ we write $G_{\leq v}$ and $G_{\geq v}$ for the subgraphs induced by all vertices up to v and from v on, respectively.

We start with $f(e) = 0$ for all $e \in E(G)$. Let

$$\alpha := \min\left\{\min\{u_f(\delta_G^-(v)) : v \in V(G) \setminus \{s\}\}, \min\{u_f(\delta_G^+(v)) : v \in V(G) \setminus \{t\}\}\right\},$$

and let v be a vertex where the minimum is attained.

We first find a v-t-flow g of value α in $G_{\geq v}$ (unless $v = t$). We do this by scanning the vertices in topological order. For each vertex w, where $w = v$ or predecessors have been processed already, set $\beta := \alpha$ if $w = v$ and $\beta := g(\delta_G^-(w))$ otherwise, let $\delta_G^+(w) = \{e_1, \ldots, e_k\}$, and set $g(e_j) := \min\{u_f(e_j), \beta - \sum_{i=1}^{j-1} g(e_i)\}$ for $j = 1, \ldots, k$. Similarly, we find an s-v-flow g' of value α in $G_{\leq v}$ (unless $v = s$) by scanning vertices in reverse topological order. We set $f(e) := f(e) + g(e) + g'(e)$ for all $e \in E(G)$. If $v \in \{s, t\}$, we stop: f is a blocking flow. If $v \notin \{s, t\}$, we delete v and its incident edges and iterate.

This algorithm stops after at most $n - 1$ iterations and is obviously correct. It can be implemented such that each edge is scanned at most once after being saturated. In each iteration it suffices to scan at most $n - 2$ edges (one at each vertex) in addition to those that are being saturated in this iteration. By always updating the current values of $u_f(\delta_G^-(v))$ and $u_f(\delta_G^+(v))$, we can find α and v in each iteration in $O(n)$ time. The running time of $O(m + n^2)$ follows. $\square$

Corollary 8.21. *There is an $O(n^3)$-time algorithm for the* MAXIMUM FLOW PROBLEM.

Proof: Use Theorem 8.20 to implement ③ of DINIC'S ALGORITHM. $\square$

Subsequent improvements are due to Cherkassky [1997], Galil [1980], Galil and Namaad [1980], Shiloach [1978], Sleator [1980], and Sleator and Tarjan [1983]. The last two references describe an $O(m \log n)$-algorithm for finding blocking flows in an acyclic network using a data structure called dynamic trees. Using this as a subroutine of DINIC'S ALGORITHM one has an $O(mn \log n)$-algorithm for the MAXIMUM FLOW PROBLEM. However, we do not describe any of the above-mentioned algorithms here (see Tarjan [1983]), because an even faster network flow algorithm will be the subject of the next section.

We close this section by describing the weakly polynomial algorithm by Fujishige [2003], mainly because of its simplicity:

FUJISHIGE'S ALGORITHM

Input: A network (G, u, s, t) with $u : E(G) \to \mathbb{Z}_+$.

Output: An s-t-flow f of maximum value.

①   Set $f(e) := 0$ for all $e \in E(G)$. Set $\alpha := \max\{u(e) : e \in E(G)\}$.

②   Set $i := 1$, $v_1 := s$, $X := \emptyset$, and $b(v) := 0$ for all $v \in V(G)$.

③   **For** $e = (v_i, w) \in \delta_{G_f}^+(v_i)$ with $w \notin \{v_1, \ldots, v_i\}$ **do:**
      Set $b(w) := b(w) + u_f(e)$. **If** $b(w) \geq \alpha$ **then** set $X := X \cup \{w\}$.

④   **If** $X = \emptyset$ **then:**
      Set $\alpha := \lfloor \frac{\alpha}{2} \rfloor$. **If** $\alpha = 0$ **then stop else** go to ②.

⑤   Set $i := i + 1$. Choose $v_i \in X$ and set $X := X \setminus \{v_i\}$.
      If $v_i \neq t$ **then** go to ③.

⑥   Set $\beta(t) := \alpha$ and $\beta(v) := 0$ for all $v \in V(G) \setminus \{t\}$.
      While $i > 1$ **do:**
          For $e = (p, v_i) \in \delta_{G_f}^-(v_i)$ with $p \in \{v_1, \ldots, v_{i-1}\}$ **do:**
              Set $\beta' := \min\{\beta(v_i), u_f(e)\}$.
              Augment f along e by β'.
              Set $\beta(v_i) := \beta(v_i) - \beta'$ and $\beta(p) := \beta(p) + \beta'$.
          Set $i := i - 1$.

⑦   **Go to** ②.

Theorem 8.22. FUJISHIGE'S ALGORITHM *correctly solves the* MAXIMUM FLOW PROBLEM *for simple digraphs G and integral capacities $u : E(G) \to \mathbb{Z}_+$ in $O(mn \log u_{\max})$ time, where $n := |V(G)|$, $m := |E(G)|$ and $u_{\max} := \max\{u(e) : e \in E(G)\}$.*

Proof: Let us call an iteration a sequence of steps ending with ④ or ⑦. In ②–⑤, $v_1, \ldots, v_i$ is always an order of a subset of vertices such that $b(v_j) = u_f(E^+(\{v_1, \ldots, v_{j-1}\}, \{v_j\})) \geq \alpha$ for $j = 2, \ldots, i$. In ⑥ the flow f is augmented with the invariant $\sum_{v \in V(G)} \beta(v) = \alpha$, and by the above the result is an s-t-flow whose value is α units larger.

Thus after at most $n - 1$ iterations, α will be decreased for the first time. When we decrease α to $\alpha' = \lfloor \frac{\alpha}{2} \rfloor \geq \frac{\alpha}{3}$ in ④, we have an s-t-cut $\delta_{G_f}^+(\{v_1, \ldots, v_i\})$ in G_f of capacity less than $\alpha(|V(G)| - i)$ because $b(v) = u_f(E^+(\{v_1, \ldots, v_i\}, \{v\})) < \alpha$ for all $v \in V(G) \setminus \{v_1, \ldots, v_i\}$. By Lemma 8.3(b), a maximum s-t-flow in G_f has value less than $\alpha(n - i) < 3\alpha'n$. Hence after less than $3n$ iterations, α will be decreased again. If α is decreased from 1 to 0, we have an s-t-cut of capacity 0 in G_f, so f is maximum.

As α is decreased at most $1 + \log u_{\max}$ times before it reaches 0, and each iteration takes $O(m)$ time, the overall running time is $O(mn \log u_{\max})$.   □

Such a scaling technique is useful in many contexts and will reappear in Chapter 9. Fujishige [2003] also described a variant of his algorithm without scaling, where v_i in ⑤ is chosen as a vertex attaining $\max\{b(v) : v \in V(G) \setminus \{v_1, \ldots, v_{i-1}\}\}$. The resulting order is called MA order and will reappear in Section 8.7. The running time of this variant is slightly higher than the above and not strongly polynomial either (Shioura [2004]). See Exercise 25.

8.5 The Goldberg-Tarjan Algorithm

In this section we shall describe the PUSH-RELABEL ALGORITHM due to Goldberg and Tarjan [1988]. We shall derive an $O(n^2 \sqrt{m})$ bound for the running time.

Sophisticated implementations using dynamic trees (see Sleator and Tarjan [1983]) result in network flow algorithms with running time $O\left(nm \ \log \frac{n^2}{m}\right)$ (Goldberg and Tarjan [1988]) and $O\left(nm \ \log\left(\frac{n}{m}\sqrt{\log u_{\max}} + 2\right)\right)$, where $u_{\max}$ is the maximum (integral) edge capacity (Ahuja, Orlin and Tarjan [1989]). The best known bounds today are $O(nm)$ (Orlin [2013]),

$$O\left(\min\{m^{1/2}, n^{2/3}\}m \log\left(\frac{n^2}{m}\right) \log u_{\max}\right)$$

(Goldberg and Rao [1998]), and $O(m\sqrt{n} \log^{O(1)}(n) \log^2 u_{\max})$ (Lee and Sidford [2014]).

By definition and Theorem 8.5, a flow f is a maximum s-t-flow if and only if the following conditions hold:

- $\mathrm{ex}_f(v) = 0$ for all $v \in V(G) \setminus \{s, t\}$;
- There is no f-augmenting path.

In the algorithms discussed so far, the first condition is always satisfied, and the algorithms stop when the second condition is satisfied. The PUSH-RELABEL ALGORITHM starts with an f satisfying the second condition and maintains it throughout. Naturally it stops when the first condition is satisfied as well. So f will not be an s-t-flow during the algorithm (except at termination), but an s-t-preflow (cf. Definition 8.19).

Definition 8.23. *Let (G, u, s, t) be a network and f an s-t-preflow. A **distance labeling** is a function $\psi : V(G) \to \mathbb{Z}_+$ such that $\psi(t) = 0$, $\psi(s) = n := |V(G)|$ and $\psi(v) \leq \psi(w) + 1$ for all $(v, w) \in E(G_f)$. An edge $e = (v, w) \in E(\overleftrightarrow{G})$ is called **admissible** if $e \in E(G_f)$ and $\psi(v) = \psi(w) + 1$.*

If ψ is a distance labeling, $\psi(v)$ (for $v \neq s$) must be a lower bound on the distance to t (number of edges in a shortest v-t-path) in G_f.

The PUSH-RELABEL ALGORITHM to be described below always works with an s-t-preflow f and a distance labeling ψ. It starts with the preflow that is equal to the capacity on each edge leaving s and zero on all other edges. The initial distance labeling is $\psi(s) = n$ and $\psi(v) = 0$ for all $v \in V(G) \setminus \{s\}$.

Then the algorithm performs the update operations PUSH (updating f) and RELABEL (updating ψ) in any order.

PUSH-RELABEL ALGORITHM

Input: A network (G, u, s, t).

Output: A maximum s-t-flow f.

---

①   Set $f(e) := u(e)$ for each $e \in \delta^+(s)$.
  Set $f(e) := 0$ for each $e \in E(G) \setminus \delta^+(s)$.

②   Set $\psi(s) := n := |V(G)|$ and $\psi(v) := 0$ for all $v \in V(G) \setminus \{s\}$.

③   **While** there exists an active vertex **do**:
  Let v be an active vertex.
  If no $e \in \delta_{G_f}^+(v)$ is admissible
    then RELABEL(v),
    else let $e \in \delta_{G_f}^+(v)$ be an admissible edge and PUSH(e).

PUSH(e)

①   Set $\gamma := \min\{\mathrm{ex}_f(v), u_f(e)\}$, where v is the tail of e.

②   Augment f along e by γ.

RELABEL(v)

①   Set $\psi(v) := \min\{\psi(w) + 1 : (v, w) \in \delta_{G_f}^+(v)\}$.

Proposition 8.24. *During the execution of the* PUSH-RELABEL ALGORITHM f *is always an s-t-preflow and ψ is always a distance labeling with respect to f. For each $v \in V(G)$, $\psi(v)$ is strictly increased by every* RELABEL(v).

Proof: We have to show that the procedures PUSH and RELABEL preserve these properties. It is clear that after a PUSH operation, f is still an s-t-preflow. A RELABEL operation does not even change f.

If RELABEL(v) is called and ψ was a distance labeling before, then $\psi(v)$ is strictly increased (as no $e \in \delta_{G_f}^+(v)$ was admissible), and ψ remains a distance labeling.

We finally show that after a PUSH operation, ψ is still a distance labeling with respect to the new preflow. We have to check $\psi(a) \le \psi(b) + 1$ for all new edges (a, b) in G_f. But if we apply PUSH(e) for some $e = (v, w)$, the only possible new edge in G_f is the reverse edge of e, and here we have $\psi(w) = \psi(v) - 1$, since e is admissible. $\square$

Lemma 8.25. *If f is an s-t-preflow and ψ is a distance labeling with respect to f, then:*

(a) *s is reachable from any active vertex v in G_f.*
(b) *If w is reachable from v in G_f for some $v, w \in V(G)$, then $\psi(v) \le \psi(w)+n-1$.*
(c) *t is not reachable from s in G_f.*

Proof: (a): Let v be an active vertex, and let R be the set of vertices reachable from v in G_f. Then $f(e) = 0$ for all $e \in \delta_G^-(R)$. So

$$\sum_{w \in R} \mathrm{ex}_f(w) \;=\; \sum_{e \in \delta_G^-(R)} f(e) - \sum_{e \in \delta_G^+(R)} f(e) \;\le\; 0.$$

But v is active, meaning $\mathrm{ex}_f(v) > 0$, and therefore there must exist a vertex $w \in R$ with $\mathrm{ex}_f(w) < 0$. Since f is an s-t-preflow, this vertex must be s.

(b): Suppose there is a v-w-path in G_f, say with vertices $v = v_0, v_1, \ldots, v_k = w$. Since ψ is a distance labeling with respect to f, $\psi(v_i) \le \psi(v_{i+1}) + 1$ for $i = 0, \ldots, k - 1$. So $\psi(v) \le \psi(w) + k$. Note that $k \le n - 1$.

(c): follows from (b) as $\psi(s) = n$ and $\psi(t) = 0$. $\square$

Part (c) helps us to prove the following:

Theorem 8.26. *When the algorithm terminates, f is a maximum s-t-flow.*

Proof: f is an s-t-flow because there are no active vertices. Lemma 8.25(c) implies that there is no augmenting path. Then by Theorem 8.5 we know that f is maximum. $\square$

The question now is how many PUSH and RELABEL operations are performed.

Lemma 8.27.

(a) *For each $v \in V(G)$, $\psi(v)$ never decreases, and $\psi(v) \leq 2n - 1$ at any stage of the algorithm.*

(b) *No vertex is relabelled more than $2n - 1$ times. The total increase of $\sum_{v \in V(G)} \psi(v)$ during the algorithm is at most $2n^2 - n$.*

Proof: Recall from Proposition 8.24 that $\psi(v)$ is strictly increased by every RELABEL(v). Moreover, we only change $\psi(v)$ by RELABEL(v) if v is active. By Lemma 8.25(a) and (b), $\psi(v) \leq \psi(s) + n - 1 = 2n - 1$. This implies (a) and (b). $\square$

We shall now analyse the number of PUSH operations. We distinguish between **saturating** pushes (where $u_f(e) = 0$ after the push) and **nonsaturating** pushes. As usual we denote $m := |E(G)|$ (and $n := |V(G)|$).

Lemma 8.28. *The number of saturating pushes is at most $2mn$.*

Proof: After each saturating push from v to w, another such push cannot occur until $\psi(w)$ increases by at least 2, a push from w to v occurs, and $\psi(v)$ increases by at least 2. Together with Lemma 8.27(a), this proves that there are at most n saturating pushes on each edge $(v, w) \in E(\overset{\leftrightarrow}{G})$. $\square$

The number of nonsaturating pushes can be in the order of $n^2 m$ in general (Exercise 26). By choosing an active vertex v with $\psi(v)$ maximum in ③ we can prove a better bound. We may assume $n \leq m \leq n^2$.

Lemma 8.29. *If we always choose v to be an active vertex with $\psi(v)$ maximum in ③ of the PUSH-RELABEL ALGORITHM, the number of nonsaturating pushes is at most $8n^2 \sqrt{m}$.*

Proof: Call a phase the time between two consecutive changes of $\psi^* := \max\{\psi(v) : v$ active$\}$. As ψ^* can increase only by relabeling, its total increase is less than $2n^2$. As $\psi^* = 0$ initially, it decreases less than $2n^2$ times, and the number of phases is less than $4n^2$.

Call a phase cheap if it contains at most $\sqrt{m}$ nonsaturating pushes and expensive otherwise. Clearly there are at most $4n^2 \sqrt{m}$ nonsaturating pushes in cheap phases. Let

$$\Phi := \sum_{v \in V(G): v \text{ active}} |\{w \in V(G) : \psi(w) \leq \psi(v)\}|.$$

Initially $\Phi \leq n^2$. A relabeling step may increase Φ by at most n. A saturating push may increase Φ by at most n. A nonsaturating push does not increase Φ. Since $\Phi = 0$ at termination, the total decrease of Φ is at most $n^2 + n(2n^2 - n) + n(2mn) \leq 4mn^2$.

Now consider the nonsaturating pushes in an expensive phase. Each of them pushes flow along an edge (v, w) with $\psi(v) = \psi^* = \psi(w) + 1$, deactivating v and possibly activating w.

As the phase ends by deactivating the last active vertex v with $\psi(v) = \psi^*$ or by relabeling, the set of vertices w with $\psi(w) = \psi^*$ remains constant during the phase, and it contains more than $\sqrt{m}$ vertices as the phase is expensive. Hence each nonsaturating push in an expensive phase decreases Φ by at least $\sqrt{m}$. Thus the total number of nonsaturating pushes in expensive phases is at most $\frac{4mn^2}{\sqrt{m}} = 4n^2\sqrt{m}$.

$\square$

This proof is due to Cheriyan and Mehlhorn [1999]. We finally get:

Theorem 8.30. (Goldberg and Tarjan [1988], Cheriyan and Maheshwari [1989], Tunçel [1994]) *The* PUSH-RELABEL ALGORITHM *solves the* MAXIMUM FLOW PROBLEM *correctly and can be implemented to run in* $O(n^2\sqrt{m})$ *time.*

Proof: The correctness follows from Theorem 8.26.

As in Lemma 8.29 we always choose v in ③ to be an active vertex with $\psi(v)$ maximum. To make this easy we keep track of doubly-linked lists $L_0, \ldots, L_{2n-1}$, where L_i contains the active vertices v with $\psi(v) = i$. These lists can be updated during each PUSH and RELABEL operation in constant time.

We can then start by scanning L_i for $i = 0$. When a vertex is relabelled, we increase i accordingly. When we find a list L_i for the current i empty (after deactivating the last active vertex at that level), we decrease i until L_i is nonempty. As we increase i at most $2n^2$ times by Lemma 8.27(b), we also decrease i at most $2n^2$ times.

As a second data structure, we store a doubly-linked list A_v containing the admissible edges leaving v for each vertex v. They can also be updated in each PUSH operation in constant time, and in each RELABEL operation in time proportional to the total number of edges incident to the relabelled vertex.

So RELABEL(v) takes a total of $O(|\delta_G(v)|)$ time, and by Lemma 8.27(b) the overall time for relabelling is $O(mn)$. Each PUSH takes constant time, and by Lemma 8.28 and Lemma 8.29 the total number of pushes is $O(n^2\sqrt{m})$. $\square$

8.6 Gomory-Hu Trees

Any algorithm for the MAXIMUM FLOW PROBLEM also implies a solution to the following problem:

MINIMUM CAPACITY CUT PROBLEM

Instance: A network (G, u, s, t).

Task: An s-t-cut in G with minimum capacity.

Proposition 8.31.   *The* MINIMUM CAPACITY CUT PROBLEM *can be solved in the same running time as the* MAXIMUM FLOW PROBLEM, *in particular in* $O(n^2\sqrt{m})$ *time.*

Proof: For a network (G, u, s, t) we compute a maximum s-t-flow f and define X to be the set of all vertices reachable from s in G_f. X can be computed with the GRAPH SCANNING ALGORITHM in linear time (Proposition 2.17). By Lemma 8.3 and Theorem 8.5, $\delta_G^+(X)$ constitutes a minimum capacity s-t-cut. The $O(n^2 \sqrt{m})$ running time follows from Theorem 8.30 (and is not best possible). $\qquad \square$

In this section we consider the problem of finding a minimum capacity s-t-cut for each pair of vertices s, t in an undirected graph G with capacities $u : E(G) \to \mathbb{R}_+$.

This problem can be reduced to the above one: For all pairs $s, t \in V(G)$ we solve the MINIMUM CAPACITY CUT PROBLEM for (G', u', s, t), where (G', u') arises from (G, u) by replacing each undirected edge $\{v, w\}$ by two oppositely directed edges (v, w) and (w, v) with $u'((v, w)) = u'((w, v)) = u(\{v, w\})$. In this way we obtain minimum s-t-cuts for all s, t after $\binom{n}{2}$ flow computations.

This section is devoted to the elegant method of Gomory and Hu [1961], which requires only $n - 1$ flow computations. We shall see some applications in Sections 12.3 and 20.4.

Definition 8.32. *Let G be an undirected graph and $u : E(G) \to \mathbb{R}_+$ a capacity function. For two vertices $s, t \in V(G)$ we denote by λ_{st} their **local edge-connectivity**, i.e. the minimum capacity of a cut separating s and t.*

The edge-connectivity of a graph is obviously the minimum local edge-connectivity with respect to unit capacities.

Lemma 8.33. *For all vertices $i, j, k \in V(G)$ we have $\lambda_{ik} \geq \min\{\lambda_{ij}, \lambda_{jk}\}$.*

Proof: Let $\delta(A)$ be a cut with $i \in A$, $k \in V(G) \setminus A$ and $u(\delta(A)) = \lambda_{ik}$. If $j \in A$ then $\delta(A)$ separates j and k, so $u(\delta(A)) \geq \lambda_{jk}$. If $j \in V(G) \setminus A$ then $\delta(A)$ separates i and j, so $u(\delta(A)) \geq \lambda_{ij}$. We conclude that $\lambda_{ik} = u(\delta(A)) \geq \min\{\lambda_{ij}, \lambda_{jk}\}$. $\qquad \square$

Indeed, this condition is not only necessary but also sufficient for numbers $(\lambda_{ij})_{1 \leq i, j \leq n}$ with $\lambda_{ij} = \lambda_{ji}$ to be local edge-connectivities of some graph (Exercise 32).

Definition 8.34. *Let G be an undirected graph and $u : E(G) \to \mathbb{R}_+$ a capacity function. A tree T is called a **Gomory-Hu tree** for (G, u) if $V(T) = V(G)$ and*

$$\lambda_{st} = \min_{e \in E(P_{st})} u(\delta_G(C_e)) \quad \text{for all } s, t \in V(G),$$

where P_{st} is the (unique) s-t-path in T and, for $e \in E(T)$, C_e and $V(G) \setminus C_e$ are the connected components of $T - e$ (i.e. $\delta_G(C_e)$ is the fundamental cut of e with respect to T).

We shall see that every undirected graph possesses a Gomory-Hu tree. This implies that for any undirected graph G there is a list of $n - 1$ cuts such that for each pair $s, t \in V(G)$ a minimum s-t-cut belongs to the list. This is not true for digraphs: for each $n \in \mathbb{N}$, Jelinek and Mayeda [1963] constructed a digraph G

with n vertices and capacities $u : E(G) \to \mathbb{R}_+$ such that the set $\{\min\{u(\delta^+(X))\} :$ $s \in X \subseteq V(G) \setminus \{t\}\} : s, t \in V(G), s \neq t\}$ contains $(n + 2)(n - 1)/2$ different numbers.

In general, a Gomory-Hu tree cannot be chosen as a subgraph of G. For example, consider $G = K_{3,3}$ and $u \equiv 1$. Here $\lambda_{st} = 3$ for all $s, t \in V(G)$. It is easy to see that the Gomory-Hu trees for (G, u) are exactly the stars with five edges.

The main idea of the algorithm for constructing a Gomory-Hu tree is as follows. First we choose any $s, t \in V(G)$ and find some minimum s-t-cut, say $\delta(A)$. Let $B := V(G) \setminus A$. Then we contract A (or B) to a single vertex, choose any $s', t' \in B$ (or $s', t' \in A$, respectively) and look for a minimum s'-t'-cut in the contracted graph G'. We continue this process, always choosing a pair s', t' of vertices not separated by any cut obtained so far. At each step, we contract – for each cut $E(A', B')$ obtained so far – A' or B', depending on which part does not contain s' and t'.

Eventually each pair of vertices is separated. We have obtained a total of $n - 1$ cuts. The crucial observation is that a minimum s'-t'-cut in the contracted graph G' is also a minimum s'-t'-cut in G. This is the subject of the following lemma. Note that when contracting a set A of vertices in (G, u), the capacity of each edge in G' is the capacity of the corresponding edge in G.

Lemma 8.35. *Let G be an undirected graph and $u : E(G) \to \mathbb{R}_+$ a capacity function. Let $s, t \in V(G)$, and let $\delta(A)$ be a minimum s-t-cut in (G, u). Let now $s', t' \in V(G) \setminus A$, and let (G', u') arise from (G, u) by contracting A to a single vertex. Then for any minimum s'-t'-cut $\delta(K \cup \{A\})$ in (G', u'), $\delta(K \cup A)$ is a minimum s'-t'-cut in (G, u).*

Proof: Let s, t, A, s', t', G', u' be as above. W.l.o.g. $s \in A$. It suffices to prove that there is a minimum s'-t'-cut $\delta(A')$ in (G, u) such that $A \subset A'$. So let $\delta(C)$ be any minimum s'-t'-cut in (G, u). W.l.o.g. $s \in C$.

Since $u(\delta(\cdot))$ is submodular (cf. Lemma 2.1(c)), we have $u(\delta(A)) + u(\delta(C)) \geq u(\delta(A \cap C)) + u(\delta(A \cup C))$. But $\delta(A \cap C)$ is an s-t-cut, so $u(\delta(A \cap C)) \geq \lambda_{st} = u(\delta(A))$. Therefore $u(\delta(A \cup C)) \leq u(\delta(C)) = \lambda_{s't'}$ proving that $\delta(A \cup C)$ is a minimum s'-t'-cut. (See Figure 8.3.)   □

Now we describe the algorithm which constructs a Gomory-Hu tree. Note that the vertices of the intermediate trees T will be vertex sets of the original graph; indeed they form a partition of $V(G)$. At the beginning, the only vertex of T is $V(G)$. In each iteration, a vertex of T containing at least two vertices of G is chosen and split into two.

GOMORY-HU ALGORITHM

Input: An undirected graph G and a capacity function $u : E(G) \to \mathbb{R}_+$.

Output: A Gomory-Hu tree T for (G, u).

①   Set $V(T) := \{V(G)\}$ and $E(T) := \emptyset$.

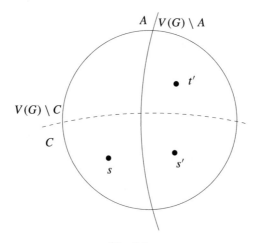

Fig. 8.3.

②   Choose some $X \in V(T)$ with $|X| \geq 2$. **If** no such X exists **then go to** ⑥.

③   Choose $s, t \in X$ with $s \neq t$.
     For each connected component C of $T - X$ **do**: Let $S_C := \bigcup_{Y \in V(C)} Y$.
     Let (G', u') arise from (G, u) by contracting S_C to a single vertex v_C for
       each connected component C of $T - X$.
     (So $V(G') = X \cup \{v_C : C$ is a connected component of $T - X\}$.)

④   Find a minimum s-t-cut $\delta(A')$ in (G', u'). Let $B' := V(G') \setminus A'$.

     Set $A := \left(\bigcup_{v_C \in A' \setminus X} S_C \right) \cup (A' \cap X)$ and $B := \left(\bigcup_{v_C \in B' \setminus X} S_C \right) \cup (B' \cap X)$.

⑤   Set $V(T) := (V(T) \setminus \{X\}) \cup \{A \cap X, B \cap X\}$.
     For each edge $e = \{X, Y\} \in E(T)$ incident to the vertex X **do**:
       If $Y \subseteq A$ **then** set $e' := \{A \cap X, Y\}$ **else** set $e' := \{B \cap X, Y\}$.
       Set $E(T) := (E(T) \setminus \{e\}) \cup \{e'\}$ and $w(e') := w(e)$.
     Set $E(T) := E(T) \cup \{\{A \cap X, B \cap X\}\}$.
     Set $w(\{A \cap X, B \cap X\}) := u'(\delta_{G'}(A'))$.
     Go to ②.

⑥   Replace all $\{x\} \in V(T)$ by x and all $\{\{x\}, \{y\}\} \in E(T)$ by $\{x, y\}$. **Stop.**

Figure 8.4 illustrates the modification of T in ⑤. To prove the correctness of
this algorithm, we first show the following lemma:

Lemma 8.36. *Each time at the end of* ④ *we have*

(a) $A \stackrel{.}{\cup} B = V(G)$
(b) $E(A, B)$ *is a minimum s-t-cut in* (G, u).

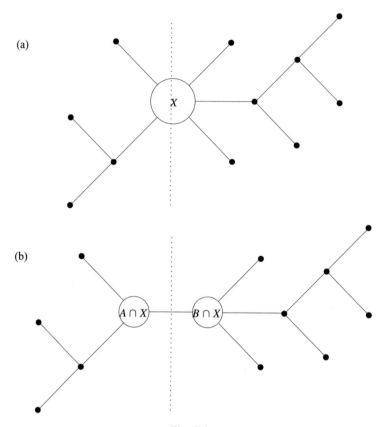

(a)

(b)

Fig. 8.4.

Proof: The elements of $V(T)$ are always nonempty subsets of $V(G)$, indeed $V(T)$ constitutes a partition of $V(G)$. From this, (a) follows easily.

We now prove (b). The claim is trivial for the first iteration (since here $G' = G$). We show that the property is preserved in each iteration.

Let $C_1, \ldots, C_k$ be the connected components of $T - X$. Let us contract them one by one; for $i = 0, \ldots, k$ let (G_i, u_i) arise from (G, u) by contracting each of $S_{C_1}, \ldots, S_{C_i}$ to a single vertex. So (G_k, u_k) is the graph which is denoted by (G', u') in ③ of the algorithm.

Claim: For any minimum s-t-cut $\delta(A_i)$ in (G_i, u_i), $\delta(A_{i-1})$ is a minimum s-t-cut in (G_{i-1}, u_{i-1}), where

$$A_{i-1} := \begin{cases} (A_i \setminus \{v_{C_i}\}) \cup S_{C_i} & \text{if } v_{C_i} \in A_i \\ A_i & \text{if } v_{C_i} \notin A_i \end{cases}.$$

Applying this claim successively for $k, k-1, \ldots, 1$ implies (b).

To prove the claim, let $\delta(A_i)$ be a minimum s-t-cut in (G_i, u_i). By our assumption that (b) is true for the previous iterations, $\delta(S_{C_i})$ is a minimum s_i-t_i-cut

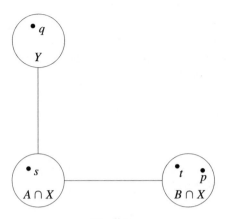

Fig. 8.5.

in (G, u) for some appropriate $s_i, t_i \in V(G)$. Furthermore, $s, t \in V(G) \setminus S_{C_i}$. So applying Lemma 8.35 completes the proof.   □

Lemma 8.37. *At any stage of the algorithm (until* ⑥ *is reached) for all* $e \in E(T)$

$$w(e) \;=\; u\left(\delta_G\left(\bigcup_{Z \in C_e} Z\right)\right),$$

where C_e *and* $V(T) \setminus C_e$ *are the connected components of* $T - e$. *Moreover for all* $e = \{P, Q\} \in E(T)$ *there are vertices* $p \in P$ *and* $q \in Q$ *with* $\lambda_{pq} = w(e)$.

Proof: Both statements are trivial at the beginning of the algorithm when T contains no edges; we show that they are never violated. So let X be the vertex of T chosen in ② in some iteration of the algorithm. Let s, t, A', B', A, B be as determined in ③ and ④ next. W.l.o.g. assume $s \in A'$.

Edges of T not incident to X are not affected by ⑤. For the new edge $\{A \cap X, B \cap X\}$, $w(e)$ is clearly set correctly, and we have $\lambda_{st} = w(e)$, $s \in A \cap X$, $t \in B \cap X$.

So let us consider an edge $e = \{X, Y\}$ that is replaced by e' in ⑤. We assume w.l.o.g. $Y \subseteq A$, so $e' = \{A \cap X, Y\}$. Assuming that the assertions were true for e we claim that they remain true for e'. This is trivial for the first assertion, because $w(e) = w(e')$ and $u\left(\delta_G\left(\bigcup_{Z \in C_e} Z\right)\right)$ does not change.

To show the second statement, we assume that there are $p \in X, q \in Y$ with $\lambda_{pq} = w(e)$. If $p \in A \cap X$ then we are done. So henceforth assume that $p \in B \cap X$ (see Figure 8.5).

We claim that $\lambda_{sq} = \lambda_{pq}$. Since $\lambda_{pq} = w(e) = w(e')$ and $s \in A \cap X$, this will conclude the proof.

By Lemma 8.33,

$$\lambda_{sq} \;\geq\; \min\{\lambda_{st}, \lambda_{tp}, \lambda_{pq}\}.$$

Since by Lemma 8.36(b) $E(A, B)$ is a minimum s-t-cut, and since $s, q \in A$, we may conclude from Lemma 8.35 that λ_{sq} does not change if we contract B. Since $t, p \in B$, this means that adding an edge $\{t, p\}$ with arbitrary high capacity does not change λ_{sq}. Hence

$$\lambda_{sq} \geq \min\{\lambda_{st}, \lambda_{pq}\}.$$

Now observe that $\lambda_{st} \geq \lambda_{pq}$ because the minimum s-t-cut $E(A, B)$ also separates p and q. So we have

$$\lambda_{sq} \geq \lambda_{pq}.$$

To prove equality, observe that $w(e)$ is the capacity of a cut separating X and Y, and thus s and q. Hence

$$\lambda_{sq} \leq w(e) = \lambda_{pq}.$$

This completes the proof.   □

Theorem 8.38. (Gomory and Hu [1961]) *The GOMORY-HU ALGORITHM works correctly. Every undirected graph possesses a Gomory-Hu tree, and such a tree is found in* $O(n^3 \sqrt{m})$ *time.*

Proof: The complexity of the algorithm is clearly determined by $n - 1$ times the complexity of finding a minimum s-t-cut, since everything else can be implemented in $O(n^3)$ time. By Proposition 8.31 we obtain the $O(n^3 \sqrt{m})$ bound.

We prove that the output T of the algorithm is a Gomory-Hu tree for (G, u). It should be clear that T is a tree with $V(T) = V(G)$. Now let $s, t \in V(G)$. Let P_{st} be the (unique) s-t-path in T and, for $e \in E(T)$, let C_e and $V(G) \setminus C_e$ be the connected components of $T - e$.

Since $\delta(C_e)$ is an s-t-cut for each $e \in E(P_{st})$,

$$\lambda_{st} \leq \min_{e \in E(P_{st})} u(\delta(C_e)).$$

On the other hand, a repeated application of Lemma 8.33 yields

$$\lambda_{st} \geq \min_{\{v,w\} \in E(P_{st})} \lambda_{vw}.$$

Hence applying Lemma 8.37 to the situation before execution of ⑥ (where each vertex X of T is a singleton) yields

$$\lambda_{st} \geq \min_{e \in E(P_{st})} u(\delta(C_e)),$$

so equality holds.   □

A similar algorithm for the same task (which might be easier to implement) was suggested by Gusfield [1990]. For digraphs, Cheung, Lau and Leung [2011] showed how to compute the minimum cardinality of an s-t-cut for all pairs $s, t \in V(G)$ in $O(m^{2.38})$ time.

8.7 The Minimum Capacity of a Cut in an Undirected Graph

If we are only interested in a minimum capacity cut in an undirected graph G with capacities $u : E(G) \rightarrow \mathbb{R}_+$, there is a simpler method using $n - 1$ flow computations: just compute a minimum s-t-cut for some fixed vertex s and each $t \in V(G) \setminus \{s\}$. However, there are more efficient algorithms.

Hao and Orlin [1994] found an $O(nm \log \frac{n^2}{m})$-algorithm for determining a minimum capacity cut. They use a modified version of the PUSH-RELABEL ALGORITHM.

If we just want to compute the edge-connectivity of a given graph (i.e. unit capacities), the currently fastest algorithms are due to Henzinger, Rao and Wang [2017] and Gabow [1995] with near-linear running times. We remark that the MAXIMUM FLOW PROBLEM in undirected graphs with unit capacities can also be solved faster than in general (Karger and Levine [1998]).

Nagamochi and Ibaraki [1992] found a completely different algorithm to determine a minimum capacity cut in an undirected graph. Their algorithm does not use max-flow computations at all. In this section we present this algorithm in a simplified form due to Stoer and Wagner [1997] and independently to Frank [1994]. We start with an easy definition.

Definition 8.39.   *Given an undirected graph G with capacities $u : E(G) \rightarrow \mathbb{R}_+$, we call an order $v_1, \ldots, v_n$ of the vertices an* **MA (maximum adjacency) order** *if for all $i \in \{2, \ldots, n\}$:*

$$\sum_{e \in E(\{v_1, \ldots, v_{i-1}\}, \{v_i\})} u(e) \;=\; \max_{j \in \{i, \ldots, n\}} \sum_{e \in E(\{v_1, \ldots, v_{i-1}\}, \{v_j\})} u(e).$$

Proposition 8.40.   *Given an undirected graph G with capacities $u : E(G) \rightarrow \mathbb{R}_+$, an MA order can be found in $O(m + n \log n)$ time.*

Proof: Consider the following algorithm. First set $\alpha(v) := 0$ for all $v \in V(G)$. Then for $i := 1$ to n do the following: choose v_i from among $V(G) \setminus \{v_1, \ldots, v_{i-1}\}$ such that it has maximum α-value (breaking ties arbitrarily), and set $\alpha(v) := \alpha(v) + \sum_{e \in E(\{v_i\}, \{v\})} u(e)$ for all $v \in V(G) \setminus \{v_1, \ldots, v_i\}$.

The correctness of this algorithm is obvious. By implementing it with a Fibonacci heap, storing each vertex v with key $-\alpha(v)$ until it is selected, we get a running time of $O(m + n \log n)$ by Theorem 6.7 as there are n INSERT-, n DELETEMIN- and (at most) m DECREASEKEY-operations. $\square$

Lemma 8.41.   (Stoer and Wagner [1997], Frank [1994])   *Let G be an undirected graph with $n := |V(G)| \geq 2$, capacities $u : E(G) \rightarrow \mathbb{R}_+$ and an MA order $v_1, \ldots, v_n$. Then*

$$\lambda_{v_{n-1} v_n} \;=\; \sum_{e \in \delta(v_n)} u(e).$$

Proof: Of course we only have to show "$\geq$". We shall use induction on $|V(G)| + |E(G)|$. For $|V(G)| < 3$ the statement is trivial. We may assume that there is no edge $e = \{v_{n-1}, v_n\} \in E(G)$, because otherwise we would delete it (both left-hand side and right-hand side decrease by $u(e)$) and apply the induction hypothesis.

Denote the right-hand side by R. Of course $v_1, \ldots, v_{n-1}$ is an MA order in $G - v_n$. So by induction,

$$\lambda^{G-v_n}_{v_{n-2}v_{n-1}} = \sum_{e \in E(\{v_{n-1}\}, \{v_1, \ldots, v_{n-2}\})} u(e) \geq \sum_{e \in E(\{v_n\}, \{v_1, \ldots, v_{n-2}\})} u(e) = R.$$

Here the inequality holds because $v_1, \ldots, v_n$ was an MA order for G. The last equality is true because $\{v_{n-1}, v_n\} \notin E(G)$. So $\lambda^G_{v_{n-2}v_{n-1}} \geq \lambda^{G-v_n}_{v_{n-2}v_{n-1}} \geq R$.

On the other hand $v_1, \ldots, v_{n-2}, v_n$ is an MA order in $G - v_{n-1}$. So by induction,

$$\lambda^{G-v_{n-1}}_{v_{n-2}v_n} = \sum_{e \in E(\{v_n\}, \{v_1, \ldots, v_{n-2}\})} u(e) = R,$$

again because $\{v_{n-1}, v_n\} \notin E(G)$. So $\lambda^G_{v_{n-2}v_n} \geq \lambda^{G-v_{n-1}}_{v_{n-2}v_n} = R$.

Now by Lemma 8.33 $\lambda_{v_{n-1}v_n} \geq \min\{\lambda_{v_{n-1}v_{n-2}}, \lambda_{v_{n-2}v_n}\} \geq R$.   □

Note that the existence of two vertices x, y with $\lambda_{xy} = \sum_{e \in \delta(x)} u(e)$ was already shown by Mader [1972], and follows easily from the existence of a Gomory-Hu tree (Exercise 34).

Theorem 8.42. (Nagamochi and Ibaraki [1992], Stoer and Wagner [1997]) *A minimum capacity cut in an undirected graph with nonnegative capacities can be found in $O(mn + n^2 \log n)$ time.*

Proof: We may assume that the given graph G is simple since we can unite parallel edges. Denote by $\lambda(G)$ the minimum capacity of a cut in G. The algorithm proceeds as follows:

Let $G_0 := G$. In the i-th step ($i = 1, \ldots, n-1$) choose vertices $x, y \in V(G_{i-1})$ with

$$\lambda^{G_{i-1}}_{xy} = \sum_{e \in \delta_{G_{i-1}}(x)} u(e).$$

By Proposition 8.40 and Lemma 8.41 this can be done in $O(m + n \log n)$ time. Set $\gamma_i := \lambda^{G_{i-1}}_{xy}$, $z_i := x$, and let G_i result from G_{i-1} by contracting $\{x, y\}$. Observe that

$$\lambda(G_{i-1}) = \min\{\lambda(G_i), \gamma_i\}, \tag{8.1}$$

because a minimum cut in G_{i-1} either separates x and y (in this case its capacity is γ_i) or does not (in this case contracting $\{x, y\}$ does not change anything).

After arriving at G_{n-1} which has only one vertex, we choose a $k \in \{1, \ldots, n-1\}$ for which γ_k is minimum. We claim that $\delta(X)$ is a minimum capacity cut in G, where X is the vertex set in G whose contraction resulted in the vertex z_k of G_{k-1}.

But this is easy to see, since by (8.1) $\lambda(G) = \min\{\gamma_1, \ldots, \gamma_{n-1}\} = \gamma_k$ and γ_k is the capacity of the cut $\delta(X)$. $\square$

A randomized contraction algorithm for finding a minimum cut (with high probability) is discussed in Exercise 39. Moreover, we mention that the vertex-connectivity of a graph can be computed by $O(n^2)$ flow computations in a graph with unit capacities (Exercise 41).

In this section we have shown how to minimize $f(X) := u(\delta(X))$ over $\emptyset \neq X \subset V(G)$. Note that this $f : 2^{V(G)} \to \mathbb{R}_+$ is submodular and symmetric (i.e. $f(A) = f(V(G) \setminus A)$ for all A). The algorithm presented here has been generalized by Queyranne [1998] to minimize general symmetric submodular functions; see Section 14.5.

The problem of finding a maximum cut is much harder and will be discussed in Section 16.2.

Exercises

1. Let (G, u, s, t) be a network, and let $\delta^+(X)$ and $\delta^+(Y)$ be minimum s-t-cuts in (G, u). Show that $\delta^+(X \cap Y)$ and $\delta^+(X \cup Y)$ are also minimum s-t-cuts in (G, u).
2. Show that in case of irrational capacities, the FORD-FULKERSON ALGORITHM may not terminate at all.
   Hint: Consider the following network (Figure 8.6):
   All lines represent edges in both directions. All edges have capacity $S = \frac{1}{1-\sigma}$ except

$$u((x_1, y_1)) = 1, \quad u((x_2, y_2)) = \sigma, \quad u((x_3, y_3)) = u((x_4, y_4)) = \sigma^2$$

   where $\sigma = \frac{\sqrt{5}-1}{2}$. Note that $\sigma^n = \sigma^{n+1} + \sigma^{n+2}$.
   (Ford and Fulkerson [1962])

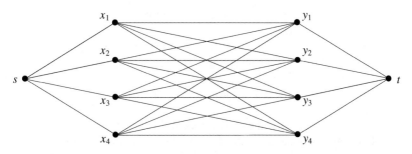

Fig. 8.6.

* 3. Let G be a digraph and M the incidence matrix of G. Prove that for all $c, l, u \in \mathbb{Z}^{E(G)}$ with $l \le u$:

$$\max \left\{ cx : x \in \mathbb{Z}^{E(G)}, l \le x \le u, Mx = 0 \right\} =$$

$$\min \left\{ y'u - y''l : y', y'' \in \mathbb{Z}_+^{E(G)}, zM + y' - y'' = c \text{ for some } z \in \mathbb{Z}^{V(G)} \right\}.$$

Show how this implies Theorem 8.6 and Corollary 8.7.

4. Prove Hoffman's circulation theorem: Given a digraph G and lower and upper capacities $l, u : E(G) \to \mathbb{R}_+$ with $l(e) \le u(e)$ for all $e \in E(G)$, there is circulation f with $l(e) \le f(e) \le u(e)$ for all $e \in E(G)$ if and only if

$$\sum_{e \in \delta^-(X)} l(e) \le \sum_{e \in \delta^+(X)} u(e) \quad \text{for all } X \subseteq V(G).$$

Note: Hoffman's circulation theorem in turn quite easily implies the Max-Flow-Min-Cut Theorem.
(Hoffman [1960])

5. Consider a network (G, u, s, t), a maximum s-t-flow f and the residual graph G_f. Form a digraph H from G_f by contracting the set S of vertices reachable from s to a vertex v_S, contracting the set T of vertices from which t is reachable to a vertex v_T, and contracting each strongly connected component X of $G_f - (S \cup T)$ to a vertex v_X. Observe that H is acyclic. Prove that there is a one-to-one correspondence between the sets $X \subseteq V(G)$ for which $\delta_G^+(X)$ is a minimum s-t-cut in (G, u) and the sets $Y \subseteq V(H)$ for which $\delta_H^+(Y)$ is a directed v_T-v_S-cut in H (i.e. a directed cut in H separating v_T and v_S).
Note: This statement also holds for G_f without any contraction instead of H. However, we shall use the statement in the above form in Section 20.5.
(Picard and Queyranne [1980])

6. Let G be a digraph and $c, c' : E(G) \to \mathbb{R}$. We look for a set $X \subset V(G)$ with $s \in X$ and $t \notin X$ such that $\sum_{e \in \delta^+(X)} c(e) - \sum_{e \in \delta^-(X)} c'(e)$ is minimum.
   (a) Show how to reduce this problem to the MINIMUM CAPACITY CUT PROBLEM.
   (b) Now consider the special case where $c = c'$. Can you solve this problem in linear time?

* 7. Let G be an acyclic digraph with mappings $\sigma, \tau, c : E(G) \to \mathbb{R}_+$, and a number $C \in \mathbb{R}_+$. We look for a mapping $x : E(G) \to \mathbb{R}_+$ such that $\sigma(e) \le x(e) \le \tau(e)$ for all $e \in E(G)$ and $\sum_{e \in E(G)} (\tau(e) - x(e)) c(e) \le C$. Among the feasible solutions we want to minimize the length (with respect to x) of the longest path in G.
The meaning behind the above is the following. The edges correspond to jobs, $\sigma(e)$ and $\tau(e)$ stand for the minimum and maximum completion time of job e, and $c(e)$ is the cost of reducing the completion time of job e by one unit. If there are two jobs $e = (i, j)$ and $e' = (j, k)$, job e has to be finished before job e' can be processed. We have a fixed budget C and want to minimize the total completion time.

Show how to solve this problem using network flow techniques. (This application is known as PERT, program evaluation and review technique, or CPM, critical path method. The problem is also known as the budget version of the time-cost tradeoff problem.)

Hint: Introduce one source s and one sink t. Start with $x = \tau$ and successively reduce the length of the longest s-t-path (with respect to x) at the minimum possible cost. Use Exercise 8 of Chapter 7, Exercise 9 of Chapter 3, and Exercise 6.

(Phillips and Dessouky [1977])

* 8. Let (G, c, s, t) be a network such that G is planar even when an edge $e = (s, t)$ is added. Consider the following algorithm. Start with the flow $f \equiv 0$ and let $G' := G_f$. At each step consider the boundary B of a face of $G' + e$ containing e (with respect to some fixed planar embedding). Augment f along $B - e$. Let G' consist of the forward edges of G_f only and iterate as long as t is reachable from s in G'.

   Prove that this algorithm computes a maximum s-t-flow. Use Theorem 2.40 to show that it can be implemented to run in $O(n^2)$ time.

   (Ford and Fulkerson [1956], Hu [1969])

   Note: This problem can be solved in $O(n)$ time. For general planar networks an $O(n \log n)$-algorithm exists; see Weihe [1997] and Borradaile and Klein [2009].

9. Show that the directed edge-disjoint version of Menger's Theorem 8.9 also follows directly from Theorem 6.18.

10. Let G be an undirected graph. Prove that one can compute an orientation G' of G in linear time such that for each $v, w \in V(G)$ the following holds: if G has two edge-disjoint v-w-paths, then G' has a (directed) v-w-path.

    Hint: Use DFS.

    (Tarjan [1972])

11. Let G be a digraph with conservative weights $c : E(G) \to \mathbb{R}$ and two vertices $s, t \in V(G)$ such that t is reachable from s. Suppose that for every edge $e \in E(G)$ we have $\text{dist}_{(G-e,c)}(s, t) = \text{dist}_{(G,c)}(s, t)$. Prove that then there are two edge-disjoint shortest s-t-paths in (G, c).

12. Consider an undirected graph G with edge-connectivity $k \in \mathbb{N}$ and (not necessarily distinct) vertices $v_0, v_1, \ldots, v_k \in V(G)$. Prove that there are edge-disjoint paths $P_1, \ldots, P_k$ such that P_i is a v_0-v_i-path $(i = 1, \ldots, k)$.

13. Let G be a graph (directed or undirected), x, y, z three vertices, and $\alpha, \beta \in \mathbb{N}$ with $\alpha \le \lambda_{xy}$, $\beta \le \lambda_{xz}$ and $\alpha + \beta \le \max\{\lambda_{xy}, \lambda_{xz}\}$. Prove that there are α x-y-paths and β x-z-paths such that these $\alpha + \beta$ paths are edge-disjoint.

14. Let G be a digraph that contains k edge-disjoint s-t-paths for any two vertices s and t (such a graph is called strongly k-edge-connected).

    Let H be any digraph with $V(H) = V(G)$ and $|E(H)| = k$. Prove that the instance (G, H) of the DIRECTED EDGE-DISJOINT PATHS PROBLEM has a solution.

    (Mader [1981] and Shiloach [1979])

15. Let G be a digraph with at least k edges. Prove: G contains k edge-disjoint s-t-paths for any two vertices s and t if and only if for any k distinct edges $e_1 = (x_1, y_1), \ldots, e_k = (x_k, y_k)$, $G - \{e_1, \ldots, e_k\}$ contains k edge-disjoint spanning arborescences $T_1, \ldots, T_k$ such that T_i is rooted at y_i $(i = 1, \ldots, k)$. *Note:* This generalizes Exercise 14. *Hint:* Use Theorem 6.18.
    (Su [1997])

16. Let G be a digraph with capacities $c : E(G) \to \mathbb{R}_+$ and $r \in V(G)$. Can one determine an r-cut with minimum capacity in polynomial time? Can one determine a directed cut with minimum capacity in polynomial time (or decide that G is strongly connected)?
    Note: The answer to the first question solves the SEPARATION PROBLEM for the MINIMUM WEIGHT ROOTED ARBORESCENCE PROBLEM; see Corollary 6.15.

17. An airline wants to conduct a given set of scheduled flights with as few airplanes as possible. All available airplanes are of the same type. For each flight we know the departure time and the flight time. We also know, for any pair of flights i and j, how much time an airplane needs after finishing flight i until it can begin flight j (this time will depend in particular on where i ends and where j begins). Show how to compute efficiently a feasible schedule with as few airplanes as possible.

18. Let G be an undirected graph, $s, t \in V(G)$, and $k \in \mathbb{N}$. How can you find in polynomial time k edge-disjoint shortest s-t-paths, or decide that such paths do not exist?

19. Prove that the value of a blocking s-t-flow in a network (G, u, s, t) with an acyclic digraph G is at least $\frac{1}{|V(G)|}$ times the value of a maximum s-t-flow. Show that this bound is sharp up to a constant factor.

20. Show how to find a blocking flow in an acyclic network in $O(nm)$ time by successively augmenting along a path of non-saturated edges and using DEPTH-FIRST SEARCH to find such a path. Show how to obtain a running time of $O(m)$ if all edges have capacity 1.

* 21. Let (G, u, s, t) be a network with $u(e) = 1$ for all edges $e \in E(G)$.
    (a) Show that then a maximum s-t-flow can be computed in $O(mn^{2/3})$ time.
    (b) Let in addition G have the property that for each $v \in V(G) \setminus \{s, t\}$ we have $|\delta^-(v)| = 1$ or $|\delta^+(v)| = 1$. Show that then a maximum s-t-flow can be computed in $O(m\sqrt{n})$ time.
    Hint: Consider DINIC'S ALGORITHM and the situation when no augmenting path has length less than $\lceil n^{2/3} \rceil$ in (a) and $\lceil \sqrt{n} \rceil$ in (b). Bound the number of remaining iterations and use the second part of Exercise 20.
    (Karzanov [1973], Even and Tarjan [1975])

22. An s-t-preflow f is called *maximum* if $\text{ex}_f(t)$ is maximum.
    (a) Show that for any maximum preflow f there exists a maximum flow f' with $f'(e) \le f(e)$ for all $e \in E(G)$.
    (b) Show how a maximum preflow can be converted into a maximum flow in $O(nm)$ time.

23. Let (G, u, s, t) be a network such that $G - t$ is an arborescence. Show how to find a maximum s-t-flow in linear time.
    Hint: Use DFS.

* 24. Let (G, u, s, t) be a network such that the underlying undirected graph of $G - \{s, t\}$ is a forest. Show how to find a maximum s-t-flow in linear time. (Vygen [2002])

25. Consider a modified version of FUJISHIGE'S ALGORITHM where in ⑤ we choose $v_i \in V(G) \setminus \{v_1, \ldots, v_{i-1}\}$ such that $b(v_i)$ is maximum, step ④ is replaced by stopping if $b(v) = 0$ for all $v \in V(G) \setminus \{v_1, \ldots, v_i\}$, and in the beginning of ⑥ we set $\beta(t) := \min_{j=2}^{i} b(j)$. Then X and α are not needed anymore.
    (a) Show that this variant of the algorithm works correctly.
    (b) Let α_k be the number $\min_{j=2}^{i} b(j)$ in iteration k (or zero if the algorithm stops before iteration k). Show that $\min_{l=k+1}^{k+2n} \alpha_l \le \frac{1}{2}\alpha_k$ for all k. Conclude that the number of iterations is $O(n \log u_{\max})$.
    (c) Show how to implement one iteration in $O(m + n \log n)$ time.

26. Prove that the PUSH-RELABEL ALGORITHM performs $O(n^2 m)$ nonsaturating pushes, independent of the choice of v in ③.

27. Let (G, u, s, t) be a network, f an s-t-preflow, and ψ a distance labeling with respect to f with $\psi(v) \le 2n$ for $v \in V(G)$. Define $\psi'(v) := \min\{\text{dist}_{G_f}(v, t), n + \text{dist}_{G_f}(v, s), 2n\}$ for $v \in V(G)$. Show that ψ' is a distance labeling with respect to f, and $\psi \le \psi'$.
    Note: Replacing ψ by ψ' from time to time, e.g. after every n RELABEL operations, improves the performance of the PUSH-RELABEL ALGORITHM in practice.

28. Given an acyclic digraph G with weights $c : E(G) \to \mathbb{R}_+$, find a maximum weight directed cut in G. Show how this problem can be reduced to the MINIMUM CAPACITY CUT PROBLEM.
    Hint: Use Exercise 6.

29. Let G be an acyclic digraph with weights $c : E(G) \to \mathbb{R}_+$. We look for the maximum weight edge set $F \subseteq E(G)$ such that no path in G contains more than one edge of F. Show that this problem is equivalent to looking for the maximum weight directed cut in G (and thus can be solved in $O(n^3)$ time by Exercise 28).

30. Let G be a digraph and $p : V(G) \to \mathbb{R}$. Show how to find a set $X \subseteq V(G)$ with $\delta^+(X) = \emptyset$ such that $p(X)$ is maximum.
    Note: This was used to model open-pit mining, where $p(v)$ is the (possibly negative) profit of mining v, and an edge (v, w) models the constraint that we cannot mine v unless we mine w.

31. Given an undirected graph G with capacities $u : E(G) \to \mathbb{R}_+$ and a set $T \subseteq V(G)$ with $|T| \ge 2$. We look for a set $X \subset V(G)$ with $T \cap X \ne \emptyset$ and $T \setminus X \ne \emptyset$ such that $\sum_{e \in \delta(X)} u(e)$ is minimum. Show how to solve this problem in $O(n^4)$ time, where $n = |V(G)|$.

32. Let λ_{ij}, $1 \leq i, j \leq n$, be nonnegative numbers with $\lambda_{ij} = \lambda_{ji}$ and $\lambda_{ik} \geq \min\{\lambda_{ij}, \lambda_{jk}\}$ for any three distinct indices $i, j, k \in \{1, \ldots, n\}$. Show that there exists a graph G with $V(G) = \{1, \ldots, n\}$ and capacities $u : E(G) \to \mathbb{R}_+$ such that the local edge-connectivities are precisely the λ_{ij}.
    Hint: Consider a maximum weight spanning tree in (K_n, c), where $c(\{i, j\}) := \lambda_{ij}$.
    (Gomory and Hu [1961])

33. Let G be an undirected graph with capacities $u : E(G) \to \mathbb{R}_+$, and let $T \subseteq V(G)$ with $|T|$ even. A T-cut in G is a cut $\delta(X)$ with $|X \cap T|$ odd. Construct a polynomial time algorithm for finding a T-cut of minimum capacity in (G, u).
    Hint: Use a Gomory-Hu tree.
    (A solution of this exercise can be found in Section 12.3.)

34. Let G be a simple undirected graph with at least two vertices. Suppose the degree of each vertex of G is at least k. Prove that there are two vertices s and t such that at least k edge-disjoint s-t-paths exist. What if there is exactly one vertex with degree less than k?
    Hint: Consider a Gomory-Hu tree for G.

35. Consider the problem of determining the edge-connectivity $\lambda(G)$ of an undirected graph (with unit capacities). Section 8.7 shows how to solve this problem in $O(mn)$ time, provided that we can find an MA order of an undirected graph with unit capacities in $O(m + n)$ time. How can this be done?

36. Let $k \in \mathbb{N}$ and G an undirected graph in which the degree of every vertex is at least k. Prove that there exist two vertices $s \neq t$ and k edge-disjoint s-t-paths.

$*$ 37. Let G be a simple undirected graph with an MA order $v_1, \ldots, v_n$. Let κ_{uv}^G denote the maximum number of internally disjoint u-v-paths in G. Prove $\kappa_{v_{n-1}v_n}^G = |\delta_G(v_n)|$ (the vertex-disjoint counterpart of Lemma 8.41).
    Hint: Prove by induction that $\kappa_{v_i v_j}^{G_{ij}} = |\delta_{G_{ij}}(v_j)|$ for all $1 \leq i < j \leq n$, where $G_{ij} = G[\{v_1, \ldots, v_i\} \cup \{v_j\}]$. To do this, assume w.l.o.g. that $\{v_i, v_j\} \notin E(G)$, choose a minimal set $Z \subseteq \{v_1, \ldots, v_{i-1}\}$ separating v_j and v_i (Menger's Theorem 8.10), and let $h \leq i$ be the maximum number such that $v_h \notin Z$ and v_h is adjacent to v_i or v_j.
    (Frank [unpublished])

$*$ 38. An undirected graph is called chordal if it has no circuit of length at least four as an induced subgraph. An order $v_1, \ldots, v_n$ of an undirected graph G is called simplicial if $\{v_i, v_j\}, \{v_i, v_k\} \in E(G)$ implies $\{v_j, v_k\} \in E(G)$ for $i < j < k$.
    (a) Prove that a graph with a simplicial order must be chordal.
    (b) Let G be a chordal graph, and let $v_1, \ldots, v_n$ be an MA order. Prove that $v_n, v_{n-1}, \ldots, v_1$ is a simplicial order.
        Hint: Use Exercise 37 and Menger's Theorem 8.10.
    Note: The fact that a graph is chordal if and only if it has a simplicial order is due to Rose [1970].

39. Let G be an undirected graph with capacities $u : E(G) \to \mathbb{R}_+$. Let $\emptyset \neq A \subset V(G)$ such that $\delta(A)$ is a minimum capacity cut in G.

   (a) Show that $u(\delta(A)) \leq \frac{2}{n} u(E(G))$. (*Hint:* Consider the trivial cuts $\delta(x)$, $x \in V(G)$.)

   (b) Assume w.l.o.g. that $u(\delta(A)) > 0$ and consider the following procedure. We randomly choose an edge and contract it; each edge e is chosen with probability $\frac{u(e)}{u(E(G))}$. We repeat this operation until there are only two vertices. Prove that the probability that we never contract an edge of $\delta(A)$ is at least $\frac{2}{(n-1)n}$.

   (c) Conclude that running the randomized algorithm in (b) kn^2 times yields $\delta(A)$ with probability at least $1 - e^{-2k}$. (Such an algorithm with a positive probability of a correct answer is called a Monte Carlo algorithm.)

   (Karger and Stein [1996] ; see also Karger [2000])

40. Conclude from Exercise 39 that there are at most $\binom{n}{2}$ cuts with minimum capacity in an undirected graph with n vertices and nonnegative edge capacities. Show that this bound is tight.

   Note: Nagamochi, Nishimura and Ibaraki [1997] showed that there are at most $\binom{n}{2}$ cuts whose capacity exceeds the minimum by at most a factor $\frac{4}{3}$.

41. Show how the vertex-connectivity of an undirected graph can be determined in $O(n^{2.5}m)$ time.

   Hint: Recall the proof of Menger's Theorem and use Exercise 21.

   Note: Faster algorithms were proposed by Henzinger, Rao and Gabow [2000], and by Gabow [2006].

42. Let G be a connected undirected graph with capacities $u : E(G) \to \mathbb{R}_+$. We are looking for a minimum capacity 3-cut, i.e. an edge set whose deletion splits G into at least three connected components.

   Let $n := |V(G)| \geq 4$. Let $\delta(X_1), \delta(X_2), \ldots$ be a list of the cuts ordered by nondecreasing capacities: $u(\delta(X_1)) \leq u(\delta(X_2)) \leq \cdots$. Assume that we know the first $2n - 2$ elements of this list (note: they can be computed in polynomial time by a method of Vazirani and Yannakakis [1992]).

   (a) Show that for some indices $i, j \in \{1, \ldots, 2n - 2\}$ all sets $X_i \setminus X_j$, $X_j \setminus X_i$, $X_i \cap X_j$ and $V(G) \setminus (X_i \cup X_j)$ are nonempty.

   (b) Show that there is a 3-cut of capacity at most $\frac{3}{2} u(\delta(X_{2n-2}))$.

   (c) For each $i = 1, \ldots, 2n - 2$ consider $\delta(X_i)$ plus a minimum capacity cut of $G - X_i$, and also $\delta(X_i)$ plus a minimum capacity cut of $G[X_i]$. This yields a list of at most $4n - 4$ 3-cuts. Prove that one of them is optimum.

   (Nagamochi and Ibaraki [2000])

   Note: This was generalized to k-cuts (for any fixed k) by Kamidoi, Yoshida and Nagamochi [2007]; see also Thorup [2008]. The problem of finding the optimum 3-cut separating three given vertices is much harder; see Dahlhaus et al. [1994] and Cheung, Cunningham and Tang [2006].

43. Let G be an undirected graph with capacities $u : E(G) \to \mathbb{Z}_+$.

   (a) Show that if $\delta(X)$ and $\delta(Y)$ are two minimum capacity cuts with $X \cap Y \neq \emptyset$ and $X \cup Y \neq V(G)$, then $\delta(X \setminus Y) \cap \delta(Y \setminus X) = \emptyset$.

(b) Suppose that the minimum capacity of a cut is odd. Show that then the family of vertex sets X for which $\delta(X)$ is a minimum capacity cut is cross-free, and hence there are at most $n - 1$ minimum capacity cuts.

Note: Dinitz, Karzanov and Lomonosov [1976] showed that there are at most $\binom{n}{2}$ minimum capacity cuts in general, which follows also from Exercise 39(b). They can be described by a so-called cactus representation, generalizing tree-representations. See also Frank [2011].

References

General Literature

Ahuja, R.K., Magnanti, T.L., and Orlin, J.B. [1993]: Network Flows. Prentice-Hall, Englewood Cliffs 1993

Cook, W.J., Cunningham, W.H., Pulleyblank, W.R., and Schrijver, A. [1998]: Combinatorial Optimization. Wiley, New York 1998, Chapter 3

Cormen, T.H., Leiserson, C.E., Rivest, R.L., and Stein, C. [2009]: Introduction to Algorithms. Third Edition. MIT Press, Cambridge 2009, Chapter 26

Ford, L.R., and Fulkerson, D.R. [1962]: Flows in Networks. Princeton University Press, Princeton 1962

Frank, A. [1995]: Connectivity and network flows. In: Handbook of Combinatorics; Vol. 1 (R.L. Graham, M. Grötschel, L. Lovász, eds.), Elsevier, Amsterdam, 1995

Frank, A. [2011]: Connections in Combinatorial Optimization. Oxford University Press, Oxford 2011

Goldberg, A.V., Tardos, É., and Tarjan, R.E. [1990]: Network flow algorithms. In: Paths, Flows, and VLSI-Layout (B. Korte, L. Lovász, H.J. Prömel, A. Schrijver, eds.), Springer, Berlin 1990, pp. 101–164

Gondran, M., and Minoux, M. [1984]: Graphs and Algorithms. Wiley, Chichester 1984, Chapter 5

Jungnickel, D. [2013]: Graphs, Networks and Algorithms. Fourth Edition. Springer, Berlin 2013

Phillips, D.T., and Garcia-Diaz, A. [1981]: Fundamentals of Network Analysis. Prentice-Hall, Englewood Cliffs 1981

Ruhe, G. [1991]: Algorithmic Aspects of Flows in Networks. Kluwer Academic Publishers, Dordrecht 1991

Schrijver, A. [2003]: Combinatorial Optimization: Polyhedra and Efficiency. Springer, Berlin 2003, Chapters 9,10,13–15

Tarjan, R.E. [1983]: Data Structures and Network Algorithms. SIAM, Philadelphia 1983, Chapter 8

Thulasiraman, K., and Swamy, M.N.S. [1992]: Graphs: Theory and Algorithms. Wiley, New York 1992, Chapter 12

Cited References

Ahuja, R.K., Orlin, J.B., and Tarjan, R.E. [1989]: Improved time bounds for the maximum flow problem. SIAM Journal on Computing 18 (1989), 939–954

Borradaile, G. and Klein, P. [2009]: An $O(n \log n)$ algorithm for maximum st-flow in a directed planar graph. Journal of the ACM 56 (2009), Article 9

Cheriyan, J., and Maheshwari, S.N. [1989]: Analysis of preflow push algorithms for maximum network flow. SIAM Journal on Computing 18 (1989), 1057–1086

Cheriyan, J., and Mehlhorn, K. [1999]: An analysis of the highest-level selection rule in the preflow-push max-flow algorithm. Information Processing Letters 69 (1999), 239–242

Cherkassky, B.V. [1977]: Algorithm of construction of maximal flow in networks with complexity of $O(V^2 \sqrt{E})$ operations. Mathematical Methods of Solution of Economical Problems 7 (1977), 112–125 [in Russian]

Cheung, K.K.H., Cunningham, W.H., and Tang, L. [2006]: Optimal 3-terminal cuts and linear programming. Mathematical Programming 106 (2006), 1–23

Cheung, H.Y., Lau, L.C., and Leung, K.M. [2011]: Graph connectivities, network coding, and expander graphs. Proceedings of the 52nd Annual IEEE Symposium on Foundations of Computer Science (2011), 197–206

Dahlhaus, E., Johnson, D.S., Papadimitriou, C.H., Seymour, P.D., and Yannakakis, M. [1994]: The complexity of multiterminal cuts. SIAM Journal on Computing 23 (1994), 864–894

Dantzig, G.B., and Fulkerson, D.R. [1956]: On the max-flow min-cut theorem of networks. In: Linear Inequalities and Related Systems (H.W. Kuhn, A.W. Tucker, eds.), Princeton University Press, Princeton 1956, pp. 215–221

Dinic, E.A. [1970]: Algorithm for solution of a problem of maximum flow in a network with power estimation. Soviet Mathematics Doklady 11 (1970), 1277–1280

Dinitz, E.A., Karzanov, A.V., and Lomonosov, M.V. [1976]: On the structure of the system of minimum edge cuts of a graph. Issledovaniya po Diskretnoĭ Optimizatsii (A.A. Fridman, ed.), Nauka, Moscow, 1976, pp. 290–306 [in Russian]

Edmonds, J., and Karp, R.M. [1972]: Theoretical improvements in algorithmic efficiency for network flow problems. Journal of the ACM 19 (1972), 248–264

Elias, P., Feinstein, A., and Shannon, C.E. [1956]: Note on maximum flow through a network. IRE Transactions on Information Theory, IT-2 (1956), 117–119

Even, S., and Tarjan, R.E. [1975]: Network flow and testing graph connectivity. SIAM Journal on Computing 4 (1975), 507–518

Ford, L.R., and Fulkerson, D.R. [1956]: Maximal Flow Through a Network. Canadian Journal of Mathematics 8 (1956), 399–404

Ford, L.R., and Fulkerson, D.R. [1957]: A simple algorithm for finding maximal network flows and an application to the Hitchcock problem. Canadian Journal of Mathematics 9 (1957), 210–218

Frank, A. [1994]: On the edge-connectivity algorithm of Nagamochi and Ibaraki. Laboratoire Artemis, IMAG, Université J. Fourier, Grenoble, 1994

Fujishige, S. [2003]: A maximum flow algorithm using MA ordering. Operations Research Letters 31 (2003), 176–178

Gabow, H.N. [1995]: A matroid approach to finding edge connectivity and packing arborescences. Journal of Computer and System Sciences 50 (1995), 259–273

Gabow, H.N. [2006]: Using expander graphs to find vertex-connectivity. Journal of the ACM 53 (2006), 800–844

Galil, Z. [1980]: An $O(V^{\frac{5}{3}}E^{\frac{2}{3}})$ algorithm for the maximal flow problem. Acta Informatica 14 (1980), 221–242

Galil, Z., and Namaad, A. [1980]: An $O(EV \log^2 V)$ algorithm for the maximal flow problem. Journal of Computer and System Sciences 21 (1980), 203–217

Gallai, T. [1958]: Maximum-minimum Sätze über Graphen. Acta Mathematica Academiae Scientiarum Hungaricae 9 (1958), 395–434

Goldberg, A.V., and Rao, S. [1998]: Beyond the flow decomposition barrier. Journal of the ACM 45 (1998), 783–797

Goldberg, A.V., and Tarjan, R.E. [1988]: A new approach to the maximum flow problem. Journal of the ACM 35 (1988), 921–940

Gomory, R.E., and Hu, T.C. [1961]: Multi-terminal network flows. Journal of SIAM 9 (1961), 551–570

Gusfield, D. [1990]: Very simple methods for all pairs network flow analysis. SIAM Journal on Computing 19 (1990), 143–155

Hao, J., and Orlin, J.B. [1994]: A faster algorithm for finding the minimum cut in a directed graph. Journal of Algorithms 17 (1994), 409–423

Henzinger, M.R., Rao, S., and Wang, D. [2017]: Local flow partitioning for faster edge connectivity. Proceedings of the 28th Annual ACM-SIAM Symposium on Discrete Algorithms (2017), 1919–1938

Henzinger, M.R., Rao, S., and Gabow, H.N. [2000]: Computing vertex connectivity: new bounds from old techniques. Journal of Algorithms 34 (2000), 222–250

Hoffman, A.J. [1960]: Some recent applications of the theory of linear inequalities to extremal combinatorial analysis. In: Combinatorial Analysis (R.E. Bellman, M. Hall, eds.), AMS, Providence 1960, pp. 113–128

Hu, T.C. [1969]: Integer Programming and Network Flows. Addison-Wesley, Reading 1969

Jelinek, F., and Mayeda, W. [1963]: On the maximum number of different entries in the terminal capacity matrix of oriented communication nets. IEEE Transactions on Circuit Theory 10 (1963), 307–308

Kamidoi, Y., Yoshida, N., and Nagamochi, H. [2007]: A deterministic algorithm for finding all minimum k-way cuts. SIAM Journal on Computing 36 (2007), 1329–1341

Karger, D.R. [2000]: Minimum cuts in near-linear time. Journal of the ACM 47 (2000), 46–76

Karger, D.R., and Levine, M.S. [1998]: Finding maximum flows in undirected graphs seems easier than bipartite matching. Proceedings of the 30th Annual ACM Symposium on Theory of Computing (1998), 69–78

Karger, D.R., and Stein, C. [1996]: A new approach to the minimum cut problem. Journal of the ACM 43 (1996), 601–640

Karzanov, A.V. [1973]: On finding a maximum flow in a network with special structure and some applications. In: Matematicheskie Voprosy Upravleniya Proizvodstvom 5 (L.A. Lyusternik, ed.), Moscow State University Press, Moscow, 1973, pp. 81–94 [in Russian]

Karzanov, A.V. [1974]: Determining a maximal flow in a network by the method of preflows. Soviet Mathematics Doklady 15 (1974), 434–437

Lee, Y.T., and Sidford, A. [2014]: Path finding methods for linear programming. Proceedings of the 55th Annual IEEE Symposium on Foundations of Computer Science (2014), 424–433

Mader, W. [1972]: Über minimal n-fach zusammenhängende, unendliche Graphen und ein Extremalproblem. Arch. Math. 23 (1972), 553–560

Mader, W. [1981]: On a property of n edge-connected digraphs. Combinatorica 1 (1981), 385–386

Malhotra, V.M., Kumar, M.P., and Maheshwari, S.N. [1978]: An $O(|V|^3)$ algorithm for finding maximum flows in networks. Information Processing Letters 7 (1978), 277–278

Menger, K. [1927]: Zur allgemeinen Kurventheorie. Fundamenta Mathematicae 10 (1927), 96–115

Nagamochi, H., and Ibaraki, T. [1992]: Computing edge-connectivity in multigraphs and capacitated graphs. SIAM Journal on Discrete Mathematics 5 (1992), 54–66

Nagamochi, H., and Ibaraki, T. [2000]: A fast algorithm for computing minimum 3-way and 4-way cuts. Mathematical Programming 88 (2000), 507–520

Nagamochi, H., Nishimura, K., and Ibaraki, T. [1997]: Computing all small cuts in an undirected network. SIAM Journal on Discrete Mathematics 10 (1997), 469–481

Orlin, J.B. [2013]: Max flows in $O(nm)$ time, or better. Proceedings of the 45th Annual ACM Symposium on Theory of Computing (2013), 765–774

Phillips, S., and Dessouky, M.I. [1977]: Solving the project time/cost tradeoff problem using the minimal cut concept. Management Science 24 (1977), 393–400

Picard, J., and Queyranne, M. [1980]: On the structure of all minimum cuts in a network and applications. Mathematical Programming Study 13 (1980), 8–16

Queyranne, M. [1998]: Minimizing symmetric submodular functions. Mathematical Programming B 82 (1998), 3–12

Rose, D.J. [1970]: Triangulated graphs and the elimination process. Journal of Mathematical Analysis and Applications 32 (1970), 597–609

Shiloach, Y. [1978]: An $O(nI\log^2 I)$ maximum-flow algorithm. Technical Report STAN-CS-78-802, Computer Science Department, Stanford University, 1978

Shiloach, Y. [1979]: Edge-disjoint branching in directed multigraphs. Information Processing Letters 8 (1979), 24–27

Shioura, A. [2004]: The MA ordering max-flow algorithm is not strongly polynomial for directed networks. Operations Research Letters 32 (2004), 31–35

Sleator, D.D. [1980]: An $O(nm\log n)$ algorithm for maximum network flow. Technical Report STAN-CS-80-831, Computer Science Department, Stanford University, 1978

Sleator, D.D., and Tarjan, R.E. [1983]: A data structure for dynamic trees. Journal of Computer and System Sciences 26 (1983), 362–391

Su, X.Y. [1997]: Some generalizations of Menger's theorem concerning arc-connected digraphs. Discrete Mathematics 175 (1997), 293–296

Stoer, M., and Wagner, F. [1997]: A simple min cut algorithm. Journal of the ACM 44 (1997), 585–591

Tarjan, R.E. [1972]: Depth first search and linear graph algorithms. SIAM Journal on Computing 1 (1972), 146–160

Thorup, M. [2008]: Minimum k-way cuts via deterministic greedy tree packing. Proceedings of the 40th Annual ACM Symposium on Theory of Computing (2008), 159–165

Tunçel, L. [1994]: On the complexity preflow-push algorithms for maximum flow problems. Algorithmica 11 (1994), 353–359

Vazirani, V.V., and Yannakakis, M. [1992]: Suboptimal cuts: their enumeration, weight, and number. In: Automata, Languages and Programming; Proceedings of the 19th ICALP conference; LNCS 623 (W. Kuich, ed.), Springer, Berlin 1992, pp. 366–377

Vygen, J. [2002]: On dual minimum cost flow algorithms. Mathematical Methods of Operations Research 56 (2002), 101–126

Weihe, K. [1997]: Maximum (s, t)-flows in planar networks in $O(|V| \log |V|)$ time. Journal of Computer and System Sciences 55 (1997), 454–475

Whitney, H. [1932]: Congruent graphs and the connectivity of graphs. American Journal of Mathematics 54 (1932), 150–168

9 Minimum Cost Flows

In this chapter we show how we can take edge costs into account. For example, in our application of the MAXIMUM FLOW PROBLEM to the JOB ASSIGNMENT PROBLEM mentioned in the introduction of Chapter 8 one could introduce edge costs to model that the employees have different salaries; our goal is to meet a deadline when all jobs must be finished at a minimum cost. Of course, there are many more applications.

A second generalization, allowing several sources and sinks, is more due to technical reasons. We introduce the general problem and an important special case in Section 9.1. In Section 9.2 we prove optimality criteria that are the basis of the minimum cost flow algorithms presented in Sections 9.3, 9.4, 9.5 and 9.6. Most of these use algorithms of Chapter 7 for finding a minimum mean cycle or a shortest path as a subroutine. Section 9.7 concludes this chapter with an application to time-dependent flows.

9.1 Problem Formulation

We are again given a digraph G with capacities $u : E(G) \rightarrow \mathbb{R}_+$, but in addition numbers $c : E(G) \rightarrow \mathbb{R}$ indicating the cost of each edge. Furthermore, we allow several sources and sinks:

Definition 9.1. *Given a digraph G, capacities $u : E(G) \rightarrow \mathbb{R}_+$, and numbers $b : V(G) \rightarrow \mathbb{R}$ with $\sum_{v \in V(G)} b(v) = 0$, a **b-flow** in (G, u) is a function $f : E(G) \rightarrow \mathbb{R}_+$ with $f(e) \leq u(e)$ for all $e \in E(G)$ and $\sum_{e \in \delta^+(v)} f(e) - \sum_{e \in \delta^-(v)} f(e) = b(v)$ for all $v \in V(G)$.*

Thus a b-flow with $b \equiv 0$ is a circulation. $b(v)$ is called the **balance** of vertex v. $|b(v)|$ is sometimes called the **supply** (if $b(v) > 0$) or the **demand** (if $b(v) < 0$) of v. Vertices v with $b(v) > 0$ are called **sources**, those with $b(v) < 0$ **sinks**.

Bernhard Korte and Jens Vygen, *Combinatorial Optimization*, Algorithms and Combinatorics 21, https://doi.org/10.1007/978-3-662-56039-6_9

Theorem 9.2. (Gale [1957]) *Let G be a digraph with capacities $u : E(G) \to \mathbb{R}_+$, and let $b : V(G) \to \mathbb{R}$ with $\sum_{v \in V(G)} b(v) = 0$. There exists a b-flow if and only if*

$$\sum_{e \in \delta^+(X)} u(e) \geq \sum_{v \in X} b(v) \quad \text{for all } X \subseteq V(G).$$

We can find one or decide that none exists in $O(n^2 \sqrt{m})$ time, where $n := |V(G)|$ and $m := |E(G)|$.

Proof: Let G' result from G by adding two vertices s and t and edges (s, v), (v, t) with capacities $u((s, v)) := \max\{0, b(v)\}$ and $u((v, t)) := \max\{0, -b(v)\}$ for all $v \in V(G)$. Let $B := \sum_{v \in V(G)} u((s, v)) = \sum_{v \in V(G)} u((v, t))$. Then the b-flows in (G, u) are precisely the restrictions of the s-t-flows with value B in (G', u) to $E(G)$. By the Max-Flow-Min-Cut Theorem 8.6, an s-t-flow of value B exists in (G', u) if and only if $\sum_{e \in \delta_{G'}^+(\{s\} \cup X)} u(e) \geq B$ for every $X \subseteq V(G)$. Note that $\sum_{e \in \delta_{G'}^+(\{s\} \cup X)} u(e) = \sum_{e \in \delta_G^+(X)} u(e) + B + \sum_{v \in X}(\max\{0, -b(v)\} - \max\{0, b(v)\}) = \sum_{e \in \delta_G^+(X)} u(e) + B - \sum_{v \in X} b(v)$, implying the first claim.

We can find a b-flow or decide that none exists by solving the MAXIMUM FLOW PROBLEM in (G', u, s, t). The running time follows from Theorem 8.30. $\square$

The problem that we study in this chapter is to find a minimum cost b-flow:

MINIMUM COST FLOW PROBLEM

Instance: A digraph G, capacities $u : E(G) \to \mathbb{R}_+$, numbers $b : V(G) \to \mathbb{R}$ with $\sum_{v \in V(G)} b(v) = 0$, and weights $c : E(G) \to \mathbb{R}$.

Task: Find a b-flow f whose cost $c(f) := \sum_{e \in E(G)} f(e)c(e)$ is minimum (or decide that none exists).

Sometimes one also allows infinite capacities. In this case an instance can be unbounded, but this can be checked in advance easily; see Exercise 5. Note that (in contrast to many other problems, e.g. the MAXIMUM FLOW PROBLEM) we cannot assume without loss of generality that the input graph is simple when considering the MINIMUM COST FLOW PROBLEM.

The MINIMUM COST FLOW PROBLEM is quite general and has a couple of interesting special cases. The uncapacitated case ($u \equiv \infty$) is sometimes called the transshipment problem. An even more restricted problem, also known as the transportation problem, has been formulated quite early by Hitchcock [1941] and others:

HITCHCOCK PROBLEM

Instance: A digraph G with $V(G) = A \cup B$ and $E(G) \subseteq A \times B$. Supplies $b(v) \geq 0$ for $v \in A$ and demands $-b(v) \geq 0$ for $v \in B$ with $\sum_{v \in V(G)} b(v) = 0$. Weights $c : E(G) \to \mathbb{R}$.

Task: Find a b-flow f in (G, ∞) of minimum cost (or decide that none exists).

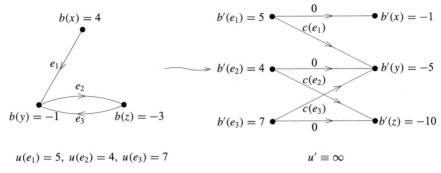

$u(e_1) = 5, \; u(e_2) = 4, \; u(e_3) = 7$

$u' \equiv \infty$

Fig. 9.1.

In the HITCHCOCK PROBLEM it causes no loss of generality to assume that c is nonnegative: Adding a constant α to each weight increases the cost of each b-flow by the same amount, namely by $\alpha \sum_{v \in A} b(v)$. Often only the special case where c is nonnegative and $E(G) = A \times B$ is considered.

Obviously, any instance of the HITCHCOCK PROBLEM can be written as an instance of the MINIMUM COST FLOW PROBLEM on a bipartite graph with infinite capacities. It is less obvious that any instance of the MINIMUM COST FLOW PROBLEM can be transformed to an equivalent (but larger) instance of the HITCHCOCK PROBLEM:

Lemma 9.3.   (Orden [1956], Wagner [1959])   *An instance of the* MINIMUM COST FLOW PROBLEM *with n vertices and m edges can be transformed to an equivalent instance of the* HITCHCOCK PROBLEM *with $n + m$ vertices and $2m$ edges.*

Proof: Let (G, u, b, c) be an instance of the MINIMUM COST FLOW PROBLEM. We define an equivalent instance (G', A', B', b', c') of the HITCHCOCK PROBLEM as follows:

Let $A' := E(G)$, $B' := V(G)$ and $G' := (A' \cup B', E_1 \cup E_2)$, where $E_1 := \{((x, y), x) : (x, y) \in E(G)\}$ and $E_2 := \{((x, y), y) : (x, y) \in E(G)\}$. Let $c'((e, x)) := 0$ for $(e, x) \in E_1$ and $c'((e, y)) := c(e)$ for $(e, y) \in E_2$. Finally let $b'(e) := u(e)$ for $e \in E(G)$ and

$$b'(x) \; := \; b(x) - \sum_{e \in \delta_G^+(x)} u(e) \quad \text{for } x \in V(G).$$

For an example, see Figure 9.1.

We prove that both instances are equivalent. Let f be a b-flow in (G, u). Define $f'((e, y)) := f(e)$ and $f'((e, x)) := u(e) - f(e)$ for $e = (x, y) \in E(G)$. Obviously f' is a b'-flow in G' with $c'(f') = c(f)$.

Conversely, if f' is a b'-flow in G', then $f((x, y)) := f'(((x, y), y))$ defines a b-flow in G with $c(f) = c'(f')$. $\square$

The above proof is due to Ford and Fulkerson [1962].

9.2 An Optimality Criterion

In this section we prove some simple results, in particular an optimality criterion, which will be the basis for the algorithms in the subsequent sections. We again use the concepts of residual graphs and augmenting paths. We extend the weights c to $\overleftrightarrow{G}$ by defining $c(\overleftarrow{e}) := -c(e)$ for each edge $e \in E(G)$. Our definition of a residual graph has the advantage that the weight of an edge in a residual graph G_f is independent of the flow f.

The following simple observation will prove useful:

Proposition 9.4. *Let (G, u, b, c) be an instance of the* MINIMUM COST FLOW PROBLEM. *Let f and f' be b-flows in (G, u). Then $g : E(\overleftrightarrow{G}) \to \mathbb{R}_+$ defined by $g(e) := \max\{0, f'(e) - f(e)\}$ and $g(\overleftarrow{e}) := \max\{0, f(e) - f'(e)\}$ for $e \in E(G)$ is a circulation in $\overleftrightarrow{G}$. Furthermore, $g(e) = 0$ for all $e \notin E(G_f)$ and $c(g) = c(f') - c(f)$.*

Proof: At each vertex $v \in V(\overleftrightarrow{G})$ we have

$$\sum_{e \in \delta^+_{\overleftrightarrow{G}}(v)} g(e) \;-\; \sum_{e \in \delta^-_{\overleftrightarrow{G}}(v)} g(e) \;=\; \sum_{e \in \delta^+_G(v)} (f'(e) - f(e)) \;-\; \sum_{e \in \delta^-_G(v)} (f'(e) - f(e))$$

$$=\; b(v) - b(v) \;=\; 0,$$

so g is a circulation in $\overleftrightarrow{G}$.

For $e \in E(\overleftrightarrow{G}) \setminus E(G_f)$ we consider two cases: If $e \in E(G)$ then $f(e) = u(e)$ and thus $f'(e) \le f(e)$, implying $g(e) = 0$. If $e = \overleftarrow{e_0}$ for some $e_0 \in E(G)$ then $f(e_0) = 0$ and thus $g(\overleftarrow{e_0}) = 0$.

The last statement is easily verified:

$$c(g) \;=\; \sum_{e \in E(\overleftrightarrow{G})} c(e) g(e) \;=\; \sum_{e \in E(G)} c(e) f'(e) - \sum_{e \in E(G)} c(e) f(e) \;=\; c(f') - c(f).$$

$\square$

Just as Eulerian graphs can be partitioned into circuits, circulations can be decomposed into flows on single circuits:

Proposition 9.5. (Ford and Fulkerson [1962]) *For any circulation f in a digraph G there is a family $\mathcal{C}$ of at most $|E(G)|$ circuits in G and positive numbers $h(C)$ $(C \in \mathcal{C})$ such that $f(e) = \sum_{C \in \mathcal{C}, e \in E(C)} h(C)$ for all $e \in E(G)$.*

Proof: This is a special case of Theorem 8.8. $\square$

Definition 9.6. *Given a digraph G with capacities $u : E(G) \to \mathbb{R}_+$ and a b-flow f in (G, u), an f-augmenting cycle is a circuit in G_f.*

Now we can prove an optimality criterion:

Theorem 9.7. (Klein [1967]) *Let (G, u, b, c) be an instance of the* MINIMUM COST FLOW PROBLEM. *A b-flow f is of minimum cost if and only if there is no f-augmenting cycle with negative total weight.*

Proof: If there is an f-augmenting cycle C with weight $\gamma < 0$, we can augment f along C by some $\varepsilon > 0$ and get a b-flow f' with cost decreased by $-\gamma\varepsilon$. So f is not a minimum cost flow.

If f is not a minimum cost b-flow, there is another b-flow f' with smaller cost. Consider g as defined in Proposition 9.4. Then g is a circulation with $c(g) < 0$. By Proposition 9.5, g can be decomposed into flows on single circuits. Since $g(e) = 0$ for all $e \notin E(G_f)$, all these circuits are f-augmenting. At least one of them must have negative total weight, proving the theorem. □

This result goes back essentially to Tolstoĭ [1930] and has been rediscovered several times in different forms. One equivalent formulation is the following:

Corollary 9.8. (Ford and Fulkerson [1962]) *Let (G, u, b, c) be an instance of the* MINIMUM COST FLOW PROBLEM. *A b-flow f is of minimum cost if and only if there exists a feasible potential for (G_f, c).*

Proof: By Theorem 9.7 f is a minimum cost b-flow if and only if G_f contains no negative circuit. By Theorem 7.7 there is no negative circuit in (G_f, c) if and only if there exists a feasible potential. □

Feasible potentials can also be regarded as solutions of the linear programming dual of the MINIMUM COST FLOW PROBLEM. This is shown by the following different proof of the above optimality criterion:

Second Proof of Corollary 9.8:   We write the MINIMUM COST FLOW PROBLEM as a maximization problem and consider the LP

$$
\max \quad \sum_{e \in E(G)} -c(e)x_e
$$

$$
\text{s.t.} \quad \sum_{e \in \delta^+(v)} x_e - \sum_{e \in \delta^-(v)} x_e \;=\; b(v) \qquad (v \in V(G)) \qquad (9.1)
$$

$$
x_e \;\leq\; u(e) \qquad (e \in E(G))
$$

$$
x_e \;\geq\; 0 \qquad (e \in E(G))
$$

and its dual

$$
\min \quad \sum_{v \in V(G)} b(v)y_v + \sum_{e \in E(G)} u(e)z_e
$$

$$
\qquad\qquad\qquad\qquad\qquad\qquad\qquad\qquad\qquad (9.2)
$$

$$
\text{s.t.} \quad y_v - y_w + z_e \;\geq\; -c(e) \qquad (e = (v, w) \in E(G))
$$

$$
z_e \;\geq\; 0 \qquad (e \in E(G))
$$

Let x be any b-flow, i.e. any feasible solution of (9.1). By Corollary 3.23 x is optimum if and only if there exists a feasible dual solution (y, z) of (9.2) such that x and (y, z) satisfy the complementary slackness conditions

$$z_e(u(e) - x_e) = 0 \quad \text{and} \quad x_e(c(e) + z_e + y_v - y_w) = 0 \quad \text{for all } e = (v, w) \in E(G).$$

So x is optimum if and only if there exists a pair of vectors (y, z) with

$$0 = -z_e \leq c(e) + y_v - y_w \qquad \text{for } e = (v, w) \in E(G) \text{ with } x_e < u(e) \quad \text{and}$$
$$c(e) + y_v - y_w = -z_e \leq 0 \qquad \text{for } e = (v, w) \in E(G) \text{ with } x_e > 0.$$

This is equivalent to the existence of a vector y such that $c(e) + y_v - y_w \geq 0$ for all residual edges $e = (v, w) \in E(G_x)$, i.e. to the existence of a feasible potential y for (G_x, c). $\qquad\square$

9.3 Minimum Mean Cycle-Cancelling Algorithm

Note that Klein's Theorem 9.7 already suggests an algorithm: first find any b-flow (using a max-flow algorithm as described in Theorem 9.2), and then successively augment along negative weight augmenting cycles until no more exist. We must however be careful in choosing the cycle if we want to have polynomial running time (see Exercise 7). A good strategy is to choose an augmenting cycle with minimum mean weight each time:

MINIMUM MEAN CYCLE-CANCELLING ALGORITHM

Input: A digraph G, capacities $u : E(G) \to \mathbb{R}_+$, numbers $b : V(G) \to \mathbb{R}$ with $\sum_{v \in V(G)} b(v) = 0$, and weights $c : E(G) \to \mathbb{R}$.

Output: A minimum cost b-flow f.

---

①   Find a b-flow f.

②   Find a circuit C in G_f whose mean weight is minimum.
      If C has nonnegative total weight (or G_f is acyclic) **then stop**.

③   Compute $\gamma := \min_{e \in E(C)} u_f(e)$. Augment f along C by γ.

      Go to ②.

As described in the proof of Theorem 9.2, ① can be implemented with any algorithm for the MAXIMUM FLOW PROBLEM. ② can be implemented with the algorithm presented in Section 7.3. We shall now prove that this algorithm terminates after a polynomial number of iterations. The proof will be similar to the one in Section 8.3. Let $\mu(f)$ denote the minimum mean weight of a circuit in G_f. Then Theorem 9.7 says that a b-flow f is optimum if and only if $\mu(f) \geq 0$.

We first show that $\mu(f)$ is non-decreasing throughout the algorithm. Moreover, we can show that it is strictly increasing with every $|E(G)|$ iterations. As usual we denote by n and m the number of vertices and edges of G, respectively.

Lemma 9.9. *Let (G, u, b, c) be an instance of the* MINIMUM COST FLOW PROB-LEM. *Let $f_1, f_2, \ldots, f_t$ be a sequence of b-flows such that for $i = 1, \ldots, t - 1$ we have $\mu(f_i) < 0$ and f_{i+1} results from f_i by augmenting along C_i, where C_i is a circuit of minimum mean weight in G_{f_i}. Then*

(a) $\mu(f_k) \leq \mu(f_{k+1})$ *for all k.*
(b) $\mu(f_k) \leq \frac{n}{n-2} \mu(f_l)$ *for all $k < l$ such that $C_k \cup C_l$ contains a pair of reverse edges.*

Proof: (a): Let f_k, f_{k+1} be two consecutive flows in this sequence. Consider the Eulerian graph H resulting from $(V(G), E(C_k) \dot\cup E(C_{k+1}))$ by deleting pairs of reverse edges. (Edges appearing both in C_k and C_{k+1} are counted twice.) Every simple subgraph of H is a subgraph of G_{f_k}, because each edge in $E(G_{f_{k+1}}) \setminus E(G_{f_k})$ must be the reverse of an edge in $E(C_k)$. Since H is Eulerian, it can be decomposed into circuits, and each of these circuits has mean weight at least $\mu(f_k)$. So $c(E(H)) \geq \mu(f_k)|E(H)|$.

Since the total weight of each pair of reverse edges is zero,

$$c(E(H)) \;=\; c(E(C_k)) + c(E(C_{k+1})) \;=\; \mu(f_k)|E(C_k)| + \mu(f_{k+1})|E(C_{k+1})|.$$

Since $|E(H)| \leq |E(C_k)| + |E(C_{k+1})|$, we conclude

$$
\begin{aligned}
\mu(f_k)(|E(C_k)| + |E(C_{k+1})|) \;&\leq\; \mu(f_k)|E(H)| \\
&\leq\; c(E(H)) \\
&=\; \mu(f_k)|E(C_k)| + \mu(f_{k+1})|E(C_{k+1})|,
\end{aligned}
$$

implying $\mu(f_{k+1}) \geq \mu(f_k)$.

(b): By (a) it is enough to prove the statement for those k, l such that for $k < i < l$, $C_i \cup C_l$ contains no pair of reverse edges.

As in the proof of (a), consider the Eulerian graph H resulting from $(V(G), E(C_k) \dot\cup E(C_l))$ by deleting pairs of reverse edges. Every simple subgraph of H is a subgraph of G_{f_k} because any edge in $E(C_l) \setminus E(G_{f_k})$ must be the reverse of an edge in one of $C_k, C_{k+1}, \ldots, C_{l-1}$. But – due to the choice of k and l – only C_k among these contains the reverse of an edge of C_l.

So as in (a) we have $c(E(H)) \geq \mu(f_k)|E(H)|$ and

$$c(E(H)) \;=\; \mu(f_k)|E(C_k)| + \mu(f_l)|E(C_l)|.$$

Since $|E(H)| \leq |E(C_k)| + \frac{n-2}{n}|E(C_l)|$ (we deleted at least two edges) we get

$$
\begin{aligned}
\mu(f_k)\left(|E(C_k)| + \frac{n-2}{n}|E(C_l)|\right) \;&\leq\; \mu(f_k)|E(H)| \\
&\leq\; c(E(H)) \\
&=\; \mu(f_k)|E(C_k)| + \mu(f_l)|E(C_l)|,
\end{aligned}
$$

implying $\mu(f_k) \leq \frac{n}{n-2} \mu(f_l)$. $\square$

Corollary 9.10. *During the execution of the* MINIMUM MEAN CYCLE-CANCELLING ALGORITHM, $|\mu(f)|$ *decreases by at least a factor of 2 with every mn iterations.*

Proof: Let $C_k, C_{k+1}, \ldots, C_{k+m}$ be the augmenting cycles in consecutive iterations of the algorithm. Since each of these circuits contains one edge as a bottleneck edge (an edge removed afterwards from the residual graph), there must be two of these circuits, say C_i and C_j ($k \le i < j \le k + m$) whose union contains a pair of reverse edges. By Lemma 9.9 we then have

$$\mu(f_k) \;\le\; \mu(f_i) \;\le\; \frac{n}{n-2}\mu(f_j) \;\le\; \frac{n}{n-2}\mu(f_{k+m}).$$

So $|\mu(f)|$ decreases by at least a factor of $\frac{n}{n-2}$ with every m iterations. The corollary follows from this because of $\left(\frac{n}{n-2}\right)^n > e^2 > 2$. $\qquad\square$

This already proves that the algorithm runs in polynomial time provided that all edge costs are integral: $|\mu(f)|$ is at most $|c_{\min}|$ at the beginning, where $c_{\min}$ is the minimum cost of any edge, and decreases by at least a factor of 2 with every mn iterations (or $\mu(f)$ becomes positive and the algorithms stops). So after $O(mn\log(n|c_{\min}|))$ iterations, $\mu(f)$ is greater than $-\frac{1}{n}$. If the edge costs are integral, this implies $\mu(f) \ge 0$ and the algorithm stops. So by Corollary 7.13, the running time is $O\left(m^2n^2\log(n|c_{\min}|)\right)$.

Even better, we can also derive a strongly polynomial running time for the MINIMUM COST FLOW PROBLEM (first obtained by Tardos [1985]):

Theorem 9.11. (Goldberg and Tarjan [1989]) *The* MINIMUM MEAN CYCLE-CANCELLING ALGORITHM *runs in* $O\left(m^3n^2\log n\right)$ *time.*

Proof: We show that every $mn(\lceil \log n\rceil + 1)$ iterations at least one edge is fixed, i.e. the flow on this edge changes for the last time. Therefore there are at most $O\left(m^2n\log n\right)$ iterations. Using Theorem 9.2 for ① and Corollary 7.13 for ② then proves the theorem.

Let f be the flow at some iteration, and let f' be the flow $mn(\lceil\log n\rceil + 1)$ iterations later. Define weights c' by $c'(e) := c(e) - \mu(f')$ ($e \in E(G_{f'})$). Let π be a feasible potential of $(G_{f'}, c')$ (which exists by Theorem 7.7). We have $0 \le c'_\pi(e) = c_\pi(e) - \mu(f')$, so

$$c_\pi(e) \;\ge\; \mu(f') \quad \text{for all } e \in E(G_{f'}). \tag{9.3}$$

Now let C be the circuit of minimum mean weight in G_f that is chosen in the algorithm to augment f. Since by Corollary 9.10

$$\mu(f) \;\le\; 2^{\lceil\log n\rceil+1}\mu(f') \;\le\; 2n\mu(f')$$

Fig. 9.2.

(see Figure 9.2), we have

$$\sum_{e\in E(C)} c_\pi(e) = \sum_{e\in E(C)} c(e) = \mu(f)|E(C)| \leq 2n\mu(f')|E(C)|.$$

So let $e_0 \in E(C)$ with $c_\pi(e_0) \leq 2n\mu(f')$. By (9.3) we have $e_0 \notin E(G_{f'})$.

Claim: For any b-flow f'' with $e_0 \in E(G_{f''})$ we have $\mu(f'') < \mu(f')$.

By Lemma 9.9(a) the claim implies that e_0 will never be in the residual graph anymore, i.e. e_0 and $\overleftarrow{e_0}$ are fixed $mn(\lceil \log n \rceil + 1)$ iterations after e_0 is used in C. This completes the proof.

To prove the claim, let f'' be a b-flow with $e_0 \in E(G_{f''})$. We apply Proposition 9.4 to f' and f'' and obtain a circulation g with $g(e) = 0$ for all $e \notin E(G_{f'})$ and $g(\overleftarrow{e_0}) > 0$ (because $e_0 \in E(G_{f''}) \setminus E(G_{f'})$).

By Proposition 9.5, g can be written as the sum of flows on f'-augmenting cycles. One of these circuits, say W, must contain $\overleftarrow{e_0}$. By using $c_\pi(\overleftarrow{e_0}) = -c_\pi(e_0) \geq -2n\mu(f')$ and applying (9.3) to all $e \in E(W) \setminus \{\overleftarrow{e_0}\}$ we obtain a lower bound for the total weight of W:

$$c(E(W)) = \sum_{e\in E(W)} c_\pi(e) \geq -2n\mu(f') + (n-1)\mu(f') > -n\mu(f').$$

But the reverse of W is an f''-augmenting cycle (this can be seen by exchanging the roles of f' and f''), and its total weight is less than $n\mu(f')$. This means that $G_{f''}$ contains a circuit whose mean weight is less than $\mu(f')$, and so the claim is proved.   □

The MINIMUM MEAN CYCLE-CANCELLING ALGORITHM was generalized in various ways by Karzanov and McCormick [1997].

9.4 Successive Shortest Path Algorithm

The following theorem gives rise to another algorithm:

Theorem 9.12. (Jewell [1958], Iri [1960], Busacker and Gowen [1961]) *Let (G, u, b, c) be an instance of the* MINIMUM COST FLOW PROBLEM, *and let f be a minimum cost b-flow. Let P be a shortest (with respect to c) s-t-path in G_f (for some s and t). Let f' be a flow obtained when augmenting f along P by at most the minimum residual capacity on P. Then f' is a minimum cost b'-flow (for some b').*

Proof: f' is a b'-flow for some b'. Suppose f' is not a minimum cost b'-flow. Then by Theorem 9.7 there is a circuit C in $G_{f'}$ with negative total weight. Consider the graph H resulting from $(V(G), E(C) \cup E(P))$ by deleting pairs of reverse edges. (Again, edges appearing both in C and P are taken twice.)

For any edge $e \in E(G_{f'}) \setminus E(G_f)$, the reverse of e must be in $E(P)$. Therefore every simple subgraph of H is a subgraph of G_f. As f is a minimum cost b-flow, no circuit in G_f, and hence in H, has negative total weight.

We have $c(E(H)) = c(E(C)) + c(E(P)) < c(E(P))$. Furthermore, H is the union of an s-t-path and some circuits. None of these circuits has negative weight. Therefore H, and thus G_f, contains an s-t-path of less weight than P, contradicting the choice of P. □

If the weights are conservative, we can start with $f \equiv 0$ as an optimum circulation (b-flow with $b \equiv 0$). Otherwise we can initially saturate all edges of negative cost, i.e., set $f(e) := u(e)$ for $e \in F := \{e' \in E(G) : c(e') < 0\}$ and $f(e) := 0$ for $e \in E(G) \setminus F$, and find a minimum cost b'-flow in (G_f, u_f), where $b'(v) = b(v) + \mathrm{ex}_f(v)$ for $v \in V(G)$. For instances with infinite capacities more work is needed; see Exercise 5.

SUCCESSIVE SHORTEST PATH ALGORITHM

Input: A digraph G, capacities $u : E(G) \to \mathbb{R}_+$, numbers $b : V(G) \to \mathbb{R}$
with $\sum_{v \in V(G)} b(v) = 0$, and conservative weights $c : E(G) \to \mathbb{R}$.

Output: A minimum cost b-flow f.

①   Set $b' := b$ and $f(e) := 0$ for all $e \in E(G)$.

②   **If** $b' = 0$ **then stop, else:**
Choose a vertex s with $b'(s) > 0$.
Choose a vertex t with $b'(t) < 0$ such that t is reachable from s in G_f.
If there is no such t **then stop.** (There exists no b-flow.)

③   Find an s-t-path P in G_f of minimum weight.

④   Compute $\gamma := \min \left\{ \min_{e \in E(P)} u_f(e), b'(s), -b'(t) \right\}$.
Set $b'(s) := b'(s) - \gamma$ and $b'(t) := b'(t) + \gamma$. Augment f along P by γ.
Go to ②.

If we allow arbitrary capacities, we have the same problems as with the FORD-FULKERSON ALGORITHM (see Exercise 2 of Chapter 8; set all costs to zero). So henceforth we assume u and b to be integral. Then we have:

Theorem 9.13. *For integer u and b, the* SUCCESSIVE SHORTEST PATH ALGO-RITHM *is correct and terminates after at most $B := \frac{1}{2} \sum_{v \in V(G)} |b(v)|$ iterations.*

Proof: The bound on the number of iterations is obvious. The algorithm maintains the invariant $b(v) = b'(v) - \mathrm{ex}_f(v)$ for all $v \in V(G)$. By Theorem 9.12, the resulting flow is optimum if the initial zero flow is optimum. This is true if and only if c is conservative.

If the algorithm decides that there is no b-flow, this decision is indeed correct: let R be the set of vertices reachable from a vertex s with $b'(s) > 0$ in G_f, and suppose that $b'(v) \geq 0$ for all $v \in R$. Then $b(R) = \sum_{v \in R}(b'(v) - \text{ex}_f(v)) > f(\delta^+(R)) - f(\delta^-(R)) = u(\delta^+(R))$, and there is no b-flow by Gale's Theorem 9.2. $\qquad\square$

Each augmentation requires a shortest path computation. Since negative weights occur, we have to use the MOORE-BELLMAN-FORD ALGORITHM whose running time is $O(nm)$ (Theorem 7.5), so the overall running time will be $O(Bnm)$. However, as in the proof of Theorem 7.8, it can be arranged that (except at the beginning) the shortest paths are computed in a graph with nonnegative weights:

Theorem 9.14. (Tomizawa [1971], Edmonds and Karp [1972]) *If u and b are integral, the* SUCCESSIVE SHORTEST PATH ALGORITHM *can be implemented with a running time of* $O\left(nm + B(m + n \log n)\right)$*, where* $B = \frac{1}{2}\sum_{v \in V(G)}|b(v)|$.

Proof: We introduce a potential $\pi_i : V(G) \to \mathbb{R}$ for each iteration i of the SUCCESSIVE SHORTEST PATH ALGORITHM. We start with any feasible potential π_0 of (G, c). By Corollary 7.7, this exists and can be computed in $O(mn)$ time.

Now let f_{i-1} be the flow before iteration i. Then the shortest path computation in iteration i is done with the reduced costs $c_{\pi_{i-1}}$ instead of c. Moreover, we extend $G_{f_{i-1}}$ to $G'_{f_{i-1}}$ by adding edges (t, v) with reduced cost zero for all $v \in V(G)\backslash\{s, t\}$ (this is to ensure that all vertices are reachable from s). For $v \in V(G)$, let $l_i(v)$ denote the length of a shortest s-v-path in $G'_{f_{i-1}}$ with respect to the weights $c_{\pi_{i-1}}$. Then we set $\pi_i(v) := \pi_{i-1}(v) + l_i(v)$.

We prove by induction on i that π_i is a feasible potential for (G_{f_i}, c). This is clear for $i = 0$. For $i > 0$ and any edge $e \in E(\overset{\leftrightarrow}{G})$ we have

$$c_{\pi_i}(e) = c(e) + \pi_i(x) - \pi_i(y) = c(e) + \pi_{i-1}(x) + l_i(x) - \pi_{i-1}(y) - l_i(y)$$
$$= c_{\pi_{i-1}}(e) + l_i(x) - l_i(y).$$

For $e = (x, y) \in E(G_{f_{i-1}})$ we have (by definition of l_i and the induction hypothesis) $l_i(y) \leq l_i(x) + c_{\pi_{i-1}}(e)$ and hence $c_{\pi_i}(e) \geq 0$. For any edge $e = (x, y) \in P_i$ (where P_i is the augmenting path in iteration i) we have $l_i(y) = l_i(x) + c_{\pi_{i-1}}(e)$ (because P_i is a shortest path) and hence $c_{\pi_i}(\overset{\leftarrow}{e}) = -c_{\pi_i}(e) = 0$. Since each edge in $E(G_{f_i}) \backslash E(G_{f_{i-1}})$ is the reverse of an edge in P_i, c_{π_i} is indeed a nonnegative weight function on $E(G_{f_i})$.

We observe that in iteration i the shortest s-t-paths in $G_{f_{i-1}}$ with respect to c are precisely the shortest s-t-paths in $G'_{f_{i-1}}$ with respect to $c_{\pi_{i-1}}$, because the added edges leaving t cannot occur and $c_{\pi_{i-1}}(P) - c(P) = \pi_{i-1}(s) - \pi_{i-1}(t)$ for any s-t-path P.

Hence we can use DIJKSTRA'S ALGORITHM – which runs in $O(m + n \log n)$ time when implemented with a Fibonacci heap by Theorem 7.4 – for all shortest path computations except the initial one. Since we have at most B iterations, we obtain an overall running time of $O(nm + B(m + n \log n))$.   □

The running time of Theorem 9.14 is still exponential unless B is known to be small. If $B = O(n)$, this is the fastest algorithm known. For an application, see Section 11.1.

In the rest of this section we show how to modify the algorithm in order to reduce the number of shortest path computations. We only consider the case when all capacities are infinite. By Lemma 9.3 each instance of the MINIMUM COST FLOW PROBLEM can be transformed to an equivalent instance with infinite capacities.

The basic idea – due to Edmonds and Karp [1972] – is the following. In early iterations we consider only augmenting paths where γ – the amount of flow that can be pushed – is large. We start with $\gamma = 2^{\lfloor \log b_{\max} \rfloor}$ and reduce γ by a factor of two if no more augmentations by γ can be done. After $\lfloor \log b_{\max} \rfloor + 1$ iterations we have $\gamma = 1$ and stop (we again assume b to be integral). Such a scaling technique has proved useful for many algorithms (see also Exercise 13). A detailed description of the first scaling algorithm reads as follows:

CAPACITY SCALING ALGORITHM

Input: A digraph G with infinite capacities $u(e) = \infty$ ($e \in E(G)$), numbers $b : V(G) \to \mathbb{Z}$ with $\sum_{v \in V(G)} b(v) = 0$, and conservative weights $c : E(G) \to \mathbb{R}$.

Output: A minimum cost b-flow f.

①   Set $b' := b$ and $f(e) := 0$ for all $e \in E(G)$.
      Set $\gamma = 2^{\lfloor \log b_{\max} \rfloor}$, where $b_{\max} = \max\{b(v) : v \in V(G)\}$.

②   **If $b' = 0$ then stop, else:**
      Choose vertices s and t with $b'(s) \geq \gamma$ and $b'(t) \leq -\gamma$ such that t is reachable from s in G_f.
      If there is no such pair (s, t) then go to ⑤.

③   Find an s-t-path P in G_f of minimum weight.

④   Set $b'(s) := b'(s) - \gamma$ and $b'(t) := b'(t) + \gamma$. Augment f along P by γ.
      Go to ②.

⑤   **If $\gamma = 1$ then stop.** (There exists no b-flow.)
                  Else set $\gamma := \frac{\gamma}{2}$ and **go to ②.**

Theorem 9.15. (Edmonds and Karp [1972]) *The CAPACITY SCALING ALGORITHM correctly solves the MINIMUM COST FLOW PROBLEM for integral b, infinite capacities and conservative weights. It can be implemented to run in $O(n(m + n \log n) \log(2 + b_{\max}))$ time, where $b_{\max} = \max\{b(v) : v \in V(G)\}$.*

Proof: As above, the correctness follows directly from Theorem 9.12. Note that at any time, the residual capacity of any edge is either infinite or an integral multiple of γ.

To establish the running time, we call the period in which γ remains constant a *phase*. We prove that there are at most n augmentations within each phase. Let γ, f, and b' be the values at the beginning of some phase. Let $S := \{v \in V(G) : b'(v) \geq \gamma\}$, $S^+ := \{v \in V(G) : b'(v) \geq 2\gamma\}$, $T := \{v \in V(G) : b'(v) \leq -\gamma\}$, and $T^+ := \{v \in V(G) : b'(v) \leq -2\gamma\}$. Let R be the set of vertices that are reachable from some element of S^+ in G_f. Note that $S^+ \subseteq R$. We also have $R \cap T^+ = \emptyset$, because otherwise the previous phase would not have ended.

Let (s_i, t_i), $i = 1, \ldots, k$, be the pairs (s, t) in the considered phase; repetitions are of course possible. As $\delta^+_{G_f}(R) = \emptyset$, we have $|\{i : s_i \in R, t_i \notin R\}| \leq |\{i : s_i \notin R, t_i \in R\}|$. Thus we can bound the number k of iterations in this phase by $k \leq |\{i : s_i, t_i \in R\}| + 2|\{i : s_i \notin R, t_i \in R\}| + |\{i : s_i, t_i \notin R\}| = |\{i : t_i \in R\}| + |\{i : s_i \notin R\}| \leq |T \cap R| + |S \setminus R| \leq n$, where the second to last inequality holds because $T^+ \cap R = \emptyset$ and $S^+ \setminus R = \emptyset$.

This means that the total number of shortest path computations is $O(n \log(2 + b_{\max}))$. Combining this with the technique of Theorem 9.14 we obtain the $O(mn + n \log(2 + b_{\max})(m + n \log n))$ bound. $\qquad \square$

This was the first polynomial-time algorithm for the MINIMUM COST FLOW PROBLEM. By some further modifications we can even obtain a strongly polynomial running time. This is the subject of the next section.

9.5 Orlin's Algorithm

The CAPACITY SCALING ALGORITHM of the previous section can be improved further. A basic idea is that if an edge carries more than $2n\gamma$ units of flow at any stage of the CAPACITY SCALING ALGORITHM, it may be contracted. Namely, observe that such an edge will always keep a positive flow (and thus zero reduced cost with respect to any feasible potential in the residual graph): there are at most n more augmentations by γ, another n by $\frac{\gamma}{2}$ and so on; hence the total amount of flow moved in the rest of the algorithm is less than $2n\gamma$.

We shall describe ORLIN'S ALGORITHM without explicitly using contraction. This simplifies the description, especially from the point of view of implementing the algorithm. A set F keeps track of the edges (and their reverse edges) that can be contracted. $(V(G), F)$ will always arise from a forest by orienting each edge in both directions. A representative is chosen out of each connected component of $(V(G), F)$. The algorithm maintains the property that the representative of a connected component is its only non-balanced vertex. This will lead to more augmentations in each phase, but still by less than $4n\gamma$.

For any vertex x, $r(x)$ denotes the representative of the connected component of $(V(G), F)$ containing x. G' denotes the subgraph of G containing the edges in F and all edges (x, y) with $r(x) \neq r(y)$. Edges outside G' will not be used anymore.

ORLIN'S ALGORITHM does not require that b is integral. However, it can deal with uncapacitated problems only (but recall Lemma 9.3).

ORLIN'S ALGORITHM

Input:	A digraph G with infinite capacities $u(e) = \infty$ ($e \in E(G)$), numbers $b : V(G) \to \mathbb{R}$ with $\sum_{v \in V(G)} b(v) = 0$, and conservative weights $c : E(G) \to \mathbb{R}$.
Output:	A minimum cost b-flow f.

---

①   Set $b' := b$ and $f(e) := 0$ for all $e \in E(G)$.
Set $r(v) := v$ for all $v \in V(G)$. Set $F := \emptyset$ and $G' := G$.
Set $\gamma = \max_{v \in V(G)} |b'(v)|$.

②   **If $b' = 0$ then stop.**

③   Choose a vertex s with $b'(s) > \frac{n-1}{n}\gamma$.
If there is no such s then go to ④.
Choose a vertex t with $b'(t) < -\frac{1}{n}\gamma$ such that t is reachable from s in G_f.
If there is no such t then stop. (There exists no b-flow.)
Go to ⑤.

④   Choose a vertex t with $b'(t) < -\frac{n-1}{n}\gamma$.
If there is no such t then go to ⑥.
Choose a vertex s with $b'(s) > \frac{1}{n}\gamma$ such that t is reachable from s in G_f.
If there is no such s then stop. (There exists no b-flow.)

⑤   Find an s-t-path P in G'_f of minimum weight.
Set $b'(s) := b'(s) - \gamma$ and $b'(t) := b'(t) + \gamma$. Augment f along P by γ.
Go to ②.

⑥   **If $f(e) = 0$ for all $e \in E(G') \setminus F$ then** set $\gamma := \min\left\{\frac{\gamma}{2}, \max_{v \in V(G)} |b'(v)|\right\}$,

                                     else set $\gamma := \frac{\gamma}{2}$.

⑦   **While** there is an $e = (x, y) \in E(G') \setminus F$ with $f(e) > 6n\gamma$ **do:**
         Set $F := F \cup \{e, \overleftarrow{e}\}$.
         Let $x' := r(x)$ and $y' := r(y)$. Let Q be the x'-y'-path in F.
         If $b'(x') > 0$ then augment f along Q by $b'(x')$,
                            else augment f along the reverse of Q by $-b'(x')$.
         Set $b'(y') := b'(y') + b'(x')$ and $b'(x') := 0$.
         For all $e' = (v, w) \in E(G') \setminus F$ with $\{r(v), r(w)\} = \{x', y'\}$ **do:**
                 Set $E(G') := E(G') \setminus \{e'\}$.
         Set $r(z) := y'$ for all vertices z reachable from y' in F.

⑧   **Go to ②.**

This algorithm is due to Orlin [1993]. See also (Plotkin and Tardos [1990]). We first prove its correctness. Let us call the time between two changes of γ a **phase**. We call a vertex v **important** if $|b'(v)| > \frac{n-1}{n}\gamma$. A phase ends when there is no important vertex.

Lemma 9.16. *The number of augmentations in* ⑤ *during one phase is at most the number of important vertices at the beginning of this phase plus the number of augmentations in* ⑦ *at the beginning of this phase.*

Proof: Each augmentation in ⑤ decreases $\Phi := \sum_{v \in V(G)} \left\lceil \frac{|b'(v)|}{\gamma} - \frac{n-1}{n} \right\rceil$ by at least one, while an augmentation in ⑦ cannot increase Φ by more than one. Moreover, Φ is the number of important vertices at the beginning of each phase.
$\square$

Lemma 9.17. ORLIN'S ALGORITHM *solves the uncapacitated* MINIMUM COST FLOW PROBLEM *with conservative weights correctly. At any stage f is a minimum cost $(b - b')$-flow.*

Proof: We first prove that f is always a $(b - b')$-flow. In particular, we have to show that f is always nonnegative. To this end, we first observe that at any time the flow on each edge in $E(G') \setminus F$, and hence also the residual capacity of the reverse edge, is an integral multiple of γ.

Moreover we claim that every edge $e \in F$ always has positive residual capacity. To see this, first observe that after e has become a member of F in the γ-phase there are at most $n - 1$ augmentations in ⑦, adding up to less than $2\gamma(n - 1)$. Moreover, any phase has at most $2n - 1$ augmentations by γ in ⑤ (cf. Lemma 9.16); hence the total amount of flow moved after e has become a member of F in the γ-phase is less than $2n\gamma$ in ⑦, less than $2n\gamma$ in ⑤ during this phase, less than $2n\frac{\gamma}{2}$ in ⑤ during the next phase, and so on, thus less than $6n\gamma$ overall. Hence we always have positive flow on every edge in $F \cap E(G)$.

Hence f is always nonnegative and thus it is always a $(b - b')$-flow. We now prove that f is always a minimum cost $(b - b')$-flow. This is true initially because c is conservative. While f is an optimum $(b - b')$-flow, each v-w-path in $(V(G), F)$ is a shortest v-w-path in G_f (a shorter path would result in a negative circuit in G_f). Now the claim follows from Theorem 9.12: P in ⑤ and Q in ⑦ are shortest paths in G'_f and $(V(G), F)$, respectively, and hence in G_f (note that for any $(v, w) \in E(G_f) \setminus E(G'_f)$ there is a v-w-path in $(V(G), F)$).

We finally show that if the algorithm stops in ③ or ④ with $b' \neq 0$, then there is indeed no b-flow. Suppose the algorithm stops in ③, implying that there is a vertex s with $b'(s) > \frac{n-1}{n}\gamma$, but that no vertex t with $b'(t) < -\frac{1}{n}\gamma$ is reachable from s in G_f. Then let R be the set of vertices reachable from s in G_f. Since f is a $(b - b')$-flow, $\sum_{x \in R}(b(x) - b'(x)) = 0$. Therefore we have

$$\sum_{x \in R} b(x) \;=\; \sum_{x \in R} b'(x) \;=\; b'(s) + \sum_{x \in R \setminus \{s\}} b'(x) \;>\; 0$$

but $\delta_G^+(R) = \emptyset$. This proves that no b-flow exists (cf. Theorem 9.2). An analogous proof applies in the case that the algorithm stops in ④.
$\square$

We now analyse the running time.

Lemma 9.18. (Plotkin and Tardos [1990]) *If a vertex z is important at the beginning of a phase at some stage of the algorithm, then the connected component of $(V(G), F)$ containing z increases during the next $\lceil 2 \log n + \log m \rceil + 3$ phases.*

Proof: Let $|b'(z)| > \frac{n-1}{n} \gamma_1$ for a vertex z at the beginning of some phase of the algorithm where $\gamma = \gamma_1$. Let γ_0 be the γ-value in the preceding phase (and $\gamma_0 = 2\gamma_1$ if there is none), and γ_2 the γ-value $\lceil 2 \log n + \log m \rceil + 2$ phases later. We have $\frac{1}{2}\gamma_0 \geq \gamma_1 \geq 4n^2 m \gamma_2$. Let b_1' and f_1 be the b' and f at the beginning of the γ_1-phase, respectively, and let b_2' and f_2 be the b' and f at the end of the γ_2-phase, respectively.

Let Z be the connected component of $(V(G), F)$ containing z in the γ_1-phase, and suppose that this remains unchanged for the $\lceil 2 \log n + \log m \rceil + 3$ phases considered. Note that ⑦ guarantees $b'(v) = 0$ for all vertices v with $r(v) \neq v$. Hence $b'(v) = 0$ for all $v \in Z \setminus \{z\}$ and

$$\sum_{x \in Z} b(x) - b_1'(z) = \sum_{x \in Z} (b(x) - b_1'(x)) = \sum_{e \in \delta^+(Z)} f_1(e) - \sum_{e \in \delta^-(Z)} f_1(e). \quad (9.4)$$

The right-hand side is an integral multiple of γ_0, and

$$\frac{1}{n}\gamma_1 \leq \frac{n-1}{n}\gamma_1 < |b_1'(z)| \leq \frac{n-1}{n}\gamma_0 < \gamma_0 - \frac{1}{n}\gamma_1. \quad (9.5)$$

Hence

$$\left| \sum_{x \in Z} b(x) \right| > \frac{1}{n}\gamma_1. \quad (9.6)$$

As in (9.4), we have $\sum_{e \in \delta^+(Z)} f_2(e) - \sum_{e \in \delta^-(Z)} f_2(e) = \sum_{x \in Z} b(x) - b_2'(z)$. Using (9.6) and $|b_2'(z)| \leq \frac{n-1}{n}\gamma_2$ we obtain

$$\sum_{e \in \delta^+(Z) \cup \delta^-(Z)} |f_2(e)| \geq \left| \sum_{x \in Z} b(x) \right| - |b_2'(z)| > \frac{1}{n}\gamma_1 - \frac{n-1}{n}\gamma_2$$

$$> (4nm - 1)\gamma_2 > m\left(6n\frac{\gamma_2}{2}\right).$$

Thus there exists at least one edge e with exactly one end in Z and $f_2(e) > 6n\frac{\gamma_2}{2}$. By ⑦ of the algorithm, this means that Z is increased at the beginning of the $\frac{\gamma_2}{2}$-phase. $\square$

Theorem 9.19. (Orlin [1993]) ORLIN'S ALGORITHM *solves the uncapacitated* MINIMUM COST FLOW PROBLEM *with conservative weights correctly in* $O(n \log n (m + n \log n))$ *time.*

Proof: The correctness has been proved above (Lemma 9.17). In each phase, ⑦ takes $O(m(i + 1))$ time, where i is the number of iterations in the while-loop. Note that the total number of iterations in this while-loop is at most $n - 1$ as the number of connected components of $(V(G), F)$ decreases each time.

By ⑥ and ⑦, there can be at most $\lceil \log n \rceil + 3$ consecutive phases before we have an important vertex or add a pair of edges to F. Therefore, Lemma 9.18 implies that the total number of phases is $O(n \log m)$.

By Lemma 9.16, the total number of augmentations in ⑤ is at most $n - 1$ plus the number of pairs (γ, z), where z is important at the beginning of the γ-phase. By Lemma 9.18, and since all vertices v with $r(v) \neq v$ have $b'(v) = 0$ at any time, the number of these pairs is $O(\log m)$ times the number of sets that are at some stage of the algorithm a connected component of $(V(G), F)$. Since the family of these sets is laminar, there are at most $2n - 1$ such sets (Corollary 2.15), and thus $O(n \log m)$ augmentations in ⑤ altogether.

Using the technique of Theorem 9.14, we obtain an overall running time of $O(mn \log m + (n \log m)(m + n \log n))$. We may assume $m = O(n^2)$, and thus $\log m = O(\log n)$, because among a set of uncapacitated parallel edges we need only a cheapest one.   □

This is the best known running time for the uncapacitated MINIMUM COST FLOW PROBLEM.

Theorem 9.20. (Orlin [1993]) *The general* MINIMUM COST FLOW PROBLEM *can be solved in* $O(m \log m(m + n \log n))$ *time, where* $n = |V(G)|$ *and* $m = |E(G)|$.

Proof: We apply the construction given in Lemma 9.3. Thus we have to solve an uncapacitated MINIMUM COST FLOW PROBLEM on a bipartite graph H with $V(H) = A' \dot{\cup} B'$, where $A' = E(G)$ and $B' = V(G)$. Since H is acyclic, an initial feasible potential can be computed in $O(|E(H)|) = O(m)$ time. As shown above (Theorem 9.19), the overall running time is dominated by $O(m \log m)$ shortest path computations in a subgraph of $\overset{\leftrightarrow}{H}$ with nonnegative weights.

Before we call DIJKSTRA'S ALGORITHM we apply the following operation to each vertex $a \in A'$ that is not an endpoint of the path we are looking for: add an edge (b, b') for each pair of edges $(b, a), (a, b')$ and set its weight to the sum of the weights of (b, a) and (a, b'); finally delete a. Clearly the resulting instance of the SHORTEST PATH PROBLEM is equivalent. Since each vertex in A' has four incident edges in $\overset{\leftrightarrow}{H}$, the resulting graph has $O(m)$ edges and at most $n + 2$ vertices. The preprocessing takes constant time per vertex, i.e. $O(m)$. The same holds for the transformation of the shortest path to the corresponding path in $\overset{\leftrightarrow}{H}$, and for computing the distance labels of the deleted vertices. We get an overall running time of $O((m \log m)(m + n \log n))$.   □

This is the fastest known strongly polynomial algorithm for the general MINIMUM COST FLOW PROBLEM, except if there are significantly more than n^2 edges (see Matsui and Scheifele [2016] for details). An algorithm which achieves the same running time as Orlin's but works directly on capacitated instances has been described by Vygen [2002]. A fast weakly polynomial algorithm is due to Lee and Sidford [2014].

9.6 The Network Simplex Algorithm

The MINIMUM COST FLOW PROBLEM is a special case of LINEAR PROGRAMMING. By applying the SIMPLEX ALGORITHM and exploiting the special structure we arrive at what is called the NETWORK SIMPLEX ALGORITHM. To make the connection clear, we first characterize the set of basic solutions (although we will not need this to prove correctness).

Definition 9.21. *Let* (G, u, b, c) *be an instance of the* MINIMUM COST FLOW PROBLEM. *A b-flow* f *in* (G, u) *is a* **spanning tree solution** *if* $(V(G), \{e \in E(G) : 0 < f(e) < u(e)\})$ *contains no undirected circuit.*

Proposition 9.22. *An instance of the* MINIMUM COST FLOW PROBLEM *either has an optimum solution which is a spanning tree solution or has no optimum solution at all.*

Proof: Given an optimum solution f and an undirected circuit C in $(V(G), \{e \in E(G) : 0 < f(e) < u(e)\})$, we have two directed circuits C' and C'' in G_f with the same underlying undirected graph as C. Let ϵ be the minimum residual capacity in $E(C') \cup E(C'')$. We can obtain two other feasible solutions f' and f'' by augmenting f by ϵ along C' and C'', respectively. As $2c(f) = c(f') + c(f'')$, both f' and f'' are also optimum solutions. At least one of them has fewer edges e with $0 < f(e) < u(e)$ than f, so after less than $|E(G)|$ steps we arrive at an optimum spanning tree solution.   □

Corollary 9.23. *Let* (G, u, b, c) *be an instance of the* MINIMUM COST FLOW PROBLEM. *Then the basic solutions of*

$$\left\{ x \in \mathbb{R}^{E(G)} \; : \; 0 \le x_e \le u(e) \; (e \in E(G)), \right.$$

$$\left. \sum_{e \in \delta^+(v)} x_e - \sum_{e \in \delta^-(v)} x_e = b(v) \; (v \in V(G)) \right\}$$

are precisely the spanning tree solutions of (G, u, b, c).

Proof: Proposition 9.22 shows that each basic solution is a spanning tree solution.

For a spanning tree solution f we consider the inequalities $x_e \ge 0$ for $e \in E(G)$ with $f(e) = 0$, $x_e \le u(e)$ for $e \in E(G)$ with $f(e) = u(e)$, and $\sum_{e \in \delta^+(v)} x_e - \sum_{e \in \delta^-(v)} x_e = b(v)$ for all v except one vertex of each connected component of $(V(G), \{e \in E(G) : 0 < f(e) < u(e)\})$. These $|E(G)|$ inequalities are all satisfied by f with equality, and the submatrix corresponding to these inequalities is nonsingular. Thus f is a basic solution.   □

In a spanning tree solution there are three types of edges: those with zero flow, those with saturated capacity, and those whose flow is positive but less than the capacity. Assuming that G is connected, we can extend the last set of edges to a connected spanning subgraph without undirected circuits (i.e., an oriented spanning tree; hence the name "spanning tree solution").

Definition 9.24. *Let (G, u, b, c) be an instance of the* MINIMUM COST FLOW PROBLEM *where G is connected. A* **spanning tree structure** *is a quadruple (r, T, L, U) where $r \in V(G)$, $E(G) = T \mathbin{\dot{\cup}} L \mathbin{\dot{\cup}} U$, $|T| = |V(G)| - 1$, and $(V(G), T)$ contains no undirected circuit.*

The b-flow **associated with the spanning tree structure** *(r, T, L, U) is defined by*

- $f(e) := 0$ *for* $e \in L$,
- $f(e) := u(e)$ *for* $e \in U$,
- $f(e) := \sum_{v \in C_e} b(v) + \sum_{e' \in U \cap \delta^-(C_e)} u(e') - \sum_{e' \in U \cap \delta^+(C_e)} u(e')$ *for* $e \in T$, *where for $e = (v, w)$ we denote by C_e the connected component of $(V(G), T \setminus \{e\})$ containing v.*

(r, T, L, U) is called **feasible** *if $0 \leq f(e) \leq u(e)$ for all $e \in T$.*

Call an edge (v, w) in T downward *if v belongs to the undirected r-w-path in T, otherwise* upward*. (r, T, L, U) is called* **strongly feasible** *if $0 < f(e) \leq u(e)$ for each downward edge $e \in T$ and $0 \leq f(e) < u(e)$ for each upward edge $e \in T$.*

The unique function $\pi : V(G) \to \mathbb{R}$ with $\pi(r) = 0$ and $c_\pi(e) = 0$ for all $e \in T$ is called the **potential associated with the spanning tree structure** *(r, T, L, U).*

Evidently the b-flow f associated with a spanning tree structure satisfies $\sum_{e \in \delta^+(v)} f(e) - \sum_{e \in \delta^-(v)} f(e) = b(v)$ for all $v \in V(G)$ (though it is not always a feasible b-flow). Moreover we note:

Proposition 9.25. *Given an instance (G, u, b, c) of the* MINIMUM COST FLOW PROBLEM *and a spanning tree structure (r, T, L, U), the b-flow f and the potential π associated with it can be computed in $O(m)$ and $O(n)$ time, respectively.*

Moreover, f is integral whenever b and u are integral, and π is integral whenever c is integral.

Proof: The potential associated with (r, T, L, U) can be computed simply by applying the GRAPH SCANNING ALGORITHM to the edges of T and their reverse edges. The b-flow associated with (r, T, L, U) can be computed in linear time by scanning the vertices in an order of nonincreasing distance from r. The integrality properties follow immediately from the definition. $\qquad\square$

The NETWORK SIMPLEX ALGORITHM maintains a strongly feasible spanning tree structure and works towards optimality. Note that the optimality criterion in Corollary 9.8 immediately implies:

Proposition 9.26. *Let (r, T, L, U) be a feasible spanning tree structure and let π be the potential associated with it. Suppose that*

- $c_\pi(e) \geq 0$ *for all $e \in L$, and*
- $c_\pi(e) \leq 0$ *for all $e \in U$.*

Then (r, T, L, U) is associated with an optimum b-flow. $\qquad\square$

Note that $\pi(v)$ is the length of the r-v-path in $\overset{\leftrightarrow}{G}$ that contains only edges of T or their reverse edges. For an edge $e = (v, w) \in E(\overset{\leftrightarrow}{G})$ we define the *fundamental circuit* C of e to consist of e and the w-v-path that contains only edges of T and their reverse edges. The vertex of C that is closest to r in T is called its *peak*.

Hence for $e = (v, w) \notin T$, $c_\pi(e) = c(e) + \pi(v) - \pi(w)$ is the cost of sending one unit of flow along the fundamental circuit of e.

There are several ways to obtain an initial feasible spanning tree structure. For example, one could compute any b-flow (by solving a MAXIMUM FLOW PROBLEM), apply the procedure in the proof of Proposition 9.22, choose r arbitrarily, and define T, L, U according to the flow (adding suitable edges to T if necessary). Alternatively, one could apply "phase one" of the SIMPLEX METHOD.

However, this may not lead to a strongly feasible spanning tree structure. The simplest possibility is to introduce very expensive auxiliary edges with sufficient capacity between r and each other vertex: For each sink $v \in V(G)\backslash\{r\}$ we introduce an edge (r, v) with capacity $-b(v)$, and for each other vertex $v \in V(G) \backslash \{r\}$ we introduce an edge (v, r) with capacity $b(v) + 1$. The cost of each auxiliary edge should be high enough so that they never occur in an optimum solution, for example $1 + (|V(G)| - 1) \max_{e \in E(G)} |c(e)|$ (Exercise 19). Then we can choose T as the set of all auxiliary edges, L as the set of all original edges, and $U := \emptyset$ to obtain an initial strongly feasible spanning tree structure.

NETWORK SIMPLEX ALGORITHM

Input: An instance (G, u, b, c) of the MINIMUM COST FLOW PROBLEM and a strongly feasible spanning tree structure (r, T, L, U).

Output: An optimum solution f.

---

① Compute the b-flow f and the potential π associated with (r, T, L, U).

② Let $e \in L$ with $c_\pi(e) < 0$ or $e \in U$ with $c_\pi(e) > 0$.
   If such an edge e does not exist **then stop**.

③ Let C be the fundamental circuit of e (if $e \in L$) or of $\overleftarrow{e}$ (if $e \in U$).
   Let $\gamma := c_\pi(e)$.

④ Let $\delta := \min_{e' \in E(C)} u_f(e')$, and let e' be the last edge where the minimum is attained when traversing C in its orientation, starting at its peak.
   Let $e_0 \in E(G)$ such that e' is e_0 or $\overleftarrow{e_0}$.

⑤ Remove e from L or U.
   Set $T := (T \cup \{e\}) \backslash \{e_0\}$.
   If $e' = e_0$ **then** insert e_0 into U **else** insert e_0 into L.

⑥ Augment f by δ along C.
   Let X be the connected component of $(V(G), T \backslash \{e\})$ containing r.
   If $e \in \delta^+(X)$ **then** set $\pi(v) := \pi(v) + \gamma$ for $v \in V(G) \backslash X$.
   If $e \in \delta^-(X)$ **then** set $\pi(v) := \pi(v) - \gamma$ for $v \in V(G) \backslash X$.
   Go to ②.

We remark that ⑥ could be replaced simply by going back to ①, as f and π computed in ⑥ are associated with the new spanning tree structure. Also note that $e = e_0$ is possible; in this case $X = V(G)$, and T and π do not change, but e moves from L to U or vice versa, and $\delta = u(e)$.

Theorem 9.27. (Dantzig [1951], Cunningham [1976]) *The* NETWORK SIMPLEX ALGORITHM *stops after finitely many iterations and returns an optimum solution.*

Proof: First observe that ⑥ maintains the property that f and π are the b-flow and potential associated with (r, T, L, U).

Next we prove that the spanning tree structure is always strongly feasible. By the choice of δ we maintain the condition $0 \leq f(e) \leq u(e)$ for all e, and hence the spanning tree structure remains feasible.

As the edges of the subpath of C from the head of e' to the peak did not attain the minimum in ④, they will continue to have positive residual capacity after the augmentation.

For the edges of the subpath of C from the peak to the tail of e' we have to make sure that their reverse edges have positive residual capacity after the augmentation. This is clear if $\delta > 0$. Otherwise (if $\delta = 0$), the fact that the spanning tree structure was strongly feasible before implies that neither e nor $\overleftarrow{e}$ can belong to this subpath (i.e. $e = e_0$ or $\delta^-(X) \cap E(C) \cap \{e, \overleftarrow{e}\} \neq \emptyset$), and that the reverse edges of the subpath of C from the peak to tail of e or $\overleftarrow{e}$ had positive residual capacity.

By Proposition 9.26 the computed flow f is optimum when the algorithm terminates. We show that there are no two iterations with the same pair (f, π), and hence each spanning tree structure occurs at most once.

In each iteration the cost of the flow is reduced by $|\gamma|\delta$. As $\gamma \neq 0$, we only need to consider iterations with $\delta = 0$. Here the cost of the flow remains constant. If $e \neq e_0$, then $e \in L \cap \delta^-(X)$ or $e \in U \cap \delta^+(X)$, and hence $\sum_{v \in V(G)} \pi(v)$ strictly increases (by at least $|\gamma|$). Finally, if $\delta = 0$ and $e = e_0$, then $u(e) = 0$, $X = V(G)$, π remains constant, and $|\{e \in L : c_\pi(e) < 0\}| + |\{e \in U : c_\pi(e) > 0\}|$ strictly decreases. This shows that no two iterations feature the same spanning tree structure.   □

Although the NETWORK SIMPLEX ALGORITHM is not a polynomial-time algorithm, it is quite efficient in practice. Orlin [1997] proposed a variant that runs in polynomial time. Polynomial-time dual network simplex algorithms were found by Orlin, Plotkin and Tardos [1993], and Armstrong and Jin [1997].

9.7 Flows Over Time

We now consider flows over time (also sometimes called dynamic flows); i.e. the flow value on each edge may change over time, and flow entering an edge arrives at the endvertex after a specified delay:

Definition 9.28. *Let (G, u, s, t) be a network with transit times $l : E(G) \to \mathbb{R}_+$ and a time horizon $T \in \mathbb{R}_+$. Then an **s-t-flow over time** f consists of a Lebesgue-measurable function $f_e : [0, T] \to \mathbb{R}_+$ for each $e \in E(G)$ with $f_e(\tau) \le u(e)$ for all $\tau \in [0, T]$ and $e \in E(G)$ and*

$$\mathrm{ex}_f(v, a) := \sum_{e \in \delta^-(v)} \int_0^{\max\{0, a - l(e)\}} f_e(\tau) \, d\tau - \sum_{e \in \delta^+(v)} \int_0^a f_e(\tau) \, d\tau \ge 0 \quad (9.7)$$

for all $v \in V(G) \setminus \{s\}$ and $a \in [0, T]$.

$f_e(\tau)$ is called the rate of flow entering e at time τ (and leaving this edge $l(e)$ time units later). (9.7) allows intermediate storage at vertices, like in s-t-preflows. It is natural to maximize the flow arriving at sink t:

MAXIMUM FLOW OVER TIME PROBLEM

Instance: A network (G, u, s, t). Transit times $l : E(G) \to \mathbb{R}_+$ and a time horizon $T \in \mathbb{R}_+$.

Task: Find an s-t-flow over time f such that value $(f) := \mathrm{ex}_f(t, T)$ is maximum.

Following Ford and Fulkerson [1958], we show that this problem can be reduced to the MINIMUM COST FLOW PROBLEM.

Theorem 9.29. *The* MAXIMUM FLOW OVER TIME PROBLEM *can be solved in $O(m \log m(m + n \log n))$ time.*

Proof: Given an instance (G, u, s, t, l, T) as above, define a new edge $e' = (t, s)$ and $G' := G + e'$. Set $u(e') := u(E(G)) + 1$, $c(e') := -T$ and $c(e) := l(e)$ for $e \in E(G)$. Consider the instance $(G', u, 0, c)$ of the MINIMUM COST FLOW PROBLEM. Let f' be an optimum solution, i.e. a minimum cost (with respect to c) circulation in (G', u). By Proposition 9.5, f' can be decomposed into flows on circuits, i.e. there is a set $\mathcal{C}$ of circuits in G' and positive numbers $h : \mathcal{C} \to \mathbb{R}_+$ such that $f'(e) = \sum\{h(C) : C \in \mathcal{C}, e \in E(C)\}$. We have $c(E(C)) \le 0$ for all $C \in \mathcal{C}$ as f' is a minimum cost circulation.

Let $C \in \mathcal{C}$ with $c(E(C)) < 0$. C must contain e'. For $e = (v, w) \in E(C) \setminus \{e'\}$, let d_e^C be the distance from s to v in (C, c). Set

$$f_e^*(\tau) := \sum \left\{ h(C) : C \in \mathcal{C}, c(E(C)) < 0, e \in E(C), d_e^C \le \tau \le d_e^C - c(E(C)) \right\}$$

for $e \in E(G)$ and $\tau \in [0, T]$. This defines an s-t-flow over time without intermediate storage (i.e. $\mathrm{ex}_f(v, a) = 0$ for all $v \in V(G) \setminus \{s, t\}$ and all $a \in [0, T]$). Moreover,

$$
\begin{aligned}
\text{value} (f^*) &= \sum_{e \in \delta^-(t)} \int_0^{T-l(e)} f_e^*(\tau) \, d\tau \\
&= \sum_{C \in \mathcal{C}} h(C)(-c(E(C))) \\
&= - \sum_{e \in E(G')} c(e) f'(e).
\end{aligned}
$$

Let $\pi(v) := T + \min\{0, \mathrm{dist}_{(G'_{f'}, c)}(t, v)\}$ for $v \in V(G)$. We have $\pi(t) = T$, $\pi(s) = 0$, and $0 \le \pi(v) \le T$ for all $v \in V(G)$.

Claim: π is a feasible potential in $(G'_{f'}, c)$.

To prove this, first observe that for all $e = (x, y) \in E(G')$ with $f'(e) > 0$ we have $\mathrm{dist}_{(G'_{f'}, c)}(t, y) \le 0$ and hence $\pi(y) = T + \mathrm{dist}_{(G'_{f'}, c)}(t, y)$. Hence, for any $e = (v, w) \in E(G'_{f'})$ we have either $\pi(v) = T + \mathrm{dist}_{(G'_{f'}, c)}(t, v)$ or $(\pi(v) = T$ and $c(e) \ge 0)$. In the first case, we conclude $c_\pi(e) = c(e) + T + \mathrm{dist}_{(G'_{f'}, c)}(t, v) - \pi(w) \ge c(e) + T + \mathrm{dist}_{(G'_{f'}, c)}(t, v) - (T + \mathrm{dist}_{(G'_{f'}, c)}(t, w)) \ge 0$, because $G'_{f'}$ contains no negative circuit (cf. Theorem 9.7). In the second case, $c_\pi(e) = c(e) + T - \pi(w) \ge c(e) + T - T = c(e) \ge 0$. The Claim is proved.

We finally prove that f^* is optimum. To see this, let f be any s-t-flow over time, and set $f_e(\tau) := 0$ for $e \in E(G)$ and $\tau \notin [0, T]$. Using (9.7) we get

$$
\begin{aligned}
\text{value} (f) &= \mathrm{ex}_f(t, T) \\
&\le \sum_{v \in V(G)} \mathrm{ex}_f(v, \pi(v)) \\
&= \sum_{e=(v,w) \in E(G)} \left(\int_0^{\pi(w)-l(e)} f_e(\tau) \, d\tau - \int_0^{\pi(v)} f_e(\tau) \, d\tau \right) \\
&\le \sum_{e=(v,w) \in E(G): \pi(w)-l(e) > \pi(v)} (\pi(w) - l(e) - \pi(v)) \, u(e) \\
&= \sum_{e=(v,w) \in E(G)} (\pi(w) - l(e) - \pi(v)) f'(e).
\end{aligned}
$$

Here the last equation follows since π is a feasible potential in $(G'_{f'}, c)$.

Hence we conclude

$$
\begin{aligned}
\text{value} (f) &\le \sum_{e=(v,w) \in E(G)} (\pi(w) - l(e) - \pi(v)) f'(e) \\
&= \sum_{e=(v,w) \in E(G')} (\pi(w) - c(e) - \pi(v)) f'(e)
\end{aligned}
$$

$$= - \sum_{e=(v,w)\in E(G')} c(e)f'(e)$$

$$= \text{value}(f^*).$$

□

The proof also shows that there always exists an optimum flow over time with piecewise constant functions and with excess constantly zero at any vertex except s and t.

Other flow over time problems are significantly more difficult. Hoppe and Tardos [2000] solved the so-called quickest transshipment problem (with several sources and sinks) with integral transit times using submodular function minimization (see Chapter 14). Finding minimum cost flows over time is *NP*-hard (Klinz and Woeginger [2004]). See Fleischer and Skutella [2007] for approximation algorithms and more information.

Exercises

1. Show that the MAXIMUM FLOW PROBLEM can be regarded as a special case of the MINIMUM COST FLOW PROBLEM.
2. Let G be a digraph with lower and upper capacities $l, u : E(G) \to \mathbb{R}_+$, where $l(e) \le u(e)$ for all $e \in E(G)$, and let $b_1, b_2 : V(G) \to \mathbb{R}$ with $b_1(v) \le b_2(v)$ for all $v \in V(G)$. Prove that there exists a flow f with $l(e) \le f(e) \le u(e)$ for all $e \in E(G)$ and

$$b_1(v) \le \sum_{e\in\delta^+(v)} f(e) - \sum_{e\in\delta^-(v)} f(e) \le b_2(v) \quad \text{for all } v \in V(G)$$

if and only if

$$\sum_{e\in\delta^+(X)} u(e) \ge \max\left\{\sum_{v\in X} b_1(v), -\sum_{v\in V(G)\setminus X} b_2(v)\right\} + \sum_{e\in\delta^-(X)} l(e)$$

for all $X \subseteq V(G)$.
Note: This generalizes Exercise 4 of Chapter 8 and Theorem 9.2.
(Hoffman [1960])
3. Prove the following theorem of Ore [1956]. Given a digraph G and nonnegative integers $a(x), b(x)$ for each $x \in V(G)$, then G has a spanning subgraph H with $|\delta_H^+(x)| = a(x)$ and $|\delta_H^-(x)| = b(x)$ for all $x \in V(G)$ if and only if

$$\sum_{x\in V(G)} a(x) = \sum_{x\in V(G)} b(x) \quad \text{and}$$

$$\sum_{x\in X} a(x) \le \sum_{y\in V(G)} \min\{b(y), |\delta_G^+(X \setminus \{y\}) \cap \delta_G^-(y)|\} \quad \text{for all } X \subseteq V(G).$$

(Ford and Fulkerson [1962])

4. Let (G, u, c, b) be an instance of the MINIMUM COST FLOW PROBLEM with $c(e) \geq 0$ for all $e \in E(G)$. Let F be the set of edges $e \in E(G)$ for which an optimum solution f with $f(e) > 0$ exists. Prove that every circuit in $(V(G), F)$ consists only of edges e with $c(e) = 0$.

5. Consider the MINIMUM COST FLOW PROBLEM where infinite capacities ($u(e) = \infty$ for some edges e) are allowed.
   (a) Show that an instance is unbounded if and only if it is feasible and there is a negative circuit all whose edges have infinite capacity.
   (b) Show how to decide in $O(n^3 + m)$ time whether an instance is unbounded.
   (c) Show that for an instance that is not unbounded each infinite capacity can be equivalently replaced by a finite capacity.

* 6. Let (G, u, c, b) be an instance of the MINIMUM COST FLOW PROBLEM. We call a function $\pi : V(G) \to \mathbb{R}$ an optimal potential if there exists a minimum cost b-flow f such that π is a feasible potential with respect to (G_f, c).
   (a) Prove that a function $\pi : V(G) \to \mathbb{R}$ is an optimal potential if and only if for all $X \subseteq V(G)$:

$$b(X) + \sum_{e \in \delta^-(X): c_\pi(e) < 0} u(e) \leq \sum_{e \in \delta^+(X): c_\pi(e) \leq 0} u(e).$$

   (b) Given $\pi : V(G) \to \mathbb{R}$, show how to find a set X violating the condition in (a) or determine that none exists.
   (c) Suppose an optimal potential is given; show how to find a minimum cost b-flow in $O(n^3)$ time.
   Note: This leads to so-called cut cancelling algorithms for the MINIMUM COST FLOW PROBLEM.
   (Hassin [1983])

7. Consider the following algorithm scheme for the MINIMUM COST FLOW PROBLEM: first find any b-flow, then as long as there is a negative augmenting cycle, augment the flow along it (by the maximum possible amount). We have seen in Section 9.3 that we obtain a strongly polynomial running time if we always choose a circuit of minimum mean weight. Prove that without this specification one cannot guarantee that the algorithm terminates.
   (Use the construction in Exercise 2 of Chapter 8.)

8. Consider the problem as described in Exercise 2 with a weight function $c : E(G) \to \mathbb{R}$. Can one find a minimum cost flow that satisfies the constraints of Exercise 2? (Reduce this problem to a standard MINIMUM COST FLOW PROBLEM.)

9. The DIRECTED CHINESE POSTMAN PROBLEM can be formulated as follows: given a strongly connected simple digraph G with weights $c : E(G) \to \mathbb{R}_+$, find $f : E(G) \to \mathbb{N}$ such that the graph which contains $f(e)$ copies of each edge $e \in E(G)$ is Eulerian and $\sum_{e \in E(G)} c(e) f(e)$ is minimum. How can this problem be solved by a polynomial-time algorithm?
   (For the UNDIRECTED CHINESE POSTMAN PROBLEM, see Section 12.2.)

* 10. The FRACTIONAL b-MATCHING PROBLEM is defined as follows: Given an undirected graph G, capacities $u : E(G) \to \mathbb{R}_+$, numbers $b : V(G) \to \mathbb{R}_+$ and weights $c : E(G) \to \mathbb{R}$, we are looking for an $f : E(G) \to \mathbb{R}_+$ with $f(e) \le u(e)$ for all $e \in E(G)$ and $\sum_{e \in \delta(v)} f(e) \le b(v)$ for all $v \in V(G)$ such that $\sum_{e \in E(G)} c(e) f(e)$ is maximum.

    (a) Show how to solve this problem by reducing it to a MINIMUM COST FLOW PROBLEM.

    (b) Suppose now b and u are integral. Show that then the FRACTIONAL b-MATCHING PROBLEM always has a half-integral solution f (i.e., $2f(e) \in \mathbb{Z}$ for all $e \in E(G)$).

    Note: The (integral) MAXIMUM WEIGHT b-MATCHING PROBLEM is the subject of Section 12.1.

* 11. Find a combinatorial polynomial-time algorithm for the interval packing problem defined in Exercise 16 of Chapter 5.
    (Arkin and Silverberg [1987])

12. Consider a linear program $\max\{cx : Ax \le b\}$ where all entries of A are -1, 0, or 1, and each column of A contains at most one 1 and at most one -1. Show that such an LP is equivalent to an instance of the MINIMUM COST FLOW PROBLEM.

13. The scaling technique introduced in Sections 8.4 and 9.4 can be considered in a quite general setting: Let Ψ be a family of set systems each of which contains the empty set. Suppose that there is an algorithm which solves the following problem: given an $(E, \mathcal{F}) \in \Psi$, weights $c : E \to \mathbb{Z}_+$ and a set $X \in \mathcal{F}$; find a $Y \in \mathcal{F}$ with $c(Y) > c(X)$ or assert that no such Y exists. Suppose this algorithm has a running time which is polynomial in size(c). Prove that then there is an algorithm for finding a maximum weight set $X \in \mathcal{F}$ for a given $(E, \mathcal{F}) \in \Psi$ and $c : E \to \mathbb{Z}_+$, whose running time is polynomial in size(c).
    (Grötschel and Lovász [1995]; see also Schulz, Weismantel and Ziegler [1995], and Schulz and Weismantel [2002])

14. Prove that ORLIN'S ALGORITHM can be rendered more precisely so that it always computes a spanning tree solution.

15. Prove that in ⑦ of ORLIN'S ALGORITHM one can replace the $6n\gamma$-bound by $5n\gamma$.

16. Consider the shortest path computations with nonnegative weights (using DIJKSTRA'S ALGORITHM) in the algorithms of Section 9.4 and 9.5. Show that even for graphs with parallel edges each of these computations can be performed in $O(n^2)$ time, provided that we have the incidence list of G sorted by edge costs. Conclude that ORLIN'S ALGORITHM runs in $O(mn^2 \log m)$ time.

* 17. The PUSH-RELABEL ALGORITHM (Section 8.5) can be generalized to the MINIMUM COST FLOW PROBLEM. For an instance (G, u, b, c) with integral costs c, we look for a b-flow f and a feasible potential π in (G_f, c). We start by setting $\pi := 0$ and saturating all edges e with negative cost. Then we apply ③ of the PUSH-RELABEL ALGORITHM with the following modifications. An

edge e is admissible if $e \in E(G_f)$ and $c_\pi(e) < 0$. A vertex v is active if $b(v) + \mathrm{ex}_f(v) > 0$. RELABEL($v$) consists of setting $\pi(v) := \max\{\pi(w) - c(e) - 1 : e = (v, w) \in E(G_f)\}$. In PUSH($e$) for $e \in \delta^+(v)$ we set $\gamma := \min\{b(v) + \mathrm{ex}_f(v), u_f(e)\}$.

(a) Prove that the number of RELABEL operations is $O(n^2 |c_{\max}|)$, where $c_{\max} = \max_{e \in E(G)} c(e)$.

Hint: Some vertex w with $b(w) + \mathrm{ex}_f(w) < 0$ must be reachable in G_f from any active vertex v. Note that $b(w)$ has never changed and recall the proofs of Lemmata 8.25 and 8.27.

(b) Show that the overall running time is $O(n^2 m c_{\max})$.

(c) Prove that the algorithm computes an optimum solution.

(d) Apply scaling to obtain an $O(n^2 m \log c_{\max})$-algorithm for the MINIMUM COST FLOW PROBLEM with integral costs c.

(Goldberg and Tarjan [1990])

18. Let (G, u, b, c) be an instance of the MINIMUM COST FLOW PROBLEM, where u and b are integral. Prove the following statement using (a) the MINIMUM MEAN CYCLE-CANCELLING ALGORITHM or (b) the SUCCESSIVE SHORTEST PATH ALGORITHM, or (c) total unimodularity. If there exists a b-flow in (G, u), then there exists a minimum cost b-flow which is integral.

19. Let (G, u, c, b) be an instance of the MINIMUM COST FLOW PROBLEM. Let $\bar{e} \in E(G)$ with $c(\bar{e}) > (|V(G)| - 1) \max_{e \in E(G) \setminus \{\bar{e}\}} |c(e)|$. Prove: If there exists a b-flow f in (G, u) with $f(\bar{e}) = 0$, then $f(\bar{e}) = 0$ holds for every optimum solution f.

20. Given a network (G, u, s, t) with integral transit times $l : E(G) \to \mathbb{Z}_+$, a time horizon $T \in \mathbb{N}$, a value $V \in \mathbb{R}_+$, and costs $c : E(G) \to \mathbb{R}_+$. We look for an s-t-flow over time f with value$(f) = V$ and minimum cost $\sum_{e \in E(G)} c(e) \int_0^T f_e(\tau) \, d\tau$. Show how to solve this in polynomial time if T is a constant.

Hint: Consider a time-expanded network with a copy of G for each discrete time step.

References

General Literature

Ahuja, R.K., Magnanti, T.L., and Orlin, J.B. [1993]: Network Flows. Prentice-Hall, Englewood Cliffs 1993

Cook, W.J., Cunningham, W.H., Pulleyblank, W.R., and Schrijver, A. [1998]: Combinatorial Optimization. Wiley, New York 1998, Chapter 4

Goldberg, A.V., Tardos, É., and Tarjan, R.E. [1990]: Network flow algorithms. In: Paths, Flows, and VLSI-Layout (B. Korte, L. Lovász, H.J. Prömel, A. Schrijver, eds.), Springer, Berlin 1990, pp. 101–164

Gondran, M., and Minoux, M. [1984]: Graphs and Algorithms. Wiley, Chichester 1984, Chapter 5

Jungnickel, D. [2013]: Graphs, Networks and Algorithms. Fourth Edition. Springer, Berlin 2013, Chapters 10 and 11

Lawler, E.L. [1976]: Combinatorial Optimization: Networks and Matroids. Holt, Rinehart and Winston, New York 1976, Chapter 4

Ruhe, G. [1991]: Algorithmic Aspects of Flows in Networks. Kluwer Academic Publishers, Dordrecht 1991

Skutella, M. [2009]: An introduction to network flows over time. In: Research Trends in Combinatorial Optimization (W.J. Cook, L. Lovász, J. Vygen, eds.), Springer, Berlin 2009, pp. 451–482

Cited References

Arkin, E.M., and Silverberg, E.B. [1987]: Scheduling jobs with fixed start and end times. Discrete Applied Mathematics 18 (1987), 1–8

Armstrong, R.D., and Jin, Z. [1997]: A new strongly polynomial dual network simplex algorithm. Mathematical Programming 78 (1997), 131–148

Busacker, R.G., and Gowen, P.J. [1961]: A procedure for determining a family of minimum-cost network flow patterns. ORO Technical Paper 15, Operational Research Office, Johns Hopkins University, Baltimore 1961

Cunningham, W.H. [1976]: A network simplex method. Mathematical Programming 11 (1976), 105–116

Dantzig, G.B. [1951]: Application of the simplex method to a transportation problem. In: Activity Analysis and Production and Allocation (T.C. Koopmans, Ed.), Wiley, New York 1951, pp. 359–373

Edmonds, J., and Karp, R.M. [1972]: Theoretical improvements in algorithmic efficiency for network flow problems. Journal of the ACM 19 (1972), 248–264

Fleischer, L., and Skutella, M. [2007]: Quickest flows over time. SIAM Journal on Computing 36 (2007), 1600–1630

Ford, L.R., and Fulkerson, D.R. [1958]: Constructing maximal dynamic flows from static flows. Operations Research 6 (1958), 419–433

Ford, L.R., and Fulkerson, D.R. [1962]: Flows in Networks. Princeton University Press, Princeton 1962

Gale, D. [1957]: A theorem on flows in networks. Pacific Journal of Mathematics 7 (1957), 1073–1082

Goldberg, A.V., and Tarjan, R.E. [1989]: Finding minimum-cost circulations by cancelling negative cycles. Journal of the ACM 36 (1989), 873–886

Goldberg, A.V., and Tarjan, R.E. [1990]: Finding minimum-cost circulations by successive approximation. Mathematics of Operations Research 15 (1990), 430–466

Grötschel, M., and Lovász, L. [1995]: Combinatorial optimization. In: Handbook of Combinatorics; Vol. 2 (R.L. Graham, M. Grötschel, L. Lovász, eds.), Elsevier, Amsterdam 1995

Hassin, R. [1983]: The minimum cost flow problem: a unifying approach to dual algorithms and a new tree-search algorithm. Mathematical Programming 25 (1983), 228–239

Hitchcock, F.L. [1941]: The distribution of a product from several sources to numerous localities. Journal of Mathematical Physics 20 (1941), 224–230

Hoffman, A.J. [1960]: Some recent applications of the theory of linear inequalities to extremal combinatorial analysis. In: Combinatorial Analysis (R.E. Bellman, M. Hall, eds.), AMS, Providence 1960, pp. 113–128

Hoppe, B., and Tardos, É. [2000]: The quickest transshipment problem. Mathematics of Operations Research 25 (2000), 36–62

Iri, M. [1960]: A new method for solving transportation-network problems. Journal of the Operations Research Society of Japan 3 (1960), 27–87

Jewell, W.S. [1958]: Optimal flow through networks. Interim Technical Report 8, MIT 1958

Karzanov, A.V., and McCormick, S.T. [1997]: Polynomial methods for separable convex optimization in unimodular linear spaces with applications. SIAM Journal on Computing 26 (1997), 1245–1275

Klein, M. [1967]: A primal method for minimum cost flows, with applications to the assignment and transportation problems. Management Science 14 (1967), 205–220

Klinz, B., and Woeginger, G.J. [2004]: Minimum cost dynamic flows: the series-parallel case. Networks 43 (2004), 153–162

Lee, Y.T., and Sidford, A. [2014]: Path finding methods for linear programming. Proceedings of the 55th Annual IEEE Symposium on Foundations of Computer Science (2014), 424–433

Matsui, T., and Scheifele, R. [2016]: A linear time algorithm for the unbalanced Hitchcock transportation problem. Networks 67 (2016), 170–182

Orden, A. [1956]: The transshipment problem. Management Science 2 (1956), 276–285

Ore, O. [1956]: Studies on directed graphs I. Annals of Mathematics 63 (1956), 383–406

Orlin, J.B. [1993]: A faster strongly polynomial minimum cost flow algorithm. Operations Research 41 (1993), 338–350

Orlin, J.B. [1997]: A polynomial time primal network simplex algorithm for minimum cost flows. Mathematical Programming 78 (1997), 109–129

Orlin, J.B., Plotkin, S.A., and Tardos, É. [1993]: Polynomial dual network simplex algorithms. Mathematical Programming 60 (1993), 255–276

Plotkin, S.A., and Tardos, É. [1990]: Improved dual network simplex. Proceedings of the 1st Annual ACM-SIAM Symposium on Discrete Algorithms (1990), 367–376

Schulz, A.S., Weismantel, R., and Ziegler, G.M. [1995]: 0/1-Integer Programming: optimization and augmentation are equivalent. In: Algorithms – ESA '95; LNCS 979 (P. Spirakis, ed.), Springer, Berlin 1995, pp. 473–483

Schulz, A.S., and Weismantel, R. [2002]: The complexity of generic primal algorithms for solving general integer problems. Mathematics of Operations Research 27 (2002), 681–692

Tardos, É. [1985]: A strongly polynomial minimum cost circulation algorithm. Combinatorica 5 (1985), 247–255

Tolstoĭ, A.N. [1930]: Metody nakhozhdeniya naimen'shego summovogo kilometrazha pri planirovanii perevozok v prostanstve. In: Planirovanie Perevozok, Sbornik pervyĭ, Transpechat' NKPS, Moskow 1930, pp. 23–55. (See A. Schrijver, On the history of the transportation and maximum flow problems, Mathematical Programming 91 (2002), 437–445)

Tomizawa, N. [1971]: On some techniques useful for solution of transportation network problems. Networks 1 (1971), 173–194

Vygen, J. [2002]: On dual minimum cost flow algorithms. Mathematical Methods of Operations Research 56 (2002), 101–126

Wagner, H.M. [1959]: On a class of capacitated transportation problems. Management Science 5 (1959), 304–318

10 Maximum Matchings

Matching theory is one of the classical and most important topics in combinatorial theory and optimization. All the graphs in this chapter are undirected. Recall that a matching is a set of pairwise disjoint edges. Our main problem is:

CARDINALITY MATCHING PROBLEM

Instance: An undirected graph G.

Task: Find a maximum cardinality matching in G.

Since the weighted version of this problem is significantly more difficult we postpone it to Chapter 11. But already the above cardinality version has applications: Suppose in the JOB ASSIGNMENT PROBLEM each job has the same processing time, say one hour, and we ask whether we can finish all the jobs within one hour. In other words: given a bipartite graph G with bipartition $V(G) = A \,\dot\cup\, B$, we look for numbers $x : E(G) \to \mathbb{R}_+$ with $\sum_{e \in \delta(a)} x(e) = 1$ for each job $a \in A$ and $\sum_{e \in \delta(b)} x(e) \leq 1$ for each employee $b \in B$. We can write this as a linear inequality system $x \geq 0, Mx \leq \mathbb{1}, M'x \geq \mathbb{1}$, where the rows of M and M' are rows of the node-edge incidence matrix of G. These matrices are totally unimodular by Theorem 5.26. From Theorem 5.21 we conclude that if there is any solution x, then there is also an integral solution. Now observe that the integral solutions to the above linear inequality system are precisely the incidence vectors of the matchings in G covering A.

Definition 10.1. *Let G be a graph and M a matching in G. We say that a vertex v is **covered** by M if $v \in e$ for some $e \in M$; otherwise v is **exposed** (by M). M is called a **perfect matching** if all vertices are covered by M.*

In Section 10.1 we consider matchings in bipartite graphs. Algorithmically this problem can be reduced to the MAXIMUM FLOW PROBLEM. The Max-Flow-Min-Cut Theorem as well as the concept of augmenting paths have nice interpretations in our context.

Matching in general, non-bipartite graphs does not reduce directly to network flows. We introduce two necessary and sufficient conditions for a general graph to have a perfect matching in Sections 10.2 and 10.3. In Section 10.4 we consider factor-critical graphs which have a matching covering all vertices but v, for each $v \in$

© Springer-Verlag GmbH Germany 2018
Bernhard Korte and Jens Vygen, *Combinatorial Optimization*, Algorithms and Combinatorics 21, https://doi.org/10.1007/978-3-662-56039-6_10

$V(G)$. These play an important role in Edmonds' algorithm for the CARDINALITY MATCHING PROBLEM, described in Section 10.5, and its weighted version which we postpone to Sections 11.2 and 11.3.

10.1 Bipartite Matching

Since the CARDINALITY MATCHING PROBLEM is easier if G is bipartite, we shall deal with this case first. In this section, a bipartite graph G is always assumed to have the bipartition $V(G) = A \mathbin{\dot\cup} B$. Since we may assume that G is connected, we can regard this bipartition as unique (Exercise 26 of Chapter 2).

For a graph G, let $\nu(G)$ denote the maximum cardinality of a matching in G, while $\tau(G)$ is the minimum cardinality of a vertex cover in G.

Theorem 10.2. (König [1931]) *If G is bipartite, then $\nu(G) = \tau(G)$.*

Proof: Consider the graph $G' = (V(G) \mathbin{\dot\cup} \{s, t\}, \ E(G) \cup \{\{s, a\} : a \in A\} \cup \{\{b, t\} : b \in B\})$. Then $\nu(G)$ is the maximum number of internally disjoint s-t-paths, while $\tau(G)$ is the minimum number of vertices whose deletion makes t unreachable from s. The theorem now immediately follows from Menger's Theorem 8.10. $\square$

$\nu(G) \le \tau(G)$ evidently holds for any graph (bipartite or not), but we do not have equality in general (as the triangle K_3 shows).

Several statements are equivalent to König's Theorem. Hall's Theorem is probably the best-known version.

Theorem 10.3. (Hall [1935]) *Let G be a bipartite graph with bipartition $V(G) = A \mathbin{\dot\cup} B$. Then G has a matching covering A if and only if*

$$|\Gamma(X)| \ge |X| \qquad \text{for all } X \subseteq A. \tag{10.1}$$

Proof: The necessity of the condition is obvious. To prove the sufficiency, assume that G has no matching covering A, i.e. $\nu(G) < |A|$. By Theorem 10.2 this implies $\tau(G) < |A|$.

Let $A' \subseteq A$, $B' \subseteq B$ such that $A' \cup B'$ covers all the edges and $|A' \cup B'| < |A|$. Obviously $\Gamma(A \setminus A') \subseteq B'$. Therefore $|\Gamma(A \setminus A')| \le |B'| < |A| - |A'| = |A \setminus A'|$, and the Hall condition (10.1) is violated. $\square$

It is worthwhile to mention that it is not too difficult to prove Hall's Theorem directly. The following proof is due to Halmos and Vaughan [1950]:

Second Proof of Theorem 10.3: We show that any G satisfying the Hall condition (10.1) has a matching covering A. We use induction on $|A|$, the cases $|A| = 0$ and $|A| = 1$ being trivial.

If $|A| \ge 2$, we consider two cases: If $|\Gamma(X)| > |X|$ for every nonempty proper subset X of A, then we take any edge $\{a, b\}$ ($a \in A$, $b \in B$), delete its two vertices and apply induction. The smaller graph satisfies the Hall condition because $|\Gamma(X)| - |X|$ can have decreased by at most one for any $X \subseteq A \setminus \{a\}$.

Now assume that there is a nonempty proper subset X of A with $|\Gamma(X)| = |X|$. By induction there is a matching covering X in $G[X \cup \Gamma(X)]$. We claim that we can extend this to a matching in G covering A. Again by the induction hypothesis, we have to show that $G[(A \setminus X) \cup (B \setminus \Gamma(X))]$ satisfies the Hall condition. To check this, observe that for any $Y \subseteq A \setminus X$ we have (in the original graph G):

$$|\Gamma(Y) \setminus \Gamma(X)| = |\Gamma(X \cup Y)| - |\Gamma(X)| \geq |X \cup Y| - |X| = |Y|.$$

$\square$

A special case of Hall's Theorem is the so-called "Marriage Theorem":

Theorem 10.4.   (Frobenius [1917])   *Let G be a bipartite graph with bipartition $V(G) = A \mathbin{\dot{\cup}} B$. Then G has a perfect matching if and only if $|A| = |B|$ and $|\Gamma(X)| \geq |X|$ for all $X \subseteq A$.* $\square$

The variety of applications of Hall's Theorem is indicated by Exercises 4–7.

The proof of König's Theorem 10.2 shows how to solve the bipartite matching problem algorithmically:

Theorem 10.5.   *The CARDINALITY MATCHING PROBLEM for bipartite graphs G can be solved in $O(nm)$ time, where $n = |V(G)|$ and $m = |E(G)|$.*

Proof: Let G be a bipartite graph with bipartition $V(G) = A \mathbin{\dot{\cup}} B$. Add a vertex s and connect it to all vertices of A, and add another vertex t connected to all vertices of B. Orient the edges from s to A, from A to B, and from B to t. Let the capacities be 1 everywhere. Then a maximum integral s-t-flow corresponds to a maximum cardinality matching (and vice versa).

So we apply the FORD-FULKERSON ALGORITHM and find a maximum s-t-flow (and thus a maximum matching) after at most n augmentations. Since each augmentation takes $O(m)$ time, we are done. $\square$

This result is essentially due to Kuhn [1955]. In fact, one can use the concept of shortest augmenting paths again (cf. the EDMONDS-KARP ALGORITHM). In this way one obtains the $O\left(\sqrt{n}(m + n)\right)$-algorithm of Hopcroft and Karp [1973] and Karzanov [1973]. This algorithm will be discussed in Exercises 11 and 12. Slight improvements of the HOPCROFT-KARP ALGORITHM yield running times of $O\left(n\sqrt{\frac{mn}{\log n}}\right)$ (Alt et al. [1991]) and $O\left(m\sqrt{n}\frac{\log(n^2/m)}{\log n}\right)$ (Feder and Motwani [1995]). The latter bound is the best known for dense graphs. For sparse graphs, Mądry [2013] found a faster algorithm which is completely different and has running time $O(m^{10/7} \log^{O(1)} m)$.

Let us reformulate the augmenting path concept in our context.

Definition 10.6.   *Let G be a graph (bipartite or not), and let M be some matching in G. A path P is an M-alternating path if $E(P) \setminus M$ is a matching. An M-alternating path is M-augmenting if it has positive length and its endpoints are exposed by M.*

One immediately checks that M-augmenting paths must have odd length.

Theorem 10.7. (Petersen [1891], Berge [1957]) *Let G be a graph (bipartite or not) with some matching M. Then M is maximum if and only if there is no M-augmenting path.*

Proof: If there is an M-augmenting path P, the symmetric difference $M \triangle E(P)$ is a matching and has greater cardinality than M, so M is not maximum. On the other hand, if there is a matching M' such that $|M'| > |M|$, the symmetric difference $M \triangle M'$ is the vertex-disjoint union of alternating circuits and paths, where at least one path must be M-augmenting.   □

This was already observed by Petersen and rediscovered by Berge; it is commonly known as Berge's Theorem. In the bipartite case it of course also follows from Theorem 8.5.

10.2 The Tutte Matrix

We now consider maximum matchings from an algebraic point of view. Let G be a simple undirected graph, and let G' be the directed graph resulting from G by arbitrarily orienting the edges. For any vector $x = (x_e)_{e \in E(G)}$ of variables, we define the **Tutte matrix**

$$T_G(x) = (t^x_{vw})_{v,w \in V(G)}$$

by

$$t^x_{vw} := \begin{cases} x_{\{v,w\}} & \text{if } (v, w) \in E(G') \\ -x_{\{v,w\}} & \text{if } (w, v) \in E(G') \ . \\ 0 & \text{otherwise} \end{cases}$$

Such a matrix M, where $M = -M^\top$, is called **skew-symmetric**. The matrix $T_G(x)$, but not its rank, depends on the orientation G' of G. Note that $\det T_G(x)$ is a polynomial in the variables x_e ($e \in E(G)$).

Theorem 10.8. (Tutte [1947]) *G has a perfect matching if and only if $\det T_G(x)$ is not identically zero.*

Proof: Let $V(G) = \{v_1, \ldots, v_n\}$, and let S_n be the set of all permutations on $\{1, \ldots, n\}$. By definition of the determinant,

$$\det T_G(x) = \sum_{\pi \in S_n} \text{sgn}(\pi) \prod_{i=1}^{n} t^x_{v_i, v_{\pi(i)}}.$$

Let $S'_n := \left\{ \pi \in S_n : \prod_{i=1}^{n} t^x_{v_i, v_{\pi(i)}} \neq 0 \right\}$. Each permutation $\pi \in S_n$ corresponds to a directed graph $H_\pi := (V(G), \{(v_i, v_{\pi(i)}) : i = 1, \ldots, n\})$ where each vertex x has $|\delta^-_{H_\pi}(x)| = |\delta^+_{H_\pi}(x)| = 1$. For permutations $\pi \in S'_n$, H_π is a subgraph of $\overset{\leftrightarrow}{G'}$.

If there exists a permutation $\pi \in S'_n$ such that H_π consists of even circuits only, then by taking every second edge of each circuit (and ignoring the orientations) we obtain a perfect matching in G.

Otherwise, for each $\pi \in S'_n$ there is a permutation $r(\pi) \in S'_n$ such that $H_{r(\pi)}$ is obtained by reversing the first odd circuit in H_π, i.e. the odd circuit containing the vertex with minimum index. Of course $r(r(\pi)) = \pi$.

Observe that $\text{sgn}(\pi) = \text{sgn}(r(\pi))$, i.e. the two permutations have the same sign: if the first odd circuit consists of the vertices $v_{i_1}, \ldots, v_{i_{2k+1}}$ with $\pi(i_j) = i_{j+1}$ ($j = 1, \ldots, 2k$) and $\pi(i_{2k+1}) = i_1$, then we obtain $r(\pi)$ by $2k$ transpositions: for $j = 1, \ldots, k$ exchange $\pi(i_{2j-1})$ with $\pi(i_{2k})$ and then $\pi(i_{2j})$ with $\pi(i_{2k+1})$.

Moreover, $\prod_{i=1}^n t^x_{v_i, v_{\pi(i)}} = -\prod_{i=1}^n t^x_{v_i, v_{r(\pi)(i)}}$. So the two corresponding terms in the sum

$$\det T_G(x) = \sum_{\pi \in S'_n} \text{sgn}(\pi) \prod_{i=1}^n t^x_{v_i, v_{\pi(i)}}$$

cancel each other. Since this holds for all pairs $\pi, r(\pi) \in S'_n$, we conclude that $\det T_G(x)$ is identically zero.

So if G has no perfect matching, $\det T_G(x)$ is identically zero. On the other hand, if G has a perfect matching M, consider the permutation defined by $\pi(i) := j$ and $\pi(j) := i$ for all $\{v_i, v_j\} \in M$. The corresponding term $\prod_{i=1}^n t^x_{v_i, v_{\pi(i)}} = \prod_{e \in M} \left(-x_e^2\right)$ cannot cancel out with any other term, so $\det T_G(x)$ is not identically zero. $\qquad\square$

Originally, Tutte used Theorem 10.8 to prove his main theorem on matchings, Theorem 10.13. Theorem 10.8 does not provide a good characterization of the property that a graph has a perfect matching. The problem is that the determinant is easy to compute if the entries are numbers (Theorem 4.10) but difficult to compute if the entries are variables. However, the theorem suggests a randomized algorithm for the CARDINALITY MATCHING PROBLEM:

Corollary 10.9. (Lovász [1979]) *Let $x = (x_e)_{e \in E(G)}$ be a random vector whose coordinates are independent and uniformly distributed in $[0, 1]$. Then with probability 1 the rank of $T_G(x)$ is exactly twice the size of a maximum matching.*

Proof: Suppose the rank of $T_G(x)$ is k, say the first k rows are linearly independent. We write $T_G(x) = \begin{pmatrix} A & B \\ -B^\top & C \end{pmatrix}$, where A is a skew-symmetric $(k \times k)$-matrix and $\begin{pmatrix} A & B \end{pmatrix}$ has rank k. Then there is a matrix D with $D \begin{pmatrix} A & B \end{pmatrix} = \begin{pmatrix} -B^\top & C \end{pmatrix}$. We have $AD^\top = -(DA)^\top = B$, and hence A has rank k. So the principal submatrix A of $T_G(x)$ is nonsingular, and by Theorem 10.8 the subgraph induced by the corresponding vertices has a perfect matching. In particular, k is even and G has a matching of cardinality $\frac{k}{2}$.

On the other hand, if G has a matching of cardinality k, the determinant of the principal submatrix T' whose rows and columns correspond to the $2k$ vertices covered by M is not identically zero by Theorem 10.8. The set of vectors x for which $\det T'(x) = 0$ must then have measure zero. So with probability one, the rank of $T_G(x)$ is at least $2k$.   □

Of course it is not possible to choose random numbers from $[0, 1]$ with a digital computer. However, it can be shown that it suffices to choose random integers from the finite set $\{1, 2, \ldots, N\}$. For sufficiently large N, the probability of error will become arbitrarily small (see Lovász [1979]). Lovász' algorithm can be used to determine a maximum matching (not only its cardinality). See Rabin and Vazirani [1989], Mulmuley, Vazirani and Vazirani [1987], and Mucha and Sankowski [2004] for further randomized algorithms for finding a maximum matching in a graph. Moreover we note that Geelen [2000] has shown how to derandomize Lovász' algorithm. Although its running time is worse than that of Edmonds' matching algorithm (see Section 10.5), it is important for some generalizations of the CARDINALITY MATCHING PROBLEM (e.g., see Geelen and Iwata [2005]).

10.3 Tutte's Theorem

We now consider the CARDINALITY MATCHING PROBLEM in general graphs. A necessary condition for a graph to have a perfect matching is that every connected component is even (i.e. has an even number of vertices). This condition is not sufficient, as the graph $K_{1,3}$ (Figure 10.1(a)) shows.

The reason that $K_{1,3}$ has no perfect matching is that there is one vertex (the black one) whose deletion produces three odd connected components. The graph shown in Figure 10.1(b) is more complicated. Does this graph have a perfect matching? If we delete the three black vertices, we get five odd connected components (and one even connected component). If there were a perfect matching, at least one vertex of each odd connected component would have to be connected to one of the black

(a)   (b)

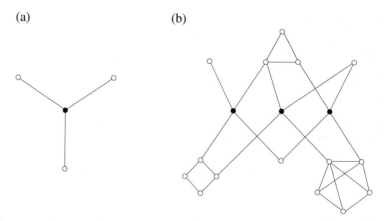

Fig. 10.1.

vertices. This is impossible because the number of odd connected components exceeds the number of black vertices.

More generally, for $X \subseteq V(G)$ let $q_G(X)$ denote the number of odd connected components in $G - X$. Then a graph for which $q_G(X) > |X|$ holds for some $X \subseteq V(G)$ cannot have a perfect matching: otherwise there must be, for each odd connected component in $G - X$, at least one matching edge connecting this connected component with X, which is impossible if there are more odd connected components than elements of X. Tutte's Theorem says that the above necessary condition is also sufficient:

Definition 10.10. *A graph G satisfies the* **Tutte condition** *if $q_G(X) \leq |X|$ for all $X \subseteq V(G)$. A nonempty vertex set $X \subseteq V(G)$ is a* **barrier** *if $q_G(X) = |X|$.*

To prove the sufficiency of the Tutte condition we shall need an easy observation and an important definition:

Proposition 10.11. *For any graph G and any $X \subseteq V(G)$ we have*

$$q_G(X) - |X| \equiv |V(G)| \pmod 2.$$

$\square$

Definition 10.12. *A graph G is called* **factor-critical** *if $G - v$ has a perfect matching for each $v \in V(G)$. A matching is called* **near-perfect** *if it covers all vertices but one.*

Now we can prove Tutte's Theorem:

Theorem 10.13. *(Tutte [1947]) A graph G has a perfect matching if and only if it satisfies the Tutte condition:*

$$q_G(X) \leq |X| \qquad \text{for all } X \subseteq V(G).$$

Proof: We have already seen the necessity of the Tutte condition. We now prove the sufficiency by induction on $|V(G)|$ (the case $|V(G)| \leq 2$ being trivial).

Let G be a graph satisfying the Tutte condition. $|V(G)|$ cannot be odd since otherwise the Tutte condition is violated because $q_G(\emptyset) \geq 1$.

So by Proposition 10.11, $|X| - q_G(X)$ must be even for every $X \subseteq V(G)$. Since $|V(G)|$ is even and the Tutte condition holds, every singleton is a barrier.

We choose a maximal barrier X. $G - X$ has $|X|$ odd connected components. $G - X$ cannot have any even connected components because otherwise $X \cup \{v\}$, where v is a vertex of some even connected component, is a barrier ($G - (X \cup \{v\})$ has $|X| + 1$ odd connected components), contradicting the maximality of X.

We now claim that each odd connected component of $G - X$ is factor-critical. To prove this, let C be some odd connected component of $G - X$ and $v \in V(C)$. If $C - v$ has no perfect matching, by the induction hypothesis there is some

$Y \subseteq V(C) \setminus \{v\}$ such that $q_{C-v}(Y) > |Y|$. By Proposition 10.11, $q_{C-v}(Y) - |Y|$ must be even, so

$$q_{C-v}(Y) \geq |Y| + 2.$$

Since X, Y and $\{v\}$ are pairwise disjoint, we have

$$\begin{aligned}
q_G(X \cup Y \cup \{v\}) &= q_G(X) - 1 + q_C(Y \cup \{v\}) \\
&= |X| - 1 + q_{C-v}(Y) \\
&\geq |X| - 1 + |Y| + 2 \\
&= |X \cup Y \cup \{v\}|.
\end{aligned}$$

So $X \cup Y \cup \{v\}$ is a barrier, contradicting the maximality of X.

We now consider the bipartite graph G' with bipartition $V(G') = X \dot\cup Z$ which arises when we delete edges with both ends in X and contract the odd connected components of $G - X$ to single vertices (forming the set Z).

It remains to show that G' has a perfect matching. If not, then by Frobenius' Theorem 10.4 there is some $A \subseteq Z$ such that $|\Gamma_{G'}(A)| < |A|$. This implies $q_G(\Gamma_{G'}(A)) \geq |A| > |\Gamma_{G'}(A)|$, a contradiction.   □

This proof is due to Anderson [1971]. The Tutte condition provides a good characterization of the perfect matching problem: either a graph has a perfect matching or it has a so-called **Tutte set** X proving that it has no perfect matching. An important consequence of Tutte's Theorem is the so-called Berge-Tutte formula:

Theorem 10.14. (Berge [1958])

$$2\nu(G) + \max_{X \subseteq V(G)} (q_G(X) - |X|) = |V(G)|.$$

Proof: For any $X \subseteq V(G)$, any matching must leave at least $q_G(X) - |X|$ vertices uncovered. Therefore $2\nu(G) + q_G(X) - |X| \leq |V(G)|$.

To prove the reverse inequality, let

$$k := \max_{X \subseteq V(G)} (q_G(X) - |X|).$$

We construct a new graph H by adding k new vertices to G, each of which is connected to all the old vertices.

If we can prove that H has a perfect matching, then

$$2\nu(G) + k \geq 2\nu(H) - k = |V(H)| - k = |V(G)|,$$

and the theorem is proved.

Suppose H has no perfect matching, then by Tutte's Theorem there is a set $Y \subseteq V(H)$ such that $q_H(Y) > |Y|$. By Proposition 10.11, k has the same parity as $|V(G)|$, implying that $|V(H)|$ is even. Therefore $Y \neq \emptyset$ and thus $q_H(Y) > 1$. But then Y contains all the new vertices, so

$$q_G(Y \cap V(G)) = q_H(Y) > |Y| = |Y \cap V(G)| + k,$$

contradicting the definition of k. $\square$

Let us close this section with a proposition for later use.

Proposition 10.15. *Let G be a graph and $X \subseteq V(G)$ with $|V(G)| - 2\nu(G) = q_G(X) - |X|$. Then any maximum matching of G contains a perfect matching in each even connected component of $G - X$, a near-perfect matching in each odd connected component of $G - X$, and matches all the vertices in X to vertices of distinct odd connected components of $G - X$.* $\square$

Later we shall see (Theorem 10.32) that X can be chosen such that each odd connected component of $G - X$ is factor-critical.

10.4 Ear-Decompositions of Factor-Critical Graphs

This section contains some results on factor-critical graphs which we shall need later. In Exercise 22 of Chapter 2 we have seen that the graphs having an ear-decomposition are exactly the 2-edge-connected graphs. Here we are interested in odd ear-decompositions only.

Definition 10.16. *An ear-decomposition is called **odd** if every ear has odd length.*

Theorem 10.17. (Lovász [1972]) *A graph is factor-critical if and only if it has an odd ear-decomposition. Furthermore, the initial vertex of the ear-decomposition can be chosen arbitrarily.*

Proof: Let G be a graph with a fixed odd ear-decomposition. We prove that G is factor-critical by induction on the number of ears. Let P be the last ear in the odd ear-decomposition, say P goes from x to y, and let G' be the graph before adding P. We have to show for any vertex $v \in V(G)$ that $G - v$ contains a perfect matching. If v is not an inner vertex of P this is clear by induction (add every second edge of P to the perfect matching in $G' - v$). If v is an inner vertex of P, then exactly one of $P_{[v,x]}$ and $P_{[v,y]}$ must be even, say $P_{[v,x]}$. By induction there is a perfect matching in $G' - x$. By adding every second edge of $P_{[y,v]}$ and of $P_{[v,x]}$ we obtain a perfect matching in $G - v$.

We now prove the reverse direction. Choose the initial vertex z of the ear-decomposition arbitrarily, and let M be a near-perfect matching in G covering $V(G) \setminus \{z\}$. Suppose we already have an odd ear-decomposition of a subgraph G' of G such that $z \in V(G')$ and $M \cap E(G')$ is a near-perfect matching in G'. If $G = G'$, we are done.

If not, then – since G is connected – there must be an edge $e = \{x, y\} \in E(G) \setminus E(G')$ with $x \in V(G')$. If $y \in V(G')$, e is the next ear. Otherwise let N be a near-perfect matching in G covering $V(G) \setminus \{y\}$. $M \triangle N$ obviously contains the edges of a y-z-path P. Let w be the first vertex of P (when traversed from y) that belongs to $V(G')$. The last edge of $P' := P_{[y,w]}$ cannot belong to M (because no edge of M leaves $V(G')$), and the first edge cannot belong to N. Since P' is M-N-alternating, $|E(P')|$ must be even, so together with e it forms the next ear.

$\square$

In fact, we have constructed a special type of odd ear-decomposition:

Definition 10.18. *Given a factor-critical graph G and a near-perfect matching M, an **M-alternating ear-decomposition** of G is an odd ear-decomposition such that each ear is an M-alternating path or a circuit C with $|E(C) \cap M| + 1 = |E(C) \setminus M|$.*

It is clear that the initial vertex of an M-alternating ear-decomposition must be the vertex exposed by M. The proof of Theorem 10.17 immediately yields:

Corollary 10.19. *For any factor-critical graph G and any near-perfect matching M in G there exists an M-alternating ear-decomposition.* $\square$

From now on, we shall only be interested in M-alternating ear-decompositions. An interesting way to store an M-alternating ear-decomposition efficiently is due to Lovász and Plummer [1986]:

Definition 10.20. *Let G be a factor-critical graph and M a near-perfect matching in G. Let $r, P_1, \ldots, P_k$ be an M-alternating ear-decomposition of G and $\mu, \varphi : V(G) \to V(G)$ two functions. We say that μ and φ are **associated with the ear-decomposition** $r, P_1, \ldots, P_k$ if*

- $\mu(x) = y$ *if* $\{x, y\} \in M$,
- $\varphi(x) = y$ *if* $\{x, y\} \in E(P_i) \setminus M$ *and* $x \notin \{r\} \cup V(P_1) \cup \cdots \cup V(P_{i-1})$,
- $\mu(r) = \varphi(r) = r$.

If M is fixed, we also say that φ is associated with $r, P_1, \ldots, P_k$.

If M is some fixed near-perfect matching and μ, φ are associated with two M-alternating ear-decompositions, they are the same up to the order of the ears. Moreover, an explicit list of the ears can be obtained in linear time:

EAR-DECOMPOSITION ALGORITHM

Input: A factor-critical graph G, functions μ, φ associated with an M-alternating ear-decomposition.

Output: An M-alternating ear-decomposition $r, P_1, \ldots, P_k$.

---

① Let initially be $X := \{r\}$, where r is the vertex with $\mu(r) = r$. Let $k := 0$, and let the stack be empty.

②   **If** $X = V(G)$ **then go to** ⑤.
      If the stack is nonempty
          then let $v \in V(G) \setminus X$ be an endpoint of the topmost element of the
           stack,
          else choose $v \in V(G) \setminus X$ arbitrarily.

③   Set $x := v$, $y := \mu(v)$ and $P := (\{x, y\}, \{\{x, y\}\})$.
      While $\varphi(\varphi(x)) = x$ **do**:
          Set $P := P + \{x, \varphi(x)\} + \{\varphi(x), \mu(\varphi(x))\}$ and $x := \mu(\varphi(x))$.
      While $\varphi(\varphi(y)) = y$ **do**:
          Set $P := P + \{y, \varphi(y)\} + \{\varphi(y), \mu(\varphi(y))\}$ and $y := \mu(\varphi(y))$.
      Set $P := P + \{x, \varphi(x)\} + \{y, \varphi(y)\}$. P is the ear containing y as an inner
      vertex. Put P on top of the stack.

④   **While** both endpoints of the topmost element P of the stack are in X **do**:
          Delete P from the stack, set $k := k+1$, $P_k := P$ and $X := X \cup V(P)$.
      Go to ②.

⑤   **For** all $\{y, z\} \in E(G) \setminus (E(P_1) \cup \cdots \cup E(P_k))$ **do**:
      Set $k := k + 1$ and $P_k := (\{y, z\}, \{\{y, z\}\})$.

Proposition 10.21. *Let G be a factor-critical graph and μ, φ functions associated with an M-alternating ear-decomposition. Then this ear-decomposition is unique up to the order of the ears. The* EAR-DECOMPOSITION ALGORITHM *correctly determines an explicit list of these ears; it runs in linear time.*

Proof: Let $\mathcal{D}$ be an M-alternating ear-decomposition associated with μ and φ. The uniqueness of $\mathcal{D}$ as well as the correctness of the algorithm follows from the obvious fact that P as computed in ③ is indeed an ear of $\mathcal{D}$. The running time of ① – ④ is evidently $O(|V(G)|)$, while ⑤ takes $O(|E(G)|)$ time. $\square$

The most important property of the functions associated with an alternating ear-decomposition is the following:

Lemma 10.22. *Let G be a factor-critical graph and μ, φ two functions associated with an M-alternating ear-decomposition. Let r be the vertex exposed by M. Then the maximal path given by an initial subsequence of*

$$x, \mu(x), \varphi(\mu(x)), \mu(\varphi(\mu(x))), \varphi(\mu(\varphi(\mu(x)))), \ldots$$

defines an M-alternating x-r-path of even length for all $x \in V(G)$.

Proof: Let $x \in V(G) \setminus \{r\}$, and let P_i be the first ear containing x. Clearly some initial subsequence of

$$x, \mu(x), \varphi(\mu(x)), \mu(\varphi(\mu(x))), \varphi(\mu(\varphi(\mu(x)))), \ldots$$

must be a subpath Q of P_i from x to y, where $y \in \{r\} \cup V(P_1) \cup \cdots \cup V(P_{i-1})$. Because we have an M-alternating ear-decomposition, the last edge of Q does not

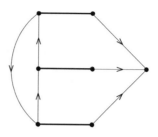

Fig. 10.2.

belong to M; hence Q has even length. If $y = r$, we are done, otherwise we apply induction on i. $\square$

The converse of Lemma 10.22 is not true: In the counterexample in Figure 10.2 (bold edges are matching edges, edges directed from u to v indicate $\varphi(u) = v$), μ and φ also define alternating paths to the vertex exposed by the matching. However, μ and φ are not associated with any alternating ear-decomposition.

For the WEIGHTED MATCHING ALGORITHM (Section 11.3) we shall need a fast routine for updating an alternating ear-decomposition when the matching changes. Although the proof of Theorem 10.17 is algorithmic (provided that we can find a maximum matching in a graph), this is far too inefficient. We make use of the old ear-decomposition:

Lemma 10.23. *Given a factor-critical graph G, two near-perfect matchings M and M', and functions μ, φ associated with an M-alternating ear-decomposition. Then functions μ', φ' associated with an M'-alternating ear-decomposition can be found in $O(|V(G)|)$ time.*

Proof: Let v be the vertex exposed by M, and let v' be the vertex exposed by M'. Let P be the v'-v-path in $M \triangle M'$, say $P = x_0, x_1, \ldots, x_k$ with $x_0 = v'$ and $x_k = v$.

An explicit list of the ears of the old ear-decomposition can be obtained from μ and φ by the EAR-DECOMPOSITION ALGORITHM in linear time (Proposition 10.21). Indeed, since we do not have to consider ears of length one, we can omit ⑤: then the total number of edges considered is at most $\frac{3}{2}(|V(G)| - 1)$ (cf. Exercise 21).

Suppose we have already constructed an M'-alternating ear-decomposition of a spanning subgraph of $G[X]$ for some $X \subseteq V(G)$ with $v' \in X$ (initially $X := \{v'\}$). Of course no M'-edge leaves X. Let $p := \max\{i \in \{0, \ldots, k\} : x_i \in X\}$ (illustrated in Figure 10.3). At each stage we keep track of p and of the edge set $\delta(X) \cap M$. Their update when extending X is clearly possible in linear total time.

Now we show how to extend the ear-decomposition. We shall add one or more ears in each step. The time needed for each step will be proportional to the total number of edges in new ears.

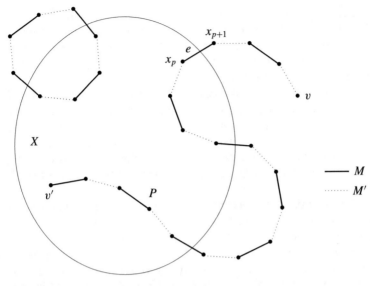

Fig. 10.3.

Case 1: $|\delta(X) \cap M| \geq 2$. Let $f \in \delta(X) \cap M$ with $x_p \notin f$. Evidently, f belongs to an M-M'-alternating path which can be added as the next ear. The time needed to find this ear is proportional to its length.

Case 2: $|\delta(X) \cap M| = 1$. Then $v \notin X$, and $e = \{x_p, x_{p+1}\}$ is the only edge in $\delta(X) \cap M$. Let R' be the x_{p+1}-v-path determined by μ and φ (cf. Lemma 10.22). The first edge of R' is e. Let q be the minimum index $i \in \{p+2, p+4, \ldots, k\}$ with $x_i \in V(R')$ and $V(R'_{[x_{p+1}, x_i]}) \cap \{x_{i+1}, \ldots, x_k\} = \emptyset$ (cf. Figure 10.4). Let $R := R'_{[x_p, x_q]}$. So R has vertices x_p, $\varphi(x_p)$, $\mu(\varphi(x_p))$, $\varphi(\mu(\varphi(x_p)))$, $\ldots$, x_q, and can be traversed in time proportional to its length.

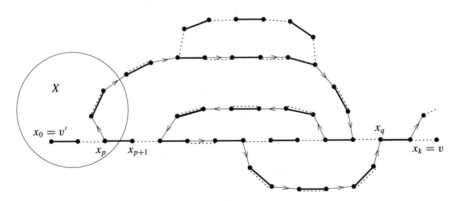

Fig. 10.4.

Let $S := E(R) \setminus E(G[X])$, $D := (M \triangle M') \setminus (E(G[X]) \cup E(P_{[x_q,v]}))$, and $Z := S \triangle D$. S and D consist of M-alternating paths and circuits. Observe that every vertex outside X has degree 0 or 2 with respect to Z. Moreover, for every vertex outside X with two incident edges of Z, one of them belongs to M'. (Here the choice of q is essential.)

Hence all connected components C of $(V(G), Z)$ with $E(C) \cap \delta(X) \neq \emptyset$ can be added as next ears, and after these ears have been added, $S \setminus Z = S \cap (M \triangle M')$ is the vertex-disjoint union of paths each of which can then be added as an ear. Since $e \in D \setminus S \subseteq Z$, we have $Z \cap \delta(X) \neq \emptyset$, so at least one ear is added.

It remains to show that the time needed for the above construction is proportional to the total number of edges in new ears. Obviously, it suffices to find S in $O(|S|)$ time.

This is difficult because of the subpaths of R inside X. However, we do not really care what they look like. So we would like to shortcut these paths whenever possible. To achieve this, we modify the φ-variables.

Namely, in each application of Case 2, let $R_{[a,b]}$ be a maximal subpath of R inside X with $a \neq b$. Let $y := \mu(b)$; y is the predecessor of b on R. We set $\varphi(x) := y$ for all vertices x on $R_{[a,y]}$ where $R_{[x,y]}$ has odd length. It does not matter whether x and y are joined by an edge. See Figure 10.5 for an illustration.

The time required for updating the φ-variables is proportional to the number of edges examined. Note that these changes of φ do not destroy the property of Lemma 10.22, and the φ-variables are not used anymore except for finding M-alternating paths to v in Case 2.

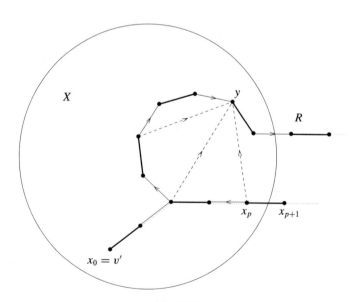

Fig. 10.5.

Now it is guaranteed that the time required for finding the subpaths of R inside X is proportional to the number of subpaths plus the number of edges examined for the first time inside X. Since the number of subpaths inside X is less than or equal to the number of new ears in this step, we obtain an overall linear running time.

Case 3: $\delta(X) \cap M = \emptyset$. Then $v \in X$. We consider the ears of the (old) M-alternating ear-decomposition in their order. Let R be the first ear with $V(R)\backslash X \neq \emptyset$.

Similar to Case 2, let $S := E(R) \setminus E(G[X])$, $D := (M\triangle M') \setminus E(G[X])$, and $Z := S\triangle D$. Again, all connected components C of $(V(G), Z)$ with $E(C)\cap\delta(X) \neq \emptyset$ can be added as next ears, and after these ears have been added, $S \setminus Z$ is the vertex-disjoint union of paths each of which can then be added as an ear. The total time needed for Case 3 is obviously linear. $\square$

10.5 Edmonds' Matching Algorithm

Recall Berge's Theorem 10.7: A matching in a graph is maximum if and only if there is no augmenting path. Since this holds for non-bipartite graphs as well, our matching algorithm will again be based on augmenting paths.

However, it is not at all clear how to find an augmenting path (or decide that there is none). In the bipartite case (Theorem 10.5) it was sufficient to mark the vertices that are reachable from an exposed vertex via an alternating edge progression. Since there were no odd circuits, vertices reachable by an alternating edge progression were also reachable by an alternating path. This is no longer the case when dealing with general graphs.

Consider the example in Figure 10.6 (the bold edges constitute a matching M). When starting at v_1, we have an alternating edge progression $v_1, v_2, v_3, v_4, v_5, v_6, v_7, v_5, v_4, v_8$, but this is not a path. We have run through an odd circuit, namely v_5, v_6, v_7. Note that in our example there exists an augmenting path $(v_1, v_2, v_3, v_7, v_6, v_5, v_4, v_8)$ but it is not clear how to find it.

The question arises what to do if we encounter an odd circuit. Surprisingly, it suffices to get rid of it by shrinking it to a single vertex. It turns out that the smaller graph has a perfect matching if and only if the original graph has one.

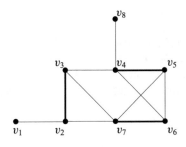

Fig. 10.6.

This is the general idea of EDMONDS' CARDINALITY MATCHING ALGORITHM. We formulate this idea in Lemma 10.25 after giving the following definition:

Definition 10.24. *Let G be a graph and M a matching in G. A* **blossom** *in G with respect to M is a factor-critical subgraph C of G with* $|M \cap E(C)| = \frac{|V(C)|-1}{2}$. *The vertex of C exposed by M* $\cap$ *E(C) is called the* **base** *of C.*

The blossom we have encountered in the above example (Figure 10.6) is induced by $\{v_5, v_6, v_7\}$. Note that this example contains other blossoms. Any single vertex is also a blossom in terms of our definition. Now we can formulate the Blossom Shrinking Lemma:

Lemma 10.25. *Let G be a graph, M a matching in G, and C a blossom in G (with respect to M). Suppose there is an M-alternating v-r-path Q of even length from a vertex v exposed by M to the base r of C, where* $E(Q) \cap E(C) = \emptyset$.

Let G' and M' result from G and M by shrinking V(C) to a single vertex. Then M is a maximum matching in G if and only if M' is a maximum matching in G'.

Proof: Suppose that M is not a maximum matching in G. $N := M \triangle E(Q)$ is a matching of the same cardinality, so it is not maximum either. By Berge's Theorem 10.7 there then exists an N-augmenting path P in G. Note that N does not cover r.

At least one of the endpoints of P, say x, does not belong to C. If P and C are disjoint, let y be the other endpoint of P. Otherwise let y be the first vertex on P – when traversed from x – belonging to C. Let P' result from $P_{[x,y]}$ when shrinking $V(C)$ in G. The endpoints of P' are exposed by N' (the matching in G' corresponding to N). Hence P' is an N'-augmenting path in G'. So N' is not a maximum matching in G', and nor is M' (which has the same cardinality).

To prove the converse, suppose that M' is not a maximum matching in G'. Let N' be a larger matching in G'. N' corresponds to a matching N_0 in G which covers at most one vertex of C in G. Since C is factor-critical, N_0 can be extended by $k := \frac{|V(C)|-1}{2}$ edges to a matching N in G, where

$$|N| \; = \; |N_0| + k \; = \; |N'| + k \; > \; |M'| + k \; = \; |M|,$$

proving that M is not a maximum matching in G.   □

It is necessary to require that the base of the blossom is reachable from a vertex exposed by M by an M-alternating path of even length which is disjoint from the blossom. For example, the blossom induced by $\{v_4, v_6, v_7, v_2, v_3\}$ in Figure 10.6 cannot be shrunk without destroying the only augmenting path.

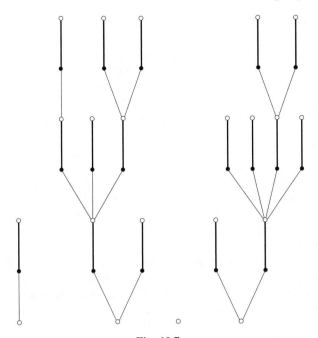

Fig. 10.7.

When looking for an augmenting path, we shall build up an alternating forest:

Definition 10.26. *Given a graph G and a matching M in G. An* **alternating forest** *with respect to M in G is a forest F in G with the following properties:*

(a) *V(F) contains all the vertices exposed by M. Each connected component of F contains exactly one exposed vertex, its* **root**.

(b) *We call a vertex v ∈ V(F) an* **outer (inner)** *vertex if it has even (odd) distance to the root of the connected component containing v. (In particular, the roots are outer vertices.) All inner vertices have degree 2 in F.*

(c) *For any v ∈ V(F), the unique path from v to the root of the connected component containing v is M-alternating.*

Figure 10.7 shows an alternating forest. The bold edges belong to the matching. The black vertices are inner, the white vertices outer.

Proposition 10.27. *In any alternating forest the number of outer vertices that are not a root equals the number of inner vertices.*

Proof: Each outer vertex that is not a root has exactly one neighbour which is an inner vertex and whose distance to the root is smaller. This is obviously a bijection between the outer vertices that are not a root and the inner vertices.   □

Informally, EDMONDS' CARDINALITY MATCHING ALGORITHM works as follows. Given some matching M, we build up an M-alternating forest F. We start with the set S of exposed vertices and no edges.

At any stage of the algorithm we consider a neighbour y of an outer vertex x. Let $P(x)$ denote the unique path in F from x to a root. There are three interesting cases, corresponding to three operations ("grow", "augment", and "shrink"):

Case 1: $y \notin V(F)$. Then the forest will grow when we add $\{x, y\}$ and the matching edge covering y.

Case 2: y is an outer vertex in a different connected component of F. Then we augment M along $P(x) \cup \{x, y\} \cup P(y)$.

Case 3: y is an outer vertex in the same connected component of F (with root q). Let r be the first vertex of $P(x)$ (starting at x) also belonging to $P(y)$. (r can be one of x, y.) If r is not a root, it must have degree at least 3. So r is an outer vertex. Therefore $C := P(x)_{[x,r]} \cup \{x, y\} \cup P(y)_{[y,r]}$ is a blossom with at least three vertices. We shrink C.

If none of the cases applies, all the neighbours of outer vertices are inner. We claim that M is maximum. Let X be the set of inner vertices, $s := |X|$, and let t be the number of outer vertices. $G - X$ has t odd components (each outer vertex is isolated in $G - X$), so $q_G(X) - |X| = t - s$. Hence by the trivial part of the Berge-Tutte formula, any matching must leave at least $t - s$ vertices uncovered. But on the other hand, the number of vertices exposed by M, i.e. the number of roots of F, is exactly $t - s$ by Proposition 10.27. Hence M is indeed maximum.

Since this is not at all a trivial task, we shall spend some time on implementation details. The difficult question is how to perform the shrinking efficiently so that the original graph can be recovered afterwards. Of course, several shrinking operations may involve the same vertex. Our presentation is based on the one given by Lovász and Plummer [1986].

Rather than actually performing the shrinking operation, we allow our forest to contain blossoms.

Definition 10.28. *Given a graph G and a matching M in G. A subgraph F of G is a* **general blossom forest** *(with respect to M) if there exists a partition $V(F) = V_1 \cup V_2 \cup \cdots \cup V_k$ such that $F_i := F[V_i]$ is a maximal factor-critical subgraph of F with $|M \cap E(F_i)| = \frac{|V_i|-1}{2}$ $(i = 1, \ldots, k)$ and after contracting each of $V_1, \ldots, V_k$ we obtain an alternating forest F'.*

F_i is called an **outer blossom (inner blossom)** *if V_i is an outer (inner) vertex in F'. All the vertices of an outer (inner) blossom are called* **outer (inner)**. *A general blossom forest where each inner blossom is a single vertex is a* **special blossom forest**.

Figure 10.8 shows a connected component of a special blossom forest with five nontrivial outer blossoms. This corresponds to one of the connected components of the alternating forest in Figure 10.7. The orientations of the edges will be explained later. All vertices of G not belonging to the special blossom forest are called **out-of-forest**.

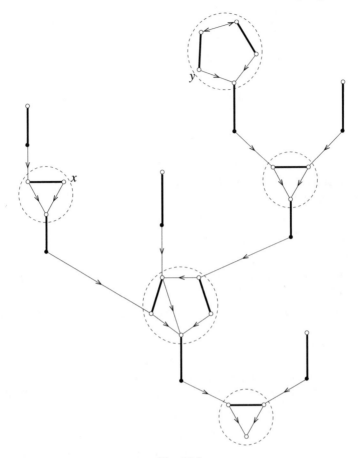

Fig. 10.8.

Note that the Blossom Shrinking Lemma 10.25 applies to outer blossoms only. However, in this section we shall deal only with special blossom forests. General blossom forests will appear only in the WEIGHTED MATCHING ALGORITHM in Chapter 11.

To store a special blossom forest F, we introduce the following data structures. For each vertex $x \in V(G)$ we have three variables $\mu(x)$, $\varphi(x)$, and $\rho(x)$ with the following properties:

$$\mu(x) \;=\; \begin{cases} x & \text{if } x \text{ is exposed by } M \\ y & \text{if } \{x, y\} \in M \end{cases} \tag{10.2}$$

$$\varphi(x) \;=\; \begin{cases} x & \text{if } x \notin V(F) \text{ or } x \text{ is the base of an outer blossom} \\ y & \text{for } \{x,y\} \in E(F) \setminus M \text{ if } x \text{ is an inner vertex} \\ y & \text{such that } \{x,y\} \in E(F) \setminus M, \text{ and } \mu \text{ and } \varphi \text{ are} \\ & \text{associated with an } M\text{-alternating ear-decomposition} \\ & \text{of the blossom containing } x, \text{ if } x \text{ is an outer vertex} \end{cases} \qquad (10.3)$$

$$\rho(x) \;=\; \begin{cases} x & \text{if } x \text{ is not an outer vertex} \\ y & \text{if } x \text{ is an outer vertex and } y \text{ is the base of} \\ & \text{the outer blossom in } F \text{ containing } x \end{cases} \qquad (10.4)$$

For each outer vertex v we define $P(v)$ to be the maximal path given by an initial subsequence of

$$v, \mu(v), \varphi(\mu(v)), \mu(\varphi(\mu(v))), \varphi(\mu(\varphi(\mu(v)))), \dots$$

We have the following properties:

Proposition 10.29. *Let F be a special blossom forest with respect to a matching M, and let $\mu, \varphi : V(G) \to V(G)$ be functions satisfying (10.2) and (10.3). Then we have:*

(a) *For each outer vertex v, $P(v)$ is an alternating v-q-path, where q is the root of the tree of F containing v.*

(b) *A vertex x is*
- *outer iff either $\mu(x) = x$ or $\varphi(\mu(x)) \neq \mu(x)$;*
- *inner iff $\varphi(\mu(x)) = \mu(x)$ and $\varphi(x) \neq x$;*
- *out-of-forest iff $\mu(x) \neq x$ and $\varphi(x) = x$ and $\varphi(\mu(x)) = \mu(x)$.*

Proof: (a): By (10.3) and Lemma 10.22, some initial subsequence of

$$v, \mu(v), \varphi(\mu(v)), \mu(\varphi(\mu(v))), \varphi(\mu(\varphi(\mu(v)))), \dots$$

must be an M-alternating path of even length to the base r of the blossom containing v. If r is not the root of the tree containing v, then r is covered by M. Hence the above sequence continues with the matching edge $\{r, \mu(r)\}$ and also with $\{\mu(r), \varphi(\mu(r))\}$, because $\mu(r)$ is an inner vertex. But $\varphi(\mu(r))$ is an outer vertex again, and so we are done by induction.

(b): If a vertex x is outer, then it is either a root (i.e. $\mu(x) = x$) or $P(x)$ is a path of length at least two, i.e. $\varphi(\mu(x)) \neq \mu(x)$.

If x is inner, then $\mu(x)$ is the base of an outer blossom, so by (10.3) $\varphi(\mu(x)) = \mu(x)$. Furthermore, $P(\mu(x))$ is a path of length at least 2, so $\varphi(x) \neq x$.

If x is out-of-forest, then by definition x is covered by M, so by (10.2) $\mu(x) \neq x$. Of course $\mu(x)$ is also out-of-forest, so by (10.3) we have $\varphi(x) = x$ and $\varphi(\mu(x)) = \mu(x)$.

Since each vertex is either outer or inner or out-of-forest, and each vertex satisfies exactly one of the three right-hand side conditions, the proof is complete.

$\square$

In Figure 10.8, an edge is oriented from u to v if $\varphi(u) = v$. We are now ready for a detailed description of the algorithm.

EDMONDS' CARDINALITY MATCHING ALGORITHM

Input: A graph G.

Output: A maximum matching in G given by the edges $\{x, \mu(x)\}$.

---

①   Set $\mu(v) := v$, $\varphi(v) := v$, $\rho(v) := v$ and *scanned*$(v) := false$ for all
     $v \in V(G)$.

②   **If** all outer vertices are scanned
         then stop,
         else let x be an outer vertex with *scanned*$(x) = false$.

③   Let y be a neighbour of x such that y is out-of-forest or (y is outer and
     $\rho(y) \neq \rho(x)$).
     If there is no such y **then** set *scanned*$(x) := true$ and **go to** ②.

④   ("grow")
     If y is out-of-forest **then** set $\varphi(y) := x$ and **go to** ③.

⑤   ("augment")
     If $P(x)$ and $P(y)$ are vertex-disjoint **then**
         Set $\mu(\varphi(v)) := v$, $\mu(v) := \varphi(v)$ for all $v \in V(P(x)) \cup V(P(y))$
             with odd distance from x or y on $P(x)$ or $P(y)$, respectively.
         Set $\mu(x) := y$.
         Set $\mu(y) := x$.
         Set $\varphi(v) := v$, $\rho(v) := v$, *scanned*$(v) := false$ for all $v \in V(G)$.
         Go to ②.

⑥   ("shrink")
     Let r be the first vertex on $V(P(x)) \cap V(P(y))$ with $\rho(r) = r$.
     For $v \in V(P(x)_{[x,r]}) \cup V(P(y)_{[y,r]})$ with odd distance from x or y on
         $P(x)_{[x,r]}$ or $P(y)_{[y,r]}$, respectively, and $\rho(\varphi(v)) \neq r$ **do:**
         Set $\varphi(\varphi(v)) := v$.
     If $\rho(x) \neq r$ **then** set $\varphi(x) := y$.
     If $\rho(y) \neq r$ **then** set $\varphi(y) := x$.
     For all $v \in V(G)$ with $\rho(v) \in V(P(x)_{[x,r]}) \cup V(P(y)_{[y,r]})$ **do:**
         Set $\rho(v) := r$.
     Go to ③.

For an illustration of the effect of shrinking on the φ-values, see Figure 10.9,
where ⑥ of the algorithm has been applied to x and y in Figure 10.8.

Lemma 10.30. *The following statements hold at any stage of* EDMONDS' CAR-
DINALITY MATCHING ALGORITHM:

(a) *The edges* $\{x, \mu(x)\}$ *form a matching M;*
(b) *The edges* $\{x, \mu(x)\}$ *and* $\{x, \varphi(x)\}$ *form a special blossom forest F with respect
    to M (plus some isolated matching edges);*
(c) *The properties (10.2), (10.3) and (10.4) are satisfied with respect to F.*

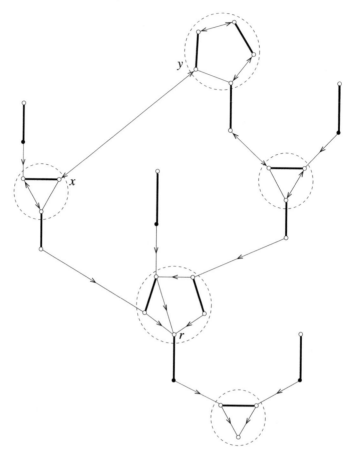

Fig. 10.9.

Proof: (a): The only place where μ is changed is ⑤, where the augmentation is obviously done correctly.

(b): Since after ① and ⑤ we trivially have a blossom forest without any edges and ④ correctly grows the blossom forest by two edges, we only have to check ⑥. r either is a root or must have degree at least three, so it must be outer. The new blossom consists of the vertices $B := \{v \in V(G) : \rho(v) \in V(P(x)_{[x,r]}) \cup V(P(y)_{[y,r]})\}$. Consider an edge $\{u, v\}$ of the blossom forest with $u \in B$ and $v \notin B$. Since $F[B]$ contains a near-perfect matching, $\{u, v\}$ is a matching edge only if it is $\{r, \mu(r)\}$. Moreover, u has been outer before applying ⑥. The fact that $F[B]$ is factor-critical follows from the existence of an M-alternating ear-decomposition (see (c)) and Theorem 10.17. This implies that F continues to be a special blossom forest.

(c): Here the only nontrivial fact is that, after shrinking, μ and φ are associated with an alternating ear-decomposition of the new blossom. So let x and y be two outer vertices in the same connected component of the special blossom forest, and let r be the first vertex of $V(P(x)) \cap V(P(y))$ for which $\rho(r) = r$. As above let B be the vertex set of the new blossom.

We note that $\varphi(v)$ is not changed for any $v \in B$ with $\rho(v) = r$. So the ear-decomposition of the old blossom $B' := \{v \in V(G) : \rho(v) = r\}$ is the starting point of the ear-decomposition of B. The next ear consists of $P(x)_{[x,x']}$, $P(y)_{[y,y']}$, and the edge $\{x, y\}$, where x' and y' is the first vertex on $P(x)$ and $P(y)$, respectively, that belongs to B'. Finally, for each ear Q of an old outer blossom $B'' \subseteq B$, $Q \setminus (E(P(x)) \cup E(P(y)))$ is an ear of the new ear-decomposition of B. $\square$

Theorem 10.31. (Edmonds [1965]) EDMONDS' CARDINALITY MATCHING AL-GORITHM *correctly determines a maximum matching in* $O(n^3)$ *time, where* $n = |V(G)|$.

Proof: Lemma 10.30 and Proposition 10.29 show that the algorithm works correctly. Consider the situation when the algorithm terminates. Let M and F be the matching and the special blossom forest according to Lemma 10.30(a) and (b). It is clear that any neighbour of an outer vertex x is either inner or a vertex y belonging to the same blossom (i.e. $\rho(y) = \rho(x)$).

To show that M is a maximum matching, let X denote the set of inner vertices, while B is the set of vertices that are the base of some outer blossom in F. Then every unmatched vertex belongs to B, and the matched vertices of B are matched with elements of X:

$$|B| = |X| + |V(G)| - 2|M|. \tag{10.5}$$

On the other hand, the outer blossoms in F are odd connected components in $G - X$. Therefore any matching must leave at least $|B| - |X|$ vertices uncovered. By (10.5), M leaves exactly $|B| - |X|$ vertices uncovered and thus is maximum.

We now consider the running time. By Proposition 10.29(b), the status of each vertex (inner, outer, or out-of-forest) can be checked in constant time. Each of ④, ⑤, ⑥ can be done in $O(n)$ time. Between two augmentations, ④ or ⑥ are executed at most $O(n)$ times, since the number of fixed points of φ decreases each time. Moreover, between two augmentations no vertex is scanned twice. Thus the time spent between two augmentations is $O(n^2)$, yielding an $O(n^3)$ total running time. $\square$

Micali and Vazirani [1980] improved the running time to $O\left(\sqrt{n}\, m\right)$. They used the results of Exercise 11, but the existence of blossoms makes the search for a maximal set of vertex-disjoint minimum length augmenting paths more difficult than in the bipartite case (which was solved earlier by Hopcroft and Karp [1973] and Karzanov [1973], see Exercise 12). See also Vazirani [1994]. The currently best known time complexity for the CARDINALITY MATCHING PROBLEM is $O\left(m\sqrt{n}\frac{\log(n^2/m)}{\log n}\right)$, just as in the bipartite case. This was obtained by Goldberg and Karzanov [2004] and by Fremuth-Paeger and Jungnickel [2003].

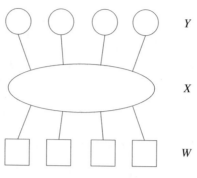

Y

X

W

Fig. 10.10.

With the matching algorithm we can easily prove the Gallai-Edmonds Structure Theorem. This was first proved by Gallai, but EDMONDS' CARDINALITY MATCHING ALGORITHM turns out to be a constructive proof thereof.

Theorem 10.32. (Gallai [1964]) *Let G be any graph. Denote by Y the set of vertices exposed by at least one maximum matching, by X the neighbours of Y in $V(G) \setminus Y$, and by W all other vertices. Then:*

(a) *Any maximum matching in G contains a perfect matching of $G[W]$ and near-perfect matchings of the connected components of $G[Y]$, and matches all vertices in X to distinct connected components of $G[Y]$;*
(b) *The connected components of $G[Y]$ are factor-critical;*
(c) $2\nu(G) = |V(G)| - q_G(X) + |X|$.

We call W, X, Y the **Gallai-Edmonds decomposition** *of G (see Figure 10.10).*

Proof: We apply EDMONDS' CARDINALITY MATCHING ALGORITHM and consider the matching M and the special blossom forest F at termination. Let X' be the set of inner vertices, Y' the set of outer vertices, and W' the set of out-of-forest vertices. We first prove that X', Y', W' satisfy (a)–(c), and then observe that $X = X'$, $Y = Y'$, and $W = W'$.

The proof of Theorem 10.31 shows that $2\nu(G) = |V(G)| - q_G(X') + |X'|$. We apply Proposition 10.15 to X'. Since the odd connected components of $G - X'$ are exactly the outer blossoms in F, (a) holds for X', Y', W'. Since the outer blossoms are factor-critical, (b) also holds.

Since part (a) holds for X', Y', and W', we know that any maximum matching covers all the vertices in $V(G) \setminus Y'$. In other words, $Y \subseteq Y'$. We claim that $Y' \subseteq Y$ also holds. Let v be an outer vertex in F. Then $M \triangle E(P(v))$ is a maximum matching M', and M' does not cover v. So $v \in Y$.

Hence $Y = Y'$. This implies $X = X'$ and $W = W'$, and the theorem is proved. $\square$

Exercises

1. Let G be a graph and M_1, M_2 two maximal matchings in G. Prove that $|M_1| \leq 2|M_2|$.

2. Let $\alpha(G)$ denote the size of a maximum stable set in G, and $\zeta(G)$ the minimum cardinality of an edge cover. Prove:

   (a) $\alpha(G) + \tau(G) = |V(G)|$ for any graph G.

   (b) $\nu(G) + \zeta(G) = |V(G)|$ for any graph G with no isolated vertices.

   (c) $\zeta(G) = \alpha(G)$ for any bipartite graph G with no isolated vertices.

   (König [1933], Gallai [1959])

3. Prove that a k-regular bipartite graph has k pairwise disjoint perfect matchings. Deduce from this that the edge set of a bipartite graph of maximum degree k can be partitioned into k matchings.

   (König [1916]; see Rizzi [1998] or Theorem 16.16)

* 4. A partially ordered set (or poset) is defined to be a set S together with a partial order on S, i.e. a relation $R \subseteq S \times S$ that is reflexive $((x, x) \in R$ for all $x \in S)$, anti-symmetric (if $(x, y) \in R$ and $(y, x) \in R$ then $x = y$), and transitive (if $(x, y) \in R$ and $(y, z) \in R$ then $(x, z) \in R$). Two elements $x, y \in S$ are called comparable if $(x, y) \in R$ or $(y, x) \in R$, otherwise they are incomparable. A chain (an antichain) is a subset of pairwise comparable (incomparable) elements of S. Use König's Theorem 10.2 to prove the following theorem of Dilworth [1950]:

   In a finite poset the maximum size of an antichain equals the minimum number of chains into which the poset can be partitioned.

   Hint: Take two copies v' and v'' of each $v \in S$ and consider the graph with an edge $\{v', w''\}$ for each $(v, w) \in R$.

   (Fulkerson [1956])

5. (a) Let $S = \{1, 2, \ldots, n\}$ and $0 \leq k < \frac{n}{2}$. Let A and B be the collection of all k-element and $(k + 1)$-element subsets of S, respectively. Construct a bipartite graph

   $$G = (A \,\dot\cup\, B, \{\{a, b\} : a \in A, \ b \in B, a \subseteq b\}).$$

   Prove that G has a matching covering A.

* (b) Prove Sperner's Lemma: the maximum number of subsets of an n-element set such that none is contained in any other is $\binom{n}{\lfloor \frac{n}{2} \rfloor}$.

   (Sperner [1928])

6. Let $(U, \mathcal{S})$ be a set system. An injective function $\Phi : \mathcal{S} \to U$ such that $\Phi(S) \in S$ for all $S \in \mathcal{S}$ is called a system of distinct representatives of $\mathcal{S}$. Prove:

   (a) $\mathcal{S}$ has a system of distinct representatives if and only if the union of any k of the sets in $\mathcal{S}$ has cardinality at least k.

   (Hall [1935])

   (b) For $u \in U$ let $r(u) := |\{S \in \mathcal{S} : u \in S\}|$. Let $n := |\mathcal{S}|$ and $N := \sum_{S \in \mathcal{S}} |S| = \sum_{u \in U} r(u)$. Suppose $|S| < \frac{N}{n-1}$ for $S \in \mathcal{S}$ and $r(u) < \frac{N}{n-1}$

for $u \in U$. Then S has a system of distinct representatives.
(Mendelsohn and Dulmage [1958])

7. Let G be a bipartite graph with bipartition $V(G) = A \,\dot{\cup}\, B$. Suppose that $S \subseteq A$, $T \subseteq B$, and there is a matching covering S and a matching covering T. Prove that then there is a matching covering $S \cup T$.
(Mendelsohn and Dulmage [1958])

8. A set of students applies for a set of seminars. Each student chooses exactly three seminars that he or she prefers. Two seminars are chosen by 40 students, all others by fewer.
   (a) Prove that each student can be assigned to a seminar that he or she chose, without assigning more than 13 students to any seminar.
   (b) Show how to compute such an assignment in $O(n^2)$ time, where n is the number of seminars.

9. How can you check whether a given undirected graph G has an orientation so that no vertex has in-degree more than a given threshold k? Give a polynomial-time algorithm and a necessary and sufficient condition ("for all $X \subseteq V(G)\dots$").

10. Show that any simple graph on n vertices with minimum degree k has a matching of cardinality $\min\{k, \lfloor \frac{n}{2} \rfloor\}$.
    Hint: Use Berge's Theorem 10.7.

11. Let G be a graph and M a matching in G that is not maximum.
    (a) Show that there are $\nu(G) - |M|$ vertex-disjoint M-augmenting paths in G.
        Hint: Recall the proof of Berge's Theorem 10.7.
    (b) Prove that there exists an M-augmenting path of length at most $\frac{\nu(G)+|M|}{\nu(G)-|M|}$ in G.
    (c) Let P be a shortest M-augmenting path in G, and P' an $(M \triangle E(P))$-augmenting path. Then $|E(P')| \geq |E(P)| + |E(P \cap P')|$.
    Consider the following generic algorithm. We start with the empty matching and in each iteration augment the matching along a shortest augmenting path. Let $P_1, P_2, \dots$ be the sequence of augmenting paths chosen. By (c), $|E(P_k)| \leq |E(P_{k+1})|$ for all k.
    (d) Show that if $|E(P_i)| = |E(P_j)|$ for $i \neq j$ then P_i and P_j are vertex-disjoint.
    (e) Use (b) to prove that the sequence $|E(P_1)|, |E(P_2)|, \dots$ contains at most $2\sqrt{\nu(G)} + 2$ different numbers.
    (Hopcroft and Karp [1973], Karzanov [1973])

* 12. Let G be a bipartite graph and consider the generic algorithm of Exercise 11.
    (a) Prove that – given a matching M – the union of all shortest M-augmenting paths in G can be found in $O(n + m)$ time.
        Hint: Use a kind of breadth-first search with matching edges and non-matching edges alternating.
    (b) Consider a sequence of iterations of the algorithm where the length of the augmenting path remains constant. Show that the time needed for the whole sequence is no more than $O(n + m)$.

Hint: First apply (a) and then find the paths successively by DFS. Mark vertices already visited.

(c) Combine (b) with Exercise 11(e) to obtain an $O\left(\sqrt{n}(m+n)\right)$-algorithm for the CARDINALITY MATCHING PROBLEM in bipartite graphs.

Note: This can be viewed as a special case of Exercise 21(b) of Chapter 8.

(Hopcroft and Karp [1973], Karzanov [1973])

13. Let G be a bipartite graph with bipartition $V(G) = A \mathbin{\dot\cup} B$, $A = \{a_1, \ldots, a_k\}$, $B = \{b_1, \ldots, b_k\}$. For any vector $x = (x_e)_{e \in E(G)}$ we define a matrix $M_G(x) = (m_{ij}^x)_{1 \le i,j \le k}$ by

$$m_{ij}^x := \begin{cases} x_e & \text{if } e = \{a_i, b_j\} \in E(G) \\ 0 & \text{otherwise} \end{cases}.$$

Its determinant $\det M_G(x)$ is a polynomial in $x = (x_e)_{e \in E(G)}$. Prove that G has a perfect matching if and only if $\det M_G(x)$ is not identically zero.

14. The **permanent** of a square matrix $M = (m_{ij})_{1 \le i,j \le n}$ is defined by

$$\text{per}(M) := \sum_{\pi \in S_n} \prod_{i=1}^{n} m_{i,\pi(i)},$$

where S_n is the set of permutations of $\{1, \ldots, n\}$. Prove that a simple bipartite graph G has exactly $\text{per}(M_G(\mathbb{1}))$ perfect matchings, where $M_G(x)$ is defined as in the previous exercise.

15. A **doubly stochastic matrix** is a nonnegative square matrix whose column sums and row sums are all 1. Integral doubly stochastic matrices are called **permutation matrices**.

Falikman [1981] and Egoryčev [1980] proved that for a doubly stochastic $n \times n$-matrix M,

$$\text{per}(M) \ge \frac{n!}{n^n},$$

and equality holds if and only if every entry of M is $\frac{1}{n}$. (This was a famous conjecture of van der Waerden; see also Schrijver [1998].)

Brègman [1973] proved that for a 0-1-matrix M with row sums $r_1, \ldots, r_n$,

$$\text{per}(M) \le (r_1!)^{\frac{1}{r_1}} \cdot \ldots \cdot (r_n!)^{\frac{1}{r_n}}.$$

Use these results and Exercise 14 to prove the following. Let G be a simple k-regular bipartite graph on $2n$ vertices, and let $\Phi(G)$ be the number of perfect matchings in G. Then

$$n! \left(\frac{k}{n}\right)^n \le \Phi(G) \le (k!)^{\frac{n}{k}}.$$

16. Prove that every 3-regular graph with at most two bridges has a perfect matching. Is there a 3-regular graph without a perfect matching?

Hint: Use Tutte's Theorem 10.13.

(Petersen [1891])

* 17. Let G be a graph, $n := |V(G)|$ even, and for any set $X \subseteq V(G)$ with $|X| \leq \frac{3}{4}n$ we have

$$\left| \bigcup_{x \in X} \Gamma(x) \right| \geq \frac{4}{3}|X|.$$

Prove that G has a perfect matching.

Hint: Let S be a set violating the Tutte condition. Prove that the number of connected components in $G - S$ with just one element is at most $\max \left\{ 0, \frac{4}{3}|S| - \frac{1}{3}n \right\}$. Consider the cases $|S| \geq \frac{n}{4}$ and $|S| < \frac{n}{4}$ separately.
(Anderson [1971])

18. Prove that an undirected graph G is factor-critical if and only if G is connected and $\nu(G) = \nu(G - v)$ for all $v \in V(G)$.

19. Prove that the number of ears in any two odd ear-decompositions of a factor-critical graph G is the same.

* 20. For a 2-edge-connected graph G let $\varphi(G)$ be the minimum number of even ears in an ear-decomposition of G (cf. Exercise 22(a) of Chapter 2). Show that for any edge $e \in E(G)$ we have either $\varphi(G/e) = \varphi(G)+1$ or $\varphi(G/e) = \varphi(G)-1$.
Note: The function $\varphi(G)$ has been studied by Frank [1993], Szigeti [1996], and Szegedy and Szegedy [2006].

21. Prove that a minimal factor-critical graph G (i.e. after the deletion of any edge the graph is no longer factor-critical) has at most $\frac{3}{2}(|V(G)| - 1)$ edges. Show that this bound is tight.

22. Show how EDMONDS' CARDINALITY MATCHING ALGORITHM finds a maximum matching in the graph shown in Figure 10.1(b).

23. Given an undirected graph, can one find an edge cover of minimum cardinality in polynomial time?

* 24. Given an undirected graph G, an edge is called unmatchable if it is not contained in any perfect matching. How can one determine the set of unmatchable edges in $O(n^3)$ time?
Hint: First determine a perfect matching in G. Then determine for each vertex v the set of unmatchable edges incident to v.

25. Let G be a graph, M a maximum matching in G, and F_1 and F_2 two special blossom forests with respect to M, each with the maximum possible number of edges. Show that the set of inner vertices in F_1 and F_2 is the same.

26. Let G be a k-connected graph with $2\nu(G) < |V(G)| - 1$. Prove:
    (a) $\nu(G) \geq k$;
    (b) $\tau(G) \leq 2\nu(G) - k$.
    Hint: Use the Gallai-Edmonds Theorem 10.32.
    (Erdős and Gallai [1961])

References

General Literature

Gerards, A.M.H. [1995]: Matching. In: Handbooks in Operations Research and Management Science; Volume 7: Network Models (M.O. Ball, T.L. Magnanti, C.L. Monma, G.L. Nemhauser, eds.), Elsevier, Amsterdam 1995, pp. 135–224

Lawler, E.L. [1976]: Combinatorial Optimization; Networks and Matroids. Holt, Rinehart and Winston, New York 1976, Chapters 5 and 6

Lovász, L., and Plummer, M.D. [1986]: Matching Theory. Akadémiai Kiadó, Budapest 1986, and North-Holland, Amsterdam 1986

Papadimitriou, C.H., and Steiglitz, K. [1982]: Combinatorial Optimization; Algorithms and Complexity. Prentice-Hall, Englewood Cliffs 1982, Chapter 10

Pulleyblank, W.R. [1995]: Matchings and extensions. In: Handbook of Combinatorics; Vol. 1 (R.L. Graham, M. Grötschel, L. Lovász, eds.), Elsevier, Amsterdam 1995

Schrijver, A. [2003]: Combinatorial Optimization: Polyhedra and Efficiency. Springer, Berlin 2003, Chapters 16 and 24

Tarjan, R.E. [1983]: Data Structures and Network Algorithms. SIAM, Philadelphia 1983, Chapter 9

Cited References

Alt, H., Blum, N., Mehlhorn, K., and Paul, M. [1991]: Computing a maximum cardinality matching in a bipartite graph in time $O\left(n^{1.5}\sqrt{m/\log n}\right)$. Information Processing Letters 37 (1991), 237–240

Anderson, I. [1971]: Perfect matchings of a graph. Journal of Combinatorial Theory B 10 (1971), 183–186

Berge, C. [1957]: Two theorems in graph theory. Proceedings of the National Academy of Science of the U.S. 43 (1957), 842–844

Berge, C. [1958]: Sur le couplage maximum d'un graphe. Comptes Rendus Hebdomadaires des Séances de l'Académie des Sciences (Paris) Sér. I Math. 247 (1958), 258–259

Brègman, L.M. [1973]: Certain properties of nonnegative matrices and their permanents. Doklady Akademii Nauk SSSR 211 (1973), 27–30 [in Russian]. English translation: Soviet Mathematics Doklady 14 (1973), 945–949

Dilworth, R.P. [1950]: A decomposition theorem for partially ordered sets. Annals of Mathematics 51 (1950), 161–166

Edmonds, J. [1965]: Paths, trees, and flowers. Canadian Journal of Mathematics 17 (1965), 449–467

Egoryčev, G.P. [1980]: Solution of the van der Waerden problem for permanents. Soviet Mathematics Doklady 23 (1982), 619–622

Erdős, P., and Gallai, T. [1961]: On the minimal number of vertices representing the edges of a graph. Magyar Tudományos Akadémia; Matematikai Kutató Intézetének Közleményei 6 (1961), 181–203

Falikman, D.I. [1981]: A proof of the van der Waerden conjecture on the permanent of a doubly stochastic matrix. Matematicheskie Zametki 29 (1981), 931–938 [in Russian]. English translation: Math. Notes of the Acad. Sci. USSR 29 (1981), 475–479

Feder, T., and Motwani, R. [1995]: Clique partitions, graph compression and speeding-up algorithms. Journal of Computer and System Sciences 51 (1995), 261–272

Frank, A. [1993]: Conservative weightings and ear-decompositions of graphs. Combinatorica 13 (1993), 65–81

Fremuth-Paeger, C., and Jungnickel, D. [2003]: Balanced network flows VIII: a revised theory of phase-ordered algorithms and the $O(\sqrt{n}m \log(n^2/m)/\log n)$ bound for the nonbipartite cardinality matching problem. Networks 41 (2003), 137–142

Frobenius, G. [1917]: Über zerlegbare Determinanten. Sitzungsbericht der Königlich Preussischen Akademie der Wissenschaften XVIII (1917), 274–277

Fulkerson, D.R. [1956]: Note on Dilworth's decomposition theorem for partially ordered sets. Proceedings of the AMS 7 (1956), 701–702

Gallai, T. [1959]: Über extreme Punkt- und Kantenmengen. Annales Universitatis Scientiarum Budapestinensis de Rolando Eötvös Nominatae; Sectio Mathematica 2 (1959), 133–138

Gallai, T. [1964]: Maximale Systeme unabhängiger Kanten. Magyar Tudományos Akadémia; Matematikai Kutató Intézetének Közleményei 9 (1964), 401–413

Geelen, J.F. [2000]: An algebraic matching algorithm. Combinatorica 20 (2000), 61–70

Geelen, J. and Iwata, S. [2005]: Matroid matching via mixed skew-symmetric matrices. Combinatorica 25 (2005), 187–215

Goldberg, A.V., and Karzanov, A.V. [2004]: Maximum skew-symmetric flows and matchings. Mathematical Programming A 100 (2004), 537–568

Hall, P. [1935]: On representatives of subsets. Journal of the London Mathematical Society 10 (1935), 26–30

Halmos, P.R., and Vaughan, H.E. [1950]: The marriage problem. American Journal of Mathematics 72 (1950), 214–215

Hopcroft, J.E., and Karp, R.M. [1973]: An $n^{5/2}$ algorithm for maximum matchings in bipartite graphs. SIAM Journal on Computing 2 (1973), 225–231

Karzanov, A.V. [1973]: Tochnaya otsenka algoritma nakhozhdeniya maksimal'nogo potoka, primenennogo k zadache "o predstavitelyakh". In: Voprosy Kibernetiki, Trudy Seminara po Kombinatornoĭ Matematike, Sovetskoe Radio, Moscow 1973, pp. 66–70 [in Russian]

König, D. [1916]: Über Graphen und ihre Anwendung auf Determinantentheorie und Mengenlehre. Mathematische Annalen 77 (1916), 453–465

König, D. [1931]: Graphs and matrices. Matematikaiés Fizikai Lapok 38 (1931), 116–119 [in Hungarian]

König, D. [1933]: Über trennende Knotenpunkte in Graphen (nebst Anwendungen auf Determinanten und Matrizen). Acta Litteratum ac Scientiarum Regiae Universitatis Hungaricae Francisco-Josephinae (Szeged). Sectio Scientiarum Mathematicarum 6 (1933), 155–179

Kuhn, H.W. [1955]: The Hungarian method for the assignment problem. Naval Research Logistics Quarterly 2 (1955), 83–97

Lovász, L. [1972]: A note on factor-critical graphs. Studia Scientiarum Mathematicarum Hungarica 7 (1972), 279–280

Lovász, L. [1979]: On determinants, matchings and random algorithms. In: Fundamentals of Computation Theory (L. Budach, ed.), Akademie-Verlag, Berlin 1979, pp. 565–574

Mądry, A. [2013]: Navigating central path with electrical flows: from flows to matchings, and back. Proceedings of the 54th Annual IEEE Symposium on Foundations of Computer Science (2013), 253–262

Mendelsohn, N.S., and Dulmage, A.L. [1958]: Some generalizations of the problem of distinct representatives. Canadian Journal of Mathematics 10 (1958), 230–241

Micali, S., and Vazirani, V.V. [1980]: An $O(V^{1/2}E)$ algorithm for finding maximum matching in general graphs. Proceedings of the 21st Annual IEEE Symposium on Foundations of Computer Science (1980), 17–27

Mucha, M., and Sankowski, P. [2004]: Maximum matchings via Gaussian elimination. Proceedings of the 45th Annual IEEE Symposium on Foundations of Computer Science (2004), 248–255

Mulmuley, K., Vazirani, U.V., and Vazirani, V.V. [1987]: Matching is as easy as matrix inversion. Combinatorica 7 (1987), 105–113

Petersen, J. [1891]: Die Theorie der regulären Graphs. Acta Mathematica 15 (1891), 193–220

Rabin, M.O., and Vazirani, V.V. [1989]: Maximum matchings in general graphs through randomization. Journal of Algorithms 10 (1989), 557–567

Rizzi, R. [1998]: König's edge coloring theorem without augmenting paths. Journal of Graph Theory 29 (1998), 87

Schrijver, A. [1998]: Counting 1-factors in regular bipartite graphs. Journal of Combinatorial Theory B 72 (1998), 122–135

Sperner, E. [1928]: Ein Satz über Untermengen einer endlichen Menge. Mathematische Zeitschrift 27 (1928), 544–548

Szegedy, B., and Szegedy, C. [2006]: Symplectic spaces and ear-decomposition of matroids. Combinatorica 26 (2006), 353–377

Szigeti, Z. [1996]: On a matroid defined by ear-decompositions. Combinatorica 16 (1996), 233–241

Tutte, W.T. [1947]: The factorization of linear graphs. Journal of the London Mathematical Society 22 (1947), 107–111

Vazirani, V.V. [1994]: A theory of alternating paths and blossoms for proving correctness of the $O(\sqrt{V}E)$ general graph maximum matching algorithm. Combinatorica 14 (1994), 71–109

11 Weighted Matching

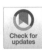

Nonbipartite weighted matching appears to be one of the "hardest" combinatorial optimization problems that can be solved in polynomial time. We shall extend EDMONDS' CARDINALITY MATCHING ALGORITHM to the weighted case and shall again obtain an $O(n^3)$-implementation. This algorithm has many applications, some of which are mentioned in the exercises and in Section 12.2. There are two basic formulations of the weighted matching problem:

MAXIMUM WEIGHT MATCHING PROBLEM

Instance: An undirected graph G and weights $c : E(G) \to \mathbb{R}$.

Task: Find a maximum weight matching in G.

MINIMUM WEIGHT PERFECT MATCHING PROBLEM

Instance: An undirected graph G and weights $c : E(G) \to \mathbb{R}$.

Task: Find a minimum weight perfect matching in G or decide that G has no perfect matching.

Proposition 11.1. *The* MAXIMUM WEIGHT MATCHING PROBLEM *and the* MINIMUM WEIGHT PERFECT MATCHING PROBLEM *are equivalent.*

Proof: Given an instance (G, c) of the MINIMUM WEIGHT PERFECT MATCHING PROBLEM, we set $c'(e) := K - c(e)$ for all $e \in E(G)$, where $K := 1 + \sum_{e \in E(G)} |c(e)|$. Then any maximum weight matching in (G, c') is a maximum cardinality matching, and hence gives a solution of the MINIMUM WEIGHT PERFECT MATCHING PROBLEM (G, c).

Conversely, let (G, c) be an instance of the MAXIMUM WEIGHT MATCHING PROBLEM. Let H be the graph with $V(H) := \{(v, i) : v \in V(G), i \in \{1, 2\}\}$, and $E(H) := \{\{(v, i), (w, i)\} : \{v, w\} \in E(G), i \in \{1, 2\}\} \cup \{\{(v, 1), (v, 2)\} : v \in V(G)\}$. So H consists of two copies of G, and there is an edge joining the two copies of each vertex. H has a perfect matching. Let $c'(\{(v, 1), (w, 1)\}) := -c(e)$

© Springer-Verlag GmbH Germany 2018
Bernhard Korte and Jens Vygen, *Combinatorial Optimization*, Algorithms and Combinatorics 21, https://doi.org/10.1007/978-3-662-56039-6_11

for all $e = \{v, w\} \in E(G)$, and let $c'(e) := 0$ for all other edges $e \in E(H)$. Then a minimum weight perfect matching M in (H, c') yields a maximum weight matching in (G, c), simply by taking $\{\{v, w\} \in E(G) : \{(v, 1), (w, 1)\} \in M\}$. $\square$

So in the following we consider only the MINIMUM WEIGHT PERFECT MATCH-ING PROBLEM. As in the previous chapter, we start by considering bipartite graphs in Section 11.1. After an outline of the weighted matching algorithm in Section 11.2 we spend some effort on implementation details in Section 11.3 in order to obtain an $O(n^3)$ running time. Sometimes one is interested in solving many matching problems that differ only on a few edges; in such a case it is not necessary to solve the problem from scratch each time as is shown in Section 11.4. Finally, in Section 11.5 we discuss the matching polytope, i.e. the convex hull of the incidence vectors of matchings. We use a description of the related perfect matching polytope already for designing the weighted matching algorithm; in turn, this algorithm will directly imply that this description is complete.

11.1 The Assignment Problem

The ASSIGNMENT PROBLEM is just another name for the MINIMUM WEIGHT PERFECT MATCHING PROBLEM in bipartite graphs. It is one of the classical combinatorial optimization problems; its history probably started with the work of Monge [1784].

As in the proof of Theorem 10.5, we can reduce the assignment problem to a network flow problem:

Theorem 11.2. *The* ASSIGNMENT PROBLEM *can be solved in* $O(nm + n^2 \log n)$ *time.*

Proof: Let G be a bipartite graph with bipartition $V(G) = A \mathbin{\dot{\cup}} B$. We assume $|A| = |B| = n$. Add a vertex s and connect it to all vertices of A, and add another vertex t connected to all vertices of B. Orient the edges from s to A, from A to B, and from B to t. Let the capacities be 1 everywhere, and let the new edges have zero cost.

Then any integral s-t-flow of value n corresponds to a perfect matching with the same cost, and vice versa. So we have to solve a MINIMUM COST FLOW PROBLEM. We do this by applying the SUCCESSIVE SHORTEST PATH ALGORITHM (see Section 9.4). The total demand is n. So by Theorem 9.14, the running time is $O(nm + n^2 \log n)$. $\square$

This is the fastest known algorithm. It is essentially equivalent to the "Hungarian method" by Kuhn [1955] and Munkres [1957], the oldest polynomial-time algorithm for the ASSIGNMENT PROBLEM (cf. Exercise 9).

It is worthwhile looking at the linear programming formulation of the ASSIGN-MENT PROBLEM. It turns out that in the integer programming formulation

$$\min\left\{\sum_{e\in E(G)} c(e)x_e : x_e \in \{0, 1\}\ (e \in E(G)),\ \sum_{e\in\delta(v)} x_e = 1\ (v \in V(G))\right\}$$

the integrality constraints can be omitted (replace $x_e \in \{0, 1\}$ by $x_e \geq 0$). The dual LP is

$$\max\left\{\sum_{v\in V(G)} z_v : z_v + z_w \leq c(e)\ \text{for all}\ e = \{v, w\} \in E(G)\right\}.$$

Based on this we have the following reformulation of the Hungarian method.

Proposition 11.3. *Let G be a graph and $c : E(G) \to \mathbb{R}$. Let $z \in \mathbb{R}^{V(G)}$ with $z_v + z_w \leq c(e)$ for all $e = \{v, w\} \in E(G)$. Let $G_z := (V(G), \{e = \{v, w\} \in E(G) : z_v + z_w = c(e)\})$. Let M be a matching in G_z, and let F be a maximal alternating forest in G_z with respect to M. Let X and Y denote the set of inner and outer vertices, respectively.*

(a) *If M is a perfect matching, then it is a minimum weight perfect matching in G.*

(b) *If $\Gamma_G(y) \subseteq X$ for all $y \in Y$, then M is a maximum matching in G.*

(c) *Otherwise let*

$$\varepsilon := \min\big\{\min\{(c(e) - z_v - z_w)/2 : e = \{v, w\} \in E(G[Y])\},$$
$$\min\{c(e) - z_v - z_w : e = \{v, w\} \in \delta_G(Y) \cap \delta_G(V(F))\}\big\}$$

and set $z'_v := z_v - \varepsilon$ for $v \in X$, $z'_v := z_v + \varepsilon$ for $v \in Y$, and $z'_v := z_v$ for $v \in V(G) \setminus V(F)$. Then $z'_v + z'_w \leq c(e)$ for all $e = \{v, w\} \in E(G)$, $M \cup E(F) \subseteq E(G_{z'})$, and $\Gamma_{G_{z'}}(y) \setminus X \neq \emptyset$ for some $y \in Y$.

Proof: (a) follows from complementary slackness: for any other perfect matching M' we have $\sum_{e\in M'} c(e) = \sum_{v\in V(G)} z_v + \sum_{e=\{v,w\}\in M'}(c(e) - z_v - z_w) \geq \sum_{v\in V(G)} z_v = \sum_{v\in V(G)} z_v + \sum_{e=\{v,w\}\in M}(c(e) - z_v - z_w) = \sum_{e\in M} c(e)$.

(b) follows from the Berge-Tutte formula (Theorem 10.14): each outer vertex is an odd connected component of $G - X$, and M leaves exactly $|Y| - |X|$ vertices uncovered.

(c) follows from the choice of ε.   □

Note that this result also holds for nonbipartite graphs. If G is bipartite, we can grow the alternating forest or augment the matching after each dual change in case (c). Eventually we will end up with case (a), i.e., a minimum weight perfect matching, which is also an optimum LP solution because the incidence vector of M and z satisfy the complementary slackness conditions (cf. Corollary 3.21). Hence Q in the following theorem is integral. We now give a different proof.

Theorem 11.4. *Let G be a graph, and let*

$$
P := \left\{ x \in \mathbb{R}_+^{E(G)} : \sum_{e \in \delta(v)} x_e \le 1 \ \text{for all } v \in V(G) \right\} \ \text{and}
$$

$$
Q := \left\{ x \in \mathbb{R}_+^{E(G)} : \sum_{e \in \delta(v)} x_e = 1 \ \text{for all } v \in V(G) \right\}
$$

be the **fractional matching polytope** *and the* **fractional perfect matching polytope** *of G. If G is bipartite, then P and Q are both integral.*

Proof: If G is bipartite, then the incidence matrix of G is totally unimodular due to Theorem 5.26. Hence by the Hoffman-Kruskal Theorem 5.21, P is integral. Q is a face of P and thus it is also integral. $\square$

There is a nice corollary concerning doubly-stochastic matrices. A **doubly stochastic matrix** is a nonnegative square matrix such that the sum of the entries in each row and each column is 1. Integral doubly stochastic matrices are called **permutation matrices**.

Corollary 11.5. (Birkhoff [1946], von Neumann [1953]) *Any doubly stochastic matrix M can be written as a convex combination of permutation matrices $P_1, \ldots, P_k$ (i.e. $M = c_1 P_1 + \ldots + c_k P_k$ for nonnegative $c_1, \ldots, c_k$ with $c_1 + \ldots + c_k = 1$).*

Proof: Let $M = (m_{ij})_{i,j \in \{1,\ldots,n\}}$ be a doubly stochastic $n \times n$-matrix, and let $K_{n,n}$ be the complete bipartite graph with bipartition $\{a_1, \ldots, a_n\} \cup \{b_1, \ldots, b_n\}$. For $e = \{a_i, b_j\} \in E(K_{n,n})$ let $x_e = m_{ij}$. Since M is doubly stochastic, x is in the fractional perfect matching polytope Q of $K_{n,n}$. By Theorem 11.4 and Corollary 3.32, x can be written as a convex combination of integral vertices of Q. These obviously correspond to permutation matrices. $\square$

This corollary, often called the Birkhoff-von-Neumann Theorem, can also be proved directly (Exercise 3).

Several more general assignment problems have been studied; see Burkard, Dell'Amico and Martello [2009].

11.2 Outline of the Weighted Matching Algorithm

The purpose of this and the next section is to describe a polynomial-time algorithm for the general MINIMUM WEIGHT PERFECT MATCHING PROBLEM. This algorithm was developed by Edmonds [1965] and uses the concepts of his algorithm for the CARDINALITY MATCHING PROBLEM (Section 10.5).

Let us first outline the main ideas without considering the implementation. Given a graph G with weights $c : E(G) \to \mathbb{R}$, the MINIMUM WEIGHT PERFECT MATCHING PROBLEM can be formulated as the integer linear program

$$\min \left\{ \sum_{e \in E(G)} c(e) x_e : x_e \in \{0, 1\} \ (e \in E(G)), \ \sum_{e \in \delta(v)} x_e = 1 \ (v \in V(G)) \right\}.$$

If A is a subset of $V(G)$ with odd cardinality, any perfect matching must contain an odd number of edges in $\delta(A)$, in particular at least one. So adding the constraint

$$\sum_{e \in \delta(A)} x_e \geq 1$$

does not change anything. Throughout this chapter we use the notation $\mathcal{A} := \{A \subseteq V(G) : |A| \text{ odd}\}$. Now consider the LP relaxation:

$$
\begin{array}{lllr}
\min & \displaystyle\sum_{e \in E(G)} c(e) x_e & & \\
\text{s.t.} & x_e \geq 0 & (e \in E(G)) & \\
& \displaystyle\sum_{e \in \delta(v)} x_e = 1 & (v \in V(G)) & (11.1) \\
& \displaystyle\sum_{e \in \delta(A)} x_e \geq 1 & (A \in \mathcal{A}, |A| > 1) &
\end{array}
$$

We shall prove later that the polytope described by (11.1) is integral; hence this LP describes the MINIMUM WEIGHT PERFECT MATCHING PROBLEM (this will be Theorem 11.15, a major result of this chapter). In the following we do not need this fact, but will rather use the LP formulation as a motivation.

To formulate the dual of (11.1), we introduce a variable z_A for each primal constraint, i.e. for each $A \in \mathcal{A}$. The dual linear program is:

$$
\begin{array}{lllr}
\max & \displaystyle\sum_{A \in \mathcal{A}} z_A & & \\
\text{s.t.} & z_A \geq 0 & (A \in \mathcal{A}, |A| > 1) & (11.2) \\
& \displaystyle\sum_{A \in \mathcal{A} : e \in \delta(A)} z_A \leq c(e) & (e \in E(G)) &
\end{array}
$$

Note that the dual variables $z_{\{v\}}$ for $v \in V(G)$ are not restricted to be non-negative. Edmonds' algorithm is a primal-dual algorithm. It starts with the empty matching ($x_e = 0$ for all $e \in E(G)$) and the feasible dual solution

$$
z_A := \begin{cases} \frac{1}{2} \min\{c(e) : e \in \delta(A)\} & \text{if } |A| = 1 \\ 0 & \text{otherwise} \end{cases}.
$$

At any stage of the algorithm, z will be a feasible dual solution, and we have

$$
\begin{aligned}
x_e > 0 \quad &\Rightarrow \quad \sum_{A \in \mathcal{A}:e \in \delta(A)} z_A = c(e); \\
z_A > 0 \quad &\Rightarrow \quad \sum_{e \in \delta(A)} x_e \leq 1.
\end{aligned}
\tag{11.3}
$$

The algorithm stops when x is the incidence vector of a perfect matching (i.e. we have primal feasibility). Due to the complementary slackness conditions (11.3) (Corollary 3.23) we then have the optimality of the primal and dual solutions. As x is optimal for (11.1) and integral, it is the incidence vector of a minimum weight perfect matching.

Given a feasible dual solution z, we call an edge e **tight** if the corresponding dual constraint is satisfied with equality, i.e. if

$$
\sum_{A \in \mathcal{A}:e \in \delta(A)} z_A = c(e).
$$

At any stage, the current matching will consist of tight edges only. The algorithm proceeds similar to Proposition 11.3 (which also holds for nonbipartite graphs). However, if G is nonbipartite, blossoms may occur. If we contract them, we can continue but obtain dual variables $z_B > 0$ for $B \in \mathcal{A}$.

Let henceforth G_z be the graph which results from G by deleting all edges that are not tight and contracting each set B with $z_B > 0$ to a single vertex. The family $\mathcal{B} := \{B \in \mathcal{A} : |B| = 1 \text{ or } z_B > 0\}$ will be laminar at any stage, and for each element of $B \in \mathcal{B}$ there is a factor-critical subgraph with vertex set B and only tight edges. Initially $\mathcal{B}$ consists of the singletons.

One iteration of the algorithm roughly proceeds as follows. We first find a maximum cardinality matching M in G_z, using EDMONDS' CARDINALITY MATCHING ALGORITHM. If M is a perfect matching, we are done: we can complete M to a perfect matching in G using tight edges only. Since the conditions (11.3) are satisfied, the matching is optimal.

Otherwise we consider the Gallai-Edmonds decomposition W, X, Y of G_z (cf. Theorem 10.32). For each vertex v of G_z let $B(v) \in \mathcal{B}$ be the vertex set whose contraction resulted in v. We modify the dual solution as follows (see Figure 11.1 for an illustration). For each $v \in X$ we decrease $z_{B(v)}$ by some positive constant ε. For each connected component C of $G_z[Y]$ we increase z_A by ε, where $A = \bigcup_{v \in C} B(v)$.

Note that tight matching edges remain tight, since by Theorem 10.32 all matching edges with one endpoint in X have the other endpoint in Y. (Indeed, all edges of the alternating forest we are working with remain tight).

We choose ε maximum possible while preserving dual feasibility. Since the current graph contains no perfect matching, the number of connected components of $G_z[Y]$ is greater than $|X|$. Hence the above dual change increases the dual objective function value $\sum_{A \in \mathcal{A}} z_A$ by at least ε. If ε can be chosen arbitrarily large, the dual LP (11.2) is unbounded, hence the primal LP (11.1) is infeasible (Theorem 3.27) and G has no perfect matching.

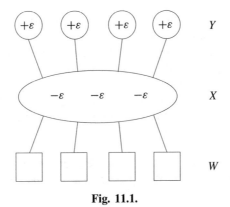

Fig. 11.1.

Due to the change of the dual solution the graph G_z will also change: new edges may become tight, new vertex sets may be contracted (corresponding to the components of Y that are not singletons), and some contracted sets may be "unpacked" (non-singletons whose dual variables become zero, corresponding to vertices of X).

The above is iterated until a perfect matching is found. We shall show later that this procedure is finite. This will follow from the fact that between two augmentations, each step (grow, shrink, unpack) increases the number of outer vertices.

11.3 Implementation of the Weighted Matching Algorithm

After this informal description we now turn to the implementation details. As with EDMONDS' CARDINALITY MATCHING ALGORITHM we do not explicitly shrink blossoms but rather store their ear-decomposition. However, there are several difficulties.

The "shrink"-step of EDMONDS' CARDINALITY MATCHING ALGORITHM produces an outer blossom. By the "augment"-step two connected components of the blossom forest become out-of-forest. Since the dual solution remains unchanged, we must retain the blossoms: we get so-called out-of-forest blossoms. The "grow"-step may involve out-of-forest blossoms which then become either inner or outer blossoms. Hence we have to deal with general blossom forests.

Another problem is that we must be able to recover nested blossoms one by one. Namely, if z_A becomes zero for some inner blossom A, there may be subsets $A' \subseteq A$ with $|A'| > 1$ and $z_{A'} > 0$. Then we have to unpack the blossom A, but not the smaller blossoms inside A (except if they remain inner and their dual variables are also zero).

Throughout the algorithm we have a laminar family $\mathcal{B} \subseteq \mathcal{A}$, containing at least all singletons. All elements of $\mathcal{B}$ are blossoms. We have $z_A = 0$ for all $A \notin \mathcal{B}$. The set B is laminar and is stored by a tree-representation (cf. Proposition 2.14). For easy reference, a number is assigned to each blossom in B that is not a singleton.

We store ear-decompositions of all blossoms in B at any stage of the algorithm. The variables $\mu(x)$ for $x \in V(G)$ again encode the current matching M. We denote by $b^1(x), \ldots, b^{k_x}(x)$ the blossoms in B containing x, without the singleton. $b^{k_x}(x)$ is the outermost blossom. We have variables $\rho^i(x)$ and $\varphi^i(x)$ for each $x \in V(G)$ and $i = 1, \ldots, k_x$. $\rho^i(x)$ is the base of the blossom $b^i(x)$. $\mu(x)$ and $\varphi^j(x)$, for all x and j with $b^j(x) = i$, are associated with an M-alternating ear-decomposition of blossom i.

Of course, we must update the blossom structures (φ and ρ) after each augmentation. Updating ρ is easy. Updating φ can also be done in linear time by Lemma 10.23.

For inner blossoms we need, in addition to the base, the vertex nearest to the root of the tree in the general blossom forest, and the neighbour in the next outer blossom. These two vertices are denoted by $\sigma(x)$ and $\chi(\sigma(x))$ for each base x of an inner blossom. See Figure 11.2 for an illustration.

With these variables, the alternating paths to the root of the tree can be determined. Since the blossoms are retained after an augmentation, we must choose the augmenting path such that each blossom still contains a near-perfect matching afterwards.

Figure 11.2 shows that we must be careful: There are two nested inner blossoms, induced by $\{x_3, x_4, x_5\}$ and $\{x_1, x_2, x_3, x_4, x_5\}$. If we just consider the ear-decomposition of the outermost blossom to find an alternating path from x_0 to the root x_6, we will end up with $(x_0, x_1, x_4, x_5 = \sigma(x_1), x_6 = \chi(x_5))$. After augmenting along $(y_6, y_5, y_4, y_3, y_2, y_1, y_0, x_0, x_1, x_4, x_5, x_6)$, the factor-critical subgraph induced by $\{x_3, x_4, x_5\}$ no longer contains a near-perfect matching.

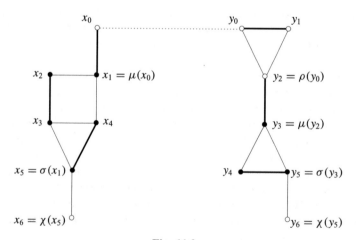

Fig. 11.2.

Thus we must find an alternating path within each blossom which contains an even number of edges within each sub-blossom. This is accomplished by the following procedure:

BLOSSOMPATH

Input: A vertex x_0.

Output: An M-alternating path $Q(x_0)$ from x_0 to $\rho^{k_{x_0}}(x_0)$.

①   Set $h := 0$ and $B := \{b^j(x_0) : j = 1, \ldots, k_{x_0}\}$.

②   **While** $x_{2h} \neq \rho^{k_{x_0}}(x_0)$ **do:**
        Set $x_{2h+1} := \mu(x_{2h})$ and $x_{2h+2} := \varphi^i(x_{2h+1})$, where
        $i = \min\{j \in \{1, \ldots, k_{x_{2h+1}}\} : b^j(x_{2h+1}) \in B\}$.
        Add all blossoms of B to B that contain x_{2h+2} but not x_{2h+1}.
        Delete all blossoms from B whose base is x_{2h+2}.
        Set $h := h + 1$.

③   Let $Q(x_0)$ be the path with vertices $x_0, x_1, \ldots, x_{2h}$.

Proposition 11.6. *The procedure* BLOSSOMPATH *can be implemented in* $O(n)$ *time.* $M \triangle E(Q(x_0))$ *contains a near-perfect matching within each blossom.*

Proof: Let us first check that the procedure indeed computes a path. In fact, if a blossom of B is left, it is never entered again. This follows from the fact that contracting the maximal sub-blossoms of any blossom in B results in a circuit (a property which will be maintained).

At the beginning of each iteration, B is the list of all blossoms that either contain x_0 or have been entered via a non-matching edge and have not been left yet. The constructed path leaves any blossom in B via a matching edge. So the number of edges within each blossom is even, proving the second statement of the proposition.

When implementing the procedure in $O(n)$ time, the only nontrivial task is the update of B. We store B as a sorted list. Using the tree-representation of B and the fact that each blossom is entered and left at most once, we get a running time of $O(n + |B|)$. Note that $|B| = O(n)$, because B is laminar. $\qquad\square$

Now determining an augmenting path consists of applying the procedure BLOSSOMPATH within blossoms, and using μ and χ between blossoms. When we find adjacent outer vertices x, y in different trees of the general blossom forest, we apply the following procedure to both x and y. The union of the two paths together with the edge $\{x, y\}$ will be the augmenting path.

TREEPATH

Input: An outer vertex v.

Output: An alternating path $P(v)$ from v to the root of the tree in the blossom
             forest.

①   Let initially $P(v)$ consist of v only. Let $x := v$.

②   Let $y := \rho^{k_x}(x)$. Let $Q(x) := \text{BLOSSOMPATH}(x)$. Append $Q(x)$ to $P(v)$.
      If $\mu(y) = y$ then stop.

③   Set $P(v) := P(v) + \{y, \mu(y)\}$.
      Let $Q(\sigma(\mu(y))) := \text{BLOSSOMPATH}(\sigma(\mu(y)))$.
      Append the reverse of $Q(\sigma(\mu(y)))$ to $P(v)$.
      Let $P(v) := P(v) + \{\sigma(\mu(y)), \chi(\sigma(\mu(y)))\}$.
      Set $x := \chi(\sigma(\mu(y)))$ and **go to** ②.

The second main problem is how to determine ε efficiently. The general blossom forest, after all possible grow-, shrink- and augment-steps are done, yields the Gallai-Edmonds decomposition W, X, Y of G_z. W contains the out-of-forest blossoms, X contains the inner blossoms, and Y consists of the outer blossoms.

For a simpler notation, let us define $c(\{v, w\}) := \infty$ if $\{v, w\} \notin E(G)$. Moreover, we use the abbreviation

$$slack(v, w) := c(\{v, w\}) - \sum_{A \in \mathcal{A}, \{v,w\} \in \delta(A)} z_A.$$

So $\{v, w\}$ is a tight edge if and only if $slack(v, w) = 0$. Then let

$$\varepsilon_1 := \min\{z_A : A \text{ is a maximal inner blossom}, |A| > 1\};$$

$$\varepsilon_2 := \min\{slack(x, y) : x \text{ outer}, y \text{ out-of-forest}\};$$

$$\varepsilon_3 := \frac{1}{2}\min\{slack(x, y) : x, y \text{ outer, belonging to different blossoms}\};$$

$$\varepsilon := \min\{\varepsilon_1, \varepsilon_2, \varepsilon_3\}.$$

This ε is the maximum number such that the dual change by ε preserves dual feasibility. If $\varepsilon = \infty$, (11.2) is unbounded and so (11.1) is infeasible. In this case G has no perfect matching.

Obviously, ε can be computed in finite time. However, in order to obtain an $O(n^3)$ overall running time we must be able to compute ε in $O(n)$ time. This is easy as far as ε_1 is concerned, but requires additional data structures for ε_2 and ε_3.

For $A \in \mathcal{B}$ let

$$\zeta_A := \sum_{B \in \mathcal{B}: A \subseteq B} z_B.$$

We shall update these values whenever changing the dual solution; this can easily be done in linear time (using the tree-representation of $\mathcal{B}$). Then

$$\varepsilon_2 = \min\{c(\{x, y\}) - \zeta_{\{x\}} - \zeta_{\{y\}} : x \text{ outer}, y \text{ out-of-forest}\},$$

$$\varepsilon_3 = \frac{1}{2}\min\{c(\{x, y\}) - \zeta_{\{x\}} - \zeta_{\{y\}} : x, y \text{ outer}, \{x, y\} \not\subseteq B \text{ for } B \in \mathcal{B}\}.$$

We introduce variables t_v^A and τ_v^A for each outer vertex v and each $A \in \mathcal{B}$, unless there is a $B \in \mathcal{B}$ with $A \cup \{v\} \subseteq B$. τ_v^A is a vertex in A minimizing

$slack(v, \tau_v^A)$, and $t_v^A := slack(v, \tau_v^A) + \Delta + \zeta_A$, where Δ denotes the sum of the ε-values in all dual changes. Observe that t_v^A does not change as long as v remains outer and $A \in \mathcal{B}$. Finally, we write $t^A := \min\{t_v^A : v \notin A, v \text{ outer}\}$. We have

$$\varepsilon_2 = \min\left\{slack(v, \tau_v^A) : v \text{ outer, } A \in \mathcal{B} \text{ maximal out-of-forest}\right\}$$

$$= \min\left\{t^A - \Delta - \zeta_A : A \in \mathcal{B} \text{ maximal out-of-forest}\right\},$$

and, similarly,

$$\varepsilon_3 = \frac{1}{2}\min\left\{t^A - \Delta - \zeta_A : A \in \mathcal{B} \text{ maximal outer}\right\}.$$

Although when computing ε_2 and ε_3 we are interested only in the values t_v^A for maximal out-of-forest and maximal outer blossoms of $\mathcal{B}$, we update these variables also for inner blossoms and those that are not maximal, because they may become relevant later. Blossoms that are outer but not maximal will not become maximal outer before an augmentation takes place. After each augmentation, however, all these variables are recomputed.

In the beginning, after each augmentation, and when a vertex v becomes outer which was not outer before, we have to compute τ_v^A and t_v^A, and possibly update t^A, for all $A \in \mathcal{B}$ (except those that are outer but not maximal). This can be done as follows:

UPDATE

Input: An outer vertex v.

Output: Updated values of τ_v^A, t_v^A and t^A for all $A \in \mathcal{B}$ and τ_w for all out-of-forest vertices w.

①   **For each** $x \in V(G)$ **do:** Set $\tau_v^{\{x\}} := x$ and $t_v^{\{x\}} := c(\{v, x\}) - \zeta_{\{v\}} + \Delta$.

②   **For** $A \in \mathcal{B}$ with $|A| > 1$ (in an order of nondecreasing cardinality) **do:**
       Set $\tau_v^A := \tau_v^{A'}$ and $t_v^A := t_v^{A'} - \zeta_{A'} + \zeta_A$, where A' is a maximal proper subset of A in $\mathcal{B}$ for which $t_v^{A'} - \zeta_{A'}$ is minimum.

③   **For** $A \in \mathcal{B}$ with $v \notin A$, except those that are outer but not maximal, **do:**
       Set $t^A := \min\{t^A, t_v^A\}$.

Obviously this computation coincides with the above definition of τ_v^A and t_v^A. It is important that this procedure runs in linear time:

Lemma 11.7. *If $\mathcal{B}$ is laminar, the procedure* UPDATE *can be implemented with* $O(n)$ *time.*

Proof: By Proposition 2.15, a laminar family of subsets of $V(G)$ has cardinality at most $2|V(G)| = O(n)$. If $\mathcal{B}$ is stored by its tree-representation, then a linear-time implementation is easy.   □

We can now go ahead with the formal description of the algorithm. Instead of identifying inner and outer vertices by the μ-, ϕ- and ρ-values, we directly mark each vertex with its status (inner, outer or out-of-forest).

WEIGHTED MATCHING ALGORITHM

Input: A graph G, weights $c : E(G) \to \mathbb{R}$.

Output: A minimum weight perfect matching in G, given by the edges $\{x, \mu(x)\}$, or the answer that G has no perfect matching.

---

①   Set $\mathcal{B} := \{\{v\} : v \in V(G)\}$ and $K := 0$. Set $\Delta := 0$.
Set $z_{\{v\}} := \frac{1}{2} \min\{c(e) : e \in \delta(v)\}$ and $\zeta_{\{v\}} := z_{\{v\}}$ for all $v \in V(G)$.
Set $k_v := 0$, $\mu(v) := v$, $\rho^0(v) := v$, and $\varphi^0(v) := v$ for all $v \in V(G)$.
Mark all vertices as outer.

②   Set $t^A := \infty$ for all $A \in \mathcal{B}$.
For all outer vertices v **do**: UPDATE(v).

③   ("dual change")
Set $\varepsilon_1 := \min\{z_A : A \text{ maximal inner element of } \mathcal{B}, |A| > 1\}$.
Set $\varepsilon_2 := \min\{t^A - \Delta - \zeta_A : A \text{ maximal out-of-forest element of } \mathcal{B}\}$.
Set $\varepsilon_3 := \min\{\frac{1}{2}(t^A - \Delta - \zeta_A) : A \text{ maximal outer element of } \mathcal{B}\}$.
Set $\varepsilon := \min\{\varepsilon_1, \varepsilon_2, \varepsilon_3\}$. **If** $\varepsilon = \infty$, **then stop** (G has no perfect matching).
For each maximal outer element A of $\mathcal{B}$ **do**:
     Set $z_A := z_A + \varepsilon$ and $\zeta_{A'} := \zeta_{A'} + \varepsilon$ for all $A' \in \mathcal{B}$ with $A' \subseteq A$.
For each maximal inner element A of $\mathcal{B}$ **do**:
     Set $z_A := z_A - \varepsilon$ and $\zeta_{A'} := \zeta_{A'} - \varepsilon$ for all $A' \in \mathcal{B}$ with $A' \subseteq A$.
Set $\Delta := \Delta + \varepsilon$.

④   **If** $\varepsilon = \varepsilon_1$ **then go to** ⑧.
If $\varepsilon = \varepsilon_2$ and $t_x^A - \Delta - \zeta_A = slack(x, y) = 0$, x outer, $y \in A$ out-of-forest
then go to ⑤.
If $\varepsilon = \varepsilon_3$ and $t_x^A - \Delta - \zeta_A = slack(x, y) = 0$, x, y outer, A maximal outer
element of $\mathcal{B}$, $x \notin A$, $y \in A$ **then**:
     Let $P(x) := \text{TREEPATH}(x)$ be given by $(x = x_0, x_1, x_2, \ldots, x_{2h})$.
     Let $P(y) := \text{TREEPATH}(y)$ be given by $(y = y_0, y_1, y_2, \ldots, y_{2j})$.
     If $P(x)$ and $P(y)$ are vertex-disjoint **then go to** ⑥, **else go to** ⑦.

⑤   ("grow")
Set $\sigma(\rho^{k_y}(y)) := y$ and $\chi(y) := x$.
Mark all vertices v with $\rho^{k_v}(v) = \rho^{k_y}(y)$ as inner.
Mark all vertices v with $\mu(\rho^{k_v}(v)) = \rho^{k_y}(y)$ as outer.
For each new outer vertex v **do**: UPDATE(v).
Go to ③.

⑥   ("augment")

For $i := 0$ **to** $h - 1$ **do**: Set $\mu(x_{2i+1}) := x_{2i+2}$ and $\mu(x_{2i+2}) := x_{2i+1}$.

For $i := 0$ **to** $j - 1$ **do**: Set $\mu(y_{2i+1}) := y_{2i+2}$ and $\mu(y_{2i+2}) := y_{2i+1}$.

Set $\mu(x) := y$ and $\mu(y) := x$.

Mark all vertices v for which the endpoint of TREEPATH(v) is either x_{2h} or y_{2j} as out-of-forest.

Update all values $\varphi^i(v)$ and $\rho^i(v)$ (using Lemma 10.23).

If $\mu(v) \neq v$ for all v **then stop, else go to** ②.

⑦   ("shrink")

Let $r = x_{2h'} = y_{2j'}$ be the first outer vertex of $V(P(x)) \cap V(P(y))$ with $\rho^{k_r}(r) = r$.

Let $A := \{v \in V(G) : \rho^{k_v}(v) \in V(P(x)_{[x,r]}) \cup V(P(y)_{[y,r]})\}$.

Set $K := K + 1$, $\mathcal{B} := \mathcal{B} \cup \{A\}$, $z_A := 0$ and $\zeta_A := 0$.

For all $v \in A$ **do**:

Set $k_v := k_v + 1$, $b^{k_v}(v) := K$, $\rho^{k_v}(v) := r$, $\varphi^{k_v}(v) := \varphi^{k_v-1}(v)$.

For $i := 1$ **to** h' **do**:

If $\rho^{k_{x_{2i}}-1}(x_{2i}) \neq r$ **then set** $\varphi^{k_{x_{2i}}}(x_{2i}) := x_{2i-1}$.

If $\rho^{k_{x_{2i-1}}-1}(x_{2i-1}) \neq r$ **then set** $\varphi^{k_{x_{2i-1}}}(x_{2i-1}) := x_{2i}$.

For $i := 1$ **to** j' **do**:

If $\rho^{k_{y_{2i}}-1}(y_{2i}) \neq r$ **then set** $\varphi^{k_{y_{2i}}}(y_{2i}) := y_{2i-1}$.

If $\rho^{k_{y_{2i-1}}-1}(y_{2i-1}) \neq r$ **then set** $\varphi^{k_{y_{2i-1}}}(y_{2i-1}) := y_{2i}$.

If $\rho^{k_x-1}(x) \neq r$ **then set** $\varphi^{k_x}(x) := y$.

If $\rho^{k_y-1}(y) \neq r$ **then set** $\varphi^{k_y}(y) := x$.

For each outer vertex $v \notin A$ **do**:

Set $t^A_v := t^{A'}_v - \zeta_{A'}$ and $\tau^A_v := \tau^{A'}_v$, where A' is a maximal proper subset of A in $\mathcal{B}$ for which $t^{A'}_v - \zeta_{A'}$ is minimum.

Set $t^A := \min\{t^A_v : v$ outer, there is no $\bar{A} \in \mathcal{B}$ with $A \cup \{v\} \subseteq \bar{A}\}$.

Mark all $v \in A$ outer. **For** each new outer vertex v **do**: UPDATE(v).

Go to ③.

⑧   ("unpack")

Let $A \in \mathcal{B}$ be a maximal inner blossom with $z_A = 0$ and $|A| > 1$.

Set $\mathcal{B} := \mathcal{B} \setminus \{A\}$.

Let $y := \sigma(\rho^{k_v}(v))$ for some $v \in A$.

Let $Q(y) := $ BLOSSOMPATH(y) be given by

$(y = r_0, r_1, r_2, \ldots, r_{2l-1}, r_{2l} = \rho^{k_y}(y))$.

Mark all $v \in A$ with $\rho^{k_v-1}(v) \notin V(Q(y))$ as out-of-forest.

Mark all $v \in A$ with $\rho^{k_v-1}(v) = r_{2i-1}$ for some i as outer.

For all $v \in A$ with $\rho^{k_v-1}(v) = r_{2i}$ for some i (v remains inner) **do**:

Set $\sigma(\rho^{k_v}(v)) := r_j$ and $\chi(r_j) := r_{j-1}$, where

$$j := \min\{j' \in \{0, \ldots, 2l\} : \rho^{k_{r_{j'}}-1}(r_{j'}) = \rho^{k_v-1}(v)\}.$$

For all $v \in A$ **do**: Set $k_v := k_v - 1$.

For each new outer vertex v **do**: UPDATE(v).

Go to ③.

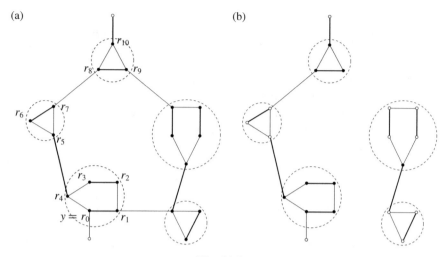

Fig. 11.3.

Note that in contrast to our previous discussion, $\varepsilon = 0$ is possible. The variables τ_v^A are not needed explicitly. The "unpack"-step ⑧ is illustrated in Figure 11.3, where a blossom with 19 vertices is unpacked. Two of the five sub-blossoms become out-of-forest, two become inner blossoms and one becomes an outer blossom.

Before analysing the algorithm, let us demonstrate its major steps by an example. Consider the graph in Figure 11.4(a). Initially, the algorithm sets $z_{\{a\}} = z_{\{d\}} = z_{\{h\}} = 2$, $z_{\{b\}} = z_{\{c\}} = z_{\{f\}} = 4$ and $z_{\{e\}} = z_{\{g\}} = 6$. In Figure 11.4(b) the slacks can be seen. So in the beginning the edges $\{a, d\}, \{a, h\}, \{b, c\}, \{b, f\}, \{c, f\}$ are tight. Therefore we will have $\epsilon = 0$ in the first iterations.

We assume that the algorithm scans the vertices in alphabetical order. So the first steps are

$$\text{augment}(a, d), \qquad \text{augment}(b, c), \qquad \text{grow}(f, b).$$

Fig. 11.4.

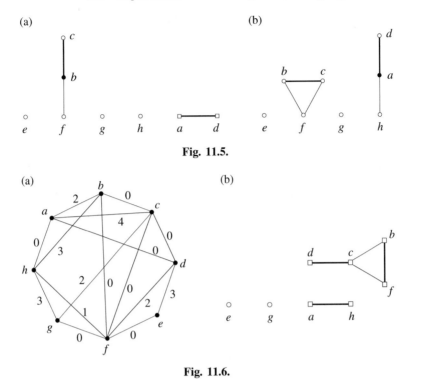

Fig. 11.5.

Fig. 11.6.

Figure 11.5(a) shows the current general blossom forest. The next steps are

$$\text{shrink}(f, c), \qquad \text{grow}(h, a),$$

resulting in the general blossom forest shown in Figure 11.5(b). Now all the tight edges are used up, so the dual variables have to change. In ③ we obtain $\varepsilon = \varepsilon_3 = 1$, say $A = \{b, c, f\}$ and $\tau_d^A = c$. The new dual variables are $z_{\{b,c,f\}} = 1$, $z_{\{a\}} = 1$, $z_{\{d\}} = z_{\{h\}} = 3$, $z_{\{b\}} = z_{\{c\}} = z_{\{f\}} = 4$, $z_{\{e\}} = z_{\{g\}} = 7$. The current slacks are shown in Figure 11.6(a). The next step is

$$\text{augment}(d, c).$$

The blossom $\{b, c, f\}$ becomes out-of-forest (Figure 11.6(b)). Now again $\varepsilon = \varepsilon_3 = 0$ in ③ as $\{e, f\}$ is tight. The next steps are

$$\text{grow}(e, f), \qquad \text{grow}(d, a).$$

We arrive at Figure 11.7(a).

No more edges incident to outer vertices are tight, and we get $\varepsilon = \varepsilon_1 = 1$ in ③ and obtain the new dual solution $z_{\{b,c,f\}} = 0$, $z_{\{a\}} = 0$, $z_{\{d\}} = z_{\{h\}} = z_{\{b\}} = z_{\{c\}} = z_{\{f\}} = 4$, $z_{\{e\}} = z_{\{g\}} = 8$. The new slacks are shown in Figure 11.7(b). Since the dual variable for the inner blossom $\{B, C, F\}$ becomes zero, we have to

Fig. 11.7.

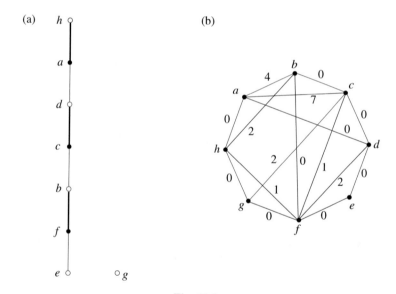

Fig. 11.8.

$$\text{unpack}(\{b, c, f\}).$$

The general blossom forest we get is shown in Figure 11.8(a). After another dual variable change with $\varepsilon = \varepsilon_3 = \frac{1}{2}$ we obtain $z_{\{a\}} = -0.5$, $z_{\{c\}} = z_{\{f\}} = 3.5$, $z_{\{b\}} = z_{\{d\}} = z_{\{h\}} = 4.5$, $z_{\{e\}} = z_{\{g\}} = 8.5$ (the slacks are shown in Figure 11.8(b)). The final steps are

$$\text{shrink}(d, e), \qquad \text{augment}(g, h),$$

and the algorithm terminates. The final matching is $M = \{\{e, f\}, \{b, c\}, \{a, d\},$ $\{g, h\}\}$. We check that M has total weight 37, equal to the sum of the dual variables.

Let us now check that the algorithm works correctly.

Proposition 11.8. *The following statements hold at any stage of the* WEIGHTED MATCHING ALGORITHM:

(a) *For each* $j \in \{1, \ldots, K\}$ *let* $X(j) := \{v \in V(G) : j \in \{b^1(v), \ldots, b^{k_v}(v)\}\}$. *Then* $\mathcal{B} = \{X(j) : j = 1, \ldots, K\} \cup \{\{v\} : v \in V(G)\}$ *is a laminar family. The sets* $V_r := \{v : \rho^{k_v}(v) = r\}$ *for* $r \in V(G)$ *with* $\rho^{k_r}(r) = r$ *are exactly the maximal elements of* $\mathcal{B}$. *The vertices in each* V_r *are marked either all outer or all inner or all out-of-forest. Each* $(V_r, \{\{v, \varphi^{k_v}(v)\} : v \in V_r \setminus \{r\}\} \cup \{\{v, \mu(v)\} : v \in V_r \setminus \{r\}\})$ *is a blossom with base* r.
(b) *The edges* $\{x, \mu(x)\}$ *form a matching* M. *M contains a near-perfect matching within each element of* $\mathcal{B}$.
(c) *For* $b = 1, \ldots, K$ *the variables* $\mu(v)$ *and* $\varphi^i(v)$, *for those* v *and* i *with* $b^i(v) = b$, *are associated with an* M-alternating ear-decomposition in $G[X(b)]$.
(d) *The edges* $\{x, \mu(x)\}$ *and* $\{x, \varphi^i(x)\}$ *for all* x *and* i, *and the edges* $\{\sigma(x), \chi(\sigma(x))\}$ *for all bases* x *of maximal inner blossoms, are all tight.*
(e) *The edges* $\{x, \mu(x)\}$, $\{x, \varphi^{k_x}(x)\}$ *for all inner or outer* x, *together with the edges* $\{\sigma(x), \chi(\sigma(x))\}$ *for all bases* x *of maximal inner blossoms, form a general blossom forest* F *with respect to* M. *The vertex marks (inner, outer, out-of-forest) are consistent with* F.
(f) *Contracting the maximal sub-blossoms of any blossom in* $B \in \mathcal{B}$ *with* $|B| > 1$ *results in a circuit.*
(g) *For each outer vertex* v, *the procedure* TREEPATH *gives an* M-alternating v-r-path, where r is the root of the tree in F containing v.

Proof: The properties clearly hold at the beginning (after ② is executed the first time). We show that they are maintained throughout the algorithm. This is easily seen for (a) by considering ⑦ and ⑧. For (b), this follows from Proposition 11.6 and the assumption that (f) and (g) hold before augmenting.

The proof that (c) continues to hold after shrinking is the same as in the non-weighted case (see Lemma 10.30 (c)). The φ-values are recomputed after augmenting and not changed elsewhere. (d) is guaranteed by ③.

It is easy to see that (e) is maintained by ⑤: The blossom containing y was out-of-forest, and setting $\chi(y) := x$ and $\sigma(v) := y$ for the base v of the blossom makes it inner. The blossom containing $\mu(\rho^{k_y}(y))$ was also out-of-forest, and becomes outer.

In ⑥, two connected components of the general blossom forest become out-of-forest, so (e) is maintained. In ⑦, the vertices in the new blossom become outer because r was outer before. In ⑧, for the vertices $v \in A$ with $\rho^{k_v-1}(v) \notin V(Q(y))$ we also have $\mu(\rho^{k_v}(v)) \notin V(Q(y))$, so they become out-of-forest. For each other $v \in A$ we have $\rho^{k_v-1}(v) = r_k$ for some k. Since $\{r_i, r_{i+1}\} \in M$ iff i is even, v becomes outer iff k is odd.

(f) holds as any new blossom arises from an odd circuit in ⑦. To see that (g) is maintained, it suffices to observe that $\sigma(x)$ and $\chi(\sigma(x))$ are set correctly for all bases x of maximal inner blossoms. This is easily checked for both ⑤ and ⑧. □

Proposition 11.8(a) justifies calling the maximal elements of $\mathcal{B}$ inner, outer or out-of-forest in ③ and ⑧ of the algorithm.

Next we show that the algorithm maintains a feasible dual solution.

Lemma 11.9. *At any stage of the algorithm, z is a feasible dual solution. If $\varepsilon = \infty$ then G has no perfect matching.*

Proof: We always have $z_A = 0$ for all $A \in \mathcal{A} \setminus \mathcal{B}$. z_A is decreased only for those $A \in \mathcal{B}$ that are maximal in $\mathcal{B}$ and inner. So the choice of ε_1 guarantees that z_A continues to be nonnegative for all A with $|A| > 1$.

How can the constraints $\sum_{A \in \mathcal{A}:e \in \delta(A)} z_A \leq c(e)$ be violated? If $\sum_{A \in \mathcal{A}:e \in \delta(A)} z_A$ increases in ③, e must either connect an outer and an out-of-forest vertex or two different outer blossoms. So the maximal ε such that the new z still satisfies $\sum_{A \in \mathcal{A}:e \in \delta(A)} z_A \leq c(e)$ is $slack(e)$ in the first case and $\frac{1}{2} slack(e)$ in the second case.

We thus have to prove that ε_2 and ε_3 are computed correctly:

$$\varepsilon_2 = \min\{slack(v, w) : v \text{ outer}, w \text{ out-of-forest}\}$$

and

$$\varepsilon_3 = \frac{1}{2} \min\left\{slack(v, w) : v, w \text{ outer}, \rho^{k_v}(v) \neq \rho^{k_w}(w)\right\}.$$

We claim that at any stage of the algorithm the following holds for any outer vertex v and any $A \in \mathcal{B}$ such that there is no $\bar{A} \in \mathcal{B}$ with $A \cup \{v\} \subseteq \bar{A}$:

(a) $\tau_v^A \in A$.
(b) $slack(v, \tau_v^A) = \min\{slack(v, u) : u \in A\}$.
(c) $\zeta_A = \sum_{B \in \mathcal{B}:A \subseteq B} z_B$. Δ is the sum of the ε-values in all dual changes so far.
(d) $slack(v, \tau_v^A) = t_v^A - \Delta - \zeta_A$.
(e) $t^A = \min\{t_v^A : v \text{ outer and there is no } \bar{A} \in \mathcal{B} \text{ with } A \cup \{v\} \subseteq \bar{A}\}$.

(a), (c), and (e) are easily seen to be true. (b) and (d) hold when τ_v^A is defined (in ⑦ or in UPDATE(v)), and afterwards $slack(v, u)$ decreases exactly by the amount that $\Delta + \zeta_A$ increases (due to (c)). Now (a), (b), (d), and (e) imply that ε_3 is computed correctly.

Now suppose $\varepsilon = \infty$, i.e. ε can be chosen arbitrarily large without destroying dual feasibility. Since the dual objective $\mathbb{1}z$ increases by at least ε in ③, we conclude that the dual LP (11.2) is unbounded. Hence by Theorem 3.27 the primal LP (11.1) is infeasible. □

Now the correctness of the algorithm follows:

Theorem 11.10. *If the algorithm terminates in ⑥, the edges $\{x, \mu(x)\}$ form a minimum weight perfect matching in G.*

Proof: Let x be the incidence vector of M (the matching consisting of the edges $\{x, \mu(x)\}$). The complementary slackness conditions

$$x_e > 0 \quad \Rightarrow \quad \sum_{A \in \mathcal{A}: e \in \delta(A)} z_A = c(e)$$

$$z_A > 0 \quad \Rightarrow \quad \sum_{e \in \delta(A)} x_e = 1$$

are satisfied: the first one holds since all the matching edges are tight (Proposition 11.8(d)), and the second one follows from Proposition 11.8(b).

Since we have feasible primal and dual solutions (Lemma 11.9), both must be optimal (Corollary 3.23). So x is optimal for the LP (11.1) and integral, proving that M is a minimum weight perfect matching. $\square$

Until now we have not proved that the algorithm terminates.

Theorem 11.11.   *The running time of the* WEIGHTED MATCHING ALGORITHM *between two augmentations is* $O(n^2)$. *The overall running time is* $O(n^3)$.

Proof: By Lemma 11.7 and Proposition 11.8(a), the UPDATE procedure runs in linear time. Both ② and ⑥ take $O(n^2)$ time, once per augmentation. ③ and ④ take $O(n)$ time. Moreover, each of ⑤, ⑦, and ⑧ can be done in $O(nk)$ time, where k is the number of new outer vertices. (In ⑦, the number of maximal proper subsets A' of A to be considered is at most $2k + 1$: every second sub-blossom of a new blossom must have been inner.)

Since an outer vertex continues to be outer until the next augmentation, the total time spent by ⑤, ⑦, and ⑧ between two augmentations is $O(n^2)$. Moreover, each call of ⑤, ⑦, and ⑧ creates at least one new outer vertex. Since at least one of ⑤, ⑥, ⑦, ⑧ is called in each iteration, the number of iterations between two augmentations is $O(n)$. This proves the $O(n^2)$ running time between two augmentations. Since there are only $\frac{n}{2}$ augmentations, the total running time is $O(n^3)$. $\square$

Corollary 11.12.   *The* MINIMUM WEIGHT PERFECT MATCHING PROBLEM *can be solved in* $O(n^3)$ *time.*

Proof: This follows from Theorems 11.10 and 11.11. $\square$

The first $O(n^3)$-implementation of Edmonds' algorithm for the MINIMUM WEIGHT PERFECT MATCHING PROBLEM was due to Gabow [1973] (see also Gabow [1976] and Lawler [1976]). The theoretically best running time, namely $O(mn + n^2 \log n)$, has also been obtained by Gabow [1990]. For planar graphs a minimum weight perfect matching can be found in $O\left(n^{\frac{3}{2}} \log n\right)$ time, as Lipton and Tarjan [1979, 1980] showed by a divide and conquer approach, using the fact that planar graphs have small "separators". For Euclidean instances (a set of points in the plane defining a complete graph whose edge weights are given by the Euclidean distances) Varadarajan [1998] found an $O\left(n^{\frac{3}{2}} \log^5 n\right)$ algorithm.

Efficient implementations are described by Mehlhorn and Schaefer [2000], Cook and Rohe [1999], and Kolmogorov [2009]. They solve matching problems with millions of vertices optimally. A "primal" weighted matching algorithm – always maintaining a perfect matching and obtaining a feasible dual solution only at termination – was described by Cunningham and Marsh [1978]. A matching whose weight is at least $1 - \epsilon$ times the maximum weight of any matching can be computed in linear time, for any fixed $\epsilon > 0$ (Duan and Pettie [2014]).

11.4 Postoptimality

In this section we prove a postoptimality result which we shall need in Section 12.2. We add two vertices to an instance that has already been solved:

Lemma 11.13. *Let (G, c) be an instance of the* MINIMUM WEIGHT PERFECT MATCHING PROBLEM, *and let $s, t \in V(G)$ be two vertices. Suppose we have run the* WEIGHTED MATCHING ALGORITHM *for the instance $(G - \{s, t\}, c)$. Then a minimum weight perfect matching with respect to (G, c) can be determined in $O(n^2)$ time.*

Proof: The addition of two vertices requires the initialization of the data structures. In particular, for each $v \in \{s, t\}$ we mark v as outer, set $\mu(v) := v$, add $\{v\}$ to $\mathcal{B}$, set $k_v := 0$, $\rho^0(v) := v$, $\varphi^0(v) := v$, and $\zeta_{\{v\}} := z_v := \min\{\frac{1}{2}c(\{s, t\}), \min\{\{c(\{v, w\}) - \zeta_{\{w\}} : w \in V(G) \setminus \{s, t\}\}\}$, where we used the notation $c(e) := \infty$ for $e \notin E(G)$. Then we start the WEIGHTED MATCHING ALGORITHM with ②. By Theorem 11.11 the algorithm terminates after $O(n^2)$ steps with an augmentation, yielding a minimum weight perfect matching in (G, c). □

We also get a second postoptimality result:

Lemma 11.14. (Weber [1981], Ball and Derigs [1983]) *Suppose we have run the* WEIGHTED MATCHING ALGORITHM *for an instance (G, c). Let $s \in V(G)$, and let $c' : E(G) \to \mathbb{R}$ with $c'(e) = c(e)$ for all $e \notin \delta(s)$. Then a minimum weight perfect matching with respect to (G, c') can be determined in $O(n^2)$ time.*

Proof: Let G' result from G by adding two vertices x, y, an edge $\{s, x\}$, and an edge $\{v, y\}$ for each edge $\{v, s\} \in E(G)$. We set $c(\{v, y\}) := c'(\{v, s\})$ for these new edges. The weight of $\{s, x\}$ can be chosen arbitrarily. We then find a minimum weight perfect matching in (G', c), using Lemma 11.13. Removing the edge $\{s, x\}$ and replacing the matching edge $\{v, y\}$ by $\{v, s\}$ yields a minimum weight perfect matching with respect to (G, c'). □

The same result for a "primal" weighted matching algorithm was found by Cunningham and Marsh [1978].

11.5 The Matching Polytope

The correctness of the WEIGHTED MATCHING ALGORITHM also yields Edmonds' characterization of the perfect matching polytope as a by-product. We again use the notation $\mathcal{A} := \{A \subseteq V(G) : |A| \text{ odd}\}$.

Theorem 11.15. (Edmonds [1965]) *Let G be an undirected graph. The **perfect matching polytope** of G, i.e. the convex hull of the incidence vectors of all perfect matchings in G, is the set of vectors x satisfying*

$$
\begin{aligned}
x_e &\geq 0 &&(e \in E(G)) \\
\sum_{e \in \delta(v)} x_e &= 1 &&(v \in V(G)) \\
\sum_{e \in \delta(A)} x_e &\geq 1 &&(A \in \mathcal{A})
\end{aligned}
$$

Proof: By Corollary 3.32 it suffices to show that all vertices of the polytope described above are integral. By Theorem 5.14 this is true if the minimization problem has an integral optimum solution for any weight function. But our WEIGHTED MATCHING ALGORITHM finds such a solution for any weight function (cf. the proof of Theorem 11.10). $\square$

An alternative proof will be given in Section 12.3 (see the remark after Theorem 12.18).

We can also describe the **matching polytope**, i.e. the convex hull of the incidence vectors of all matchings in an undirected graph G:

Theorem 11.16. (Edmonds [1965]) *Let G be a graph. The matching polytope of G is the set of vectors $x \in \mathbb{R}_+^{E(G)}$ satisfying*

$$
\sum_{e \in \delta(v)} x_e \leq 1 \text{ for all } v \in V(G) \quad \text{and} \quad \sum_{e \in E(G[A])} x_e \leq \frac{|A| - 1}{2} \text{ for all } A \in \mathcal{A}.
$$

Proof: Since the incidence vector of any matching obviously satisfies these inequalities, we only have to prove one direction. Let $x \in \mathbb{R}_+^{E(G)}$ be a vector with $\sum_{e \in \delta(v)} x_e \leq 1$ for $v \in V(G)$ and $\sum_{e \in E(G[A])} x_e \leq \frac{|A|-1}{2}$ for $A \in \mathcal{A}$. We prove that x is a convex combination of incidence vectors of matchings.

We use the same construction as in the proof of Proposition 11.1: let H be the graph with $V(H) := \{(v, i) : v \in V(G), i \in \{1, 2\}\}$, and $E(H) := \{\{(v, i), (w, i)\} : \{v, w\} \in E(G), i \in \{1, 2\}\} \cup \{\{(v, 1), (v, 2)\} : v \in V(G)\}$. So H consists of two copies of G, and there is an edge joining the two copies of each vertex. Let $y_{\{(v,i),(w,i)\}} := x_e$ for each $e = \{v, w\} \in E(G)$ and $i \in \{1, 2\}$, and let $y_{\{(v,1),(v,2)\}} := 1 - \sum_{e \in \delta_G(v)} x_e$ for each $v \in V(G)$. We claim that y belongs to the perfect matching polytope of H. Considering the subgraph induced by $\{(v, 1) : v \in V(G)\}$, which is isomorphic to G, we then get that x is a convex combination of incidence vectors of matchings in G.

$V(G)$ $\{(v, 1) : v \in V(G)\}$ $\{(v, 2) : v \in V(G)\}$

$/\!/\!/ : X$

Fig. 11.9.

Obviously, $y \in \mathbb{R}_+^{E(H)}$ and $\sum_{e \in \delta_H(v)} y_e = 1$ for all $v \in V(H)$. To show that y belongs to the perfect matching polytope of H, we use Theorem 11.15. So let $X \subseteq V(H)$ with $|X|$ odd. We prove that $\sum_{e \in \delta_H(X)} y_e \geq 1$. Let $A := \{v \in V(G) : (v, 1) \in X, (v, 2) \notin X\}$, $B := \{v \in V(G) : (v, 1) \in X, (v, 2) \in X\}$ and $C := \{v \in V(G) : (v, 1) \notin X, (v, 2) \in X\}$. Since $|X|$ is odd, either A or C must have odd cardinality, w.l.o.g. $|A|$ is odd. We write $A_i := \{(a, i) : a \in A\}$ and $B_i := \{(b, i) : b \in B\}$ for $i = 1, 2$ (see Figure 11.9). Then

$$
\sum_{e \in \delta_H(X)} y_e \geq \sum_{v \in A_1} \sum_{e \in \delta_H(v)} y_e - 2 \sum_{e \in E(H[A_1])} y_e - \sum_{e \in E_H(A_1, B_1)} y_e + \sum_{e \in E_H(B_2, A_2)} y_e
$$

$$
= \sum_{v \in A_1} \sum_{e \in \delta_H(v)} y_e - 2 \sum_{e \in E(G[A])} x_e
$$

$$
\geq |A_1| - (|A| - 1) = 1.
$$

$\square$

Indeed, we can prove the following stronger result:

Theorem 11.17. (Cunningham and Marsh [1978]) *For any undirected graph G the linear inequality system*

$$
\begin{aligned}
x_e &\geq 0 & (e \in E(G)) \\
\sum_{e \in \delta(v)} x_e &\leq 1 & (v \in V(G)) \\
\sum_{e \subseteq A} x_e &\leq \frac{|A| - 1}{2} & (A \in \mathcal{A}, |A| > 1)
\end{aligned}
$$

is TDI.

Proof: For $c : E(G) \to \mathbb{Z}$ we consider the LP max $\sum_{e \in E(G)} c(e) x_e$ subject to the above constraints. The dual LP is:

$$\min \quad \sum_{v \in V(G)} y_v + \sum_{A \in \mathcal{A}, |A| > 1} \frac{|A| - 1}{2} z_A$$

$$\text{s.t.} \quad \sum_{v \in e} y_v + \sum_{A \in \mathcal{A}, e \subseteq A} z_A \geq c(e) \qquad (e \in E(G))$$

$$y_v \geq 0 \qquad (v \in V(G))$$

$$z_A \geq 0 \qquad (A \in \mathcal{A}, |A| > 1)$$

Let (G, c) be the smallest counterexample, i.e. there is no integral optimum dual solution and $|V(G)| + |E(G)| + \sum_{e \in E(G)} |c(e)|$ is minimum. Then $c(e) \geq 1$ for all e (otherwise we can delete any edge of nonpositive weight), and G has no isolated vertices (otherwise we can delete them).

Moreover, for any optimum solution y, z we claim that $y = 0$. To prove this, suppose $y_v > 0$ for some $v \in V(G)$. Then by complementary slackness (Corollary 3.23) $\sum_{e \in \delta(v)} x_e = 1$ for every primal optimum solution x. But then decreasing $c(e)$ by one for each $e \in \delta(v)$ yields a smaller instance (G, c'), whose optimum LP value is one less (here we use primal integrality, i.e. Theorem 11.16). Since (G, c) is the smallest counterexample, there exists an integral optimum dual solution y', z' for (G, c'). Increasing y'_v by one yields an integral optimum dual solution for (G, c), a contradiction.

Now let $y = 0$ and z be an optimum dual solution for which

$$\sum_{A \in \mathcal{A}, |A| > 1} |A|^2 z_A \tag{11.4}$$

is as large as possible. We claim that $\mathcal{F} := \{A : z_A > 0\}$ is laminar. To see this, suppose there are sets $X, Y \in \mathcal{F}$ with $X \setminus Y \neq \emptyset$, $Y \setminus X \neq \emptyset$ and $X \cap Y \neq \emptyset$. Let $\epsilon := \min\{z_X, z_Y\} > 0$.

If $|X \cap Y|$ is odd, then $|X \cup Y|$ is also odd. Set $z'_X := z_X - \epsilon$, $z'_Y := z_Y - \epsilon$, $z'_{X \cap Y} := z_{X \cap Y} + \epsilon$ (unless $|X \cap Y| = 1$), $z'_{X \cup Y} := z_{X \cup Y} + \epsilon$ and $z'_A := z_A$ for all other sets A. y, z' is also a feasible dual solution; moreover it is optimum as well. This is a contradiction since (11.4) is larger.

If $|X \cap Y|$ is even, then $|X \setminus Y|$ and $|Y \setminus X|$ are odd. Set $z'_X := z_X - \epsilon$, $z'_Y := z_Y - \epsilon$, $z'_{X \setminus Y} := z_{X \setminus Y} + \epsilon$ (unless $|X \setminus Y| = 1$), $z'_{Y \setminus X} := z_{Y \setminus X} + \epsilon$ (unless $|Y \setminus X| = 1$) and $z'_A := z_A$ for all other sets A. Set $y'_v := \epsilon$ for $v \in X \cap Y$ and $y'_v := 0$ for $v \notin X \cap Y$. Then (y', z') is a feasible dual solution with

$$\sum_{v \in V(G)} y'_v + \sum_{A \in \mathcal{A}, |A| > 1} \frac{|A| - 1}{2} z'_A = \epsilon |X \cap Y| + \sum_{A \in \mathcal{A}, |A| > 1} \frac{|A| - 1}{2} z_A$$

$$+ \epsilon \left(\frac{|X \setminus Y| - 1}{2} + \frac{|Y \setminus X| - 1}{2} - \frac{|X| - 1}{2} - \frac{|Y| - 1}{2} \right)$$

$$= \sum_{A \in \mathcal{A}, |A| > 1} \frac{|A| - 1}{2} z_A,$$

hence it is also optimum. This contradicts the fact that any optimum dual solution (y', z') must have $y' = 0$.

Now let $A \in \mathcal{F}$ with $z_A \notin \mathbb{Z}$ and A maximal. Set $\epsilon := z_A - \lfloor z_A \rfloor > 0$. Let $A_1, \ldots, A_k$ be the maximal proper subsets of A in $\mathcal{F}$; they must be pairwise disjoint because $\mathcal{F}$ is laminar. Setting $z'_A := z_A - \epsilon$ and $z'_{A_i} := z_{A_i} + \epsilon$ for $i = 1, \ldots, k$ (and $z'_D := z_D$ for all other $D \in \mathcal{A}$) yields another feasible dual solution $y = 0, z'$ (since c is integral). We have

$$\sum_{B \in \mathcal{A}, |B| > 1} \frac{|B| - 1}{2} z'_B \; < \; \sum_{B \in \mathcal{A}, |B| > 1} \frac{|B| - 1}{2} z_B,$$

contradicting the optimality of the original dual solution $y = 0, z$. $\square$

This proof is due to Schrijver [1983a]. For different proofs, see Lovász [1979] and Schrijver [1983b]. The latter does not use Theorem 11.16. Moreover, replacing $\sum_{e \in \delta(v)} x_e \leq 1$ by $\sum_{e \in \delta(v)} x_e = 1$ for $v \in V(G)$ in Theorem 11.17 yields an alternative description of the perfect matching polytope, which is also TDI (by Theorem 5.19). Theorem 11.15 can easily be derived from this; however, the linear inequality system of Theorem 11.15 is not TDI in general (K_4 is a counterexample). Theorem 11.17 also implies the Berge-Tutte formula (Theorem 10.14; see Exercise 16). Generalizations will be discussed in Section 12.1.

Exercises

1. Use Theorem 11.4 to prove a weighted version of König's Theorem 10.2. (Egerváry [1931])
2. Describe the convex hull of the incidence vectors of all
   (a) vertex covers,
   (b) stable sets,
   (c) edge covers,
   in a bipartite graph G. Show how Theorem 10.2 and the statement of Exercise 2(c) of Chapter 10 follow.
   Hint: Use Theorem 5.26 and Corollary 5.22.
3. Prove the Birkhoff-von-Neumann Theorem 11.5 directly.
4. Let G be a graph and P the fractional perfect matching polytope of G. Prove that the vertices of P are exactly the vectors x with

$$x_e = \begin{cases} \frac{1}{2} & \text{if } e \in E(C_1) \cup \cdots \cup E(C_k) \\ 1 & \text{if } e \in M \\ 0 & \text{otherwise} \end{cases},$$

where $C_1, \ldots, C_k$ are vertex-disjoint odd circuits and M is a perfect matching in $G - (V(C_1) \cup \cdots \cup V(C_k))$.
(Balinski [1972]; see Lovász [1979])

5. Let G be a bipartite graph with bipartition $V = A \cup B$ and $A = \{a_1, \ldots, a_p\}$, $B = \{b_1, \ldots, b_q\}$. Let $c : E(G) \to \mathbb{R}$ be weights on the edges. We look for the maximum weight order-preserving matching M, i.e. for any two edges $\{a_i, b_j\}, \{a_{i'}, b_{j'}\} \in M$ with $i < i'$ we require $j < j'$. Solve this problem with an $O(n^3)$-algorithm.

   Hint: Use dynamic programming.

6. Prove that, at any stage of the WEIGHTED MATCHING ALGORITHM, $|\mathcal{B}| \leq \frac{3}{2}n$.

7. Let G be a graph with nonnegative weights $c : E(G) \to \mathbb{R}_+$. Let M be the matching at any intermediate stage of the WEIGHTED MATCHING ALGO-RITHM. Let X be the set of vertices covered by M. Show that any matching covering X is at least as expensive as M.

   (Ball and Derigs [1983])

8. A graph with integral weights on the edges is said to have the even circuit property if the total weight of every circuit is even. Show that the WEIGHTED MATCHING ALGORITHM applied to a graph with the even circuit property maintains this property (with respect to the slacks) and also maintains a dual solution that is integral. Conclude that for any graph there exists an optimum dual solution z that is half-integral (i.e. $2z$ is integral).

9. When the WEIGHTED MATCHING ALGORITHM is restricted to bipartite graphs, it becomes much simpler. Show which parts are necessary even in the bipartite case and which are not.

   Note: One arrives at what is called the Hungarian method for the ASSIGNMENT PROBLEM (Kuhn [1955]). This algorithm can also be regarded as an equivalent description of the procedure proposed in the proof of Theorem 11.2.

10. Suppose G is a complete graph and $c : E(G) \to \mathbb{R}_+$ satisfies the triangle inequality, i.e., $c(\{x, z\}) \leq c(\{x, y\}) + c(\{y, z\})$ for all $x, y, z \in V(G)$. Show that then there exists an optimum solution z to (11.2) with $z \geq 0$.

* 11. Show that after each augment-step of the WEIGHTED MATCHING ALGORITHM, the current matching M has minimum weight among all matchings of cardi-nality $|M|$.

    Hint: Extend the dual solution to an instance with $|V(G)| - 2|M|$ added aux-iliary nodes.

    Note: This also gives an alternative solution to the MAXIMUM WEIGHT MATCHING PROBLEM, which could be more efficient than the one in the proof of Proposition 11.1.

    (Lawler [1976], Walter and Kaibel [2017])

12. How can the bottleneck matching problem (find a perfect matching M such that $\max\{c(e) : e \in M\}$ is minimum) be solved in $O(n^3)$ time?

13. Show how to solve the MINIMUM WEIGHT EDGE COVER PROBLEM in poly-nomial time: given an undirected graph G and weights $c : E(G) \to \mathbb{R}$, find a minimum weight edge cover.

14. Given an undirected graph G with weights $c : E(G) \to \mathbb{R}_+$ and two vertices s and t, we look for a shortest s-t-path with an even (or with an odd) number of edges. Reduce this to a MINIMUM WEIGHT PERFECT MATCHING PROBLEM.

Hint: Take two copies of G, connect each vertex with its copy by an edge of zero weight and delete s and t (or s and the copy of t).
(Grötschel and Pulleyblank [1981])

15. Let G be a k-regular and $(k-1)$-edge-connected graph with an even number of vertices, and let $c : E(G) \to \mathbb{R}_+$. Prove that there exists a perfect matching M in G with $c(M) \geq \frac{1}{k} c(E(G))$.
   Hint: Show that $\frac{1}{k} \mathbb{1}$ is in the perfect matching polytope.
   (Naddef and Pulleyblank [1981])

* 16. Show that Theorem 11.17 implies:
   (a) the Berge-Tutte formula (Theorem 10.14);
   (b) Theorem 11.15;
   (c) the existence of an optimum half-integral dual solution to the dual LP (11.2) (cf. Exercise 8).
   Hint: Use Theorem 5.19.

17. The fractional perfect matching polytope Q of G is identical to the perfect matching polytope if G is bipartite (Theorem 11.4). Consider the first Gomory-Chvátal-truncation Q' of Q (Definition 5.30). Prove that Q' is always identical to the perfect matching polytope.

18. Show that, given a graph G with weights $c : V(G) \to \mathbb{R}_+$ such that G is the line graph of a graph H, one can find a maximum weight stable set in $O(n^3)$ time, where $n = |V(G)|$.
   Note: Faenza, Oriolo and Stauffer [2014] found an $O(mn + n^2 \log n)$-time algorithm for *claw-free* graphs (graphs that do not contain $K_{1,3}$ as induced subgraph). Note that line graphs are claw-free, so this generalizes weighted matching.

References

General Literature

Gerards, A.M.H. [1995]: Matching. In: Handbooks in Operations Research and Management Science; Volume 7: Network Models (M.O. Ball, T.L. Magnanti, C.L. Monma, G.L. Nemhauser, eds.), Elsevier, Amsterdam 1995, pp. 135–224

Lawler, E.L. [1976]: Combinatorial Optimization; Networks and Matroids. Holt, Rinehart and Winston, New York 1976, Chapters 5 and 6

Papadimitriou, C.H., and Steiglitz, K. [1982]: Combinatorial Optimization; Algorithms and Complexity. Prentice-Hall, Englewood Cliffs 1982, Chapter 11

Pulleyblank, W.R. [1995]: Matchings and extensions. In: Handbook of Combinatorics; Vol. 1 (R.L. Graham, M. Grötschel, L. Lovász, eds.), Elsevier, Amsterdam 1995

Cited References

Balinski, M.L. [1972]: Establishing the matching polytope. Journal of Combinatorial Theory 13 (1972), 1–13

Ball, M.O., and Derigs, U. [1983]: An analysis of alternative strategies for implementing matching algorithms. Networks 13 (1983), 517–549

Birkhoff, G. [1946]: Tres observaciones sobre el algebra lineal. Revista Universidad Nacional de Tucumán, Series A 5 (1946), 147–151

Burkard, R., Dell'Amico, M., and Martello, S. [2009]: Assignment Problems. SIAM, Philadelphia 2009

Cook, W., and Rohe, A. [1999]: Computing minimum-weight perfect matchings. INFORMS Journal of Computing 11 (1999), 138–148

Cunningham, W.H., and Marsh, A.B. [1978]: A primal algorithm for optimum matching. Mathematical Programming Study 8 (1978), 50–72

Duan, R. and Pettie, S. [2014]: Linear-time approximation for maximum weight matching. Journal of the ACM 61 (2014), 1–23

Edmonds, J. [1965]: Maximum matching and a polyhedron with (0,1) vertices. Journal of Research of the National Bureau of Standards B 69 (1965), 125–130

Egervary, E. [1931]: Matrixok kombinatorikus tulajdonságairol. Matematikai és Fizikai Lapok 38 (1931), 16–28 [in Hungarian]

Faenza, Y., Oriolo, G., and Stauffer, G. [2014]: Solving the weighted stable set problem in claw-free graphs via decomposition. Journal of the ACM 61 (2014), Article 20

Gabow, H.N. [1973]: Implementation of algorithms for maximum matching on non-bipartite graphs. Ph.D. Thesis, Stanford University, Dept. of Computer Science, 1973

Gabow, H.N. [1976]: An efficient implementation of Edmonds' algorithm for maximum matching on graphs. Journal of the ACM 23 (1976), 221–234

Gabow, H.N. [1990]: Data structures for weighted matching and nearest common ancestors with linking. Proceedings of the 1st Annual ACM-SIAM Symposium on Discrete Algorithms (1990), 434–443. Full version in arXiv:1611.07541

Grötschel, M., and Pulleyblank, W.R. [1981]: Weakly bipartite graphs and the max-cut problem. Operations Research Letters 1 (1981), 23–27

Kolmogorov, V. [2009]: Blossom V: a new implementation of a minimum cost perfect matching algorithm. Mathematical Programming Computation 1 (2009), 43–67

Kuhn, H.W. [1955]: The Hungarian method for the assignment problem. Naval Research Logistics Quarterly 2 (1955), 83–97

Lipton, R.J., and Tarjan, R.E. [1979]: A separator theorem for planar graphs. SIAM Journal on Applied Mathematics 36 (1979), 177–189

Lipton, R.J., and Tarjan, R.E. [1980]: Applications of a planar separator theorem. SIAM Journal on Computing 9 (1980), 615–627

Lovász, L. [1979]: Graph theory and integer programming. In: Discrete Optimization I; Annals of Discrete Mathematics 4 (P.L. Hammer, E.L. Johnson, B.H. Korte, eds.), North-Holland, Amsterdam 1979, pp. 141–158

Mehlhorn, K., and Schäfer, G. [2000]: Implementation of $O(nm \log n)$ weighted matchings in general graphs: the power of data structures. In: Algorithm Engineering; WAE-2000; LNCS 1982 (S. Näher, D. Wagner, eds.), pp. 23–38; also electronically in The ACM Journal of Experimental Algorithmics 7 (2002)

Monge, G. [1784]: Mémoire sur la théorie des déblais et des remblais. Histoire de l'Académie Royale des Sciences 2 (1784), 666–704

Munkres, J. [1957]: Algorithms for the assignment and transportation problems. Journal of the Society for Industrial and Applied Mathematics 5 (1957), 32–38

Naddef, D., and Pulleyblank, W.R. [1981]: Matchings in regular graphs. Discrete Mathematics 34 (1981), 283–291

von Neumann, J. [1953]: A certain zero-sum two-person game equivalent to the optimal assignment problem. In: Contributions to the Theory of Games II; Ann. of Math. Stud. 28 (H.W. Kuhn, ed.), Princeton University Press, Princeton 1953, pp. 5–12

Schrijver, A. [1983a]: Short proofs on the matching polyhedron. Journal of Combinatorial Theory B 34 (1983), 104–108

Schrijver, A. [1983b]: Min-max results in combinatorial optimization. In: Mathematical Programming; The State of the Art – Bonn 1982 (A. Bachem, M. Grötschel, B. Korte, eds.), Springer, Berlin 1983, pp. 439–500

Varadarajan, K.R. [1998]: A divide-and-conquer algorithm for min-cost perfect matching in the plane. Proceedings of the 39th Annual IEEE Symposium on Foundations of Computer Science (1998), 320–329

Walter, M., and Kaibel, V. [2017]: A note on matchings constructed during Edmonds' weighted perfect matching algorithm. arXiv:1703.09505

Weber, G.M. [1981]: Sensitivity analysis of optimal matchings. Networks 11 (1981), 41–56

12 *b*-Matchings and *T*-Joins

Check for updates

In this chapter we introduce two more combinatorial optimization problems, the MAXIMUM WEIGHT *b*-MATCHING PROBLEM in Section 12.1 and the MINIMUM WEIGHT *T*-JOIN PROBLEM in Section 12.2. Both can be regarded as generalizations of the MINIMUM WEIGHT PERFECT MATCHING PROBLEM and also include other important problems. On the other hand, both problems can be reduced to the MINIMUM WEIGHT PERFECT MATCHING PROBLEM. They have combinatorial polynomial-time algorithms as well as polyhedral descriptions. Since in both cases the SEPARATION PROBLEM turns out to be solvable in polynomial time, we obtain another polynomial-time algorithm for these generalized matching problems (using the ELLIPSOID METHOD; see Section 4.6). In fact, the SEPARATION PROBLEM can be reduced to finding a minimum capacity *T*-cut in both cases; see Sections 12.3 and 12.4. This problem, finding a minimum capacity cut $\delta(X)$ such that $|X \cap T|$ is odd for a specified vertex set T, can be solved with network flow techniques.

12.1 *b*-Matchings

Definition 12.1. *Let G be an undirected graph with integral edge capacities* $u : E(G) \to \mathbb{N} \cup \{\infty\}$ *and numbers* $b : V(G) \to \mathbb{N}$. *Then a **b-matching** in* (G, u) *is a function* $f : E(G) \to \mathbb{Z}_+$ *with* $f(e) \leq u(e)$ *for all* $e \in E(G)$ *and* $\sum_{e \in \delta(v)} f(e) \leq b(v)$ *for all* $v \in V(G)$. *In the case* $u \equiv 1$ *we speak of a **simple** b-matching in G. A b-matching f is called **perfect** if* $\sum_{e \in \delta(v)} f(e) = b(v)$ *for all* $v \in V(G)$.

In the case $b \equiv 1$ the capacities are irrelevant, and we are back to ordinary matchings. A simple perfect *b*-matching is sometimes also called a *b*-factor. It can be regarded as a subset of edges. In Chapter 21 we shall be interested in perfect simple 2-matchings, i.e. subsets of edges such that each vertex is incident to exactly two of them.

MAXIMUM WEIGHT *b*-MATCHING PROBLEM

Instance: A graph G, capacities $u : E(G) \to \mathbb{N} \cup \{\infty\}$, weights $c : E(G) \to \mathbb{R}$, and numbers $b : V(G) \to \mathbb{N}$.

Task: Find a *b*-matching f in (G, u) whose weight $\sum_{e \in E(G)} c(e) f(e)$ is maximum.

© Springer-Verlag GmbH Germany 2018
Bernhard Korte and Jens Vygen, *Combinatorial Optimization*, Algorithms and Combinatorics 21, https://doi.org/10.1007/978-3-662-56039-6_12

Edmonds' WEIGHTED MATCHING ALGORITHM can be extended to solve this problem (Marsh [1979]). We shall not describe this algorithm here, but shall rather give a polyhedral description and show that the SEPARATION PROBLEM can be solved in polynomial time. This yields a polynomial-time algorithm via the EL-LIPSOID METHOD (cf. Corollary 3.33).

The **b-matching polytope** of (G, u) is defined to be the convex hull of all b-matchings in (G, u). We first consider the uncapacitated case ($u \equiv \infty$):

Theorem 12.2. (Edmonds [1965]) *Let G be an undirected graph and $b : V(G) \to \mathbb{N}$. The b-matching polytope of (G, ∞) is the set of vectors $x \in \mathbb{R}_+^{E(G)}$ satisfying*

$$\sum_{e \in \delta(v)} x_e \leq b(v) \qquad (v \in V(G));$$

$$\sum_{e \in E(G[X])} x_e \leq \left\lfloor \tfrac{1}{2} \sum_{v \in X} b(v) \right\rfloor \qquad (X \subseteq V(G)).$$

Proof: Since any b-matching obviously satisfies these constraints, we only have to show one direction. So let $x \in \mathbb{R}_+^{E(G)}$ with $\sum_{e \in \delta(v)} x_e \leq b(v)$ for all $v \in V(G)$ and $\sum_{e \in E(G[X])} x_e \leq \left\lfloor \tfrac{1}{2} \sum_{v \in X} b(v) \right\rfloor$ for all $X \subseteq V(G)$. We show that x is a convex combination of b-matchings.

We define a new graph H by splitting up each vertex v into $b(v)$ copies: we define $X_v := \{(v, i) : i \in \{1, \ldots, b(v)\}\}$ for $v \in V(G)$, $V(H) := \bigcup_{v \in V(G)} X_v$ and $E(H) := \{\{v', w'\} : \{v, w\} \in E(G), v' \in X_v, w' \in X_w\}$. Let $y_e := \frac{1}{b(v)b(w)} x_{\{v,w\}}$ for each edge $e = \{v', w'\} \in E(H)$, $v' \in X_v, w' \in X_w$. We claim that y is a convex combination of incidence vectors of matchings in H. By contracting the sets X_v ($v \in V(G)$) in H we then return to G and x, and conclude that x is a convex combination of b-matchings in G.

To prove that y is in the matching polytope of H we use Theorem 11.16. $\sum_{e \in \delta(v)} y_e \leq 1$ obviously holds for each $v \in V(H)$. Let $C \subseteq V(H)$ with $|C|$ odd. We show that $\sum_{e \in E(H[C])} y_e \leq \frac{1}{2}(|C| - 1)$.

If $X_v \subseteq C$ or $X_v \cap C = \emptyset$ for each $v \in V(G)$, this follows directly from the inequalities assumed for x. Otherwise let $a, b \in X_v$, $a \in C$, $b \notin C$. Then

$$
\begin{aligned}
2 \sum_{e \in E(H[C])} y_e &= \sum_{c \in C \setminus \{a\}} \sum_{e \in E(\{c\}, C \setminus \{c\})} y_e + \sum_{e \in E(\{a\}, C \setminus \{a\})} y_e \\
&\leq \sum_{c \in C \setminus \{a\}} \sum_{e \in \delta(c) \setminus \{\{c,b\}\}} y_e + \sum_{e \in E(\{a\}, C \setminus \{a\})} y_e \\
&= \sum_{c \in C \setminus \{a\}} \sum_{e \in \delta(c)} y_e - \sum_{e \in E(\{b\}, C \setminus \{a\})} y_e + \sum_{e \in E(\{a\}, C \setminus \{a\})} y_e \\
&= \sum_{c \in C \setminus \{a\}} \sum_{e \in \delta(c)} y_e \\
&\leq |C| - 1.
\end{aligned}
$$

$\square$

Note that this construction yields an algorithm which, however, in general has an exponential running time. But we note that in the special case $\sum_{v \in V(G)} b(v) = O(n)$ we can solve the uncapacitated MAXIMUM WEIGHT b-MATCHING PROBLEM in $O(n^3)$ time (using the WEIGHTED MATCHING ALGORITHM; cf. Corollary 11.12). Pulleyblank [1973, 1980] described the facets of this polytope and showed that the linear inequality system in Theorem 12.2 is TDI. The following generalization allows finite capacities:

Theorem 12.3.   (Edmonds and Johnson [1970])   *Let G be an undirected graph, $u : E(G) \to \mathbb{N} \cup \{\infty\}$ and $b : V(G) \to \mathbb{N}$. The b-matching polytope of (G, u) is the set of vectors $x \in \mathbb{R}_+^{E(G)}$ satisfying*

$$
\begin{array}{rcll}
x_e & \leq & u(e) & (e \in E(G)); \\
\displaystyle\sum_{e \in \delta(v)} x_e & \leq & b(v) & (v \in V(G)); \\
\displaystyle\sum_{e \in E(G[X])} x_e + \sum_{e \in F} x_e & \leq & \left\lfloor \dfrac{1}{2} \left(\displaystyle\sum_{v \in X} b(v) + \sum_{e \in F} u(e) \right) \right\rfloor & \begin{array}{l} (X \subseteq V(G), \\ F \subseteq \delta(X)). \end{array}
\end{array}
$$

Proof: First observe that every b-matching x satisfies the constraints. This is clear except for the last one; but note that any vector $x \in \mathbb{R}_+^{E(G)}$ with $x_e \leq u(e)$ $(e \in E(G))$ and $\sum_{e \in \delta(v)} x_e \leq b(v)$ $(v \in V(G))$ satisfies

$$
\begin{aligned}
\sum_{e \in E(G[X])} x_e + \sum_{e \in F} x_e &= \frac{1}{2} \left(\sum_{v \in X} \sum_{e \in \delta(v)} x_e + \sum_{e \in F} x_e - \sum_{e \in \delta(X) \setminus F} x_e \right) \\
&\leq \frac{1}{2} \left(\sum_{v \in X} b(v) + \sum_{e \in F} u(e) \right).
\end{aligned}
$$

If x is integral, the left-hand side is an integer, so we may round down the right-hand side.

Now let $x \in \mathbb{R}_+^{E(G)}$ be a vector with $x_e \leq u(e)$ for all $e \in E(G)$, $\sum_{e \in \delta(v)} x_e \leq b(v)$ for all $v \in V(G)$ and

$$
\sum_{e \in E(G[X])} x_e + \sum_{e \in F} x_e \leq \left\lfloor \frac{1}{2} \left(\sum_{v \in X} b(v) + \sum_{e \in F} u(e) \right) \right\rfloor
$$

for all $X \subseteq V(G)$ and $F \subseteq \delta(X)$. We show that x is a convex combination of b-matchings in (G, u).

Let H be the graph resulting from G by subdividing each edge $e = \{v, w\}$ with $u(e) \neq \infty$ by means of two new vertices (e, v), (e, w). (Instead of e, H now contains the edges $\{v, (e, v)\}$, $\{(e, v), (e, w)\}$ and $\{(e, w), w\}$.) Set $b((e, v)) := b((e, w)) := u(e)$ for the new vertices.

For each subdivided edge $e = \{v, w\}$ set $y_{\{v,(e,v)\}} := y_{\{(e,w),w\}} := x_e$ and $y_{\{(e,v),(e,w)\}} := u(e) - x_e$. For each original edge e with $u(e) = \infty$ set $y_e := x_e$. We claim that y is in the b-matching polytope P of (H, ∞).

We use Theorem 12.2. Obviously $y \in \mathbb{R}_+^{E(H)}$ and $\sum_{e \in \delta(v)} y_e \leq b(v)$ for all $v \in V(H)$. Suppose there is a set $A \subseteq V(H)$ with

$$\sum_{e \in E(H[A])} y_e > \left\lfloor \frac{1}{2} \sum_{a \in A} b(a) \right\rfloor. \tag{12.1}$$

Let $B := A \cap V(G)$. For each $e = \{v, w\} \in E(G[B])$ we may assume $(e, v), (e, w) \in A$, for otherwise the addition of (e, v) and (e, w) does not destroy (12.1). On the other hand, we may assume that $(e, v) \in A$ implies $v \in A$: If $(e, v), (e, w) \in A$ but $v \notin A$, we can delete (e, v) and (e, w) from A without destroying (12.1). If $(e, v) \in A$ but $v, (e, w) \notin A$, we can just delete (e, v) from A. Figure 12.1 shows the remaining possible edge types.

Let $F := \{e = \{v, w\} \in E(G) : |A \cap \{(e, v), (e, w)\}| = 1\}$. We have

$$\sum_{e \in E(G[B])} x_e + \sum_{e \in F} x_e = \sum_{e \in E(H[A])} y_e - \sum_{\substack{e \in E(G[B]), \\ u(e) < \infty}} u(e)$$

$$> \left\lfloor \frac{1}{2} \sum_{a \in A} b(a) \right\rfloor - \sum_{\substack{e \in E(G[B]), \\ u(e) < \infty}} u(e)$$

$$= \left\lfloor \frac{1}{2} \left(\sum_{v \in B} b(v) + \sum_{e \in F} u(e) \right) \right\rfloor,$$

contradicting our assumption. So $y \in P$, and in fact y belongs to the face

$$\left\{ z \in P : \sum_{e \in \delta(v)} z_e = b(v) \text{ for all } v \in V(H) \setminus V(G) \right\}$$

of P. Since the vertices of this face are also vertices of P, y is a convex combination of b-matchings $f_1, \ldots, f_m$ in (H, ∞), each of which satisfies $\sum_{e \in \delta(v)} f_i(e) = b(v)$ for all $v \in V(H) \setminus V(G)$. This implies $f_i(\{v, (e, v)\}) = f_i(\{(e, w), w\}) \leq u(e)$ for

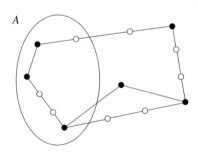

Fig. 12.1.

each subdivided edge $e = \{v, w\} \in E(G)$. By returning from H to G we obtain that x is a convex combination of b-matchings in (G, u). □

The constructions in the proofs of Theorems 12.2 and 12.3 are both due to Tutte [1954]. They can also be used to prove a generalization of Tutte's Theorem 10.13 (Exercise 4):

Theorem 12.4. (Tutte [1952]) *Let G be a graph, $u : E(G) \to \mathbb{N} \cup \{\infty\}$ and $b : V(G) \to \mathbb{N}$. Then (G, u) has a perfect b-matching if and only if for any two disjoint subsets $X, Y \subseteq V(G)$ the number of connected components C in $G - X - Y$ for which $\sum_{c \in V(C)} b(c) + \sum_{e \in E_G(V(C), Y)} u(e)$ is odd does not exceed*

$$\sum_{v \in X} b(v) + \sum_{y \in Y} \left(\sum_{e \in \delta(y)} u(e) - b(y) \right) - \sum_{e \in E_G(X, Y)} u(e).$$

12.2 Minimum Weight T-Joins

Consider the following problem: A postman has to deliver the mail within his district. To do this, he must start at the post office, walk along each street at least once, and finally return to the post office. The problem is to find a postman's tour of minimum length. This is known as the CHINESE POSTMAN PROBLEM (Guan [1962]).

Of course we model the street map as a graph which we assume to be connected. (Otherwise the postman has to use streets that do not belong to his district, in which case the problem becomes *NP*-hard; see Exercise 17(d) of Chapter 15.) By Euler's Theorem 2.24 we know that there is a postman's tour using each edge exactly once (i.e. an Eulerian walk) if and only if every vertex has even degree.

If the graph is not Eulerian, we have to use some edges several times. Knowing Euler's Theorem, we can formulate the CHINESE POSTMAN PROBLEM as follows: given a graph G with weights $c : E(G) \to \mathbb{R}_+$, find a function $n : E(G) \to \mathbb{N}$ such that G', the graph which arises from G by taking $n(e)$ copies of each edge $e \in E(G)$, is Eulerian and $\sum_{e \in E(G)} n(e)c(e)$ is minimum.

All this is true in the directed and undirected case. In the directed case, the problem can be solved with network flow techniques (Exercise 9 of Chapter 9). Hence from now on we shall deal with undirected graphs only. Here we need the WEIGHTED MATCHING ALGORITHM.

Of course it makes no sense to walk through an edge e more than twice, because then we may subtract 2 from some $n(e)$ and obtain a solution that cannot be worse. So the problem is to find a minimum weight $J \subseteq E(G)$ such that $(V(G), E(G) \,\dot\cup\, J)$ (the graph we obtain by doubling the edges in J) is Eulerian. In this section, we solve a generalization of this problem.

Definition 12.5. *Given an undirected graph G and a set $T \subseteq V(G)$. A set $J \subseteq E(G)$ is a **T-join** if $|J \cap \delta(x)|$ is odd if and only if $x \in T$.*

MINIMUM WEIGHT *T*-JOIN PROBLEM

Instance: An undirected graph G, weights $c : E(G) \to \mathbb{R}$, and a set $T \subseteq V(G)$.

Task: Find a minimum weight T-join in G or decide that none exists.

The MINIMUM WEIGHT *T*-JOIN PROBLEM generalizes several combinatorial optimization problems:

- If c is nonnegative and T is the set of vertices having odd degree in G, then we have the UNDIRECTED CHINESE POSTMAN PROBLEM (if G is connected).
- If $T = \emptyset$, the T-joins are exactly the Eulerian subgraphs. So the empty set is a minimum weight $\emptyset$-join if and only if c is conservative.
- If $|T| = 2$, say $T = \{s, t\}$, each T-join is the union of an s-t-path and possibly some circuits. So if c is conservative, the MINIMUM WEIGHT *T*-JOIN PROBLEM is equivalent to the SHORTEST PATH PROBLEM in undirected graphs. (Note that we were not able to solve the SHORTEST PATH PROBLEM in undirected graphs in Chapter 7, except for nonnegative weights.)
- If $T = V(G)$, the T-joins of cardinality $\frac{|V(G)|}{2}$ are exactly the perfect matchings. So the MINIMUM WEIGHT PERFECT MATCHING PROBLEM can be reduced to the MINIMUM WEIGHT *T*-JOIN PROBLEM by adding a large constant to each edge weight.

Let us start with a simple proposition:

Proposition 12.6. *Let G be a graph, $T, T' \subseteq V(G)$. Let J be a T-join and let J' be a T'-join. Then $J \triangle J'$ is a $(T \triangle T')$-join.*

Proof: For each $v \in V(G)$ we have

$$
\begin{aligned}
|\delta_{J \triangle J'}(v)| &\equiv |\delta_J(v)| + |\delta_{J'}(v)| \\
&\equiv |\{v\} \cap T| + |\{v\} \cap T'| \\
&\equiv |\{v\} \cap (T \triangle T')| \qquad \pmod 2.
\end{aligned}
$$

$\square$

The main purpose of this section is to give a polynomial-time algorithm for the MINIMUM WEIGHT *T*-JOIN PROBLEM. The question whether a T-join exists at all can be answered easily:

Proposition 12.7. *Let G be a graph and $T \subseteq V(G)$. There exists a T-join in G if and only if $|V(C) \cap T|$ is even for each connected component C of G.*

Proof: If J is a T-join, then for each connected component C of G we have that $\sum_{v \in V(C)} |J \cap \delta(v)| = 2|J \cap E(C)|$, so $|J \cap \delta(v)|$ is odd for an even number of vertices $v \in V(C)$. Since J is a T-join, this means that $|V(C) \cap T|$ is even.

Conversely, let $|V(C) \cap T|$ be even for each connected component C of G. Then T can be partitioned into pairs $\{v_1, w_1\}, \ldots, \{v_k, w_k\}$ with $k = \frac{|T|}{2}$, such that v_i and w_i are in the same connected component for $i = 1, \ldots, k$. Let P_i be some v_i-w_i-path ($i = 1, \ldots, k$), and let $J := E(P_1) \triangle E(P_2) \triangle \cdots \triangle E(P_k)$. By Proposition 12.6 J is a T-join. $\square$

We call a T-join optimum if it has minimum weight. A simple optimality criterion is:

Proposition 12.8.   *A T-join J in a graph G with weights $c : E(G) \rightarrow \mathbb{R}$ is optimum if and only if $c(J \cap E(C)) \leq c(E(C) \setminus J)$ for each circuit C in G.*

Proof: If $c(J \cap E(C)) > c(E(C) \setminus J)$, then $J \triangle E(C)$ is a T-join whose weight is less than the weight of J. On the other hand, if J' is a T-join with $c(J') < c(J)$, $J' \triangle J$ is Eulerian, i.e. the union of circuits, where for at least one circuit C we have $c(J \cap E(C)) > c(J' \cap E(C)) = c(E(C) \setminus J)$. $\square$

This proposition can be regarded as a special case of Theorem 9.7. We now solve the MINIMUM WEIGHT T-JOIN PROBLEM with nonnegative weights by reducing it to the MINIMUM WEIGHT PERFECT MATCHING PROBLEM. The main idea is contained in the following lemma:

Lemma 12.9.   *Let G be a graph, $c : E(G) \rightarrow \mathbb{R}_+$, and $T \subseteq V(G)$. Every optimum T-join in G is the disjoint union of the edge sets of $\frac{|T|}{2}$ paths whose ends are distinct and in T, and possibly some zero-weight circuits.*

Proof: By induction on $|T|$. The case $T = \emptyset$ is trivial since the minimum weight of an $\emptyset$-join is zero.

Let J be any optimum T-join in G; w.l.o.g. J contains no zero-weight circuit. By Proposition 12.8 J contains no circuit of positive weight. As c is nonnegative, J thus forms a forest. Let x, y be two leaves of the same connected component of this forest, i.e. $|J \cap \delta(x)| = |J \cap \delta(y)| = 1$, and let P be the x-y-path with edges in J. We have $x, y \in T$, and $J \setminus E(P)$ is a minimum cost $(T \setminus \{x, y\})$-join (a cheaper $(T \setminus \{x, y\})$-join J' would imply a T-join $J' \triangle E(P)$ that is cheaper than J). The assertion now follows from the induction hypothesis. $\square$

Theorem 12.10.   (Edmonds and Johnson [1973])   *In the case of nonnegative weights, the MINIMUM WEIGHT T-JOIN PROBLEM can be solved in $O(n^3)$ time.*

Proof: Let (G, c, T) be an instance. We first solve an ALL PAIRS SHORTEST PATHS PROBLEM in (G, c); more precisely: in the graph resulting by replacing each edge by a pair of oppositely directed edges with the same weight. By Theorem 7.8 this takes $O(mn + n^2 \log n)$ time. In particular, we obtain the metric closure $(\bar{G}, \bar{c})$ of (G, c) (cf. Corollary 7.10).

Now we find a minimum weight perfect matching M in $(\bar{G}[T], \bar{c})$. By Corollary 11.12, this takes $O(n^3)$ time. By Lemma 12.9, $\bar{c}(M)$ is at most the minimum weight of a T-join.

We consider the shortest x-y-path in G for each $\{x, y\} \in M$ (which we have already computed). Let J be the symmetric difference of the edge sets of all these paths. Evidently, J is a T-join in G. Moreover, $c(J) \leq \bar{c}(M)$, so J is optimum. □

This method no longer works if we allow negative weights, because we would introduce negative circuits. However, we can reduce the MINIMUM WEIGHT T-JOIN PROBLEM with arbitrary weights to that with nonnegative weights:

Theorem 12.11. *Let G be a graph with weights $c : E(G) \rightarrow \mathbb{R}$ and $T \subseteq V(G)$. Let E^- be the set of edges with negative weight, V^- the set of vertices that are incident with an odd number of negative edges, and $d : E(G) \rightarrow \mathbb{R}_+$ with $d(e) := |c(e)|$.*

Then $J \triangle E^-$ is a minimum c-weight T-join if and only if J is a minimum d-weight $(T \triangle V^-)$-join.

Proof: As E^- is a V^--join, Proposition 12.6 implies that $J \triangle E^-$ is a T-join if and only if J is a $(T \triangle V^-)$-join. Moreover, for any subset J of $E(G)$ we have

$$
\begin{aligned}
c(J \triangle E^-) &= c(J \setminus E^-) + c(E^- \setminus J) \\
&= d(J \setminus E^-) + c(E^- \setminus J) + c(J \cap E^-) + d(J \cap E^-) \\
&= d(J) + c(E^-) \, .
\end{aligned}
$$

Since $c(E^-)$ is constant, the theorem follows. □

Corollary 12.12. *The MINIMUM WEIGHT T-JOIN PROBLEM can be solved in $O(n^3)$ time.*

Proof: This follows directly from Theorems 12.10 and 12.11. □

We are finally able to solve the SHORTEST PATH PROBLEM in undirected graphs:

Corollary 12.13. *The problem of finding a shortest path between two specified vertices in an undirected graph with conservative weights can be solved in $O(n^3)$ time.*

Proof: Let s and t be the two specified vertices. Set $T := \{s, t\}$ and apply Corollary 12.12. After deleting zero-weight circuits, the resulting T-join is the edge set of a shortest s-t-path. □

Of course this also implies an $O(mn^3)$-algorithm for finding a circuit of minimum total weight in an undirected graph with conservative weights (and in particular to compute the girth). If we are interested in the ALL PAIRS SHORTEST PATHS PROBLEM in undirected graphs, we do not have to do $\binom{n}{2}$ independent weighted matching computations (which would give a running time of $O(n^5)$). Using the postoptimality results of Section 11.4 we can prove:

Theorem 12.14. *The problem of finding shortest paths for all pairs of vertices in an undirected graph G with conservative weights $c : E(G) \to \mathbb{R}$ can be solved in $O(n^4)$ time.*

Proof: By Theorem 12.11 and the proof of Corollary 12.13 we have to compute an optimum $(\{s, t\} \triangle V^-)$-join with respect to the weights $d(e) := |c(e)|$ for all $s, t \in V(G)$, where V^- is the set of vertices incident to an odd number of negative edges. Let $\bar{d}(\{x, y\}) := \operatorname{dist}_{(G,d)}(x, y)$ for $x, y \in V(G)$, and let H_X be the complete graph on $X \triangle V^-$ ($X \subseteq V(G)$). By the proof of Theorem 12.10 it is sufficient to compute a minimum weight perfect matching in $(H_{\{s,t\}}, \bar{d})$ for all s and t.

Our $O(n^4)$-algorithm proceeds as follows. We first compute $\bar{d}$ (cf. Corollary 7.10) and run the WEIGHTED MATCHING ALGORITHM for the instance $(H_\emptyset, \bar{d})$. Up to now we have spent $O(n^3)$ time.

We show that we can now compute a minimum weight perfect matching of $(H_{\{s,t\}}, \bar{d})$ in $O(n^2)$ time, for any s and t. There are four cases:

Case 1: $s, t \notin V^-$. Then we add these two vertices and reoptimize by Lemma 11.13. In $O(n^2)$ time we obtain a minimum weight perfect matching in $(H_{\{s,t\}}, \bar{d})$.

Case 2: $s, t \in V^-$. Then we construct H' by adding two auxiliary vertices s', t' and two edges $\{s, s'\}, \{t, t'\}$ with arbitrary weight. We reoptimize by Lemma 11.13 and delete the two new edges from the resulting minimum weight perfect matching in H'.

Case 3: $s \in V^-$ and $t \notin V^-$. Then we construct H' by adding t, an auxiliary vertex s', and an edge $\{s, s'\}$ (with arbitrary cost) in addition to the edges incident to t. We reoptimize by Lemma 11.13 and delete the edge $\{s, s'\}$ from the resulting minimum weight perfect matching in H'.

Case 4: $s \notin V^-$ and $t \in V^-$. Symmetric to Case 3. $\square$

Gabow [1983] improved the running time to $O(\min\{n^3, nm \log n\})$.

12.3   *T*-Joins and *T*-Cuts

In this section we shall derive a polyhedral description of the MINIMUM WEIGHT T-JOIN PROBLEM. In contrast to the description of the perfect matching polytope (Theorem 11.15), where we had a constraint for each cut $\delta(X)$ with $|X|$ odd, we now need a constraint for each T-cut. A **T-cut** is a cut $\delta(X)$ with $|X \cap T|$ odd. The following simple observation is very useful:

Proposition 12.15. *Let G be an undirected graph and $T \subseteq V(G)$ with $|T|$ even. Then for any T-join J and any T-cut C we have $J \cap C \neq \emptyset$.*

Proof: Suppose $C = \delta(X)$, then $|X \cap T|$ is odd. So the number of edges in $J \cap C$ must be odd, in particular nonzero. $\square$

A stronger statement can be found in Exercise 12.

Proposition 12.15 implies that the minimum cardinality of a T-join is not less than the maximum number of pairwise edge-disjoint T-cuts. In general, we do not have equality: consider $G = K_4$ and $T = V(G)$. However, for bipartite graphs equality holds:

Theorem 12.16. (Seymour [1981]) *Let G be a bipartite graph and $T \subseteq V(G)$ such that a T-join in G exists. Then the minimum cardinality of a T-join equals the maximum number of pairwise edge-disjoint T-cuts.*

Proof: (Sebő [1987])   We only have to prove "$\leq$". We use induction on $|V(G)|$. If $T = \emptyset$ (in particular if $|V(G)| = 1$), the statement is trivial. So we assume $|V(G)| \geq |T| \geq 2$. Denote by $\tau(G, T)$ the minimum cardinality of a T-join in G (and ∞ if no T-join exists). We will use Proposition 12.8 (for unit weights) and Proposition 12.6 several times. We may assume that there exists a T-join in G because otherwise there is an empty T-cut that we can list an infinite number of times.

Choose $a, b \in V(G)$ such that $\tau(G, T \triangle \{a\} \triangle \{b\})$ is minimum. Let $T' := T \triangle \{a\} \triangle \{b\}$. As one possible choice of a and b consists of the endpoints of an edge in a minimum T-join (note that $T \neq \emptyset$), we have $\tau(G, T') < \tau(G, T)$ and hence $a \neq b$ and $T' = T \triangle \{a, b\}$.

Claim: For any minimum T-join J in G we have $|J \cap \delta(a)| = |J \cap \delta(b)| = 1$.

To prove this claim, let J be a minimum T-join and J' a minimum T'-join. $J \triangle J'$ is the edge-disjoint union of an a-b-path P and some circuits $C_1, \ldots, C_k$. We have $|E(C_i) \cap J| = |E(C_i) \cap J'|$ for each i, because both J and J' are minimum. So $|J \triangle E(P)| = |J'|$, and $J'' := J \triangle E(P)$ is also a minimum T'-join. Now $J'' \cap \delta(a) = J'' \cap \delta(b) = \emptyset$, because if, say, $\{b, b'\} \in J''$, $J'' \setminus \{\{b, b'\}\}$ is a $(T \triangle \{a\} \triangle \{b'\})$-join, and we have $\tau(G, T \triangle \{a\} \triangle \{b'\}) < |J''| = |J'| = \tau(G, T')$, contradicting the choice of a and b. We conclude that $|J \cap \delta(a)| = |J \cap \delta(b)| = 1$, and the claim is proved.

In particular, $a, b \in T$. Now let J be a minimum T-join in G. Contract $B := \{b\} \cup \Gamma(b)$ to a single vertex v_B, and let the resulting graph be G^*. G^* is also bipartite. Let $T^* := T \setminus B$ if $|T \cap B|$ is even and $T^* := (T \setminus B) \cup \{v_B\}$ otherwise. The set J^*, resulting from J by the contraction of B, is obviously a T^*-join in G^*. Since $\Gamma(b)$ is a stable set in G (as G is bipartite), the claim implies that $|J| = |J^*| + 1$.

It suffices to prove that J^* is a minimum T^*-join in G^*, because then we have $\tau(G, T) = |J| = |J^*| + 1 = \tau(G^*, T^*) + 1$, and the theorem follows by induction (observe that $\delta(b)$ is a T-cut in G disjoint from $E(G^*)$).

So suppose that J^* is not a minimum T^*-join in G^*. Then by Proposition 12.8 there is a circuit C^* in G^* with $|J^* \cap E(C^*)| > |E(C^*) \setminus J^*|$. Since G^* is bipartite, $|J^* \cap E(C^*)| \geq |E(C^*) \setminus J^*| + 2$. $E(C^*)$ corresponds to an edge set Q in G. Q cannot be a circuit, because $|J \cap Q| > |Q \setminus J|$ and J is a minimum T-join. Hence Q is an x-y-path in G for some $x, y \in \Gamma(b)$ with $x \neq y$. Let C be the circuit in G

formed by Q together with $\{x, b\}$ and $\{b, y\}$. Since J is a minimum T-join in G,

$$|J \cap E(C)| \le |E(C) \setminus J| \le |E(C^*) \setminus J^*| + 2 \le |J^* \cap E(C^*)| \le |J \cap E(C)|.$$

Thus we must have equality throughout, in particular $\{x, b\}, \{b, y\} \notin J$ and $|J \cap E(C)| = |E(C) \setminus J|$. So $\bar{J} := J \triangle E(C)$ is also a minimum T-join and $|\bar{J} \cap \delta(b)| = 3$. But this is impossible by the claim. □

Corollary 12.17. *Let G be a graph, $c : E(G) \to \mathbb{Z}_+$, and $T \subseteq V(G)$ such that a T-join in G exists. Let k be the minimum cost of a T-join in G. Then there are T-cuts $C_1, \ldots, C_{2k}$ such that each edge $e \in E(G)$ is contained in at most $2c(e)$ of them.*

Proof: Let E_0 be the set of edges with weight zero. We construct a bipartite graph G' by contracting the connected components of $(V(G), E_0)$ and replacing each edge e by a path of length $2c(e)$. Let T' be the set of vertices in G' corresponding to connected components X of $(V(G), E_0)$ with $|X \cap T|$ odd.
Claim: The minimum cardinality of a T'-join in G' is $2k$.

To prove this, first note that it cannot be more than $2k$ as every T-join J in G corresponds to a T'-join in G' of cardinality at most $2c(J)$. Conversely, let J' be a T'-join in G'. It corresponds to an edge set J in G. Let $\bar{T} := T \triangle \{v \in V(G) : |\delta(v) \cap J| \text{ odd}\}$. Then each connected component X of $(V(G), E_0)$ contains an even number of vertices of $\bar{T}$ (as $|\delta(X) \cap J| \equiv |X \cap T| \pmod 2$). By Proposition 12.7 $(V(G), E_0)$ has a $\bar{T}$-join $\bar{J}$, and $J \cup \bar{J}$ is a T-join in G with weight $c(J) = \frac{|J'|}{2}$. The claim is proved.

By Theorem 12.16, there are $2k$ pairwise edge-disjoint T'-cuts in G'. Back in G, this yields a list of $2k$ T-cuts in G such that every edge e is contained in at most $2c(e)$ of these. □

Karzanov [1986] described an efficient algorithm for finding such a T-cut packing. T-cuts are also essential in the following description of the T-join polyhedron:

Theorem 12.18.   (Edmonds and Johnson [1973])   *Let G be an undirected graph, $c : E(G) \to \mathbb{R}_+$, and $T \subseteq V(G)$ such that a T-join in G exists. Then the incidence vector of each minimum weight T-join is an optimum solution of the LP*

$$\min \left\{ cx : x \ge 0, \sum_{e \in C} x_e \ge 1 \text{ for all } T\text{-cuts } C \right\}.$$

*(This polyhedron is called the **T-join polyhedron** of G.)*

Proof: By Proposition 12.15, the incidence vector of a T-join satisfies the constraints. Moreover, by Proposition 12.7, every 0-1-vector satisfying the constraints is the incidence vector of an edge set containing a T-join. Therefore, by (g) ⇒(f) of Theorem 5.14, it suffices to show that the LP value is an integer for every integral c for which the minimum is finite, i.e., for every $c : E(G) \to \mathbb{Z}_+$.

So let $c : E(G) \rightarrow \mathbb{Z}_+$, and let k be the minimum weight (with respect to c) of a T-join in G. By Corollary 12.17 there are T-cuts $C_1, \ldots, C_{2k}$ in G such that every edge e is contained in at most $2c(e)$ of these.

So for any feasible solution x of the above LP we have

$$2cx \geq \sum_{i=1}^{2k} \sum_{e \in C_i} x_e \geq 2k,$$

proving that the optimum value of the LP is k.   □

This implies Theorem 11.15: let G be a graph with a perfect matching and $T := V(G)$. Then Theorem 12.18 implies that

$$\min \left\{ cx : x \geq 0, \sum_{e \in C} x_e \geq 1 \text{ for all } T\text{-cuts } C \right\}$$

is an integer for each $c \in \mathbb{Z}^{E(G)}$ for which the minimum is finite. By Theorem 5.14, the polyhedron is integral, and so is its face

$$\left\{ x \in \mathbb{R}_+^{E(G)} : \sum_{e \in C} x_e \geq 1 \text{ for all } T\text{-cuts } C, \sum_{e \in \delta(v)} x_e = 1 \text{ for all } v \in V(G) \right\}.$$

One can also derive a description of the convex hull of the incidence vectors of all T-joins (Exercise 16). Theorems 12.18 and 4.21 (along with Corollary 3.33) imply another polynomial-time algorithm for the MINIMUM WEIGHT T-JOIN PROBLEM if we can solve the SEPARATION PROBLEM for the above description. This is obviously equivalent to checking whether there exists a T-cut with capacity less than one (here x serves as capacity vector). So it suffices to solve the following problem:

MINIMUM CAPACITY T-CUT PROBLEM

Instance: A graph G, capacities $u : E(G) \rightarrow \mathbb{R}_+$, and a nonempty set $T \subseteq V(G)$ of even cardinality.

Task: Find a minimum capacity T-cut in G.

Note that the MINIMUM CAPACITY T-CUT PROBLEM also solves the SEPARATION PROBLEM for the perfect matching polytope (Theorem 11.15; $T := V(G)$). The following theorem solves the MINIMUM CAPACITY T-CUT PROBLEM: it suffices to consider the fundamental cuts of a Gomory-Hu tree. Recall that we can find a Gomory-Hu tree for an undirected graph with capacities in $O(n^4)$ time (Theorem 8.38).

Theorem 12.19. (Padberg and Rao [1982]) *Let G be an undirected graph with capacities $u : E(G) \rightarrow \mathbb{R}_+$. Let H be a Gomory-Hu tree for (G, u). Let $T \subseteq V(G)$ with $|T| \geq 2$ even. Then there is a minimum capacity T-cut among the fundamental cuts of H. Hence a minimum capacity T-cut can be found in $O(n^4)$ time.*

Proof: Let $\delta_G(X)$ be a minimum capacity T-cut in (G, u). Let J be the set of edges e of H for which $|C_e \cap T|$ is odd, where C_e is a connected component of $H - e$. As $|\delta_J(x)| \equiv \sum_{e \in \delta_H(x)} |C_e \cap T| \equiv |\{x\} \cap T|$ (mod 2) for all $x \in V(G)$, J is a T-join in H. By Proposition 12.15, there exists an edge $f \in \delta_H(X) \cap J$. We have

$$u(\delta_G(X)) \geq \min\{u(\delta_G(Y)) : |Y \cap f| = 1\} = u(\delta_G(C_f)),$$

showing that $\delta_G(C_f)$ is a minimum capacity T-cut. $\square$

12.4 The Padberg-Rao Theorem

Theorem 12.19 was generalized by Letchford, Reinelt and Theis [2008]:

Lemma 12.20.   *Let G be an undirected graph with at least one edge, $T \subseteq V(G)$ with $|T|$ even, and $c, c' : E(G) \to \mathbb{R}_+ \cup \{\infty\}$. Then there is an $O(n^4)$-algorithm which finds sets $X \subseteq V(G)$ and $F \subseteq \delta(X)$ such that $|X \cap T| + |F|$ is odd and $\sum_{e \in \delta(X) \setminus F} c(e) + \sum_{e \in F} c'(e)$ is minimum.*

Proof: Since adding edges e with $c(e) = 0$ and $c'(e) = \infty$ does not change anything, we may assume that G is connected. Let $d(e) := \min\{c(e), c'(e)\}$ ($e \in E(G)$). Let $E' := \{e \in E(G) : c'(e) < c(e)\}$ and $V' := \{v \in V(G) : |\delta_{E'}(v)|$ odd$\}$. Let $T' := T \triangle V'$. Note: for $X \subseteq V(G)$ we have $|X \cap T| + |\delta(X) \cap E'| \equiv |X \cap T| + |X \cap V'| \equiv |X \cap T'|$ (mod 2).

The algorithm first computes a Gomory-Hu tree H for (G, d). For each $f \in E(H)$, let X_f be the vertex set of a connected component of $H - f$. Let $g_f \in \delta_G(X_f)$ with $|c'(g_f) - c(g_f)|$ minimum. Then let $F_f := \delta_G(X_f) \cap E'$ if $|X_f \cap T'|$ is odd and $F_f := (\delta_G(X_f) \cap E') \triangle \{g_f\}$ otherwise. Finally, we choose an $f \in E(H)$ such that $\sum_{e \in \delta_G(X_f) \setminus F_f} c(e) + \sum_{e \in F_f} c'(e)$ is minimum, and output $X := X_f$ and $F := F_f$.

The total running time is clearly dominated by computing the Gomory-Hu tree.

Let $X^* \subseteq V(G)$ and $F^* \subseteq \delta(X^*)$ be optimum sets, i.e. $|X^* \cap T| + |F^*|$ is odd and $\sum_{e \in \delta_G(X^*) \setminus F^*} c(e) + \sum_{e \in F^*} c'(e)$ is minimum.

Case 1: $|X^* \cap T'|$ is odd. Then the set of $f \in E(H)$ such that $|X_f \cap T'|$ is odd is a T'-join in H, and thus has nonempty intersection with the T'-cut $\delta_H(X^*)$. Let $f \in \delta_H(X^*)$ with $|X_f \cap T'|$ odd. By definition of the Gomory-Hu tree, $d(\delta_G(X_f)) \leq d(\delta_G(X^*))$ and $\sum_{e \in \delta_G(X_f) \setminus F_f} c(e) + \sum_{e \in F_f} c'(e) = d(\delta_G(X_f))$.

Case 2: $|X^* \cap T'|$ is even. Let $g^* \in \delta_G(X^*)$ with $|c'(g^*) - c(g^*)|$ minimum. The unique circuit in $H + g^*$ contains an edge $f \in \delta_H(X^*)$. Then $\sum_{e \in \delta_G(X^*) \setminus F^*} c(e) + \sum_{e \in F^*} c'(e) = d(\delta_G(X^*)) + |c'(g^*) - c(g^*)| \geq d(\delta_G(X_f)) + |c'(g^*) - c(g^*)| \geq \sum_{e \in \delta_G(X_f) \setminus F_f} c(e) + \sum_{e \in F_f} c'(e)$. Here the first inequality follows from the definition of a Gomory-Hu tree (note that $f \in \delta_H(X^*)$), and the second inequality follows from $g^* \in \delta_G(X_f)$. $\square$

With this we can solve the SEPARATION PROBLEM for the b-matching polytope (Theorem 12.3) in polynomial time. This result is known as Padberg-Rao Theorem. Letchford, Reinelt and Theis [2008] simplified the proof and improved the running time:

Theorem 12.21. (Padberg and Rao [1982], Letchford, Reinelt and Theis [2008]) *For undirected graphs* G, $u : E(G) \rightarrow \mathbb{N} \cup \{\infty\}$ *and* $b : V(G) \rightarrow \mathbb{N}$, *the* SEPARATION PROBLEM *for the b-matching polytope of* (G, u) *can be solved in* $O(n^4)$ *time.*

Proof: Given a vector $x \in \mathbb{R}_+^{E(G)}$ with $x_e \leq u(e)$ for all $e \in E(G)$ and $\sum_{e \in \delta_G(v)} x_e \leq b(v)$ for all $v \in V(G)$ (these trivial inequalities can be checked in linear time), we have to check the last set of inequalities in Theorem 12.3. In the proof of Theorem 12.3 we saw that these inequalities are automatically satisfied whenever $b(X) + u(F)$ is even. They are violated iff

$$b(X) - 2 \sum_{e \in E(G[X])} x_e + \sum_{e \in F} (u(e) - 2x_e) \; < \; 1$$

for some $X \subseteq V(G)$ and $F \subseteq \delta(X)$ with $b(X) + u(F)$ odd.

We extend G to a graph $\bar{G}$ by adding a new vertex z and edges $\{z, v\}$ for all $v \in V(G)$. Let $T := \{v \in V(\bar{G}) : b(v) \text{ odd}\}$, where $b(z) := \sum_{v \in V(G)} b(v)$.

Let $E' := \{e \in E(G) : u(e) \text{ finite and odd}\}$. We define $c(e) := x_e$ and $c'(e) := u(e) - x_e$ for $e \in E'$, $c(e) := \min\{x_e, u(e) - x_e\}$ and $c'(e) := \infty$ for $e \in E(G) \setminus E'$, and $c(\{z, v\}) := b(v) - \sum_{e \in \delta_G(v)} x_e$ and $c'(\{z, v\}) := \infty$ for $v \in V(G)$.

For $X \subseteq V(G)$ we write $D_X := \{e \in \delta_G(X) \setminus E' : u(e) < 2x_e\}$.

For each $X \subseteq V(G)$ and $F \subseteq \delta_G(X) \cap E'$ we have:

$$|X \cap T| + |F| \; \equiv \; b(X) + u(F \cup D_X) \quad (\text{mod } 2)$$

and

$$
\begin{aligned}
c(\delta_{\bar{G}}(X) \setminus F) + c'(F) \; &= \; \sum_{v \in X} \left(b(v) - \sum_{e \in \delta_G(v)} x_e \right) + \sum_{e \in (\delta_G(X) \cap E') \setminus F} x_e \\
&\quad + \sum_{e \in \delta_G(X) \setminus E'} \min\{x_e, u(e) - x_e\} + \sum_{e \in F} (u(e) - x_e) \\
&= \; b(X) - 2 \sum_{e \in E(G[X])} x_e + \sum_{e \in F \cup D_X} (u(e) - 2x_e).
\end{aligned}
$$

We conclude: if there are sets $X \subseteq V(\bar{G})$ and $F \subseteq \delta_{\bar{G}}(X)$ with $c(\delta_{\bar{G}}(X) \setminus F) + c'(F) < 1$, then we have $F \subseteq E'$ and w.l.o.g. $z \notin X$ (otherwise take the complement), and hence $b(X) - 2 \sum_{e \in E(G[X])} x_e + \sum_{e \in F \cup D_X} (u(e) - 2x_e) < 1$.

Conversely, if $b(X) - 2 \sum_{e \in E(G[X])} x_e + \sum_{e \in F} (u(e) - 2x_e) < 1$ for some $X \subseteq V(G)$ and $F \subseteq \delta_G(X)$, then w.l.o.g. $D_X \subseteq F \subseteq D_X \cup E'$, and hence $c(\delta_{\bar{G}}(X) \setminus (F \setminus D_X)) + c'(F \setminus D_X) < 1$.

Hence the separation problem reduces to finding sets $X \subseteq V(\bar{G})$ and $F \subseteq \delta_{\bar{G}}(X)$ with $|X \cap T| + |F|$ odd and $c(\delta_{\bar{G}}(X) \setminus F) + c'(F)$ minimum. This can be done with Lemma 12.20. $\qquad\square$

A generalization of this result has been found by Caprara and Fischetti [1996]. The Padberg-Rao Theorem implies:

Corollary 12.22. *The* MAXIMUM WEIGHT b-MATCHING PROBLEM *can be solved in polynomial time.*

Proof: By Corollary 3.33 we have to solve the LP given in Theorem 12.3. By Theorem 4.21 it suffices to have a polynomial-time algorithm for the SEPARATION PROBLEM. Such an algorithm is provided by Theorem 12.21. $\qquad\square$

Marsh [1979] extended Edmonds' WEIGHTED MATCHING ALGORITHM to the MAXIMUM WEIGHT b-MATCHING PROBLEM. This combinatorial algorithm is of course more practical than using the ELLIPSOID METHOD. But Theorem 12.21 is also interesting for other purposes (see e.g. Section 21.4). See also Gerards [1995]. Combinatorial algorithms with strongly polynomial running time were described by Anstee [1987], and for the unweighted case by Goldberg and Karzanov [2004].

Exercises

1. Show that a minimum weight perfect simple 2-matching in an undirected graph G can be found in $O(n^6)$ time.

* 2. Let G be an undirected graph and $b_1, b_2 : V(G) \to \mathbb{N}$. Describe the convex hull of functions $f : E(G) \to \mathbb{Z}_+$ with $b_1(v) \le \sum_{e \in \delta(v)} f(e) \le b_2(v)$.
   Hint: For $X, Y \subseteq V(G)$ with $X \cap Y = \emptyset$ consider the constraint

$$\sum_{e \in E(G[X])} f(e) - \sum_{e \in E(G[Y]) \cup E(Y,Z)} f(e) \le \left\lfloor \frac{1}{2} \left(\sum_{x \in X} b_2(x) - \sum_{y \in Y} b_1(y) \right) \right\rfloor,$$

   where $Z := V(G) \setminus (X \cup Y)$. Use Theorem 12.3.
   (Schrijver [1983])

* 3. Can one generalize the result of Exercise 2 further by introducing lower and upper capacities on the edges?
   Note: This can be regarded as an undirected version of the problem in Exercise 2 of Chapter 9. For a common generalization of both problems and also the MINIMUM WEIGHT T-JOIN PROBLEM see the papers of Edmonds and Johnson [1973], and Schrijver [1983]. Even here a description of the polytope that is TDI is known.

* 4. Prove Theorem 12.4.
   Hint: For the sufficiency, use Tutte's Theorem 10.13 and the constructions in the proofs of Theorems 12.2 and 12.3.

5. The subgraph degree polytope of a graph G is defined to be the convex hull of all vectors $b \in \mathbb{Z}_+^{V(G)}$ such that G has a perfect simple b-matching. Prove that its dimension is $|V(G)| - k$, where k is the number of connected components of G that are bipartite.

* 6. Given an undirected graph, an odd cycle cover is defined to be a subset of edges containing at least one edge of each odd circuit. Show how to find in polynomial time a minimum weight odd cycle cover in a planar graph with nonnegative weights on the edges. Can you also solve the problem for general weights?

   Hint: Consider the UNDIRECTED CHINESE POSTMAN PROBLEM in the planar dual graph and use Theorem 2.26 and Corollary 2.45.

7. Consider the MAXIMUM WEIGHT CUT PROBLEM in planar graphs: Given an undirected planar graph G with weights $c : E(G) \to \mathbb{R}_+$, we look for a maximum weight cut. Can one solve this problem in polynomial time?

   Hint: Use Exercise 6.

   Note: For general graphs this problem is NP-hard; see Theorem 16.6.
   (Hadlock [1975])

8. Given a graph G with weights $c : E(G) \to \mathbb{R}_+$ and a set $T \subseteq V(G)$ with $|T|$ even. We construct a new graph G' by setting

$$
\begin{aligned}
V(G') \ := \ & \{(v, e) : v \in e \in E(G)\} \ \cup \\
& \{\bar{v} : v \in V(G), |\delta_G(v)| + |\{v\} \cap T| \text{ odd}\}, \\
E(G') \ := \ & \{\{(v, e), (w, e)\} : e = \{v, w\} \in E(G)\} \ \cup \\
& \{\{(v, e), (v, f)\} : v \in V(G), e, f \in \delta_G(v), e \neq f\} \ \cup \\
& \{\{\bar{v}, (v, e)\} : v \in e \in E(G), \bar{v} \in V(G')\},
\end{aligned}
$$

   and define $c'(\{(v, e), (w, e)\}) := c(e)$ for $e = \{v, w\} \in E(G)$ and $c'(e') = 0$ for all other edges e' of G'.

   Show that a minimum weight perfect matching in G' corresponds to a minimum weight T-join in G. Is this reduction preferable to the one used in the proof of Theorem 12.10?

* 9. The following problem combines simple perfect b-matchings and T-joins. We are given an undirected graph G with weights $c : E(G) \to \mathbb{R}$, a partition of the vertex set $V(G) = R \,\dot\cup\, S \,\dot\cup\, T$, and a function $b : R \to \mathbb{Z}_+$. We ask for a minimum weight subset of edges $J \subseteq E(G)$ such that $|J \cap \delta(v)| = b(v)$ for $v \in R$, $|J \cap \delta(v)|$ is even for $v \in S$, and $|J \cap \delta(v)|$ is odd for $v \in T$. Show how to reduce this problem to a MINIMUM WEIGHT PERFECT MATCHING PROBLEM.

   Hint: Consider the constructions in Section 12.1 and Exercise 8.

10. We consider the UNDIRECTED MINIMUM MEAN CYCLE PROBLEM: Given an undirected graph G and weights $c : E(G) \to \mathbb{R}$, find a circuit C in G whose mean weight $\frac{c(E(C))}{|E(C)|}$ is minimum. Let (G, c) be an instance. Let $\mathcal{D}$ be

the set of $\emptyset$-joins in G. For $a \in \mathbb{R}$ and $X \in \mathcal{D}$ let $c^a(X) := c(X) - a|X|$ and $\bar{c}^a(X) := \frac{c(X)}{|X|} - a$. Prove:

(a) For any $a \in \mathbb{R}$, we can find an element $X \in \mathcal{D}$ with $c^a(X)$ minimum in $O(n^3)$ time.

(b) Let $a \in \mathbb{R}$ such that $c^a(Y) \leq 0$ for all $Y \in \mathcal{D}$, and let $X \in \mathcal{D}$ with $c^a(X)$ minimum. Then there exists a $Y \in \mathcal{D}$ with $c^{a+b}(Y) \leq 0$, where $b = \bar{c}^a(X)$.

(c) Consider the following procedure. Start with $a = \max\{c(e) : e \in E(G)\}$. Compute an element $X \in \mathcal{D}$ with $c^a(X)$ minimum. Stop if $c^a(X) = 0$. Otherwise replace a by $a + \bar{c}^a(X)$ and iterate. Show that $|X|$ is getting smaller in each iteration.

(d) Show that the algorithm in (c) computes an optimum solution in $O(n^5)$ time.

Note: The running time can be improved to $O(n^2 m + n^3 \log n)$. Essentially the same algorithm works also for other minimum ratio problems.

(Karzanov [1985], Barahona [1993], Babenko and Karzanov [2009])

11. Given a graph G and a set $T \subseteq V(G)$, describe a linear-time algorithm to find a T-join in G or decide that none exists.

Hint: Consider a maximal forest in G.

12. Let G be an undirected graph, $T \subseteq V(G)$ with $|T|$ even, and $F \subseteq E(G)$. Prove: F has nonempty intersection with every T-join if and only if F contains a T-cut. F has nonempty intersection with every T-cut if and only if F contains a T-join.

13. Let G be a graph and $F \subseteq E(G)$. Prove that there exists a set A with $F \subseteq A \subseteq E(G)$ such that $(V(G), A)$ is Eulerian if and only if G contains no cut B with $B \subseteq F$ and $|B|$ odd.

* 14. Let G be a planar 2-connected graph with a fixed embedding, let C be the circuit bounding the outer face, and let T be an even cardinality subset of $V(C)$. Prove that the minimum cardinality of a T-join equals the maximum number of pairwise edge-disjoint T-cuts.

Hint: Colour the edges of C red and blue such that, when traversing C, colours change precisely at the vertices in T. Consider the planar dual graph, split the vertex representing the outer face into a red and a blue vertex, and apply Menger's Theorem 8.9.

15. Prove Theorem 12.18 using Theorem 11.15 and the construction of Exercise 8.

(Edmonds and Johnson [1973])

16. Let G be an undirected graph and $T \subseteq V(G)$ with $|T|$ even. Prove that the convex hull of the incidence vectors of all T-joins in G is the set of all vectors $x \in [0, 1]^{E(G)}$ satisfying

$$\sum_{e \in \delta_G(X) \setminus F} x_e + \sum_{e \in F}(1 - x_e) \geq 1$$

for all $X \subseteq V(G)$ and $F \subseteq \delta_G(X)$ with $|X \cap T| + |F|$ odd.

Hint: Use Theorems 12.18 and 12.11.

17. Let G be an undirected graph. Show that the cone generated by the incidence vectors of all circuits in G is

$$\left\{ x \in \mathbb{R}^{E(G)} : x \geq 0, \quad \sum_{e' \in C \setminus \{e\}} x(e') \geq x(e) \text{ for all cuts } C \text{ in } G \text{ and } e \in C \right\}.$$

Hint: Use Exercise 16.

18. Describe the UNDIRECTED MINIMUM MEAN CYCLE PROBLEM as a linear program with polynomially many variables and constraints.
    Hint: Use Exercise 17, add the constraint $x(E(G)) = 1$, and use the Max-Flow-Min-Cut Theorem.
    (Barahona [1993])
    Note: Since we need extra variables, this is called an extended formulation (cf. Yannakakis [1991]). Barahona [1993] also showed how to solve the MINIMUM WEIGHT T-JOIN PROBLEM by starting with any T-join and successively applying Proposition 12.8, looking for a minimum mean weight circuit, where weights of the current T-join are multiplied by -1. The number of iterations can be bounded by $O(m^2 \log n)$. Hence one can solve the MINIMUM WEIGHT T-JOIN PROBLEM (and hence the MINIMUM WEIGHT PERFECT MATCHING PROBLEM) by solving a polynomial number of polynomial-size LPs. Nevertheless, any LP description for weighted matching has exponential size (Rothvoß [2017]).

19. Let G be an undirected graph and $T \subseteq V(G)$ with $|T| = 2k$ even. Prove that the minimum cardinality of a T-cut in G equals the maximum of $\min_{i=1}^{k} \lambda_{s_i, t_i}$ over all pairings $T = \{s_1, t_1, s_2, t_2, \ldots, s_k, t_k\}$. ($\lambda_{s,t}$ denotes the maximum number of pairwise edge-disjoint s-t-paths.) Can you think of a weighted version of this min-max formula?
    Hint: Use Theorem 12.19.
    (Rizzi [2002])

20. This exercise gives an algorithm for the MINIMUM CAPACITY T-CUT PROBLEM without using Gomory-Hu trees. The algorithm is recursive and – given G, u and T – proceeds as follows:
    1. First we find a set $X \subseteq V(G)$ with $T \cap X \neq \emptyset$ and $T \setminus X \neq \emptyset$, such that $u(X) := \sum_{e \in \delta_G(X)} u(e)$ is minimum (cf. Exercise 31 of Chapter 8). If $|T \cap X|$ happens to be odd, we are done (return X).
    2. Otherwise we apply the algorithm recursively first to G, u and $T \cap X$, and then to G, u and $T \setminus X$. We obtain a set $Y \subseteq V(G)$ with $|(T \cap X) \cap Y|$ odd and $u(Y)$ minimum and a set $Z \subseteq V(G)$ with $|(T \setminus X) \cap Z|$ odd and $u(Z)$ minimum. W.l.o.g. $T \setminus X \not\subseteq Y$ and $X \cap T \not\subseteq Z$ (otherwise replace Y by $V(G) \setminus Y$ and/or Z by $V(G) \setminus Z$).
    3. If $u(X \cap Y) < u(Z \setminus X)$ then return $X \cap Y$ else return $Z \setminus X$.
    Show that this algorithm works correctly and that its running time is $O(n^5)$, where $n = |V(G)|$.

21. Show how to solve the MAXIMUM WEIGHT b-MATCHING PROBLEM for the special case when $b(v)$ is even for all $v \in V(G)$ in strongly polynomial time. *Hint:* Reduction to a MINIMUM COST FLOW PROBLEM as in Exercise 10 of Chapter 9.

References

General Literature

Cook, W.J., Cunningham, W.H., Pulleyblank, W.R., and Schrijver, A. [1998]: Combinatorial Optimization. Wiley, New York 1998, Sections 5.4 and 5.5

Frank, A. [1996]: A survey on T-joins, T-cuts, and conservative weightings. In: Combinatorics, Paul Erdős is Eighty; Volume 2 (D. Miklós, V.T. Sós, T. Szőnyi, eds.), Bolyai Society, Budapest 1996, pp. 213–252

Gerards, A.M.H. [1995]: Matching. In: Handbooks in Operations Research and Management Science; Volume 7: Network Models (M.O. Ball, T.L. Magnanti, C.L. Monma, G.L. Nemhauser, eds.), Elsevier, Amsterdam 1995, pp. 135–224

Lovász, L., and Plummer, M.D. [1986]: Matching Theory. Akadémiai Kiadó, Budapest 1986, and North-Holland, Amsterdam 1986

Schrijver, A. [1983]: Min-max results in combinatorial optimization; Section 6. In: Mathematical Programming; The State of the Art – Bonn 1982 (A. Bachem, M. Grötschel, B. Korte, eds.), Springer, Berlin 1983, pp. 439–500

Schrijver, A. [2003]: Combinatorial Optimization: Polyhedra and Efficiency. Springer, Berlin 2003, Chapters 29–33

Cited References

Anstee, R.P. [1987]: A polynomial algorithm for b-matchings: an alternative approach. Information Processing Letters 24 (1987), 153–157

Babenko, M.A. and Karzanov, A.V. [2009]: Minimum mean cycle problem in bidirected and skew-symmetric graphs. Discrete Optimization 6 (2009), 92–97

Barahona, F. [1993]: Reducing matching to polynomial size linear programming. SIAM Journal on Optimization 3 (1993), 688–695

Caprara, A., and Fischetti, M. [1996]: $\{0, \frac{1}{2}\}$-Chvátal-Gomory cuts. Mathematical Programming 74 (1996), 221–235

Edmonds, J. [1965]: Maximum matching and a polyhedron with (0,1) vertices. Journal of Research of the National Bureau of Standards B 69 (1965), 125–130

Edmonds, J., and Johnson, E.L. [1970]: Matching: A well-solved class of integer linear programs. In: Combinatorial Structures and Their Applications; Proceedings of the Calgary International Conference on Combinatorial Structures and Their Applications 1969 (R. Guy, H. Hanani, N. Sauer, J. Schönheim, eds.), Gordon and Breach, New York 1970, pp. 69–87

Edmonds, J., and Johnson, E.L. [1973]: Matching, Euler tours and the Chinese postman. Mathematical Programming 5 (1973), 88–124

Gabow, H.N. [1983]: An efficient reduction technique for degree-constrained sub-graph and bidirected network flow problems. Proceedings of the 15th Annual ACM Symposium on Theory of Computing (1983), 448–456

Goldberg, A.V., and Karzanov, A.V. [2004]: Maximum skew-symmetric flows and matchings. Mathematical Programming A 100 (2004), 537–568

Guan, M. [1962]: Graphic programming using odd and even points. Chinese Mathematics 1 (1962), 273–277

Hadlock, F. [1975]: Finding a maximum cut of a planar graph in polynomial time. SIAM Journal on Computing 4 (1975), 221–225

Karzanov, A.V. [1985]: Minimum mean weight cuts and cycles in directed graphs. In: Kachestvennye i Priblizhennye Metody Issledovaniya Operatornykh Uravneniĭ (V.S. Klimov, ed.), Yaroslavl State University Press, Yaroslavl 1985, pp. 72–83 [in Russian]. English translation: American Mathematical Society Translations Ser. 2, Vol. 158 (1994), 47–55

Karzanov, A.V. [1986]: An algorithm for determining a maximum packing of odd-terminus cuts and its applications. In: Isslidovaniya po Prikladnoĭ Teorii Grafov (A.S. Alekseev, ed.), Nauka Siberian Dept., Novosibirsk, 1986, pp. 126–140 [in Russian]. English translation: American Mathematical Society Translations Ser. 2, Vol. 158 (1994), 57–70

Letchford, A.N., Reinelt, G., and Theis, D.O. [2008]: Odd minimum cut sets and b-matchings revisited. SIAM Journal on Discrete Mathematics 22 (2008), 1480–1487

Marsh, A.B. [1979]: Matching algorithms. Ph.D. thesis, Johns Hopkins University, Baltimore 1979

Padberg, M.W., and Rao, M.R. [1982]: Odd minimum cut-sets and b-matchings. Mathematics of Operations Research 7 (1982), 67–80

Pulleyblank, W.R. [1973]: Faces of matching polyhedra. Ph.D. thesis, University of Waterloo, 1973

Pulleyblank, W.R. [1980]: Dual integrality in b-matching problems. Mathematical Programming Study 12 (1980), 176–196

Rizzi, R. [2002]: Minimum T-cuts and optimal T-pairings. Discrete Mathematics 257 (2002), 177–181

Rothvoß, T. [2017]: The matching polytope has exponential extension complexity. Journal of the ACM 64 (2017), Article 41

Sebő, A. [1987]: A quick proof of Seymour's theorem on T-joins. Discrete Mathematics 64 (1987), 101–103

Seymour, P.D. [1981]: On odd cuts and multicommodity flows. Proceedings of the London Mathematical Society (3) 42 (1981), 178–192

Tutte, W.T. [1952]: The factors of graphs. Canadian Journal of Mathematics 4 (1952), 314–328

Tutte, W.T. [1954]: A short proof of the factor theorem for finite graphs. Canadian Journal of Mathematics 6 (1954), 347–352

Yannakakis, M. [1991]: Expressing combinatorial optimization problems by linear programs. Journal of Computer and System Sciences 43 (1991), 441–466

13 Matroids

Many combinatorial optimization problems can be formulated as follows. Given a set system $(E, \mathcal{F})$, i.e. a finite set E and some $\mathcal{F} \subseteq 2^E$, and a cost function $c : \mathcal{F} \to \mathbb{R}$, find an element of $\mathcal{F}$ whose cost is minimum or maximum. In the following we consider modular functions c, i.e. assume that $c(X) = c(\emptyset) + \sum_{x \in X}(c(\{x\}) - c(\emptyset))$ for all $X \subseteq E$; equivalently we are given a function $c : E \to \mathbb{R}$ and write $c(X) = \sum_{e \in X} c(e)$.

In this chapter we restrict ourselves to those combinatorial optimization problems where $\mathcal{F}$ describes an independence system (i.e. is closed under subsets) or even a matroid. The results of this chapter generalize several results obtained in previous chapters.

In Section 13.1 we introduce independence systems and matroids and show that many combinatorial optimization problems can be described in this context. There are several equivalent axiom systems for matroids (Section 13.2) and an interesting duality relation discussed in Section 13.3. The main reason why matroids are important is that a simple greedy algorithm can be used for optimization over matroids. We analyze greedy algorithms in Section 13.4 before turning to the problem of optimizing over the intersection of two matroids. As shown in Sections 13.5 and 13.7 this problem can be solved in polynomial time. This also solves the problem of covering a matroid by independent sets as discussed in Section 13.6.

13.1 Independence Systems and Matroids

Definition 13.1. *A set system* $(E, \mathcal{F})$ *is an* **independence system** *if*

(M1) $\emptyset \in \mathcal{F}$;
(M2) *If* $X \subseteq Y \in \mathcal{F}$ *then* $X \in \mathcal{F}$.

The elements of $\mathcal{F}$ *are called* **independent**, *the elements of* $2^E \setminus \mathcal{F}$ **dependent**. *Minimal dependent sets are called* **circuits**, *maximal independent sets are called* **bases**. *For* $X \subseteq E$, *the maximal independent subsets of* X *are called bases of* X.

© Springer-Verlag GmbH Germany 2018
Bernhard Korte and Jens Vygen, *Combinatorial Optimization*, Algorithms and Combinatorics 21, https://doi.org/10.1007/978-3-662-56039-6_13

Definition 13.2. *Let $(E, \mathcal{F})$ be an independence system. For $X \subseteq E$ we define the* **rank** *of X by $r(X) := \max\{|Y| : Y \subseteq X, Y \in \mathcal{F}\}$. Moreover, we define the* **closure** *of X by $\sigma(X) := \{y \in E : r(X \cup \{y\}) = r(X)\}$.*

Throughout this chapter, $(E, \mathcal{F})$ will be an independence system, and $c : E \to \mathbb{R}$ will be a cost function. We shall concentrate on the following two problems:

MAXIMIZATION PROBLEM FOR INDEPENDENCE SYSTEMS

Instance: An independence system $(E, \mathcal{F})$ and $c : E \to \mathbb{R}$.

Task: Find an $X \in \mathcal{F}$ such that $c(X) := \sum_{e \in X} c(e)$ is maximum.

MINIMIZATION PROBLEM FOR INDEPENDENCE SYSTEMS

Instance: An independence system $(E, \mathcal{F})$ and $c : E \to \mathbb{R}$.

Task: Find a basis B such that $c(B)$ is minimum.

The instance specification is somewhat vague. The set E and the cost function c are given explicitly as usual. However, the set $\mathcal{F}$ is usually not given by an explicit list of its elements. Rather one assumes an oracle which – given a subset $F \subseteq E$ – decides whether $F \in \mathcal{F}$. We shall return to this question in Section 13.4.

The following list shows that many combinatorial optimization problems actually have one of the above two forms:

(1) MAXIMUM WEIGHT STABLE SET PROBLEM
   Given a graph G and weights $c : V(G) \to \mathbb{R}$, find a stable set X in G of maximum weight.
   Here $E = V(G)$ and $\mathcal{F} = \{F \subseteq E : F \text{ is stable in } G\}$.
(2) TSP
   Given a complete undirected graph G and weights $c : E(G) \to \mathbb{R}_+$, find a minimum weight Hamiltonian circuit in G.
   Here $E = E(G)$ and $\mathcal{F} = \{F \subseteq E : F \text{ is a subset of edges of a Hamiltonian circuit in } G\}$.
(3) SHORTEST PATH PROBLEM
   Given a graph G (directed or undirected), $c : E(G) \to \mathbb{R}$ and $s, t \in V(G)$ such that t is reachable from s, find a shortest s-t-path in G with respect to c.
   Here $E = E(G)$ and $\mathcal{F} = \{F \subseteq E : F \text{ is a subset of edges of an } s\text{-}t\text{-path}\}$.
(4) KNAPSACK PROBLEM
   Given $n \in \mathbb{N}$ and nonnegative numbers c_i, w_i $(1 \le i \le n)$ and W, find a subset $S \subseteq \{1, \ldots, n\}$ such that $\sum_{j \in S} w_j \le W$ and $\sum_{j \in S} c_j$ is maximum.
   Here $E = \{1, \ldots, n\}$ and $\mathcal{F} = \left\{F \subseteq E : \sum_{j \in F} w_j \le W\right\}$.
(5) MINIMUM SPANNING TREE PROBLEM
   Given a connected undirected graph G and weights $c : E(G) \to \mathbb{R}$, find a minimum weight spanning tree in G.
   Here $E = E(G)$ and the elements of $\mathcal{F}$ are the edge sets of the forests in G.

(6) MAXIMUM WEIGHT FOREST PROBLEM

Given an undirected graph G and weights $c : E(G) \rightarrow \mathbb{R}$, find a maximum weight forest in G.

Here $(E, \mathcal{F})$ is as in (5).

(7) STEINER TREE PROBLEM

Given a connected undirected graph G, weights $c : E(G) \rightarrow \mathbb{R}_+$, and a set $T \subseteq V(G)$ of terminals, find a Steiner tree for T, i.e. a tree S with $T \subseteq V(S)$ and $E(S) \subseteq E(G)$, such that $c(E(S))$ is minimum.

Here $E = E(G)$ and $\mathcal{F}$ contains all subsets of edges of Steiner trees for T.

(8) MAXIMUM WEIGHT BRANCHING PROBLEM

Given a digraph G and weights $c : E(G) \rightarrow \mathbb{R}$, find a maximum weight branching in G.

Here $E = E(G)$ and $\mathcal{F}$ contains the edge sets of the branchings in G.

(9) MAXIMUM WEIGHT MATCHING PROBLEM

Given an undirected graph G and weights $c : E(G) \rightarrow \mathbb{R}$, find a maximum weight matching in G.

Here $E = E(G)$ and $\mathcal{F}$ is the set of matchings in G.

This list contains NP-hard problems ((1),(2),(4),(7)) as well as polynomially solvable problems ((5),(6),(8),(9)). Problem (3) is NP-hard in the above form but polynomially solvable for nonnegative weights. (NP-hardness is introduced in Chapter 15.)

Definition 13.3. *An independence system is a* **matroid** *if*

(M3) *If $X, Y \in \mathcal{F}$ and $|X| > |Y|$, then there is an $x \in X \setminus Y$ with $Y \cup \{x\} \in \mathcal{F}$.*

The name matroid points out that the structure is a generalization of matrices. This will become clear by our first example:

Proposition 13.4. *The following independence systems $(E, \mathcal{F})$ are matroids:*

(a) *E is the set of columns of a matrix A over some field, and*
   $\mathcal{F} := \{F \subseteq E : \text{The columns in } F \text{ are linearly independent over that field}\}$.
(b) *E is the set of edges of some undirected graph G and*
   $\mathcal{F} := \{F \subseteq E : (V(G), F) \text{ is a forest}\}$.
(c) *E is a finite set, k a nonnegative integer, and $\mathcal{F} := \{F \subseteq E : |F| \leq k\}$.*
(d) *E is the set of edges of some undirected graph G, S a stable set in G, $k_s \in \mathbb{Z}_+$ $(s \in S)$, and $\mathcal{F} := \{F \subseteq E : |\delta_F(s)| \leq k_s \text{ for all } s \in S\}$.*
(e) *E is the set of edges of some digraph G, $S \subseteq V(G)$, $k_s \in \mathbb{Z}_+$ $(s \in S)$, and $\mathcal{F} := \{F \subseteq E : |\delta_F^-(s)| \leq k_s \text{ for all } s \in S\}$.*

Proof: In all cases it is obvious that $(E, \mathcal{F})$ is indeed an independence system. So it remains to show that (M3) holds. For (a) this is well-known from linear algebra, for (c) it is trivial.

To prove (M3) for (b), let $X, Y \in \mathcal{F}$ and suppose $Y \cup \{x\} \notin \mathcal{F}$ for all $x \in X \setminus Y$. We show that $|X| \leq |Y|$. For each edge $x = \{v, w\} \in X$, v and w are in the same connected component of $(V(G), Y)$. Hence each connected component $Z \subseteq V(G)$ of $(V(G), X)$ is a subset of a connected component of $(V(G), Y)$. So the number p of connected components of the forest $(V(G), X)$ is greater than or equal to the number q of connected components of the forest $(V(G), Y)$. But then $|V(G)| - |X| = p \geq q = |V(G)| - |Y|$, implying $|X| \leq |Y|$.

To verify (M3) for (d), let $X, Y \in \mathcal{F}$ with $|X| > |Y|$. Let $S' := \{s \in S : |\delta_Y(s)| = k_s\}$. As $|X| > |Y|$ and $|\delta_X(s)| \leq k_s$ for all $s \in S'$, there exists an $e \in X \setminus Y$ with $e \notin \delta(s)$ for $s \in S'$. Then $Y \cup \{e\} \in \mathcal{F}$.

For (e) the proof is identical except for replacing δ by δ^-. $\square$

Some of these matroids have special names: The matroid in (a) is called the **vector matroid** of A. Let $\mathcal{M}$ be a matroid. If there is a matrix A over the field F such that $\mathcal{M}$ is the vector matroid of A, then $\mathcal{M}$ is called **representable over** F. There are matroids that are not representable over any field.

The matroid in (b) is called the **cycle matroid of** G and will sometimes be denoted by $\mathcal{M}(G)$. A matroid that is the cycle matroid of some graph, which may contain loops, is called a **graphic matroid**.

The matroids in (c) are called **uniform matroids**.

In our list of independence systems at the beginning of this section, the only matroids are the graphic matroids in (5) and (6). To check that all the other independence systems in the above list are not matroids in general is easily proved with the help of the following theorem (Exercise 1):

Theorem 13.5. *Let $(E, \mathcal{F})$ be an independence system. Then the following statements are equivalent:*

(M3) *If $X, Y \in \mathcal{F}$ and $|X| > |Y|$, then there is an $x \in X \setminus Y$ with $Y \cup \{x\} \in \mathcal{F}$.*
(M3') *If $X, Y \in \mathcal{F}$ and $|X| = |Y|+1$, then there is an $x \in X \setminus Y$ with $Y \cup \{x\} \in \mathcal{F}$.*
(M3'') *For each $X \subseteq E$, all bases of X have the same cardinality.*

Proof: Trivially, (M3)$\Leftrightarrow$(M3') and (M3)$\Rightarrow$(M3''). To prove (M3'')$\Rightarrow$(M3), let $X, Y \in \mathcal{F}$ and $|X| > |Y|$. By (M3''), Y cannot be a basis of $X \cup Y$. So there must be an $x \in (X \cup Y) \setminus Y = X \setminus Y$ such that $Y \cup \{x\} \in \mathcal{F}$. $\square$

Sometimes it is useful to have a second rank function:

Definition 13.6. *Let $(E, \mathcal{F})$ be an independence system. For $X \subseteq E$ we define the **lower rank** by*

$$\rho(X) := \min\{|Y| : Y \subseteq X, Y \in \mathcal{F} \text{ and } Y \cup \{x\} \notin \mathcal{F} \text{ for all } x \in X \setminus Y\}.$$

*The **rank quotient** of $(E, \mathcal{F})$ is defined by*

$$q(E, \mathcal{F}) := \min_{F \subseteq E} \frac{\rho(F)}{r(F)}.$$

Proposition 13.7. *Let $(E, \mathcal{F})$ be an independence system. Then $q(E, \mathcal{F}) \leq 1$. Furthermore, $(E, \mathcal{F})$ is a matroid if and only if $q(E, \mathcal{F}) = 1$.*

Proof: $q(E, \mathcal{F}) \leq 1$ follows from the definition. $q(E, \mathcal{F}) = 1$ is obviously equivalent to (M3″). □

To estimate the rank quotient, the following statement can be used:

Theorem 13.8. (Hausmann, Jenkyns and Korte [1980]) *Let $(E, \mathcal{F})$ be an independence system. If, for any $A \in \mathcal{F}$ and $e \in E$, $A \cup \{e\}$ contains at most p circuits, then $q(E, \mathcal{F}) \geq \frac{1}{p}$.*

Proof: Let $F \subseteq E$ and J, K two bases of F. We show $\frac{|J|}{|K|} \geq \frac{1}{p}$.
Let $J \setminus K = \{e_1, \ldots, e_t\}$. We construct a sequence $K = K_0, K_1, \ldots, K_t$ of independent subsets of $J \cup K$ such that $J \cap K \subseteq K_i$, $K_i \cap \{e_1, \ldots, e_t\} = \{e_1, \ldots, e_i\}$ and $|K_{i-1} \setminus K_i| \leq p$ for $i = 1, \ldots, t$.
Since $K_i \cup \{e_{i+1}\}$ contains at most p circuits and each such circuit must meet $K_i \setminus J$ (because J is independent), there is an $X \subseteq K_i \setminus J$ such that $|X| \leq p$ and $(K_i \setminus X) \cup \{e_{i+1}\} \in \mathcal{F}$. We set $K_{i+1} := (K_i \setminus X) \cup \{e_{i+1}\}$.
Now $J \subseteq K_t \in \mathcal{F}$. Since J is a basis of F, $J = K_t$. We conclude that

$$|K \setminus J| \;=\; \sum_{i=1}^{t} |K_{i-1} \setminus K_i| \;\leq\; pt \;=\; p \, |J \setminus K|,$$

proving $|K| \leq p \, |J|$. □

This shows that in example (9) we have $q(E, \mathcal{F}) \geq \frac{1}{2}$ (see also Exercise 1 of Chapter 10). In fact $q(E, \mathcal{F}) = \frac{1}{2}$ iff G contains a path of length 3 as a subgraph (otherwise $q(E, \mathcal{F}) = 1$). For the independence system in example (1) of our list, the rank quotient can become arbitrarily small (choose G to be a star). In Exercise 6, the rank quotients for other independence systems will be discussed.

13.2 Other Matroid Axioms

In this section we consider other axiom systems defining matroids. They characterize fundamental properties of the family of bases, the rank function, the closure operator and the family of circuits of a matroid.

Theorem 13.9. *Let E be a finite set and $\mathcal{B} \subseteq 2^E$. $\mathcal{B}$ is the set of bases of some matroid $(E, \mathcal{F})$ if and only if the following holds:*

(B1) $\mathcal{B} \neq \emptyset$;
(B2) *For any $B_1, B_2 \in \mathcal{B}$ and $x \in B_1 \setminus B_2$ there exists a $y \in B_2 \setminus B_1$ with $(B_1 \setminus \{x\}) \cup \{y\} \in \mathcal{B}$.*

Proof: The set of bases of a matroid satisfies (B1) (by (M1)) and (B2): For bases B_1, B_2 and $x \in B_1 \setminus B_2$ we have that $B_1 \setminus \{x\}$ is independent. By (M3) there is some $y \in B_2 \setminus B_1$ such that $(B_1 \setminus \{x\}) \cup \{y\}$ is independent. Indeed, it must be a basis, because all bases of a matroid have the same cardinality.

On the other hand, let B satisfy (B1) and (B2). We first show that all elements of B have the same cardinality: Otherwise let $B_1, B_2 \in B$ with $|B_1| > |B_2|$ such that $|B_1 \cap B_2|$ is maximum. Let $x \in B_1 \setminus B_2$. By (B2) there is a $y \in B_2 \setminus B_1$ with $(B_1 \setminus \{x\}) \cup \{y\} \in B$, contradicting the maximality of $|B_1 \cap B_2|$.

Now let

$$\mathcal{F} := \{F \subseteq E : \text{there exists a } B \in B \text{ with } F \subseteq B\}.$$

$(E, \mathcal{F})$ is an independence system, and B is the family of its bases. To show that $(E, \mathcal{F})$ satisfies (M3), let $X, Y \in \mathcal{F}$ with $|X| > |Y|$. Let $X \subseteq B_1 \in B$ and $Y \subseteq B_2 \in B$, where B_1 and B_2 are chosen such that $|B_1 \cap B_2|$ is maximum. If $B_2 \cap (X \setminus Y) \neq \emptyset$, we are done because we can augment Y.

We claim that the other case, $B_2 \cap (X \setminus Y) = \emptyset$, is impossible. Namely, with this assumption we get

$$|B_1 \cap B_2| + |Y \setminus B_1| + |(B_2 \setminus B_1) \setminus Y| = |B_2| = |B_1| \geq |B_1 \cap B_2| + |X \setminus Y|.$$

Since $|X \setminus Y| > |Y \setminus X| \geq |Y \setminus B_1|$, this implies $(B_2 \setminus B_1) \setminus Y \neq \emptyset$. So let $y \in (B_2 \setminus B_1) \setminus Y$. By (B2) there exists an $x \in B_1 \setminus B_2$ with $(B_2 \setminus \{y\}) \cup \{x\} \in B$, contradicting the maximality of $|B_1 \cap B_2|$. $\square$

See Exercise 8 for a similar statement. A very important property of matroids is that the rank function is submodular:

Theorem 13.10.   *Let E be a finite set and $r : 2^E \to \mathbb{Z}_+$. Then the following statements are equivalent:*

(a) *r is the rank function of a matroid $(E, \mathcal{F})$ (and $\mathcal{F} = \{F \subseteq E : r(F) = |F|\}$).*
(b) *For all $X, Y \subseteq E$:*
  (R1) *$r(X) \leq |X|$;*
  (R2) *If $X \subseteq Y$ then $r(X) \leq r(Y)$;*
  (R3) *$r(X \cup Y) + r(X \cap Y) \leq r(X) + r(Y)$.*
(c) *For all $X \subseteq E$ and $x, y \in E$:*
  (R1') *$r(\emptyset) = 0$;*
  (R2') *$r(X) \leq r(X \cup \{y\}) \leq r(X) + 1$;*
  (R3') *If $r(X \cup \{x\}) = r(X \cup \{y\}) = r(X)$ then $r(X \cup \{x, y\}) = r(X)$.*

Proof: (a)$\Rightarrow$(b): If r is a rank function of an independence system $(E, \mathcal{F})$, (R1) and (R2) evidently hold. If $(E, \mathcal{F})$ is a matroid, we can also show (R3):

Let $X, Y \subseteq E$, and let A be a basis of $X \cap Y$. By (M3), A can be extended to a basis $A \dot\cup B$ of X and to a basis $(A \cup B) \dot\cup C$ of $X \cup Y$. Then $A \cup C$ is an independent subset of Y, so

$$
\begin{aligned}
r(X) + r(Y) &\geq |A \cup B| + |A \cup C| \\
&= 2|A| + |B| + |C| \\
&= |A \cup B \cup C| + |A| \\
&= r(X \cup Y) + r(X \cap Y).
\end{aligned}
$$

(b)$\Rightarrow$(c): (R1$'$) is implied by (R1). $r(X) \leq r(X \cup \{y\})$ follows from (R2). By (R3) and (R1),

$$
r(X \cup \{y\}) \;\leq\; r(X) + r(\{y\}) - r(X \cap \{y\}) \;\leq\; r(X) + r(\{y\}) \;\leq\; r(X) + 1,
$$

proving (R2$'$).

(R3$'$) is trivial for $x = y$. For $x \neq y$ we have, by (R2) and (R3),

$$
2r(X) \;\leq\; r(X) + r(X \cup \{x, y\}) \;\leq\; r(X \cup \{x\}) + r(X \cup \{y\}),
$$

implying (R3$'$).

(c)$\Rightarrow$(a): Let $r : 2^E \to \mathbb{Z}_+$ be a function satisfying (R1$'$)–(R3$'$). Let

$$
\mathcal{F} := \{F \subseteq E : r(F) = |F|\}.
$$

We claim that $(E, \mathcal{F})$ is a matroid. (M1) follows from (R1$'$). (R2$'$) implies $r(X) \leq |X|$ for all $X \subseteq E$. If $Y \in \mathcal{F}$, $y \in Y$ and $X := Y \setminus \{y\}$, we have

$$
|X| + 1 \;=\; |Y| \;=\; r(Y) \;=\; r(X \cup \{y\}) \;\leq\; r(X) + 1 \;\leq\; |X| + 1,
$$

so $X \in \mathcal{F}$. This implies (M2).

Now let $X, Y \in \mathcal{F}$ and $|X| = |Y| + 1$. Let $X \setminus Y = \{x_1, \ldots, x_k\}$. Suppose that (M3$'$) is violated, i.e. $r(Y \cup \{x_i\}) = |Y|$ for $i = 1, \ldots, k$. Then by (R3$'$) $r(Y \cup \{x_1, x_i\}) = r(Y)$ for $i = 2, \ldots, k$. Repeated application of this argument yields $r(Y) = r(Y \cup \{x_1, \ldots, x_k\}) = r(X \cup Y) \geq r(X)$, a contradiction.

So $(E, \mathcal{F})$ is indeed a matroid. To show that r is the rank function of this matroid, we have to prove that $r(X) = \max\{|Y| : Y \subseteq X, r(Y) = |Y|\}$ for all $X \subseteq E$. So let $X \subseteq E$, and let Y a maximum subset of X with $r(Y) = |Y|$. For all $x \in X \setminus Y$ we have $r(Y \cup \{x\}) < |Y| + 1$, so by (R2$'$) $r(Y \cup \{x\}) = |Y|$. Repeated application of (R3$'$) implies $r(X) = |Y|$. $\square$

Theorem 13.11. *Let E be a finite set and $\sigma : 2^E \to 2^E$ a function. σ is the closure operator of a matroid $(E, \mathcal{F})$ if and only if the following conditions hold for all $X, Y \subseteq E$ and $x, y \in E$:*

(S1) $X \subseteq \sigma(X)$;

(S2) $X \subseteq Y \subseteq E$ *implies* $\sigma(X) \subseteq \sigma(Y)$;

(S3) $\sigma(X) = \sigma(\sigma(X))$;

(S4)   *If* $y \notin \sigma(X)$ *and* $y \in \sigma(X \cup \{x\})$ *then* $x \in \sigma(X \cup \{y\})$.

Proof: If σ is the closure operator of a matroid, then (S1) holds trivially.
For $X \subseteq Y$ and $z \in \sigma(X)$ we have by (R3) and (R2)

$$
\begin{aligned}
r(X) + r(Y) &= r(X \cup \{z\}) + r(Y) \\
&\geq r((X \cup \{z\}) \cap Y) + r(X \cup \{z\} \cup Y) \\
&\geq r(X) + r(Y \cup \{z\}),
\end{aligned}
$$

implying $z \in \sigma(Y)$ and thus proving (S2).

By repeated application of (R3′) we have $r(\sigma(X)) = r(X)$ for all X, which implies (S3).

To prove (S4), suppose that there are X, x, y with $y \notin \sigma(X)$, $y \in \sigma(X \cup \{x\})$ and $x \notin \sigma(X \cup \{y\})$. Then $r(X \cup \{y\}) = r(X) + 1$, $r(X \cup \{x, y\}) = r(X \cup \{x\})$ and $r(X \cup \{x, y\}) = r(X \cup \{y\}) + 1$. Thus $r(X \cup \{x\}) = r(X) + 2$, contradicting (R2′).

To show the converse, let $\sigma : 2^E \to 2^E$ be a function satisfying (S1)–(S4). Let

$$
\mathcal{F} := \{X \subseteq E : x \notin \sigma(X \setminus \{x\}) \text{ for all } x \in X\}.
$$

We claim that $(E, \mathcal{F})$ is a matroid.

(M1) is trivial. For $X \subseteq Y \in \mathcal{F}$ and $x \in X$ we have $x \notin \sigma(Y \setminus \{x\}) \supseteq \sigma(X \setminus \{x\})$, so $X \in \mathcal{F}$ and (M2) holds. To prove (M3) we need the following statement:

Claim: For $X \in \mathcal{F}$ and $Y \subseteq E$ with $|X| > |Y|$ we have $X \not\subseteq \sigma(Y)$.

We prove the claim by induction on $|Y \setminus X|$. If $Y \subset X$, then let $x \in X \setminus Y$. Since $X \in \mathcal{F}$ we have $x \notin \sigma(X \setminus \{x\}) \supseteq \sigma(Y)$ by (S2). Hence $x \in X \setminus \sigma(Y)$ as required.

If $|Y \setminus X| > 0$, then let $y \in Y \setminus X$. By the induction hypothesis there exists an $x \in X \setminus \sigma(Y \setminus \{y\})$. If $x \notin \sigma(Y)$, then we are done. Otherwise $x \notin \sigma(Y \setminus \{y\})$ but $x \in \sigma(Y) = \sigma((Y \setminus \{y\}) \cup \{y\})$, so by (S4) $y \in \sigma((Y \setminus \{y\}) \cup \{x\})$. By (S1) we get $Y \subseteq \sigma((Y \setminus \{y\}) \cup \{x\})$ and thus $\sigma(Y) \subseteq \sigma((Y \setminus \{y\}) \cup \{x\})$ by (S2) and (S3). Applying the induction hypothesis to X and $(Y \setminus \{y\}) \cup \{x\}$ (note that $x \neq y$) yields $X \not\subseteq \sigma((Y \setminus \{y\}) \cup \{x\})$, so $X \not\subseteq \sigma(Y)$ as required.

Having proved the claim we can easily verify (M3). Let $X, Y \in \mathcal{F}$ with $|X| > |Y|$. By the claim there exists an $x \in X \setminus \sigma(Y)$. Now for each $z \in Y \cup \{x\}$ we have $z \notin \sigma(Y \setminus \{z\})$, because $Y \in \mathcal{F}$ and $x \notin \sigma(Y) = \sigma(Y \setminus \{x\})$. By (S4) $z \notin \sigma(Y \setminus \{z\})$ and $x \notin \sigma(Y)$ imply $z \notin \sigma((Y \setminus \{z\}) \cup \{x\}) \supseteq \sigma((Y \cup \{x\}) \setminus \{z\})$. Hence $Y \cup \{x\} \in \mathcal{F}$.

So (M3) indeed holds and $(E, \mathcal{F})$ is a matroid, say with rank function r and closure operator σ'. It remains to prove that $\sigma = \sigma'$.

By definition, $\sigma'(X) = \{y \in E : r(X \cup \{y\}) = r(X)\}$ and

$$
r(X) = \max\{|Y| : Y \subseteq X, \, y \notin \sigma(Y \setminus \{y\}) \text{ for all } y \in Y\}
$$

for all $X \subseteq E$.

Let $X \subseteq E$. To show $\sigma'(X) \subseteq \sigma(X)$, let $z \in \sigma'(X) \setminus X$. Let Y be a basis of X. Since $r(Y \cup \{z\}) \leq r(X \cup \{z\}) = r(X) = |Y| < |Y \cup \{z\}|$ we have $y \in \sigma((Y \cup \{z\}) \setminus \{y\})$ for some $y \in Y \cup \{z\}$. If $y = z$, then we have $z \in \sigma(Y)$. Otherwise

(S4) and $y \notin \sigma(Y \setminus \{y\})$ also yield $z \in \sigma(Y)$. Hence by (S2) $z \in \sigma(X)$. Together with (S1) this implies $\sigma'(X) \subseteq \sigma(X)$.

Now let $z \notin \sigma'(X)$, i.e. $r(X \cup \{z\}) > r(X)$. Let now Y be a basis of $X \cup \{z\}$. Then $z \in Y$ and $|Y \setminus \{z\}| = |Y| - 1 = r(X \cup \{z\}) - 1 = r(X)$. Therefore $Y \setminus \{z\}$ is a basis of X, implying $X \subseteq \sigma'(Y \setminus \{z\}) \subseteq \sigma(Y \setminus \{z\})$, and thus $\sigma(X) \subseteq \sigma(Y \setminus \{z\})$. As $z \notin \sigma(Y \setminus \{z\})$, we conclude that $z \notin \sigma(X)$. □

Theorem 13.12. *Let E be a finite set and $\mathcal{C} \subseteq 2^E$. $\mathcal{C}$ is the set of circuits of an independence system $(E, \mathcal{F})$, where $\mathcal{F} = \{F \subset E : \text{there exists no } C \in \mathcal{C} \text{ with } C \subseteq F\}$, if and only if the following conditions hold:*

(C1) $\emptyset \notin \mathcal{C}$;
(C2) *For any $C_1, C_2 \in \mathcal{C}$, $C_1 \subseteq C_2$ implies $C_1 = C_2$.*

Moreover, if $\mathcal{C}$ is the set of circuits of an independence system $(E, \mathcal{F})$, then the following statements are equivalent:

(a) *$(E, \mathcal{F})$ is a matroid.*
(b) *For any $X \in \mathcal{F}$ and $e \in E$, $X \cup \{e\}$ contains at most one circuit.*
(C3) *For any $C_1, C_2 \in \mathcal{C}$ with $C_1 \neq C_2$ and $e \in C_1 \cap C_2$ there exists a $C_3 \in \mathcal{C}$ with $C_3 \subseteq (C_1 \cup C_2) \setminus \{e\}$.*
(C3′) *For any $C_1, C_2 \in \mathcal{C}$, $e \in C_1 \cap C_2$ and $f \in C_1 \setminus C_2$ there exists a $C_3 \in \mathcal{C}$ with $f \in C_3 \subseteq (C_1 \cup C_2) \setminus \{e\}$.*

Proof: By definition, the family of circuits of any independence system satisfies (C1) and (C2). If $\mathcal{C}$ satisfies (C1), then $(E, \mathcal{F})$ is an independence system. If $\mathcal{C}$ also satisfies (C2), it is the set of circuits of this independence system.

(a)⇒(C3′): Let $\mathcal{C}$ be the family of circuits of a matroid, and let $C_1, C_2 \in \mathcal{C}$, $e \in C_1 \cap C_2$ and $f \in C_1 \setminus C_2$. By applying (R3) twice we have

$$
\begin{aligned}
&|C_1| - 1 + r((C_1 \cup C_2) \setminus \{e, f\}) + |C_2| - 1 \\
=\ & r(C_1) + r((C_1 \cup C_2) \setminus \{e, f\}) + r(C_2) \\
\geq\ & r(C_1) + r((C_1 \cup C_2) \setminus \{f\}) + r(C_2 \setminus \{e\}) \\
\geq\ & r(C_1 \setminus \{f\}) + r(C_1 \cup C_2) + r(C_2 \setminus \{e\}) \\
=\ & |C_1| - 1 + r(C_1 \cup C_2) + |C_2| - 1.
\end{aligned}
$$

So $r((C_1 \cup C_2) \setminus \{e, f\}) = r(C_1 \cup C_2)$. Let B be a basis of $(C_1 \cup C_2) \setminus \{e, f\}$. Then $B \cup \{f\}$ contains a circuit C_3, with $f \in C_3 \subseteq (C_1 \cup C_2) \setminus \{e\}$ as required.

(C3′)⇒(C3): trivial.

(C3)⇒(b): If $X \in \mathcal{F}$ and $X \cup \{e\}$ contains two circuits C_1, C_2, (C3) implies $(C_1 \cup C_2) \setminus \{e\} \notin \mathcal{F}$. However, $(C_1 \cup C_2) \setminus \{e\}$ is a subset of X.

(b)⇒(a): Follows from Theorem 13.8 and Proposition 13.7. □

Especially property (b) will be used often. For $X \in \mathcal{F}$ and $e \in E$ such that $X \cup \{e\} \notin \mathcal{F}$ we write $C(X, e)$ for the unique circuit in $X \cup \{e\}$. If $X \cup \{e\} \in \mathcal{F}$ we write $C(X, e) := \emptyset$.

13.3 Duality

Another basic concept in matroid theory is duality.

Definition 13.13. *Let $(E, \mathcal{F})$ be an independence system. We define the **dual** of $(E, \mathcal{F})$ by $(E, \mathcal{F}^*)$, where*

$$\mathcal{F}^* = \{F \subseteq E : \text{ there is a basis } B \text{ of } (E, \mathcal{F}) \text{ such that } F \cap B = \emptyset\}.$$

It is obvious that the dual of an independence system is again an independence system.

Proposition 13.14. $(E, \mathcal{F}^{**}) = (E, \mathcal{F}).$

Proof: $F \in \mathcal{F}^{**} \Leftrightarrow$ there is a basis B^* of $(E, \mathcal{F}^*)$ such that $F \cap B^* = \emptyset \Leftrightarrow$ there is a basis B of $(E, \mathcal{F})$ such that $F \cap (E \setminus B) = \emptyset \Leftrightarrow F \in \mathcal{F}$. $\quad\square$

Theorem 13.15. *Let $(E, \mathcal{F})$ be an independence system, $(E, \mathcal{F}^*)$ its dual, and let r and r^* be the corresponding rank functions.*

(a) $(E, \mathcal{F})$ *is a matroid if and only if $(E, \mathcal{F}^*)$ is a matroid.* (Whitney [1935])
(b) *If $(E, \mathcal{F})$ is a matroid, then $r^*(F) = |F| + r(E \setminus F) - r(E)$ for $F \subseteq E$.*

Proof: Due to Proposition 13.14 we have to show only one direction of (a). So let $(E, \mathcal{F})$ be a matroid. We define $q : 2^E \to \mathbb{Z}_+$ by $q(F) := |F| + r(E \setminus F) - r(E)$. We claim that q satisfies (R1), (R2) and (R3). By this claim and Theorem 13.10, q is the rank function of a matroid. Since obviously $q(F) = |F|$ if and only if $F \in \mathcal{F}^*$, we conclude that $q = r^*$, and (a) and (b) are proved.

Now we prove the above claim: q satisfies (R1) because r satisfies (R2). To check that q satisfies (R2), let $X \subseteq Y \subseteq E$. Since $(E, \mathcal{F})$ is a matroid, (R3) holds for r, so

$$r(E \setminus X) + 0 = r((E \setminus Y) \cup (Y \setminus X)) + r(\emptyset) \leq r(E \setminus Y) + r(Y \setminus X).$$

We conclude that

$$r(E \setminus X) - r(E \setminus Y) \leq r(Y \setminus X) \leq |Y \setminus X| = |Y| - |X|$$

(note that r satisfies (R1)), so $q(X) \leq q(Y)$.

It remains to show that q satisfies (R3). Let $X, Y \subseteq E$. Using the fact that r satisfies (R3) we have

$$
\begin{aligned}
& q(X \cup Y) + q(X \cap Y) \\
= {} & |X \cup Y| + |X \cap Y| + r(E \setminus (X \cup Y)) + r(E \setminus (X \cap Y)) - 2r(E) \\
= {} & |X| + |Y| + r((E \setminus X) \cap (E \setminus Y)) + r((E \setminus X) \cup (E \setminus Y)) - 2r(E) \\
\leq {} & |X| + |Y| + r(E \setminus X) + r(E \setminus Y) - 2r(E) \\
= {} & q(X) + q(Y).
\end{aligned}
$$

$\quad\square$

For any graph G we have introduced the cycle matroid $\mathcal{M}(G)$ which of course has a dual. For an embedded planar graph G there is also a planar dual G^* (which in general depends on the embedding of G). It is interesting that the two concepts of duality coincide:

Theorem 13.16. *Let G be a connected planar graph with an arbitrary planar embedding, and G^* the planar dual. Then*

$$\mathcal{M}(G^*) = (\mathcal{M}(G))^* .$$

(Formally, the two matroids are isomorphic: they are identical up to renaming the elements of the ground set.)

Proof: For $T \subseteq E(G)$ we write $\overline{T}^* := \{e^* : e \in E(G) \setminus T\}$, where e^* is the dual of edge e. We have to prove the following:

Claim: T is the edge set of a spanning tree in G iff $\overline{T}^*$ is the edge set of a spanning tree in G^*.

Since $(G^*)^* = G$ (by Proposition 2.42) and $\overline{(\overline{T}^*)}^* = T$ it suffices to prove one direction of the claim.

So let $T \subseteq E(G)$, where $\overline{T}^*$ is the edge set of a spanning tree in G^*. $(V(G), T)$ must be connected, for otherwise a connected component would define a cut, the dual of which contains a circuit in $\overline{T}^*$ (Theorem 2.43). On the other hand, if $(V(G), T)$ contains a circuit, then the dual edge set is a cut and $(V(G^*), \overline{T}^*)$ is disconnected. Hence $(V(G), T)$ is indeed a spanning tree in G. □

This implies that if G is planar then $(\mathcal{M}(G))^*$ is a graphic matroid. If, for any graph G, $(\mathcal{M}(G))^*$ is a graphic matroid, say $(\mathcal{M}(G))^* = \mathcal{M}(G')$, then G' is evidently an abstract dual of G. By Exercise 40 of Chapter 2, the converse is also true: G is planar if and only if G has an abstract dual (Whitney [1933]). This implies that $(\mathcal{M}(G))^*$ is graphic if and only if G is planar.

Note that Theorem 13.16 quite directly implies Euler's formula (Theorem 2.32): Let G be a connected planar graph with a planar embedding, and let $\mathcal{M}(G)$ be the cycle matroid of G. By Theorem 13.15 (b), $r(E(G)) + r^*(E(G)) = |E(G)|$. Since $r(E(G)) = |V(G)| - 1$ (the number of edges in a spanning tree) and $r^*(E(G)) = |V(G^*)| - 1$ (by Theorem 13.16), we obtain that the number of faces of G is $|V(G^*)| = |E(G)| - |V(G)| + 2$, Euler's formula. (However, we used Euler's formula in the proof of Proposition 2.42, which was used in our proof of Theorem 13.16.)

Duality of independence systems has also some nice applications in polyhedral combinatorics. A set system $(E, \mathcal{F})$ is called a **clutter** if $X \not\subseteq Y$ for all $X, Y \in \mathcal{F}$. If $(E, \mathcal{F})$ is a clutter, then we define its **blocking clutter** by

$$BL(E, \mathcal{F}) := (E, \{X \subseteq E : X \cap Y \neq \emptyset \text{ for all } Y \in \mathcal{F},$$
$$X \text{ minimal with this property}\}).$$

For an independence system $(E, \mathcal{F})$ and its dual $(E, \mathcal{F}^*)$ let $\mathcal{B}$ and $\mathcal{B}^*$ be the family of bases, and $\mathcal{C}$ and $\mathcal{C}^*$ the family of circuits, respectively. (Every clutter arises in both of these ways except for $\mathcal{F} = \emptyset$ or $\mathcal{F} = \{\emptyset\}$.) It follows immediately from the definitions that $(E, \mathcal{B}^*) = BL(E, \mathcal{C})$ and $(E, \mathcal{C}^*) = BL(E, \mathcal{B})$. Together with Proposition 13.14 this implies $BL(BL(E, \mathcal{F})) = (E, \mathcal{F})$ for every clutter $(E, \mathcal{F})$. We give some examples for clutters $(E, \mathcal{F})$ and their blocking clutters $(E, \mathcal{F}')$. In each case $E = E(G)$ for some graph G:

(1) $\mathcal{F}$ is the set of spanning trees, $\mathcal{F}'$ is the set of minimal cuts;
(2) $\mathcal{F}$ is the set of arborescences rooted at r, $\mathcal{F}'$ is the set of minimal r-cuts;
(3) $\mathcal{F}$ is the set of s-t-paths, $\mathcal{F}'$ is the set of minimal cuts separating s and t (this example works in undirected graphs and in digraphs);
(4) $\mathcal{F}$ is the set of circuits in an undirected graph, $\mathcal{F}'$ is the set of complements of maximal forests;
(5) $\mathcal{F}$ is the set of circuits in a digraph, $\mathcal{F}'$ is the set of minimal feedback edge sets (a feedback edge set is a set of edges whose removal makes the digraph acyclic);
(6) $\mathcal{F}$ is the set of minimal edge sets whose contraction makes the digraph strongly connected, $\mathcal{F}'$ is the set of minimal directed cuts;
(7) $\mathcal{F}$ is the set of minimal T-joins, $\mathcal{F}'$ is the set of minimal T-cuts.

All these blocking relations can be verified easily: (1) and (2) follow directly from Theorems 2.4 and 2.5, (3), (4) and (5) are trivial, (6) follows from Corollary 2.7, and (7) from Proposition 12.7.

In some cases, the blocking clutter gives a polyhedral characterization of the MINIMIZATION PROBLEM FOR INDEPENDENCE SYSTEMS for nonnegative cost functions:

Definition 13.17. *Let* $(E, \mathcal{F})$ *be a clutter,* $(E, \mathcal{F}')$ *its blocking clutter and* P *the convex hull of the incidence vectors of the elements of* $\mathcal{F}$. *We say that* $(E, \mathcal{F})$ *has the* **Max-Flow-Min-Cut property** *if*

$$\left\{ x + y : x \in P, \ y \in \mathbb{R}_+^E \right\} = \left\{ x \in \mathbb{R}_+^E : \sum_{e \in B} x_e \geq 1 \text{ for all } B \in \mathcal{F}' \right\}.$$

Examples are (2) and (7) of our list above (by Theorems 6.15 and 12.18), but also (3) and (6) (see Exercise 11). The following theorem relates the above covering-type formulation to a packing formulation of the dual problem and allows to derive certain min-max theorems from others:

Theorem 13.18. (Fulkerson [1971], Lehman [1979]) *Let* $(E, \mathcal{F})$ *be a clutter and* $(E, \mathcal{F}')$ *its blocking clutter. Then the following statements are equivalent:*

(a) $(E, \mathcal{F})$ *has the Max-Flow-Min-Cut property;*
(b) $(E, \mathcal{F}')$ *has the Max-Flow-Min-Cut property;*
(c) $\min\{c(A) : A \in \mathcal{F}\} = \max\{\mathbb{1}y : y \in \mathbb{R}_+^{\mathcal{F}'}, \sum_{B \in \mathcal{F}':e \in B} y_B \leq c(e)$
    for all $e \in E\}$ *for every* $c : E \to \mathbb{R}_+$.

Proof: Since $BL(E, \mathcal{F}') = BL(BL(E, \mathcal{F})) = (E, \mathcal{F})$ it suffices to prove (a)$\Rightarrow$(c)$\Rightarrow$(b). The other implication (b)$\Rightarrow$(a) then follows by exchanging the roles of $\mathcal{F}$ and $\mathcal{F}'$.

(a)$\Rightarrow$(c): By Corollary 3.33 we have for every $c : E \to \mathbb{R}_+$

$$\min\{c(A) : A \in \mathcal{F}\} = \min\{cx : x \in P\} = \min\left\{c(x + y) : x \in P, \ y \in \mathbb{R}_+^E\right\},$$

where P is the convex hull of the incidence vectors of elements of $\mathcal{F}$. From this, the Max-Flow-Min-Cut property and the LP Duality Theorem 3.20 we get (c).

(c)$\Rightarrow$(b): Let P' denote the convex hull of the incidence vectors of the elements of $\mathcal{F}'$. We have to show that

$$\left\{x + y : x \in P', \ y \in \mathbb{R}_+^E\right\} = \left\{x \in \mathbb{R}_+^E : \sum_{e \in A} x_e \geq 1 \text{ for all } A \in \mathcal{F}\right\}.$$

Since "$\subseteq$" is trivial from the definition of blocking clutters we only show the other inclusion. So let $c \in \mathbb{R}_+^E$ be a vector with $\sum_{e \in A} c_e \geq 1$ for all $A \in \mathcal{F}$. By (c) we have

$$
\begin{aligned}
1 &\leq \min\{c(A) : A \in \mathcal{F}\} \\
&= \max\left\{\mathbb{1}y : y \in \mathbb{R}_+^{\mathcal{F}'}, \sum_{B \in \mathcal{F}' : e \in B} y_B \leq c(e) \text{ for all } e \in E\right\},
\end{aligned}
$$

so let $y \in \mathbb{R}_+^{\mathcal{F}'}$ be a vector with $\mathbb{1}y = 1$ and $\sum_{B \in \mathcal{F}' : e \in B} y_B \leq c(e)$ for all $e \in E$. Then $x_e := \sum_{B \in \mathcal{F}' : e \in B} y_B$ $(e \in E)$ defines a vector $x \in P'$ with $x \leq c$, proving that $c \in \left\{x + y : x \in P', \ y \in \mathbb{R}_+^E\right\}$. $\square$

For example, this theorem implies the Max-Flow-Min-Cut Theorem 8.6 quite directly: Let (G, u, s, t) be a network. By Exercise 1 of Chapter 7 the minimum length of an s-t-path in (G, u) equals the maximum number of s-t-cuts such that each edge e is contained in at most $u(e)$ of them. Hence the clutter of s-t-paths (example (3) in the above list) has the Max-Flow-Min-Cut Property, and so has its blocking clutter. Now (c) applied to the clutter of minimal s-t-cuts implies the Max-Flow-Min-Cut Theorem.

Note however that Theorem 13.18 does not guarantee an integral vector attaining the maximum in (c), even if c is integral. The clutter of T-joins for $G = K_4$ and $T = V(G)$ shows that this does not exist in general.

13.4 The Greedy Algorithm

Again, let $(E, \mathcal{F})$ be an independence system and $c : E \to \mathbb{R}_+$. We consider the MAXIMIZATION PROBLEM for $(E, \mathcal{F}, c)$ and formulate two "greedy algorithms". We do not have to consider negative weights since elements with negative weight never appear in an optimum solution.

We assume that $(E, \mathcal{F})$ is given by an oracle. For the first algorithm we simply assume an **independence oracle**, i.e. an oracle which, given a set $F \subseteq E$, decides whether $F \in \mathcal{F}$ or not.

BEST-IN-GREEDY ALGORITHM

Input: An independence system $(E, \mathcal{F})$, given by an independence oracle.
          Weights $c : E \rightarrow \mathbb{R}_+$.

Output: A set $F \in \mathcal{F}$.

---

①   Sort $E = \{e_1, e_2, \ldots, e_n\}$ such that $c(e_1) \geq c(e_2) \geq \cdots \geq c(e_n)$.

②   Set $F := \emptyset$.

③   **For** $i := 1$ **to** n **do: If** $F \cup \{e_i\} \in \mathcal{F}$ **then** set $F := F \cup \{e_i\}$.

The second algorithm requires a more complicated oracle. Given a set $F \subseteq E$, this oracle decides whether F contains a basis. Let us call such an oracle a **basis-superset oracle**.

WORST-OUT-GREEDY ALGORITHM

Input: An independence system $(E, \mathcal{F})$, given by a basis-superset oracle.
          Weights $c : E \rightarrow \mathbb{R}_+$.

Output: A basis F of $(E, \mathcal{F})$.

---

①   Sort $E = \{e_1, e_2, \ldots, e_n\}$ such that $c(e_1) \leq c(e_2) \leq \cdots \leq c(e_n)$.

②   Set $F := E$.

③   **For** $i := 1$ **to** n **do: If** $F \setminus \{e_i\}$ contains a basis **then** set $F := F \setminus \{e_i\}$.

Before we analyse these algorithms, let us take a closer look at the oracles required. It is an interesting question whether such oracles are polynomially equivalent, i.e. whether one can be simulated by a polynomial-time oracle algorithm using the other. The independence oracle and the basis-superset oracle do not seem to be polynomially equivalent:

If we consider the independence system for the TSP (example (2) of the list in Section 13.1), it is easy (and the subject of Exercise 14) to decide whether a set of edges is independent, i.e. the subset of a Hamiltonian circuit (recall that we are working with a complete graph). On the other hand, it is a difficult problem to decide whether a set of edges contains a Hamiltonian circuit (this is *NP*-complete; cf. Theorem 15.25).

Conversely, in the independence system for the SHORTEST PATH PROBLEM (example (3)), it is easy to decide whether a set of edges contains an s-t-path. Here it is not known how to decide whether a given set is independent (i.e. subset of an s-t-path) in polynomial time (Korte and Monma [1979] proved *NP*-completeness).

For matroids, both oracles are polynomially equivalent. Other equivalent oracles are the **rank oracle** and **closure oracle**, which return the rank and the closure of a given subset of E, respectively (Exercise 17).

However, even for matroids there are other natural oracles that are not polynomially equivalent. For example, the oracle deciding whether a given set is a basis is weaker than the independence oracle. The oracle which for a given $F \subseteq E$ returns the minimum cardinality of a dependent subset of F is stronger than the independence oracle (Hausmann and Korte [1981]).

One can analogously formulate both greedy algorithms for the MINIMIZATION PROBLEM. It is easy to see that the BEST-IN-GREEDY for the MAXIMIZATION PROBLEM for $(E, \mathcal{F}, c)$ corresponds to the WORST-OUT-GREEDY for the MINIMIZATION PROBLEM for $(E, \mathcal{F}^*, c)$: adding an element to F in the BEST-IN-GREEDY corresponds to removing an element from F in the WORST-OUT-GREEDY. Observe that KRUSKAL'S ALGORITHM (see Section 6.1) is a BEST-IN-GREEDY algorithm for the MINIMIZATION PROBLEM in a cycle matroid.

The rest of this section contains some results concerning the quality of a solution found by the greedy algorithms.

Theorem 13.19.   (Jenkyns [1976], Korte and Hausmann [1978]) *Let $(E, \mathcal{F})$ be an independence system. For $c : E \to \mathbb{R}_+$ we denote by $G(E, \mathcal{F}, c)$ the cost of some solution found by the* BEST-IN-GREEDY *for the* MAXIMIZATION PROBLEM, *and by* $\mathrm{OPT}(E, \mathcal{F}, c)$ *the cost of an optimum solution. Then*

$$q(E, \mathcal{F}) \le \frac{G(E, \mathcal{F}, c)}{\mathrm{OPT}(E, \mathcal{F}, c)} \le 1$$

for all $c : E \to \mathbb{R}_+$. There is a cost function where the lower bound is attained.

Proof: Let $E = \{e_1, e_2, \ldots, e_n\}$, $c : E \to \mathbb{R}_+$, and $c(e_1) \ge c(e_2) \ge \ldots \ge c(e_n)$. Let G_n be the solution found by the BEST-IN-GREEDY (when sorting E like this), while O_n is an optimum solution. We define $E_j := \{e_1, \ldots, e_j\}$, $G_j := G_n \cap E_j$ and $O_j := O_n \cap E_j$ $(j = 0, \ldots, n)$. Set $d_n := c(e_n)$ and $d_j := c(e_j) - c(e_{j+1})$ for $j = 1, \ldots, n - 1$.

Since $O_j \in \mathcal{F}$, we have $|O_j| \le r(E_j)$. Since G_j is a basis of E_j, we have $|G_j| \ge \rho(E_j)$. With these two inequalities we conclude that

$$
\begin{aligned}
c(G_n) &= \sum_{j=1}^{n} (|G_j| - |G_{j-1}|)\, c(e_j) \\
&= \sum_{j=1}^{n} |G_j|\, d_j \\
&\ge \sum_{j=1}^{n} \rho(E_j)\, d_j \\
&\ge q(E, \mathcal{F}) \sum_{j=1}^{n} r(E_j)\, d_j
\end{aligned}
\tag{13.1}
$$

$$\geq q(E, \mathcal{F}) \sum_{j=1}^{n} |O_j| d_j$$

$$= q(E, \mathcal{F}) \sum_{j=1}^{n} (|O_j| - |O_{j-1}|) c(e_j)$$

$$= q(E, \mathcal{F}) c(O_n).$$

Finally we show that the lower bound is tight. Choose $F \subseteq E$ and bases B_1, B_2 of F such that

$$\frac{|B_1|}{|B_2|} = q(E, \mathcal{F}).$$

Define

$$c(e) := \begin{cases} 1 & \text{for } e \in F \\ 0 & \text{for } e \in E \setminus F \end{cases}$$

and sort $e_1, \ldots, e_n$ such that $c(e_1) \geq c(e_2) \geq \ldots \geq c(e_n)$ and $B_1 = \{e_1, \ldots, e_{|B_1|}\}$. Then $G(E, \mathcal{F}, c) = |B_1|$ and $\text{OPT}(E, \mathcal{F}, c) = |B_2|$, and the lower bound is attained. $\square$

In particular we have the so-called Edmonds-Rado Theorem:

Theorem 13.20. (Rado [1957], Edmonds [1971]) *An independence system* $(E, \mathcal{F})$ *is a matroid if and only if the* BEST-IN-GREEDY *finds an optimum solution for the* MAXIMIZATION PROBLEM *for* $(E, \mathcal{F}, c)$ *for all cost functions* $c : E \to \mathbb{R}_+$.

Proof: By Theorem 13.19 we have $q(E, \mathcal{F}) < 1$ if and only if there exists a cost function $c : E \to \mathbb{R}_+$ for which the BEST-IN-GREEDY does not find an optimum solution. By Proposition 13.7 we have $q(E, \mathcal{F}) < 1$ if and only if $(E, \mathcal{F})$ is not a matroid. $\square$

This is one of the rare cases where we can define a structure by its algorithmic behaviour. We also obtain a polyhedral description:

Theorem 13.21. (Edmonds [1970]) *Let* $(E, \mathcal{F})$ *be a matroid and* $r : 2^E \to \mathbb{Z}_+$ *its rank function. Then the* **matroid polytope** *of* $(E, \mathcal{F})$, *i.e. the convex hull of the incidence vectors of all elements of* $\mathcal{F}$, *is equal to*

$$\left\{ x \in \mathbb{R}^E : x \geq 0, \sum_{e \in A} x_e \leq r(A) \text{ for all } A \subseteq E \right\}.$$

Proof: Obviously, this polytope contains all incidence vectors of independent sets. By Corollary 3.32 it remains to show that all vertices of this polytope are integral. By Theorem 5.14 this is equivalent to showing that

$$\max \left\{ cx : x \geq 0, \sum_{e \in A} x_e \leq r(A) \text{ for all } A \subseteq E \right\} \tag{13.2}$$

has an integral optimum solution for any $c : E \to \mathbb{R}$. W.l.o.g. $c(e) \geq 0$ for all e, since for $e \in E$ with $c(e) < 0$ any optimum solution x of (13.2) has $x_e = 0$.

Let x be an optimum solution of (13.2). In (13.1) we replace $|O_j|$ by $\sum_{e \in E_j} x_e$ ($j = 0, \ldots, n$). We obtain $c(G_n) \geq \sum_{e \in E} c(e) x_e$. So the BEST-IN-GREEDY produces a solution whose incidence vector is another optimum solution of (13.2). $\square$

When applied to graphic matroids, this also yields Theorem 6.13. As in this special case, we also have total dual integrality in general. A generalization of this result will be proved in Section 14.2.

The above observation that the BEST-IN-GREEDY for the MAXIMIZATION PROBLEM for $(E, \mathcal{F}, c)$ corresponds to the WORST-OUT-GREEDY for the MINIMIZATION PROBLEM for $(E, \mathcal{F}^*, c)$ suggests the following dual counterpart of Theorem 13.19:

Theorem 13.22. (Korte and Monma [1979]) *Let $(E, \mathcal{F})$ be an independence system. For $c : E \to \mathbb{R}_+$ let $G(E, \mathcal{F}, c)$ denote a solution found by the* WORST-OUT-GREEDY *for the* MINIMIZATION PROBLEM. *Then*

$$1 \leq \frac{G(E, \mathcal{F}, c)}{\text{OPT}(E, \mathcal{F}, c)} \leq \max_{F \subseteq E} \frac{|F| - \rho^*(F)}{|F| - r^*(F)} \tag{13.3}$$

for all $c : E \to \mathbb{R}_+$, where ρ^* and r^* are the rank functions of the dual independence system $(E, \mathcal{F}^*)$. There is a cost function where the upper bound is attained.

Proof: We use the same notation as in the proof of Theorem 13.19. By construction, $G_j \cup (E \setminus E_j)$ contains a basis of E, but $(G_j \cup (E \setminus E_j)) \setminus \{e\}$ does not contain a basis of E for any $e \in G_j$ ($j = 1, \ldots, n$). In other words, $E_j \setminus G_j$ is a basis of E_j with respect to $(E, \mathcal{F}^*)$, so $|E_j| - |G_j| \geq \rho^*(E_j)$.

Since $O_n \subseteq E \setminus (E_j \setminus O_j)$ and O_n is a basis, $E_j \setminus O_j$ is independent in $(E, \mathcal{F}^*)$, so $|E_j| - |O_j| \leq r^*(E_j)$.

We conclude that

$$\begin{aligned} |G_j| &\leq |E_j| - \rho^*(E_j) \qquad \text{and} \\ |O_j| &\geq |E_j| - r^*(E_j). \end{aligned}$$

Now the same calculation as (13.1) provides the upper bound. To see that this bound is tight, consider

$$c(e) := \begin{cases} 1 & \text{for } e \in F \\ 0 & \text{for } e \in E \setminus F \end{cases},$$

where $F \subseteq E$ is a set where the maximum in (13.3) is attained. Let B_1 be a basis of F with respect to $(E, \mathcal{F}^*)$, with $|B_1| = \rho^*(F)$. If we sort $e_1, \ldots, e_n$ such that $c(e_1) \geq c(e_2) \geq \ldots \geq c(e_n)$ and $B_1 = \{e_1, \ldots, e_{|B_1|}\}$, we have $G(E, \mathcal{F}, c) = |F| - |B_1|$ and $\text{OPT}(E, \mathcal{F}, c) = |F| - r^*(F)$. $\square$

Fig. 13.1.

If we apply the WORST-OUT-GREEDY to the MAXIMIZATION PROBLEM or the BEST-IN-GREEDY to the MINIMIZATION PROBLEM, there is no positive lower/finite upper bound for $\frac{G(E,\mathcal{F},c)}{\text{OPT}(E,\mathcal{F},c)}$. To see this, consider the problem of finding a minimal vertex cover of maximum weight or a maximal stable set of minimum weight in the simple graph shown in Figure 13.1.

However in the case of matroids, it does not matter whether we use the BEST-IN-GREEDY or the WORST-OUT-GREEDY: since all bases have the same cardinality, the MINIMIZATION PROBLEM for $(E, \mathcal{F}, c)$ is equivalent to the MAXIMIZATION PROBLEM for $(E, \mathcal{F}, c')$, where $c'(e) := M - c(e)$ for all $e \in E$ and $M := 1 + \max\{c(e) : e \in E\}$. Therefore KRUSKAL'S ALGORITHM (Section 6.1) solves the MINIMUM SPANNING TREE PROBLEM optimally.

The Edmonds-Rado Theorem 13.20 also yields the following characterization of optimum k-element solutions of the MAXIMIZATION PROBLEM.

Theorem 13.23. *Let $(E, \mathcal{F})$ be a matroid, $c : E \to \mathbb{R}$, $k \in \mathbb{N}$ and $X \in \mathcal{F}$ with $|X| = k$. Then $c(X) = \max\{c(Y) : Y \in \mathcal{F}, |Y| = k\}$ if and only if the following two conditions hold:*

(a) *For all $y \in E \setminus X$ with $X \cup \{y\} \notin \mathcal{F}$ and all $x \in C(X, y)$ we have $c(x) \geq c(y)$;*
(b) *For all $y \in E \setminus X$ with $X \cup \{y\} \in \mathcal{F}$ and all $x \in X$ we have $c(x) \geq c(y)$.*

Proof: The necessity is trivial: if one of the conditions is violated for some y and x, the k-element set $X' := (X \cup \{y\}) \setminus \{x\} \in \mathcal{F}$ has greater weight than X.

To see the sufficiency, let $\mathcal{F}' := \{F \in \mathcal{F} : |F| \leq k\}$ and $c'(e) := c(e) + M$ for all $e \in E$, where $M = \max\{|c(e)| : e \in E\}$. Sort $E = \{e_1, \ldots, e_n\}$ such that $c'(e_1) \geq \cdots \geq c'(e_n)$ and, for any i, $c'(e_i) = c'(e_{i+1})$ and $e_{i+1} \in X$ imply $e_i \in X$ (i.e. elements of X come first among those of equal weight).

Let X' be the solution found by the BEST-IN-GREEDY for the instance $(E, \mathcal{F}', c')$ where ① sorts the elements as above. Since $(E, \mathcal{F})$ is a matroid, the Edmonds-Rado Theorem 13.20 implies:

$$c(X') + kM = c'(X') = \max\{c'(Y) : Y \in \mathcal{F}'\}$$
$$= \max\{c(Y) : Y \in \mathcal{F}, |Y| = k\} + kM.$$

We conclude the proof by showing that $X = X'$. We know that $|X| = k = |X'|$. So suppose $X \neq X'$, and let $e_i \in X' \setminus X$ with i minimum. Then $X \cap \{e_1, \ldots, e_{i-1}\} = X' \cap \{e_1, \ldots, e_{i-1}\}$. Now if $X \cup \{e_i\} \notin \mathcal{F}$, then (a) implies $C(X, e_i) \subseteq X'$, a contradiction. If $X \cup \{e_i\} \in \mathcal{F}$, then (b) implies $X \subseteq X'$ which is also impossible. $\square$

We shall need this theorem in Section 13.7. The special case that $(E, \mathcal{F})$ is a graphic matroid and $k = r(E)$ is part of Theorem 6.3.

13.5 Matroid Intersection

Definition 13.24. *Given two independence systems* $(E, \mathcal{F}_1)$ *and* $(E, \mathcal{F}_2)$, *we define their* **intersection** *by* $(E, \mathcal{F}_1 \cap \mathcal{F}_2)$.

The intersection of a finite number of independence systems is defined analogously. It is clear that the result is again an independence system.

Proposition 13.25. *Any independence system* $(E, \mathcal{F})$ *is the intersection of a finite number of matroids.*

Proof: Each circuit C of $(E, \mathcal{F})$ defines a matroid $(E, \{F \subseteq E : C \setminus F \neq \emptyset\})$ by Theorem 13.12 (this also follows easily from the definition). The intersection of all these matroids is of course $(E, \mathcal{F})$. $\square$

Since the intersection of matroids is not a matroid in general, we cannot hope to get an optimum common independent set by a greedy algorithm. However, the following result, together with Theorem 13.19, implies a bound for the solution found by the BEST-IN-GREEDY:

Proposition 13.26. *If* $(E, \mathcal{F})$ *is the intersection of* p *matroids, then* $q(E, \mathcal{F}) \geq \frac{1}{p}$.

Proof: By Theorem 13.12(b), $X \cup \{e\}$ contains at most p circuits for any $X \in \mathcal{F}$ and $e \in E$. The statement now follows from Theorem 13.8. $\square$

Of particular interest are independence systems that are the intersection of two matroids. The prime example here is the matching problem in a bipartite graph $G = (A \,\dot\cup\, B, E)$. If $\mathcal{F} := \{F \subseteq E : F \text{ is a matching in } G\}$, then $(E, \mathcal{F})$ is the intersection of two matroids. Namely, let

$$\mathcal{F}_1 := \{F \subseteq E : |\delta_F(x)| \leq 1 \text{ for all } x \in A\} \quad \text{and}$$
$$\mathcal{F}_2 := \{F \subseteq E : |\delta_F(x)| \leq 1 \text{ for all } x \in B\}.$$

$(E, \mathcal{F}_1), (E, \mathcal{F}_2)$ are matroids by Proposition 13.4(d). Clearly, $\mathcal{F} = \mathcal{F}_1 \cap \mathcal{F}_2$.

A second example is the independence system consisting of all branchings in a digraph G (Example 8 of the list at the beginning of Section 13.1). Here one matroid contains all sets of edges such that each vertex has at most one entering edge (see Proposition 13.4(e)), while the second matroid is the cycle matroid $\mathcal{M}(G)$ of the underlying undirected graph.

We shall now describe Edmonds' algorithm for the following problem:

MATROID INTERSECTION PROBLEM

Instance: Two matroids $(E, \mathcal{F}_1), (E, \mathcal{F}_2)$, given by independence oracles.

Task: Find a set $F \in \mathcal{F}_1 \cap \mathcal{F}_2$ such that $|F|$ is maximum.

We start with the following lemma. Recall that, for $X \in \mathcal{F}$ and $e \in E$, $C(X, e)$ denotes the unique circuit in $X \cup \{e\}$ if $X \cup \{e\} \notin \mathcal{F}$, and $C(X, e) = \emptyset$ otherwise.

Lemma 13.27. (Frank [1981]) *Let $(E, \mathcal{F})$ be a matroid and $X \in \mathcal{F}$. Let $x_1, \ldots, x_s \in X$ and $y_1, \ldots, y_s \notin X$ with*

(a) $x_k \in C(X, y_k)$ *for* $k = 1, \ldots, s$ *and*
(b) $x_j \notin C(X, y_k)$ *for* $1 \le j < k \le s$.

Then $(X \setminus \{x_1, \ldots, x_s\}) \cup \{y_1, \ldots, y_s\} \in \mathcal{F}$.

Proof: Let $X_r := (X \setminus \{x_1, \ldots, x_r\}) \cup \{y_1, \ldots, y_r\}$. We show that $X_r \in \mathcal{F}$ for all r by induction. For $r = 0$ this is trivial. Let us assume that $X_{r-1} \in \mathcal{F}$ for some $r \in \{1, \ldots, s\}$. If $X_{r-1} \cup \{y_r\} \in \mathcal{F}$ then we immediately have $X_r \in \mathcal{F}$. Otherwise $X_{r-1} \cup \{y_r\}$ contains a unique circuit C (by Theorem 13.12(b)). Since $C(X, y_r) \subseteq X_{r-1} \cup \{y_r\}$ (by (b)), we must have $C = C(X, y_r)$. But then by (a) $x_r \in C(X, y_r) = C$, so $X_r = (X_{r-1} \cup \{y_r\}) \setminus \{x_r\} \in \mathcal{F}$.   □

The idea behind EDMONDS' MATROID INTERSECTION ALGORITHM is the following. Starting with $X = \emptyset$, we augment X by one element in each iteration. Since in general we cannot hope for an element e such that $X \cup \{e\} \in \mathcal{F}_1 \cap \mathcal{F}_2$, we shall look for "alternating paths". To make this convenient, we define an auxiliary graph. We apply the notion $C(X, e)$ to $(E, \mathcal{F}_i)$ and write $C_i(X, e)$ $(i = 1, 2)$.

Given a set $X \in \mathcal{F}_1 \cap \mathcal{F}_2$, we define a directed auxiliary graph G_X by

$$A_X^{(1)} := \{ (x, y) : y \in E \setminus X, \, x \in C_1(X, y) \setminus \{y\} \},$$
$$A_X^{(2)} := \{ (y, x) : y \in E \setminus X, \, x \in C_2(X, y) \setminus \{y\} \},$$
$$G_X := (E, A_X^{(1)} \cup A_X^{(2)}).$$

We set

$$S_X := \{ y \in E \setminus X : X \cup \{y\} \in \mathcal{F}_1 \},$$
$$T_X := \{ y \in E \setminus X : X \cup \{y\} \in \mathcal{F}_2 \}$$

(see Figure 13.2) and look for a shortest path from S_X to T_X. Such a path will enable us to augment the set X. (If $S_X \cap T_X \ne \emptyset$, we have a path of length zero and we can augment X by any element in $S_X \cap T_X$.)

Lemma 13.28. *Let $X \in \mathcal{F}_1 \cap \mathcal{F}_2$. Let $y_0, x_1, y_1, \ldots, x_s, y_s$ be the vertices of a shortest y_0-y_s-path in G_X (in this order), with $y_0 \in S_X$ and $y_s \in T_X$. Then*

$$X' := (X \cup \{y_0, \ldots, y_s\}) \setminus \{x_1, \ldots, x_s\} \in \mathcal{F}_1 \cap \mathcal{F}_2.$$

Proof: First we show that $X \cup \{y_0\}, x_1, \ldots, x_s$ and $y_1, \ldots, y_s$ satisfy the requirements of Lemma 13.27 with respect to $\mathcal{F}_1$. Observe that $X \cup \{y_0\} \in \mathcal{F}_1$ because $y_0 \in S_X$. (a) is satisfied because $(x_j, y_j) \in A_X^{(1)}$ for all j, and (b) is satisfied because otherwise the path could be shortcut. We conclude that $X' \in \mathcal{F}_1$.

Secondly, we show that $X \cup \{y_s\}, x_s, x_{s-1}, \ldots, x_1$ and $y_{s-1}, \ldots, y_1, y_0$ satisfy the requirements of Lemma 13.27 with respect to $\mathcal{F}_2$. Observe that $X \cup \{y_s\} \in \mathcal{F}_2$

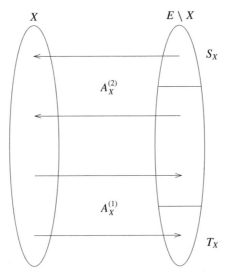

Fig. 13.2.

because $y_s \in T_X$. (a) is satisfied because $(y_{j-1}, x_j) \in A_X^{(2)}$ for all j, and (b) is satisfied because otherwise the path could be shortcut. We conclude that $X' \in \mathcal{F}_2$.   □

We shall now prove that if there exists no S_X-T_X-path in G_X, then X is already maximum. We need the following simple fact:

Proposition 13.29.   *Let* $(E, \mathcal{F}_1)$ *and* $(E, \mathcal{F}_2)$ *be two matroids with rank functions* r_1 *and* r_2. *Then for any* $F \in \mathcal{F}_1 \cap \mathcal{F}_2$ *and any* $Q \subseteq E$ *we have*

$$|F| \leq r_1(Q) + r_2(E \setminus Q).$$

Proof: $F \cap Q \in \mathcal{F}_1$ implies $|F \cap Q| \leq r_1(Q)$. Similarly $F \setminus Q \in \mathcal{F}_2$ implies $|F \setminus Q| \leq r_2(E \setminus Q)$. Adding the two inequalities completes the proof.   □

Lemma 13.30. $X \in \mathcal{F}_1 \cap \mathcal{F}_2$ *is maximum if and only if there is no* S_X-T_X-*path in* G_X.

Proof: If there is an S_X-T_X-path, there is also a shortest one. We apply Lemma 13.28 and obtain a set $X' \in \mathcal{F}_1 \cap \mathcal{F}_2$ of greater cardinality.

Otherwise let R be the set of vertices reachable from S_X in G_X (see Figure 13.3). We have $R \cap T_X = \emptyset$. Let r_1 and r_2 be the rank function of $\mathcal{F}_1$ and $\mathcal{F}_2$, respectively.

We claim that $r_2(R) = |X \cap R|$. If not, there would be a $y \in R \setminus X$ with $(X \cap R) \cup \{y\} \in \mathcal{F}_2$. Since $X \cup \{y\} \notin \mathcal{F}_2$ (because $y \notin T_X$), the circuit $C_2(X, y)$ must contain an element $x \in X \setminus R$. But then $(y, x) \in A_X^{(2)}$ means that there is an edge leaving R. This contradicts the definition of R.

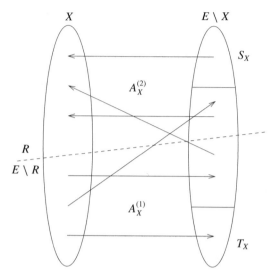

Fig. 13.3.

Next we prove that $r_1(E \setminus R) = |X \setminus R|$. If not, there would be a $y \in (E \setminus R) \setminus X$ with $(X \setminus R) \cup \{y\} \in \mathcal{F}_1$. Since $X \cup \{y\} \notin \mathcal{F}_1$ (because $y \notin S_X$), the circuit $C_1(X, y)$ must contain an element $x \in X \cap R$. But then $(x, y) \in A_X^{(1)}$ means that there is an edge leaving R. This contradicts the definition of R.

Altogether we have $|X| = r_2(R) + r_1(E \setminus R)$. By Proposition 13.29, this implies optimality. $\square$

The last paragraph of this proof yields the following min-max-equality:

Theorem 13.31. (Edmonds [1970]) *Let* $(E, \mathcal{F}_1)$ *and* $(E, \mathcal{F}_2)$ *be two matroids with rank functions* r_1 *and* r_2. *Then*

$$\max \{|X| : X \in \mathcal{F}_1 \cap \mathcal{F}_2\} = \min \{r_1(Q) + r_2(E \setminus Q) : Q \subseteq E\}.$$

$\square$

We are now ready for a detailed description of the algorithm.

EDMONDS' MATROID INTERSECTION ALGORITHM

Input: Two matroids $(E, \mathcal{F}_1)$ and $(E, \mathcal{F}_2)$, given by independence oracles.

Output: A set $X \in \mathcal{F}_1 \cap \mathcal{F}_2$ of maximum cardinality.

①   Set $X := \emptyset$.

②   **For** each $y \in E \setminus X$ and $i \in \{1, 2\}$ **do**: Compute
       $C_i(X, y) := \{x \in X \cup \{y\} : X \cup \{y\} \notin \mathcal{F}_i, (X \cup \{y\}) \setminus \{x\} \in \mathcal{F}_i\}$.

③   Compute S_X, T_X, and G_X as defined above.

④   Apply BFS to find a shortest S_X-T_X-path P in G_X.
       If none exists **then stop**.

⑤   Set $X := X \triangle V(P)$ and **go to** ②.

Theorem 13.32. EDMONDS' MATROID INTERSECTION ALGORITHM *correctly solves the* MATROID INTERSECTION PROBLEM *in* $O(|E|^3\theta)$ *time, where θ is the maximum complexity of the two independence oracles.*

Proof: The correctness follows from Lemmata 13.28 and 13.30. ② and ③ can be done in $O(|E|^2\theta)$, ④ in $O(|E|^2)$ time. Since there are at most $|E|$ augmentations, the overall complexity is $O(|E|^3\theta)$.   □

Faster matroid intersection algorithms are discussed by Cunningham [1986] and Gabow and Xu [1996]. We remark that the problem of finding a maximum cardinality set in the intersection of three matroids is an *NP*-hard problem; see Exercise 17(c) of Chapter 15.

13.6 Matroid Partitioning

Instead of the intersection of matroids we now consider their union which is defined as follows:

Definition 13.33. *Let $(E, \mathcal{F}_1), \ldots, (E, \mathcal{F}_k)$ be k matroids. A set $X \subseteq E$ is called* **partitionable** *if there exists a partition $X = X_1 \;\dot{\cup}\; \cdots \;\dot{\cup}\; X_k$ with $X_i \in \mathcal{F}_i$ for $i = 1, \ldots, k$. Let $\mathcal{F}$ be the family of partitionable subsets of E. Then $(E, \mathcal{F})$ is called the* **union** *or* **sum** *of $(E, \mathcal{F}_1), \ldots, (E, \mathcal{F}_k)$.*

We shall prove that the union of matroids is a matroid again. Moreover, we solve the following problem via matroid intersection:

MATROID PARTITIONING PROBLEM

Instance: A number $k \in \mathbb{N}$, k matroids $(E, \mathcal{F}_1), \ldots, (E, \mathcal{F}_k)$, given by independence oracles.

Task: Find a partitionable set $X \subseteq E$ of maximum cardinality.

The main theorem with respect to matroid partitioning is:

Theorem 13.34. (Nash-Williams [1967]) *Let $(E, \mathcal{F}_1), \ldots, (E, \mathcal{F}_k)$ be matroids with rank functions $r_1, \ldots, r_k$, and let $(E, \mathcal{F})$ be their union. Then $(E, \mathcal{F})$ is a matroid, and its rank function r is given by $r(X) = \min_{A \subseteq X} \left(|X \setminus A| + \sum_{i=1}^{k} r_i(A) \right)$.*

Proof: $(E, \mathcal{F})$ is obviously an independence system. Let $X \subseteq E$. We first prove
$$r(X) = \min_{A \subseteq X} \left(|X \setminus A| + \sum_{i=1}^{k} r_i(A) \right).$$

For any $Y \subseteq X$ such that Y is partitionable, i.e. $Y = Y_1 \dot\cup \cdots \dot\cup Y_k$ with $Y_i \in \mathcal{F}_i$ $(i = 1, \ldots, k)$, and any $A \subseteq X$ we have

$$|Y| = |Y \setminus A| + |Y \cap A| \leq |X \setminus A| + \sum_{i=1}^{k} |Y_i \cap A| \leq |X \setminus A| + \sum_{i=1}^{k} r_i(A),$$

so $r(X) \leq \min_{A \subseteq X} \left(|X \setminus A| + \sum_{i=1}^{k} r_i(A) \right)$.

On the other hand, let $X' := X \times \{1, \ldots, k\}$. We define two matroids on X'. For $Q \subseteq X'$ and $i \in \{1, \ldots, k\}$ we write $Q_i := \{e \in X : (e, i) \in Q\}$. Let

$$\mathcal{I}_1 := \{Q \subseteq X' : Q_i \in \mathcal{F}_i \text{ for all } i = 1, \ldots, k\}$$

and

$$\mathcal{I}_2 := \{Q \subseteq X' : Q_i \cap Q_j = \emptyset \text{ for all } i \neq j\}.$$

Evidently, both $(X', \mathcal{I}_1)$ and $(X', \mathcal{I}_2)$ are matroids, and their rank functions are given by $s_1(Q) := \sum_{i=1}^{k} r_i(Q_i)$ and $s_2(Q) := \left| \bigcup_{i=1}^{k} Q_i \right|$ for $Q \subseteq X'$.

Now the family of partitionable subsets of X can be written as

$$\{A \subseteq X : \text{there is a function } f : A \to \{1, \ldots, k\}$$
$$\text{with } \{(e, f(e)) : e \in A\} \in \mathcal{I}_1 \cap \mathcal{I}_2\}.$$

So the maximum cardinality of a partitionable set is the maximum cardinality of a common independent set in $\mathcal{I}_1$ and $\mathcal{I}_2$. By Theorem 13.31 this maximum cardinality equals $\min \{s_1(Q) + s_2(X' \setminus Q) : Q \subseteq X'\}$. If $Q \subseteq X'$ attains this minimum, then for $A := Q_1 \cap \cdots \cap Q_k$ we have

$$r(X) = s_1(Q) + s_2(X' \setminus Q) = \sum_{i=1}^{k} r_i(Q_i) + \left| X \setminus \bigcap_{i=1}^{k} Q_i \right| \geq \sum_{i=1}^{k} r_i(A) + |X \setminus A|.$$

So we have found a set $A \subseteq X$ with $\sum_{i=1}^{k} r_i(A) + |X \setminus A| \leq r(X)$.

Having proved the formula for the rank function r, we finally show that r is submodular. By Theorem 13.10, this implies that $(E, \mathcal{F})$ is a matroid. To show the submodularity, let $X, Y \subseteq E$, and let $A \subseteq X$, $B \subseteq Y$ with $r(X) = |X \setminus A| + \sum_{i=1}^{k} r_i(A)$ and $r(Y) = |Y \setminus B| + \sum_{i=1}^{k} r_i(B)$. Then

$$r(X) + r(Y)$$
$$= |X \setminus A| + |Y \setminus B| + \sum_{i=1}^{k} (r_i(A) + r_i(B))$$
$$\geq |(X \cup Y) \setminus (A \cup B)| + |(X \cap Y) \setminus (A \cap B)| + \sum_{i=1}^{k} (r_i(A \cup B) + r_i(A \cap B))$$
$$\geq r(X \cup Y) + r(X \cap Y).$$

$\square$

The construction in the above proof (Edmonds [1970]) reduces the MATROID PARTITIONING PROBLEM to the MATROID INTERSECTION PROBLEM. A reduction in the other direction is also possible (Exercise 22), so both problems can be regarded as equivalent.

Note that we can find a maximum independent set in the union of an arbitrary number of matroids efficiently, while the intersection of more than two matroids is intractable.

13.7 Weighted Matroid Intersection

Now we consider the more general weighted version of matroid intersection.

WEIGHTED MATROID INTERSECTION PROBLEM

Instance: Two matroids $(E, \mathcal{F}_1)$ and $(E, \mathcal{F}_2)$, given by independence oracles. Weights $c : E \to \mathbb{R}$.

Task: Find a set $X \in \mathcal{F}_1 \cap \mathcal{F}_2$ whose weight $c(X)$ is maximum.

We shall describe a primal-dual algorithm due to Frank [1981] for this problem. It generalizes EDMONDS' MATROID INTERSECTION ALGORITHM. Again we start with $X := X_0 = \emptyset$ and increase the cardinality in each iteration by one. We obtain sets $X_0, \ldots, X_m \in \mathcal{F}_1 \cap \mathcal{F}_2$ with $|X_k| = k$ $(k = 0, \ldots, m)$ and $m = \max\{|X| : X \in \mathcal{F}_1 \cap \mathcal{F}_2\}$. Each X_k will be optimum, i.e.

$$c(X_k) = \max\{c(X) : X \in \mathcal{F}_1 \cap \mathcal{F}_2, |X| = k\}. \tag{13.4}$$

Hence at the end we just choose an optimum set among $X_0, \ldots, X_m$.

The main idea is to split up the weight function. At any stage we have two functions $c_1, c_2 : E \to \mathbb{R}$ with $c_1(e) + c_2(e) = c(e)$ for all $e \in E$. For each k we shall guarantee

$$c_i(X_k) = \max\{c_i(X) : X \in \mathcal{F}_i, |X| = k\} \quad (i = 1, 2). \tag{13.5}$$

This condition obviously implies (13.4). To obtain (13.5) we use the optimality criterion of Theorem 13.23. Instead of G_X, S_X and T_X only a subgraph $\bar{G}$ and subsets $\bar{S}$, $\bar{T}$ are considered.

WEIGHTED MATROID INTERSECTION ALGORITHM

Input: Two matroids $(E, \mathcal{F}_1)$ and $(E, \mathcal{F}_2)$, given by independence oracles. Weights $c : E \to \mathbb{R}$.

Output: A set $X \in \mathcal{F}_1 \cap \mathcal{F}_2$ of maximum weight.

---

①   Set $k := 0$ and $X_0 := \emptyset$. Set $c_1(e) := c(e)$ and $c_2(e) := 0$ for all $e \in E$.

②   **For each** $y \in E \setminus X_k$ and $i \in \{1, 2\}$ **do**: Compute
$$C_i(X_k, y) := \{x \in X_k \cup \{y\} : X_k \cup \{y\} \notin \mathcal{F}_i, (X_k \cup \{y\}) \setminus \{x\} \in \mathcal{F}_i\}.$$

③   Compute

$$A^{(1)} := \{(x, y) : y \in E \setminus X_k, \, x \in C_1(X_k, y) \setminus \{y\}\},$$
$$A^{(2)} := \{(y, x) : y \in E \setminus X_k, \, x \in C_2(X_k, y) \setminus \{y\}\},$$
$$S := \{y \in E \setminus X_k : X_k \cup \{y\} \in \mathcal{F}_1\},$$
$$T := \{y \in E \setminus X_k : X_k \cup \{y\} \in \mathcal{F}_2\}.$$

④   Compute

$$m_1 := \max\{c_1(y) : y \in S\},$$
$$m_2 := \max\{c_2(y) : y \in T\},$$
$$\bar{S} := \{y \in S : c_1(y) = m_1\},$$
$$\bar{T} := \{y \in T : c_2(y) = m_2\},$$
$$\bar{A}^{(1)} := \{(x, y) \in A^{(1)} : c_1(x) = c_1(y)\},$$
$$\bar{A}^{(2)} := \{(y, x) \in A^{(2)} : c_2(x) = c_2(y)\},$$
$$\bar{G} := (E, \bar{A}^{(1)} \cup \bar{A}^{(2)}).$$

⑤   Apply BFS to compute the set R of vertices reachable from $\bar{S}$ in $\bar{G}$.

⑥   **If** $R \cap \bar{T} \neq \emptyset$ **then:** Find an $\bar{S}$-$\bar{T}$-path P in $\bar{G}$ with a minimum number of edges, set $X_{k+1} := X_k \triangle V(P)$ and $k := k + 1$ and **go to** ②.

⑦   Compute

$$\varepsilon_1 := \min\{c_1(x) - c_1(y) : (x, y) \in \delta^+_{A^{(1)}}(R)\},$$
$$\varepsilon_2 := \min\{c_2(x) - c_2(y) : (y, x) \in \delta^+_{A^{(2)}}(R)\},$$
$$\varepsilon_3 := \min\{m_1 - c_1(y) : y \in S \setminus R\},$$
$$\varepsilon_4 := \min\{m_2 - c_2(y) : y \in T \cap R\},$$
$$\varepsilon := \min\{\varepsilon_1, \varepsilon_2, \varepsilon_3, \varepsilon_4\}$$

(where $\min \emptyset := \infty$).

⑧   **If** $\varepsilon < \infty$ **then:**
        Set $c_1(x) := c_1(x) - \varepsilon$ and $c_2(x) := c_2(x) + \varepsilon$ for all $x \in R$. **Go to** ④.
    If $\varepsilon = \infty$ **then:**
        Among $X_0, X_1, \ldots, X_k$, let X be the one with maximum weight. **Stop**.

See Edmonds [1979] and Lawler [1976] for earlier versions of this algorithm. To prove correctness, we need the following generalization of Lemma 13.27:

Lemma 13.35. (Frank [1981]) *Let* $(E, \mathcal{F})$ *be a matroid,* $c : E \to \mathbb{R}$, *and* $X \in \mathcal{F}$. *Let* $x_1, \ldots, x_l \in X$ *and* $y_1, \ldots, y_l \notin X$ *with*

(a) $x_j \in C(X, y_j)$ *and* $c(x_j) = c(y_j)$ *for* $j = 1, \ldots, l$, *and*
(b) $x_i \notin C(X, y_j)$ *or* $c(x_i) > c(y_j)$ *for* $1 \leq i < j \leq l$, *and*

(c) $x_i \notin C(X, y_j)$ or $c(x_i) \geq c(y_j)$ for $1 \leq j < i \leq l$.

Then $(X \setminus \{x_1, \ldots, x_l\}) \cup \{y_1, \ldots, y_l\} \in \mathcal{F}$.

Proof: We again use induction on l. The case $l = 1$ is trivial by (a). Let $\mu = \min_{i=1}^{l} c(x_i)$, and let h be the smallest index for which the minimum is attained. Let $X' := (X \setminus \{x_h\}) \cup \{y_h\}$. By (a), $X' \in \mathcal{F}$. We show $C(X', y_j) = C(X, y_j)$ for all $j \neq h$; then (a) and (b) hold also for X' and the remaining indices $\{1, \ldots, l\} \setminus \{h\}$, and we are done by induction.

Let $j \neq h$, and suppose that $C(X', y_j) \neq C(X, y_j)$. Then $x_h \in C(X, y_j)$. Then, if $h < j$, (a) and (b) imply $\mu = c(x_h) > c(y_j) = c(x_j)$, a contradiction to the definition of μ. If $h > j$, (a) and (c) imply $\mu = c(x_h) \geq c(y_j) = c(x_j)$, a contradiction to the choice of h. □

Theorem 13.36. (Frank [1981]) *The* WEIGHTED MATROID INTERSECTION AL-GORITHM *correctly solves the* WEIGHTED MATROID INTERSECTION PROBLEM *in* $O(|E|^4 + |E|^3 \theta)$ *time, where* θ *is the maximum complexity of the two independence oracles.*

Proof: Let m be the final value of k (we will show later that the algorithm terminates). The algorithm computes sets $X_0, X_1, \ldots, X_m$. When the algorithm stops, we have $\varepsilon_1 = \varepsilon_2 = \varepsilon_3 = \varepsilon_4 = \infty$, so T is not reachable from S in $(E, A^{(1)} \cup A^{(2)})$. Then Lemma 13.30 implies $m = |X_m| = \max\{|X| : X \in \mathcal{F}_1 \cap \mathcal{F}_2\}$.

To prove correctness, we show that we have at any stage $X_k \in \mathcal{F}_1 \cap \mathcal{F}_2$ and (13.5). Since $c = c_1 + c_2$ always holds, this implies $c(X_k) = \max\{c(X) : X \in \mathcal{F}_1 \cap \mathcal{F}_2, |X| = k\}$ for all $k = 0, \ldots, m$, and hence the correctness.

Initially, $k = 0$, $X_k = \emptyset \in \mathcal{F}_1 \cap \mathcal{F}_2$, and (13.5) is trivial. We now show that each iteration of the algorithm preserves the properties.

We first note that (13.5) and Theorem 13.23(a) imply

(i) $c_1(x) > c_1(y)$ for all $(x, y) \in A^{(1)} \setminus \bar{A}^{(1)}$, and
(ii) $c_2(x) > c_2(y)$ for all $(y, x) \in A^{(2)} \setminus \bar{A}^{(2)}$.

We first consider an iteration in which $R \cap \bar{T} \neq \emptyset$. Then we find a path P in ⑤, say with vertices $y_0, x_1, y_1, \ldots, x_l, y_l$ in this order.

Similarly as in Lemma 13.28, we show that $(E, \mathcal{F}_1)$, c_1, $X \cup \{y_0\}$, $x_1, \ldots, x_l$, and $y_1, \ldots, y_l$ satisfy the requirements of Lemma 13.35: $X \cup \{y_0\} \in \mathcal{F}_1$ follows from $y_0 \in S$. Condition (a) of Lemma 13.35 follows from $(x_j, y_j) \in \bar{A}^{(1)}$ for $j = 1, \ldots, l$, (c) follows from (i), and (b) follows from (i) and the fact that P is a shortest path in $\bar{G}$. Hence $X_{k+1} \in \mathcal{F}_1$.

Analogously, $(E, \mathcal{F}_2)$, c_2, $X_k \cup \{y_l\}$, $x_l, \ldots, x_1$, and $y_{l-1}, \ldots, y_0$ satisfy the requirements of Lemma 13.35. Here $X_k \cup \{y_l\} \in \mathcal{F}_2$ follows from $y_l \in T$. Condition (a) of Lemma 13.35 follows from $(y_{j-1}, x_j) \in \bar{A}^{(2)}$ for $j = 1, \ldots, l$, (c) follows from (ii), and (b) again follows from (ii) and the fact that P is a shortest path in $\bar{G}$. Hence $X_{k+1} \in \mathcal{F}_2$.

So we have shown $X_{k+1} \in \mathcal{F}_1 \cap \mathcal{F}_2$. To show that X_{k+1} satisfies (13.5) we use Theorem 13.23.

By definition of $\bar{G}$ we have $c_1(X_{k+1}) = c_1(X_k) + c_1(y_0)$ and $c_2(X_{k+1}) = c_2(X_k) + c_2(y_s)$. Since X_k satisfies (13.5), conditions (a) and (b) of Theorem 13.23 must hold with respect to X_k and each of $\mathcal{F}_1$ and $\mathcal{F}_2$.

By definition of $\bar{S}$ and $y_0 \in \bar{S}$, both conditions continue to hold for $X_k \cup \{y_0\}$ and $\mathcal{F}_1$. Therefore $c_1(X_{k+1}) = c_1(X_k \cup \{y_0\}) = \max\{c_1(Y) : Y \in \mathcal{F}_1, |Y| = k+1\}$. Moreover, by definition of $\bar{T}$ and $y_l \in \bar{T}$, (a) and (b) of Theorem 13.23 continue to hold for $X_k \cup \{y_l\}$ and $\mathcal{F}_2$, implying $c_2(X_{k+1}) = c_2(X_k \cup \{y_l\}) = \max\{c_2(Y) : Y \in \mathcal{F}_2, |Y| = k+1\}$. In other words, (13.5) indeed holds for X_{k+1}.

Now consider an iteration in which $R \cap \bar{T} = \emptyset$, so we change c_1 and c_2 in ⑧. We first show that $\varepsilon > 0$. Since by the definition of R we have $\delta_{\bar{G}}^+(R) = \emptyset$, (i) and (ii) directly imply $\varepsilon_1 > 0$ and $\varepsilon_2 > 0$.

$m_1 \geq c_1(y)$ holds for all $y \in S$. If in addition $y \notin R$ then $y \notin \bar{S}$, so $m_1 > c_1(y)$. Therefore $\varepsilon_3 > 0$. Similarly, $\varepsilon_4 > 0$ (using $\bar{T} \cap R = \emptyset$). We conclude that $\varepsilon > 0$.

We can now prove that ⑧ preserves (13.5). Let c_1' be the modified c_1, i.e.

$$c_1'(x) := \begin{cases} c_1(x) - \varepsilon & \text{if } x \in R \\ c_1(x) & \text{if } x \notin R \end{cases}.$$

We prove that X_k and c_1' satisfy the conditions of Theorem 13.23 with respect to $\mathcal{F}_1$.

To prove (a), let $y \in E \setminus X_k$ and $x \in C_1(X_k, y) \setminus \{y\}$. Suppose $c_1'(x) < c_1'(y)$. Since $c_1(x) \geq c_1(y)$ and $\varepsilon > 0$, we must have $x \in R$ and $y \notin R$. Since also $(x, y) \in A^{(1)}$, we have $\varepsilon \leq \varepsilon_1 \leq c_1(x) - c_1(y) = (c_1'(x) + \varepsilon) - c_1'(y)$, a contradiction.

To prove (b), let $x \in X_k$ and $y \in E \setminus X_k$ with $X_k \cup \{y\} \in \mathcal{F}_1$. Now suppose $c_1'(y) > c_1'(x)$. Since $c_1(y) \leq m_1 \leq c_1(x)$, we must have $x \in R$ and $y \notin R$. Since $y \in S$ we have $\varepsilon \leq \varepsilon_3 \leq m_1 - c_1(y) \leq c_1(x) - c_1(y) = (c_1'(x) + \varepsilon) - c_1'(y)$, a contradiction.

Let c_2' be the modified c_2, i.e.

$$c_2'(x) := \begin{cases} c_2(x) + \varepsilon & \text{if } x \in R \\ c_2(x) & \text{if } x \notin R \end{cases}.$$

We show that X_k and c_2' satisfy the conditions of Theorem 13.23 with respect to $\mathcal{F}_2$.

To prove (a), let $y \in E \setminus X_k$ and $x \in C_2(X_k, y) \setminus \{y\}$. Suppose $c_2'(x) < c_2'(y)$. Since $c_2(x) \geq c_2(y)$, we must have $y \in R$ and $x \notin R$. Since also $(y, x) \in A^{(2)}$, we have $\varepsilon \leq \varepsilon_2 \leq c_2(x) - c_2(y) = c_2'(x) - (c_2'(y) - \varepsilon)$, a contradiction.

To prove (b), let $x \in X_k$ and $y \in E \setminus X_k$ with $X_k \cup \{y\} \in \mathcal{F}_2$. Now suppose $c_2'(y) > c_2'(x)$. Since $c_2(y) \leq m_2 \leq c_2(x)$, we must have $y \in R$ and $x \notin R$. Since $y \in T$ we have $\varepsilon \leq \varepsilon_4 \leq m_2 - c_2(y) \leq c_2(x) - c_2(y) = c_2'(x) - (c_2'(y) - \varepsilon)$, a contradiction.

So we have proved that (13.5) is not violated during ⑧, and thus the algorithm works correctly.

We now consider the running time. Observe that after an update of the weights in ⑧, the new sets $\bar{S}$, $\bar{T}$, and R, as computed subsequently in ④ and ⑤, are supersets of the old $\bar{S}$, $\bar{T}$, and R, respectively. If $\varepsilon = \varepsilon_4 < \infty$, an augmentation (increase of k) follows. Otherwise the cardinality of R increases immediately (in ⑤) by at least one. So ④ – ⑧ are repeated less than $|E|$ times between two augmentations.

Since the running time of ④ – ⑧ is $O(|E|^2)$, the total running time between two augmentations is $O(|E|^3)$ plus $O(|E|^2)$ oracle calls (in ②). Since there are $m \leq |E|$ augmentations, the stated overall running time follows.   □

The running time can easily be improved to $O(|E|^3\theta)$ (Exercise 24).

Exercises

1. Prove that all the independence systems apart from (5) and (6) in the list at the beginning of Section 13.1 are – in general – not matroids.

2. Show that the uniform matroid with four elements and rank 2 is not a graphic matroid.

3. Prove that every graphic matroid is representable over every field.

4. Let G be an undirected graph, $K \in \mathbb{N}$, and let $\mathcal{F}$ contain those subsets of $E(G)$ that are the union of K forests. Prove that $(E(G), \mathcal{F})$ is a matroid.

5. Let G be a bipartite graph with bipartition $V(G) = A \cup B$ and $(A, \mathcal{F})$ a matroid. Prove that then $(B, \{Y \subseteq B : Y = \emptyset \text{ or } \exists X \in \mathcal{F} : G[X \cup Y] \text{ has a perfect matching}\})$ is a matroid.

6. Compute tight lower bounds for the rank quotients of the independence systems listed at the beginning of Section 13.1.

7. Let S be a finite family of finite sets. A set T is a transversal of S if there is a bijection $\Phi : T \to S$ with $t \in \Phi(t)$ for all $t \in T$. (For a necessary and sufficient condition for the existence of a transversal, see Exercise 6 of Chapter 10.) Assume that S has a transversal. Prove that the family of transversals of S is the family of bases of a matroid.

8. Let E be a finite set and $\mathcal{B} \subseteq 2^E$. Show that $\mathcal{B}$ is the set of bases of some matroid $(E, \mathcal{F})$ if and only if the following holds:
   (B1) $\mathcal{B} \neq \emptyset$;
   (B2) For any $B_1, B_2 \in \mathcal{B}$ and $y \in B_2 \setminus B_1$ there exists an $x \in B_1 \setminus B_2$ with $(B_1 \setminus \{x\}) \cup \{y\} \in \mathcal{B}$.

9. Let G be a graph. Let $\mathcal{F}$ be the family of sets $X \subseteq V(G)$, for which a maximum matching exists that covers no vertex in X. Prove that $(V(G), \mathcal{F})$ is a matroid. What is the dual matroid?

10. Show that $\mathcal{M}(G^*) = (\mathcal{M}(G))^*$ also holds for disconnected graphs G, extending Theorem 13.16.
    Hint: Use Exercise 37(a) of Chapter 2.

11. Show that the clutters in (3) and (6) in the list of Section 13.3 have the Max-Flow-Min-Cut property. (Use Theorem 19.17.) Show that the clutters in (1), (4) and (5) do not have the Max-Flow-Min-Cut property in general.

* 12. A clutter $(E, \mathcal{F})$ is called binary if for all $X_1, \ldots, X_k \in \mathcal{F}$ with k odd there exists a $Y \in \mathcal{F}$ with $Y \subseteq X_1 \triangle \cdots \triangle X_k$. Prove that the clutter of minimal T-joins and the clutter of minimal T-cuts (example (7) of the list in Section 13.3) are binary. Prove that a clutter is binary if and only if $|A \cap B|$ is odd for all $A \in \mathcal{F}$ and all $B \in \mathcal{F}'$, where $(E, \mathcal{F}')$ is the blocking clutter. Conclude that a clutter is binary if and only if its blocking clutter is binary.
  Note: Seymour [1977] classified the binary clutters with the Max-Flow-Min-Cut property.

* 13. Let P be a polyhedron of blocking type, i.e. we have $x + y \in P$ for all $x \in P$ and $y \geq 0$. The blocking polyhedron of P is defined to be $B(P) := \{z : z^\top x \geq 1$ for all $x \in P\}$. Prove that $B(P)$ is again a polyhedron of blocking type and that $B(B(P)) = P$.
  Note: Compare this with Theorem 4.22.

14. How can one check (in polynomial time) whether a given set of edges of a complete graph G is a subset of some Hamiltonian circuit in G?

15. Prove that if $(E, \mathcal{F})$ is a matroid, then the BEST-IN-GREEDY maximizes any bottleneck function $c(F) = \min\{c_e : e \in F\}$ over the bases.

16. Let $(E, \mathcal{F})$ be a matroid and $c : E \to \mathbb{R}$ such that $c(e) \neq c(e')$ for all $e \neq e'$ and $c(e) \neq 0$ for all e. Prove that both the MAXIMIZATION and the MINIMIZATION PROBLEM for $(E, \mathcal{F}, c)$ have a unique optimum solution.

* 17. Prove that for matroids the independence, basis-superset, closure and rank oracles are polynomially equivalent.
  Hint: To show that the rank oracle reduces to the independence oracle, use the BEST-IN-GREEDY. To show that the independence oracle reduces to the basis-superset oracle, use the WORST-OUT-GREEDY.
  (Hausmann and Korte [1981])

18. Let $(E, \mathcal{F})$ be a matroid with rank function r, E' a finite set, and $f : E \to E'$. Let $\mathcal{F}' := \{f(F) : F \in \mathcal{F}\}$. Prove:
  (a) $(E', \mathcal{F}')$ is a matroid.
  (b) The rank function r' of $(E', \mathcal{F}')$ is given by

$$r(X) = \min\left\{|X \setminus W| + r\left(f^{-1}(W)\right) : W \subseteq X\right\}$$

  for all $X \subseteq E'$.
  Hint: Use Theorem 13.31.
  (Nash-Williams [1967])

19. Given an undirected graph G, we wish to colour the edges with a minimum number of colours such that for any circuit C of G, the edges of C do not all have the same colour. Show that there is a polynomial-time algorithm for this problem.

20. Let $(E, \mathcal{F}_1), \ldots, (E, \mathcal{F}_k)$ be matroids with rank functions $r_1, \ldots, r_k$. Prove that a set $X \subseteq E$ is partitionable if and only if $|A| \leq \sum_{i=1}^{k} r_i(A)$ for all $A \subseteq X$. Show that Theorem 6.20 is a special case.
  (Edmonds and Fulkerson [1965])

21. Let $(E, \mathcal{F})$ be a matroid with rank function r. Prove (using Theorem 13.34):

(a) $(E, \mathcal{F})$ has k pairwise disjoint bases if and only if $kr(A) + |E \setminus A| \geq kr(E)$ for all $A \subseteq E$.

(b) $(E, \mathcal{F})$ has k independent sets whose union is E if and only if $kr(A) \geq |A|$ for all $A \subseteq E$.

Show that Theorem 6.20 and Theorem 6.17 are special cases.

22. Let $(E, \mathcal{F}_1)$ and $(E, \mathcal{F}_2)$ be two matroids. Let X be a maximal partitionable subset with respect to $(E, \mathcal{F}_1)$ and $(E, \mathcal{F}_2^*)$: $X = X_1 \cup X_2$ with $X_1 \in \mathcal{F}_1$ and $X_2 \in \mathcal{F}_2^*$. Let $B_2 \supseteq X_2$ be a basis of $\mathcal{F}_2^*$. Prove that then $X \setminus B_2$ is a maximum-cardinality set in $\mathcal{F}_1 \cap \mathcal{F}_2$.
(Edmonds [1970])

23. Let $(E, \mathcal{S})$ be a set system, and let $(E, \mathcal{F})$ be a matroid with rank function r. Show that $\mathcal{S}$ has a transversal that is independent in $(E, \mathcal{F})$ if and only if $r\left(\bigcup_{B \in \mathcal{B}} B\right) \geq |\mathcal{B}|$ for all $\mathcal{B} \subseteq \mathcal{S}$.
Hint: First describe the rank function of the matroid whose independent sets are all transversals (Exercise 7), using Theorem 13.34. Then apply Theorem 13.31. (Rado [1942])

24. Show that the running time of the WEIGHTED MATROID INTERSECTION ALGORITHM (cf. Theorem 13.36) can be improved to $O(|E|^3 \theta)$.

25. Let $(E, \mathcal{F}_1)$ and $(E, \mathcal{F}_2)$ be two matroids, and $c : E \to \mathbb{R}$. Let $X_0, \ldots, X_m \in \mathcal{F}_1 \cap \mathcal{F}_2$ with $|X_k| = k$ and $c(X_k) = \max\{c(X) : X \in \mathcal{F}_1 \cap \mathcal{F}_2, |X| = k\}$ for all k. Prove that for $k = 1, \ldots, m - 2$

$$c(X_{k+1}) - c(X_k) \leq c(X_k) - c(X_{k-1}).$$

(Krogdahl [unpublished])

26. Consider the following problem. Given a digraph G with edge weights, a vertex $s \in V(G)$, and a number k, find a minimum weight subgraph H of G containing k edge-disjoint paths from s to each other vertex. Show that this reduces to the WEIGHTED MATROID INTERSECTION PROBLEM.
Hint: See Exercise 26 of Chapter 6 and Exercise 4 of this chapter. (Edmonds [1970]; Frank and Tardos [1989]; Gabow [1995])

27. Let A and B be two finite sets of cardinality $n \in \mathbb{N}$, $G = (A \cup B, \{\{a, b\} : a \in A, b \in B\})$ the complete bipartite graph, $\bar{a} \in A$, and $c : E(G) \to \mathbb{R}$ a cost function. Let $\mathcal{T}$ be the family of edge sets of all spanning trees T in G with $|\delta_T(a)| = 2$ for all $a \in A \setminus \{\bar{a}\}$. Show that a minimum cost element of $\mathcal{T}$ can be computed in $O(n^7)$ time. How many edges will be incident to $\bar{a}$?

References

General Literature

Bixby, R.E., and Cunningham, W.H. [1995]: Matroid optimization and algorithms. In: Handbook of Combinatorics; Vol. 1 (R.L. Graham, M. Grötschel, L. Lovász, eds.), Elsevier, Amsterdam, 1995

Cook, W.J., Cunningham, W.H., Pulleyblank, W.R., and Schrijver, A. [1998]: Combinatorial Optimization. Wiley, New York 1998, Chapter 8

Faigle, U. [1987]: Matroids in combinatorial optimization. In: Combinatorial Geometries (N. White, ed.), Cambridge University Press, 1987

Frank, A. [2011]: Connections in Combinatorial Optimization. Oxford University Press, Oxford 2011

Gondran, M., and Minoux, M. [1984]: Graphs and Algorithms. Wiley, Chichester 1984, Chapter 9

Lawler, E.L. [1976]: Combinatorial Optimization; Networks and Matroids. Holt, Rinehart and Winston, New York 1976, Chapters 7 and 8

Oxley, J.G. [1992]: Matroid Theory. Oxford University Press, Oxford 1992

von Randow, R. [1975]: Introduction to the Theory of Matroids. Springer, Berlin 1975

Recski, A. [1989]: Matroid Theory and its Applications. Springer, Berlin 1989

Schrijver, A. [2003]: Combinatorial Optimization: Polyhedra and Efficiency. Springer, Berlin 2003, Chapters 39–42

Welsh, D.J.A. [1976]: Matroid Theory. Academic Press, London 1976

Cited References

Cunningham, W.H. [1986] : Improved bounds for matroid partition and intersection algorithms. SIAM Journal on Computing 15 (1986), 948–957

Edmonds, J. [1970]: Submodular functions, matroids and certain polyhedra. In: Combinatorial Structures and Their Applications; Proceedings of the Calgary International Conference on Combinatorial Structures and Their Applications 1969 (R. Guy, H. Hanani, N. Sauer, J. Schönheim, eds.), Gordon and Breach, New York 1970, pp. 69–87

Edmonds, J. [1971]: Matroids and the greedy algorithm. Mathematical Programming 1 (1971), 127–136

Edmonds, J. [1979]: Matroid intersection. In: Discrete Optimization I; Annals of Discrete Mathematics 4 (P.L. Hammer, E.L. Johnson, B.H. Korte, eds.), North-Holland, Amsterdam 1979, pp. 39–49

Edmonds, J., and Fulkerson, D.R. [1965]: Transversals and matroid partition. Journal of Research of the National Bureau of Standards B 69 (1965), 67–72

Frank, A. [1981]: A weighted matroid intersection algorithm. Journal of Algorithms 2 (1981), 328–336

Frank, A., and Tardos, É. [1989]: An application of submodular flows. Linear Algebra and Its Applications 114/115 (1989), 329–348

Fulkerson, D.R. [1971]: Blocking and anti-blocking pairs of polyhedra. Mathematical Programming 1 (1971), 168–194

Gabow, H.N. [1995]: A matroid approach to finding edge connectivity and packing arborescences. Journal of Computer and System Sciences 50 (1995), 259–273

Gabow, H.N., and Xu, Y. [1996]: Efficient theoretic and practical algorithms for linear matroid intersection problems. Journal of Computer and System Sciences 53 (1996), 129–147

Hausmann, D., Jenkyns, T.A., and Korte, B. [1980]: Worst case analysis of greedy type algorithms for independence systems. Mathematical Programming Study 12 (1980), 120–131

Hausmann, D., and Korte, B. [1981]: Algorithmic versus axiomatic definitions of matroids. Mathematical Programming Study 14 (1981), 98–111

Jenkyns, T.A. [1976]: The efficiency of the greedy algorithm. Proceedings of the 7th S-E Conference on Combinatorics, Graph Theory, and Computing, Utilitas Mathematica, Winnipeg 1976, pp. 341–350

Korte, B., and Hausmann, D. [1978]: An analysis of the greedy algorithm for independence systems. In: Algorithmic Aspects of Combinatorics; Annals of Discrete Mathematics 2 (B. Alspach, P. Hell, D.J. Miller, eds.), North-Holland, Amsterdam 1978, pp. 65–74

Korte, B., and Monma, C.L. [1979]: Some remarks on a classification of oracle-type algorithms. In: Numerische Methoden bei graphentheoretischen und kombina- torischen Problemen; Band 2 (L. Collatz, G. Meinardus, W. Wetterling, eds.), Birkhäuser, Basel 1979, pp. 195–215

Lehman, A. [1979]: On the width-length inequality. Mathematical Programming 17 (1979), 403–417

Nash-Williams, C.S.J.A. [1967]: An application of matroids to graph theory. In: Theory of Graphs; Proceedings of an International Symposium in Rome 1966 (P. Rosenstiehl, ed.), Gordon and Breach, New York, 1967, pp. 263–265

Rado, R. [1942]: A theorem on independence relations. Quarterly Journal of Math. Oxford 13 (1942), 83–89

Rado, R. [1957]: Note on independence functions. Proceedings of the London Mathematical Society 7 (1957), 300–320

Seymour, P.D. [1977]: The matroids with the Max-Flow Min-Cut property. Journal of Combinatorial Theory B 23 (1977), 189–222

Whitney, H. [1933]: Planar graphs. Fundamenta Mathematicae 21 (1933), 73–84

Whitney, H. [1935]: On the abstract properties of linear dependence. American Journal of Mathematics 57 (1935), 509–533

14 Generalizations of Matroids

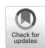

There are several interesting generalizations of matroids. We have already seen independence systems in Section 13.1, which arose from dropping the axiom (M3). In Section 14.1 we consider greedoids, arising by dropping (M2) instead. Moreover, certain polytopes related to matroids and to submodular functions, called polymatroids, lead to strong generalizations of important theorems; we shall discuss them in Section 14.2. In Sections 14.3 and 14.4 we consider two approaches to the problem of minimizing an arbitrary submodular function: one using the ELLIPSOID METHOD, and one with a combinatorial algorithm. For the important special case of symmetric submodular functions we mention a simpler algorithm in Section 14.5. Finally, in Section 14.6, we discuss the problem of maximizing a submodular function.

14.1 Greedoids

By definition, set systems $(E, \mathcal{F})$ are matroids if and only if they satisfy

(M1) $\emptyset \in \mathcal{F}$;
(M2) If $X \subseteq Y \in \mathcal{F}$ then $X \in \mathcal{F}$;
(M3) If $X, Y \in \mathcal{F}$ and $|X| > |Y|$, then there is an $x \in X \setminus Y$ with $Y \cup \{x\} \in \mathcal{F}$.

If we drop (M3), we obtain independence systems, discussed in Sections 13.1 and 13.4. Now we drop (M2) instead:

Definition 14.1. *A* **greedoid** *is a set system* $(E, \mathcal{F})$ *satisfying* (M1) *and* (M3).

Instead of the subclusiveness (M2) we have accessibility: We call a set system $(E, \mathcal{F})$ **accessible** if $\emptyset \in \mathcal{F}$ and for any $X \in \mathcal{F} \setminus \{\emptyset\}$ there exists an $x \in X$ with $X \setminus \{x\} \in \mathcal{F}$. Greedoids are accessible (accessibility follows directly from (M1) and (M3)). Though more general than matroids, they comprise a rich structure and, on the other hand, generalize many different, seemingly unrelated concepts. We start with the following result:

© Springer-Verlag GmbH Germany 2018
Bernhard Korte and Jens Vygen, *Combinatorial Optimization*, Algorithms and Combinatorics 21, https://doi.org/10.1007/978-3-662-56039-6_14

Theorem 14.2. *Let $(E, \mathcal{F})$ be an accessible set system. The following statements are equivalent:*

(a) *For any $X \subseteq Y \subset E$ and $z \in E \setminus Y$ with $X \cup \{z\} \in \mathcal{F}$ and $Y \in \mathcal{F}$ we have $Y \cup \{z\} \in \mathcal{F}$;*
(b) *$\mathcal{F}$ is closed under union.*

Proof: (a) $\Rightarrow$(b): Let $X, Y \in \mathcal{F}$; we show that $X \cup Y \in \mathcal{F}$. Let Z be a maximal set with $Z \in \mathcal{F}$ and $X \subseteq Z \subseteq X \cup Y$. Suppose $Y \setminus Z \neq \emptyset$. By repeatedly applying accessibility to Y we get a set $Y' \in \mathcal{F}$ with $Y' \subseteq Z$ and an element $y \in Y \setminus Z$ with $Y' \cup \{y\} \in \mathcal{F}$. We apply (a) to Z, Y' and y and obtain $Z \cup \{y\} \in \mathcal{F}$, contradicting the choice of Z.

(b) $\Rightarrow$(a) is trivial. $\qquad\square$

If the conditions in Theorem 14.2 hold, then $(E, \mathcal{F})$ is called an **antimatroid**.

Proposition 14.3. *Every antimatroid is a greedoid.*

Proof: Let $(E, \mathcal{F})$ be an antimatroid, i.e. accessible and closed under union. To prove (M3), let $X, Y \in \mathcal{F}$ with $|X| > |Y|$. Since $(E, \mathcal{F})$ is accessible there is an order $X = \{x_1, \ldots, x_n\}$ with $\{x_1, \ldots, x_i\} \in \mathcal{F}$ for $i = 0, \ldots, n$. Let $i \in \{1, \ldots, n\}$ be the minimum index with $x_i \notin Y$; then $Y \cup \{x_i\} = Y \cup \{x_1, \ldots, x_i\} \in \mathcal{F}$ (since $\mathcal{F}$ is closed under union). $\qquad\square$

Another equivalent definition of antimatroids is by a closure operator:

Proposition 14.4. *Let $(E, \mathcal{F})$ be a set system such that $\mathcal{F}$ is closed under union and $\emptyset \in \mathcal{F}$. Define*

$$\tau(A) := \bigcap\{X \subseteq E : A \subseteq X, \ E \setminus X \in \mathcal{F}\}$$

Then τ is a closure operator, i.e. satisfies (S1)–(S3) of Theorem 13.11.

Proof: Let $X \subseteq Y \subseteq E$. $X \subseteq \tau(X) \subseteq \tau(Y)$ is trivial. To prove (S3), assume that there exists a $y \in \tau(\tau(X)) \setminus \tau(X)$. Then $y \in Y$ for all $Y \subseteq E$ with $\tau(X) \subseteq Y$ and $E \setminus Y \in \mathcal{F}$, but there exists a $Z \subseteq E \setminus \{y\}$ with $X \subseteq Z$ and $E \setminus Z \in \mathcal{F}$. This implies $\tau(X) \not\subseteq Z$, a contradiction. $\qquad\square$

Theorem 14.5. *Let $(E, \mathcal{F})$ be a set system such that $\mathcal{F}$ is closed under union and $\emptyset \in \mathcal{F}$. Then $(E, \mathcal{F})$ is accessible if and only if the closure operator τ of Proposition 14.4 satisfies the anti-exchange property: if $X \subseteq E$, $y, z \in E \setminus \tau(X)$, $y \neq z$ and $z \in \tau(X \cup \{y\})$, then $y \notin \tau(X \cup \{z\})$.*

Proof: If $(E, \mathcal{F})$ is accessible, then (M3) holds by Proposition 14.3. To show the anti-exchange property, let $X \subseteq E$, $B := E \setminus \tau(X)$, and $y, z \in B$ with $z \notin A := E \setminus \tau(X \cup \{y\})$. Observe that $A \in \mathcal{F}$, $B \in \mathcal{F}$ and $A \subseteq B \setminus \{y, z\}$.

By applying (M3) to A and B we get an element $b \in B \setminus A \subseteq E \setminus (X \cup A)$ with $A \cup \{b\} \in \mathcal{F}$. $A \cup \{b\}$ cannot be a subset of $E \setminus (X \cup \{y\})$ (otherwise $\tau(X \cup \{y\}) \subseteq E \setminus (A \cup \{b\})$, contradicting $\tau(X \cup \{y\}) = E \setminus A$). Hence $b = y$. So we have $A \cup \{y\} \in \mathcal{F}$ and thus $\tau(X \cup \{z\}) \subseteq E \setminus (A \cup \{y\})$. We have proved $y \notin \tau(X \cup \{z\})$.

To show the converse, let $A \in \mathcal{F} \setminus \{\emptyset\}$ and let $X := E \setminus A$. We have $\tau(X) = X$. Let $a \in A$ such that $|\tau(X \cup \{a\})|$ is minimum. We claim that $\tau(X \cup \{a\}) = X \cup \{a\}$, i.e. $A \setminus \{a\} \in \mathcal{F}$.

Suppose, on the contrary, that $b \in \tau(X \cup \{a\}) \setminus (X \cup \{a\})$. By the anti-exchange property we have $a \notin \tau(X \cup \{b\})$. Moreover,

$$\tau(X \cup \{b\}) \subseteq \tau(\tau(X \cup \{a\}) \cup \{b\}) = \tau(\tau(X \cup \{a\})) = \tau(X \cup \{a\}).$$

Hence $\tau(X \cup \{b\})$ is a proper subset of $\tau(X \cup \{a\})$, contradicting the choice of a. $\quad\square$

The anti-exchange property of Theorem 14.5 is different from (S4). While (S4) of Theorem 13.11 is a property of linear hulls in $\mathbb{R}^n$, this is a property of convex hulls in $\mathbb{R}^n$: if $y \neq z$, $z \notin \text{conv}(X)$ and $z \in \text{conv}(X \cup \{y\})$, then clearly $y \notin \text{conv}(X \cup \{z\})$. So for any finite set $E \subset \mathbb{R}^n$, $(E, \{X \subseteq E : X \cap \text{conv}(E \setminus X) = \emptyset\})$ is an antimatroid.

Greedoids generalize matroids and antimatroids, but they also contain other interesting structures. One example is the blossom structure we used in EDMONDS' CARDINALITY MATCHING ALGORITHM (Exercise 1). Another basic example is:

Proposition 14.6. *Let G be a graph (directed or undirected) and $r \in V(G)$. Let $\mathcal{F}$ be the family of all edge sets of arborescences in G rooted at r, or trees in G containing r (not necessarily spanning). Then $(E(G), \mathcal{F})$ is a greedoid.*

Proof: (M1) is trivial. We prove (M3) for the directed case; the same argument applies to the undirected case. Let (X_1, F_1) and (X_2, F_2) be two arborescences in G rooted at r with $|F_1| > |F_2|$. Then $|X_1| = |F_1| + 1 > |F_2| + 1 = |X_2|$, so let $x \in X_1 \setminus X_2$. The r-x-path in (X_1, F_1) contains an edge (v, w) with $v \in X_2$ and $w \notin X_2$. This edge can be added to (X_2, F_2), proving that $F_2 \cup \{(v, w)\} \in \mathcal{F}$. $\quad\square$

This greedoid is called the directed (undirected) branching greedoid of G.

The problem of finding a maximum weight spanning tree in a connected graph G with nonnegative weights is the MAXIMIZATION PROBLEM for the cycle matroid $\mathcal{M}(G)$. The BEST-IN-GREEDY ALGORITHM is in this case nothing but KRUSKAL'S ALGORITHM. Now we have a second formulation of the same problem: we are looking for a maximum weight set F with $F \in \mathcal{F}$, where $(E(G), \mathcal{F})$ is the undirected branching greedoid of G.

We now formulate a general greedy algorithm for greedoids. In the special case of matroids, it is exactly the BEST-IN-GREEDY ALGORITHM discussed in Section 13.4. If we have an undirected branching greedoid with a modular cost function c, it is PRIM'S ALGORITHM:

GREEDY ALGORITHM FOR GREEDOIDS

Input: A greedoid $(E, \mathcal{F})$ and a function $c : 2^E \rightarrow \mathbb{R}$, given by an oracle
which for any given $X \subseteq E$ says whether $X \in \mathcal{F}$ and returns $c(X)$.

Output: A set $F \in \mathcal{F}$.

①   Set $F := \emptyset$.

②   Let $e \in E \setminus F$ such that $F \cup \{e\} \in \mathcal{F}$ and $c(F \cup \{e\})$ is maximum;
      if no such e exists **then stop**.

③   Set $F := F \cup \{e\}$ and **go to** ②.

Even for modular cost functions c this algorithm does not always provide an optimal solution. At least we can characterize those greedoids where it works:

Theorem 14.7. *Let $(E, \mathcal{F})$ be a greedoid. The* GREEDY ALGORITHM FOR GREE-DOIDS *finds a set $F \in \mathcal{F}$ of maximum weight for each modular weight function $c : 2^E \rightarrow \mathbb{R}_+$ if and only if $(E, \mathcal{F})$ has the so-called strong exchange property: For all $A \in \mathcal{F}$, B maximal in $\mathcal{F}$, $A \subseteq B$ and $x \in E \setminus B$ with $A \cup \{x\} \in \mathcal{F}$ there exists a $y \in B \setminus A$ such that $A \cup \{y\} \in \mathcal{F}$ and $(B \setminus \{y\}) \cup \{x\} \in \mathcal{F}$.*

Proof: Suppose $(E, \mathcal{F})$ is a greedoid with the strong exchange property. Let $c : E \rightarrow \mathbb{R}_+$, and let $A = \{a_1, \ldots, a_l\}$ be the solution found by the GREEDY AL-GORITHM FOR GREEDOIDS, where the elements are chosen in the order $a_1, \ldots, a_l$.

Let $B = \{a_1, \ldots, a_k\} \cup B'$ be an optimum solution such that k is maximum, and suppose that $k < l$. Then we apply the strong exchange property to $\{a_1, \ldots, a_k\}$, B and a_{k+1}. We conclude that there exists a $y \in B'$ with $\{a_1, \ldots, a_k, y\} \in \mathcal{F}$ and $(B \setminus y) \cup \{a_{k+1}\} \in \mathcal{F}$. By the choice of a_{k+1} in ② of the GREEDY ALGORITHM FOR GREEDOIDS we have $c(a_{k+1}) \geq c(y)$ and thus $c((B \setminus y) \cup \{a_{k+1}\}) \geq c(B)$, contradicting the choice of B.

Conversely, let $(E, \mathcal{F})$ be a greedoid that does not have the strong exchange property. Let $A \in \mathcal{F}$, B maximal in $\mathcal{F}$, $A \subseteq B$ and $x \in E \setminus B$ with $A \cup \{x\} \in \mathcal{F}$ such that for all $y \in B \setminus A$ with $A \cup \{y\} \in \mathcal{F}$ we have $(B \setminus y) \cup \{x\} \notin \mathcal{F}$.

Let $Y := \{y \in B \setminus A : A \cup \{y\} \in \mathcal{F}\}$. We set $c(e) := 2$ for $e \in B \setminus Y$, and $c(e) := 1$ for $e \in Y \cup \{x\}$ and $c(e) := 0$ for $e \in E \setminus (B \cup \{x\})$. Then the GREEDY ALGORITHM FOR GREEDOIDS might choose the elements of A first (they have weight 2) and then might choose x. It will eventually end up with a set $F \in \mathcal{F}$ that cannot be optimal, since $c(F) \leq c(B \cup \{x\}) - 2 < c(B \cup \{x\}) - 1 = c(B)$ and $B \in \mathcal{F}$. $\square$

Indeed, optimizing modular functions over general greedoids is *NP*-hard. This follows from the following observation (together with Corollary 15.24):

Proposition 14.8. *The problem of deciding, given an undirected graph G and $k \in \mathbb{N}$, whether G has a vertex cover of cardinality k, linearly reduces to the following problem: Given a greedoid $(E, \mathcal{F})$ (by a membership oracle) and a function $c : E \rightarrow \mathbb{R}_+$, find an $F \in \mathcal{F}$ with $c(F)$ maximum.*

Proof: Let G be any undirected graph and $k \in \mathbb{N}$. Let $D := V(G) \,\dot\cup\, E(G)$ and

$$\mathcal{F} := \{X \subseteq D : \text{ for all } e = \{v, w\} \in E(G) \cap X \text{ we have } v \in X \text{ or } w \in X\}.$$

$(D, \mathcal{F})$ is an antimatroid: it is accessible and closed under union. In particular, by Proposition 14.3, it is a greedoid.

Now consider $\mathcal{F}' := \{X \in \mathcal{F} : |X| \leq |E(G)| + k\}$. Since (M1) and (M3) are preserved, $(D, \mathcal{F}')$ is also a greedoid. Set $c(e) := 1$ for $e \in E(G)$ and $c(v) := 0$ for $v \in V(G)$. Then there exists a set $F \in \mathcal{F}'$ with $c(F) = |E(G)|$ if and only if G contains a vertex cover of size k. $\quad\square$

On the other hand, there are interesting functions that can be maximized over arbitrary greedoids, for example bottleneck functions $c(F) := \min\{c'(e) : e \in F\}$ for some $c' : E \to \mathbb{R}_+$ over the maximal elements of $\mathcal{F}$ (Exercise 2). See the book by Korte, Lovász and Schrader [1991] for more results in this area.

14.2 Polymatroids

From Theorem 13.10 we know the tight connection between matroids and submodular functions. Submodular functions define the following interesting class of polyhedra:

Definition 14.9. *A* **polymatroid** *is a polytope of type*

$$P(f) := \left\{ x \in \mathbb{R}^E : x \geq 0, \sum_{e \in A} x_e \leq f(A) \text{ for all } A \subseteq E \right\}$$

where E is a finite set and $f : 2^E \to \mathbb{R}_+$ is a submodular function.

It is not hard to see that for any polymatroid f can be chosen such that $f(\emptyset) = 0$ and f is monotone (Exercise 6; a function $f : 2^E \to \mathbb{R}$ is called **monotone** if $f(X) \leq f(Y)$ for $X \subseteq Y \subseteq E$). Edmonds' original definition was different; see Exercise 7. Moreover, we mention that the term polymatroid is sometimes not used for the polytope but for the pair (E, f).

If f is the rank function of a matroid, $P(f)$ is the convex hull of the incidence vectors of the independent sets of this matroid (Theorem 13.21). We know that the BEST-IN-GREEDY optimizes any linear function over a matroid polytope. A similar greedy algorithm also works for general polymatroids. We assume that f is monotone:

POLYMATROID GREEDY ALGORITHM

Input: A finite set E and a submodular, monotone function $f : 2^E \to \mathbb{R}_+$ with $f(\emptyset) \geq 0$ (given by an oracle). A vector $c \in \mathbb{R}^E$.

Output: A vector $x \in P(f)$ with cx maximum.

---

① Sort $E = \{e_1, \ldots, e_n\}$ such that $c(e_1) \geq \cdots \geq c(e_k) > 0 \geq c(e_{k+1}) \geq \cdots \geq c(e_n)$.

② **If** $k \geq 1$ **then** set $x(e_1) := f(\{e_1\})$.
    Set $x(e_i) := f(\{e_1, \ldots, e_i\}) - f(\{e_1, \ldots, e_{i-1}\})$ for $i = 2, \ldots, k$.
    Set $x(e_i) := 0$ for $i = k + 1, \ldots, n$.

Proposition 14.10. *Let* $E = \{e_1, \ldots, e_n\}$ *and* $f : 2^E \to \mathbb{R}$ *be a submodular function with* $f(\emptyset) \geq 0$. *Let* $b : E \to \mathbb{R}$ *with* $b(e_1) \leq f(\{e_1\})$ *and* $b(e_i) \leq f(\{e_1, \ldots, e_i\}) - f(\{e_1, \ldots, e_{i-1}\})$ *for* $i = 2, \ldots, n$. *Then* $\sum_{a \in A} b(a) \leq f(A)$ *for all* $A \subseteq E$.

Proof: Induction on $i = \max\{j : e_j \in A\}$. The assertion is trivial for $A = \emptyset$ and $A = \{e_1\}$. If $i \geq 2$, then $\sum_{a \in A} b(a) = \sum_{a \in A \setminus \{e_i\}} b(a) + b(e_i) \leq f(A \setminus \{e_i\}) + b(e_i) \leq f(A \setminus \{e_i\}) + f(\{e_1, \ldots, e_i\}) - f(\{e_1, \ldots, e_{i-1}\}) \leq f(A)$, where the first inequality follows from the induction hypothesis and the third one from submodularity. □

Theorem 14.11. *The* POLYMATROID GREEDY ALGORITHM *correctly finds an* $x \in P(f)$ *with* cx *maximum. If* f *is integral, then* x *is also integral.*

Proof: Let $x \in \mathbb{R}^E$ be the output of the POLYMATROID GREEDY ALGORITHM for E, f and c. By definition, if f is integral, then x is also integral. We have $x \geq 0$ since f is monotone, and thus $x \in P(f)$ by Proposition 14.10.

Now let $y \in \mathbb{R}_+^E$ with $cy > cx$. Similarly as in the proof of Theorem 13.19 we set $d_j := c(e_j) - c(e_{j+1})$ ($j = 1, \ldots, k-1$) and $d_k := c(e_k)$, and we have

$$\sum_{j=1}^{k} d_j \sum_{i=1}^{j} x(e_i) = cx < cy \leq \sum_{j=1}^{k} c(e_j) y(e_j) = \sum_{j=1}^{k} d_j \sum_{i=1}^{j} y(e_i).$$

Since $d_j \geq 0$ for all j there is an index $j \in \{1, \ldots, k\}$ with $\sum_{i=1}^{j} y(e_i) > \sum_{i=1}^{j} x(e_i)$; however, since $\sum_{i=1}^{j} x(e_i) = f(\{e_1, \ldots, e_j\})$ this means that $y \notin P(f)$. □

As with matroids, we can also handle the intersection of two polymatroids. The following polymatroid intersection theorem has many implications:

Theorem 14.12. (Edmonds [1970, 1979]) *Let* E *be a finite set, and let* $f, g : 2^E \to \mathbb{R}_+$ *be submodular functions. Then the system*

$$
\begin{aligned}
x &\geq 0 \\
\sum_{e \in A} x_e &\leq f(A) \qquad (A \subseteq E) \\
\sum_{e \in A} x_e &\leq g(A) \qquad (A \subseteq E)
\end{aligned}
$$

is TDI.

Proof: Consider the primal-dual pair of LPs

$$\max\left\{cx : x \geq 0, \sum_{e \in A} x_e \leq f(A) \text{ and } \sum_{e \in A} x_e \leq g(A) \text{ for all } A \subseteq E\right\}$$

and

$$\min\left\{\sum_{A \subseteq E} (f(A)y_A + g(A)z_A) : y, z \geq 0, \sum_{A \subseteq E, e \in A} (y_A + z_A) \geq c_e \text{ for all } e \in E\right\}.$$

To show total dual integrality, we use Lemma 5.24.

Let $c : E(G) \to \mathbb{Z}$, and let y, z be an optimum dual solution for which

$$\sum_{A \subseteq E} (y_A + z_A)|A||E \setminus A| \tag{14.1}$$

is as small as possible. We claim that $\mathcal{F} := \{A \subseteq E : y_A > 0\}$ is a chain, i.e. for any $A, B \in \mathcal{F}$ either $A \subseteq B$ or $B \subseteq A$.

To see this, suppose $A, B \in \mathcal{F}$ with $A \cap B \neq A$ and $A \cap B \neq B$. Let $\epsilon := \min\{y_A, y_B\}$. Set $y'_A := y_A - \epsilon$, $y'_B := y_B - \epsilon$, $y'_{A \cap B} := y_{A \cap B} + \epsilon$, $y'_{A \cup B} := y_{A \cup B} + \epsilon$, and $y'(S) := y(S)$ for all other $S \subseteq E$. Since y', z is a feasible dual solution, it is also optimum (f is submodular) and contradicts the choice of y, because (14.1) is smaller for y', z.

By the same argument, $\mathcal{F}' := \{A \subseteq E : z_A > 0\}$ is a chain. Now let M and M' be the matrices whose columns are indexed with the elements of E and whose rows are the incidence vectors of the elements of $\mathcal{F}$ and $\mathcal{F}'$, respectively. By Lemma 5.24, it suffices to show that $\binom{M}{M'}$ is totally unimodular.

Here we use Ghouila-Houri's Theorem 5.25. Let $\mathcal{R}$ be a set of rows, say $\mathcal{R} = \{A_1, \ldots, A_p, B_1, \ldots, B_q\}$ with $A_1 \supseteq \cdots \supseteq A_p$ and $B_1 \supseteq \cdots \supseteq B_q$. Let $\mathcal{R}_1 := \{A_i : i \text{ odd}\} \cup \{B_i : i \text{ even}\}$ and $\mathcal{R}_2 := \mathcal{R} \setminus \mathcal{R}_1$. Since for any $e \in E$ we have $\{R \in \mathcal{R} : e \in R\} = \{A_1, \ldots, A_{p_e}\} \cup \{B_1, \ldots, B_{q_e}\}$ for some $p_e \in \{0, \ldots, p\}$ and $q_e \in \{0, \ldots, q\}$, the sum of the rows in $\mathcal{R}_1$ minus the sum of the rows in $\mathcal{R}_2$ is a vector with entries $-1, 0, 1$ only. So the criterion of Theorem 5.25 is satisfied. $\square$

One can optimize linear functions over the intersection of two polymatroids. However, this is not as easy as with a single polymatroid. But we can use the ELLIPSOID METHOD if we can solve the SEPARATION PROBLEM for each polymatroid. We return to this question in Section 14.3.

Corollary 14.13. (Edmonds [1970]) *Let $(E, \mathcal{M}_1)$ and $(E, \mathcal{M}_2)$ be two matroids with rank functions r_1 and r_2. Then the convex hull of the incidence vectors of the elements of $\mathcal{M}_1 \cap \mathcal{M}_2$ is the polytope*

$$\left\{x \in \mathbb{R}_+^E : \sum_{e \in A} x_e \leq \min\{r_1(A), r_2(A)\} \text{ for all } A \subseteq E\right\}.$$

Proof: As r_1 and r_2 are nonnegative and submodular (by Theorem 13.10), the above inequality system is TDI (by Theorem 14.12). Since r_1 and r_2 are integral, the polytope is integral (by Corollary 5.16). Since $r_1(A) \leq |A|$ for all $A \subseteq E$, the vertices (the convex hull of which the polytope is by Corollary 3.32) are 0-1-vectors, and thus incidence vectors of common independent sets (elements of $\mathcal{M}_1 \cap \mathcal{M}_2$). On the other hand, each such incidence vector satisfies the inequalities (by definition of the rank function). $\qquad\square$

Of course, the description of the matroid polytope (Theorem 13.21) follows from this by setting $\mathcal{M}_1 = \mathcal{M}_2$. Theorem 14.12 has some further consequences:

Corollary 14.14. (Edmonds [1970]) *Let E be a finite set, and let $f, g : 2^E \to \mathbb{R}_+$ be submodular and monotone functions with $f(\emptyset) = g(\emptyset) = 0$. Then*

$$\max\{\mathbb{1}x : x \in P(f) \cap P(g)\} = \min_{A \subseteq E}(f(A) + g(E \setminus A)).$$

Moreover, if f and g are integral, there exists an integral x attaining the maximum.

Proof: By Theorem 14.12, the dual to

$$\max\{\mathbb{1}x : x \in P(f) \cap P(g)\},$$

which is

$$\min\left\{\sum_{A \subseteq E}(f(A)y_A + g(A)z_A) : y, z \geq 0, \sum_{A \subseteq E, e \in A}(y_A + z_A) \geq 1 \text{ for all } e \in E\right\},$$

has an integral optimum solution y, z. Let $B := \bigcup_{A:y_A \geq 1} A$ and $C := \bigcup_{A:z_A \geq 1} A$. We have $B \cup C = E$. Since f and g are submodular and nonnegative, and $f(\emptyset) = g(\emptyset) = 0$,

$$\sum_{A \subseteq E}(f(A)y_A + g(A)z_A) \geq f(B) + g(C).$$

Since $E \setminus B \subseteq C$ and g is monotone, this is at least $f(B) + g(E \setminus B)$, proving "$\geq$".

The other inequality "$\leq$" is trivial, because for any $A \subseteq E$ we obtain a feasible dual solution y, z by setting $y_A := 1$, $z_{E \setminus A} := 1$ and all other components to zero.

The integrality follows directly from Theorem 14.12 and Corollary 5.16. $\qquad\square$

Theorem 13.31 is a special case. Moreover we obtain:

Corollary 14.15. (Frank [1982]) *Let E be a finite set and $f, g : 2^E \to \mathbb{R}$ such that f is supermodular, g is submodular and $f \leq g$. Then there exists a modular function $h : 2^E \to \mathbb{R}$ with $f \leq h \leq g$. If f and g are integral, h can be chosen integral.*

Proof: Without loss of generality, $f(\emptyset) = g(\emptyset)$ and $f(E) = g(E)$. Let $M :=$ $2\max\{|f(A)| + |g(A)| : A \subseteq E\}$. Let $f'(A) := g(E) - f(E \setminus A) + M|A|$ and $g'(A) := g(A) - f(\emptyset) + M|A|$ for all $A \subseteq E$. f' and g' are nonnegative, submodular and monotone. An application of Corollary 14.14 yields

$$
\begin{aligned}
& \max\{1\!\!1 x : x \in P(f') \cap P(g')\} \\
= {} & \min_{A \subseteq E}(f'(A) + g'(E \setminus A)) \\
= {} & \min_{A \subseteq E}(g(E) - f(E \setminus A) + M|A| + g(E \setminus A) - f(\emptyset) + M|E \setminus A|) \\
\geq {} & g(E) - f(\emptyset) + M|E|.
\end{aligned}
$$

So let $x \in P(f') \cap P(g')$ with $1\!\!1 x = g(E) - f(\emptyset) + M|E|$. If f and g are integral, x can be chosen integral. Let $h'(A) := \sum_{e \in A} x_e$ and $h(A) := h'(A) + f(\emptyset) - M|A|$ for all $A \subseteq E$. The function h is modular. Moreover, for all $A \subseteq E$ we have $h(A) \leq g'(A) + f(\emptyset) - M|A| = g(A)$ and $h(A) = 1\!\!1 x - h'(E \setminus A) + f(\emptyset) - M|A| \geq g(E) + M|E| - M|A| - f'(E \setminus A) = f(A)$. $\square$

The analogy to convex and concave functions is obvious; see also Exercise 10.

14.3 Minimizing Submodular Functions

The SEPARATION PROBLEM for a polymatroid $P(f)$ and a vector x asks for a set A with $f(A) < \sum_{e \in A} x(e)$. So this problem reduces to finding a set A minimizing $g(A)$, where $g(A) := f(A) - \sum_{e \in A} x(e)$. Note that if f is submodular, then g is also submodular. Therefore it is an interesting problem to minimize submodular functions.

Another motivation might be that submodular functions can be regarded as the discrete analogue of convex functions (Corollary 14.15 and Exercise 10). We have already solved a special case in Section 8.7: finding the minimum cut in an undirected graph can be regarded as minimizing a certain symmetric submodular function $f : 2^U \to \mathbb{R}_+$ over $2^U \setminus \{\emptyset, U\}$. Before returning to this special case we first show how to minimize general submodular functions. For simplicity we restrict ourselves to integer-valued submodular functions:

SUBMODULAR FUNCTION MINIMIZATION PROBLEM

Instance: A finite set U. A submodular function $f : 2^U \to \mathbb{R}$ (given by an oracle).

Task: Find a subset $X \subseteq U$ with $f(X)$ minimum.

Grötschel, Lovász and Schrijver [1981] showed how this problem can be solved with the help of the ELLIPSOID METHOD. The idea is to determine the minimum by binary search; this will reduce the problem to the SEPARATION PROBLEM for a polymatroid. Using the equivalence of separation and optimization (Section 4.6), it thus suffices to optimize linear functions over polymatroids. However, this can be

done easily by the POLYMATROID GREEDY ALGORITHM. We first need an upper
bound on $|f(S)|$ for $S \subseteq U$:

Proposition 14.16. *For any submodular function $f : 2^U \to \mathbb{R}$ and any $S \subseteq U$
we have*

$$f(U) - \sum_{u \in U} \max\{0, f(\{u\}) - f(\emptyset)\} \leq f(S) \leq f(\emptyset) + \sum_{u \in U} \max\{0, f(\{u\}) - f(\emptyset)\}.$$

*In particular, a number B with $|f(S)| \leq B$ for all $S \subseteq U$ can be computed in
linear time, with $|U| + 2$ oracle calls to f.*

Proof: By repeated application of submodularity we get for $\emptyset \neq S \subseteq U$ (let
$x \in S$):

$$f(S) \leq -f(\emptyset) + f(S \setminus \{x\}) + f(\{x\}) \leq \cdots \leq -|S| f(\emptyset) + f(\emptyset) + \sum_{x \in S} f(\{x\}),$$

and for $S \subset U$ (let $y \in U \setminus S$):

$$\begin{aligned} f(S) &\geq -f(\{y\}) + f(S \cup \{y\}) + f(\emptyset) \geq \cdots \\ &\geq -\sum_{y \in U \setminus S} f(\{y\}) + f(U) + |U \setminus S| f(\emptyset). \end{aligned} \qquad \square$$

For the rest of this section, we restrict to integer-valued functions and show
how to solve the SUBMODULAR FUNCTION MINIMIZATION PROBLEM using the
equivalence of optimization and separation in the unusual direction.

Proposition 14.17. *The following problem can be solved in polynomial time:
Given a finite set U, a submodular and monotone function $f : 2^U \to \mathbb{Z}_+$ (by an
oracle) with $f(S) > 0$ for $S \neq \emptyset$, a number $B \in \mathbb{N}$ with $f(S) \leq B$ for all $S \subseteq U$,
and a vector $x \in \mathbb{Z}_+^U$, decide if $x \in P(f)$ and otherwise return a set $S \subseteq U$ with
$\sum_{v \in S} x(v) > f(S)$.*

Proof: This is the SEPARATION PROBLEM for the polymatroid $P(f)$. We will
use Theorem 4.23, because we have already solved the optimization problem for
$P(f)$: the POLYMATROID GREEDY ALGORITHM maximizes any linear function
over $P(f)$ (Theorem 14.11).

We have to check the prerequisites of Theorem 4.23. Since the zero vector
and the unit vectors are all in $P(f)$, we can take $x_0 := \epsilon \mathbb{1}$ as a point in the
interior, where $\epsilon = \frac{1}{|U|+1}$. We have $\text{size}(x_0) = O(|U| \log |U|)$. Moreover, each
vertex of $P(f)$ is produced by the POLYMATROID GREEDY ALGORITHM (for
some objective function; cf. Theorem 14.11) and thus has size $O(|U|(2 + \log B))$.
We conclude that the SEPARATION PROBLEM can be solved in polynomial time.
By Theorem 4.23, we get a facet-defining inequality of $P(f)$ violated by x if
$x \notin P(f)$. This corresponds to a set $S \subseteq U$ with $\sum_{v \in S} x(v) > f(S)$. $\qquad \square$

If f is not monotone, we cannot apply this result directly. Instead we consider a different function:

Proposition 14.18. *Let* $f : 2^U \to \mathbb{R}$ *be a submodular function and* $\beta \in \mathbb{R}$. *Then* $g : 2^U \to \mathbb{R}$, *defined by*

$$g(X) := f(X) - \beta + \sum_{e \in X} (f(U \setminus \{e\}) - f(U)),$$

is submodular and monotone.

Proof: The submodularity of g follows directly from the submodularity of f. To show that g is monotone, let $X \subset U$ and $e \in U \setminus X$. We have $g(X \cup \{e\}) - g(X) = f(X \cup \{e\}) - f(X) + f(U \setminus \{e\}) - f(U) \geq 0$ since f is submodular. □

Theorem 14.19. *Integer-valued submodular functions* $f : 2^U \to \mathbb{Z}$ *(given by an oracle) can be minimized in time polynomial in* $|U| + \log \max\{|f(S)| : S \subseteq U\}$.

Proof: Let U be a finite set; suppose we are given f by an oracle. First compute a number $B \in \mathbb{N}$ with $\max\{|f(S)| : S \subseteq U\} \leq B \leq (2|U|+1) \max\{|f(S)| : S \subseteq U\}$ (cf. Proposition 14.16). Since f is submodular, we have for each $e \in U$ and for each $X \subseteq U \setminus \{e\}$:

$$f(\{e\}) - f(\emptyset) \geq f(X \cup \{e\}) - f(X) \geq f(U) - f(U \setminus \{e\}). \tag{14.2}$$

If, for some $e \in U$, $f(\{e\}) - f(\emptyset) \leq 0$, then by (14.2) there is an optimum set S containing e. In this case we consider the instance (U', B, f') defined by $U' := U \setminus \{e\}$ and $f'(X) := f(X \cup \{e\})$ for $X \subseteq U \setminus \{e\}$, find a set $S' \subseteq U'$ with $f'(S')$ minimum and output $S := S' \cup \{e\}$.

Similarly, if $f(U) - f(U \setminus \{e\}) \geq 0$, then by (14.2) there is an optimum set S not containing e. In this case we simply minimize f restricted to $U \setminus \{e\}$. In both cases we have reduced the size of the ground set.

So we may assume that $f(\{e\}) - f(\emptyset) > 0$ and $f(U \setminus \{e\}) - f(U) > 0$ for all $e \in U$. Let $x(e) := f(U \setminus \{e\}) - f(U)$. For each integer β with $-B \leq \beta \leq f(\emptyset)$ we define $g(X) := f(X) - \beta + \sum_{e \in X} x(e)$. By Proposition 14.18, g is submodular and monotone. Furthermore we have $g(\emptyset) = f(\emptyset) - \beta \geq 0$ and $g(\{e\}) = f(\{e\}) - \beta + x(e) > 0$ for all $e \in U$, and thus $g(X) > 0$ for all $\emptyset \neq X \subseteq U$. Now we apply Proposition 14.17 and check if $x \in P(g)$. If yes, we have $f(X) \geq \beta$ for all $X \subseteq U$. Otherwise we get a set S with $f(S) < \beta$.

Now we apply binary search: by choosing β appropriately each time, we need $O(\log(2B))$ iterations to find the number $\beta^* \in \{-B, -B+1, \ldots, f(\emptyset)\}$ for which $f(X) \geq \beta^*$ for all $X \subseteq U$ but $f(S) < \beta^* + 1$ for some $S \subseteq U$. This set S minimizes f. □

The first strongly polynomial-time algorithm has been designed by Grötschel, Lovász and Schrijver [1988], also based on the ellipsoid method. Combinatorial algorithms to solve the SUBMODULAR FUNCTION MINIMIZATION PROBLEM in strongly polynomial time have been found by Schrijver [2000] and independently by Iwata, Fleischer and Fujishige [2001], both based on earlier work by Cunningham [1984]. In the next section we describe Schrijver's algorithm.

14.4 Schrijver's Algorithm

For a finite set U and a submodular function $f : 2^U \to \mathbb{R}$, assume w.l.o.g. that $U = \{1, \ldots, n\}$ and $f(\emptyset) = 0$. At each stage, Schrijver's [2000] algorithm maintains a point x in the so-called **base polyhedron** of f, defined by

$$\left\{ x \in \mathbb{R}^U : \sum_{u \in A} x(u) \le f(A) \text{ for all } A \subseteq U, \sum_{u \in U} x(u) = f(U) \right\}.$$

We mention that the set of vertices of this base polyhedron is precisely the set of vectors $b^\prec$ for all total orders $\prec$ of U, where we define

$$b^\prec(u) := f(\{v \in U : v \preceq u\}) - f(\{v \in U : v \prec u\})$$

$(u \in U)$. This fact, which we will not need here, can be proved in a similar way as Theorem 14.11 (Exercise 14).

The point x is always written as an explicit convex combination $x = \lambda_1 b^{\prec_1} + \cdots + \lambda_k b^{\prec_k}$ of these vertices. Initially, one can choose $k = 1$ and any total order. For a total order $\prec$ and $s, u \in U$, we denote by $\prec^{s,u}$ the total order that results from $\prec$ by moving u just before s. Moreover, let χ^u denote the incidence vector of u $(u \in U)$.

SCHRIJVER'S ALGORITHM

Input: A finite set $U = \{1, \ldots, n\}$. A submodular function $f : 2^U \to \mathbb{R}$ with $f(\emptyset) = 0$ (given by an oracle).

Output: A subset $X \subseteq U$ with $f(X)$ minimum.

①   Set $k := 1$, let $\prec_1$ be any total order on U, and set $x := b^{\prec_1}$.

②   Set $D := (U, A)$, where $A = \{(u, v) : u \prec_i v$ for some $i \in \{1, \ldots, k\}\}$.

③   Let $P := \{v \in U : x(v) > 0\}$ and $N := \{v \in U : x(v) < 0\}$, and let X be the set of vertices not reachable from P in the digraph D.
    If $N \subseteq X$, **then stop else** let $d(v)$ denote the distance from P to v in D.

④   Choose the vertex $t \in N$ reachable from P with $(d(t), t)$ lexicographically maximum.
    Choose the maximal vertex s with $(s, t) \in A$ and $d(s) = d(t) - 1$.
    Let $i \in \{1, \ldots, k\}$ such that $\alpha := |\{v : s \prec_i v \preceq_i t\}|$ is maximum (the number of indices attaining this maximum will be denoted by β).

⑤   Compute a number ϵ with $0 \le \epsilon \le -x(t)$ and write $x' := x + \epsilon(\chi^t - \chi^s)$ as an explicit convex combination of at most n vectors, chosen among $b^{\prec_1}, \ldots, b^{\prec_k}$ and $b_i^{\prec^{s,u}}$ for all $u \in U$ with $s \prec_i u \preceq_i t$, with the additional property that $b^{\prec_i}$ does not occur if $x'(t) < 0$.

⑥   Set $x := x'$, rename the vectors in the convex combination of x as $b^{\prec_1}, \ldots, b^{\prec_{k'}}$, set $k := k'$, and **go to** ②.

Theorem 14.20. (Schrijver [2000]) SCHRIJVER'S ALGORITHM *works correctly.*

Proof: The algorithm terminates if D contains no path from P to N and outputs the set X of vertices not reachable from P. Clearly $N \subseteq X \subseteq U \setminus P$, so $\sum_{u \in X} x(u) \leq \sum_{u \in W} x(u)$ for each $W \subseteq U$. Moreover, no edge enters X, so either $X = \emptyset$ or for each $j \in \{1, \ldots, k\}$ there exists a $v \in X$ with $X = \{u \in U : u \preceq_j v\}$. Hence, by definition, $\sum_{u \in X} b^{\prec_j}(u) = f(X)$ for all $j \in \{1, \ldots, k\}$. Moreover, by Proposition 14.10, $\sum_{u \in W} b^{\prec_j}(u) \leq f(W)$ for all $W \subseteq U$ and $j \in \{1, \ldots, k\}$. Therefore, for each $W \subseteq U$,

$$
f(W) \;\geq\; \sum_{j=1}^{k} \lambda_j \sum_{u \in W} b^{\prec_j}(u) = \sum_{u \in W} \sum_{j=1}^{k} \lambda_j b^{\prec_j}(u) = \sum_{u \in W} x(u)
$$

$$
\geq\; \sum_{u \in X} x(u) = \sum_{u \in X} \sum_{j=1}^{k} \lambda_j b^{\prec_j}(u) = \sum_{j=1}^{k} \lambda_j \sum_{u \in X} b^{\prec_j}(u) = f(X),
$$

proving that X is an optimum solution. $\square$

Lemma 14.21. (Schrijver [2000]) *Each iteration can be performed in $O(n^3 + \gamma n^2)$ time, where γ is the time for an oracle call.*

Proof: It suffices to show that ⑤ can be done in $O(n^3 + \gamma n^2)$ time. Let $x = \lambda_1 b^{\prec_1} + \cdots + \lambda_k b^{\prec_k}$ and $s \prec_i t$. We first show:

Claim: $\delta(\chi^t - \chi^s)$, for some $\delta \geq 0$, can be written as a convex combination of the vectors $b^{\prec_i^{s,v}} - b^{\prec_i}$ for $s \prec_i v \preceq_i t$ in $O(\gamma n^2)$ time.

To prove this, we need some preliminaries. Let $s \prec_i v \preceq_i t$. By definition, $b^{\prec_i^{s,v}}(u) = b^{\prec_i}(u)$ for $u \prec_i s$ or $u \succ_i v$. As f is submodular, we have for $s \preceq_i u \prec_i v$:

$$
b^{\prec_i^{s,v}}(u) = f(\{w \in U : w \preceq_i^{s,v} u\}) - f(\{w \in U : w \prec_i^{s,v} u\})
$$
$$
\leq f(\{w \in U : w \preceq_i u\}) - f(\{w \in U : w \prec_i u\}) = b^{\prec_i}(u).
$$

Moreover, for $u = v$ we have:

$$
b^{\prec_i^{s,v}}(v) = f(\{w \in U : w \preceq_i^{s,v} v\}) - f(\{w \in U : w \prec_i^{s,v} v\})
$$
$$
= f(\{w \in U : w \prec_i s\} \cup \{v\}) - f(\{w \in U : w \prec_i s\})
$$
$$
\geq f(\{w \in U : w \preceq_i v\}) - f(\{w \in U : w \prec_i v\})
$$
$$
= b^{\prec_i}(v).
$$

Finally, observe that $\sum_{u \in U} b^{\prec_i^{s,v}}(u) = f(U) = \sum_{u \in U} b^{\prec_i}(u)$.

As the claim is trivial if $b^{\prec_i^{s,v}} = b^{\prec_i}$ for some $s \prec_i v \preceq_i t$, we may assume $b^{\prec_i^{s,v}}(v) > b^{\prec_i}(v)$ for all $s \prec_i v \preceq_i t$. We recursively set

$$
\kappa_v := \frac{\chi_v^t - \sum_{v \prec_i w \preceq_i t} \kappa_w (b^{\prec_i^{s,w}}(v) - b^{\prec_i}(v))}{b^{\prec_i^{s,v}}(v) - b^{\prec_i}(v)} \geq 0
$$

for $s \prec_i v \preceq_i t$, and obtain $\sum_{s \prec_i v \preceq_i t} \kappa_v(b^{\prec_i^{s,v}} - b^{\prec_i}) = \chi^t - \chi^s$, because $\sum_{s \prec_i v \preceq_i t} \kappa_v(b^{\prec_i^{s,v}}(u) - b^{\prec_i}(u)) = \sum_{u \preceq_i v \preceq_i t} \kappa_v(b^{\prec_i^{s,v}}(u) - b^{\prec_i}(u)) = \chi_u^t$ for all $s \prec_i u \preceq_i t$, and the sum over all components is zero.

By setting $\delta := \frac{1}{\sum_{s \prec_i v \preceq_i t} \kappa_v}$ and multiplying each κ_u by δ, we see that the claim follows.

Now consider $\epsilon := \min\{\lambda_i \delta, -x(t)\}$ and $x' := x + \epsilon(\chi^t - \chi^s)$. If $\epsilon = \lambda_i \delta \le -x(t)$, then we have $x' = \sum_{j=1}^k \lambda_j b^{\prec_j} + \lambda_i \sum_{s \prec_i v \preceq_i t} \kappa_v(b^{\prec_i^{s,v}} - b^{\prec_i})$, i.e. we have written x' as a convex combination of $b^{\prec_j}$ ($j \in \{1, \ldots, k\} \setminus \{i\}$) and $b^{\prec_i^{s,v}}$ ($s \prec_i v \preceq_i t$). If $\epsilon = -x(t)$, we may additionally use $b^{\prec_i}$ in the convex combination.

We finally reduce this convex combination to at most n vectors in $O(n^3)$ time, as shown in Exercise 5 of Chapter 4. $\qquad\Box$

Lemma 14.22. (Vygen [2003]) SCHRIJVER'S ALGORITHM *terminates after* $O(n^5)$ *iterations.*

Proof: If an edge (v, w) is introduced after a new vector $b^{\prec_i^{s,v}}$ has been added in ⑤ of an iteration, then $s \preceq_i w \prec_i v \preceq_i t$ in this iteration. Thus $d(w) \le d(s) + 1 = d(t) \le d(v) + 1$ in this iteration, and the introduction of the new edge cannot make the distance from P to any $u \in U$ smaller. As ⑤ makes sure that no element is ever added to P, the distance $d(u)$ never decreases for any $u \in U$.

Call a *block* a sequence of iterations where the pair (s, t) remains constant. Note that each block has $O(n^2)$ iterations, because (α, β) decreases lexicographically in each iteration within each block. It remains to prove that there are $O(n^3)$ blocks.

A block can end only because of at least one of the following reasons (by the choice of t and s, since an iteration with $t = t^*$ does not add any edge whose head is t^*, and since a vertex v can enter N only if $v = s$ and hence $d(v) < d(t)$):

(a) the distance $d(v)$ increases for some $v \in U$.
(b) t is removed from N.
(c) (s, t) is removed from A.

We now count the number of blocks of these three types. Clearly there are $O(n^2)$ blocks of type (a).

Now consider type (b). We claim that for each $t^* \in U$ there are $O(n^2)$ iterations with $t = t^*$ and $x'(t) = 0$. This is easy to see: between every two such iterations, $d(v)$ must change for some $v \in U$, and this can happen $O(n^2)$ times as d-values can only increase. Thus there are $O(n^3)$ blocks of type (b).

We finally show that there are $O(n^3)$ blocks of type (c). It suffices to show that $d(t)$ will change before the next such block with the pair (s, t).

For $s, t \in U$, we say that s is *t-boring* if $(s, t) \notin A$ or $d(t) \le d(s)$. Let $s^*, t^* \in U$, and consider the time period after a block with $s = s^*$ and $t = t^*$ ending because (s^*, t^*) is removed from A, until the subsequent change of $d(t^*)$. We prove that each $v \in \{s^*, \ldots, n\}$ is t^*-boring throughout this period. Applying this for $v = s^*$ concludes the proof.

At the beginning of the period, each $v \in \{s^* + 1, \ldots, n\}$ is t^*-boring due to the choice of $s = s^*$ in the iteration immediately preceding the period. s^* is also t^*-boring as (s^*, t^*) is removed from A. As $d(t^*)$ remains constant within the considered time period and $d(v)$ never decreases for any v, we only have to check the introduction of new edges.

Suppose that, for some $v \in \{s^*, \ldots, n\}$, the edge (v, t^*) is added to A after an iteration that chooses the pair (s, t). Then, by the initial remarks of this proof, $s \preceq_i t^* \prec_i v \preceq_i t$ in this iteration, and thus $d(t^*) \leq d(s) + 1 = d(t) \leq d(v) + 1$. Now we distinguish two cases: If $s > v$, then we have $d(t^*) \leq d(s)$: either because $t^* = s$, or as s was t^*-boring and $(s, t^*) \in A$. If $s < v$, then we have $d(t) \leq d(v)$: either because $t = v$, or by the choice of s and since $(v, t) \in A$. In both cases we conclude that $d(t^*) \leq d(v)$, and v remains t^*-boring. $\square$

Theorem 14.20, Lemma 14.21 and Lemma 14.22 imply:

Theorem 14.23. *The* SUBMODULAR FUNCTION MINIMIZATION PROBLEM *can be solved in* $O(n^8 + \gamma n^7)$ *time, where γ is the time for an oracle call.* $\square$

Iwata [2002] described a fully combinatorial algorithm (using only additions, subtractions, comparisons and oracle calls, but no multiplication or division). He also improved the running time (Iwata [2003]). A faster strongly polynomial-time algorithm was found by Orlin [2007]; it runs in $O(n^6 + \gamma n^5)$ time.

Recently, Lee, Sidford and Wong [2015] found a strongly polynomial-time algorithm that minimizes a submodular function in $O(\gamma n^3 \log^2 n + n^4 \log^{O(1)} n)$ time.

14.5 Symmetric Submodular Functions

A submodular function $f : 2^U \to \mathbb{R}$ is called **symmetric** if $f(A) = f(U \setminus A)$ for all $A \subseteq U$. In this special case the SUBMODULAR FUNCTION MINIMIZATION PROBLEM is trivial, since $2f(\emptyset) = f(\emptyset) + f(U) \leq f(A) + f(U \setminus A) = 2f(A)$ for all $A \subseteq U$, implying that the empty set is optimal. Hence the problem is interesting only if this trivial case is excluded: one looks for a nonempty proper subset A of U such that $f(A)$ is minimum.

Generalizing the algorithm of Section 8.7, Queyranne [1998] has found a relatively simple combinatorial algorithm for this problem using only $O(n^3)$ oracle calls. The following lemma is a generalization of Lemma 8.41 (Exercise 16):

Lemma 14.24. *Given a symmetric submodular function $f : 2^U \to \mathbb{R}$ with $n := |U| \geq 2$, we can find two elements $x, y \in U$ with $x \neq y$ and $f(\{x\}) = \min\{f(X) : x \in X \subseteq U \setminus \{y\}\}$ in $O(n^2 \theta)$ time, where θ is the time bound of the oracle for f.*

Proof: We construct an order $U = \{u_1, \ldots, u_n\}$ by doing the following for $k = 1, \ldots, n - 1$. Suppose that $u_1, \ldots, u_{k-1}$ are already constructed; let $U_{k-1} :=$

$\{u_1, \ldots, u_{k-1}\}$. For $C \subseteq U$ we define

$$w_k(C) := f(C) - \frac{1}{2}(f(C \setminus U_{k-1}) + f(C \cup U_{k-1}) - f(U_{k-1})).$$

Note that w_k is also symmetric. Let u_k be an element of $U \setminus U_{k-1}$ that maximizes $w_k(\{u_k\})$.

Finally, let u_n be the only element in $U \setminus \{u_1, \ldots, u_{n-1}\}$. Obviously the construction of the order $u_1, \ldots, u_n$ can be done in $O(n^2\theta)$ time.

Claim: For all $k = 1, \ldots, n-1$ and all $x, y \in U \setminus U_{k-1}$ with $x \neq y$ and $w_k(\{x\}) \leq w_k(\{y\})$ we have

$$w_k(\{x\}) = \min\{w_k(C) : x \in C \subseteq U \setminus \{y\}\}.$$

We prove the claim by induction on k. For $k = 1$ the assertion is trivial since $w_1(C) = \frac{1}{2}f(\emptyset)$ for all $C \subseteq U$.

Let now $k > 1$ and $x, y \in U \setminus U_{k-1}$ with $x \neq y$ and $w_k(\{x\}) \leq w_k(\{y\})$. Moreover, let $Z \subseteq U$ with $u_{k-1} \notin Z$, and let $z \in Z \setminus U_{k-1}$. By the choice of u_{k-1} we have $w_{k-1}(\{z\}) \leq w_{k-1}(\{u_{k-1}\})$; thus by the induction hypothesis we get $w_{k-1}(\{z\}) \leq w_{k-1}(Z)$. Furthermore, the submodularity of f implies

$$(w_k(Z) - w_{k-1}(Z)) - (w_k(\{z\}) - w_{k-1}(\{z\}))$$
$$= \frac{1}{2}(f(Z \cup U_{k-2}) - f(Z \cup U_{k-1}) - f(U_{k-2}) + f(U_{k-1}))$$
$$\quad - \frac{1}{2}(f(\{z\} \cup U_{k-2}) - f(\{z\} \cup U_{k-1}) - f(U_{k-2}) + f(U_{k-1}))$$
$$= \frac{1}{2}(f(Z \cup U_{k-2}) + f(\{z\} \cup U_{k-1}) - f(Z \cup U_{k-1}) - f(\{z\} \cup U_{k-2}))$$
$$\geq 0.$$

Hence $w_k(Z) - w_k(\{z\}) \geq w_{k-1}(Z) - w_{k-1}(\{z\}) \geq 0$.

To conclude the proof of the claim, let $C \subseteq U$ with $x \in C$ and $y \notin C$. There are two cases:

Case 1: $u_{k-1} \notin C$. Then the above result for $Z = C$ and $z = x$ yields $w_k(C) \geq w_k(\{x\})$ as required.

Case 2: $u_{k-1} \in C$. Then we apply the above to $Z = U \setminus C$ and $z = y$ and get $w_k(C) = w_k(U \setminus C) \geq w_k(\{y\}) \geq w_k(\{x\})$.

This completes the proof of the claim. Applying it to $k = n-1$, $x = u_n$ and $y = u_{n-1}$ we get

$$w_{n-1}(\{u_n\}) = \min\{w_{n-1}(C) : u_n \in C \subseteq U \setminus \{u_{n-1}\}\}.$$

Since $w_{n-1}(C) = f(C) - \frac{1}{2}(f(\{u_n\}) + f(U \setminus \{u_{n-1}\}) - f(U_{n-2}))$ for all $C \subseteq U$ with $u_n \in C$ and $u_{n-1} \notin C$, the lemma follows (set $x := u_n$ and $y := u_{n-1}$). $\quad\square$

The above proof is due to Fujishige [1998]. Now we can proceed analogously to the proof of Theorem 8.42:

Theorem 14.25. (Queyranne [1998]) *Given a symmetric submodular function* f : $2^U \to \mathbb{R}$, *a nonempty proper subset* A *of* U *such that* $f(A)$ *is minimum can be found in* $O(n^3 \theta)$ *time where* θ *is the time bound of the oracle for* f.

Proof: If $|U| = 1$, the problem is trivial. Otherwise we apply Lemma 14.24 and find two elements $x, y \in U$ with $f(\{x\}) = \min\{f(X) : x \in X \subseteq U \setminus \{y\}\}$ in $O(n^2 \theta)$ time. Next we recursively find a nonempty proper subset of $U \setminus \{x\}$ minimizing the function $f' : 2^{U \setminus \{x\}} \to \mathbb{R}$, defined by $f'(X) := f(X)$ if $y \notin X$ and $f'(X) := f(X \cup \{x\})$ if $y \in X$. One readily observes that f' is symmetric and submodular.

Let $\emptyset \neq Y \subset U \setminus \{x\}$ be a set minimizing f'; w.l.o.g. $y \in Y$ (as f' is symmetric). We claim that either $\{x\}$ or $Y \cup \{x\}$ minimizes f (over all nonempty proper subsets of U). To see this, consider any $C \subset U$ with $x \in C$. If $y \notin C$, then we have $f(\{x\}) \leq f(C)$ by the choice of x and y. If $y \in C$, then $f(C) = f'(C \setminus \{x\}) \geq f'(Y) = f(Y \cup \{x\})$. Hence $f(C) \geq \min\{f(\{x\}), f(Y \cup \{x\})\}$ for all nonempty proper subsets C of U.

To achieve the asserted running time we of course cannot compute f' explicitly. Rather we store a partition of U, initially consisting of the singletons. At each step of the recursion we build the union of those two sets of the partition that contain x and y. In this way f' can be computed efficiently (using the oracle for f). $\square$

This result has been further generalized by Nagamochi and Ibaraki [1998] and by Rizzi [2000].

14.6 Submodular Function Maximization

The problem of maximizing a submodular function (again given by an oracle) cannot be solved with a polynomial number of oracle calls, as was shown by Feige, Mirrokni and Vondrák [2011]; see Exercise 18. However, there are efficient algorithms that find an approximate solution, i.e., given a nonnegative submodular function $f : 2^U \to \mathbb{R}_+$ find a set $A \subseteq U$ with $f(A) \geq \frac{1}{k} \max\{f(W) : W \subseteq U\}$. This is called a k-approximation. Already the expected value of a set chosen uniformly at random is a 4-approximation (cf. Exercise 17). But one can do better. Consider the following simple randomized algorithm. It starts with $A = \emptyset$ and $B = U$ and, in iteration i, decides to include i (and adds it to A) or to not include i (and remove it from B). At the end, A and B coincide.

SIMPLE SUBMODULAR FUNCTION MAXIMIZATION ALGORITHM

Input: A finite set $U = \{1, \ldots, n\}$. A submodular function $f : 2^U \to \mathbb{R}_+$ (given by an oracle).

Output: A subset $A \subseteq U$.

---

①   Set $A := \emptyset$ and $B := U$.

②   **For** $i := 1, \ldots, n$ **do**:

       $\Delta_A := f(A \cup \{i\}) - f(A)$.

       $\Delta_B := f(B \setminus \{i\}) - f(B)$.

       Set $A := A \cup \{i\}$ with probability x and $B := B \setminus \{i\}$ otherwise, where

$$x = \begin{cases} 1 & \text{if } \Delta_B \leq 0 \\ 0 & \text{if } \Delta_A < 0 \\ \frac{\Delta_A}{\Delta_A + \Delta_B} & \text{if } \Delta_A \geq 0 \text{ and } \Delta_B > 0 \end{cases} \tag{14.3}$$

③   **Output** A.

Theorem 14.26. (Buchbinder et al. [2015]) *The expected value of $f(A)$, where A is the output of the* SIMPLE SUBMODULAR FUNCTION MAXIMIZATION AL-GORITHM, *is at least* $\frac{1}{2} \max\{f(W) : W \subseteq U\}$.

Proof: First note that $\Delta_A + \Delta_B = f(A \cup \{i\}) - f(A) + f(B \setminus \{i\}) - f(B) \geq 0$ in every iteration because $A = B \cap \{1, \ldots, i-1\}$, $B = A \cup \{i, \ldots, n\}$, and f is submodular.

Let $A_0 := \emptyset$, $B_0 := U$, and let A_i, B_i be the sets A, B after i iterations. Let $O_0 \subseteq U$ be a set with $f(O_0) = \max\{f(W) : W \subseteq U\}$, and $O_i \subseteq U$ defined by $O_i \cap \{i+1, \ldots, n\} = O_0 \cap \{i+1, \ldots, n\}$ and $O_i \cap \{1, \ldots, i\} = A_i = B_i \cap \{1, \ldots, i\}$. In other words, O_i agrees to the decisions made in the first i iterations, and agrees to the optimum solution O_0 on the elements $i+1, \ldots, n$. Now consider the change in iteration i. We claim:

$$\text{Exp}[f(O_{i-1}) - f(O_i)] \leq \max\{x \Delta_B, (1-x) \Delta_A\}, \tag{14.4}$$

where x is the probability that we add i to A, independently of how x is chosen in the algorithm.

To show (14.4), we consider two cases. If $i \notin O_0$, then $\text{Exp}[f(O_{i-1}) - f(O_i)] = x(f(O_{i-1}) - f(O_{i-1} \cup \{i\})) + (1-x)(f(O_{i-1}) - f(O_{i-1})) = x(f(O_{i-1}) - f(O_{i-1} \cup \{i\})) \leq x(f(B_{i-1} \setminus \{i\}) - f(B_{i-1})) = x \Delta_B$, where the inequality follows from submodularity and $O_{i-1} \subseteq B_{i-1} \setminus \{i\}$.

If $i \in O_0$, then $\text{Exp}[f(O_{i-1}) - f(O_i)] = x(f(O_{i-1}) - f(O_{i-1})) + (1-x)(f(O_{i-1}) - f(O_{i-1} \setminus \{i\})) = (1-x)(f(O_{i-1}) - f(O_{i-1} \setminus \{i\})) \leq (1-x)(f(A_{i-1} \cup \{i\}) - f(A_{i-1})) = (1-x) \Delta_A$, where the inequality follows from submodularity and $O_{i-1} \setminus \{i\} \supseteq A_{i-1}$.

Next we show that our choice of x implies that

$$\max\{x \Delta_B, (1-x) \Delta_A\} \leq \frac{1}{2} \big(\text{Exp}[f(A_i) - f(A_{i-1})] + \text{Exp}[f(B_i) - f(B_{i-1})] \big). \tag{14.5}$$

Indeed, if $\Delta_A < 0$ or $\Delta_B \leq 0$, the left-hand side is 0, and $f(A_i) - f(A_{i-1}) + f(B_i) - f(B_{i-1}) = \max\{\Delta_A, \Delta_B\} \geq 0$. If $\Delta_A \geq 0$ and $\Delta_B > 0$, the inequality follows from the simple computation

$$
\begin{aligned}
\max\{x\Delta_B, (1-x)\Delta_A\} &= \frac{\Delta_A\Delta_B}{\Delta_A + \Delta_B} \\
&\leq \frac{1}{2}\frac{\Delta_A^2 + \Delta_B^2}{\Delta_A + \Delta_B} \\
&= \frac{1}{2}\left(\frac{\Delta_A}{\Delta_A + \Delta_B}\Delta_A + \frac{\Delta_B}{\Delta_A + \Delta_B}\Delta_B\right) \\
&= \frac{1}{2}\big(\mathrm{Exp}[f(A_i) - f(A_{i-1})] + \mathrm{Exp}[f(B_i) - f(B_{i-1})]\big).
\end{aligned}
$$

Together, (14.4) and (14.5) imply

$$
\mathrm{Exp}[f(O_{i-1}) - f(O_i)] \leq \frac{1}{2}\mathrm{Exp}[f(A_i) - f(A_{i-1})] + \frac{1}{2}\mathrm{Exp}[f(B_i) - f(B_{i-1})].
$$

Summing this inequality for $i = 1, \ldots, n$ yields $\mathrm{Exp}[f(O_0)] - \mathrm{Exp}[f(O_n)] \leq \frac{1}{2}\big(\mathrm{Exp}[f(A_n)] - \mathrm{Exp}[f(A_0)] + \mathrm{Exp}[f(B_n)] - \mathrm{Exp}[f(B_0)]\big) = \mathrm{Exp}[f(A_n)] - \frac{1}{2}f(\emptyset) - \frac{1}{2}f(U) \leq \mathrm{Exp}[f(A_n)]$. Using $O_n = A_n$ we get $2\,\mathrm{Exp}[f(A_n)] \geq f(O_0)$. $\square$

This algorithm can be derandomized as follows.

Theorem 14.27. (Buchbinder and Feldman [2016]) *There is a deterministic polynomial-time algorithm that computes a 2-approximation for the problem of maximizing a nonnegative submodular function.*

Proof: Instead of making random choices, we keep an explicit probability distribution on pairs (A, B). The only problem is to avoid that the support of this distribution grows exponentially. We proceed as follows. Given a set T_{i-1} of triples (p, A, B) with $p > 0$ and $A \subseteq B \subseteq U$ with $A = A \cap \{1, \ldots, i-1\} = B \cap \{1, \ldots, i-1\}$ and $\{i, \ldots, n\} \subseteq B$ and $\sum_{(p,A,B)\in T_{i-1}} p = 1$ at the beginning of iteration i (initially $T_0 = \{(1, \emptyset, U)\}$), we consider the polytope that consists of all $x \in [0, 1]^{T_{i-1}}$ satisfying

$$
\sum_{t=(p,A,B)\in T_{i-1}} p(x_t \Delta_A + (1-x_t)\Delta_B) \geq 2 \sum_{t=(p,A,B)\in T_{i-1}} px_t \Delta_B,
$$

$$
\sum_{t=(p,A,B)\in T_{i-1}} p(x_t \Delta_A + (1-x_t)\Delta_B) \geq 2 \sum_{t=(p,A,B)\in T_{i-1}} p(1-x_t)\Delta_A,
$$

where $\Delta_A := f(A \cup \{i\}) - f(A)$ and $\Delta_B := f(B \setminus \{i\}) - f(B)$ are defined as above. This polytope is nonempty because for $t = (p, A, B)$ we can choose x_t as in (14.3); see (14.5) in the previous proof.

Let x be an extreme point of this polytope. By Theorem 4.18 we can find such an extreme point in polynomial time. It satisfies $|T_{i-1}|$ linearly independent inequalities with equality (cf. Proposition 3.9), so all but at most two variables are

0 or 1. Then we can continue the next iteration with the set

$$T_i := \{(px_t, A \cup \{i\}, B) : t = (p, A, B) \in T_{i-1}, \ x_t > 0\} \cup$$
$$\{(p(1 - x_t), A, B \setminus \{i\}) : t = (p, A, B) \in T_{i-1}, \ x_t < 1\}.$$

Hence the number of triples grows by at most two in each iteration.

Now taking (A_i, B_i) as (A, B) with probability p for every $(p, A, B) \in T_i$, we get

$$\tfrac{1}{2}\big(\mathrm{Exp}[f(A_i)] - \mathrm{Exp}[f(A_{i-1})] + \mathrm{Exp}[f(B_i)] - \mathrm{Exp}[f(B_{i-1})]\big)$$

$$= \ \tfrac{1}{2} \sum_{t=(p,A,B)\in T_{i-1}} p\left(x_t(f(A \cup \{i\}) - f(A)) + (1 - x_t)(f(B \setminus \{i\}) - f(B))\right)$$

$$= \ \tfrac{1}{2} \sum_{t=(p,A,B)\in T_{i-1}} p\left(x_t \Delta_A + (1 - x_t)\Delta_B\right)$$

$$\geq \ \sum_{t=(p,A,B)\in T_{i-1}} p \max\{x_t \Delta_B, (1 - x_t)\Delta_A\},$$

where the last inequality holds since x is in the above polytope. Together with (14.4) this implies

$$\mathrm{Exp}[f(O_{i-1}) - f(O_i)] \ \leq \ \tfrac{1}{2}\mathrm{Exp}[f(A_i) - f(A_{i-1})] + \tfrac{1}{2}\mathrm{Exp}[f(B_i) - f(B_{i-1})].$$

and hence $2\,\mathrm{Exp}[f(A_n)] \geq O_0$ as in the previous proof. $\square$

In fact, it is not necessary to use the ellipsoid method to find an extreme point; instead this can be done in linear time; see Exercise 20.

Exercises

1. Let G be an undirected graph and M a maximum matching in G. Let $\mathcal{F}$ be the family of those subsets $X \subseteq E(G)$ for which there exists a special blossom forest F with respect to M with $E(F) \setminus M = X$. Prove that $(E(G) \setminus M, \mathcal{F})$ is a greedoid.
   Hint: Use Exercise 25 of Chapter 10.
2. Let $(E, \mathcal{F})$ be a greedoid and $c' : E \to \mathbb{R}_+$. We consider the bottleneck function $c(F) := \min\{c'(e) : e \in F\}$ for $F \subseteq E$. Show that the GREEDY ALGORITHM FOR GREEDOIDS, when applied to $(E, \mathcal{F})$ and c, finds a maximal element F of $\mathcal{F}$ with $c(F)$ maximum.
3. This exercise shows that greedoids can also be defined as languages (cf. Definition 15.1). Let E be a finite set. A language L over the alphabet E is called a greedoid language if
   (a) L contains the empty string;
   (b) $x_i \neq x_j$ for all $(x_1, \ldots, x_n) \in L$ and $1 \leq i < j \leq n$;

(c) $(x_1, \ldots, x_{n-1}) \in L$ for all $(x_1, \ldots, x_n) \in L$;

(d) If $(x_1, \ldots, x_n), (y_1, \ldots, y_m) \in L$ with $m < n$, then there exists an $i \in \{1, \ldots, n\}$ such that $(y_1, \ldots, y_m, x_i) \in L$.

L is called an antimatroid language if it satisfies (a), (b), (c) and

(d') If $(x_1, \ldots, x_n), (y_1, \ldots, y_m) \in L$ with $\{x_1, \ldots, x_n\} \not\subseteq \{y_1, \ldots, y_m\}$, then there exists an $i \in \{1, \ldots, n\}$ such that $(y_1, \ldots, y_m, x_i) \in L$.

Prove: A language L over the alphabet E is a greedoid language (an antimatroid language) if and only if the set system $(E, \mathcal{F})$ is a greedoid (antimatroid), where $\mathcal{F} := \{\{x_1, \ldots, x_n\} : (x_1, \ldots, x_n) \in L\}$.

4. Let U be a finite set and $f : 2^U \to \mathbb{R}$. Prove that f is submodular if and only if $f(X \cup \{y, z\}) - f(X \cup \{y\}) \leq f(X \cup \{z\}) - f(X)$ for all $X \subseteq U$ and $y, z \in U$ with $y \neq z$.

5. Let (G, u, s, t) be a network and $U := \delta^+(s)$. Let $P := \{x \in \mathbb{R}_+^U :$ there is an s-t-flow f in (G, u) with $f(e) = x_e$ for all $e \in U\}$. Prove that P is a polymatroid.

6. Let P be a polymatroid. Show that then there is a submodular and monotone function f with $f(\emptyset) = 0$ and $P = P(f)$.

$*$ 7. Prove that a nonempty compact set $P \subseteq \mathbb{R}_+^n$ is a polymatroid if and only if

(a) For all $0 \leq x \leq y \in P$ we have $x \in P$.

(b) For all $x \in \mathbb{R}_+^n$ and all $y, z \leq x$ with $y, z \in P$ that are maximal with this property (i.e. $y \leq w \leq x$ and $w \in P$ implies $w = y$, and $z \leq w \leq x$ and $w \in P$ implies $w = z$) we have $\mathbb{1}y = \mathbb{1}z$.

Note: This is the original definition of Edmonds [1970].

8. Prove that the POLYMATROID GREEDY ALGORITHM, when applied to a vector $c \in \mathbb{R}^E$ with $c(e) > 0$ for all $e \in E$ and a submodular but not necessarily monotone function $f : 2^E \to \mathbb{R}$ with $f(\emptyset) \geq 0$, solves

$$\max \left\{ cx : \sum_{e \in A} x_e \leq f(A) \text{ for all } A \subseteq E \right\}.$$

9. Prove Theorem 14.12 for the special case that f and g are rank functions of matroids by constructing an integral optimum dual solution from c_1 and c_2 as generated by the WEIGHTED MATROID INTERSECTION ALGORITHM. (Frank [1981])

$*$ 10. Let S be a finite set and $f : 2^S \to \mathbb{R}$. Define $f' : \mathbb{R}_+^S \to \mathbb{R}$ as follows. For any $x \in \mathbb{R}_+^S$ there are unique $k \in \mathbb{Z}_+$, $\lambda_1, \ldots, \lambda_k > 0$ and $\emptyset \subset T_1 \subset T_2 \subset \cdots \subset T_k \subseteq S$ such that $x = \sum_{i=1}^k \lambda_i \chi^{T_i}$, where χ^{T_i} is the incidence vector of T_i. Then $f'(x) := \sum_{i=1}^k \lambda_i f(T_i)$.

Prove that f is submodular if and only if f' is convex. (Lovász [1983])

11. Let E be a finite set and $f : 2^E \to \mathbb{R}_+$ a submodular function with $f(\{e\}) \leq 2$ for all $e \in E$. (The pair (E, f) is sometimes called a 2-polymatroid.) The POLYMATROID MATCHING PROBLEM asks for a maximum cardinality set $X \subseteq E$ with $f(X) = 2|X|$. (f is of course given by an oracle.)

Let $E_1, \ldots, E_k$ be pairwise disjoint unordered pairs and let $(E, \mathcal{F})$ be a matroid (given by an independence oracle), where $E = E_1 \cup \cdots \cup E_k$. The MATROID PARITY PROBLEM asks for a maximum cardinality set $I \subseteq \{1, \ldots, k\}$ with $\bigcup_{i \in I} E_i \in \mathcal{F}$.

(a) Show that the MATROID PARITY PROBLEM polynomially reduces to the POLYMATROID MATCHING PROBLEM.

* (b) Show that the POLYMATROID MATCHING PROBLEM polynomially reduces to the MATROID PARITY PROBLEM.

   Hint: Use an algorithm for the SUBMODULAR FUNCTION MINIMIZATION PROBLEM.

* (c) Show that there is no algorithm for the POLYMATROID MATCHING PROBLEM whose running time is polynomial in $|E|$.

   (Jensen and Korte [1982], Lovász [1981])

   (A problem polynomially reduces to another one if the former can be solved with a polynomial-time oracle algorithm using an oracle for the latter; see Chapter 15.)

   Note: A polynomial-time algorithm for an important special case (representable matroids) was given by Lovász [1980, 1981] and extended to the weighted generalization by Iwata and Kobayashi [2017] and Pap [2017+].

12. A function $f : 2^S \to \mathbb{R} \cup \{\infty\}$ is called crossing submodular if $f(X) + f(Y) \geq f(X \cup Y) + f(X \cap Y)$ for any two sets $X, Y \subseteq S$ with $X \cap Y \neq \emptyset$ and $X \cup Y \neq S$. The SUBMODULAR FLOW PROBLEM is as follows: Given a digraph G, functions $l : E(G) \to \mathbb{R} \cup \{-\infty\}$, $u : E(G) \to \mathbb{R} \cup \{\infty\}$, $c : E(G) \to \mathbb{R}$, and a crossing submodular function $b : 2^{V(G)} \to \mathbb{R} \cup \{\infty\}$. Then a feasible submodular flow is a function $f : E(G) \to \mathbb{R}$ with $l(e) \leq f(e) \leq u(e)$ for all $e \in E(G)$ and

$$\sum_{e \in \delta^-(X)} f(e) - \sum_{e \in \delta^+(X)} f(e) \leq b(X)$$

for all $X \subseteq V(G)$. The task is to decide whether a feasible flow exists and, if yes, to find one whose cost $\sum_{e \in E(G)} c(e) f(e)$ is minimum possible.

Show that this problem generalizes the MINIMUM COST FLOW PROBLEM and the problem of optimizing a linear function over the intersection of two polymatroids.

Note: The SUBMODULAR FLOW PROBLEM, introduced by Edmonds and Giles [1977], can be solved in strongly polynomial time; see Fujishige, Röck and Zimmermann [1989]. See also Fleischer and Iwata [2000].

* 13. Show that the inequality system describing a feasible submodular flow (Exercise 12) is TDI. Show that this implies Theorems 14.12 and 19.17.

   (Edmonds and Giles [1977])

14. Prove that the set of vertices of the base polyhedron of a submodular function f with $f(\emptyset) = 0$ is precisely the set of vectors $b^\prec$ for all total orders $\prec$ of

U, where

$$b^{\prec}(u) := f(\{v \in U : v \leq u\}) - f(\{v \in U : v \prec u\})$$

$(u \in U)$.

Hint: See the proof of Theorem 14.11.

15. Let $f : 2^U \to \mathbb{R}$ be a submodular function with $f(\emptyset) = 0$, and let $B(f)$ denote its base polyhedron. Prove that $\min\{f(X) : X \subseteq U\} = \max\{\sum_{u \in U} z(u) : z(A) \leq \min\{0, f(A)\}$ for all $A \subseteq U\} = \max\{\sum_{u \in U} \min\{0, y(u)\} : y \in B(f)\}$.
    Hint: Use Corollary 14.14 or the correctness of SCHRIJVER'S ALGORITHM.

16. Show that Lemma 8.41 is a special case of Lemma 14.24.

17. Let $f : 2^U \to \mathbb{R}$ be a submodular function. Let R be a random subset of U, where each element is chosen independently with probability $\frac{1}{2}$. Prove:
    (a) $\text{Exp}(f(R)) \geq \frac{1}{2}(f(\emptyset) + f(U))$.
    (b) For each $A \subseteq U$ we have $\text{Exp}(f(R)) \geq \frac{1}{4}(f(\emptyset) + f(A) + f(U \setminus A) + f(U))$.
        Hint: Apply (a) twice.
    (c) If f is nonnegative, then $\text{Exp}(f(R)) \geq \frac{1}{4} \max_{A \subseteq U} f(A)$.
    Note: Part (c) implies a randomized 4-factor approximation algorithm for (non-negative) submodular function maximization. This problem cannot be solved optimally with a polynomial number of oracle calls.
    (Feige, Mirrokni and Vondrák [2011])

18. Let $0 < \epsilon < 1$ be fixed and $n \in \mathbb{N}$ even with $\epsilon n \in \mathbb{N}$. Let $U = \{1, \ldots, n\}$. For any $C \subset U$ with $2|C| = |U|$ consider the functions $g, f_C : 2^U \to \mathbb{Z}_+$ defined as follows. For $S \subseteq U$ let $k := |S \cap C|$ and $l := |S \setminus C|$, and let $g(S) := |S||U \setminus S|$ and $f_C(S) := g(S)$ if $|k - l| \leq \epsilon n$ and $f_C(S) := n|S| - 4kl + \epsilon^2 n^2 - 2\epsilon n|k - l|$ if $|k - l| \geq \epsilon n$.
    (a) Show that the two definitions of $f_C(S)$ coincide if $|k - l| = \epsilon n$.
    (b) Show that g and f_C are submodular (*Hint:* use Exercise 4)
 * (c) Observe that an algorithm is likely to need exponentially many oracle calls to find out which of these functions (g or f_C for some C) is the input.
    (d) Show that the maximum values of f and any g_C differ by a factor more than $2(1 - \epsilon)$.
    (Feige, Mirrokni and Vondrák [2011])

19. Show that if we place the randomized step in the SIMPLE SUBMODULAR FUNCTION MAXIMIZATION ALGORITHM by setting $A := A \cup \{i\}$ if $\Delta_A \geq \Delta_B$ and $B := B \setminus \{i\}$ otherwise yields a 3-approximation (deterministically, in linear time).

20. Consider the problem of finding an extreme point in the polytope defined in the proof of Theorem 14.27.
    (a) Replace one of the two nontrivial constraints by an objective function. Use this to show that there is an extreme point in which at most one variable has a fractional value.

(b) Show how to solve the LP resulting from (a) in $O(n \log n)$ time (note that $|T_{i-1}| = O(n)$).

Remark: In fact, the LP can be solved in linear time using the results of Section 17.1.

(Buchbinder and Feldman [2016])

References

General Literature

Bixby, R.E., and Cunningham, W.H. [1995]: Matroid optimization and algorithms. In: Handbook of Combinatorics; Vol. 1 (R.L. Graham, M. Grötschel, L. Lovász, eds.), Elsevier, Amsterdam, 1995

Björner, A., and Ziegler, G.M. [1992]: Introduction to greedoids. In: Matroid Applications (N. White, ed.), Cambridge University Press, Cambridge 1992

Frank, A. [2011]: Connections in Combinatorial Optimization. Oxford University Press, Oxford 2011

Fujishige, S. [2005]: Submodular Functions and Optimization. Second Edition. Elsevier, Amsterdam 2005

Iwata, S. [2008]: Submodular function minimization. Mathematical Programming B 112 (2008), 45–64

Korte, B., Lovász, L., and Schrader, R. [1991]: Greedoids. Springer, Berlin 1991

McCormick, S.T. [2004]: Submodular function minimization. In: Discrete Optimization (K. Aardal, G.L. Nemhauser, R. Weismantel, eds.), Elsevier, Amsterdam 2005

Schrijver, A. [2003]: Combinatorial Optimization: Polyhedra and Efficiency. Springer, Berlin 2003, Chapters 44–49

Cited References

Buchbinder, N., and Feldman, M. [2016]: Deterministic algorithms for submodular maximization problems. Proceedings of the 27th Annual ACM-SIAM Symposium on Discrete Algorithms (2016), 392–403

Buchbinder, N., Feldman, M., Naor, J., and Schwartz, R. [2015]: A tight linear time (1/2)-approximation for unconstrained submodular maximization. SIAM Journal on Computing 44 (2015), 1384–1402

Cunningham, W.H. [1984]: Testing membership in matroid polyhedra. Journal of Combinatorial Theory B 36 (1984), 161–188

Edmonds, J. [1970]: Submodular functions, matroids and certain polyhedra. In: Combinatorial Structures and Their Applications; Proceedings of the Calgary International Conference on Combinatorial Structures and Their Applications 1969 (R. Guy, H. Hanani, N. Sauer, J. Schönheim, eds.), Gordon and Breach, New York 1970, pp. 69–87

Edmonds, J. [1979]: Matroid intersection. In: Discrete Optimization I; Annals of Discrete Mathematics 4 (P.L. Hammer, E.L. Johnson, B.H. Korte, eds.), North-Holland, Amsterdam 1979, pp. 39–49

Edmonds, J., and Giles, R. [1977]: A min-max relation for submodular functions on graphs. In: Studies in Integer Programming; Annals of Discrete Mathematics 1 (P.L. Hammer, E.L. Johnson, B.H. Korte, G.L. Nemhauser, eds.), North-Holland, Amsterdam 1977, pp. 185–204

Feige, U., Mirrokni, V.S., and Vondrák, J. [2011]: Maximizing non-monotone submodular functions. SIAM Journal on Computing 40 (2011), 1133–1153

Fleischer, L., and Iwata, S. [2000]: Improved algorithms for submodular function minimization and submodular flow. Proceedings of the 32nd Annual ACM Symposium on Theory of Computing (2000), 107–116

Frank, A. [1981]: A weighted matroid intersection algorithm. Journal of Algorithms 2 (1981), 328–336

Frank, A. [1982]: An algorithm for submodular functions on graphs. In: Bonn Workshop on Combinatorial Optimization; Annals of Discrete Mathematics 16 (A. Bachem, M. Grötschel, B. Korte, eds.), North-Holland, Amsterdam 1982, pp. 97–120

Fujishige, S. [1998]: Another simple proof of the validity of Nagamochi and Ibaraki's min-cut algorithm and Queyranne's extension to symmetric submodular function minimization. Journal of the Operations Research Society of Japan 41 (1998), 626–628

Fujishige, S., Röck, H., and Zimmermann, U. [1989]: A strongly polynomial algorithm for minimum cost submodular flow problems. Mathematics of Operations Research 14 (1989), 60–69

Grötschel, M., Lovász, L., and Schrijver, A. [1981]: The ellipsoid method and its consequences in combinatorial optimization. Combinatorica 1 (1981), 169–197

Grötschel, M., Lovász, L., and Schrijver, A. [1988]: Geometric Algorithms and Combinatorial Optimization. Springer, Berlin 1988

Iwata, S. [2002]: A fully combinatorial algorithm for submodular function minimization. Journal of Combinatorial Theory B 84 (2002), 203–212

Iwata, S. [2003]: A faster scaling algorithm for minimizing submodular functions. SIAM Journal on Computing 32 (2003), 833–840

Iwata, S., Fleischer, L., and Fujishige, S. [2001]: A combinatorial, strongly polynomial-time algorithm for minimizing submodular functions. Journal of the ACM 48 (2001), 761–777

Iwata, S., and Kobayashi, Y. [2017]: A weighted linear matroid parity algorithm. Proceedings of the 49th Annual ACM Symposium on Theory of Computing (2017), 264–276

Jensen, P.M., and Korte, B. [1982]: Complexity of matroid property algorithms. SIAM Journal on Computing 11 (1982), 184–190

Lee, Y.T., Sidford, A., and Wong, S.C. [2015]: A faster cutting plane method and its implications for combinatorial and convex optimization. Proceedings of the 56th Annual IEEE Symposium on Foundations of Computer Science (2015), 1049–1065

Lovász, L. [1980]: Matroid matching and some applications. Journal of Combinatorial Theory B 28 (1980), 208–236

Lovász, L. [1981]: The matroid matching problem. In: Algebraic Methods in Graph Theory; Vol. II (L. Lovász, V.T. Sós, eds.), North-Holland, Amsterdam 1981, 495–517

Lovász, L. [1983]: Submodular functions and convexity. In: Mathematical Programming: The State of the Art – Bonn 1982 (A. Bachem, M. Grötschel, B. Korte, eds.), Springer, Berlin 1983

Nagamochi, H., and Ibaraki, T. [1998]: A note on minimizing submodular functions. Information Processing Letters 67 (1998), 239–244

Orlin, J.B. [2007]: A faster strongly polynomial time algorithm for submodular function minimization. Mathematical Programming 118 (2009), 237–251

Pap, G. [2017+]. Weighted linear matroid matching. Forthcoming. Abstract in: Oberwolfach Report 53 (2011), 3046–3049

Queyranne, M. [1998]: Minimizing symmetric submodular functions. Mathematical Programming B 82 (1998), 3–12

Rizzi, R. [2000]: On minimizing symmetric set functions. Combinatorica 20 (2000), 445–450

Schrijver, A. [2000]: A combinatorial algorithm minimizing submodular functions in strongly polynomial time. Journal of Combinatorial Theory B 80 (2000), 346–355

Vygen, J. [2003]: A note on Schrijver's submodular function minimization algorithm. Journal of Combinatorial Theory B 88 (2003), 399–402

15 *NP*-Completeness

For many combinatorial optimization problems a polynomial-time algorithm is known; the most important ones are presented in this book. However, there are also many important problems for which no polynomial-time algorithm is known. Although we cannot prove that none exists we can show that a polynomial-time algorithm for one "hard" (more precisely: *NP*-hard) problem would imply a polynomial-time algorithm for almost all problems discussed in this book (more precisely: all *NP*-easy problems).

To formalize this concept and prove the above statement we need a machine model, i.e. a precise definition of a polynomial-time algorithm. Therefore we discuss Turing machines in Section 15.1. This theoretical model is not suitable to describe more complicated algorithms. However we shall argue that it is equivalent to our informal notion of algorithms: every algorithm in this book can, theoretically, be written as a Turing machine, with a loss in efficiency that is polynomially bounded. We indicate this in Section 15.2.

In Section 15.3 we introduce decision problems, and in particular the classes *P* and *NP*. While *NP* contains most decision problems appearing in this book, *P* contains only those for which there are polynomial-time algorithms. It is an open question whether $P = NP$. Although we shall discuss many problems in *NP* for which no polynomial-time algorithm is known, nobody can (so far) prove that none exists. We specify what it means that one problem reduces to another, or that one problem is at least as hard as another one. In this notion, the hardest problems in *NP* are the *NP*-complete problems; they can be solved in polynomial time if and only if $P = NP$.

In Section 15.4 we exhibit the first *NP*-complete problem, SATISFIABILITY. In Section 15.5 some more decision problems, more closely related to combinatorial optimization, are proved to be *NP*-complete. In Sections 15.6 and 15.7 we shall discuss related concepts, also extending to optimization problems.

© Springer-Verlag GmbH Germany 2018
Bernhard Korte and Jens Vygen, *Combinatorial Optimization*, Algorithms and Combinatorics 21, https://doi.org/10.1007/978-3-662-56039-6_15

15.1 Turing Machines

In this section we present a very simple model for computation: the Turing machine. It can be regarded as a sequence of simple instructions working on a string. The input and the output will be a binary string:

Definition 15.1. *An* **alphabet** *is a finite set with at least two elements, not containing the special symbol* $\sqcup$ *(which we shall use for blanks). For an alphabet A we denote by a* **string** *over A a finite sequence of elements of A, by* A^n *the set of strings of length n, and by* $A^* := \bigcup_{n \in \mathbb{Z}_+} A^n$ *the set of all strings over A. We use the convention that* A^0 *contains exactly one element, the* **empty string**. *A* **language** *over A is a subset of* A^*. *The elements of a language are often called* **words**. *If* $x \in A^n$ *we write* $\mathrm{size}(x) := n$ *for the* **length** *of the string.*

We shall often work with the alphabet $A = \{0, 1\}$ and the set $\{0, 1\}^*$ of all **0-1-strings** (or **binary strings**). The components of a 0-1-string are sometimes called its **bits**. So there is exactly one 0-1-string of zero length, the empty string.

A Turing machine gets as input a string $x \in A^*$ for some fixed alphabet A. The input is completed by blank symbols (denoted by $\sqcup$) to a two-way infinite string $s \in (A \cup \{\sqcup\})^{\mathbb{Z}}$. This string s can be regarded as a tape with a read-write head; only a single position can be read and modified at each step, and the read-write head can be moved by one position in each step.

A Turing machine consists of a set of $N + 1$ instructions numbered $0, \ldots, N$. In the beginning instruction 0 is executed and the current position of the string is position 1. Now each instruction is of the following type. Read the bit at the current position, and depending on its value do the following: overwrite the current bit by some element of $A \cup \{\sqcup\}$, possibly move the current position by one to the left or to the right, and go to an instruction which will be executed next.

There is a special instruction denoted by -1 which marks the end of the computation. The components of our infinite string s indexed by $1, 2, 3, \ldots$ up to the first $\sqcup$ then yield the output string. Formally we define a Turing machine as follows:

Definition 15.2. (Turing [1936]) *Let A be an alphabet and* $\bar{A} := A \cup \{\sqcup\}$. *A* **Turing machine** *(with alphabet A) is defined by a function*

$$\Phi : \{0, \ldots, N\} \times \bar{A} \to \{-1, \ldots, N\} \times \bar{A} \times \{-1, 0, 1\}$$

for some $N \in \mathbb{Z}_+$. *The* **computation** *of* Φ *on input x, where* $x \in A^*$, *is the finite or infinite sequence of triples* $(n^{(i)}, s^{(i)}, \pi^{(i)})$ *with* $n^{(i)} \in \{-1, \ldots, N\}$, $s^{(i)} \in \bar{A}^{\mathbb{Z}}$ *and* $\pi^{(i)} \in \mathbb{Z}$ $(i = 0, 1, 2, \ldots)$ *defined recursively as follows* $(n^{(i)}$ *denotes the current instruction,* $s^{(i)}$ *represents the string, and* $\pi^{(i)}$ *is the current position):*

$n^{(0)} := 0$. $s_j^{(0)} := x_j$ *for* $1 \leq j \leq \mathrm{size}(x)$, *and* $s_j^{(0)} := \sqcup$ *for all* $j \leq 0$ *and* $j > \mathrm{size}(x)$. $\pi^{(0)} := 1$.

If $(n^{(i)}, s^{(i)}, \pi^{(i)})$ is already defined, we distinguish two cases. If $n^{(i)} \neq -1$, then let $(m, \sigma, \delta) := \Phi\big(n^{(i)}, s^{(i)}_{\pi^{(i)}}\big)$ and set $n^{(i+1)} := m$, $s^{(i+1)}_{\pi^{(i)}} := \sigma$, $s^{(i+1)}_j := s^{(i)}_j$ for $j \in \mathbb{Z} \setminus \{\pi^{(i)}\}$, and $\pi^{(i+1)} := \pi^{(i)} + \delta$.

If $n^{(i)} = -1$, then this is the end of the sequence. We then define $\mathrm{time}(\Phi, x) := i$ and $\mathrm{output}(\Phi, x) \in A^k$, where $k := \min\{j \in \mathbb{N} : s^{(i)}_j = \sqcup\} - 1$, by $\mathrm{output}(\Phi, x)_j := s^{(i)}_j$ for $j = 1, \ldots, k$.

If this sequence is infinite (i.e. $n^{(i)} \neq -1$ for all i), then we set $\mathrm{time}(\Phi, x) := \infty$. In this case $\mathrm{output}(\Phi, x)$ is undefined.

Of course we are interested mostly in Turing machines whose computation is finite or even polynomially bounded:

Definition 15.3.   *Let A be an alphabet. A **computational problem** is a pair (X, R), where $X \subseteq A^*$ is a language and $R \subseteq X \times A^*$ is a relation such that for every $x \in X$ there exists a $y \in A^*$ with $(x, y) \in R$. Let Φ be a Turing machine with alphabet A such that $\mathrm{time}(\Phi, x) < \infty$ and $(x, \mathrm{output}(\Phi, x)) \in R$ for each $x \in X$. Then we say that Φ **computes** (X, R). If there exists a polynomial p such that for all $x \in X$ we have $\mathrm{time}(\Phi, x) \leq p(\mathrm{size}(x))$, then Φ is a **polynomial-time Turing machine** (for (X, R)).*

*If $|\{y \in A^* : (x, y) \in R\}| = 1$ for all $x \in X$, we can define $f : X \to A^*$ by $(x, f(x)) \in R$ and say that Φ **computes** f. In the case $X = A^*$ and $f : X \to \{0, 1\}$ we say that Φ **decides** the language $L := \{x \in X : f(x) = 1\}$. If there exists some polynomial-time Turing machine computing a function f (or deciding a language L), then we say that f is **computable in polynomial time** (or L is **decidable in polynomial time**, respectively).*

To make these definitions clear we give an example. The following Turing machine $\Phi : \{0, \ldots, 3\} \times \{0, 1, \sqcup\} \to \{-1, \ldots, 3\} \times \{0, 1, \sqcup\} \times \{-1, 0, 1\}$ computes the successor function $n \mapsto n + 1$ ($n \in \mathbb{N}$), where the numbers are coded by their usual binary representation.

$\Phi(0, 0)$	$=$	$(0, 0, 1)$	⓪	**While** $s_\pi \neq \sqcup$ **do** $\pi := \pi + 1$.
$\Phi(0, 1)$	$=$	$(0, 1, 1)$		
$\Phi(0, \sqcup)$	$=$	$(1, \sqcup, -1)$		Set $\pi := \pi - 1$.
$\Phi(1, 1)$	$=$	$(1, 0, -1)$	①	**While** $s_\pi = 1$ **do** $s_\pi := 0$ and $\pi := \pi - 1$.
$\Phi(1, 0)$	$=$	$(-1, 1, 0)$		**If** $s_\pi = 0$ **then** $s_\pi := 1$ and **stop**.
$\Phi(1, \sqcup)$	$=$	$(2, \sqcup, 1)$		Set $\pi := \pi + 1$.
$\Phi(2, 0)$	$=$	$(3, 1, 1)$	②	Set $s_\pi := 1$ and $\pi := \pi + 1$.
$\Phi(3, 0)$	$=$	$(3, 0, 1)$	③	**While** $s_\pi = 0$ **do** $\pi := \pi + 1$.
$\Phi(3, \sqcup)$	$=$	$(-1, 0, 0)$		Set $s_\pi := 0$ and **stop**.

Note that several values of Φ are not specified as they are never used in any computation. The comments on the right-hand side illustrate the computation. Instructions ② and ③ are used only if the input consists of 1's only, i.e. $n = 2^k - 1$ for some $k \in \mathbb{N}$. We have $\mathrm{time}(\Phi, x) \leq 3\,\mathrm{size}(x) + 3$ for all inputs x, so Φ is a polynomial-time Turing machine.

In the next section we shall show that the above definition is consistent with our informal definition of a polynomial-time algorithm in Section 1.2: each polynomial-time algorithm in this book can be simulated by a polynomial-time Turing machine.

15.2 Church's Thesis

The Turing machine is the most customary theoretical model for algorithms. Although it seems to be very restricted, it is as powerful as any other reasonable model: the set of **computable functions**, and also the set of functions computable in polynomial time, is always the same. This statement, known as Church's thesis, is of course too imprecise to be proved. However, there are strong results supporting this claim. For example, each program in a common programming language like C can be modelled by a Turing machine. In particular, all algorithms in this book can be rewritten as Turing machines. This is usually very inconvenient (thus we shall never do it), but theoretically it is possible. Moreover, any function computable in polynomial time by a C program is also computable in polynomial time by a Turing machine (and vice versa).

Since it is not a trivial task to implement more complicated programs on a Turing machine we consider as an intermediate step a Turing machine with two strings (tapes) and two independent read-write heads, one for each tape:

Definition 15.4. *Let A be an alphabet and $\bar{A} := A \cup \{\sqcup\}$. A* **two-tape Turing machine** *is defined by a function*

$$\Phi : \{0, \ldots, N\} \times \bar{A}^2 \to \{-1, \ldots, N\} \times \bar{A}^2 \times \{-1, 0, 1\}^2$$

for some $N \in \mathbb{Z}_+$. The **computation** *of Φ on input x, where $x \in A^*$, is the finite or infinite sequence of 5-tuples $(n^{(i)}, s^{(i)}, t^{(i)}, \pi^{(i)}, \rho^{(i)})$ with $n^{(i)} \in \{-1, \ldots, N\}$, $s^{(i)}, t^{(i)} \in \bar{A}^{\mathbb{Z}}$ and $\pi^{(i)}, \rho^{(i)} \in \mathbb{Z}$ ($i = 0, 1, 2, \ldots$) defined recursively as follows:*
$n^{(0)} := 0$. $s_j^{(0)} := x_j$ *for $1 \le j \le \mathrm{size}(x)$, and $s_j^{(0)} := \sqcup$ for all $j \le 0$ and $j > \mathrm{size}(x)$. $t_j^{(0)} := \sqcup$ for all $j \in \mathbb{Z}$. $\pi^{(0)} := 1$ and $\rho^{(0)} := 1$.*

If $(n^{(i)}, s^{(i)}, t^{(i)}, \pi^{(i)}, \rho^{(i)})$ is already defined, we distinguish two cases. If $n^{(i)} \ne -1$, then let $(m, \sigma, \tau, \delta, \epsilon) := \Phi\left(n^{(i)}, s_{\pi^{(i)}}^{(i)}, t_{\rho^{(i)}}^{(i)}\right)$ and set $n^{(i+1)} := m$, $s_{\pi^{(i)}}^{(i+1)} := \sigma$, $s_j^{(i+1)} := s_j^{(i)}$ for $j \in \mathbb{Z} \setminus \{\pi^{(i)}\}$, $t_{\rho^{(i)}}^{(i+1)} := \tau$, $t_j^{(i+1)} := t_j^{(i)}$ for $j \in \mathbb{Z} \setminus \{\rho^{(i)}\}$, $\pi^{(i+1)} := \pi^{(i)} + \delta$, and $\rho^{(i+1)} := \rho^{(i)} + \epsilon$.

If $n^{(i)} = -1$, then this is the end of the sequence. $\mathrm{time}(\Phi, x)$ and $\mathrm{output}(\Phi, x)$ are defined as with the one-tape Turing machine.

Turing machines with more than two tapes can be defined analogously, but we shall not need them. Before we show how to perform standard operations with a two-tape Turing machine, let us note that a two-tape Turing machine can be simulated by an ordinary (one-tape) Turing machine.

Theorem 15.5. *Let A be an alphabet, and let*

$$\Phi : \{0, \ldots, N\} \times (A \cup \{\sqcup\})^2 \to \{-1, \ldots, N\} \times (A \cup \{\sqcup\})^2 \times \{-1, 0, 1\}^2$$

be a two-tape Turing machine. Then there exists an alphabet $B \supseteq A$ and a (one-tape) Turing machine

$$\Phi' : \{0, \ldots, N'\} \times (B \cup \{\sqcup\}) \to \{-1, \ldots, N'\} \times (B \cup \{\sqcup\}) \times \{-1, 0, 1\}$$

such that $\mathrm{output}(\Phi', x) = \mathrm{output}(\Phi, x)$ *and* $\mathrm{time}(\Phi', x) = O\big((\mathrm{time}(\Phi, x))^2\big)$ *for* $x \in A^*$.

Proof: We use the letters s and t for the two strings of Φ, and denote by π and ρ the positions of the read-write heads, as in Definition 15.4. The string of Φ' will be denoted by u and its read-write head position by ψ.

We have to encode both strings s, t and both read-write head positions π, ρ in one string u. To make this possible each symbol u_j of u is a 4-tuple (s_j, p_j, t_j, r_j), where s_j and t_j are the corresponding symbols of s and t, and $p_j, r_j \in \{0, 1\}$ indicate whether the read-write heads of the first and second string currently scans position j; i.e. we have $p_j = 1$ iff $\pi = j$, and $r_j = 1$ iff $\rho = j$.

So we define $\bar{B} := (\bar{A} \times \{0, 1\} \times \bar{A} \times \{0, 1\})$; then we identify $a \in \bar{A}$ with $(a, 0, \sqcup, 0)$ to allow inputs from A^*. The first step of Φ' consists in initializing the marks p_1 and r_1 to 1:

$$\Phi'(0, (., 0, ., 0)) \;=\; (1, (., 1, ., 1), 0) \qquad \textcircled{0} \quad \text{Set } \pi := \psi \text{ and } \rho := \psi.$$

Here a dot stands for an arbitrary value (which is not modified).

Now we show how to implement a general instruction $\Phi(m, \sigma, \tau) = (m', \sigma', \tau', \delta, \epsilon)$. We first have to find the positions π and ρ. It is convenient to assume that our single read-write head ψ is already at the leftmost of the two positions π and ρ; i.e. $\psi = \min\{\pi, \rho\}$. We have to find the other position by scanning the string u to the right, we have to check whether $s_\pi = \sigma$ and $t_\rho = \tau$ and, if so, perform the operation required (write new symbols to s and t, move π and ρ, jump to the next instruction).

The following block implements one instruction $\Phi(m, \sigma, \tau) = (m', \sigma', \tau', \delta, \epsilon)$ for $m = 0$; for each m we have $|\bar{A}|^2$ such blocks, one for each choice of σ and τ. The second block for $m = 0$ starts with $\textcircled{13}$, the first block for m' with $\textcircled{M}$, where $M := 12|\bar{A}|^2 m' + 1$. All in all we get $N' = 12(N+1)|\bar{A}|^2$.

A dot again stands for an arbitrary value which is not modified. Similarly, ζ and ξ stand for an arbitrary element of $\bar{A} \setminus \{\sigma\}$ and $\bar{A} \setminus \{\tau\}$, respectively. We assume that $\psi = \min\{\pi, \rho\}$ initially; note that $\textcircled{10}$, $\textcircled{11}$ and $\textcircled{12}$ guarantee that this property also holds at the end.

$\Phi'(1, (\zeta, 1, ., .)) = (13, (\zeta, 1, ., .), 0)$ ① If $\psi = \pi$ and $s_\psi \neq \sigma$ then go to ⑬.

$\Phi'(1, (., ., \xi, 1)) = (13, (., ., \xi, 1), 0)$ If $\psi = \rho$ and $t_\psi \neq \tau$ then go to ⑬.

$\Phi'(1, (\sigma, 1, \tau, 1)) = (2, (\sigma, 1, \tau, 1), 0)$ If $\psi = \pi$ then go to ②.

$\Phi'(1, (\sigma, 1, ., 0)) = (2, (\sigma, 1, ., 0), 0)$

$\Phi'(1, (., 0, \tau, 1)) = (6, (., 0, \tau, 1), 0)$ If $\psi = \rho$ then go to ⑥.

$\Phi'(2, (., ., ., 0)) = (2, (., ., ., 0), 1)$ ② While $\psi \neq \rho$ do $\psi := \psi + 1$.

$\Phi'(2, (., ., \xi, 1)) = (12, (., ., \xi, 1), -1)$ If $t_\psi \neq \tau$ then set $\psi := \psi - 1$

    and go to ⑫.

$\Phi'(2, (., ., \tau, 1)) = (3, (., ., \tau', 0), \epsilon)$ Set $t_\psi := \tau'$ and $\psi := \psi + \epsilon$.

$\Phi'(3, (., ., ., 0)) = (4, (., ., ., 1), 1)$ ③ Set $\rho := \psi$ and $\psi := \psi + 1$.

$\Phi'(4, (., 0, ., .)) = (4, (., 0, ., .), -1)$ ④ While $\psi \neq \pi$ do $\psi := \psi - 1$.

$\Phi'(4, (\sigma, 1, ., .)) = (5, (\sigma', 0, ., .), \delta)$ Set $s_\psi := \sigma'$ and $\psi := \psi + \delta$.

$\Phi'(5, (., 0, ., .)) = (10, (., 1, ., .), -1)$ ⑤ Set $\pi := \psi$ and $\psi := \psi - 1$.

    Go to ⑩.

$\Phi'(6, (., 0, ., .)) = (6, (., 0, ., .), 1)$ ⑥ While $\psi \neq \pi$ do $\psi := \psi + 1$.

$\Phi'(6, (\zeta, 1, ., .)) = (12, (\zeta, 1, ., .), -1)$ If $s_\psi \neq \sigma$ then set $\psi := \psi - 1$

    and go to ⑫.

$\Phi'(6, (\sigma, 1, ., .)) = (7, (\sigma', 0, ., .), \delta)$ Set $s_\psi := \sigma'$ and $\psi := \psi + \delta$.

$\Phi'(7, (., 0, ., .)) = (8, (., 1, ., .), 1)$ ⑦ Set $\pi := \psi$ and $\psi := \psi + 1$.

$\Phi'(8, (., ., ., 0)) = (8, (., ., ., 0), -1)$ ⑧ While $\psi \neq \rho$ do $\psi := \psi - 1$.

$\Phi'(8, (., ., \tau, 1)) = (9, (., ., \tau', 0), \epsilon)$ Set $t_\psi := \tau'$ and $\psi := \psi + \epsilon$.

$\Phi'(9, (., ., ., 0)) = (10, (., ., ., 1), -1)$ ⑨ Set $\rho := \psi$ and $\psi := \psi - 1$.

$\Phi'(10, (., ., ., .)) = (11, (., ., ., .), -1)$ ⑩ Set $\psi := \psi - 1$.

$\Phi'(11, (., 0, ., 0)) = (11, (., 0, ., 0), 1)$ ⑪ While $\psi \notin \{\pi, \rho\}$ do $\psi := \psi + 1$.

$\Phi'(11, (., 1, ., .)) = (M, (., 1, ., .), 0)$ Go to Ⓜ.

$\Phi'(11, (., 0, ., 1)) = (M, (., 0, ., 1), 0)$

$\Phi'(12, (., 0, ., 0)) = (12, (., 0, ., 0), -1)$ ⑫ While $\psi \notin \{\pi, \rho\}$ do $\psi := \psi - 1$.

$\Phi'(12, (., 1, ., .)) = (13, (., 1, ., .), 0)$

$\Phi'(12, (., ., ., 1)) = (13, (., ., ., 1), 0)$

Any computation of Φ' passes through at most $|\bar{A}|^2$ blocks like the above for each computation step of Φ. The number of computation steps within each block is at most $2|\pi - \rho| + 10$. Since $|\bar{A}|$ is a constant and $|\pi - \rho|$ is bounded by time(Φ, x) we conclude that the whole computation of Φ is simulated by Φ' with $O\left((\text{time}(\Phi, x))^2\right)$ steps.

Finally we have to clean up the output: replace each symbol $(\sigma, ., ., .)$ by $(\sigma, 0, \sqcup, 0)$. Obviously this at most doubles the total number of steps.   □

With a two-tape Turing machine it is not too difficult to implement more complicated instructions, and thus arbitrary algorithms:

We use the alphabet $A = \{0, 1, \#\}$ and model an arbitrary number of variables by the string

$$x_0\#\#1\#x_1\#\#10\#x_2\#\#11\#x_3\#\#100\#x_4\#\#101\#x_5\#\# \ldots \qquad (15.1)$$

which we store on the first tape. Each group (except the first one) contains a binary representation of the index i followed by the value of x_i, which we assume to be a binary string. The first variable x_0 and the second tape are used only as registers for intermediate results of computation steps.

Random access to variables is not possible in constant time with a Turing machine, no matter how many tapes we have. If we simulate an arbitrary algorithm by a two-tape Turing machine, we will have to scan the first tape quite often. Moreover, if the length of the string in one variable changes, the substring to the right has to be shifted. Nevertheless each standard operation (i.e. each elementary step of an algorithm) can be simulated with $O(l^2)$ computation steps of a two-tape Turing machine, where l is the current length of the string (15.1).

We try to make this clearer with a concrete example. Consider the following instruction: Add to x_5 the value of the variable whose index is given by x_2.

To get the value of x_5 we scan the first tape for the substring ##101#. We copy the substring following this up to #, exclusively, to the second tape. This is easy since we have two separate read-write heads. Then we copy the string from the second tape to x_0. If the new value of x_0 is shorter or longer than the old one, we have to shift the rest of the string (15.1) to the left or to the right appropriately.

Next we have to search for the variable index that is given by x_2. To do this, we first copy x_2 to the second tape. Then we scan the first tape, checking each variable index (comparing it with the string on the second tape bitwise). When we have found the correct variable index, we copy the value of this variable to the second tape.

Now we add the number stored in x_0 to that on the second tape. A Turing machine for this task, using the standard method, is not hard to design. We can overwrite the number on the second tape by the result while computing it. Finally we have the result on the second tape and copy it back to x_5. If necessary we shift the substring to the right of x_5 appropriately.

All the above can be done by a two-tape Turing machine in $O(l^2)$ computation steps (in fact all but shifting the string (15.1) can be done in $O(l)$ steps). It should be clear that the same holds for all other standard operations, including multiplication and division.

By Definition 1.4 an algorithm is said to run in polynomial time if there is a $k \in \mathbb{N}$ such that the number of elementary steps is bounded by $O(n^k)$ and any number in intermediate computation can be stored with $O(n^k)$ bits, where n is the input size. Moreover, we store at most $O(n^k)$ numbers at any time. Hence we can bound the length of each of the two strings in a two-tape Turing machine simulating such an algorithm by $l = O(n^k \cdot n^k) = O(n^{2k})$, and hence its running time by $O(n^k(n^{2k})^2) = O(n^{5k})$. This is still polynomial in the input size.

Recalling Theorem 15.5 we may conclude that for any function f there is a polynomial-time algorithm computing f if and only if there is a polynomial-time Turing machine computing f. Hence we will use the terms algorithm and Turing machine interchangeably in the rest of this chapter.

Hopcroft and Ullman [1979], Lewis and Papadimitriou [1981], and van Emde Boas [1990] provide more details about the equivalence of different machine models. Another common model (which is close to our informal model of Section 1.2) is the RAM machine (cf. Exercise 3) which allows arithmetic operations on integers in constant time. Other models allow only operations on bits (or integers of fixed length) which is more realistic when dealing with large numbers. Obviously, addition and comparison of natural numbers with n bits can be done with $O(n)$ bit operations. For multiplication (and division) the obvious method takes $O(n^2)$ bit operations, but the algorithm of Schönhage and Strassen [1971] needs only $O(n \log n \log \log n)$ bit operations for multiplying two n-bit integers, and this was improved further by Fürer [2009] and Harvey, van der Hoeven and Lecerf [2016]. This of course implies algorithms for the addition and comparison of rational numbers with the same time complexity. As far as polynomial-time computability is concerned all models are equivalent, but of course the running time measures are quite different.

The model of encoding the entire input by 0-1-strings (or strings over any fixed alphabet) does not in principle exclude certain types of real numbers, e.g. algebraic numbers (if $x \in \mathbb{R}$ is the k-th smallest root of a polynomial p, then x can be coded by listing k and the degree and the coefficients of p). However, there is no way of representing arbitrary real numbers in a digital computer since there are uncountably many real numbers but only countably many 0-1-strings. We take the classical approach and restrict ourselves to rational input in this chapter.

We close this section by giving a formal definition of oracle algorithms, based on two-tape Turing machines. We may call an oracle at any stage of the computation; we use the second tape for writing the oracle's input and reading its output. We introduce a special instruction -2 for oracle calls:

Definition 15.6. *Let A be an alphabet and $\bar{A} := A \cup \{\sqcup\}$. Let (X, R) be a computational problem with $X \subseteq A^*$. An* **oracle Turing machine** *using (X, R) is a function*

$$\Phi : \{0, \dots, N\} \times \bar{A}^2 \to \{-2, \dots, N\} \times \bar{A}^2 \times \{-1, 0, 1\}^2$$

for some $N \in \mathbb{Z}_+$. Its computation is defined as for a two-tape Turing machine, but with the following differences: initially we set $\mathrm{time}^{(0)} := 0$. If, for some computation step i, $\Phi\left(n^{(i)}, s_{\pi^{(i)}}^{(i)}, t_{\rho^{(i)}}^{(i)}\right) = (-2, \sigma, \tau, \delta, \epsilon)$ for some $\sigma, \tau, \delta, \epsilon$, then consider the string on the second tape $x \in A^k$, $k := \min\left\{ j \in \mathbb{N} : t_j^{(i)} = \sqcup \right\} - 1$, given by $x_j := t_j^{(i)}$ for $j = 1, \dots, k$. If $x \in X$, then the second tape is overwritten by $t_j^{(i+1)} = y_j$ for $j = 1, \dots, \mathrm{size}(y)$ and $t_{\mathrm{size}(y)+1}^{(i+1)} = \sqcup$ for some $y \in A^*$ with $(x, y) \in R$, we set $\mathrm{time}^{(i+1)} := \mathrm{time}^{(i)} + 1 + \mathrm{size}(y)$, and the computation continues with $n^{(i+1)} := n^{(i)} + 1$. The rest remains unchanged, and in all other cases we set $\mathrm{time}^{(i+1)} := \mathrm{time}^{(i)} + 1$. As before, the computation ends when $n^{(i)} = -1$; then we set $\mathrm{time}(\Phi, x) := \mathrm{time}^{(i)}$. The output is defined as with the two-tape Turing machine.

All definitions with respect to Turing machines can be extended to oracle Turing machines. The output of an oracle is not necessarily unique; hence there can be several possible computations for the same input. When proving the correctness or estimating the running time of an oracle algorithm we have to consider all possible computations, i.e. all choices of the oracle.

By the results of this section the existence of a polynomial-time (oracle) algorithm is equivalent to the existence of a polynomial-time (oracle) Turing machine.

15.3 *P* and *NP*

Most of complexity theory is based on decision problems. Decision problems are special computational problems. Any language $L \subseteq \{0, 1\}^*$ can be interpreted as decision problem: given a 0-1-string, decide whether it belongs to L. However, we are more interested in problems like the following:

HAMILTONIAN CIRCUIT

Instance: An undirected graph G.

Question: Has G a Hamiltonian circuit?

We will always assume a fixed efficient encoding of the input as a binary string; occasionally we extend our alphabet by other symbols. For example we assume that a graph is given by an adjacency list, and such a list can be coded as a binary string of length $O(n \log m + m \log n)$, where n and m denote the number of vertices and edges. We always assume an efficient encoding, i.e. one whose length is polynomially bounded by the minimum possible encoding length.

Not all binary strings are instances of HAMILTONIAN CIRCUIT but only those representing an undirected graph. For most interesting decision problems the instances are a proper subset of the 0-1-strings. We require that we can decide in polynomial time whether an arbitrary string is an instance or not:

Definition 15.7.   *A **decision problem** is a pair $\mathcal{P} = (X, Y)$, where X is a language decidable in polynomial time and $Y \subseteq X$. The elements of X are called **instances** of $\mathcal{P}$; the elements of Y are **yes-instances**, those of $X \setminus Y$ are **no-instances**.*

A decision problem (X, Y) can be regarded as the computational problem $(X, \{(x, 1) : x \in Y\} \cup \{(x, 0) : x \in X \setminus Y\})$. Hence an algorithm for a decision problem (X, Y) is an algorithm computing the function $f : X \to \{0, 1\}$, defined by $f(x) = 1$ for $x \in Y$ and $f(x) = 0$ for $x \in X \setminus Y$.

We give two more examples, the decision problems corresponding to LINEAR PROGRAMMING and INTEGER PROGRAMMING:

LINEAR INEQUALITIES

Instance: A matrix $A \in \mathbb{Z}^{m \times n}$ and a vector $b \in \mathbb{Z}^m$.

Question: Is there a vector $x \in \mathbb{Q}^n$ such that $Ax \leq b$?

INTEGER LINEAR INEQUALITIES

Instance: A matrix $A \in \mathbb{Z}^{m \times n}$ and a vector $b \in \mathbb{Z}^m$.

Question: Is there a vector $x \in \mathbb{Z}^n$ such that $Ax \leq b$?

Definition 15.8.   *The class of all decision problems for which there is a polynomial-time algorithm is denoted by P.*

In other words, a member of P is a pair (X, Y) with $Y \subseteq X \subseteq \{0, 1\}^*$ where both X and Y are languages decidable in polynomial time. To prove that a problem is in P one usually describes a polynomial-time algorithm. By the results of Section 15.2 there is a polynomial-time Turing machine for each problem in P. By Khachiyan's Theorem 4.18, LINEAR INEQUALITIES belongs to P. It is not known whether INTEGER LINEAR INEQUALITIES or HAMILTONIAN CIRCUIT belong to P. We shall now introduce another class, called *NP*, which contains these problems, and in fact most decision problems discussed in this book.

We do not insist on a polynomial-time algorithm, but we require that for each yes-instance there is a certificate which can be checked in polynomial time. For example, for the HAMILTONIAN CIRCUIT problem a possible certificate is simply a Hamiltonian circuit. It is easy to check whether a given string is the binary encoding of a Hamiltonian circuit. Note that we do not require a certificate for no-instances. Formally we define:

Definition 15.9.   *A decision problem $\mathcal{P} = (X, Y)$ belongs to NP if there is a polynomial p and a decision problem $\mathcal{P}' = (X', Y')$ in P, where*

$$X' := \left\{ x\#c : x \in X,\ c \in \{0, 1\}^{\lfloor p(\text{size}(x)) \rfloor} \right\},$$

such that

$$Y = \left\{ y \in X : \text{there exists a string } c \in \{0, 1\}^{\lfloor p(\text{size}(y)) \rfloor} \text{ with } y\#c \in Y' \right\}.$$

*Here $x\#c$ denotes the concatenation of the string x, the symbol # and the string c. A string c with $y\#c \in Y'$ is called a **certificate** for y (since c proves that $y \in Y$). An algorithm for $\mathcal{P}'$ is called a **certificate-checking algorithm**.*

Proposition 15.10. $P \subseteq NP$.

Proof: One can choose p to be identically zero. An algorithm for $\mathcal{P}'$ just deletes the last symbol of the input "$x\#$" and then applies an algorithm for $\mathcal{P}$.   □

It is not known whether $P = NP$. In fact, this is the most important open problem in complexity theory. As an example for problems in *NP* that are not known to be in P we have:

Proposition 15.11.   HAMILTONIAN CIRCUIT *belongs to NP*.

Proof: For each yes-instance G we take any Hamiltonian circuit of G as a certificate. To check whether a given edge set is in fact a Hamiltonian circuit of a given graph is obviously possible in polynomial time. □

Proposition 15.12. INTEGER LINEAR INEQUALITIES *belongs to NP.*

Proof: As a certificate we just take a solution vector. If there exists a solution, there exists one of polynomial size by Corollary 5.8. □

The name *NP* stands for "nondeterministic polynomial". To explain this we have to define what a nondeterministic algorithm is. This is a good opportunity to define randomized algorithms in general, a concept which has already been mentioned before. The common feature of randomized algorithms is that their computation does not only depend on the input but also on some random bits.

Definition 15.13. *A* **randomized algorithm** *for computing a function $f : S \to T$ is an algorithm computing a function $g : \{s\#r : s \in S, r \in \{0, 1\}^{k(s)}\} \to T$, where $k : S \to \mathbb{Z}_+$. So for each instance $s \in S$ the algorithm can use $k(s) \in \mathbb{Z}_+$ random bits. We measure the running time dependency on* size(s) *only; thus randomized algorithms running in polynomial time can read only a polynomial number of random bits.*

Naturally we are interested in such a randomized algorithm only if f and g are related. In the ideal case, if $g(s\#r) = f(s)$ for all $s \in S$ and all $r \in \{0, 1\}^{k(s)}$, we speak of a **Las Vegas algorithm**. *A Las Vegas algorithm always computes the correct result, but the running time may vary for different runs on the same input s. Sometimes even less reliable algorithms are interesting: if there is at least a positive probability p of a correct answer, independent of the instance, i.e.*

$$p := \inf_{s \in S} \frac{|\{r \in \{0, 1\}^{k(s)} : g(s\#r) = f(s)\}|}{2^{k(s)}} > 0,$$

then we have a **Monte Carlo algorithm**.

If $T = \{0, 1\}$, and for each $s \in S$ with $f(s) = 0$ we have $g(s\#r) = 0$ for all $r \in \{0, 1\}^{k(s)}$, then we have a randomized algorithm with **one-sided error**. *If in addition for each $s \in S$ with $f(s) = 1$ there is at least one $r \in \{0, 1\}^{k(s)}$ with $g(s\#r) = 1$, then the algorithm is called a* **nondeterministic algorithm**.

Alternatively a randomized algorithm can be regarded as an oracle algorithm where the oracle produces a random bit (0 or 1) whenever called. A nondeterministic algorithm for a decision problem always answers "no" for a no-instance, and for each yes-instance there is a chance that it answers "yes". The following observation is easy:

Proposition 15.14. *A decision problem belongs to NP if and only if it has a polynomial-time nondeterministic algorithm.*

Proof: Let $P = (X, Y)$ be a decision problem in *NP*, and let $P' = (X', Y')$ be defined as in Definition 15.9. Then a polynomial-time algorithm for P' is in fact also a nondeterministic algorithm for P: the unknown certificate is simply replaced by random bits. Since the number of random bits is bounded by a polynomial in size(x), $x \in X$, so is the running time of the algorithm.

Conversely, if $P = (X, Y)$ has a polynomial-time nondeterministic algorithm using $k(x)$ random bits for instance x, then there is a polynomial p such that $k(x) \leq p(\text{size}(x))$ for each instance x. We define $X' := \left\{ x\#c : x \in X, c \in \{0, 1\}^{\lfloor p(\text{size}(x)) \rfloor} \right\}$ and $Y' := \left\{ x\#c \in X' : g(x\#r) = 1, r \text{ consists of the first } k(x) \text{ bits of } c \right\}$.

Then by the definition of nondeterministic algorithms we have $(X', Y') \in P$ and

$$Y = \left\{ y \in X : \text{there exists a string } c \in \{0, 1\}^{\lfloor p(\text{size}(y)) \rfloor} \text{ with } y\#c \in Y' \right\}.$$

$\square$

Most decision problems encountered in combinatorial optimization belong to *NP*. For many of them it is not known whether they have a polynomial-time algorithm. However, one can say that certain problems are not easier than others. To make this precise we introduce the important concept of polynomial reductions, which works for general computational problems.

Definition 15.15.   *Let P_1 and P_2 be computational problems. We say that P_1* **polynomially reduces** *to P_2 if there exists a polynomial-time oracle algorithm for P_1 using P_2.*

The following observation is the main reason for this concept:

Proposition 15.16.   *If P_1 polynomially reduces to P_2 and there is a polynomial-time algorithm for P_2, then there is a polynomial-time algorithm for P_1.*

Proof: Let A_2 be an algorithm for P_2 with time$(A_2, y) \leq p_2(\text{size}(y))$ for all instances y of P_2. Let A_1 be an oracle algorithm for P_1 using P_2 with time$(A_1, x) \leq p_1(\text{size}(x))$ for all instances x of P_1. Then replacing the oracle calls in A_1 by subroutines equivalent to A_2 yields an algorithm A_3 for P_1. For any instance x of P_1 with size$(x) = n$ we have time$(A_3, x) \leq p_1(n) \cdot p_2(p_1(n))$: there can be at most $p_1(n)$ oracle calls in A_1, and none of the instances of P_2 produced by A_1 can be longer than $p_1(n)$. Since we can choose p_1 and p_2 to be polynomials we conclude that A_3 is a polynomial-time algorithm. $\square$

The theory of *NP*-completeness is based on a special kind of polynomial-time reduction, which is defined only for decision problems:

Definition 15.17.   *Let $P_1 = (X_1, Y_1)$ and $P_2 = (X_2, Y_2)$ be decision problems. We say that P_1* **polynomially transforms** *to P_2 if there is a function $f : X_1 \to X_2$ computable in polynomial time such that $f(x_1) \in Y_2$ for all $x_1 \in Y_1$ and $f(x_1) \in X_2 \setminus Y_2$ for all $x_1 \in X_1 \setminus Y_1$.*

In other words, yes-instances are transformed to yes-instances, and no-instances are transformed to no-instances. Obviously, if a problem $\mathcal{P}_1$ polynomially transforms to $\mathcal{P}_2$, then $\mathcal{P}_1$ also polynomially reduces to $\mathcal{P}_2$. Polynomial transformations are sometimes called Karp reductions, while general polynomial reductions are also known as Turing reductions. Both are easily seen to be transitive (for polynomial reductions this follows as in the proof of Proposition 15.16).

Definition 15.18. *A decision problem $\mathcal{P} \in NP$ is called **NP-complete** if all other problems in NP polynomially transform to $\mathcal{P}$.*

By Proposition 15.16 we know that if there is a polynomial-time algorithm for any *NP*-complete problem, then $P = NP$.

Of course, the above definition would be meaningless if no *NP*-complete problems existed. The next section consists of a proof that there is an *NP*-complete problem.

15.4 Cook's Theorem

In his pioneering work, Cook [1971] proved that a certain decision problem, called SATISFIABILITY, is in fact *NP*-complete. We need some definitions:

Definition 15.19. *Let X be a finite set of **Boolean variables**. A **truth assignment** for X is a function $T : X \to \{true, false\}$. We extend T to the set $L := X \cup \{\overline{x} : x \in X\}$ by setting $T(\overline{x}) := true$ if $T(x) = false$ and vice versa ($\overline{x}$ can be regarded as the negation of x). The elements of L are called the **literals** over X.*

*A **clause** over X is a set of literals over X. A clause represents the disjunction of those literals and is **satisfied** by a truth assignment iff at least one of its members is true. A family of clauses over X is **satisfiable** iff there is some truth assignment simultaneously satisfying all of its clauses.*

Since we consider the conjunction of disjunctions of literals, we also speak of Boolean formulas (cf. Exercise 21) in conjunctive normal form. For example, the family $\{\{x_1, \overline{x_2}\}, \{\overline{x_2}, \overline{x_3}\}, \{x_1, x_2, \overline{x_3}\}, \{\overline{x_1}, x_3\}\}$ corresponds to the Boolean formula $(x_1 \vee \overline{x_2}) \wedge (\overline{x_2} \vee \overline{x_3}) \wedge (x_1 \vee x_2 \vee \overline{x_3}) \wedge (\overline{x_1} \vee x_3)$. It is satisfiable as the truth assignment $T(x_1) := true$, $T(x_2) := false$ and $T(x_3) := true$ shows. We are now ready to specify the satisfiability problem:

SATISFIABILITY

Instance: A set X of variables and a family $\mathcal{Z}$ of clauses over X.

Question: Is $\mathcal{Z}$ satisfiable?

Theorem 15.20. (Cook [1971]) SATISFIABILITY *is NP-complete.*

Proof: SATISFIABILITY belongs to *NP* because a satisfying truth assignment serves as a certificate for any yes-instance, which of course can be checked in polynomial time.

Let now $\mathcal{P} = (X, Y)$ be any other problem in *NP*. We have to show that $\mathcal{P}$ polynomially transforms to SATISFIABILITY.

By Definition 15.9 there is a polynomial p and a decision problem $\mathcal{P}' = (X', Y')$ in P, where $X' := \{x\#c : x \in X, c \in \{0, 1\}^{\lfloor p(\text{size}(x)) \rfloor}\}$ and

$$Y = \left\{ y \in X : \text{there exists a string } c \in \{0, 1\}^{\lfloor p(\text{size}(y)) \rfloor} \text{ with } y\#c \in Y' \right\}.$$

Let

$$\Phi : \{0, \ldots, N\} \times \bar{A} \;\rightarrow\; \{-1, \ldots, N\} \times \bar{A} \times \{-1, 0, 1\}$$

be a polynomial-time Turing machine for $\mathcal{P}'$ with alphabet A; let $\bar{A} := A \cup \{\sqcup\}$. Let q be a polynomial such that $\text{time}(\Phi, x\#c) \leq q(\text{size}(x\#c))$ for all instances $x\#c \in X'$. Note that $\text{size}(x\#c) = \text{size}(x) + 1 + \lfloor p(\text{size}(x)) \rfloor$.

We will now construct a collection $\mathcal{Z}(x)$ of clauses over some set $V(x)$ of Boolean variables for each $x \in X$, such that $\mathcal{Z}(x)$ is satisfiable if and only if $x \in Y$.

We abbreviate $Q := q(\text{size}(x) + 1 + \lfloor p(\text{size}(x)) \rfloor)$. Q is an upper bound on the length of any computation of Φ on input $x\#c$, for any $c \in \{0, 1\}^{\lfloor p(\text{size}(x)) \rfloor}$. $V(x)$ contains the following Boolean variables:

- a variable $v_{ij\sigma}$ for all $0 \leq i \leq Q$, $-Q \leq j \leq Q$ and $\sigma \in \bar{A}$;
- a variable w_{ijn} for all $0 \leq i \leq Q$, $-Q \leq j \leq Q$ and $-1 \leq n \leq N$.

The intended meaning is: $v_{ij\sigma}$ indicates whether at time i (i.e. after i steps of the computation) the j-th position of the string contains the symbol σ. w_{ijn} indicates whether at time i the j-th position of the string is scanned and the n-th instruction is executed.

So if $(n^{(i)}, s^{(i)}, \pi^{(i)})_{i=0,1,\ldots}$ is a computation of Φ then we intend to set $v_{ij\sigma}$ to *true* iff $s_j^{(i)} = \sigma$ and w_{ijn} to *true* iff $\pi^{(i)} = j$ and $n^{(i)} = n$.

The collection $\mathcal{Z}(x)$ of clauses to be constructed will be satisfiable if and only if there is a string c with $\text{output}(\Phi, x\#c) = 1$.

$\mathcal{Z}(x)$ contains the following clauses to model the following conditions:

At any time each position of the string contains a unique symbol:

- $\{v_{ij\sigma} : \sigma \in \bar{A}\}$ for $0 \leq i \leq Q$ and $-Q \leq j \leq Q$;
- $\{\overline{v_{ij\sigma}}, \overline{v_{ij\tau}}\}$ for $0 \leq i \leq Q$, $-Q \leq j \leq Q$ and $\sigma, \tau \in \bar{A}$ with $\sigma \neq \tau$.

At any time a unique position of the string is scanned and a single instruction is executed:

- $\{w_{ijn} : -Q \leq j \leq Q, -1 \leq n \leq N\}$ for $0 \leq i \leq Q$;
- $\{\overline{w_{ijn}}, \overline{w_{ij'n'}}\}$ for $0 \leq i \leq Q$, $-Q \leq j, j' \leq Q$ and $-1 \leq n, n' \leq N$ with $(j, n) \neq (j', n')$.

The algorithm starts correctly with input $x\#c$ for some $c \in \{0,1\}^{\lfloor p(\text{size}(x)) \rfloor}$:

- $\{v_{0,j,x_j}\}$ for $1 \leq j \leq \text{size}(x)$;
- $\{v_{0,\text{size}(x)+1,\#}\}$;
- $\{v_{0,\text{size}(x)+1+j,0},\, v_{0,\text{size}(x)+1+j,1}\}$ for $1 \leq j \leq \lfloor p(\text{size}(x)) \rfloor$;
- $\{v_{0,j,\sqcup}\}$ for $-Q \leq j \leq 0$ and $\text{size}(x)+2+\lfloor p(\text{size}(x)) \rfloor \leq j \leq Q$;
- $\{w_{010}\}$.

The algorithm works correctly:

- $\{\overline{v_{ij\sigma}}, \overline{w_{ijn}}, v_{i+1,j,\tau}\}, \{\overline{v_{ij\sigma}}, \overline{w_{ijn}}, w_{i+1,j+\delta,m}\}$ for $0 \leq i < Q$,
  $-Q \leq j \leq Q, \sigma \in \bar{A}$ and $0 \leq n \leq N$, where $\Phi(n,\sigma) = (m,\tau,\delta)$.

When the algorithm reaches instruction -1, it stops:

- $\{\overline{w_{i,j,-1}}, w_{i+1,j,-1}\}, \{\overline{w_{i,j,-1}}, \overline{v_{i,j,\sigma}}, v_{i+1,j,\sigma}\}$
  for $0 \leq i < Q, -Q \leq j \leq Q$ and $\sigma \in \bar{A}$.

Positions not being scanned remain unchanged:

- $\{\overline{v_{ij\sigma}}, \overline{w_{ij'n}}, v_{i+1,j,\sigma}\}$
  for $0 \leq i < Q, \sigma \in \bar{A}, -1 \leq n \leq N$, and $-Q \leq j, j' \leq Q$ with $j \neq j'$.

The output of the algorithm is 1:

- $\{v_{Q,1,1}\}, \{v_{Q,2,\sqcup}\}$.

The encoding length of $\mathcal{Z}(x)$ is $O(Q^3 \log Q)$: there are $O(Q^3)$ occurrences of literals, whose indices require $O(\log Q)$ space. Since Q depends polynomially on $\text{size}(x)$ we conclude that there is a polynomial-time algorithm which, given x, constructs $\mathcal{Z}(x)$. Note that p, Φ and q are fixed and not part of the input of this algorithm.

It remains to show that $\mathcal{Z}(x)$ is satisfiable if and only if $x \in Y$.

If $\mathcal{Z}(x)$ is satisfiable, consider a truth assignment T satisfying all clauses. Let $c \in \{0,1\}^{\lfloor p(\text{size}(x)) \rfloor}$ with $c_j = 1$ for all j with $T(v_{0,\text{size}(x)+1+j,1}) = \textit{true}$ and $c_j = 0$ otherwise. By the above construction the variables reflect the computation of Φ on input $x\#c$. Hence we may conclude that $\text{output}(\Phi, x\#c) = 1$. Since Φ is a certificate-checking algorithm, this implies that x is a yes-instance.

Conversely, if $x \in Y$, let c be any certificate for x. Let $(n^{(i)}, s^{(i)}, \pi^{(i)})_{i=0,1,\ldots,m}$ be the computation of Φ on input $x\#c$. Then we define $T(v_{i,j,\sigma}) := \textit{true}$ iff $s_j^{(i)} = \sigma$ and $T(w_{i,j,n}) = \textit{true}$ iff $\pi^{(i)} = j$ and $n^{(i)} = n$. For $i := m+1, \ldots, Q$ we set $T(v_{i,j,\sigma}) := T(v_{i-1,j,\sigma})$ and $T(w_{i,j,n}) := T(w_{i-1,j,n})$ for all j, n and σ. Then T is a truth assignment satisfying $\mathcal{Z}(x)$, completing the proof. $\square$

SATISFIABILITY is not the only NP-complete problem; we will encounter many others in this book. Now that we already have one NP-complete problem at hand, it is much easier to prove NP-completeness for another problem. To show that a certain decision problem $\mathcal{P}$ is NP-complete, we shall just prove that $\mathcal{P} \in NP$ and that SATISFIABILITY (or any other problem which we know already to be

NP-complete) polynomially transforms to $\mathcal{P}$. Since polynomial transformability is transitive, this will be sufficient.

The following restriction of SATISFIABILITY will prove very useful for several *NP*-completeness proofs:

3SAT

Instance: A set X of variables and a collection $\mathcal{Z}$ of clauses over X, each containing exactly three literals.

Question: Is $\mathcal{Z}$ satisfiable?

To show *NP*-completeness of 3SAT we observe that any clause can be replaced equivalently by a set of 3SAT-clauses:

Proposition 15.21. *Let X be a set of variables and Z a clause over X with k literals. Then there is a set Y of at most $\max\{k - 3, 2\}$ new variables and a family $\mathcal{Z}'$ of at most $\max\{k - 2, 4\}$ clauses over $X \,\dot{\cup}\, Y$ such that each element of $\mathcal{Z}'$ has exactly three literals, and for each family $\mathcal{W}$ of clauses over X we have that $\mathcal{W} \cup \{Z\}$ is satisfiable if and only if $\mathcal{W} \cup \mathcal{Z}'$ is satisfiable. Moreover, such a family $\mathcal{Z}'$ can be computed in $O(k)$ time.*

Proof: If Z has three literals, we set $\mathcal{Z}' := \{Z\}$. If Z has more than three literals, say $Z = \{\lambda_1, \dots, \lambda_k\}$, we choose a set $Y = \{y_1, \dots, y_{k-3}\}$ of $k - 3$ new variables and set

$$\mathcal{Z}' := \big\{ \{\lambda_1, \lambda_2, y_1\}, \{\overline{y_1}, \lambda_3, y_2\}, \{\overline{y_2}, \lambda_4, y_3\}, \dots, \\ \{\overline{y_{k-4}}, \lambda_{k-2}, y_{k-3}\}, \{\overline{y_{k-3}}, \lambda_{k-1}, \lambda_k\} \big\}.$$

If $Z = \{\lambda_1, \lambda_2\}$, we choose a new variable y_1 $(Y := \{y_1\})$ and set

$$\mathcal{Z}' := \{\{\lambda_1, \lambda_2, y_1\}, \{\lambda_1, \lambda_2, \overline{y_1}\}\}.$$

If $Z = \{\lambda_1\}$, we choose a set $Y = \{y_1, y_2\}$ of two new variables and set

$$\mathcal{Z}' := \{\{\lambda_1, y_1, y_2\}, \{\lambda_1, y_1, \overline{y_2}\}, \{\lambda_1, \overline{y_1}, y_2\}, \{\lambda_1, \overline{y_1}, \overline{y_2}\}\}.$$

Observe that in each case Z can be equivalently replaced by the clauses in $\mathcal{Z}'$ in any instance of SATISFIABILITY. $\qquad\square$

Theorem 15.22. (Cook [1971]) *3SAT is NP-complete.*

Proof: As a restriction of SATISFIABILITY, 3SAT is certainly in *NP*. We now show that SATISFIABILITY polynomially transforms to 3SAT. Consider any collection $\mathcal{Z}$ of clauses $Z_1, \dots, Z_m$. We shall construct a new collection $\mathcal{Z}'$ of clauses with three literals per clause such that $\mathcal{Z}$ is satisfiable if and only if $\mathcal{Z}'$ is satisfiable.

To do this, we replace each clause Z_i by an equivalent set of clauses, each with three literals. This is possible in linear time by Proposition 15.21. $\qquad\square$

If we restrict each clause to consist of just two literals, the problem (called 2SAT) can be solved in linear time (Exercise 8).

15.5 Some Basic *NP*-Complete Problems

Karp [1972] discovered the wealth of consequences of Cook's work for combinatorial optimization problems. As a start, we consider the following problem:

STABLE SET

Instance: A graph G and an integer k.

Question: Is there a stable set of k vertices?

Theorem 15.23.   (Karp [1972])   STABLE SET *is NP-complete.*

Proof: Obviously, STABLE SET $\in NP$. We show that SATISFIABILITY polynomially transforms to STABLE SET.

Let $\mathcal{Z}$ be a collection of clauses $Z_1, \ldots, Z_m$ with $Z_i = \{\lambda_{i1}, \ldots, \lambda_{ik_i}\}$ ($i = 1, \ldots, m$), where the λ_{ij} are literals over some set X of variables.

We shall construct a graph G such that G has a stable set of size m if and only if there is a truth assignment satisfying all m clauses.

For each clause Z_i, we introduce a clique of k_i vertices according to the literals in this clause. Vertices corresponding to different clauses are connected by an edge if and only if the literals contradict each other. Formally, let $V(G) := \{v_{ij} : 1 \leq i \leq m, \ 1 \leq j \leq k_i\}$ and

$$E(G) := \big\{\{v_{ij}, v_{kl}\} : (i = k \text{ and } j \neq l)$$
$$\text{or } (\lambda_{ij} = x \text{ and } \lambda_{kl} = \overline{x} \text{ for some } x \in X)\big\}.$$

See Figure 15.1 for an example ($m = 4$, $Z_1 = \{\overline{x_1}, x_2, x_3\}$, $Z_2 = \{x_1, \overline{x_3}\}$, $Z_3 = \{x_2, \overline{x_3}\}$ and $Z_4 = \{\overline{x_1}, \overline{x_2}, \overline{x_3}\}$).

Suppose G has a stable set of size m. Then its vertices specify pairwise compatible literals belonging to different clauses. Setting each of these literals to be *true*

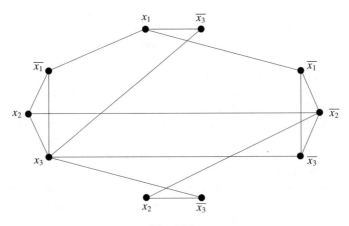

Fig. 15.1.

(and setting variables not occurring there arbitrarily) we obtain a truth assignment satisfying all m clauses.

Conversely, if some truth assignment satisfies all m clauses, then we choose a literal which is *true* out of each clause. The set of corresponding vertices then defines a stable set of size m in G.   □

It is essential that k is part of the input: for each fixed k it can be decided in $O(n^k)$ time whether a given graph with n vertices has a stable set of size k (simply by testing all vertex sets with k elements). Two interesting related problems are the following:

VERTEX COVER

Instance: A graph G and an integer k.

Question: Is there a vertex cover of cardinality k?

CLIQUE

Instance: A graph G and an integer k.

Question: Has G a clique of cardinality k?

Corollary 15.24.   (Karp [1972])   VERTEX COVER *and* CLIQUE *are NP-complete.*

Proof: By Proposition 2.2, STABLE SET polynomially transforms to both VERTEX COVER and CLIQUE.   □

We now turn to the famous Hamiltonian circuit problem (already defined in Section 15.3).

Theorem 15.25.   (Karp [1972])   HAMILTONIAN CIRCUIT *is NP-complete.*

Proof: Membership in *NP* is obvious. We prove that 3SAT polynomially transforms to HAMILTONIAN CIRCUIT. Given a collection $\mathcal{Z}$ of clauses $Z_1, \ldots, Z_m$ over $X = \{x_1, \ldots, x_n\}$, each clause containing three literals, we shall construct a graph G such that G is Hamiltonian iff $\mathcal{Z}$ is satisfiable.

We first define two gadgets which will appear several times in G. Consider the graph shown in Figure 15.2(a), which we call A. We assume that it is a subgraph of G and no vertex of A except u, u', v, v' is incident to any other edge of G. Then any Hamiltonian circuit of G must traverse A in one of the ways shown in Figure 15.3(a) and (b). So we can replace A by two edges with the additional restriction that any Hamiltonian circuit of G must contain exactly one of them (Figure 15.2(b)).

Now consider the graph B shown in Figure 15.4(a). We assume that it is a subgraph of G, and no vertex of B except u and u' is incident to any other edge of G. Then no Hamiltonian circuit of G traverses all of e_1, e_2, e_3. Moreover, one easily checks that for any $S \subset \{e_1, e_2, e_3\}$ there is a Hamiltonian path from u to

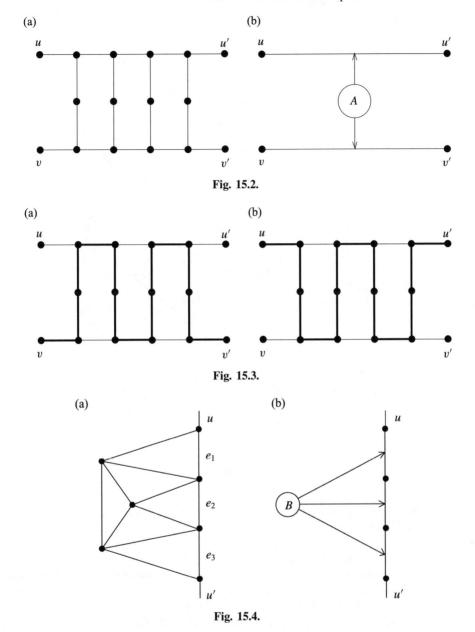

(a)   (b)

Fig. 15.2.

(a)   (b)

Fig. 15.3.

(a)   (b)

Fig. 15.4.

u' in B that contains S but none of $\{e_1, e_2, e_3\} \setminus S$. We represent B by the symbol shown in Figure 15.4(b).

We are now able to construct G. For each clause, we introduce a copy of B, joined one after another. Between the first and the last copy of B, we insert two vertices for each variable, all joined one after another. We then double the edges

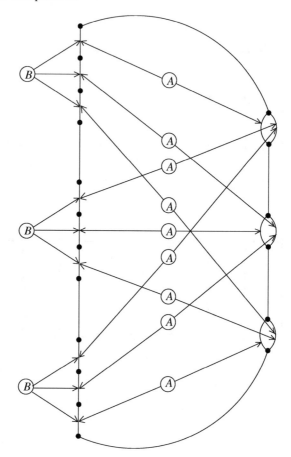

Fig. 15.5.

between the two vertices of each variable x; these two edges will correspond to x and $\overline{x}$, respectively.

The edges e_1, e_2, and e_3 in each copy of B are now connected via a copy of A to the edges corresponding to the first, second, and third literal of the corresponding clause, respectively. These constructions are done consecutively: when introducing a copy of subgraph A at an edge $e = \{u, v\}$ corresponding to a literal, the edge incident to u in Figure 15.2(a) takes the role of e: it is now the edge corresponding to that literal. The overall construction is illustrated by Figure 15.5 with the example $\{\{x_1, \overline{x_2}, \overline{x_3}\}, \{\overline{x_1}, x_2, \overline{x_3}\}, \{\overline{x_1}, \overline{x_2}, x_3\}\}$.

Now we claim that G is Hamiltonian if and only if $\mathcal{Z}$ is satisfiable. Let C be a Hamiltonian circuit. We define a truth assignment by setting a literal *true* iff C contains the corresponding edge. By the properties of the gadgets A and B each clause contains a literal that is *true*.

Conversely, any satisfying truth assignment defines a set of edges corresponding to literals that are *true*. Since each clause contains a literal that is *true* this set of edges can be completed to a tour in G.   □

This proof is essentially due to Papadimitriou and Steiglitz [1982]. The problem of deciding whether a given graph contains a Hamiltonian path is also *NP*-complete (Exercise 17(a)). Moreover, one can easily transform the undirected versions to the directed Hamiltonian circuit or Hamiltonian path problem by replacing each undirected edge by a pair of oppositely directed edges. Thus the directed versions are also *NP*-complete.

There is another fundamental *NP*-complete problem:

3-DIMENSIONAL MATCHING (3DM)

Instance: Disjoint sets U, V, W of equal cardinality and $T \subseteq U \times V \times W$.

Question: Is there a subset M of T with $|M| = |U|$ such that for distinct $(u, v, w), (u', v', w') \in M$ one has $u \neq u'$, $v \neq v'$ and $w \neq w'$?

Theorem 15.26.   (Karp [1972])   3DM *is NP-complete.*

Proof: Membership in *NP* is obvious. We shall polynomially transform SATISFIA-BILITY to 3DM. Given a collection $\mathcal{Z}$ of clauses $Z_1, \ldots, Z_m$ over $X = \{x_1, \ldots, x_n\}$, we construct an instance (U, V, W, T) of 3DM which is a yes-instance if and only if $\mathcal{Z}$ is satisfiable.

We define:

$$U := \{x_i^j, \overline{x_i}^j : i = 1, \ldots, n; \ j = 1, \ldots, m\}$$

$$V := \{a_i^j : i = 1, \ldots, n; \ j = 1, \ldots, m\} \cup \{v^j : j = 1, \ldots, m\}$$
$$\cup \ \{c_k^j : k = 1, \ldots, n-1; \ j = 1, \ldots, m\}$$

$$W := \{b_i^j : i = 1, \ldots, n; \ j = 1, \ldots, m\} \cup \{w^j : j = 1, \ldots, m\}$$
$$\cup \ \{d_k^j : k = 1, \ldots, n-1; \ j = 1, \ldots, m\}$$

$$T_1 := \{(x_i^j, a_i^j, b_i^j), (\overline{x_i}^j, a_i^{j+1}, b_i^j) : i = 1, \ldots, n; \ j = 1, \ldots, m\},$$
$$\text{where } a_i^{m+1} := a_i^1$$

$$T_2 := \{(x_i^j, v^j, w^j) : i = 1, \ldots, n; \ j = 1, \ldots, m; \ x_i \in Z_j\}$$
$$\cup \ \{(\overline{x_i}^j, v^j, w^j) : i = 1, \ldots, n; \ j = 1, \ldots, m; \ \overline{x_i} \in Z_j\}$$

$$T_3 := \{(x_i^j, c_k^j, d_k^j), (\overline{x_i}^j, c_k^j, d_k^j) : i = 1, \ldots, n; \ j = 1, \ldots, m; \ k = 1, \ldots, n-1\}$$

$$T := T_1 \cup T_2 \cup T_3.$$

For an illustration of this construction, see Figure 15.6. Here $m = 2$, $Z_1 = \{x_1, \overline{x_2}\}$, $Z_2 = \{\overline{x_1}, \overline{x_2}\}$. Each triangle corresponds to an element of $T_1 \cup T_2$. The elements c_k^j, d_k^j and the triples in T_3 are not shown.

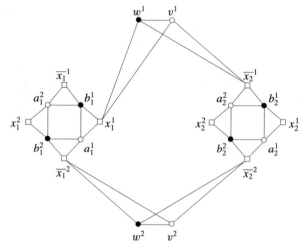

Fig. 15.6.

Suppose (U, V, W, T) is a yes-instance, so let $M \subseteq T$ be a solution. Since the a_i^j's and b_i^j's appear only in elements of T_1, for each i we have either $M \cap T_1 \supseteq \{(x_i^j, a_i^j, b_i^j) : j = 1, \ldots, m\}$ or $M \cap T_1 \supseteq \{(\overline{x_i}^j, a_i^{j+1}, b_i^j) : j = 1, \ldots, m\}$. In the first case we set x_i to *false*, in the second case to *true*.

Furthermore, for each clause Z_j we have $(\lambda^j, v^j, w^j) \in M$ for some literal $\lambda \in Z_j$. Since λ^j does not appear in any element of $M \cap T_1$ this literal is *true*; hence we have a satisfying truth assignment.

Conversely, a satisfying truth assignment suggests a set $M_1 \subseteq T_1$ of cardinality nm and a set $M_2 \subseteq T_2$ of cardinality m such that for distinct $(u, v, w), (u', v', w') \in M_1 \cup M_2$ we have $u \neq u'$, $v \neq v'$ and $w \neq w'$. It is easy to complete $M_1 \cup M_2$ by $(n - 1)m$ elements of T_3 to a solution of the 3DM instance.   □

A problem which looks simple but is not known to be solvable in polynomial time is the following:

SUBSET-SUM

Instance: Natural numbers $c_1, \ldots, c_n, K$.

Question: Is there a subset $S \subseteq \{1, \ldots, n\}$ such that $\sum_{j \in S} c_j = K$?

Corollary 15.27.   (Karp [1972])   SUBSET-SUM *is NP-complete.*

Proof: It is obvious that SUBSET-SUM is in *NP*. We prove that 3DM polynomially transforms to SUBSET-SUM. So let (U, V, W, T) be an instance of 3DM. W.l.o.g. let $U \cup V \cup W = \{u_1, \ldots, u_{3m}\}$. We write $S := \{\{a, b, c\} : (a, b, c) \in T\}$ and $S = \{s_1, \ldots, s_n\}$.

Define

$$c_j := \sum_{u_i \in s_j} (n+1)^{i-1} \qquad (j = 1, \ldots, n), \text{ and}$$

$$K := \sum_{i=1}^{3m} (n+1)^{i-1}.$$

Written in $(n + 1)$-ary form, the number c_j can be regarded as the incidence vector of s_j $(j = 1, \ldots, n)$, and K consists of 1's only. Therefore each solution to the 3DM instance corresponds to a subset R of S such that $\sum_{s_j \in R} c_j = K$, and vice versa. Moreover, $\text{size}(c_j) \leq \text{size}(K) = O(m \log n)$, so the above is indeed a polynomial transformation. $\qquad \square$

An important special case is the following problem:

PARTITION

Instance: Natural numbers $c_1, \ldots, c_n$.

Question: Is there a subset $S \subseteq \{1, \ldots, n\}$ such that $\sum_{j \in S} c_j = \sum_{j \notin S} c_j$?

Corollary 15.28. (Karp [1972]) PARTITION *is NP-complete.*

Proof: We show that SUBSET-SUM polynomially transforms to PARTITION. So let $c_1, \ldots, c_n, K$ be an instance of SUBSET-SUM. We add an element $c_{n+1} := \left| \sum_{i=1}^n c_i - 2K \right|$ (unless this number is zero) and have an instance $c_1, \ldots, c_{n+1}$ of PARTITION.

Case 1: $2K \leq \sum_{i=1}^n c_i$. Then for any $I \subseteq \{1, \ldots, n\}$ we have

$$\sum_{i \in I} c_i = K \quad \text{if and only if} \quad \sum_{i \in I \cup \{n+1\}} c_i = \sum_{i \in \{1,\ldots,n\} \setminus I} c_i.$$

Case 2: $2K > \sum_{i=1}^n c_i$. Then for any $I \subseteq \{1, \ldots, n\}$ we have

$$\sum_{i \in I} c_i = K \quad \text{if and only if} \quad \sum_{i \in I} c_i = \sum_{i \in \{1,\ldots,n+1\} \setminus I} c_i.$$

In both cases we have constructed a yes-instance of PARTITION if and only if the original instance of SUBSET-SUM is a yes-instance. $\qquad \square$

We finally note:

Theorem 15.29. INTEGER LINEAR INEQUALITIES *is NP-complete.*

Proof: We already mentioned the membership in *NP* in Proposition 15.12. Any of the above problems can easily be formulated as an instance of INTEGER LINEAR INEQUALITIES. For example a PARTITION instance $c_1, \ldots, c_n$ is a yes-instance if and only if $\{x \in \mathbb{Z}^n : 0 \leq x \leq \mathbb{1}, 2c^\top x = c^\top \mathbb{1}\}$ is nonempty. $\qquad \square$

15.6 The Class *coNP*

The definition of *NP* is not symmetric with respect to yes-instances and no-instances. For example, it is an open question whether the following problem belongs to *NP*: given a graph G, is it true that G is not Hamiltonian? We introduce the following definitions:

Definition 15.30. *For a decision problem* $\mathcal{P} = (X, Y)$ *we define its* **complement** *to be the decision problem* $(X, X \setminus Y)$. *The class coNP consists of all problems whose complements are in NP. A decision problem* $\mathcal{P} \in coNP$ *is called* **coNP-complete** *if all other problems in coNP polynomially transform to* $\mathcal{P}$.

Trivially, the complement of a problem in *P* is also in *P*. On the other hand, $NP \neq coNP$ is commonly conjectured (though not proved). For this conjecture the *NP*-complete problems play a special role:

Theorem 15.31. *A decision problem is coNP-complete if and only if its complement is NP-complete. Unless* $NP = coNP$, *no coNP-complete problem is in NP.*

Proof: The first statement follows directly from the definition.

Suppose $\mathcal{P} = (X, Y) \in NP$ is a *coNP*-complete problem. Let $\mathcal{Q} = (V, W)$ be an arbitrary problem in *coNP*. We show that $\mathcal{Q} \in NP$.

Since $\mathcal{P}$ is *coNP*-complete, $\mathcal{Q}$ polynomially transforms to $\mathcal{P}$. So there is a polynomial-time algorithm which transforms any instance v of $\mathcal{Q}$ to an instance $x = f(v)$ of $\mathcal{P}$ such that $x \in Y$ if and only if $v \in W$. Note that $\text{size}(x) \leq p(\text{size}(v))$ for some fixed polynomial p.

Since $\mathcal{P} \in NP$, there exists a polynomial q and a decision problem $\mathcal{P}' = (X', Y')$ in *P*, where $X' := \{x \# c : x \in X, c \in \{0, 1\}^{\lfloor q(\text{size}(x)) \rfloor}\}$, such that

$$Y = \left\{ y \in X : \text{there exists a string } c \in \{0, 1\}^{\lfloor q(\text{size}(y)) \rfloor} \text{ with } y \# c \in Y' \right\}$$

(cf. Definition 15.9). We define a decision problem (V', W') by $V' := \{v \# c : v \in V, c \in \{0, 1\}^{\lfloor q(p(\text{size}(v))) \rfloor}\}$, and $v \# c \in W'$ if and only if $f(v) \# c' \in Y'$ where c' consists of the first $\lfloor q(\text{size}(f(v))) \rfloor$ components of c.

Observe that $(V', W') \in P$. Therefore, by definition, $\mathcal{Q} \in NP$. We conclude $coNP \subseteq NP$ and hence, by symmetry, $NP = coNP$. $\qquad \square$

If one can show that a problem is in $NP \cap coNP$, we say that the problem has a **good characterization** (Edmonds [1965]). This means that for yes-instances as well as for no-instances there are certificates that can be checked in polynomial time. Theorem 15.31 indicates that a problem with a good characterization is probably not *NP*-complete.

To give examples, Proposition 2.9, Theorem 2.24, and Proposition 2.27 provide good characterizations for the problems of deciding whether a given graph is acyclic, whether it has an Eulerian walk, and whether it is bipartite, respectively.

Of course, this is not very interesting since all these problems can be solved easily in polynomial time. But consider the decision version of LINEAR PROGRAMMING:

Theorem 15.32.   LINEAR INEQUALITIES *is in NP ∩ coNP*.

Proof: This immediately follows from Theorem 4.4 and Corollary 3.24.   □

Of course, this theorem also follows from any polynomial-time algorithm for LINEAR PROGRAMMING, e.g. Theorem 4.18. However, before the ELLIPSOID METHOD had been discovered, Theorem 15.32 was the only theoretical evidence that LINEAR INEQUALITIES is probably not *NP*-complete. This gave hope to find a polynomial-time algorithm for LINEAR PROGRAMMING (which can be reduced to LINEAR INEQUALITIES by Proposition 4.16); a justified hope as we know today.

The following famous problem has a similar history:

PRIME

Instance: A number $n \in \mathbb{N}$ (in its binary representation).

Question: Is n a prime?

It is obvious that PRIME belongs to *coNP*. Pratt [1975] proved that PRIME also belongs to *NP*. Finally, Agrawal, Kayal and Saxena [2004] proved that PRIME $\in P$ by finding a surprisingly simple $O(\log^{7.5+\epsilon} n)$-algorithm (for any $\epsilon > 0$). Before, the best known deterministic algorithm for PRIME was due to, Adleman, Pomerance and Rumely [1983], running in $O\left((\log n)^{c \log \log \log n}\right)$ time for some constant c. Since the input size is $O(\log n)$, this is not polynomial.

We close this section by sketching the inclusions of *NP* and *coNP* (Figure 15.7). Ladner [1975] showed that, unless $P = NP$, there are problems in $NP \setminus P$ that are not *NP*-complete. However, until the $P \neq NP$ conjecture is resolved, it is still possible that all regions drawn in Figure 15.7 collapse to one.

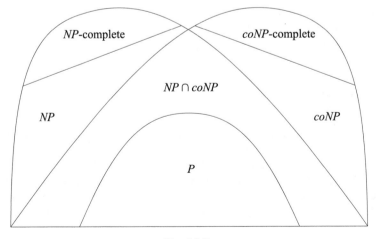

Fig. 15.7.

15.7 *NP*-Hard Problems

Now we extend our results to general computational problems, and in particular to optimization problems.

Definition 15.33. *A computational problem* $\mathcal{P}$ *is called* **NP-hard** *if all problems in NP polynomially reduce to* $\mathcal{P}$.

Note that the definition also applies to decision problems, and it is symmetric (in contrast to *NP*-completeness): a decision problem is *NP*-hard if and only if its complement is. *NP*-hard problems are at least as hard as the hardest problems in *NP*. But some may be harder than any problem in *NP*. A problem which polynomially reduces to some problem in *NP* is called **NP-easy**. A problem which is both *NP*-hard and *NP*-easy is **NP-equivalent**. In other words, a problem is *NP*-equivalent if and only if it is polynomially equivalent to SATISFIABILITY, where two problems $\mathcal{P}$ and $\mathcal{Q}$ are called **polynomially equivalent** if $\mathcal{P}$ polynomially reduces to $\mathcal{Q}$, and $\mathcal{Q}$ polynomially reduces to $\mathcal{P}$. We note:

Proposition 15.34. *Let* $\mathcal{P}$ *be an NP-equivalent computational problem. Then* $\mathcal{P}$ *has an exact polynomial-time algorithm if and only if* $P = NP$. □

Of course, all *NP*-complete problems and all *coNP*-complete problems are *NP*-equivalent. Almost all problems discussed in this book are *NP*-easy since they polynomially reduce to INTEGER PROGRAMMING; this is usually a trivial observation which we do not even mention.

We now formally define the type of optimization problems we are interested in:

Definition 15.35. *An* **NP optimization problem** *is a quadruple* $\mathcal{P} = (X, (S_x)_{x \in X}, c, \text{goal})$, *where*

- X *is a language over* $\{0, 1\}$ *decidable in polynomial time;*
- S_x *is a nonempty subset of* $\{0, 1\}^*$ *for each* $x \in X$; *there exists a polynomial* p *with* $\text{size}(y) \leq p(\text{size}(x))$ *for all* $x \in X$ *and* $y \in S_x$, *and the language* $\{(x, y) : x \in X, y \in S_x\}$ *is decidable in polynomial time;*
- $c : \{(x, y) : x \in X, y \in S_x\} \to \mathbb{Q}$ *is a function computable in polynomial time; and*
- $\text{goal} \in \{\text{max}, \text{min}\}$.

The elements of X *are called* **instances** *of* $\mathcal{P}$. *For each instance* x, *the elements of* S_x *are called* **feasible solutions** *of* x. *We write* $\text{OPT}(x) := \text{goal}\{c(x, y) : y \in S_x\}$. *An* **optimum solution** *of* x *is a feasible solution* y *of* x *with* $c(x, y) = \text{OPT}(x)$.

A **heuristic** *for* $\mathcal{P}$ *is an algorithm* A *which computes for each input* $x \in X$ *with* $S_x \neq \emptyset$ *a feasible solution* $y \in S_x$. *We sometimes write* $A(x) := c(x, y)$. *If* $A(x) = \text{OPT}(x)$ *for all* $x \in X$ *with* $S_x \neq \emptyset$, *then* A *is an* **exact algorithm** *for* $\mathcal{P}$.

Depending on the context, $c(x, y)$ is often called the cost, the weight, the profit or the length of y. If c is nonnegative, then we say that the optimization problem has nonnegative weights. The values of c are rational numbers; we assume an encoding into binary strings as usual.

Most interesting optimization problems fall into this class, but there are some exceptions (e.g., Exercise 24).

An optimization problem $(X, (S_x)_{x \in X}, c, \text{goal})$ can be regarded as the computational problem $(X, \{(x, y) : x \in X, y \in S_x, c(x, y) = \text{OPT}(x)\})$. Hence polynomial reductions also apply to optimization problems.

Theorem 15.36. *Every NP optimization problem is NP-easy.*

Proof: Let $\mathcal{P} = (X, (S_x)_{x \in X}, c, \text{goal})$ be an *NP* optimization problem. We polynomially reduce $\mathcal{P}$ to a decision problem $\mathcal{Q} \in NP$. As usual we call a string $y \in \{0, 1\}^p$, $p \in \mathbb{Z}_+$, lexicographically greater than a string $s \in \{0, 1\}^q$, $q \in \mathbb{Z}_+$, iff $y \neq s$ and $y_j > s_j$ for $j = \min\{i \in \mathbb{N} : y_i \neq s_i\}$, where $y_i := -1$ for $i > p$ and $s_i := -1$ for $i > q$.

If goal $=$ max, then $\mathcal{Q}$ is defined as follows: Given $x \in X$, $\gamma \in \mathbb{Q}$, and $s \in \{0, 1\}^*$, is there a $y \in S_x$ such that $c(x, y) \geq \gamma$ and y is equal to or lexicographically greater than s? If goal $=$ min, then $c(x, y) \geq \gamma$ is replaced by $c(x, y) \leq \gamma$.

Observe that $\mathcal{Q}$ belongs to *NP* (y serves as certificate). We polynomially reduce $\mathcal{P}$ to $\mathcal{Q}$ as follows.

As c is computable in polynomial time, there is a constant $d \in \mathbb{N}$ such that $\text{size}(c(x, y)) \leq (\text{size}(x) + p(\text{size}(x)))^d =: k(x)$ for all $x \in X$ and $y \in S_x$. Hence $\text{OPT}(x) \in [-2^{k(x)}, 2^{k(x)}]$, and $|c(x, y) - c(x, y')|$ is an integral multiple of $2^{-k(x)}$ for all $x \in X$ and $y, y' \in S_x$.

Given an instance $x \in X$, we first compute $k(x)$ and then determine $\text{OPT}(x)$ by binary search. We start with $\alpha := -2^{k(x)}$ and $\beta := 2^{k(x)}$. In each iteration we apply the oracle to (x, γ, s_0), where $\gamma = \frac{\alpha + \beta}{2}$ and s_0 is the empty string. If the answer is yes, we set $\alpha := \gamma$, otherwise $\beta := \gamma$.

After $2k(x) + 2$ iterations we have $\beta - \alpha < 2^{-k(x)}$. Then we fix $\gamma := \alpha$ and use another $2p(\text{size}(x))$ oracle calls to compute a solution $y \in S_x$ with $c(x, y) \geq \alpha$. For $i := 1, \ldots, p(\text{size}(x))$ we call the oracle to (x, α, s_{i-1}^0) and (x, α, s_{i-1}^1), where s^j results from the string s by appending the symbol $j \in \{0, 1\}$. If the both answers are yes, then we set $s_i := s_{i-1}^1$, if only the first answer is yes, then we set $s_i := s_{i-1}^0$, and if both answers are no, then we set $s_i := s_{i-1}$. We conclude that $s_{p(\text{size}(x))}$ is the lexicographically maximal string y with $y \in S_x$ and $c(x, y) = \text{OPT}(x)$. $\square$

Most problems that we discuss from now on are also *NP*-hard, and we shall usually prove this by describing a polynomial reduction from an *NP*-complete problem. As a first example we consider MAX-2SAT: given an instance of SATISFIABILITY with exactly two literals per clause, find a truth assignment that maximizes the number of satisfied clauses.

Theorem 15.37.   (Garey, Johnson and Stockmeyer [1976]) MAX-2SAT *is NP-hard.*

Proof: By reduction from 3SAT. Given an instance I of 3SAT with clauses $C_1, \ldots, C_m$, we construct an instance I' of MAX-2SAT by adding new variables $y_1, z_1, \ldots, y_m, z_m$ and replacing each clause $C_i = \{\lambda_1, \lambda_2, \lambda_3\}$ by the fourteen clauses

$$\{\lambda_1, z_i\}, \{\lambda_1, \bar{z}_i\}, \{\lambda_2, z_i\}, \{\lambda_2, \bar{z}_i\}, \{\lambda_3, z_i\}, \{\lambda_3, \bar{z}_i\}, \{y_i, z_i\}, \{y_i, \bar{z}_i\},$$
$$\{\lambda_1, \bar{y}_i\}, \{\lambda_2, \bar{y}_i\}, \{\lambda_3, \bar{y}_i\}, \{\bar{\lambda}_1, \bar{\lambda}_2\}, \{\bar{\lambda}_1, \bar{\lambda}_3\}, \{\bar{\lambda}_2, \bar{\lambda}_3\}.$$

Note that no truth assignment satisfies more than 11 of these 14 clauses. Moreover, if 11 of these clauses are satisfied, then at least one of $\lambda_1, \lambda_2, \lambda_3$ must be *true*. On the other hand, if one of $\lambda_1, \lambda_2, \lambda_3$ is *true* we can set $y_i := \lambda_1 \wedge \lambda_2 \wedge \lambda_3$ and $z_i := true$ in order to satisfy 11 of these clauses.

We conclude that I has a truth assignment satisfying all m clauses if and only if I' has a truth assignment that satisfies $11m$ clauses. $\square$

It is an open question whether each *NP*-hard decision problem $\mathcal{P} \in NP$ is *NP*-complete (recall the difference between polynomial reduction and polynomial transformation; Definitions 15.15 and 15.17). Exercises 22 and 23 discuss two *NP*-hard decision problems that appear not to be in *NP*. See also Exercise 2 of Chapter 19.

Unless $P = NP$ there is no exact polynomial-time algorithm for any *NP*-hard problem. There might, however, be a pseudopolynomial algorithm:

Definition 15.38.   *Let $\mathcal{P}$ be a decision problem or an optimization problem such that each instance x consists of a list of nonnegative integers. We denote by* largest(x) *the largest of these integers. An algorithm for $\mathcal{P}$ is called* **pseudopolynomial** *if its running time is bounded by a polynomial in* size(x) *and* largest(x).

For example there is a trivial pseudopolynomial algorithm for PRIME which divides the natural number n to be tested for primality by each integer from 2 to $\lfloor\sqrt{n}\rfloor$. Another example is:

Theorem 15.39.   *There is a pseudopolynomial algorithm for* SUBSET-SUM.

Proof: Given an instance $c_1, \ldots, c_n, K$ of SUBSET-SUM, we construct a digraph G with vertex set $\{0, \ldots, n\} \times \{0, 1, 2, \ldots, K\}$. For each $j \in \{1, \ldots, n\}$ we add edges $((j - 1, i), (j, i))$ $(i = 0, 1, \ldots, K)$ and $((j - 1, i), (j, i + c_j))$ $(i = 0, 1, \ldots, K - c_j)$.

Observe that any path from $(0, 0)$ to (j, i) corresponds to a subset $S \subseteq \{1, \ldots, j\}$ with $\sum_{k \in S} c_k = i$, and vice versa. Therefore we can solve our SUBSET-SUM instance by checking whether G contains a path from $(0, 0)$ to (n, K). With the GRAPH SCANNING ALGORITHM this can be done in $O(nK)$ time, so we have a pseudopolynomial algorithm. $\square$

The above is also a pseudopolynomial algorithm for PARTITION because $\frac{1}{2} \sum_{i=1}^{n} c_i \leq \frac{n}{2} \text{largest}(c_1, \ldots, c_n)$. We shall discuss an extension of this algorithm in Section 17.2. If the numbers are not too large, a pseudopolynomial algorithm can be quite efficient. Therefore the following definition is useful:

Definition 15.40. *For a decision problem $\mathcal{P} = (X, Y)$ or an optimization problem $\mathcal{P} = (X, (S_x)_{x \in X}, c, \text{goal})$, and a subset $X' \subseteq X$ of instances we define the* **restriction** *of $\mathcal{P}$ to X' by $\mathcal{P}' = (X', X' \cap Y)$ or $\mathcal{P}' = (X', (S_x)_{x \in X'}, c, \text{goal})$, respectively.*

Let $\mathcal{P}$ be a decision or optimization problem such that each instance consists of a list of numbers. For a polynomial p let $\mathcal{P}_p$ be the restriction of $\mathcal{P}$ to instances x consisting of nonnegative integers with $\text{largest}(x) \leq p(\text{size}(x))$. $\mathcal{P}$ is called **strongly *NP*-hard** *if there is a polynomial p such that $\mathcal{P}_p$ is NP-hard. $\mathcal{P}$ is called* **strongly *NP*-complete** *if $\mathcal{P} \in NP$ and there is a polynomial p such that $\mathcal{P}_p$ is NP-complete.*

Proposition 15.41. *Unless $P = NP$ there is no exact pseudopolynomial algorithm for any strongly NP-hard problem.* $\square$

We give some famous examples:

Theorem 15.42. INTEGER PROGRAMMING *is strongly NP-hard.*

Proof: For an undirected graph G the integer program $\max\{\mathbb{1}x : x \in \mathbb{Z}^{V(G)}, 0 \leq x \leq \mathbb{1}, x_v + x_w \leq 1 \text{ for } \{v, w\} \in E(G)\}$ has optimum value at least k if and only if G contains a stable set of cardinality k. Since $k \leq |V(G)|$ for all nontrivial instances (G, k) of STABLE SET, the result follows from Theorem 15.23. $\square$

TRAVELING SALESMAN PROBLEM (TSP)

Instance: A complete graph K_n ($n \geq 3$) and weights $c : E(K_n) \rightarrow \mathbb{R}_+$.

Task: Find a Hamiltonian circuit T whose weight $\sum_{e \in E(T)} c(e)$ is minimum.

The vertices of a TSP-instance are often called cities, the weights are also referred to as distances.

Theorem 15.43. *The* TSP *is strongly NP-hard.*

Proof: We show that the TSP is *NP*-hard even when restricted to instances where all distances are 1 or 2. We describe a polynomial reduction from the HAMILTONIAN CIRCUIT problem. Given a graph G on $n \geq 3$ vertices, we construct the following instance of TSP: Take one city for each vertex of G, and let the distances be 1 whenever the edge is in $E(G)$ and 2 otherwise. It is then obvious that G is Hamiltonian if and only if the length of an optimum TSP tour is n. $\square$

The proof also shows that the following decision problem is not easier than the TSP itself: Given an instance of the TSP and an integer k, is there a tour of length *k* or less? A similar statement is true for a large class of discrete optimization problems:

Proposition 15.44. *Let $\mathcal{F}$ and $\mathcal{F}'$ be (infinite) families of finite sets, and let $\mathcal{P}$ be the following optimization problem: given a set $E \in \mathcal{F}$ and a function $c : E \to \mathbb{Z}$, find a set $F \subseteq E$ with $F \in \mathcal{F}'$ and $c(F)$ minimum (or decide that no such F exists).*

Then $\mathcal{P}$ can be solved in polynomial time if and only if the following decision problem can be solved in polynomial time: given an instance (E, c) of $\mathcal{P}$ and an integer k, is $\mathrm{OPT}((E, c)) \leq k$? If the optimization problem is NP-hard, then so is this decision problem.

Proof: It suffices to show that there is an oracle algorithm for the optimization problem using the decision problem (the converse is trivial). Let (E, c) be an instance of $\mathcal{P}$. We first determine $\mathrm{OPT}((E, c))$ by binary search. Since there are at most $1 + \sum_{e \in E} |c(e)| \leq 2^{\mathrm{size}(c)}$ possible values we can do this with $O(\mathrm{size}(c))$ iterations, each including one oracle call.

Then we successively check for each element of E whether there exists an optimum solution without this element. This can be done by increasing its weight (say by one) and checking whether this also increases the value of an optimum solution. If so, we keep the old weight, otherwise we indeed increase the weight. After checking all elements of E, those elements whose weight we did not change constitute an optimum solution.   □

Examples where this result applies are the TSP, the MAXIMUM WEIGHT CLIQUE PROBLEM, the SHORTEST PATH PROBLEM, the KNAPSACK PROBLEM, and many others. Schulz [2009] and Orlin, Punnen and Schulz [2009] proved similar results for INTEGER PROGRAMMING.

The following chapters mostly deal with *NP*-hard combinatorial optimization problems. Since we cannot hope for a polynomial-time exact algorithm, there are essentially two alternatives: approximation algorithms that do not guarantee to find the optimum (they are discussed in the next chapter) and exact algorithms that do not have a polynomially bounded worst-case running time. We will give an example for the latter now.

Recall that VERTEX COVER is *NP*-complete. An instance is a pair (G, k); let $n := |V(G)|$ and $m = |E(G)|$. There is a trivial exact algorithm with running time $O(n^k m)$, simply by trying all n-element vertex sets. So if k is fixed rather than part of the input, this is a polynomial-time algorithm (this is often called *fixed-parameter tractable*). One can do even better by applying the following preprocessing, also called *kernelization*, before:

Theorem 15.45.   *For any VERTEX COVER instance (G, k) we can, in linear time, either decide whether G has a vertex cover of size at most k or compute an equivalent instance (G', l) with $l \leq k$ and $|V(G')| \leq |E(G')| \leq l^2$.*

Proof: Here is a recursive algorithm. If $k < 0$, output "no". If $k \geq 0$ and G has no edges, output "yes". Otherwise first delete parallel edges and isolated vertices.

Then, if G has a vertex v of degree more than k, we know that v must belong to every vertex cover of size at most k; thus call the algorithm recursively with input $(G - v, k - 1)$. Otherwise, if G has a vertex v with only one neighbour w, there is a minimum vertex cover containing w (and not v), so call the algorithm recursively with input $(G - w, k - 1)$. Otherwise, if G has no vertex of degree more than k and G has more than k^2 edges, then there is no vertex cover of size at most k; in this case output "no". If none of the reductions apply, output the instance as is. □

See Exercise 6 of Chapter 16 for an improvement. Theorem 15.45 leads to an overall running time of $O(n + m + k^{2k})$. For small k, this is a reasonable algorithm even for very large graphs.

Exercises

1. Observe that there are more languages than Turing machines. Conclude that there are languages that cannot be decided by a Turing machine.
   Turing machines can also be encoded by binary strings. Consider the famous HALTING PROBLEM: given two binary strings x and y, where x encodes a Turing machine Φ, is time$(\Phi, y) < \infty$?
   Prove that the HALTING PROBLEM is undecidable (i.e. there is no algorithm for it).
   Hint: Assuming that there is such an algorithm A, construct a Turing machine which, on input x, first runs the algorithm A on input (x, x) and then terminates if and only if output$(A, (x, x)) = 0$.

2. Describe a Turing machine which compares two strings: it should accept as input a string $a\#b$ with $a, b \in \{0, 1\}^*$ and output 1 if $a = b$ and 0 if $a \neq b$.

3. A well-known machine model is the **RAM machine**. It works with an infinite sequence of registers $x_1, x_2, \ldots$ and one special register, the accumulator Acc. Each register can store an arbitrarily large integer, possibly negative. A RAM program is a sequence of instructions. There are ten types of instructions (the meaning is illustrated on the right-hand side):

WRITE	k	$Acc := k$.
LOAD	k	$Acc := x_k$.
LOADI	k	$Acc := x_{x_k}$.
STORE	k	$x_k := Acc$.
STOREI	k	$x_{x_k} := Acc$.
ADD	k	$Acc := Acc + x_k$.
SUBTR	k	$Acc := Acc - x_k$.
HALF		$Acc := \lfloor Acc/2 \rfloor$.
IFPOS	i	**If** $Acc > 0$ **then go to** $\textcircled{i}$.
HALT		**Stop.**

A RAM program is a sequence of m instructions; each is one of the above, where $k \in \mathbb{Z}$ and $i \in \{1, \ldots, m\}$. The computation starts with instruction 1; it then proceeds as one would expect; we do not give a formal definition.

The above list of instructions may be extended. We say that a command can be simulated by a RAM program in time n if it can be substituted by RAM commands so that the total number of steps in any computation increases by at most a factor of n.

(a) Show that the following commands can be simulated by small RAM programs in constant time:

IFNEG i	**If** $Acc < 0$ **then go to** ⓘ.
IFZERO i	**If** $Acc = 0$ **then go to** ⓘ.

* (b) Show that the SUBTR and HALF commands can be simulated by RAM programs using only the other eight commands in $O(\text{size}(x_k))$ time and $O(\text{size}(Acc))$ time, respectively.

* (c) Show that the following commands can be simulated by RAM programs in $O(n)$ time, where $n = \max\{\text{size}(x_k), \text{size}(Acc)\}$:

MULT k	$Acc := Acc \cdot x_k$.
DIV k	$Acc := \lfloor Acc/x_k \rfloor$.
MOD k	$Acc := Acc \bmod x_k$.

* 4. Let $f : \{0, 1\}^* \to \{0, 1\}^*$ be a mapping. Show that if there is a Turing machine Φ computing f, then there is a RAM program (cf. Exercise 3) such that the computation on input x (in Acc) terminates after $O(\text{size}(x) + \text{time}(\Phi, x))$ steps with $Acc = f(x)$.

Show that if there is a RAM machine which, given x in Acc, computes $f(x)$ in Acc in at most $g(\text{size}(x))$ steps, then there is a Turing machine computing f with $\text{time}(\Phi, x) = O(g(\text{size}(x))^3)$.

5. Prove that the following decision problems are in *NP*:
   (a) Given two graphs G and H, is G isomorphic to a subgraph of H?
   (b) Given a natural number n (in binary encoding), is there a prime number p with $n = p^p$?
   (c) Given a matrix $A \in \mathbb{Z}^{m \times n}$ and a vector $b \in \mathbb{Z}^m$, is the polyhedron $P = \{x : Ax \leq b\}$ bounded?
   (d) Given a matrix $A \in \mathbb{Z}^{m \times n}$ and a vector $b \in \mathbb{Z}^m$, is the polyhedron $P = \{x : Ax \leq b\}$ unbounded?

6. Prove: If $\mathcal{P} \in NP$, then there exists a polynomial p such that $\mathcal{P}$ can be solved by a (deterministic) algorithm having time complexity $O(2^{p(n)})$, where n is the input size.

7. Prove that the set of decision problems in *NP* is countable.

8. Let $\mathcal{Z}$ be a 2SAT instance, i.e. a collection of clauses over X with two literals each. Consider a digraph $G(\mathcal{Z})$ as follows: $V(G)$ is the set of literals over X. There is an edge $(\lambda_1, \lambda_2) \in E(G)$ iff the clause $\{\overline{\lambda}_1, \lambda_2\}$ is a member of $\mathcal{Z}$.

(a) Show that if, for some variable x, x and $\bar{x}$ are in the same strongly connected component of $G(\mathcal{Z})$, then $\mathcal{Z}$ is not satisfiable.

(b) Show the converse of (a).

(c) Give a linear-time algorithm for 2SAT.

9. Describe a linear-time algorithm which for any instance of SATISFIABILITY finds a truth assignment satisfying at least half of the clauses.

10. Consider instances of SATISFIABILITY in which each clause has one of the forms $\{x\}$, $\{\bar{x}\}$, or $\{\bar{x}, y\}$, where x and y are variables. Given such an instance and a nonnegative weight for each clause, find (in polynomial time) a truth assignment that maximizes the total weight of the satisfied clauses.

    Hint: Reduce this to the MINIMUM CAPACITY CUT PROBLEM.

11. Consider 3-OCCURRENCE SAT, which is SATISFIABILITY restricted to instances where each clause contains at most three literals and each variable occurs in at most three clauses. Prove that even this restricted version is *NP*-complete.

12. Let $\kappa : \{0, 1\}^m \to \{0, 1\}^m$ be a (not necessarily bijective) mapping, $m \geq 2$. For $x = (x_1, \ldots, x_n) \in \{0, 1\}^m \times \cdots \times \{0, 1\}^m = \{0, 1\}^{nm}$ let $\kappa(x) := (\kappa(x_1), \ldots, \kappa(x_n))$, and for a decision problem $\mathcal{P} = (X, Y)$ with $X \subseteq \bigcup_{n \in \mathbb{Z}_+} \{0, 1\}^{nm}$ let $\kappa(\mathcal{P}) := (\{\kappa(x) : x \in X\}, \{\kappa(x) : x \in Y\})$. Prove:

    (a) For all codings κ and all $\mathcal{P} \in NP$ we have also $\kappa(\mathcal{P}) \in NP$.

    (b) If $\kappa(\mathcal{P}) \in P$ for all codings κ and all $\mathcal{P} \in P$, then $P = NP$.

    (Papadimitriou [1994])

13. Prove that STABLE SET is *NP*-complete even if restricted to graphs whose maximum degree is 4.

    Hint: Use Exercise 11.

14. Prove that the following problem, called DOMINATING SET, is *NP*-complete. Given an undirected graph G and a number $k \in \mathbb{N}$, is there a set $X \subseteq V(G)$ with $|X| \leq k$ such that $X \cup \Gamma(X) = V(G)$?

    Hint: Transformation from VERTEX COVER.

15. The decision problem CLIQUE is *NP*-complete. Is it still *NP*-complete (provided that $P \neq NP$) if restricted to

    (a) bipartite graphs,

    (b) planar graphs,

    (c) 2-connected graphs?

16. Show for each of the following decision problems either membership in P or *NP*-completeness. Given an undirected graph G, does G contain

    (a) a circuit of length at least 17?

    (b) a circuit containing at least half of the vertices?

    (c) a circuit of odd length?

    (d) a clique containing at least half of the vertices?

    (e) two cliques such that each vertex belongs to at least one of them?

17. Prove that the following problems are *NP*-complete:
    (a) HAMILTONIAN PATH and DIRECTED HAMILTONIAN PATH
        Given a graph G (directed or undirected), does G contain a Hamiltonian path?
    (b) SHORTEST PATH
        Given a graph G (directed or undirected), weights $c : E(G) \rightarrow \mathbb{Z}$, two vertices $s, t \in V(G)$, and an integer k. Is there an s-t-path of weight at most k?
    (c) 3-MATROID INTERSECTION
        Given three matroids $(E, \mathcal{F}_1), (E, \mathcal{F}_2), (E, \mathcal{F}_3)$ (by independence oracles) and a number $k \in \mathbb{N}$, decide whether there is a set $F \in \mathcal{F}_1 \cap \mathcal{F}_2 \cap \mathcal{F}_3$ with $|F| \geq k$.
    (d) CHINESE POSTMAN PROBLEM
        Given undirected graphs G and H with $V(G) = V(H)$, weights $c : E(H) \rightarrow \mathbb{Z}_+$ and an integer k. Is there a subset $F \subseteq E(H)$ with $c(F) \leq k$ such that $(V(G), E(G) \mathbin{\dot\cup} F)$ is connected and Eulerian?

18. Either find a polynomial-time algorithm or prove *NP*-completeness for the following decision problems:
    (a) Given an undirected graph G and some $T \subseteq V(G)$, is there a spanning tree in G such that all vertices in T are leaves?
    (b) Given an undirected graph G and some $T \subseteq V(G)$, is there a spanning tree in G such that all leaves are elements of T?
    (c) Given a digraph G, weights $c : E(G) \rightarrow \mathbb{R}$, a set $T \subseteq V(G)$ and a number k, is there a branching B with $|\delta_B^+(x)| \leq 1$ for all $x \in T$ and $c(B) \geq k$?

19. Prove that the following decision problem belongs to *coNP*: given a matrix $A \in \mathbb{Q}^{m \times n}$ and a vector $b \in \mathbb{Q}^n$, is the polyhedron $\{x : Ax \leq b\}$ integral?
    Hint: Use Proposition 3.9, Lemma 5.12, and Theorem 5.14.
    Note: The problem is not known to be in *NP*.

20. Prove that the following problem belongs to *coNP*: given a matrix $A \in \mathbb{Z}^{m \times n}$ and a vector $b \in \mathbb{Z}^m$, decide whether the polyhedron $P = \{x \in \mathbb{R}^n : Ax \leq b\}$ is integral.
    Note: In fact the problem is *coNP*-complete, as was shown by Papadimitriou and Yannakakis [1990].

21. Let us define *Boolean formulas*. Let X be a set of variables. Then *true* and *false* are the Boolean formulas over X of length zero, the literals are the Boolean formulas over X of length one, and the Boolean formulas over X of length $k \geq 2$ are the strings $(\psi \wedge \psi')$ and $(\psi \vee \psi')$ for all Boolean formulas ψ of length $l \in \mathbb{N}$ and ψ' of length $l' \in \mathbb{N}$ with $l + l' = k$. Given a truth assignment $T : X \rightarrow \{true, false\}$, we extend it to the Boolean formulas over X by setting $T((\psi \wedge \psi')) := T(\psi) \wedge T(\psi')$ and $T((\psi \vee \psi')) := T(\psi) \vee T(\psi')$. Two Boolean formulas ψ and ψ' over X are *equivalent* if $T(\psi) = T(\psi')$ for all truth assignments $T : X \rightarrow \{true, false\}$.

Prove that the following problem, called BOOLEAN EQUIVALENCE, is *coNP*-complete: given two Boolean formulas over a variable set X, are they equivalent?

22. Show that the following problem is *NP*-hard (it is not known to be in *NP*): given an instance of SATISFIABILITY, does the majority of all truth assignments satisfy all the clauses?

23. Show that PARTITION polynomially transforms to the following problem (which is thus *NP*-hard; it is not known to be in *NP*; cf. Haase and Kiefer [2016]):

K-TH HEAVIEST SUBSET

Instance: Integers $c_1, \ldots, c_n, K, L$.

Question: Are there K distinct subsets $S_1, \ldots, S_K \subseteq \{1, \ldots, n\}$ such that $\sum_{j \in S_i} c_j \geq L$ for $i = 1, \ldots, K$?

---

24. Prove that the following problem, called LOGIC MINIMIZATION, can be solved in polynomial time if and only if $P = NP$: given a set X of variables and a Boolean formula over X, find an equivalent Boolean formula over X with minimum length.

Hint: Use Exercise 21.

References

General Literature

Aho, A.V., Hopcroft, J.E., and Ullman, J.D. [1974]: The Design and Analysis of Computer Algorithms. Addison-Wesley, Reading 1974

Arora, S., and Barak, B. [2009]: Computational Complexity: A Modern Approach. Cambridge University Press, New York 2009

Ausiello, G., Crescenzi, P., Gambosi, G., Kann, V., Marchetti-Spaccamela, A., and Protasi, M. [1999]: Complexity and Approximation: Combinatorial Optimization Problems and Their Approximability Properties. Springer, Berlin 1999

Bovet, D.B., and Crescenzi, P. [1994]: Introduction to the Theory of Complexity. Prentice-Hall, New York 1994

Garey, M.R., and Johnson, D.S. [1979]: Computers and Intractability: A Guide to the Theory of *NP*-Completeness. Freeman, San Francisco 1979, Chapters 1–3, 5, and 7

Goldreich, O. [2008]: Computational Complexity: A Conceptual Perspective. Cambridge University Press, New York 2008

Horowitz, E., and Sahni, S. [1978]: Fundamentals of Computer Algorithms. Computer Science Press, Potomac 1978, Chapter 11

Johnson, D.S. [1981]: The *NP*-completeness column: an ongoing guide. Journal of Algorithms starting with Vol. 4 (1981)

Karp, R.M. [1975]: On the complexity of combinatorial problems. Networks 5 (1975), 45–68

Papadimitriou, C.H. [1994]: Computational Complexity. Addison-Wesley, Reading 1994

Papadimitriou, C.H., and Steiglitz, K. [1982]: Combinatorial Optimization: Algorithms and Complexity. Prentice-Hall, Englewood Cliffs 1982, Chapters 15 and 16

Wegener, I. [2005]: Complexity Theory: Exploring the Limits of Efficient Algorithms. Springer, Berlin 2005

Cited References

Adleman, L.M., Pomerance, C., and Rumely, R.S. [1983]: On distinguishing prime numbers from composite numbers. Annals of Mathematics 117 (1983), 173–206

Agrawal, M., Kayal, N., and Saxena, N. [2004]: PRIMES is in P. Annals of Mathematics 160 (2004), 781–793

Cook, S.A. [1971]: The complexity of theorem proving procedures. Proceedings of the 3rd Annual ACM Symposium on the Theory of Computing (1971), 151–158

Edmonds, J. [1965]: Minimum partition of a matroid into independent subsets. Journal of Research of the National Bureau of Standards B 69 (1965), 67–72

van Emde Boas, P. [1990]: Machine models and simulations. In: Handbook of Theoretical Computer Science; Volume A; Algorithms and Complexity (J. van Leeuwen, ed.), Elsevier, Amsterdam 1990, pp. 1–66

Fürer, M. [2009]: Faster integer multiplication. SIAM Journal on Computing 39 (2009), 979–1005

Garey, M.R., Johnson, D.S., and Stockmeyer, L. [1976]: Some simplified *NP*-complete graph problems. Theoretical Computer Science 1 (1976), 237–267

Haase, C., and Kiefer, S. [2016]: The complexity of the Kth largest subset problem and related problems. Information Processing Letters 116 (1016), 111–115

Harvey, D., van der Hoeven, J., and Lecerf, G. [2016]: Even faster integer multiplication. Journal of Complexity 36 (2016), 1–30

Hopcroft, J.E., and Ullman, J.D. [1979]: Introduction to Automata Theory, Languages, and Computation. Addison-Wesley, Reading 1979

Karp, R.M. [1972]: Reducibility among combinatorial problems. In: Complexity of Computer Computations (R.E. Miller, J.W. Thatcher, eds.), Plenum Press, New York 1972, pp. 85–103

Ladner, R.E. [1975]: On the structure of polynomial time reducibility. Journal of the ACM 22 (1975), 155–171

Lewis, H.R., and Papadimitriou, C.H. [1981]: Elements of the Theory of Computation. Prentice-Hall, Englewood Cliffs 1981

Orlin, J.B., Punnen, A.P., and Schulz, A.S. [2009]: Integer programming: optimization and evaluation are equivalent. In: Algorithms and Data Structures – Proceedings of the 11th Algorithms and Data Structures Symposium; LNCS 5664 (F. Dehne, M. Gavrilova, J.-R. Sack, C.D. Tóth, eds.), Springer, Berlin 2009, pp. 519–529

Papadimitriou, C.H., and Yannakakis, M. [1990]: On recognizing integer polyhedra. Combinatorica 10 (1990), 107–109

Pratt, V. [1975]: Every prime has a succinct certificate. SIAM Journal on Computing 4 (1975), 214–220

Schönhage, A., and Strassen, V. [1971]: Schnelle Multiplikation großer Zahlen. Computing 7 (1971), 281–292

Schulz, A.S. [2009]: On the relative complexity of 15 problems related to 0/1-integer programming. In: Research Trends in Combinatorial Optimization (W.J. Cook, L. Lovász, J. Vygen, eds.), Springer, Berlin 2009, pp. 399–428

Turing, A.M. [1936]: On computable numbers, with an application to the Entscheidungsproblem. Proceedings of the London Mathematical Society (2) 42 (1936), 230–265 and 43 (1937), 544–546

16 Approximation Algorithms

In this chapter we introduce the important concept of approximation algorithms. So far we have dealt mostly with polynomially solvable problems. In the remaining chapters we shall indicate some strategies to cope with *NP*-hard combinatorial optimization problems. Here approximation algorithms must be mentioned in the first place.

The ideal case is when the solution is guaranteed to differ from the optimum solution by a constant only:

Definition 16.1. *An* **absolute approximation algorithm** *for an optimization problem* $\mathcal{P}$ *is a polynomial-time algorithm A for* $\mathcal{P}$ *for which there exists a constant k such that*

$$|A(I) - \mathrm{OPT}(I)| \leq k$$

for all instances I of $\mathcal{P}$.

Unfortunately, an absolute approximation algorithm is known for very few classical *NP*-hard optimization problems. We shall discuss two major examples, the EDGE-COLOURING PROBLEM and the VERTEX-COLOURING PROBLEM in planar graphs in Section 16.3.

In most cases we must be satisfied with relative performance guarantees. Here we have to restrict ourselves to problems with nonnegative weights.

Definition 16.2. *Let* $\mathcal{P}$ *be an optimization problem with nonnegative weights and* $k \geq 1$. *A* **k-factor approximation algorithm** *for* $\mathcal{P}$ *is a polynomial-time algorithm A for* $\mathcal{P}$ *such that*

$$\frac{1}{k} \mathrm{OPT}(I) \leq A(I) \leq k \, \mathrm{OPT}(I)$$

for all instances I of $\mathcal{P}$. *We also say that A has* **performance ratio** (*or* **performance guarantee**) *k*.

The first inequality applies to maximization problems, the second one to minimization problems. Note that for instances I with $\mathrm{OPT}(I) = 0$ we require an exact solution. The 1-factor approximation algorithms are precisely the exact

© Springer-Verlag GmbH Germany 2018
Bernhard Korte and Jens Vygen, *Combinatorial Optimization*, Algorithms and Combinatorics 21, https://doi.org/10.1007/978-3-662-56039-6_16

polynomial-time algorithms. Sometimes the above definition is extended to the case where k is a function of the instance I, rather than a constant. We will see an example in the next section.

In Section 13.4 we saw that the BEST-IN-GREEDY ALGORITHM for the MAXIMIZATION PROBLEM for an independence system $(E, \mathcal{F})$ has performance ratio $\frac{1}{q(E,\mathcal{F})}$ (Theorem 13.19). In Section 14.6 we saw an approximation algorithm for submodular function maximization. In the following sections and chapters we shall illustrate the above definitions and analyse the approximability of various NP-hard problems. We start with covering problems.

16.1 Set Covering

In this section we focus on the following quite general problem:

MINIMUM WEIGHT SET COVER PROBLEM

Instance: A set system $(U, \mathcal{S})$ with $\bigcup_{S \in \mathcal{S}} S = U$, weights $c : \mathcal{S} \to \mathbb{R}_+$.

Task: Find a minimum weight **set cover** of $(U, \mathcal{S})$, i.e. a subfamily $\mathcal{R} \subseteq \mathcal{S}$ such that $\bigcup_{R \in \mathcal{R}} R = U$.

For $c \equiv 1$, the problem is called MINIMUM SET COVER PROBLEM. Another interesting special case arises if $|\{S \in \mathcal{S} : x \in S\}| = 2$ for all $x \in U$; this is the MINIMUM WEIGHT VERTEX COVER PROBLEM: given a graph G and $c : V(G) \to \mathbb{R}_+$, the corresponding set covering instance is defined by $U :=$ $E(G)$, $\mathcal{S} := \{\delta(v) : v \in V(G)\}$ and $c(\delta(v)) := c(v)$ for all $v \in V(G)$. As the MINIMUM WEIGHT VERTEX COVER PROBLEM is NP-hard even for unit weights (Theorem 15.24), so is the MINIMUM SET COVER PROBLEM.

Johnson [1974] and Lovász [1975] proposed a simple greedy algorithm for the MINIMUM SET COVER PROBLEM: in each iteration, pick a set which covers a maximum number of elements not already covered. Chvátal [1979] generalized this algorithm to the weighted case:

GREEDY ALGORITHM FOR SET COVER

Input: A set system $(U, \mathcal{S})$ with $\bigcup_{S \in \mathcal{S}} S = U$, weights $c : \mathcal{S} \to \mathbb{R}_+$.

Output: A set cover $\mathcal{R}$ of $(U, \mathcal{S})$.

① Set $\mathcal{R} := \emptyset$ and $W := \emptyset$.

② **While** $W \neq U$ **do**:
   Choose a set $R \in \mathcal{S} \setminus \mathcal{R}$ for which $R \setminus W \neq \emptyset$ and $\frac{c(R)}{|R \setminus W|}$ is minimum.
   Set $\mathcal{R} := \mathcal{R} \cup \{R\}$ and $W := W \cup R$.

The running time is obviously $O(|U||\mathcal{S}|)$. The following performance guarantee can be proved:

Theorem 16.3. (Chvátal [1979]) *For any instance* $(U, \mathcal{S}, c)$ *of the* MINIMUM WEIGHT SET COVER PROBLEM, *the* GREEDY ALGORITHM FOR SET COVER *finds a set cover whose weight is at most* $H(r) \operatorname{OPT}(U, \mathcal{S}, c)$, *where* $r := \max_{S \in \mathcal{S}} |S|$ *and* $H(r) = 1 + \frac{1}{2} + \cdots + \frac{1}{r}$.

Proof: Let $(U, \mathcal{S}, c)$ be an instance of the MINIMUM WEIGHT SET COVER PROBLEM, and let $\mathcal{R} = \{R_1, \ldots, R_k\}$ be the solution found by the above algorithm, where R_i is the set chosen in the i-th iteration. For $j = 0, \ldots, k$ let $W_j := \bigcup_{i=1}^{j} R_i$.

For each $e \in U$ let $j(e) := \min\{j \in \{1, \ldots, k\} : e \in R_j\}$ be the iteration where e is covered. Let

$$y(e) := \frac{c(R_{j(e)})}{|R_{j(e)} \setminus W_{j(e)-1}|}.$$

Let $S \in \mathcal{S}$ be fixed, and let $k' := \max\{j(e) : e \in S\}$. We have

$$
\begin{aligned}
\sum_{e \in S} y(e) &= \sum_{i=1}^{k'} \sum_{e \in S : j(e)=i} y(e) \\
&= \sum_{i=1}^{k'} \frac{c(R_i)}{|R_i \setminus W_{i-1}|} |S \cap (W_i \setminus W_{i-1})| \\
&= \sum_{i=1}^{k'} \frac{c(R_i)}{|R_i \setminus W_{i-1}|} (|S \setminus W_{i-1}| - |S \setminus W_i|) \\
&\leq \sum_{i=1}^{k'} \frac{c(S)}{|S \setminus W_{i-1}|} (|S \setminus W_{i-1}| - |S \setminus W_i|)
\end{aligned}
$$

by the choice of the R_i in ② (observe that $S \setminus W_{i-1} \neq \emptyset$ for $i = 1, \ldots, k'$). By writing $s_i := |S \setminus W_{i-1}|$ we get

$$
\begin{aligned}
\sum_{e \in S} y(e) &\leq c(S) \sum_{i=1}^{k'} \frac{s_i - s_{i+1}}{s_i} \\
&\leq c(S) \sum_{i=1}^{k'} \left(\frac{1}{s_i} + \frac{1}{s_i - 1} + \cdots + \frac{1}{s_{i+1} + 1} \right) \\
&= c(S) \sum_{i=1}^{k'} (H(s_i) - H(s_{i+1})) \\
&= c(S)(H(s_1) - H(s_{k'+1})) \\
&\leq c(S) H(s_1).
\end{aligned}
$$

Since $s_1 = |S| \le r$, we conclude that

$$\sum_{e \in S} y(e) \le c(S)H(r).$$

We sum over all $S \in \mathcal{O}$ for an optimum set cover $\mathcal{O}$ and obtain

$$
\begin{aligned}
c(\mathcal{O})H(r) &\ge \sum_{S \in \mathcal{O}} \sum_{e \in S} y(e) \\
&\ge \sum_{e \in U} y(e) \\
&= \sum_{i=1}^{k} \sum_{e \in U : j(e)=i} y(e) \\
&= \sum_{i=1}^{k} c(R_i) = c(\mathcal{R}).
\end{aligned}
$$

$\square$

For a slightly tighter analysis of the non-weighted case, see Slavík [1997]. Raz and Safra [1997] discovered that there exists a constant $c > 0$ such that, unless $P = NP$, no performance ratio of $c \ln |U|$ can be achieved. Indeed, a performance ratio of $c \ln |U|$ cannot be achieved for any $c < 1$ unless each problem in NP can be solved in $O\left(n^{O(\log \log n)}\right)$ time (Feige [1998]).

The MINIMUM WEIGHT EDGE COVER PROBLEM is obviously a special case of the MINIMUM WEIGHT SET COVER PROBLEM. Here we have $r = 2$ in Theorem 16.3, hence the above algorithm is a $\frac{3}{2}$-factor approximation algorithm in this special case. However, the problem can also be solved optimally in polynomial time; cf. Exercise 13 of Chapter 11.

For the MINIMUM VERTEX COVER PROBLEM, the above algorithm reads as follows:

GREEDY ALGORITHM FOR VERTEX COVER

Input: A graph G.

Output: A vertex cover R of G.

---

①   Set $R := \emptyset$.

②   **While** $E(G) \neq \emptyset$ **do**:
　　Choose a vertex $v \in V(G) \setminus R$ with maximum degree.
　　Set $R := R \cup \{v\}$ and delete all edges incident to v.

This algorithm looks reasonable, so one might ask for which k it is a k-factor approximation algorithm. It may be surprising that there is no such k. Indeed, the bound given in Theorem 16.3 is almost best possible:

Theorem 16.4. (Johnson [1974], Papadimitriou and Steiglitz [1982]) *For all* $n \ge 3$ *there is an instance G of the* MINIMUM VERTEX COVER PROBLEM *such that*

$nH(n-1)+2 \leq |V(G)| \leq nH(n-1)+n$, *the maximum degree of G is* $n-1$, *OPT(G) = n, and the above algorithm can find a vertex cover containing all but n vertices.*

Proof: For each $n \geq 3$ and $i \leq n$ we define $A_n^i := \sum_{j=2}^{i} \lfloor \frac{n}{j} \rfloor$ and

$$V(G_n) := \left\{ a_1, \ldots, a_{A_n^{n-1}}, b_1, \ldots, b_n, c_1, \ldots, c_n \right\}.$$

$$E(G_n) := \{\{b_i, c_i\} : i = 1, \ldots, n\} \cup$$

$$\bigcup_{i=2}^{n-1} \bigcup_{j=A_n^{i-1}+1}^{A_n^i} \left\{ \{a_j, b_k\} : (j - A_n^{i-1} - 1)i + 1 \leq k \leq (j - A_n^{i-1})i \right\}.$$

Observe that $|V(G_n)| = 2n + A_n^{n-1}$, $A_n^{n-1} \leq nH(n-1)-n$ and $A_n^{n-1} \geq nH(n-1) - n - (n-2)$. Figure 16.1 shows G_6.

If we apply our algorithm to G_n, it may first choose vertex $a_{A_n^{n-1}}$ (because it has maximum degree), and subsequently the vertices $a_{A_n^{n-1}-1}, a_{A_n^{n-1}-2}, \ldots, a_1$. After this there are n pairwise disjoint edges left, so n more vertices are needed. Hence the constructed vertex cover consists of $A_n^{n-1} + n$ vertices, while the optimum vertex cover $\{b_1, \ldots, b_n\}$ has size n.   □

There are, however, 2-factor approximation algorithms for the MINIMUM VER-TEX COVER PROBLEM. The simplest one is due to Gavril (see Garey and Johnson [1979]): just find any maximal matching M and take the ends of all edges in M. This is obviously a vertex cover and contains $2|M|$ vertices. Since any vertex cover must contain $|M|$ vertices (no vertex covers two edges of M), this is a 2-factor approximation algorithm.

This performance guarantee is tight: simply think of a graph consisting of many pairwise disjoint edges. It may be surprising that the above is the best known approximation algorithm for the MINIMUM VERTEX COVER PROBLEM. Later we

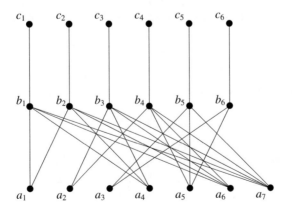

Fig. 16.1.

shall show that there is a number $k > 1$ such that no k-factor approximation algorithm exists unless $P = NP$ (Theorem 16.46). Indeed, a 1.36-factor approximation algorithm does not exist unless $P = NP$ (Dinur and Safra [2002]). See also Khot and Regev [2008].

At least Gavril's algorithm can be extended to the weighted case. We present the algorithm of Bar-Yehuda and Even [1981], which is applicable to the general MINIMUM WEIGHT SET COVER PROBLEM:

BAR-YEHUDA-EVEN ALGORITHM

Input: A set system $(U, \mathcal{S})$ with $\bigcup_{S \in \mathcal{S}} S = U$, weights $c : \mathcal{S} \to \mathbb{R}_+$.

Output: A set cover $\mathcal{R}$ of $(U, \mathcal{S})$.

①   Set $\mathcal{R} := \emptyset$ and $W := \emptyset$. Set $y(e) := 0$ for all $e \in U$.
     Set $c'(S) := c(S)$ for all $S \in \mathcal{S}$.

②   **While** $W \neq U$ **do:**
     Choose an element $e \in U \setminus W$.
     Let $R \in \mathcal{S}$ with $e \in R$ and $c'(R)$ minimum. Set $y(e) := c'(R)$.
     Set $c'(S) := c'(S) - y(e)$ for all $S \in \mathcal{S}$ with $e \in S$.
     Set $\mathcal{R} := \mathcal{R} \cup \{R\}$ and $W := W \cup R$.

Theorem 16.5. (Bar-Yehuda and Even [1981]) *For any instance $(U, \mathcal{S}, c)$ of the MINIMUM WEIGHT SET COVER PROBLEM, the BAR-YEHUDA-EVEN ALGORITHM finds a set cover whose weight is at most $p\,\mathrm{OPT}(U, \mathcal{S}, c)$, where $p := \max_{e \in U} |\{S \in \mathcal{S} : e \in S\}|$.*

Proof: The MINIMUM WEIGHT SET COVER PROBLEM can be written as the integer linear program

$$\min \left\{ cx : Ax \geq \mathbb{1}, \ x \in \{0, 1\}^{\mathcal{S}} \right\},$$

where A is the matrix whose rows correspond to the elements of U and whose columns are the incidence vectors of the sets in $\mathcal{S}$. The optimum of the LP relaxation

$$\min \{cx : Ax \geq \mathbb{1}, \ x \geq 0\}$$

is a lower bound for $\mathrm{OPT}(U, \mathcal{S}, c)$ (the omission of the constraints $x \leq \mathbb{1}$ does not change the optimum value of this LP). Hence, by Proposition 3.13, the optimum of the dual LP

$$\max\{y\mathbb{1} : yA \leq c, \ y \geq 0\}$$

is also a lower bound for $\mathrm{OPT}(U, \mathcal{S}, c)$.

Now observe that $c'(S) \geq 0$ for all $S \in \mathcal{S}$ at any stage of the algorithm. Let $\bar{y}$ be the vector y at termination. We have $\bar{y} \geq 0$ and $\sum_{e \in S} \bar{y}(e) \leq c(S)$ for all $S \in \mathcal{S}$, i.e. $\bar{y}$ is a feasible solution of the dual LP and

$$\bar{y}\mathbb{1} \ \leq \ \max\{y\mathbb{1} : yA \leq c, \ y \geq 0\} \ \leq \ \mathrm{OPT}(U, \mathcal{S}, c).$$

Finally observe that

$$
\begin{aligned}
c(\mathcal{R}) &= \sum_{R \in \mathcal{R}} c(R) \\
&= \sum_{R \in \mathcal{R}} \sum_{e \in R} \bar{y}(e) \\
&\leq \sum_{e \in U} p \bar{y}(e) \\
&= p \bar{y} \mathbb{1} \\
&\leq p \, \mathrm{OPT}(U, \mathcal{S}, c). \qquad \square
\end{aligned}
$$

Since we have $p = 2$ in the vertex cover case, this is a 2-factor approximation algorithm for the MINIMUM WEIGHT VERTEX COVER PROBLEM. The first 2-factor approximation algorithm was due to Hochbaum [1982]. She proposed finding an optimum solution y of the dual LP in the above proof and taking all sets S with $\sum_{e \in S} y(e) = c(S)$. Alternatively, one could find an optimum solution x of the primal LP and take all sets S with $x_S \geq \frac{1}{p}$.

The advantage of the BAR-YEHUDA-EVEN ALGORITHM is that it does not explicitly use linear programming. In fact it can easily be implemented with $O\left(\sum_{S \in \mathcal{S}} |S|\right)$ time. It is our first example of a primal-dual approximation algorithm; more complicated examples will follow in Sections 20.5 and 22.3.

16.2 The Max-Cut Problem

In this section we consider another basic problem:

MAXIMUM WEIGHT CUT PROBLEM

Instance: An undirected graph G and weights $c : E(G) \to \mathbb{R}_+$.

Task: Find a cut in G with maximum total weight.

This problem is often called MAX-CUT for short. In contrast to minimum weight cuts, discussed in Section 8.7, this is a difficult problem. It is strongly *NP*-hard; even the special case where $c \equiv 1$ (the MAXIMUM CUT PROBLEM) is hard:

Theorem 16.6.   (Garey, Johnson and Stockmeyer [1976])   *The* MAXIMUM CUT PROBLEM *is NP-hard.*

Proof: By reduction from MAX-2SAT (cf. Theorem 15.37). Given an instance of MAX-2SAT with n variables and m clauses, we construct a graph G whose vertices are the literals plus an extra vertex z. For each variable x we add $3m$ parallel edges between x and $\bar{x}$. For each clause $\{\lambda, \lambda'\}$ we add three edges $\{\lambda, \lambda'\}$, $\{\lambda, z\}$ and $\{\lambda', z\}$. So G has $2n + 1$ vertices and $3m(n + 1)$ edges.

We claim that the maximum cardinality of a cut in G is $3mn + 2t$, where t is the maximum number of clauses satisfied by any truth assignment. Indeed, given a truth assignment satisfying t clauses, let X be the set of true literals. Then $|\delta_G(X)| = 3mn + 2t$. Conversely, if there is a set $X \subseteq V(G)$ with $|\delta_G(X)| \geq 3mn + a$, then w.l.o.g. $z \notin X$ (otherwise replace X by $V(G) \setminus X$), and for each variable x we have $|X \cap \{x, \bar{x}\}| = 1$ (otherwise replace X by $X \triangle \{x\}$ and increase the cut). Thus we can set all literals in X to be *true* and get a truth assignment satisfying at least $\frac{a}{2}$ clauses.   □

It is very easy to find a 2-factor approximation algorithm for the MAXIMUM WEIGHT CUT PROBLEM: If $V(G) = \{v_1, \ldots, v_n\}$, start with $X := \{v_1\}$, and for $i = 3, \ldots, n$ add v_i to X if $\sum_{e \in E(v_i, \{v_1, \ldots, v_{i-1}\} \cap X)} c(e) < \sum_{e \in E(v_i, \{v_1, \ldots, v_{i-1}\} \setminus X)} c(e)$. (The simple analysis of this algorithm is left as Exercise 9.)

For a long time no better approximation algorithm was known. Then Goemans and Williamson [1995] found a much better one using semidefinite programming; the rest of this section is based on their paper.

Let G be an undirected graph and $c : E(G) \rightarrow \mathbb{R}_+$. W.l.o.g., $V(G) = \{1, \ldots, n\}$. For $1 \leq i, j \leq n$ let $c_{ij} := c(\{i, j\})$ if $\{i, j\} \in E(G)$ and $c_{ij} := 0$ otherwise. Then the MAXIMUM WEIGHT CUT PROBLEM consists of finding a subset $S \subseteq \{1, \ldots, n\}$ maximizing $\sum_{i \in S, j \in \{1, \ldots, n\} \setminus S} c_{ij}$. Expressing S by $y \in \{-1, 1\}^n$ with $y_i = 1$ iff $i \in S$, we can formulate the problem as follows:

$$\max \quad \frac{1}{2} \sum_{1 \leq i < j \leq n} c_{ij}(1 - y_i y_j)$$

$$\text{s.t.} \quad y_i \in \{-1, 1\} \quad (i = 1, \ldots, n)$$

The variables y_i can be regarded as one-dimensional vectors of unit norm. By relaxing them to multidimensional vectors of unit Euclidean norm we get a very interesting relaxation:

$$\max \quad \frac{1}{2} \sum_{1 \leq i < j \leq n} c_{ij}(1 - y_i^\top y_j) \qquad (16.1)$$

$$\text{s.t.} \quad y_i \in \mathcal{S}_m \quad (i = 1, \ldots, n)$$

where $m \in \mathbb{N}$ and $\mathcal{S}_m = \{x \in \mathbb{R}^m : ||x||_2 = 1\}$ denotes the unit sphere in $\mathbb{R}^m$. For example, for the triangle ($n = 3$, $c_{12} = c_{13} = c_{23} = 1$) the optimum is obtained by points on the unit sphere in $\mathbb{R}^2$ that are vertices of an equilateral triangle, e.g. $y_1 = (0, -1)$, $y_2 = (-\frac{\sqrt{3}}{2}, \frac{1}{2})$, and $y_3 = (\frac{\sqrt{3}}{2}, \frac{1}{2})$, giving an optimum value of $\frac{9}{4}$, in contrast to the maximum weight of a cut, which is 2. However, the interesting fact is that we can solve (16.1) almost optimally in polynomial time.

The trick is not to consider the variables y_i directly, not even their dimension. Instead, we consider the $n \times n$-matrix $(y_i^\top y_j)_{i,j=1,\ldots,n}$. As a matrix X is symmetric and positive semidefinite if and only if it can be written as $B^\top B$ for some matrix

B, we can write equivalently

$$\max \quad \frac{1}{2} \sum_{1 \le i < j \le n} c_{ij}(1 - x_{ij})$$

$$\text{s.t.} \quad x_{ii} = 1 \qquad (i = 1, \dots, n)$$
$$X = (x_{ij})_{1 \le i, j \le n} \text{ symmetric and positive semidefinite}$$

(16.2)

From a solution to (16.2) we can obtain a solution to (16.1) with $m \le n$ and almost the same objective function value by Cholesky factorization in $O(n^3)$ time (we have to accept an arbitrarily small rounding error; cf. Exercise 6 of Chapter 4).

Problem (16.2) is called a semidefinite programming relaxation. It can be solved approximately in polynomial time by the ELLIPSOID METHOD, applying Theorem 4.19, as we shall show now. First observe that we optimize a linear objective function over the convex set

$$P := \{ X = (x_{ij})_{1 \le i, j \le n} \in \mathbb{R}^{n \times n} : X \text{ symmetric and positive semidefinite,}$$

$$x_{ii} = 1 \ (i = 1, \dots, n) \}.$$

By projecting P to the $\frac{n^2-n}{2}$ free variables we obtain

$$P' := \{ (x_{ij})_{1 \le i < j \le n} : (x_{ij})_{1 \le i, j \le n} \in P \text{ with } x_{ii} := 1 \text{ and } x_{ji} := x_{ij} \text{ for } i < j \}.$$

Note that neither P nor P' is a polyhedron. However, P' is convex, bounded, and full-dimensional:

Proposition 16.7. P' is convex. Moreover, $B(0, \frac{1}{n}) \subseteq P' \subseteq B(0, n)$.

Proof: Convexity follows from the simple fact that convex combinations of positive semidefinite matrices are positive semidefinite.

For the first inclusion, observe that for a symmetric $n \times n$-matrix X whose diagonal entries are 1 and whose off-diagonal entries have absolute value at most $\frac{1}{n}$ we have, for any $d \in \mathbb{R}^n$,

$$d^\top X d = \sum_{i,j=1}^{n} x_{ij} d_i d_j$$

$$\ge \frac{1}{2n-2} \sum_{i \ne j} (x_{ii} d_i^2 + x_{jj} d_j^2 - (2n-2)|x_{ij}||d_i d_j|)$$

$$\ge \frac{1}{2n-2} \sum_{i \ne j}^{n} (d_i^2 + d_j^2 - 2|d_i d_j|)$$

$$= \frac{1}{2n-2} \sum_{i \ne j}^{n} (|d_i| - |d_j|)^2$$

$$\ge 0,$$

i.e. X is positive semidefinite.

For the second inclusion, note that all off-diagonal entries of a matrix in P have absolute value at most 1, and thus the Euclidean norm of the vector of upper diagonal entries is at most n.   □

It remains to show that the SEPARATION PROBLEM for P' can be solved in polynomial time. This is achieved by GAUSSIAN ELIMINATION:

Theorem 16.8. *Given a symmetric matrix $X \in \mathbb{Q}^{n \times n}$, we can decide in polynomial time whether X is positive semidefinite, and find a vector $d \in \mathbb{Q}^n$ with $d^\top X d < 0$ if one exists.*

Proof: If $x_{nn} < 0$, then we set $d = (0, \ldots, 0, 1)$ and have $d^\top X d < 0$. If $x_{nn} = 0$ and $x_{nj} \neq 0$ for some $j < n$, then we can define d by $d_j := -1$, $d_n := \frac{x_{jj}}{2x_{nj}} + x_{nj}$, and $d_i := 0$ for $i \in \{1, \ldots, n-1\} \setminus \{j\}$, and have $d^\top X d = x_{jj} - 2x_{nj}(\frac{x_{jj}}{2x_{nj}} + x_{nj}) = -2(x_{nj})^2 < 0$, again proving that X is not positive semidefinite.

In the other cases we reduce the dimension. If $x_{nj} = 0$ for all j, then the last row and column can be deleted: X is positive semidefinite if and only if $X' := (x_{ij})_{i,j=1,\ldots,n-1}$ is positive semidefinite. Moreover, if $c \in \mathbb{Q}^{n-1}$ satisfies $c^\top X' c < 0$, we set $d := \binom{c}{0}$ and have $d^\top X d < 0$.

Hence let us now assume that $x_{nn} > 0$. Then we consider $X' := (x_{ij} - \frac{x_{ni} x_{nj}}{x_{nn}})_{i,j=1,\ldots,n-1}$; this corresponds to one iteration of GAUSSIAN ELIMINATION. Note that X' is positive semidefinite if and only if X is positive semidefinite. For a vector $c \in \mathbb{Q}^{n-1}$ with $c^\top X' c < 0$ we set $d := (-\frac{1}{x_{nn}} \sum_{i=1}^{n-1} c_i x_{ni}^{\,c})$. Then

$$
\begin{aligned}
d^\top X d &= \sum_{i,j=1}^{n-1} d_i \left(x'_{ij} + \frac{x_{ni}}{x_{nn}} x_{nj} \right) d_j + 2 \sum_{j=1}^{n-1} d_n x_{nj} d_j + d_n^2 x_{nn} \\
&= c^\top X' c + \sum_{i,j=1}^{n-1} c_i \frac{x_{ni} x_{nj}}{x_{nn}} c_j (1 - 2 + 1) \\
&= c^\top X' c \\
&< 0.
\end{aligned}
$$

This defines a polynomial-time algorithm. To see that the numbers involved in the computation of d are not too large, let $X^{(n)}, X^{(n-1)}, \ldots, X^{(k)}$ be the considered matrices ($X^{(i)} \in \mathbb{Q}^{i \times i}$), and assume that we observe in iteration $n + 1 - k$ that the matrix $X^{(k)} = (y_{ij})_{i,j=1,\ldots,k}$ is not positive semidefinite (i.e. $y_{kk} < 0$ or $y_{kk} = 0$ and $y_{kj} \neq 0$ for some $j < k$). We have a vector $c \in \mathbb{Q}^k$ with $c^\top X^{(k)} c < 0$ and $\text{size}(c) \leq 2\,\text{size}(X^{(k)})$. Now a vector $d \in \mathbb{Q}^n$ with $d^\top X d < 0$ can be constructed as above; note that d is a solution of the linear equation system $Md = \binom{c}{0}$, where the j-th row of M is

- the j-th unit vector if $j \leq k$,
- the j-th unit vector if $j > k$ and the j-th row of $X^{(j)}$ is zero,
- the j-th row of $X^{(j)}$, followed by zeros, otherwise.

Hence, with Theorem 4.4, we have $\text{size}(d) \leq 4n(\text{size}(M) + \text{size}(c))$, which is polynomial by Theorem 4.10.   □

Corollary 16.9. *The* SEPARATION PROBLEM *for* P' *can be solved in polynomial time.*

Proof: Let $(y_{ij})_{1 \leq i < j \leq n}$ be given, and let $Y = (y_{ij})_{1 \leq i,j \leq n}$ be the symmetric matrix defined by $y_{ii} = 1$ for all i and $y_{ji} := y_{ij}$ for $i < j$. Apply Theorem 16.8. If Y is positive semidefinite, we are done.

Otherwise we find a vector $d \in \mathbb{Q}^n$ with $d^\top Y d < 0$. Then $-\sum_{i=1}^{n} d_i^2 > d^\top Y d - \sum_{i=1}^{n} d_i^2 = \sum_{1 \leq i < j \leq n} 2d_i d_j y_{ij}$, but $\sum_{1 \leq i < j \leq n} 2d_i d_j z_{ij} \geq -\sum_{i=1}^{n} d_i^2$ for all $z \in P'$. Thus $(d_i d_j)_{1 \leq i < j \leq n}$ constitutes a separating hyperplane.   □

We can now conclude:

Theorem 16.10. *For any instance of the* MAXIMUM WEIGHT CUT PROBLEM, *we can find a matrix* $Y = (y_{ij})_{1 \leq i,j \leq n} \in P$ *with*

$$\sum_{1 \leq i < j \leq n} c_{ij}(1 - y_{ij}) \geq \max \left\{ \sum_{1 \leq i < j \leq n} c_{ij}(1 - x_{ij}) : (x_{ij})_{1 \leq i,j \leq n} \in P \right\} - \epsilon$$

in time polynomial in n, $\text{size}((c_{ij})_{1 \leq i < j \leq n})$, *and* $\text{size}(\epsilon)$.

Proof: We apply Theorem 4.19, using Proposition 16.7 and Corollary 16.9.   □

Semidefinite programs such as (16.2) can also be solved approximately by interior point algorithms, which are more efficient than the ELLIPSOID METHOD. See Alizadeh [1995] for details. A combinatorial, even faster algorithm was suggested by Arora and Kale [2016].

As mentioned above, from an almost optimum solution to (16.2) we can derive a solution to (16.1) with almost the same objective function value by Cholesky factorization. This solution consists of a set of vectors $y_i \in \mathbb{R}^m$ ($i = 1, \ldots, n$) for some $m \leq n$. As (16.1) is a relaxation of our original problem, we have that the optimum is at most $\frac{1}{2} \sum_{1 \leq i < j \leq n} c_{ij}(1 - y_i^\top y_j) + \epsilon$.

The vectors y_i lie on a unit sphere. The idea is now to take a random hyperplane through the origin, and define S to be the set of indices i for which y_i is on one side of this hyperplane.

A random hyperplane through the origin is given by a random point on the $(m - 1)$-dimensional sphere. This can be chosen by drawing m real numbers independently from the standard normal distribution, which in turn can be done using independent random numbers uniformly distributed in $[0, 1]$. See Knuth [1969] (Section 3.4.1) for details.

The algorithm by Goemans and Williamson now reads as follows.

GOEMANS-WILLIAMSON MAX-CUT ALGORITHM

Input: A number $n \in \mathbb{N}$, numbers $c_{ij} \geq 0$ for $1 \leq i < j \leq n$.

Output: A set $S \subseteq \{1, \ldots, n\}$.

①   Solve (16.2) approximately; i.e. find a symmetric positive semidefinite
      matrix $X = (x_{ij})_{1 \le i, j \le n}$ with $x_{ii} = 1$ for $i = 1, \ldots, n$, such that
      $\sum_{1 \le i < j \le n} c_{ij}(1 - x_{ij}) \ge 0.9995 \cdot \text{OPT}(16.2)$.

②   Apply Cholesky factorization to X in order to obtain vectors
      $y_1, \ldots, y_n \in \mathbb{R}^m$ with $m \le n$ and $y_i^\top y_j \approx x_{ij}$ for all $i, j \in \{1, \ldots, n\}$.

③   Choose a random point a on the unit sphere $\{x \in \mathbb{R}^m : ||x||_2 = 1\}$.

④   Set $S := \{i \in \{1, \ldots, n\} : a^\top y_i \ge 0\}$.

Theorem 16.11. *The* GOEMANS-WILLIAMSON MAX-CUT ALGORITHM *runs in polynomial time.*

Proof: See the preceding discussion. The most difficult step, ①, can be solved in polynomial time by Theorem 16.10. Here we can choose $\epsilon = 0.00025 \sum_{1 \le i < j \le n} c_{ij}$ as $\frac{1}{2} \sum_{1 \le i < j \le n} c_{ij}$ is a lower bound on the optimum objective value (attained by randomly choosing $S \subseteq \{1, \ldots, n\}$) and thus on the optimum value of (16.2). $\square$

We now prove the performance guarantee:

Theorem 16.12. (Goemans and Williamson [1995]) *The* GOEMANS-WILLIAMSON MAX-CUT ALGORITHM *returns a set S for which the expected value of $\sum_{i \in S, j \notin S} c_{ij}$ is at least 0.878 times the maximum possible value.*

Proof: Let S_m again denote the unit sphere in $\mathbb{R}^m$, and let $H(y) := \{x \in S_m : x^\top y \ge 0\}$ be the hemisphere with pole y, for $y \in S_m$. For a subset $A \subseteq S_m$ let $\mu(A) := \frac{\text{volume}(A)}{\text{volume}(S_m)}$; this defines a probability measure on S_m. We have $|S \cap \{i, j\}| = 1$ with probability $\mu(H(y_i) \triangle H(y_j))$, where $\triangle$ denotes the symmetric difference. Note that $H(y_i) \triangle H(y_j)$ is the union of two spherical digons, each with angle $\arccos(y_i^\top y_j)$. As the volume is proportional to the angle, we have $\mu(H(y_i) \triangle H(y_j)) = \frac{1}{\pi} \arccos(y_i^\top y_j)$.

Claim: $\frac{1}{\pi} \arccos \beta \ge 0.8785 \frac{1-\beta}{2}$ for all $\beta \in [-1, 1]$.

For $\beta = 1$ we have equality. Moreover, elementary calculus yields

$$\min_{-1 \le \beta < 1} \frac{\arccos \beta}{1 - \beta} = \min_{0 < \gamma \le \pi} \frac{\gamma}{1 - \cos \gamma} = \frac{1}{\sin \gamma'},$$

where γ' is determined by $\cos \gamma' + \gamma' \sin \gamma' = 1$. We get $2.3311 < \gamma' < 2.3312$ and $\frac{1}{\sin \gamma'} > \frac{1}{\sin 2.3311} > 1.38$. As $\frac{1.38}{\pi} > \frac{0.8785}{2}$, this proves the claim.

Thus the expected value of $\sum_{i \in S, j \notin S} c_{ij}$ is

$$\sum_{1 \le i < j \le n} c_{ij} \mu(H(y_i) \triangle H(y_j)) = \sum_{1 \le i < j \le n} c_{ij} \frac{1}{\pi} \arccos(y_i^\top y_j)$$

$$\ge 0.8785 \cdot \frac{1}{2} \sum_{1 \le i < j \le n} c_{ij}(1 - y_i^\top y_j)$$

$$\approx \; 0.8785 \cdot \frac{1}{2} \sum_{1 \le i < j \le n} c_{ij} (1 - x_{ij})$$

$$\ge \; 0.8785 \cdot 0.9995 \cdot \text{OPT} \,(16.2)$$

$$> \; 0.878 \cdot \text{OPT} \,(16.2)$$

$$\ge \; 0.878 \cdot \max \left\{ \sum_{i \in S, \, j \notin S} c_{ij} : S \subseteq \{1, \ldots, n\} \right\}.$$

$\square$

Therefore we have a randomized approximation algorithm with performance ratio $\frac{1}{0.878} < 1.139$. Mahajan and Ramesh [1999] showed how to derandomize this algorithm, thus obtaining a deterministic 1.139-factor approximation algorithm. However, there is no 1.062-factor approximation algorithm unless $P = NP$ (Håstad [2001], Papadimitriou and Yannakakis [1991]).

See Lovász [2003] for other interesting connections of semidefinite programming and combinatorial optimization.

16.3 Colouring

In this section we briefly discuss two more well-known special cases of the MIN-IMUM SET COVER PROBLEM: we want to partition the vertex set of a graph into stable sets, or the edge set of a graph into matchings:

Definition 16.13. *Let G be an undirected graph. A* **vertex-colouring** *of G is a mapping $f : V(G) \to \mathbb{N}$ with $f(v) \ne f(w)$ for all $\{v, w\} \in E(G)$. An* **edge-colouring** *of G is a mapping $f : E(G) \to \mathbb{N}$ with $f(e) \ne f(e')$ for all $e, e' \in E(G)$ with $e \ne e'$ and $e \cap e' \ne \emptyset$.*

The number $f(v)$ or $f(e)$ is called the **colour** of v or e. In other words, the set of vertices or edges with the same colour (f-value) must be a stable set, or a matching, respectively. Of course we are interested in using as few colours as possible:

VERTEX-COLOURING PROBLEM

Instance: An undirected graph G.

Task: Find a vertex-colouring $f : V(G) \to \{1, \ldots, k\}$ of G with minimum k.

EDGE-COLOURING PROBLEM

Instance: An undirected graph G.

Task: Find an edge-colouring $f : E(G) \to \{1, \ldots, k\}$ of G with minimum k.

Reducing these problems to the MINIMUM SET COVER PROBLEM is not very useful: for the VERTEX-COLOURING PROBLEM we would have to list the maximal stable sets, and for the EDGE-COLOURING PROBLEM the maximal matchings; in both cases we would have to reckon with exponentially many.

The optimum value of the VERTEX-COLOURING PROBLEM (i.e. the minimum number of colours) is called the **chromatic number** of the graph. The optimum value of the EDGE-COLOURING PROBLEM is called the **edge-chromatic number** or sometimes the chromatic index. Both colouring problems are NP-hard:

Theorem 16.14. *The following decision problems are NP-complete:*

(a) (Holyer [1981]) *Decide whether a given simple graph has edge-chromatic number 3.*
(b) (Stockmeyer [1973]) *Decide whether a given planar graph has chromatic number 3.*

The problems remain NP-hard even when the graph has maximum degree three in (a), and maximum degree four in (b).

Proposition 16.15. *For any given graph we can decide in linear time whether the chromatic number is less than 3, and if so, find an optimum colouring. The same holds for the edge-chromatic number.*

Proof: A graph has chromatic number 1 iff it has no edges. By definition, the graphs with chromatic number at most 2 are precisely the bipartite graphs. By Proposition 2.27 we can check in linear time whether a graph is bipartite and in the positive case find a bipartition, i.e. a vertex-colouring with two colours.

To check whether the edge-chromatic number of a graph G is less than 3 (and, if so, find an optimum edge-colouring) we simply consider the VERTEX-COLOURING PROBLEM in the line graph of G. This is obviously equivalent. $\square$

For bipartite graphs, the EDGE-COLOURING PROBLEM can be solved, too:

Theorem 16.16. (König [1916]) *The edge-chromatic number of a bipartite graph G equals the maximum degree of a vertex in G.*

Proof: By induction on $|E(G)|$. Let G be a graph with maximum degree k, and let $e = \{v, w\}$ be an edge. By the induction hypothesis, $G - e$ has an edge-colouring f with k colours. There are colours $i, j \in \{1, \ldots, k\}$ such that $f(e') \neq i$ for all $e' \in \delta(v)$ and $f(e') \neq j$ for all $e' \in \delta(w)$. If $i = j$, we are done since we can extend f to G by giving e colour i.

The graph $H = (V(G), \{e' \in E(G) \setminus \{e\} : f(e') \in \{i, j\}\})$ has maximum degree 2, and v has degree at most 1 in H. Consider the maximal path P in H with endpoint v. The colours alternate on P; hence the other endpoint of P cannot be w. Exchange the colours i and j on P and extend the edge-colouring to G by giving e colour j. $\square$

The maximum degree of a vertex is an obvious lower bound on the edge-chromatic number of any graph. It is not always attained as the triangle K_3 shows. The following theorem shows how to find an edge-colouring of a given simple graph which needs at most one more colour than necessary:

Theorem 16.17.   (Vizing [1964])   *Let G be an undirected simple graph with maximum degree k. Then G has an edge-colouring with at most $k + 1$ colours, and such a colouring can be found in polynomial time.*

Proof: By induction on $|E(G)|$. If G has no edges, the assertion is trivial. Otherwise let $e = \{x, y_0\}$ be any edge; by the induction hypothesis there exists an edge-colouring f of $G - e$ with $k + 1$ colours. For each vertex v choose a colour $n(v) \in \{1, \ldots, k + 1\} \setminus \{f(w) : w \in \delta_{G-e}(v)\}$ missing at v.

Starting from y_0, construct a maximal sequence $y_0, y_1, \ldots, y_t$ of distinct neighbours of x such that $n(y_{i-1}) = f(\{x, y_i\})$ for $i = 1, \ldots, t$.

If no edge incident to x is coloured $n(y_t)$, then we construct an edge-colouring f' of G from f by setting $f'(\{x, y_{i-1}\}) := f(\{x, y_i\})$ $(i = 1, \ldots, t)$ and $f'(\{x, y_t\}) := n(y_t)$. So we assume that there is an edge incident to x with colour $n(y_t)$; by the maximality of t we have $f(\{x, y_s\}) = n(y_t)$ for some $s \in \{1, \ldots, t - 1\}$.

Consider the maximum path P starting at y_t in the graph $(V(G), \{e' \in E(G - e) : f(e') \in \{n(x), n(y_t)\}\})$ (this graph has maximum degree 2; cf. Figure 16.2). We distinguish three cases. In each case we construct an edge-colouring f' of G.

If P ends in x, then $\{y_s, x\}$ is the last edge of P. We construct f' from f by exchanging colours $n(x)$ and $n(y_t)$ on P, and setting $f'(\{x, y_{i-1}\}) := f(\{x, y_i\})$ $(i = 1, \ldots, s)$.

If P ends in y_{s-1}, then the last edge of P has colour $n(x)$, since colour $n(y_t) = f(\{x, y_s\}) = n(y_{s-1})$ is missing at y_{s-1}. We construct f' from f as follows: exchange colours $n(x)$ and $n(y_t)$ on P, set $f'(\{x, y_{i-1}\}) := f(\{x, y_i\})$ $(i = 1, \ldots, s - 1)$ and $f'(\{x, y_{s-1}\}) := n(x)$.

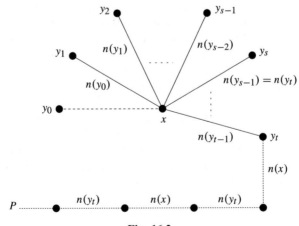

Fig. 16.2.

If P ends neither in x nor in y_{s-1}, then we can construct f' from f by exchanging colours $n(x)$ and $n(y_t)$ on P, and setting $f'(\{x, y_{i-1}\}) := f(\{x, y_i\})$ $(i = 1, \ldots, t)$ and $f'(\{x, y_t\}) := n(x)$.   □

Vizing's Theorem implies an absolute approximation algorithm for the EDGE-COLOURING PROBLEM in simple graphs. If we allow parallel edges the statement of Vizing's Theorem is no longer true: by replacing each edge of the triangle K_3 by r parallel edges we obtain a $2r$-regular graph with edge-chromatic number $3r$.

We now turn to the VERTEX-COLOURING PROBLEM. The maximum degree also gives an upper bound on the chromatic number:

Theorem 16.18. *Let G be an undirected graph with maximum degree k. Then G has an vertex-colouring with at most $k + 1$ colours, and such a colouring can be found in linear time.*

Proof: The following GREEDY COLOURING ALGORITHM obviously finds such a colouring.   □

GREEDY COLOURING ALGORITHM

Input: An undirected graph G.

Output: A vertex-colouring of G.

①   Let $V(G) = \{v_1, \ldots, v_n\}$.

②   **For** $i := 1$ **to** n **do**:
        Set $f(v_i) := \min\{k \in \mathbb{N} : k \neq f(v_j)$ for all $j < i$ with $v_j \in \Gamma(v_i)\}$.

For complete graphs and for odd circuits one evidently needs $k + 1$ colours, where k is the maximum degree. For all other connected graphs k colours suffice, as Brooks [1941] showed. However, the maximum degree is not a lower bound on the chromatic number: any star $K_{1,n}$ ($n \in \mathbb{N}$) has chromatic number 2. Therefore these results do not lead to an approximation algorithm. In fact, no algorithms for the VERTEX-COLOURING PROBLEM with a reasonable performance guarantee for general graphs are known; see Khanna, Linial and Safra [2000]. Zuckerman [2007] showed that, unless $P = NP$, no polynomial-time algorithm computes the chromatic number of any graph with n vertices up to a factor of $n^{1-\epsilon}$ for any fixed $\epsilon > 0$.

Held, Cook and Sewell [2012] showed how to compute good lower bounds on the chromatic number. A simple lower bound is the maximum size of a clique. Obviously, if a graph G contains a clique of size k, then the chromatic number of G is at least k. As the pentagon (circuit of length five) shows, the chromatic number can exceed the maximum clique size. Indeed, there are graphs with arbitrarily large chromatic number that contain no K_3. This motivates the following definition, which is due to Berge [1961, 1962]:

Definition 16.19. *A graph G is **perfect** if $\chi(H) = \omega(H)$ for every induced subgraph H of G, where $\chi(H)$ is the chromatic number and $\omega(H)$ is the maximum cardinality of a clique in H.*

It follows immediately that the decision problem whether a given perfect graph has chromatic number k has a good characterization (belongs to $NP \cap coNP$). Some examples of perfect graphs can be found in Exercise 16. A polynomial-time algorithm for recognizing perfect graphs has been found by Chudnovsky et al. [2005].

Berge [1961] conjectured that a graph is perfect if and only if it contains neither an odd circuit of length at least five nor the complement of such a circuit as an induced subgraph. This so-called strong perfect graph theorem has been proved by Chudnovsky et al. [2006]. Long before, Lovász [1972] proved the weaker assertion that a graph is perfect iff its complement is perfect. This is known as the weak perfect graph theorem; to prove it we need a lemma:

Lemma 16.20. *Let G be a perfect graph and $x \in V(G)$. Then the graph $G' := (V(G) \,\dot\cup\, \{y\}, E(G) \,\dot\cup\, \{\{y, v\} : v \in \{x\} \cup \Gamma(x)\})$, resulting from G by adding a new vertex y which is joined to x and to all neighbours of x, is perfect.*

Proof: By induction on $|V(G)|$. The case $|V(G)| = 1$ is trivial since K_2 is perfect. Now let G be a perfect graph with at least two vertices. Let $x \in V(G)$, and let G' arise by adding a new vertex y adjacent to x and all its neighbours. It suffices to prove that $\omega(G') = \chi(G')$, since for proper subgraphs H of G' this follows from the induction hypothesis: either H is a subgraph of G and thus perfect, or it arises from a proper subgraph of G by adding a vertex y as above.

Since we can colour G' with $\chi(G) + 1$ colours easily, we may assume that $\omega(G') = \omega(G)$. Then x is not contained in any maximum clique of G. Let f be a vertex-colouring of G with $\chi(G)$ colours, and let $X := \{v \in V(G) : f(v) = f(x)\}$. We have $\omega(G - X) = \chi(G - X) = \chi(G) - 1 = \omega(G) - 1$ and thus $\omega(G - (X \setminus \{x\})) = \omega(G) - 1$ (as x does not belong to any maximum clique of G). Since $(X \setminus \{x\}) \cup \{y\} = V(G') \setminus V(G - (X \setminus \{x\}))$ is a stable set, we have

$$\chi(G') = \chi(G - (X \setminus \{x\})) + 1 = \omega(G - (X \setminus \{x\})) + 1 = \omega(G) = \omega(G').$$

$\square$

Theorem 16.21. (Lovász [1972], Fulkerson [1972], Chvátal [1975]) *For a simple graph G the following statements are equivalent:*

(a) *G is perfect.*
(b) *The complement of G is perfect.*

(c) *The stable set polytope, i.e. the convex hull of the incidence vectors of the stable sets of G, is given by:*

$$\left\{ x \in \mathbb{R}_+^{V(G)} : \sum_{v \in S} x_v \le 1 \text{ for all cliques } S \text{ in } G \right\}. \tag{16.3}$$

Proof: We prove (a)$\Rightarrow$(c)$\Rightarrow$(b). This suffices, since applying (a)$\Rightarrow$(b) to the complement of G yields (b)$\Rightarrow$(a).

(a)$\Rightarrow$(c): Evidently the stable set polytope is contained in (16.3). To prove the other inclusion, let x be a rational vector in the polytope (16.3); we may write $x_v = \frac{p_v}{q}$, where $q \in \mathbb{N}$ and $p_v \in \mathbb{Z}_+$ for $v \in V(G)$. Replace each vertex v by a clique of size p_v; i.e. consider G' defined by

$$V(G') := \{(v, i) : v \in V(G), 1 \le i \le p_v\},$$
$$E(G') := \{\{(v, i), (v, j)\} : v \in V(G), 1 \le i < j \le p_v\} \cup$$
$$\{\{(v, i), (w, j)\} : \{v, w\} \in E(G), 1 \le i \le p_v, 1 \le j \le p_w\}.$$

Lemma 16.20 implies that G' is perfect. For an arbitrary clique X' in G' let $X := \{v \in V(G) : (v, i) \in X' \text{ for some } i\}$ be its projection to G (also a clique); we have

$$|X'| \le \sum_{v \in X} p_v = q \sum_{v \in X} x_v \le q.$$

So $\omega(G') \le q$. Since G' is perfect, it thus has a vertex-colouring f with at most q colours. For $v \in V(G)$ and $i = 1, \ldots, q$ let $a_{i,v} := 1$ if $f((v, j)) = i$ for some j and $a_{i,v} := 0$ otherwise. Then $\sum_{i=1}^{q} a_{i,v} = p_v$ for all $v \in V(G)$ and hence

$$x = \left(\frac{p_v}{q} \right)_{v \in V(G)} = \frac{1}{q} \sum_{i=1}^{q} a_i$$

is a convex combination of incidence vectors of stable sets, where $a_i = (a_{i,v})_{v \in V(G)}$.

(c)$\Rightarrow$(b): We show by induction on $|V(G)|$ that if (16.3) is integral then the complement of G is perfect. Since graphs with less than three vertices are perfect, let G be a graph with $|V(G)| \ge 3$ where (16.3) is integral.

We have to show that the vertex set of any induced subgraph H of G can be partitioned into $\alpha(H)$ cliques, where $\alpha(H)$ is the size of a maximum stable set in H. For proper subgraphs H this follows from the induction hypothesis, since (by Theorem 5.14) every face of the integral polytope (16.3) is integral, in particular the face defined by the supporting hyperplanes $x_v = 0$ ($v \in V(G) \setminus V(H)$).

So it remains to prove that $V(G)$ can be partitioned into $\alpha(G)$ cliques. The equation $\mathbb{1}x = \alpha(G)$ defines a supporting hyperplane of (16.3), so

$$\left\{ x \in \mathbb{R}_+^{V(G)} : \sum_{v \in S} x_v \le 1 \text{ for all cliques } S \text{ in } G, \sum_{v \in V(G)} x_v = \alpha(G) \right\} \tag{16.4}$$

is a face of (16.3). This face is contained in some facets, which cannot all be of the form $\{x \in (16.3) : x_v = 0\}$ for some v (otherwise the origin would belong to the intersection). Hence there is some clique S in G such that $\sum_{v \in S} x_v = 1$ for all x in (16.4). Hence this clique S intersects each maximum stable set of G. Now by the induction hypothesis, the vertex set of $G - S$ can partitioned into $\alpha(G - S) = \alpha(G) - 1$ cliques. Adding S concludes the proof. □

This proof is due to Lovász [1979b]. Indeed, the inequality system defining (16.3) is TDI for perfect graphs (Exercise 17). With some more work one can prove that for perfect graphs the VERTEX-COLOURING PROBLEM, the MAXIMUM WEIGHT STABLE SET PROBLEM and the MAXIMUM WEIGHT CLIQUE PROBLEM can be solved in strongly polynomial time. Although these problems are all *NP*-hard for general graphs (Theorem 15.23, Corollary 15.24, Theorem 16.14(b)), there is a number (the so-called theta-function of the complement graph, introduced by Lovász [1979a]) which is always between the maximum clique size and the chromatic number, and which can be computed in polynomial time for general graphs using the ELLIPSOID METHOD. The details are a bit involved; see Grötschel, Lovász and Schrijver [1988].

One of the best-known problems in graph theory has been the four colour problem: is it true that every planar map can be coloured with four colours such that no two countries with a common border have the same colour? If we consider the countries as regions and switch to the planar dual graph, this is equivalent to asking whether every planar graph has a vertex-colouring with four colours. Appel and Haken [1977] and Appel, Haken and Koch [1977] proved that this is indeed true: every planar graph has chromatic number at most 4. For a simpler proof of this Four Colour Theorem (which nevertheless is based on a case checking by a computer) see Robertson et al. [1997]. We prove the following weaker result, known as the Five Colour Theorem:

Theorem 16.22. (Heawood [1890]) *Any planar graph has a vertex-colouring with at most five colours, and such a colouring can be found in polynomial time.*

Proof: By induction on $|V(G)|$. We may assume that G is simple, and we fix an arbitrary planar embedding $\Phi = \left(\psi, (J_e)_{e \in E(G)}\right)$ of G. By Corollary 2.33, G has a vertex v of degree five or less. By the induction hypothesis, $G - v$ has a vertex-colouring f with at most 5 colours. We may assume that v has degree 5 and all neighbours have different colours; otherwise we can easily extend the colouring to G.

Let w_1, w_2, w_3, w_4, w_5 be the neighbours of v in the cyclic order in which the polygonal arcs $J_{\{v,w_i\}}$ leave v.

We first claim that there are no vertex-disjoint paths P from w_1 to w_3 and Q from w_2 to w_4 in $G - v$. To prove this, let P be a w_1-w_3-path, and let C be the circuit in G consisting of P and the edges $\{v, w_1\}, \{v, w_3\}$. By Theorem 2.30 $\mathbb{R}^2 \setminus \bigcup_{e \in E(C)} J_e$ splits into two connected regions, and v is on the boundary of both regions. Hence w_2 and w_4 belong to different regions of that set, implying that every w_2-w_4-path in $G - v$ must contain a vertex of C.

Let X be the connected component of the graph $G[\{x \in V(G) \setminus \{v\} : f(x) \in \{f(w_1), f(w_3)\}\}]$ which contains w_1. If X does not contain w_3, we can exchange the colours in X and afterwards extend the colouring to G by colouring v with the old colour of w_1. So we may assume that there is a w_1-w_3-path P containing only vertices coloured with $f(w_1)$ or $f(w_3)$.

Analogously, we are done if there is no w_2-w_4-path Q containing only vertices coloured with $f(w_2)$ or $f(w_4)$. But the contrary assumption means that there are vertex-disjoint paths P from w_1 to w_3 and Q from w_2 to w_4 in $G - v$, a contradiction.   □

Hence this is a second *NP*-hard problem which has an absolute approximation algorithm. Indeed, the Four Colour Theorem implies that the chromatic number of a non-bipartite planar graph can only be 3 or 4. Using the polynomial-time algorithm of Robertson et al. [1997], which colours any given planar graph with four colours, one obtains an absolute approximation algorithm which uses at most one colour more than necessary.

Fürer and Raghavachari [1994] detected a third natural problem which can be approximated up to an additive error of one: Given an undirected graph, they look for a spanning tree whose maximum degree is minimum among all the spanning trees (the problem is a generalization of the HAMILTONIAN PATH PROBLEM and thus *NP*-hard). Their algorithm also extends to a general case corresponding to the STEINER TREE PROBLEM: given a set $T \subseteq V(G)$, find a tree S in G with $V(T) \subseteq V(S)$ such that the maximum degree of S is minimum. Singh and Lau [2015] found an extension to minimum weight spanning trees with bounded degrees.

On the other hand, the following theorem tells that many problems do not have absolute approximation algorithms unless $P = NP$:

Proposition 16.23. *Let $\mathcal{I}$ be an (infinite) family of set systems, and let $\mathcal{P}$ be the following optimization problem: Given a set system $(E, \mathcal{F}) \in \mathcal{I}$ and a function $c : E \to \mathbb{Z}$, find a set $F \in \mathcal{F}$ such that $c(F)$ minimum (or decide that no such F exists).*

Then $\mathcal{P}$ has an absolute approximation algorithm if and only if $\mathcal{P}$ can be solved in polynomial time.

Proof: Suppose there is a polynomial-time algorithm A and an integer k such that

$$|A((E, \mathcal{F}, c)) - \text{OPT}((E, \mathcal{F}, c))| \leq k$$

for all instances $(E, \mathcal{F}, c)$ of $\mathcal{P}$. We show how to solve $\mathcal{P}$ exactly in polynomial time.

Given an instance $(E, \mathcal{F}, c)$ of $\mathcal{P}$, we construct a new instance $(E, \mathcal{F}, c')$, where $c'(e) := (k+1)c(e)$ for all $e \in E$. Obviously the optimum solutions remain the same. But if we now apply A to the new instance,

$$|A((E, \mathcal{F}, c')) - \text{OPT}((E, \mathcal{F}, c'))| \leq k$$

and thus $A((E, \mathcal{F}, c')) = \text{OPT}((E, \mathcal{F}, c'))$.   □

Examples are the MINIMIZATION PROBLEM FOR INDEPENDENCE SYSTEMS and the MAXIMIZATION PROBLEM FOR INDEPENDENCE SYSTEMS (multiply c by -1), and thus all problems in the list of Section 13.1.

16.4 Approximation Schemes

Recall the absolute approximation algorithm for the EDGE-COLOURING PROBLEM discussed in the previous section. This also implies a relative performance guarantee: Since one can easily decide if the edge-chromatic number is 1 or 2 (Proposition 16.15), Vizing's Theorem yields a $\frac{4}{3}$-factor approximation algorithm. On the other hand, Theorem 16.14(a) implies that no k-factor approximation algorithm exists for any $k < \frac{4}{3}$ (unless $P = NP$).

Hence the existence of an absolute approximation algorithm does not imply the existence of a k-factor approximation algorithm for all $k > 1$. We shall meet a similar situation with the BIN-PACKING PROBLEM in Chapter 18. This consideration suggests the following definition:

Definition 16.24. *Let $\mathcal{P}$ be an optimization problem with nonnegative weights. An* **asymptotic k-factor approximation algorithm** *for $\mathcal{P}$ is a polynomial-time algorithm A for $\mathcal{P}$ for which there exists a constant c such that*

$$\frac{1}{k} \operatorname{OPT}(I) - c \ \leq \ A(I) \ \leq \ k \operatorname{OPT}(I) + c$$

for all instances I of $\mathcal{P}$. We also say that A has **asymptotic performance ratio** k.

The **(asymptotic) approximation ratio** of an optimization problem $\mathcal{P}$ with nonnegative weights is defined to be the infimum of all numbers k for which there exists an (asymptotic) k-factor approximation algorithm for $\mathcal{P}$, or ∞ if there is no (asymptotic) approximation algorithm at all.

For example, the above-mentioned EDGE-COLOURING PROBLEM has approximation ratio $\frac{4}{3}$ (unless $P = NP$), but asymptotic approximation ratio 1 (not only in simple graphs; see Sanders and Steurer [2008]). Optimization problems with (asymptotic) approximation ratio 1 are of particular interest. For these problems we introduce the following notion:

Definition 16.25. *Let $\mathcal{P}$ be an optimization problem with nonnegative weights. An* **approximation scheme** *for $\mathcal{P}$ is an algorithm A accepting as input an instance I of $\mathcal{P}$ and an $\epsilon > 0$ such that, for each fixed ϵ, A is a $(1 + \epsilon)$-factor approximation algorithm for $\mathcal{P}$.*

An **asymptotic approximation scheme** *for $\mathcal{P}$ is a pair of algorithms (A, A') with the following properties: A' is a polynomial-time algorithm accepting a number $\epsilon > 0$ as input and computing a number c_ϵ. A accepts an instance I of $\mathcal{P}$ and an*

$\epsilon > 0$ *as input, and its output consists of a feasible solution for I satisfying*

$$\frac{1}{1+\epsilon} \, \text{OPT}(I) - c_\epsilon \ \leq \ A(I, \epsilon) \ \leq \ (1+\epsilon) \, \text{OPT}(I) + c_\epsilon.$$

For each fixed ϵ, the running time of A is polynomially bounded in $\text{size}(I)$.

An (asymptotic) approximation scheme is called a **fully polynomial (asymptotic) approximation scheme** *if the running time as well as the maximum size of any number occurring in the computation is bounded by a polynomial in* $\text{size}(I) + \text{size}(\epsilon) + \frac{1}{\epsilon}$, *and c_ϵ is a polynomial in $\frac{1}{\epsilon}$.*

In some other texts one finds the abbreviations PTAS for (polynomial-time) approximation scheme and FPAS or FPTAS for fully polynomial approximation scheme.

Apart from absolute approximation algorithms, a fully polynomial approximation scheme can be considered the best we may hope for when faced with an *NP*-hard optimization problem, at least if the cost of any feasible solution is a nonnegative integer (which can be assumed in many cases without loss of generality):

Proposition 16.26. *Let $\mathcal{P} = (X, (S_x)_{x \in X}, c, \text{goal})$ be an NP optimization problem where the values of c are nonnegative integers. Let A be an algorithm which, given an instance I of $\mathcal{P}$ and a number $\epsilon > 0$, computes a feasible solution of I with*

$$\frac{1}{1+\epsilon} \, \text{OPT}(I) \ \leq \ A(I, \epsilon) \ \leq \ (1+\epsilon) \, \text{OPT}(I)$$

and whose running time is bounded by a polynomial in $\text{size}(I) + \text{size}(\epsilon)$. *Then $\mathcal{P}$ can be solved exactly in polynomial time.*

Proof: Given an instance I, we first run A on $(I, 1)$. We set $\epsilon := \frac{1}{1+2A(I,1)}$ and observe that $\epsilon \, \text{OPT}(I) < 1$. Now we run A on (I, ϵ). Since $\text{size}(\epsilon)$ is polynomially bounded in $\text{size}(I)$, this procedure constitutes a polynomial-time algorithm. If $\mathcal{P}$ is a minimization problem, we have

$$A(I, \epsilon) \ \leq \ (1+\epsilon) \, \text{OPT}(I) \ < \ \text{OPT}(I) + 1,$$

which, since c is integral, implies optimality. Similarly, if $\mathcal{P}$ is a maximization problem, we have

$$A(I, \epsilon) \ \geq \ \frac{1}{1+\epsilon} \, \text{OPT}(I) \ > \ (1-\epsilon) \, \text{OPT}(I) \ > \ \text{OPT}(I) - 1.$$

$\square$

Unfortunately, a fully polynomial approximation scheme exists only for very few problems (see Theorem 17.11). Moreover we note that even the existence of a fully polynomial approximation scheme does not imply an absolute approximation algorithm; the KNAPSACK PROBLEM is an example.

In Chapters 17 and 18 we shall discuss two problems (KNAPSACK and BIN-PACKING) which have a fully polynomial approximation scheme and a fully polynomial asymptotic approximation scheme, respectively. For many problems the two types of approximation schemes coincide:

Theorem 16.27.   (Papadimitriou and Yannakakis [1993]) *Let $\mathcal{P}$ be an optimization problem with nonnegative weights. Suppose that for each constant k there is a polynomial-time algorithm which decides whether a given instance has optimum value at most k, and, if so, finds an optimum solution.*

Then $\mathcal{P}$ has an approximation scheme if and only if $\mathcal{P}$ has an asymptotic approximation scheme.

Proof: The only-if-part is trivial, so suppose that $\mathcal{P}$ has an asymptotic approximation scheme (A, A'). We describe an approximation scheme for $\mathcal{P}$.

Let a fixed $\epsilon > 0$ be given; we may assume $\epsilon < 1$. We set $\epsilon' := \frac{\epsilon - \epsilon^2}{2 + \epsilon + \epsilon^2} < \frac{\epsilon}{2}$ and first run A' on the input ϵ', yielding a constant $c_{\epsilon'}$.

For a given instance I we next test whether $\mathrm{OPT}(I)$ is at most $\frac{2c_{\epsilon'}}{\epsilon}$. This is a constant for each fixed ϵ, so we can decide this in polynomial time and find an optimum solution if $\mathrm{OPT}(I) \leq \frac{2c_{\epsilon'}}{\epsilon}$.

Otherwise we apply A to I and ϵ' and obtain a solution of value V, with

$$\frac{1}{1 + \epsilon'} \mathrm{OPT}(I) - c_{\epsilon'} \leq V \leq (1 + \epsilon') \mathrm{OPT}(I) + c_{\epsilon'}.$$

We claim that this solution is good enough. Indeed, we have $c_{\epsilon'} < \frac{\epsilon}{2} \mathrm{OPT}(I)$ which implies

$$V \leq (1 + \epsilon') \mathrm{OPT}(I) + c_{\epsilon'} < \left(1 + \frac{\epsilon}{2}\right) \mathrm{OPT}(I) + \frac{\epsilon}{2} \mathrm{OPT}(I) = (1 + \epsilon) \mathrm{OPT}(I)$$

and

$$
\begin{aligned}
V &\geq \frac{1}{(1 + \epsilon')} \mathrm{OPT}(I) - \frac{\epsilon}{2} \mathrm{OPT}(I) \\
&= \frac{2 + \epsilon + \epsilon^2}{2 + 2\epsilon} \mathrm{OPT}(I) - \frac{\epsilon}{2} \mathrm{OPT}(I) \\
&= \left(\frac{1}{1 + \epsilon} + \frac{\epsilon}{2}\right) \mathrm{OPT}(I) - \frac{\epsilon}{2} \mathrm{OPT}(I) \\
&= \frac{1}{1 + \epsilon} \mathrm{OPT}(I).
\end{aligned}
$$

$\square$

So the definition of an asymptotic approximation scheme is meaningful only for problems (such as bin-packing or colouring problems) whose restriction to a constant optimum value is still difficult. For many problems this restriction can be solved in polynomial time by some kind of complete enumeration.

16.5 Maximum Satisfiability

The SATISFIABILITY Problem was our first *NP*-complete problem. In this section we analyse the corresponding optimization problem:

MAXIMUM SATISFIABILITY (MAX-SAT)

Instance: A set X of variables, a family $\mathcal{Z}$ of clauses over X, and a weight function $c : \mathcal{Z} \to \mathbb{R}_+$.

Task: Find a truth assignment T of X such that the total weight of the clauses in $\mathcal{Z}$ that are satisfied by T is maximum.

As we shall see, approximating MAX-SAT is a nice example (and historically one of the first) for the algorithmic use of the probabilistic method.

Let us first consider the following trivial randomized algorithm: set each variable independently *true* with probability $\frac{1}{2}$. Obviously this algorithm satisfies each clause Z with probability $1 - 2^{-|Z|}$.

Let us write r for random variables which are *true* with probability $\frac{1}{2}$ and *false* otherwise, and let $R = (r, r, \ldots, r)$ be the random variable uniformly distributed over all truth assignments (several copies of r are independent of each other). If we write $c(T)$ for the total weight of the clauses satisfied by the truth assignment T, the expected total weight of the clauses satisfied by R is

$$
\begin{aligned}
\mathrm{Exp}\,(c(R)) &= \sum_{Z \in \mathcal{Z}} c(Z) \,\mathrm{Prob}(R \text{ satisfies } Z) \\
&= \sum_{Z \in \mathcal{Z}} c(Z) \left(1 - 2^{-|Z|} \right) \qquad\qquad (16.5) \\
&\geq \left(1 - 2^{-p} \right) \sum_{Z \in \mathcal{Z}} c(Z),
\end{aligned}
$$

where $p := \min_{Z \in \mathcal{Z}} |Z|$; Exp and Prob denote expectation and probability.

Since the optimum cannot exceed $\sum_{Z \in \mathcal{Z}} c(Z)$, R is expected to yield a solution within a factor $\frac{1}{1 - 2^{-p}}$ of the optimum. But what we would really like to have is a deterministic approximation algorithm. In fact, we can turn our (trivial) randomized algorithm into a deterministic algorithm while preserving the performance guarantee. This step is often called derandomization.

Let us fix the truth assignment step by step. Suppose $X = \{x_1, \ldots, x_n\}$, and we have already fixed a truth assignment T for $x_1, \ldots, x_k$ ($0 \leq k < n$). If we now set $x_{k+1}, \ldots, x_n$ randomly, setting each variable independently *true* with probability $\frac{1}{2}$, we will satisfy clauses of expected total weight $e_0 = \mathrm{Exp}(c(T(x_1), \ldots, T(x_k), r, \ldots, r))$. If we set x_{k+1} *true* (*false*), and then set $x_{k+2}, \ldots, x_n$ randomly, the satisfied clauses will have some expected total weight e_1 (e_2, respectively). e_1 and e_2 can be thought of as conditional expectations. Trivially $e_0 = \frac{e_1 + e_2}{2}$, so at least one of e_1, e_2 must be at least e_0. We set x_{k+1} to be *true* if $e_1 \geq e_2$ and *false* otherwise. This is sometimes called the method of conditional probabilities.

JOHNSON'S ALGORITHM FOR MAX-SAT

Input: A set $X = \{x_1, \ldots, x_n\}$ of variables, a family $\mathcal{Z}$ of clauses over X, and a weight function $c : \mathcal{Z} \to \mathbb{R}_+$.

Output: A truth assignment $T : X \to \{true, false\}$.

---

①   **For** $k := 1$ **to** n **do:**
     If $\mathrm{Exp}(c(T(x_1), \ldots, T(x_{k-1}), true, r, \ldots, r))$
      $\geq \mathrm{Exp}(c(T(x_1), \ldots, T(x_{k-1}), false, r, \ldots, r))$
       then set $T(x_k) := true$
       else set $T(x_k) := false$.

The expectations can be easily computed with (16.5).

Theorem 16.28. (Johnson [1974]) JOHNSON'S ALGORITHM FOR MAX-SAT *is a $\frac{1}{1-2^{-p}}$-factor approximation algorithm for* MAX-SAT, *where p is the minimum cardinality of a clause.*

Proof: Let us define the conditional expectation

$$s_k := \mathrm{Exp}(c(T(x_1), \ldots, T(x_k), r, \ldots, r))$$

for $k = 0, \ldots, n$. Observe that $s_n = c(T)$ is the total weight of the clauses satisfied by our algorithm, while $s_0 = \mathrm{Exp}(c(R)) \geq \left(1 - 2^{-p}\right) \sum_{Z \in \mathcal{Z}} c(Z)$ by (16.5).

Furthermore, $s_i \geq s_{i-1}$ by the choice of $T(x_i)$ in ① (for $i = 1, \ldots, n$). So $s_n \geq s_0 \geq \left(1 - 2^{-p}\right) \sum_{Z \in \mathcal{Z}} c(Z)$. Since the optimum is at most $\sum_{Z \in \mathcal{Z}} c(Z)$, the proof is complete. $\square$

Since $p \geq 1$, we have a 2-factor approximation algorithm. However, this is not too interesting as there is a much simpler 2-factor approximation algorithm: either set all variables *true* or all *false*, whichever is better. However, Chen, Friesen and Zheng [1999] showed that JOHNSON'S ALGORITHM FOR MAX-SAT is indeed a $\frac{3}{2}$-factor approximation algorithm.

If there are no one-element clauses ($p \geq 2$), it is a $\frac{4}{3}$-factor approximation algorithm (by Theorem 16.28), for $p \geq 3$ it is an $\frac{8}{7}$-factor approximation algorithm.

Yannakakis [1994] found a $\frac{4}{3}$-factor approximation algorithm for the general case using network flow techniques. We shall describe a simpler $\frac{4}{3}$-factor approximation algorithm due to Goemans and Williamson [1994].

It is straightforward to translate MAX-SAT into an integer linear program: If we have variables $X = \{x_1, \ldots, x_n\}$, clauses $\mathcal{Z} = \{Z_1, \ldots, Z_m\}$, and weights $c_1, \ldots, c_m$, we can write

$$\max \quad \sum_{j=1}^{m} c_j z_j$$

$$\text{s.t.} \quad z_j \leq \sum_{i:x_i \in Z_j} y_i + \sum_{i:\overline{x_i} \in Z_j} (1 - y_i) \quad (j = 1, \ldots, m)$$

$$y_i, z_j \in \{0, 1\} \quad (i = 1, \ldots, n, \ j = 1, \ldots, m).$$

Here $y_i = 1$ means that variable x_i is *true*, and $z_j = 1$ means that clause Z_j is satisfied. Now consider the LP relaxation:

$$\max \quad \sum_{j=1}^{m} c_j z_j$$

$$
\begin{array}{llll}
\text{s.t.} & z_j \leq \displaystyle\sum_{i:x_i \in Z_j} y_i + \sum_{i:\overline{x_i} \in Z_j}(1 - y_i) & (j = 1, \ldots, m) & \\
& y_i \leq 1 & (i = 1, \ldots, n) & (16.6) \\
& y_i \geq 0 & (i = 1, \ldots, n) & \\
& z_j \leq 1 & (j = 1, \ldots, m) & \\
& z_j \geq 0 & (j = 1, \ldots, m). &
\end{array}
$$

Let (y^*, z^*) be an optimum solution of (16.6). Now independently set each variable x_i *true* with probability y_i^*. This step is known as randomized rounding, a technique which has been introduced by Raghavan and Thompson [1987]. The above method constitutes another randomized algorithm for MAX-SAT, which can be derandomized as above. Let r_p be the random variable which is *true* with probability p and *false* otherwise.

GOEMANS-WILLIAMSON ALGORITHM FOR MAX-SAT

Input: A set $X = \{x_1, \ldots, x_n\}$ of variables, a family $\mathcal{Z}$ of clauses over X, and a weight function $c : \mathcal{Z} \to \mathbb{R}_+$.

Output: A truth assignment $T : X \to \{true, false\}$.

① Solve the linear program (16.6); let (y^*, z^*) be an optimum solution.

② **For** $k := 1$ **to** n **do**:
   If $\mathrm{Exp}(c(T(x_1), \ldots, T(x_{k-1}), true, r_{y_{k+1}^*}, \ldots, r_{y_n^*})$
      $\geq \mathrm{Exp}(c(T(x_1), \ldots, T(x_{k-1}), false, r_{y_{k+1}^*}, \ldots, r_{y_n^*})$
      then set $T(x_k) := true$
      else set $T(x_k) := false$.

Theorem 16.29. (Goemans and Williamson [1994]) *The* GOEMANS-WILLIAMSON ALGORITHM FOR MAX-SAT *is a* $\dfrac{1}{1 - \left(1 - \frac{1}{q}\right)^q}$-*factor approximation algorithm, where q is the maximum cardinality of a clause.*

Proof: Let us write

$$s_k := \mathrm{Exp}(c(T(x_1), \ldots, T(x_k), r_{y_{k+1}^*}, \ldots, r_{y_n^*}))$$

for $k = 0, \ldots, n$. We again have $s_i \geq s_{i-1}$ for $i = 1, \ldots, n$ and $s_n = c(T)$ is the total weight of clauses satisfied by our algorithm. So it remains to estimate $s_0 = \mathrm{Exp}(c(R_{y^*}))$, where $R_{y^*} = (r_{y_1^*}, \ldots, r_{y_n^*})$.

For $j = 1, \ldots, m$, the probability that the clause Z_j is satisfied by R_{y^*} is

$$1 - \left(\prod_{i:x_i \in Z_j} (1 - y_i^*) \right) \cdot \left(\prod_{i:\overline{x_i} \in Z_j} y_i^* \right).$$

Since the geometric mean is always less than or equal to the arithmetic mean, this probability is at least

$$1 - \left(\frac{1}{|Z_j|} \left(\sum_{i:x_i \in Z_j} (1 - y_i^*) + \sum_{i:\overline{x_i} \in Z_j} y_i^* \right) \right)^{|Z_j|}$$

$$= 1 - \left(1 - \frac{1}{|Z_j|} \left(\sum_{i:x_i \in Z_j} y_i^* + \sum_{i:\overline{x_i} \in Z_j} (1 - y_i^*) \right) \right)^{|Z_j|}$$

$$\geq 1 - \left(1 - \frac{z_j^*}{|Z_j|} \right)^{|Z_j|}$$

$$\geq \left(1 - \left(1 - \frac{1}{|Z_j|} \right)^{|Z_j|} \right) z_j^*.$$

To prove the last inequality, observe that for any $0 \leq a \leq 1$ and any $k \in \mathbb{N}$

$$1 - \left(1 - \frac{a}{k} \right)^k \geq a \left(1 - \left(1 - \frac{1}{k} \right)^k \right)$$

holds: both sides of the inequality are equal for $a \in \{0, 1\}$, and the left-hand side (as a function of a) is concave, while the right-hand side is linear.

So we have

$$s_0 = \mathrm{Exp}(c(R_{y^*})) = \sum_{j=1}^m c_j \, \mathrm{Prob}(R_{y^*} \text{ satisfies } Z_j)$$

$$\geq \sum_{j=1}^m c_j \left(1 - \left(1 - \frac{1}{|Z_j|} \right)^{|Z_j|} \right) z_j^*$$

$$\geq \left(1 - \left(1 - \frac{1}{q} \right)^q \right) \sum_{j=1}^m c_j z_j^*$$

(observe that the sequence $\left(\left(1 - \frac{1}{k} \right)^k \right)_{k \in \mathbb{N}}$ is monotonously increasing and converges to $\frac{1}{e}$). Since the optimum is less than or equal to $\sum_{j=1}^m z_j^* c_j$, the optimum value of the LP relaxation, the proof is complete. $\qquad \square$

Since $\left(1 - \frac{1}{q}\right)^q < \frac{1}{e}$, we have an $\frac{e}{e-1}$-factor approximation algorithm ($\frac{e}{e-1}$ is about 1.582).

We now have two similar algorithms that behave differently: the first one is better for long clauses, while the second is better for short clauses. Hence it is natural to combine them:

Theorem 16.30. (Goemans and Williamson [1994]) *The following is a $\frac{4}{3}$-factor approximation algorithm for* MAX-SAT: *run both* JOHNSON'S ALGORITHM FOR MAX-SAT *and the* GOEMANS-WILLIAMSON ALGORITHM FOR MAX-SAT *and choose the better of the two solutions.*

Proof: We use the notation of the above proofs. The algorithm returns a truth assignment satisfying clauses of total weight at least

$$
\max\{\mathrm{Exp}(c(R)), \mathrm{Exp}(c(R_{y^*}))\}
$$

$$
\geq \quad \frac{1}{2}\left(\mathrm{Exp}(c(R)) + \mathrm{Exp}(c(R_{y^*}))\right)
$$

$$
\geq \quad \frac{1}{2}\sum_{j=1}^{m}\left(\left(1 - 2^{-|Z_j|}\right)c_j + \left(1 - \left(1 - \frac{1}{|Z_j|}\right)^{|Z_j|}\right)z_j^* c_j\right)
$$

$$
\geq \quad \frac{1}{2}\sum_{j=1}^{m}\left(2 - 2^{-|Z_j|} - \left(1 - \frac{1}{|Z_j|}\right)^{|Z_j|}\right)z_j^* c_j
$$

$$
\geq \quad \frac{3}{4}\sum_{j=1}^{m}z_j^* c_j.
$$

For the last inequality observe that $2 - 2^{-k} - \left(1 - \frac{1}{k}\right)^k \geq \frac{3}{2}$ for all $k \in \mathbb{N}$: for $k \in \{1, 2\}$ we have equality; for $k \geq 3$ we have $2 - 2^{-k} - \left(1 - \frac{1}{k}\right)^k \geq 2 - \frac{1}{8} - \frac{1}{e} > \frac{3}{2}$. Since the optimum is at least $\sum_{j=1}^{m} z_j^* c_j$, the theorem is proved. $\square$

Slightly better approximation algorithms for MAX-SAT (using semidefinite programming) have been found; see Goemans and Williamson [1995], Mahajan and Ramesh [1999], Feige and Goemans [1995], and Asano [2006]. The currently best known algorithm achieves a performance ratio of 1.256 (Avidor, Berkovitch and Zwick [2006]).

Indeed, Bellare and Sudan [1994] showed that approximating MAX-SAT to within a factor of $\frac{74}{73}$ is NP-hard. Even for MAX-3SAT (which is MAX-SAT restricted to instances where each clause has exactly three literals) no approximation scheme exists (unless $P = NP$), as we shall show in the next section.

16.6 The *PCP* Theorem

Many non-approximability results are based on a deep theorem which gives a new characterization of the class *NP*. Recall that a decision problem belongs to *NP* if and only if there is a polynomial-time certificate-checking algorithm. Now we consider randomized certificate-checking algorithms that read the complete instance but only a small part of the certificate to be checked. They always accept yes-instances with correct certificates but sometimes also accept no-instances.

Which bits of the certificate are read is decided randomly in advance; more precisely this decision depends on the instance x and on $O(\log(\text{size}(x)))$ random bits.

We now formalize this concept. If s is a string and $t \in \mathbb{N}^k$, then s_t denotes the string of length k whose i-th component is the t_i-th component of s ($i = 1, \ldots, k$).

Definition 16.31. *A decision problem (X, Y) belongs to the class* **PCP(log *n*,1)** *if there is a polynomial p and a constant $k \in \mathbb{N}$, a function*

$$ f : \left\{ (x, r) : x \in X,\, r \in \{0, 1\}^{\lfloor \log(p(\text{size}(x))) \rfloor} \right\} \to \mathbb{N}^k $$

*computable in polynomial time, with $f(x, r) \in \{1, \ldots, \lfloor p(\text{size}(x)) \rfloor\}^k$ for all x and r, and a decision problem (X', Y') in P, where $X' := \{(x, \pi, \gamma) : x \in X,\ \pi \in \{1, \ldots, \lfloor p(\text{size}(x)) \rfloor\}^k,\ \gamma \in \{0, 1\}^k\}$, such that for any instance $x \in X$:
If $x \in Y$ then there exists a $c \in \{0, 1\}^{\lfloor p(\text{size}(x)) \rfloor}$ with $\text{Prob}\left((x, f(x, r), c_{f(x,r)}) \in Y' \right) = 1$. If $x \notin Y$ then $\text{Prob}\left((x, f(x, r), c_{f(x,r)}) \in Y' \right) < \frac{1}{2}$ for all $c \in \{0, 1\}^{\lfloor p(\text{size}(x)) \rfloor}$.
Here the probability is taken over the uniform distribution of random strings $r \in \{0, 1\}^{\lfloor \log(p(\text{size}(x))) \rfloor}$.*

The letters "*PCP*" stand for "probabilistically checkable proof". The parameters $\log n$ and 1 reflect that, for an instance of size n, $O(\log n)$ random bits are used and $O(1)$ bits of the certificate are read.

For any yes-instance there is a certificate which is always accepted; while for no-instances no string is accepted as a certificate with probability $\frac{1}{2}$ or more. Note that this error probability $\frac{1}{2}$ can be replaced equivalently by any number between zero and one (Exercise 20).

Proposition 16.32. *PCP*$(\log n, 1) \subseteq NP$.

Proof: Let $(X, Y) \in PCP(\log n, 1)$, and let $p, k, f, (X', Y')$ be given as in Definition 16.31. Let $X'' := \left\{ (x, c) : x \in X,\ c \in \{0, 1\}^{\lfloor p(\text{size}(x)) \rfloor} \right\}$, and let

$$ Y'' := \left\{ (x, c) \in X'' : \text{Prob}\left((x, f(x, r), c_{f(x,r)}) \in Y' \right) = 1 \right\}. $$

To show that $(X, Y) \in NP$ it suffices to show that $(X'', Y'') \in P$. But since there are only $2^{\lfloor \log(p(\text{size}(x))) \rfloor}$, i.e. at most $p(\text{size}(x))$ many strings $r \in \{0, 1\}^{\lfloor \log(p(\text{size}(x))) \rfloor}$, we can try them all. For each one we compute $f(x, r)$ and test whether $(x, f(x, r), c_{f(x,r)}) \in Y'$ (we use that $(X', Y') \in P$). The overall running time is polynomial in size(x). $\qquad\square$

Now the surprising result is that these randomized verifiers, which read only a constant number of bits of the certificate, are as powerful as the standard (deterministic) certificate-checking algorithms which have the full information. This is the so-called *PCP* Theorem:

Theorem 16.33.   (Arora et al. [1998])

$$NP = PCP(\log n, 1).$$

The proof of $NP \subseteq PCP(\log n, 1)$ is very difficult and beyond the scope of this book. It is based on earlier (and weaker) results of Feige et al. [1996] and Arora and Safra [1998]. For a self-contained proof of the *PCP* Theorem 16.33, see also Arora [1994], Hougardy, Prömel and Steger [1994], or Ausiello et al. [1999]. Stronger results were found subsequently by Bellare, Goldreich and Sudan [1998] and Håstad [2001]. For example, the number k in Definition 16.31 can be chosen to be 9. A new proof of the *PCP* Theorem was proposed by Dinur [2007].

We show some of its consequences for the non-approximability of combinatorial optimization problems. We start with the MAXIMUM CLIQUE PROBLEM and the MAXIMUM STABLE SET PROBLEM: given an undirected graph G, find a clique, or a stable set, of maximum cardinality in G.

Recall Proposition 2.2 (and Corollary 15.24): The problems of finding a maximum clique, a maximum stable set, or a minimum vertex cover are all equivalent. However, the 2-factor approximation algorithm for the MINIMUM VERTEX COVER PROBLEM (Section 16.1) does not imply an approximation algorithm for the MAXIMUM STABLE SET PROBLEM or the MAXIMUM CLIQUE PROBLEM.

Namely, it can happen that the algorithm returns a vertex cover C of size $n-2$, while the optimum is $\frac{n}{2}-1$ (where $n = |V(G)|$). The complement $V(G) \setminus C$ is then a stable set of cardinality 2, but the maximum stable set has cardinality $\frac{n}{2}+1$. This example shows that transferring an algorithm to another problem via a polynomial transformation does not in general preserve its performance guarantee. We shall consider a restricted type of transformation in the next section. Here we deduce a non-approximability result for the MAXIMUM CLIQUE PROBLEM from the *PCP* Theorem:

Theorem 16.34.   (Arora and Safra [1998])   *Unless $P = NP$ there is no 2-factor approximation algorithm for the* MAXIMUM CLIQUE PROBLEM.

Proof: Let $\mathcal{P} = (X, Y)$ be some *NP*-complete problem. By the *PCP* Theorem 16.33, $\mathcal{P} \in PCP(\log n, 1)$, so let p, k, f, $\mathcal{P}' := (X', Y')$ be as in Definition 16.31.

For any given $x \in X$ we construct a graph G_x as follows. Let

$$V(G_x) := \left\{ (r, a) : r \in \{0, 1\}^{\lfloor \log(p(\text{size}(x))) \rfloor}, a \in \{0, 1\}^k, (x, f(x, r), a) \in Y' \right\}$$

(representing all "accepting runs" of the randomized certificate checking algorithm). Two vertices (r, a) and (r', a') are joined by an edge if $a_i = a'_j$ whenever the

i-th component of $f(x, r)$ equals the j-th component of $f(x, r')$ (note that this implies $r \neq r'$). Since $\mathcal{P}' \in P$ and there are only a polynomial number of random strings, G_x can be computed in polynomial time (and has polynomial size).

If $x \in Y$ then by definition there exists a certificate $c \in \{0, 1\}^{\lfloor p(\text{size}(x)) \rfloor}$ such that $(x, f(x, r), c_{f(x,r)}) \in Y'$ for all $r \in \{0, 1\}^{\lfloor \log(p(\text{size}(x))) \rfloor}$. Hence there is a clique of size $2^{\lfloor \log(p(\text{size}(x))) \rfloor}$ in G_x.

On the other hand, if $x \notin Y$ then there is no clique of size $\frac{1}{2} 2^{\lfloor \log(p(\text{size}(x))) \rfloor}$ in G_x: Suppose $(r^{(1)}, a^{(1)}), \ldots, (r^{(t)}, a^{(t)})$ are the vertices of a clique. Then $r^{(1)}, \ldots, r^{(t)}$ are pairwise different. We set $c_i := a_k^{(j)}$ whenever the k-th component of $f(x, r^{(j)})$ equals i, and set the remaining components of c (if any) arbitrarily. This way we obtain a certificate c with $(x, f(x, r^{(i)}), c_{f(x,r^{(i)})}) \in Y'$ for all $i = 1, \ldots, t$. If $x \notin Y$ we have $t < \frac{1}{2} 2^{\lfloor \log(p(\text{size}(x))) \rfloor}$.

So any 2-factor approximation algorithm for the MAXIMUM CLIQUE PROBLEM is able to decide if $x \in Y$, i.e. to solve $\mathcal{P}$. Since $\mathcal{P}$ is *NP*-complete, this is possible only if $P = NP$. $\qquad \square$

The reduction in the above proof is due to Feige et al. [1996]. Since the error probability $\frac{1}{2}$ in Definition 16.31 can be replaced by any number between 0 and 1 (Exercise 20), we get that there is no ρ-factor approximation algorithm for the MAXIMUM CLIQUE PROBLEM for any $\rho \geq 1$ (unless $P = NP$).

With more effort Zuckerman [2007] showed that, unless $P = NP$, no polynomial-time algorithm computes the maximum size of a clique in any graph with n vertices up to a factor of $n^{1-\epsilon}$, for any fixed $\epsilon > 0$. The best known algorithm guarantees to find a clique of size $\frac{k \log^3 n}{n (\log \log n)^2}$ in this case (Feige [2004]). Of course, all this also holds for the MAXIMUM STABLE SET PROBLEM (by considering the complement of the given graph).

Now we turn to the following restriction of MAX-SAT:

MAX-3SAT

Instance: A set X of variables and a family $\mathcal{Z}$ of clauses over X, each with exactly three literals.

Task: Find a truth assignment T of X such that the number of clauses in $\mathcal{Z}$ that are satisfied by T is maximum.

In Section 16.5 we had a simple $\frac{8}{7}$-factor approximation algorithm for MAX-3SAT, even for the weighted form (Theorem 16.28). Håstad [2001] showed that this is best possible: no ρ-factor approximation algorithm for MAX-3SAT can exist for any $\rho < \frac{8}{7}$ unless $P = NP$. Here we prove the following weaker result:

Theorem 16.35. (Arora et al. [1998]) *Unless $P = NP$ there is no approximation scheme for* MAX-3SAT.

Proof: Let $\mathcal{P} = (X, Y)$ be some *NP*-complete problem. By the *PCP* Theorem 16.33, $\mathcal{P} \in PCP(\log n, 1)$, so let p, k, f, $\mathcal{P}' := (X', Y')$ be as in Definition 16.31.

For any given $x \in X$ we construct a 3SAT-instance J_x. Namely, for each random string $r \in \{0, 1\}^{\lfloor \log(p(\text{size}(x))) \rfloor}$ we define a family $\mathcal{Z}_r$ of 3SAT-clauses (the union of all these families will be J_x). We first construct a family $\mathcal{Z}_r'$ of clauses with an arbitrary number of literals and then apply Proposition 15.21.

So let $r \in \{0, 1\}^{\lfloor \log(p(\text{size}(x))) \rfloor}$ and $f(x, r) = (t_1, \ldots, t_k)$. Let $\{a^{(1)}, \ldots, a^{(s_r)}\}$ be the set of strings $a \in \{0, 1\}^k$ for which $(x, f(x, r), a) \in Y'$. If $s_r = 0$ then we simply set $\mathcal{Z}' := \{\{y\}, \{\bar{y}\}\}$, where y is some variable not used anywhere else.

Otherwise let $c \in \{0, 1\}^{\lfloor p(\text{size}(x)) \rfloor}$. We have that $(x, f(x, r), c_{f(x,r)}) \in Y'$ if and only if

$$\bigvee_{j=1}^{s_r} \left(\bigwedge_{i=1}^{k} \left(c_{t_i} = a_i^{(j)} \right) \right).$$

This is equivalent to

$$\bigwedge_{(i_1, \ldots, i_{s_r}) \in \{1, \ldots, k\}^{s_r}} \left(\bigvee_{j=1}^{s_r} \left(c_{t_{i_j}} = a_{i_j}^{(j)} \right) \right).$$

This conjunction of clauses can be constructed in polynomial time because $\mathcal{P}' \in P$ and k is a constant. By introducing Boolean variables $\pi_1, \ldots, \pi_{\lfloor p(\text{size}(x)) \rfloor}$ representing the bits $c_1, \ldots, c_{\lfloor p(\text{size}(x)) \rfloor}$ we obtain a family $\mathcal{Z}_r'$ of k^{s_r} clauses (each with s_r literals) such that $\mathcal{Z}_r'$ is satisfied if and only if $(x, f(x, r), c_{f(x,r)}) \in Y'$.

By Proposition 15.21, we can rewrite each $\mathcal{Z}_r'$ equivalently as a conjunction of 3SAT-clauses, where the number of clauses increases by at most a factor of $\max\{s_r - 2, 4\}$. Let this family of clauses be $\mathcal{Z}_r$. Since $s_r \leq 2^k$, each $\mathcal{Z}_r$ consists of at most $l := k^{2^k} \max\{2^k - 2, 4\}$ 3SAT-clauses.

Our 3SAT-instance J_x is the union of all the families $\mathcal{Z}_r$ for all r. Note that J_x can be computed in polynomial time.

Now if x is a yes-instance, then there exists a certificate c as in Definition 16.31. This c immediately defines a truth assignment satisfying J_x.

On the other hand, if x is a no-instance, then only $\frac{1}{2}$ of the formulas $\mathcal{Z}_r$ are simultaneously satisfiable. So in this case any truth assignment leaves at least a fraction of $\frac{1}{2l}$ of the clauses unsatisfied.

So any k-factor approximation algorithm for MAX-3SAT with $k < \frac{2l}{2l-1}$ satisfies more than a fraction of $\frac{2l-1}{2l} = 1 - \frac{1}{2l}$ of the clauses of any satisfiable instance. Hence such an algorithm can decide whether $x \in Y$ or not. Since $\mathcal{P}$ is NP-complete, such an algorithm cannot exist unless $P = NP$. $\square$

16.7 L-Reductions

Our goal is to show, for other problems than MAX-3SAT, that they have no approximation scheme unless $P = NP$. As with the NP-completeness proofs (Section 15.5), it is not necessary to have a direct proof using the definition of $PCP(\log n, 1)$ for each problem. Rather we use a certain type of reduction which preserves approximability (general polynomial transformations do not):

Definition 16.36. *Let* $\mathcal{P} = (X, (S_x)_{x \in X}, c, \text{goal})$ *and* $\mathcal{P}' = (X', (S'_x)_{x \in X'}, c', \text{goal}')$ *be two optimization problems with nonnegative weights. An* **L-reduction** *from* $\mathcal{P}$ *to* $\mathcal{P}'$ *is a pair of functions* f *and* g, *both computable in polynomial time, and two constants* $\alpha, \beta > 0$ *such that for any instance* x *of* $\mathcal{P}$:

(a) $f(x)$ *is an instance of* $\mathcal{P}'$ *with* $\text{OPT}(f(x)) \leq \alpha \, \text{OPT}(x)$;
(b) *For any feasible solution* y' *of* $f(x)$, $g(x, y')$ *is a feasible solution of* x *such that* $|c(x, g(x, y')) - \text{OPT}(x)| \leq \beta |c'(f(x), y') - \text{OPT}(f(x))|$.

We say that $\mathcal{P}$ *is* **L-reducible** *to* $\mathcal{P}'$ *if there is an L-reduction from* $\mathcal{P}$ *to* $\mathcal{P}'$.

The letter "L" in the term L-reduction stands for "linear". L-reductions were introduced by Papadimitriou and Yannakakis [1991]. The definition immediately implies that L-reductions can be composed:

Proposition 16.37. *Let* $\mathcal{P}$, $\mathcal{P}'$, $\mathcal{P}''$ *be optimization problems with nonnegative weights. If* (f, g, α, β) *is an L-reduction from* $\mathcal{P}$ *to* $\mathcal{P}'$ *and* $(f', g', \alpha', \beta')$ *is an L-reduction from* $\mathcal{P}'$ *to* $\mathcal{P}''$, *then their composition* $(f'', g'', \alpha\alpha', \beta\beta')$ *is an L-reduction from* $\mathcal{P}$ *to* $\mathcal{P}''$, *where* $f''(x) = f'(f(x))$ *and* $g''(x, y'') = g(x, g'(x', y''))$. $\quad\square$

The decisive property of L-reductions is that they preserve approximability:

Theorem 16.38. (Papadimitriou and Yannakakis [1991]) *Let* $\mathcal{P}$ *and* $\mathcal{P}'$ *be two optimization problems with nonnegative weights. Let* (f, g, α, β) *be an L-reduction from* $\mathcal{P}$ *to* $\mathcal{P}'$. *If there is an approximation scheme for* $\mathcal{P}'$, *then there is an approximation scheme for* $\mathcal{P}$.

Proof: Given an instance x of $\mathcal{P}$ and a number $0 < \epsilon < 1$, we apply the approximation scheme for $\mathcal{P}'$ to $f(x)$ and $\epsilon' := \frac{\epsilon}{2\alpha\beta}$. We obtain a feasible solution y' of $f(x)$ and finally return $y := g(x, y')$, a feasible solution of x. Since

$$
\begin{aligned}
|c(x, y) - \text{OPT}(x)| \;&\leq\; \beta |c'(f(x), y') - \text{OPT}(f(x))| \\
&\leq\; \beta \max\Big\{ (1 + \epsilon')\,\text{OPT}(f(x)) - \text{OPT}(f(x)), \\
&\qquad\qquad \text{OPT}(f(x)) - \frac{1}{1 + \epsilon'}\,\text{OPT}(f(x)) \Big\} \\
&\leq\; \beta\epsilon'\,\text{OPT}(f(x)) \\
&\leq\; \alpha\beta\epsilon'\,\text{OPT}(x) \\
&=\; \frac{\epsilon}{2}\,\text{OPT}(x)
\end{aligned}
$$

we get

$$
c(x, y) \;\leq\; \text{OPT}(x) + |c(x, y) - \text{OPT}(x)| \;\leq\; \left(1 + \frac{\epsilon}{2}\right) \text{OPT}(x)
$$

and

$$c(x, y) \geq \text{OPT}(x) - |\text{OPT}(x) - c(x, y)| \geq \left(1 - \frac{\epsilon}{2}\right) \text{OPT}(x) > \frac{1}{1 + \epsilon} \text{OPT}(x),$$

so this constitutes an approximation scheme for $\mathcal{P}$. $\square$

This theorem together with Theorem 16.35 motivates the following definition:

Definition 16.39. *An optimization problem $\mathcal{P}$ with nonnegative weights is called* **MAXSNP-hard** *if* MAX-3SAT *is L-reducible to $\mathcal{P}$.*

The name *MAXSNP* refers to a class of optimization problems introduced by Papadimitriou and Yannakakis [1991]. Here we do not need this class, so we omit its (nontrivial) definition.

Corollary 16.40. *Unless* $P = NP$ *there is no approximation scheme for any MAXSNP-hard problem.*

Proof: Directly from Theorems 16.35 and 16.38. $\square$

We shall show *MAXSNP*-hardness for several problems by describing L-reductions. We start with a restricted version of MAX-3SAT:

3-OCCURRENCE MAX-SAT PROBLEM

Instance: A set X of variables and a family $\mathcal{Z}$ of clauses over X, each with at most three literals, such that no variable occurs in more than three clauses.

Task: Find a truth assignment T of X such that the number of clauses in $\mathcal{Z}$ that are satisfied by T is maximum.

That this problem is *NP*-hard can be proved by a simple transformation from 3SAT (or MAX-3SAT), cf. Exercise 11 of Chapter 15. Since this transformation is not an L-reduction, it does not imply *MAXSNP*-hardness. We need a more complicated construction, using so-called expander graphs:

Definition 16.41. *Let G be an undirected graph and $\gamma > 0$ a constant. G is a* γ**-expander** *if for each $A \subseteq V(G)$ with $|A| \leq \frac{|V(G)|}{2}$ we have $|\Gamma(A)| \geq \gamma|A|$.*

For example, a complete graph is a 1-expander. However, one is interested in expanders with a small number of edges. We cite the following theorem without its quite complicated proof:

Theorem 16.42. (Ajtai [1994]) *There exists a positive constant γ such that for any given even integer $n \geq 4$, a 3-regular γ-expander with n vertices can be constructed in $O(n^3 \log^3 n)$ time.*

The following corollary was mentioned (and used) by Papadimitriou [1994], and a correct proof was given by Fernández-Baca and Lagergren [1998]:

Corollary 16.43. *For any given integer $n \geq 3$, a digraph G with $O(n)$ vertices and a set $S \subseteq V(G)$ of cardinality n with the following properties can be constructed in $O(n^3 \log^3 n)$ time:*
$|\delta^-(v)| + |\delta^+(v)| \leq 3$ *for each* $v \in V(G)$;
$|\delta^-(v)| + |\delta^+(v)| = 2$ *for each* $v \in S$; *and*
$|\delta^+(A)| \geq \min\{|S \cap A|, |S \setminus A|\}$ *for each* $A \subseteq V(G)$.

Proof: Let $\gamma > 0$ be the constant of Theorem 16.42, and let $k := \left\lceil \frac{1}{\gamma} \right\rceil$. We first construct a 3-regular γ-expander H with n or $n+1$ vertices, using Theorem 16.42.

We replace each edge $\{v, w\}$ by k parallel edges (v, w) and k parallel edges (w, v). Let the resulting digraph be H'. Note that for any $A \subseteq V(H')$ with $|A| \leq \frac{|V(H')|}{2}$ we have

$$|\delta^+_{H'}(A)| = k|\delta_H(A)| \geq k|\Gamma_H(A)| \geq k\gamma|A| \geq |A|.$$

Similarly we have for any $A \subseteq V(H')$ with $|A| > \frac{|V(H')|}{2}$:

$$|\delta^+_{H'}(A)| = k|\delta_H(V(H') \setminus A)| \geq k|\Gamma_H(V(H') \setminus A)|$$
$$\geq k\gamma|V(H') \setminus A| \geq |V(H') \setminus A|.$$

So in both cases we have $|\delta^+_{H'}(A)| \geq \min\{|A|, |V(H') \setminus A|\}$.

Now we split up each vertex $v \in V(H')$ into $6k+1$ vertices $x_{v,i}$ ($i = 0, \ldots, 6k$) such that each vertex except $x_{v,0}$ has degree 1. For each vertex $x_{v,i}$ we now add vertices $w_{v,i,j}$ and $y_{v,i,j}$ ($j = 0, \ldots, 6k$) connected by a path of length $12k+2$ with vertices $w_{v,i,0}, w_{v,i,1}, \ldots, w_{v,i,6k}, x_{v,i}, y_{v,i,0}, \ldots, y_{v,i,6k}$ in this order. Finally we add edges $(y_{v,i,j}, w_{v,j,i})$ for all $v \in V(H')$, all $i \in \{0, \ldots, 6k\}$ and all $j \in \{0, \ldots, 6k\} \setminus \{i\}$.

Altogether we have a vertex set Z_v of cardinality $(6k + 1)(12k + 3)$ for each $v \in V(H')$. The overall resulting graph G has $|V(H')|(6k + 1)(12k + 3) = O(n)$ vertices, each of degree two or three. By the construction, $G[Z_v]$ contains $\min\{|X_1|, |X_2|\}$ vertex-disjoint paths from X_1 to X_2 for any pair of disjoint subsets X_1, X_2 of $\{x_{v,i} : i = 0, \ldots, 6k\}$.

We choose S to be an n-element subset of $\{x_{v,0} : v \in V(H')\}$; note that each of these vertices has one entering and one leaving edge.

It remains to prove that $|\delta^+(A)| \geq \min\{|S \cap A|, |S \setminus A|\}$ for each $A \subseteq V(G)$. We prove this by induction on $|\{v \in V(H') : \emptyset \neq A \cap Z_v \neq Z_v\}|$. If this number is zero, i.e. $A = \bigcup_{v \in B} Z_v$ for some $B \subseteq V(H')$, then we have

$$|\delta^+_G(A)| = |\delta^+_{H'}(B)| \geq \min\{|B|, |V(H') \setminus B|\} \geq \min\{|S \cap A|, |S \setminus A|\}.$$

Otherwise let $v \in V(H')$ with $\emptyset \neq A \cap Z_v \neq Z_v$. Let $P := \{x_{v,i} : i = 0, \ldots, 6k\} \cap A$ and $Q := \{x_{v,i} : i = 0, \ldots, 6k\} \setminus A$. If $|P| \leq 3k$, then by the property of $G[Z_v]$ we have

$$
\begin{aligned}
|E_G^+(Z_v \cap A, Z_v \setminus A)| &\geq |P| = |P \setminus S| + |P \cap S| \\
&\geq |E_G^+(A \setminus Z_v, A \cap Z_v)| + |P \cap S|.
\end{aligned}
$$

By applying the induction hypothesis to $A \setminus Z_v$ we therefore get

$$
\begin{aligned}
|\delta_G^+(A)| &\geq |\delta_G^+(A \setminus Z_v)| - |E_G^+(A \setminus Z_v, A \cap Z_v)| + |E_G^+(Z_v \cap A, Z_v \setminus A)| \\
&\geq |\delta_G^+(A \setminus Z_v)| + |P \cap S| \\
&\geq \min\{|S \cap (A \setminus Z_v)|, |S \setminus (A \setminus Z_v)|\} + |P \cap S| \\
&\geq \min\{|S \cap A|, |S \setminus A|\}.
\end{aligned}
$$

Similarly, if $|P| \geq 3k + 1$, then $|Q| \leq 3k$ and by the property of $G[Z_v]$ we have

$$
\begin{aligned}
|E_G^+(Z_v \cap A, Z_v \setminus A)| &\geq |Q| = |Q \setminus S| + |Q \cap S| \\
&\geq |E_G^+(Z_v \setminus A, V(G) \setminus (A \cup Z_v))| + |Q \cap S|.
\end{aligned}
$$

By applying the induction hypothesis to $A \cup Z_v$ we therefore get

$$
\begin{aligned}
|\delta_G^+(A)| &\geq |\delta_G^+(A \cup Z_v)| \\
&\quad - |E_G^+(Z_v \setminus A, V(G) \setminus (A \cup Z_v))| + |E_G^+(Z_v \cap A, Z_v \setminus A)| \\
&\geq |\delta_G^+(A \cup Z_v)| + |Q \cap S| \\
&\geq \min\{|S \cap (A \cup Z_v)|, |S \setminus (A \cup Z_v)|\} + |Q \cap S| \\
&\geq \min\{|S \cap A|, |S \setminus A|\}.
\end{aligned}
$$

$\square$

Now we can prove:

Theorem 16.44. (Papadimitriou and Yannakakis [1991], Papadimitriou [1994], Fernández-Baca and Lagergren [1998]) *The 3-OCCURRENCE MAX-SAT PROBLEM is MAXSNP-hard.*

Proof: We describe an L-reduction (f, g, α, β) from MAX-3SAT. To define f, let $(X, \mathcal{Z})$ be an instance of MAX-3SAT. For each variable $x \in X$ which occurs in more than three, say in k clauses, we modify the instance as follows. We replace x by a new and distinct variable in each clause. This way we introduce new variables $x_1, \ldots, x_k$. We introduce additional constraints (and further variables) which ensure, roughly speaking, that it is favourable to assign the same truth value to all the variables $x_1, \ldots, x_k$.

We construct G and S as in Corollary 16.43 and rename the vertices such that $S = \{1, \ldots, k\}$. Now for each vertex $v \in V(G) \setminus S$ we introduce a new variable

x_v, and for each edge $(v, w) \in E(G)$ we introduce a clause $\{\overline{x_v}, x_w\}$. In total we have added at most

$$\frac{3}{2}(k + 1) \left(6 \left\lceil \frac{1}{\gamma} \right\rceil + 1 \right) \left(12 \left\lceil \frac{1}{\gamma} \right\rceil + 3 \right) \leq 315 \left\lceil \frac{1}{\gamma} \right\rceil^2 k$$

new clauses, where γ is again the constant of Theorem 16.42.

Applying the above substitution for each variable we obtain an instance $(X', \mathcal{Z}') = f(X, \mathcal{Z})$ of the 3-OCCURRENCE MAX-SAT PROBLEM with

$$|\mathcal{Z}'| \leq |\mathcal{Z}| + 315 \left\lceil \frac{1}{\gamma} \right\rceil^2 3|\mathcal{Z}| \leq 946 \left\lceil \frac{1}{\gamma} \right\rceil^2 |\mathcal{Z}|.$$

Hence

$$\mathrm{OPT}(X', \mathcal{Z}') \leq |\mathcal{Z}'| \leq 946 \left\lceil \frac{1}{\gamma} \right\rceil^2 |\mathcal{Z}| \leq 1892 \left\lceil \frac{1}{\gamma} \right\rceil^2 \mathrm{OPT}(X, \mathcal{Z}),$$

because at least half of the clauses of a MAX-SAT-instance can be satisfied (either by setting all variables *true* or all *false*). So we can set $\alpha := 1892 \left\lceil \frac{1}{\gamma} \right\rceil^2$.

To describe g, let T' be a truth assignment of X'. We first construct a truth assignment T'' of X' satisfying at least as many clauses of $\mathcal{Z}'$ as T', and satisfying all new clauses (corresponding to edges of the graphs G above). Namely, for any variable x occurring more than three times in $(X, \mathcal{Z})$, let G be the graph constructed above, and let $A := \{v \in V(G) : T'(x_v) = \mathit{true}\}$. If $|S \cap A| \geq |S \setminus A|$ then we set $T''(x_v) := \mathit{true}$ for all $v \in V(G)$, otherwise we set $T''(x_v) := \mathit{false}$ for all $v \in V(G)$. It is clear that all new clauses (corresponding to edges) are satisfied.

There are at most $\min\{|S \cap A|, |S \setminus A|\}$ old clauses satisfied by T' but not by T''. On the other hand, T' does not satisfy any of the clauses $\{\overline{x_v}, x_w\}$ for $(v, w) \in \delta_G^+(A)$. By the properties of G, the number of these clauses is at least $\min\{|S \cap A|, |S \setminus A|\}$.

Now T'' yields a truth assignment $T = g(X, \mathcal{Z}, T')$ of X in the obvious way: Set $T(x) := T''(x) = T'(x)$ for $x \in X \cap X'$ and $T(x) := T''(x_i)$ if x_i is any variable replacing x in the construction from $(X, \mathcal{Z})$ to $(X', \mathcal{Z}')$.

T violates as many clauses as T''. So if $c(X, \mathcal{Z}, T)$ denotes the number of clauses in instance $(X, \mathcal{Z})$ that are satisfied by T, we conclude

$$|\mathcal{Z}| - c(X, \mathcal{Z}, T) = |\mathcal{Z}'| - c(X', \mathcal{Z}', T'') \leq |\mathcal{Z}'| - c(X', \mathcal{Z}', T'). \quad (16.7)$$

On the other hand, any truth assignment T of X leads to a truth assignment T' of X' violating the same number of clauses (by setting the variables x_v ($v \in V(G)$) uniformly to $T(x)$ for each variable x and the corresponding graph G in the above construction). Hence

$$|\mathcal{Z}| - \mathrm{OPT}(X, \mathcal{Z}) \geq |\mathcal{Z}'| - \mathrm{OPT}(X', \mathcal{Z}'). \quad (16.8)$$

Combining (16.7) and (16.8) we get

$$
\begin{aligned}
|\,\mathrm{OPT}(X, \mathcal{Z}) - c(X, \mathcal{Z}, T)| &= (|\mathcal{Z}| - c(X, \mathcal{Z}, T)) - (|\mathcal{Z}| - \mathrm{OPT}(X, \mathcal{Z})) \\
&\leq \mathrm{OPT}(X', \mathcal{Z}') - c(X', \mathcal{Z}', T') \\
&= |\,\mathrm{OPT}(X', \mathcal{Z}') - c(X', \mathcal{Z}', T')|,
\end{aligned}
$$

where $T = g(X, \mathcal{Z}, T')$. So $(f, g, \alpha, 1)$ is indeed an L-reduction. $\qquad\square$

This result is the starting point of several *MAXSNP*-hardness proofs. For example:

Corollary 16.45. (Papadimitriou and Yannakakis [1991]) *The* MAXIMUM STABLE SET PROBLEM *restricted to graphs with maximum degree 4 is MAXSNP-hard.*

Proof: The construction of the proof of Theorem 15.23 defines an L-reduction from the 3-OCCURRENCE MAX-SAT PROBLEM to the MAXIMUM STABLE SET PROBLEM restricted to graphs with maximum degree 4: for each instance $(X, \mathcal{Z})$ a graph G is constructed such that from each truth assignment satisfying k clauses one easily obtains a stable set of cardinality k, and vice versa. $\qquad\square$

Indeed, the MAXIMUM STABLE SET PROBLEM is *MAXSNP*-hard even when restricted to 3-regular graphs (Berman and Fujito [1999]). On the other hand, a simple greedy algorithm, which in each step chooses a vertex v of minimum degree and deletes v and all its neighbours, is a $\frac{(k+2)}{3}$-factor approximation algorithm for the MAXIMUM STABLE SET PROBLEM in graphs with maximum degree k (Halldórsson and Radhakrishnan [1997]). For $k = 4$ this gives a performance ratio of 2 which is better than the ratio 8 we get from the following proof (using the 2-factor approximation algorithm for the MINIMUM VERTEX COVER PROBLEM).

Theorem 16.46. (Papadimitriou and Yannakakis [1991]) *The* MINIMUM VERTEX COVER PROBLEM *restricted to graphs with maximum degree 4 is MAXSNP-hard.*

Proof: Consider the trivial transformation from the MAXIMUM STABLE SET PROBLEM (Proposition 2.2) with $f(G) := G$ and $g(G, X) := V(G) \setminus X$ for all graphs G and all $X \subseteq V(G)$. Although this is not an L-reduction in general, it is an L-reduction if restricted to graphs with maximum degree 4, as we shall show.

If G has maximum degree 4, there exists a stable set of cardinality at least $\frac{|V(G)|}{5}$. So if we denote by $\alpha(G)$ the maximum cardinality of a stable set and by $\tau(G)$ the minimum cardinality of a vertex cover we have

$$
\alpha(G) \geq \frac{1}{4}(|V(G)| - \alpha(G)) = \frac{1}{4}\tau(G)
$$

and $\alpha(G) - |X| = |V(G) \setminus X| - \tau(G)$ for any stable set $X \subseteq V(G)$. Hence $(f, g, 4, 1)$ is an L-reduction. $\qquad\square$

See Clementi and Trevisan [1999] and Chlebík and Chlebíková [2006] for stronger statements. In particular, there is no approximation scheme for the MIN-IMUM VERTEX COVER PROBLEM (unless $P = NP$). We shall prove MAXSNP-hardness of other problems in later chapters; see also Exercise 23.

Exercises

1. Formulate a 2-factor approximation algorithm for the following problem. Given a digraph with edge weights, find an acyclic subgraph of maximum weight. *Note:* No k-factor approximation algorithm for this problem is known for $k < 2$.

2. The k-CENTER PROBLEM is defined as follows: given an undirected graph G, weights $c : E(G) \to \mathbb{R}_+$, and a number $k \in \mathbb{N}$ with $k \leq |V(G)|$, find a set $X \subseteq V(G)$ of cardinality k such that

$$\max_{v \in V(G)} \min_{x \in X} \mathrm{dist}(v, x)$$

   is minimum. As usual we denote the optimum value by $\mathrm{OPT}(G, c, k)$.
   (a) Let S be a maximal stable set in $(V(G), \{\{v, w\} : \mathrm{dist}(v, w) \leq 2R\})$. Show that then $\mathrm{OPT}(G, c, |S| - 1) > R$.
   (b) Use (a) to describe a 2-factor approximation algorithm for the k-CENTER PROBLEM.
      (Hochbaum and Shmoys [1985])
   * (c) Show that there is no r-factor approximation algorithm for the k-CENTER PROBLEM for any $r < 2$.
      Hint: Use Exercise 14 of Chapter 15.
      (Hsu and Nemhauser [1979])

3. Can one find a minimum vertex cover (or a maximum stable set) in a bipartite graph in polynomial time?

4. Show that the performance guarantee in Theorem 16.5 is tight.

5. Show that the following is a 2-factor approximation algorithm for the MINIMUM VERTEX COVER PROBLEM: compute a DFS-tree and output all its vertices with nonzero out-degree.
   (Bar-Yehuda [unpublished])

6. Show that the LP relaxation $\min\{cx : M^\top x \geq \mathbb{1}, x \geq 0\}$ of the MINIMUM WEIGHT VERTEX COVER PROBLEM, where M is the incidence matrix of an undirected graph and $c \in \mathbb{R}_+^{V(G)}$, always has a half-integral optimum solution (i.e. one with entries $0, \frac{1}{2}, 1$ only). Derive another 2-factor approximation algorithm from this fact. Moreover, use it to strengthen Theorem 15.45 to $|V(G')| \leq 2l$.

* 7. Consider the MINIMUM WEIGHT FEEDBACK VERTEX SET PROBLEM: given an undirected graph G and weights $c : V(G) \to \mathbb{R}_+$, find a vertex set $X \subseteq V(G)$ of minimum weight such that $G - X$ is a forest. Consider the following recursive algorithm A:

If $E(G) = \emptyset$, then return $A(G, c) := \emptyset$. If $|\delta_G(x)| \leq 1$ for some $x \in V(G)$, then return $A(G, c) := A(G - x, c)$. If $c(x) = 0$ for some $x \in V(G)$, then return $A(G, c) := \{x\} \cup A(G - x, c)$. Otherwise let

$$\epsilon := \min_{x \in V(G)} \frac{c(v)}{|\delta(v)|}$$

and $c'(v) := c(v) - \epsilon|\delta(v)|$ $(v \in V(G))$. Let $X := A(G, c')$. For each $x \in X$ do: If $G - (X \setminus \{x\})$ is a forest, then set $X := X \setminus \{x\}$. Return $A(G, c) := x$. Prove that this is a 2-factor approximation algorithm for the MINIMUM WEIGHT FEEDBACK VERTEX SET PROBLEM.
(Becker and Geiger [1996])

8. Show that the MAXIMUM CUT PROBLEM is NP-hard even for simple graphs.
9. Prove that the simple greedy algorithm for MAX-CUT described at the beginning of Section 16.2 is a 2-factor approximation algorithm.
10. Consider the following local search algorithm for the MAXIMUM CUT PROBLEM. Start with any nonempty proper subset S of $V(G)$. Now check iteratively if some vertex can be added to S or deleted from S such that $|\delta(S)|$ increases. Stop if no such improvement is possible.
    (a) Prove that the above is a 2-factor approximation algorithm. (Recall Exercise 14 of Chapter 2.)
    (b) Can the algorithm be extended to the MAXIMUM WEIGHT CUT PROBLEM, where we have nonnegative edge weights?
    (c) Does the algorithm always find an optimum solution for planar graphs, or for bipartite graphs? For both classes there is a polynomial-time algorithm (Exercise 7 of Chapter 12 and Proposition 2.27).
11. In the DIRECTED MAXIMUM WEIGHT CUT PROBLEM we are given a digraph G with weights $c : E(G) \to \mathbb{R}_+$, and we look for a set $X \subseteq V(G)$ such that $\sum_{e \in \delta^+(X)} c(e)$ is maximum. Show that there is a 4-factor approximation algorithm for this problem.
    Hint: Use Exercise 10.
    Note: There is a 1.165-factor but no 1.09-factor approximation algorithm unless $P = NP$ (Feige and Goemans [1995], Håstad [2001]).
12. Give a solution of (16.2) for the unweighted pentagon C_5. What are the solutions that the GOEMANS-WILLIAMSON MAX-CUT ALGORITHM can return for this instance?
13. Show that $(\frac{1}{\pi} \arccos(y_i^\top y_j))_{1 \leq i, j \leq n}$ is a convex combination of cut semimetrics δ^R, $R \subseteq \{1, \ldots, n\}$, where $\delta^R_{i,j} = 1$ if $|R \cap \{i, j\}| = 1$ and $\delta^R_{i,j} = 0$ otherwise.
    Hint: Write

$$(\mu(H(y_i) \triangle H(y_j)))_{1 \leq i, j \leq n} = \sum_{R \subseteq \{1, \ldots, n\}} \mu \left(\bigcap_{i \in R} H(y_i) \setminus \bigcup_{i \notin R} H(y_i) \right) \delta^R.$$

Note: See Deza and Laurent [1997] for a lot of related information.

14. Show that for each $n \in \mathbb{N}$ there is a bipartite graph on $2n$ vertices for which the GREEDY COLOURING ALGORITHM needs n colours. So the algorithm may give arbitrarily bad results. However, show that there always exists an order of the vertices for which the algorithm finds an optimum colouring.

15. Show that one can colour any 3-colourable graph G with at most $2\sqrt{2n}$ colours in polynomial time, where $n := |V(G)|$.
    Hint: As long as there is a vertex v of degree at least $\sqrt{2n}$, colour $\Gamma(v)$ optimally with at most two colours (not to be used again), and delete these vertices. Finally use the GREEDY COLOURING ALGORITHM.
    (Wigderson [1983])

16. Show that the following graphs are perfect:
    (a) bipartite graphs;
    (b) interval graphs: $(\{v_1, \ldots, v_n\}, \{\{v_i, v_j\} : i \neq j, [a_i, b_i] \cap [a_j, b_j] \neq \emptyset\})$, where $[a_1, b_1], \ldots, [a_n, b_n]$ is a set of closed intervals;
    (c) chordal graphs (see Exercise 38 of Chapter 8).

∗ 17. Let G be an undirected graph. Prove that the following statements are equivalent:
    (a) G is perfect.
    (b) For any weight function $c : V(G) \to \mathbb{Z}_+$ the maximum weight of a clique in G equals the minimum number of stable sets such that each vertex v is contained in $c(v)$ of them.
    (c) For any weight function $c : V(G) \to \mathbb{Z}_+$ the maximum weight of a stable set in G equals the minimum number of cliques such that each vertex v is contained in $c(v)$ of them.
    (d) The inequality system defining (16.3) is TDI.
    (e) The clique polytope of G, i.e. the convex hull of the incidence vectors of all cliques in G, is given by

    $$\left\{ x \in \mathbb{R}_+^{V(G)} : \sum_{v \in S} x_v \leq 1 \text{ for all stable sets } S \text{ in } G \right\}. \qquad (16.9)$$

    (f) The inequality system defining (16.9) is TDI.
    Note: The polytope (16.9) is called the antiblocker of the polytope (16.3).

18. An instance of MAX-SAT is called k-satisfiable if any k of its clauses can be simultaneously satisfied. Let r_k be the infimum of the fraction of clauses that one can satisfy in any k-satisfiable instance.
    (a) Prove that $r_1 = \frac{1}{2}$.
    (b) Prove that $r_2 = \frac{\sqrt{5}-1}{2}$.
        (*Hint:* Some variables occur in one-element clauses (w.l.o.g. all one-element clauses are positive), set them *true* with probability a (for some $\frac{1}{2} < a < 1$), and set the other variables *true* with probability $\frac{1}{2}$. Apply the derandomization technique and choose a appropriately.)
    (c) Prove that $r_3 \geq \frac{2}{3}$.
    (Lieberherr and Specker [1981])

19. Erdős [1967] showed the following: for each constant $k \in \mathbb{N}$, the (asymptotically) best fraction of the edges that we can guarantee to be in a maximum cut is $\frac{1}{2}$, even if we restrict attention to graphs without odd circuits of length k or less. (Compare Exercise 10(a).)
    (a) What about $k = \infty$?
    (b) Show how the MAXIMUM CUT PROBLEM can be reduced to MAX-SAT.
        Hint: Use a variable for each vertex and two clauses $\{x, y\}$, $\{\bar{x}, \bar{y}\}$ for each edge $\{x, y\}$.
    (c) Use (b) and Erdős' Theorem in order to prove that $r_k \le \frac{3}{4}$ for all k. (For a definition of r_k, see Exercise 18.)
    Note: Trevisan [2004] proved that $\lim_{k \to \infty} r_k = \frac{3}{4}$.
20. Prove that the error probability $\frac{1}{2}$ in Definition 16.31 can be replaced equivalently by any number between 0 and 1. Deduce from this (and the proof of Theorem 16.34) that there is no ρ-factor approximation algorithm for the MAXIMUM CLIQUE PROBLEM for any $\rho \ge 1$ (unless $P = NP$).
21. Prove that the MAXIMUM CLIQUE PROBLEM is L-reducible to the SET PACKING PROBLEM: given a set system $(U, \mathcal{S})$, find a maximum cardinality subfamily $\mathcal{R} \subseteq \mathcal{S}$ whose elements are pairwise disjoint.
22. Prove that the MINIMUM VERTEX COVER PROBLEM has no absolute approximation algorithm (unless $P = NP$).
23. Prove that MAX-2SAT is *MAXSNP*-hard.
    Hint: Use Corollary 16.45.
    (Papadimitriou and Yannakakis [1991])

References

General Literature

Asano, T., Iwama, K., Takada, H., and Yamashita, Y. [2000]: Designing high-quality approximation algorithms for combinatorial optimization problems. IEICE Transactions on Communications/Electronics/Information and Systems E83-D (2000), 462–478

Ausiello, G., Crescenzi, P., Gambosi, G., Kann, V., Marchetti-Spaccamela, A., and Protasi, M. [1999]: Complexity and Approximation: Combinatorial Optimization Problems and Their Approximability Properties. Springer, Berlin 1999

Garey, M.R., and Johnson, D.S. [1979]: Computers and Intractability; A Guide to the Theory of *NP*-Completeness. Freeman, San Francisco 1979, Chapter 4

Hochbaum, D.S. [1996]: Approximation Algorithms for *NP*-Hard Problems. PWS, Boston, 1996

Horowitz, E., and Sahni, S. [1978]: Fundamentals of Computer Algorithms. Computer Science Press, Potomac 1978, Chapter 12

Shmoys, D.B. [1995]: Computing near-optimal solutions to combinatorial optimization problems. In: Combinatorial Optimization; DIMACS Series in Discrete Mathematics and Theoretical Computer Science 20 (W. Cook, L. Lovász, P. Seymour, eds.), AMS, Providence 1995

Papadimitriou, C.H. [1994]: Computational Complexity, Addison-Wesley, Reading 1994, Chapter 13

Vazirani, V.V. [2001]: Approximation Algorithms. Springer, Berlin, 2001

Williamson, D.P., and Shmoys, D.B. [2011]: The Design of Approximation Algorithms. Cambridge University Press, Cambridge 2011

Cited References

Ajtai, M. [1994]: Recursive construction for 3-regular expanders. Combinatorica 14 (1994), 379–416

Alizadeh, F. [1995]: Interior point methods in semidefinite programming with applications to combinatorial optimization. SIAM Journal on Optimization 5 (1995), 13–51

Appel, K., and Haken, W. [1977]: Every planar map is four colorable; Part I; Discharging. Illinois Journal of Mathematics 21 (1977), 429–490

Appel, K., Haken, W., and Koch, J. [1977]: Every planar map is four colorable; Part II; Reducibility. Illinois Journal of Mathematics 21 (1977), 491–567

Arora, S. [1994]: Probabilistic checking of proofs and the hardness of approximation problems, Ph.D. thesis, U.C. Berkeley, 1994

Arora, S., and Kale, S. [2016]: A combinatorial, primal-dual approach to semidefinite programs. Journal of the ACM 63 (2016), Article 12

Arora, S., Lund, C., Motwani, R., Sudan, M., and Szegedy, M. [1998]: Proof verification and hardness of approximation problems. Journal of the ACM 45 (1998), 501–555

Arora, S., and Safra, S. [1998]: Probabilistic checking of proofs. Journal of the ACM 45 (1998), 70–122

Asano, T. [2006]: An improved analysis of Goemans and Williamson's LP-relaxation for MAX SAT. Theoretical Computer Science 354 (2006), 339–353

Avidor, A., Berkovitch, I., and Zwick, U. [2006]: Improved approximation algorithms for MAX NAE-SAT and MAX SAT. Approximation and Online Algorithms – Proceedings of the 3rd WAOA workshop (2005); LNCS 3879 (Erlebach, T., Persiano, G., eds.), Springer, Berlin 2006, pp. 27–40

Bar-Yehuda, R., and Even, S. [1981]: A linear-time approximation algorithm for the weighted vertex cover problem. Journal of Algorithms 2 (1981), 198–203

Becker, A., and Geiger, D. [1996]: Optimization of Pearl's method of conditioning and greedy-like approximation algorithms for the vertex feedback set problem. Artificial Intelligence Journal 83 (1996), 1–22

Bellare, M., and Sudan, M. [1994]: Improved non-approximability results. Proceedings of the 26th Annual ACM Symposium on the Theory of Computing (1994), 184–193

Bellare, M., Goldreich, O., and Sudan, M. [1998]: Free bits, PCPs and nonapproximability – towards tight results. SIAM Journal on Computing 27 (1998), 804–915

Berge, C. [1961]: Färbung von Graphen, deren sämtliche bzw. deren ungerade Kreise starr sind. Wissenschaftliche Zeitschrift, Martin Luther Universität Halle-Wittenberg, Mathematisch-Naturwissenschaftliche Reihe (1961), 114–115

Berge, C. [1962]: Sur une conjecture relative au problème des codes optimaux. Communication, 13ème assemblée générale de l'URSI, Tokyo 1962

Berman, P., and Fujito, T. [1999]: On approximation properties of the independent set problem for low degree graphs. Theory of Computing Systems 32 (1999), 115–132

Brooks, R.L. [1941]: On colouring the nodes of a network. Proceedings of the Cambridge Philosophical Society 37 (1941), 194–197

Chen, J., Friesen, D.K., and Zheng, H. [1999]: Tight bound on Johnson's algorithm for maximum satisfiability. Journal of Computer and System Sciences 58 (1999), 622–640

Chlebík, M. and Chlebíková, J. [2006]: Complexity of approximating bounded variants of optimization problems. Theoretical Computer Science 354 (2006), 320–338

Chudnovsky, M., Cornuéjols, G., Liu, X., Seymour, P., and Vušković, K. [2005]: Recognizing Berge graphs. Combinatorica 25 (2005), 143–186

Chudnovsky, M., Robertson, N., Seymour, P., and Thomas, R. [2006]: The strong perfect graph theorem. Annals of Mathematics 164 (2006), 51–229

Chvátal, V. [1975]: On certain polytopes associated with graphs. Journal of Combinatorial Theory B 18 (1975), 138–154

Chvátal, V. [1979]: A greedy heuristic for the set cover problem. Mathematics of Operations Research 4 (1979), 233–235

Clementi, A.E.F., and Trevisan, L. [1999]: Improved non-approximability results for minimum vertex cover with density constraints. Theoretical Computer Science 225 (1999), 113–128

Deza, M.M., and Laurent, M. [1997]: Geometry of Cuts and Metrics. Springer, Berlin 1997

Dinur, I. [2007]: The PCP theorem by gap amplification. Journal of the ACM 54 (2007), Article 12

Dinur, I., and Safra, S. [2002]: On the hardness of approximating minimum vertex cover. Annals of Mathematics 162 (2005), 439–485

Erdős, P. [1967]: On bipartite subgraphs of graphs. Mat. Lapok. 18 (1967), 283–288

Feige, U. [1998]: A threshold of $\ln n$ for approximating set cover. Journal of the ACM 45 (1998), 634–652

Feige, U. [2004]: Approximating maximum clique by removing subgraphs. SIAM Journal on Discrete Mathematics 18 (2004), 219–225

Feige, U., and Goemans, M.X. [1995]: Approximating the value of two prover proof systems, with applications to MAX 2SAT and MAX DICUT. Proceedings of the 3rd Israel Symposium on Theory of Computing and Systems (1995), 182–189

Feige, U., Goldwasser, S., Lovász, L., Safra, S., and Szegedy, M. [1996]: Interactive proofs and the hardness of approximating cliques. Journal of the ACM 43 (1996), 268–292

Fernández-Baca, D., and Lagergren, J. [1998]: On the approximability of the Steiner tree problem in phylogeny. Discrete Applied Mathematics 88 (1998), 129–145

Fulkerson, D.R. [1972]: Anti-blocking polyhedra. Journal of Combinatorial Theory B 12 (1972), 50–71

Fürer, M., and Raghavachari, B. [1994]: Approximating the minimum-degree Steiner tree to within one of optimal. Journal of Algorithms 17 (1994), 409–423

Garey, M.R., Johnson, D.S., and Stockmeyer, L. [1976]: Some simplified NP-complete graph problems. Theoretical Computer Science 1 (1976), 237–267

Goemans, M.X., and Williamson, D.P. [1994]: New 3/4-approximation algorithms for the maximum satisfiability problem. SIAM Journal on Discrete Mathematics 7 (1994), 656–666

Goemans, M.X., and Williamson, D.P. [1995]: Improved approximation algorithms for maximum cut and satisfiability problems using semidefinite programming. Journal of the ACM 42 (1995), 1115–1145

Grötschel, M., Lovász, L., and Schrijver, A. [1988]: Geometric Algorithms and Combinatorial Optimization. Springer, Berlin 1988

Halldórsson, M.M., and Radhakrishnan, J. [1997]: Greed is good: approximating independent sets in sparse and bounded degree graphs. Algorithmica 18 (1997), 145–163

Håstad, J. [2001]: Some optimal inapproximability results. Journal of the ACM 48 (2001), 798–859

Heawood, P.J. [1890]: Map colour theorem. Quarterly Journal of Pure Mathematics 24 (1890), 332–338

Held, S., Cook, W., and Sewell, E.C. [2012]: Maximum-weight stable sets and safe lower bounds for graph coloring. Mathematical Programming Computation 4 (2012), 363–381

Hochbaum, D.S. [1982]: Approximation algorithms for the set covering and vertex cover problems. SIAM Journal on Computing 11 (1982), 555–556

Hochbaum, D.S., and Shmoys, D.B. [1985]: A best possible heuristic for the k-center problem. Mathematics of Operations Research 10 (1985), 180–184

Holyer, I. [1981]: The NP-completeness of edge-coloring. SIAM Journal on Computing 10 (1981), 718–720

Hougardy, S., Prömel, H.J., and Steger, A. [1994]: Probabilistically checkable proofs and their consequences for approximation algorithms. Discrete Mathematics 136 (1994), 175–223

Hsu, W.L., and Nemhauser, G.L. [1979]: Easy and hard bottleneck location problems. Discrete Applied Mathematics 1 (1979), 209–216

Johnson, D.S. [1974]: Approximation algorithms for combinatorial problems. Journal of Computer and System Sciences 9 (1974), 256–278

Khanna, S., Linial, N., and Safra, S. [2000]: On the hardness of approximating the chromatic number. Combinatorica 20 (2000), 393–415

Khot, S., and Regev, O. [2008]: Vertex cover might be hard to approximate to within $2 - \epsilon$. Journal of Computer and System Sciences 74 (2008), 335–349

Knuth, D.E. [1969]: The Art of Computer Programming; Vol. 2. Seminumerical Algorithms. Addison-Wesley, Reading 1969 (third edition: 1997)

König, D. [1916]: Über Graphen und ihre Anwendung auf Determinantentheorie und Mengenlehre. Mathematische Annalen 77 (1916), 453–465

Lieberherr, K., and Specker, E. [1981]: Complexity of partial satisfaction. Journal of the ACM 28 (1981), 411–421

Lovász, L. [1972]: Normal hypergraphs and the perfect graph conjecture. Discrete Mathematics 2 (1972), 253–267

Lovász, L. [1975]: On the ratio of optimal integral and fractional covers. Discrete Mathematics 13 (1975), 383–390

Lovász, L. [1979a]: On the Shannon capacity of a graph. IEEE Transactions on Information Theory 25 (1979), 1–7

Lovász, L. [1979b]: Graph theory and integer programming. In: Discrete Optimization I; Annals of Discrete Mathematics 4 (P.L. Hammer, E.L. Johnson, B.H. Korte, eds.), North-Holland, Amsterdam 1979, pp. 141–158

Lovász, L. [2003]: Semidefinite programs and combinatorial optimization. In: Recent Advances in Algorithms and Combinatorics (B.A. Reed, C. Linhares Sales, eds.), Springer, New York 2003, pp. 137–194

Mahajan, S., and Ramesh, H. [1999]: Derandomizing approximation algorithms based on semidefinite programming. SIAM Journal on Computing 28 (1999), 1641–1663

Papadimitriou, C.H., and Steiglitz, K. [1982]: Combinatorial Optimization; Algorithms and Complexity. Prentice-Hall, Englewood Cliffs 1982, pp. 406–408

Papadimitriou, C.H., and Yannakakis, M. [1991]: Optimization, approximation, and complexity classes. Journal of Computer and System Sciences 43 (1991), 425–440

Papadimitriou, C.H., and Yannakakis, M. [1993]: The traveling salesman problem with distances one and two. Mathematics of Operations Research 18 (1993), 1–12

Raghavan, P., and Thompson, C.D. [1987]: Randomized rounding: a technique for provably good algorithms and algorithmic proofs. Combinatorica 7 (1987), 365–374

Raz, R., and Safra, S. [1997]: A sub constant error probability low degree test, and a sub constant error probability *PCP* characterization of *NP*. Proceedings of the 29th Annual ACM Symposium on Theory of Computing (1997), 475–484

Robertson, N., Sanders, D.P., Seymour, P., and Thomas, R. [1997]: The four colour theorem. Journal of Combinatorial Theory B 70 (1997), 2–44

Sanders, P., and Steurer, D. [2008]: An asymptotic approximation scheme for multigraph edge coloring. ACM Transactions on Algorithms 4 (2008), Article 21

Singh, M. and Lau, L.C. [2015]: Approximating minimum bounded degree spanning trees to within one of optimal. Journal of the ACM 62 (2015), Article 1

Slavík, P. [1997]: A tight analysis of the greedy algorithm for set cover. Journal of Algorithms 25 (1997), 237–254

Stockmeyer, L.J. [1973]: Planar 3-colorability is polynomial complete. ACM SIGACT News 5 (1973), 19–25

Trevisan, L. [2004]: On local versus global satisfiability. SIAM Journal on Discrete Mathematics 17 (2004), 541–547

Vizing, V.G. [1964]: On an estimate of the chromatic class of a p-graph. Diskret. Analiz 3 (1964), 23–30 [in Russian]

Wigderson, A. [1983]: Improving the performance guarantee for approximate graph coloring. Journal of the ACM 30 (1983), 729–735

Yannakakis, M. [1994]: On the approximation of maximum satisfiability. Journal of Algorithms 17 (1994), 475–502

Zuckerman, D. [2007]: Linear degree extractors and the inapproximability of Max Clique and Chromatic Number. Theory of Computing 3 (2007), 103–128

17 The Knapsack Problem

The MINIMUM WEIGHT PERFECT MATCHING PROBLEM and the WEIGHTED MATROID INTERSECTION PROBLEM discussed in earlier chapters are among the "hardest" problems for which a polynomial-time algorithm is known. In this chapter we deal with the following problem which turns out to be, in a sense, the "easiest" *NP*-hard problem:

KNAPSACK PROBLEM

Instance: Nonnegative integers n, $c_1, \ldots, c_n$, $w_1, \ldots, w_n$ and W.

Task: Find a subset $S \subseteq \{1, \ldots, n\}$ such that $\sum_{j \in S} w_j \leq W$ and $\sum_{j \in S} c_j$ is maximum.

Applications arise whenever we want to select an optimum subset of bounded weight from a set of elements each of which has a weight and a profit.

We start by considering the fractional version in Section 17.1, which turns out to be solvable in linear time. The integral knapsack problem is *NP*-hard as shown in Section 17.2, but a pseudopolynomial algorithm solves it optimally. Combined with a rounding technique this can be used to design a fully polynomial approximation scheme, which is the subject of Section 17.3. In Section 17.4 we discuss a multidimensional generalization. Finally, in Section 17.5 we discuss an algorithm that is efficient in practice, and explain why.

17.1 Fractional Knapsack and Weighted Median Problem

We consider the following problem:

FRACTIONAL KNAPSACK PROBLEM

Instance: Nonnegative integers n, $c_1, \ldots, c_n$, $w_1, \ldots, w_n$ and W.

Task: Find numbers $x_1, \ldots, x_n \in [0, 1]$ such that $\sum_{j=1}^{n} x_j w_j \leq W$ and $\sum_{j=1}^{n} x_j c_j$ is maximum.

© Springer-Verlag GmbH Germany 2018

Bernhard Korte and Jens Vygen, *Combinatorial Optimization*, Algorithms and Combinatorics 21, https://doi.org/10.1007/978-3-662-56039-6_17

The following observation suggests a simple algorithm which requires sorting the elements appropriately:

Proposition 17.1. (Dantzig [1957]) *Let $c_1, \ldots, c_n$, $w_1, \ldots, w_n$ and W be nonnegative integers with $\sum_{i=1}^{n} w_i > W$, $\{1 \leq i \leq n : w_i = 0\} = \{1, \ldots, h\}$, and*

$$\frac{c_{h+1}}{w_{h+1}} \geq \frac{c_{h+2}}{w_{h+2}} \geq \cdots \geq \frac{c_n}{w_n},$$

and let

$$k := \min \left\{ j \in \{1, \ldots, n\} : \sum_{i=1}^{j} w_i > W \right\}.$$

Then an optimum solution of the given instance of the FRACTIONAL KNAPSACK PROBLEM *is defined by*

$$\begin{aligned}
x_j &:= 1 & \text{for } j = 1, \ldots, k-1, \\
x_k &:= \frac{W - \sum_{j=1}^{k-1} w_j}{w_k}, \\
x_j &:= 0 & \text{for } j = k+1, \ldots, n.
\end{aligned}$$

$\square$

Sorting the elements takes $O(n \log n)$ time (Theorem 1.5), and computing k can be done in $O(n)$ time by simple linear scanning. Although this algorithm is quite fast, one can do even better. Observe that the problem reduces to a weighted median search:

Definition 17.2. *Let $n \in \mathbb{N}$, $z_1, \ldots, z_n \in \mathbb{R}$, $w_1, \ldots, w_n \in \mathbb{R}_+$ and $W \in \mathbb{R}$ with $0 < W \leq \sum_{i=1}^{n} w_i$. Then the $(w_1, \ldots, w_n; W)$-**weighted median with respect to** $(z_1, \ldots, z_n)$ is defined to be the unique number z^* for which*

$$\sum_{i:z_i < z^*} w_i < W \leq \sum_{i:z_i \leq z^*} w_i.$$

So we have to solve the following problem:

WEIGHTED MEDIAN PROBLEM

Instance: A natural number n, numbers $z_1, \ldots, z_n \in \mathbb{R}$, $w_1, \ldots, w_n \in \mathbb{R}_+$, and a number W with $0 < W \leq \sum_{i=1}^{n} w_i$.

Task: Find the $(w_1, \ldots, w_n; W)$-weighted median with respect to $(z_1, \ldots, z_n)$.

An important special case is the following:

SELECTION PROBLEM

Instance: A natural number n, numbers $z_1, \ldots, z_n \in \mathbb{R}$, and an integer $k \in \{1, \ldots, n\}$.

Task: Find the k-smallest number among $z_1, \ldots, z_n$; more precisely: an index $i \in \{1, \ldots, n\}$ with $|\{j : z_j < z_i\}| < k \leq |\{j : z_j \leq z_i\}|$.

The weighted median can be determined in $O(n)$ time: the following algorithm is a weighted version of the one by Blum et al. [1973]; see also Vygen [1997].

WEIGHTED MEDIAN ALGORITHM

Input: A natural number n, numbers $z_1, \ldots, z_n \in \mathbb{R}$, $w_1, \ldots, w_n \in \mathbb{R}_+$, and a number W with $0 < W \leq \sum_{i=1}^{n} w_i$.

Output: The $(w_1, \ldots, w_n; W)$-weighted median with respect to $(z_1, \ldots, z_n)$.

① Partition the list $z_1, \ldots, z_n$ into blocks of five elements each (the last block may contain less elements).
Find (recursively) the (non-weighted) median of each block.
Let M be the list of these $\lceil \frac{n}{5} \rceil$ median elements.

② Find (recursively) the non-weighted median of M, let it be z_m.

③ Compare each element with z_m. W.l.o.g. let $z_i < z_m$ for $i = 1, \ldots, k$,
$z_i = z_m$ for $i = k+1, \ldots, l$ and $z_i > z_m$ for $i = l+1, \ldots, n$.

④ **If** $\sum_{i=1}^{k} w_i < W \leq \sum_{i=1}^{l} w_i$ **then** stop ($z^* := z_m$).

If $\sum_{i=1}^{l} w_i < W$ **then** find recursively the $\left(w_{l+1}, \ldots, w_n; W - \sum_{i=1}^{l} w_i \right)$-weighted median with respect to $(z_{l+1}, \ldots, z_n)$. **Stop.**

If $\sum_{i=1}^{k} w_i \geq W$ **then** find recursively the $\left(w_1, \ldots, w_k; W \right)$-weighted median with respect to $(z_1, \ldots, z_k)$. **Stop.**

Theorem 17.3. *The* WEIGHTED MEDIAN ALGORITHM *works correctly and takes* $O(n)$ *time only.*

Proof: The correctness is easily checked. Let us denote the worst-case running time for n elements by $f(n)$. We obtain

$$f(n) = O(n) + f\left(\left\lceil \frac{n}{5} \right\rceil\right) + O(n) + f\left(\frac{1}{2}\left\lceil \frac{n}{5} \right\rceil 5 + \frac{1}{2}\left\lceil \frac{n}{5} \right\rceil 2\right),$$

because the recursive call in ④ omits at least three elements out of at least half of the five-element blocks. The above recursion formula yields $f(n) = O(n)$: as

$\lceil \frac{n}{5} \rceil \le \frac{9}{41}n$ for all $n \ge 37$, one obtains $f(n) \le cn + f\left(\frac{9}{41}n\right) + f\left(\frac{7}{2}\frac{9}{41}n\right)$ for a suitable c and $n \ge 37$. Given this, $f(n) \le (82c + f(36))n$ can be verified easily by induction. So indeed the overall running time is linear. $\square$

We immediately obtain the following corollaries:

Corollary 17.4.   (Blum et al. [1973])   *The* SELECTION PROBLEM *can be solved in $O(n)$ time.*

Proof: Set $w_i := 1$ for $i = 1, \ldots, n$ and $W := k$ and apply Theorem 17.3. $\square$

Corollary 17.5.   *The* FRACTIONAL KNAPSACK PROBLEM *can be solved in linear time.*

Proof: Setting $z_i := -\frac{c_i}{w_i}$ $(i = 1, \ldots, n)$ reduces the FRACTIONAL KNAPSACK PROBLEM to the WEIGHTED MEDIAN PROBLEM. $\square$

17.2 A Pseudopolynomial Algorithm

We now turn to the (integral) KNAPSACK PROBLEM. The techniques of the previous section are also of some use here:

Proposition 17.6.   *Let $c_1, \ldots, c_n, w_1, \ldots, w_n$ and W be nonnegative integers with $w_j \le W$ for $j = 1, \ldots, n$, $\sum_{i=1}^{n} w_i > W$, and*

$$\frac{c_1}{w_1} \ge \frac{c_2}{w_2} \ge \cdots \ge \frac{c_n}{w_n}.$$

Let

$$k := \min\left\{ j \in \{1, \ldots, n\} : \sum_{i=1}^{j} w_i > W \right\}.$$

Then choosing the better of the two feasible solutions $\{1, \ldots, k-1\}$ and $\{k\}$ constitutes a 2-factor approximation algorithm for the KNAPSACK PROBLEM with running time $O(n)$.

Proof: Given any instance of the KNAPSACK PROBLEM, elements $i \in \{1, \ldots, n\}$ with $w_i > W$ are of no use and can be deleted beforehand. Now if $\sum_{i=1}^{n} w_i \le W$, then $\{1, \ldots, n\}$ is an optimum solution. Otherwise we compute the number k in $O(n)$ time without sorting: this is just a WEIGHTED MEDIAN PROBLEM as above (Theorem 17.3).

By Proposition 17.1, $\sum_{i=1}^{k} c_i$ is an upper bound on the optimum value of the FRACTIONAL KNAPSACK PROBLEM, hence also for the integral KNAPSACK PROBLEM. Therefore the better of the two feasible solutions $\{1, \ldots, k-1\}$ and $\{k\}$ achieves at least half the optimum value. $\square$

But we are more interested in an exact solution of the KNAPSACK PROBLEM. However, we have to make the following observation:

Theorem 17.7. (Karp [1972]) *The* KNAPSACK PROBLEM *is NP-hard.*

Proof: We prove that the related decision problem defined as follows is *NP*-complete: given nonnegative integers n, $c_1, \ldots, c_n$, $w_1, \ldots, w_n$, W and K, is there a subset $S \subseteq \{1, \ldots, n\}$ such that $\sum_{j \in S} w_j \leq W$ and $\sum_{j \in S} c_j \geq K$?

This decision problem obviously belongs to *NP*. To show that it is *NP*-complete, we transform SUBSET-SUM (see Corollary 15.27) to it. Given an instance $c_1, \ldots, c_n$, K of SUBSET-SUM, define $w_j := c_j$ $(j = 1, \ldots, n)$ and $W := K$. Obviously this yields an equivalent instance of the above decision problem. $\square$

Since we have not shown the KNAPSACK PROBLEM to be strongly *NP*-hard there is hope for a pseudopolynomial algorithm. Indeed, the algorithm given in the proof of Theorem 15.39 can easily be generalized by introducing weights on the edges and solving a shortest path problem. This leads to an algorithm with running time $O(nW)$ (Exercise 7).

By a similar trick we can also get an algorithm with an $O(nC)$ running time, where $C := \sum_{j=1}^{n} c_j$. We describe this algorithm in a direct way, without constructing a graph and referring to shortest paths. Since the algorithm is based on simple recursion formulas we speak of a dynamic programming algorithm. It is basically due to Bellman [1956, 1957] and Dantzig [1957].

DYNAMIC PROGRAMMING KNAPSACK ALGORITHM

Input: Nonnegative integers n, $c_1, \ldots, c_n$, $w_1, \ldots, w_n$ and W.

Output: A subset $S \subseteq \{1, \ldots, n\}$ such that $\sum_{j \in S} w_j \leq W$ and $\sum_{j \in S} c_j$ is maximum.

---

①   Let C be any upper bound on the value of the optimum solution, e.g.
$C := \sum_{j=1}^{n} c_j$.

②   Set $x(0, 0) := 0$ and $x(0, k) := \infty$ for $k = 1, \ldots, C$.

③   **For** $j := 1$ **to** n **do:**
    For $k := 0$ **to** C **do:**
        Set $s(j, k) := 0$ and $x(j, k) := x(j - 1, k)$.
    For $k := c_j$ **to** C **do:**
        If $x(j - 1, k - c_j) + w_j \leq \min\{W, x(j, k)\}$ **then:**
            Set $x(j, k) := x(j - 1, k - c_j) + w_j$ and $s(j, k) := 1$.

④   Let $k = \max\{i \in \{0, \ldots, C\} : x(n, i) < \infty\}$. Set $S := \emptyset$.
    For $j := n$ **down to** 1 **do:**
        If $s(j, k) = 1$ **then** set $S := S \cup \{j\}$ and $k := k - c_j$.

Theorem 17.8. *The* DYNAMIC PROGRAMMING KNAPSACK ALGORITHM *finds an optimum solution in* $O(nC)$ *time.*

Proof: The running time is obvious.

The variable $x(j, k)$ denotes the minimum total weight of a subset $S \subseteq \{1, \ldots, j\}$ with $\sum_{i \in S} w_i \leq W$ and $\sum_{i \in S} c_i = k$. The algorithm correctly computes these values using the recursion formulas

$$
x(j, k) =
\begin{cases}
x(j-1, k-c_j) + w_j & \text{if } c_j \leq k \text{ and} \\
& x(j-1, k-c_j) + w_j \leq \min\{W, x(j-1, k)\} \\
x(j-1, k) & \text{otherwise}
\end{cases}
$$

for $j = 1, \ldots, n$ and $k = 0, \ldots, C$. The variables $s(j, k)$ indicate which of these two cases applies. So the algorithm enumerates all subsets $S \subseteq \{1, \ldots, n\}$ except those that are infeasible or those that are dominated by others: S is said to be dominated by S' if $\sum_{j \in S} c_j = \sum_{j \in S'} c_j$ and $\sum_{j \in S} w_j \geq \sum_{j \in S'} w_j$. In ④ the best feasible subset is chosen. $\square$

Of course it is desirable to have a better upper bound C than $\sum_{i=1}^{n} c_i$. For example, the 2-factor approximation algorithm of Proposition 17.6 can be run; multiplying the value of the returned solution by 2 yields an upper bound on the optimum value. We shall use this idea later.

The $O(nC)$-bound is not polynomial in the size of the input, because the input size can only be bounded by $O(n \log C + n \log W)$ (we may assume that $w_j \leq W$ for all j). But we have a pseudopolynomial algorithm which can be quite efficient if the numbers involved are not too large. If both the weights $w_1, \ldots, w_n$ and the profits $c_1, \ldots, c_n$ are small, the $O(nc_{max} w_{max})$-algorithm of Pisinger [1999] is the fastest one ($c_{max} := \max\{c_1, \ldots, c_n\}$, $w_{max} := \max\{w_1, \ldots, w_n\}$).

17.3 A Fully Polynomial Approximation Scheme

In this section we investigate approximation algorithms for the KNAPSACK PROBLEM. By Proposition 16.23, the KNAPSACK PROBLEM has no absolute approximation algorithm unless $P = NP$.

However, we shall prove that the KNAPSACK PROBLEM has a fully polynomial approximation scheme. The first such algorithm was found by Ibarra and Kim [1975], improving on the work by Sahni [1975].

Since the running time of the DYNAMIC PROGRAMMING KNAPSACK ALGORITHM depends on C, it is a natural idea to divide all numbers $c_1, \ldots, c_n$ by 2 and round them down. This will reduce the running time, but may lead to inaccurate solutions. More generally, setting

$$
\bar{c}_j := \left\lfloor \frac{c_j}{t} \right\rfloor \qquad (j = 1, \ldots, n)
$$

will reduce the running time by a factor t. Trading accuracy for running time is typical for approximation schemes. For $S \subseteq \{1, \ldots, n\}$ we write $c(S) := \sum_{i \in S} c_i$.

KNAPSACK APPROXIMATION SCHEME

Input: Nonnegative integers $n, c_1, \ldots, c_n, w_1, \ldots, w_n$ and W. A number $\epsilon > 0$.

Output: A subset $S \subseteq \{1, \ldots, n\}$ such that $\sum_{j \in S} w_j \leq W$ and $c(S) \geq \frac{1}{1+\epsilon} c(S')$ for all $S' \subseteq \{1, \ldots, n\}$ with $\sum_{j \in S'} w_j \leq W$.

---

①   Run the 2-factor approximation algorithm of Proposition 17.6. Let S_1 be the solution obtained. **If** $c(S_1) = 0$ **then** set $S := S_1$ and **stop**.

②   Set $t := \max \left\{ 1, \frac{\epsilon c(S_1)}{n} \right\}$.
Set $\bar{c}_j := \lfloor \frac{c_j}{t} \rfloor$ for $j = 1, \ldots, n$.

③   Apply the DYNAMIC PROGRAMMING KNAPSACK ALGORITHM to the instance $(n, \bar{c}_1, \ldots, \bar{c}_n, w_1, \ldots, w_n, W)$; set $C := \frac{2c(S_1)}{t}$. Let S_2 be the solution obtained.

④   **If** $c(S_1) > c(S_2)$ **then** set $S := S_1$, **else** set $S := S_2$.

Theorem 17.9. (Ibarra and Kim [1975], Gens and Levner [1979]) *The* KNAPSACK APPROXIMATION SCHEME *is a fully polynomial approximation scheme for the* KNAPSACK PROBLEM*; its running time is* $O \left(n^2 \cdot \frac{1}{\epsilon} \right)$.

Proof: If the algorithm stops in ① then S_1 is optimal by Proposition 17.6. So we now assume $c(S_1) > 0$. Let S^* be an optimum solution of the original instance. Since $2c(S_1) \geq c(S^*)$ by Proposition 17.6, C in ③ is a correct upper bound on the value of the optimum solution of the rounded instance. So by Theorem 17.8, S_2 is an optimum solution of the rounded instance. Hence we have:

$$\sum_{j \in S_2} c_j \geq \sum_{j \in S_2} t\bar{c}_j = t \sum_{j \in S_2} \bar{c}_j \geq t \sum_{j \in S^*} \bar{c}_j = \sum_{j \in S^*} t\bar{c}_j > \sum_{j \in S^*} (c_j - t)$$

$$\geq c(S^*) - nt.$$

If $t = 1$, then S_2 is optimal by Theorem 17.8. Otherwise the above inequality implies $c(S_2) \geq c(S^*) - \epsilon c(S_1)$, and we conclude that

$$(1 + \epsilon)c(S) \geq c(S_2) + \epsilon c(S_1) \geq c(S^*).$$

So we have a $(1 + \epsilon)$-factor approximation algorithm for any fixed $\epsilon > 0$. By Theorem 17.8 the running time of ③ can be bounded by

$$O(nC) = O \left(\frac{nc(S_1)}{t} \right) = O \left(n^2 \cdot \frac{1}{\epsilon} \right).$$

The other steps can easily be done in $O(n)$ time.   □

Lawler [1979] found a similar fully polynomial approximation scheme whose running time is $O \left(n \log \left(\frac{1}{\epsilon} \right) + \frac{1}{\epsilon^4} \right)$. This was improved by Kellerer and Pferschy [2004] and Chan [2018].

Unfortunately there are not many problems that have a fully polynomial approximation scheme. To state this more precisely, we consider the MAXIMIZATION PROBLEM FOR INDEPENDENCE SYSTEMS.

What we have used in our construction of the DYNAMIC PROGRAMMING KNAPSACK ALGORITHM and the KNAPSACK APPROXIMATION SCHEME is a certain dominance relation. We generalize this concept as follows:

Definition 17.10.   *Given an independence system* $(E, \mathcal{F})$, *a cost function* $c : E \to \mathbb{Z}_+$, *subsets* $S_1, S_2 \subseteq E$, *and* $\epsilon > 0$. S_1 ϵ-**dominates** S_2 *if*

$$\frac{1}{1+\epsilon} c(S_1) \ \leq \ c(S_2) \ \leq \ (1+\epsilon) c(S_1)$$

and there is a basis B_1 *with* $S_1 \subseteq B_1$ *such that for each basis* B_2 *with* $S_2 \subseteq B_2$ *we have*

$$(1+\epsilon) c(B_1) \ \geq \ c(B_2).$$

ϵ-DOMINANCE PROBLEM

Instance: An independence system $(E, \mathcal{F})$, a cost function $c : E \to \mathbb{Z}_+$, a number $\epsilon > 0$, and two subsets $S_1, S_2 \subseteq E$.

Question: Does S_1 ϵ-dominate S_2 ?

Of course the independence system is given by some oracle, e.g. an independence oracle. The DYNAMIC PROGRAMMING KNAPSACK ALGORITHM made frequent use of 0-dominance. It turns out that the existence of an efficient algorithm for the ϵ-DOMINANCE PROBLEM is essential for a fully polynomial approximation scheme.

Theorem 17.11.   (Korte and Schrader [1981])   *Let* $\mathcal{I}$ *be a family of independence systems. Let* $\mathcal{I}'$ *be the family of instances* $(E, \mathcal{F}, c)$ *of the* MAXIMIZATION PROBLEM FOR INDEPENDENCE SYSTEMS *with* $(E, \mathcal{F}) \in \mathcal{I}$ *and* $c : E \to \mathbb{Z}_+$, *and let* $\mathcal{I}''$ *be the family of instances* $(E, \mathcal{F}, c, \epsilon, S_1, S_2)$ *of the* ϵ-DOMINANCE PROBLEM *with* $(E, \mathcal{F}) \in \mathcal{I}$.

Then there exists a fully polynomial approximation scheme for the MAXIMIZATION PROBLEM FOR INDEPENDENCE SYSTEMS *restricted to* $\mathcal{I}'$ *if and only if there exists an algorithm for the* ϵ-DOMINANCE PROBLEM *restricted to* $\mathcal{I}''$ *whose running time is bounded by a polynomial in the length of the input and* $\frac{1}{\epsilon}$.

While the sufficiency is proved by generalizing the KNAPSACK APPROXIMATION SCHEME (Exercise 14), the proof of the necessity is rather involved and not presented here. The conclusion is that if a fully polynomial approximation scheme exists at all, then a modification of the KNAPSACK APPROXIMATION SCHEME does the job. See also Woeginger [2000] for a similar result.

To prove that for a certain optimization problem there is no fully polynomial approximation scheme, the following theorem is often more useful:

Theorem 17.12. (Garey and Johnson [1978]) *A strongly NP-hard optimization problem with integral objective function satisfying*

$$\text{OPT}(I) \leq p(\text{size}(I), \text{largest}(I))$$

for some polynomial p and all instances I has a fully polynomial approximation scheme only if P = NP.

Proof: Suppose it has a fully polynomial approximation scheme. Then we apply it with

$$\epsilon = \frac{1}{p(\text{size}(I), \text{largest}(I)) + 1}$$

and obtain an exact pseudopolynomial algorithm. By Proposition 15.41 this is impossible unless $P = NP$. $\square$

In the next section we consider a problem that has a pseudopolynomial algorithm but no fully polynomial approximation scheme.

17.4 Multi-Dimensional Knapsack

Throughout this section let $m \in \mathbb{N}$ be fixed. We consider the following problem.

---

m-DIMENSIONAL KNAPSACK PROBLEM

Instance: A number $n \in \mathbb{N}$ and nonnegative integers c_i, w_{ij}, and W_j for $i = 1, \ldots, n$ and $j = 1, \ldots, m$.

Task: Find a subset $S \subseteq \{1, \ldots, n\}$ such that $\sum_{i \in S} w_{ij} \leq W_j$ for all $j = 1, \ldots, m$ and $\sum_{i \in S} c_i$ is maximum.

Even for this generalization of the KNAPSACK PROBLEM (which is the case $m = 1$) there is a pseudopolynomial algorithm; see Exercise 15. However, this does not lead to a fully polynomial approximation scheme:

Proposition 17.13. (Korte and Schrader [1981]) *The m-DIMENSIONAL KNAPSACK PROBLEM has no fully polynomial approximation scheme for any $m \geq 2$, unless P = NP.*

Proof: It suffices to consider the case $m = 2$. Given an instance (P, Q, R, T) of 3DM, let $T = \{t_1, \ldots, t_n\} \subseteq P \times Q \times R$, $k := |P| = |Q| = |R|$ and w.l.o.g. $P \cup Q \cup R = \{1, \ldots, 3k\}$. We use a similar construction as in the proof of Corollary 15.27. For $i = 1, \ldots, n$ let $t_i = (p, q, r) \in P \times Q \times R$; then we set $c_i = 1$, $w_{i1} = (n + 1)^p + (n + 1)^q + (n + 1)^r$, and $w_{i2} = (n + 1)^{3k+1} - w_{i1}$. Moreover, let $W_1 = \sum_{p=1}^{3k}(n + 1)^p$ and $W_2 = k(n + 1)^{3k+1} - W_1$. This defines an instance of the 2-DIMENSIONAL KNAPSACK PROBLEM.

If there is a 3-dimensional matching $\{t_i : i \in S\}$ (where $S \subseteq \{1,\ldots,n\}$), then $|S| = k$ and $\sum_{i \in S} w_{i1} = W_1$ and hence $\sum_{i \in S} w_{i2} = W_2$. On the other hand, any feasible solution $S \subseteq \{1,\ldots,n\}$ with $|S| = k$ of this instance of the 2-DIMENSIONAL KNAPSACK PROBLEM satisfies $\sum_{i \in S} w_{i1} \leq W_1$ and

$$\sum_{i \in S} w_{i1} = k(n+1)^{3k+1} - \sum_{i \in S} w_{i2} \geq k(n+1)^{3k+1} - W_2 = W_1.$$

Therefore there is a feasible solution S with $|S| = k$ if and only if the 3DM instance (U, V, W, T) is a yes-instance.

We conclude that the optimum value of this instance of the 2-DIMENSIONAL KNAPSACK PROBLEM is k if I is a yes-instance and at most $k - 1$ otherwise. If we had a fully polynomial approximation scheme, we could get a $(1 + \epsilon)$-approximation for $\epsilon = \frac{1}{k}$ in polynomial time, and thus distinguish between the two cases. Thus we could solve 3DM in polynomial time. As 3DM is NP-complete (Theorem 15.26), this would imply $P = NP$.   □

At least there is an approximation scheme:

Theorem 17.14. (Frieze and Clarke [1984]) *For any fixed $m \in \mathbb{N}$ and $\epsilon > 0$, there is a $(1 + \epsilon)$-factor approximation algorithm for the m-DIMENSIONAL KNAPSACK PROBLEM.*

Proof: We set $k := \lceil \frac{m}{\epsilon} \rceil$; note that k is a constant. We enumerate all subsets of $\{1,\ldots,n\}$ with less than k elements. Moreover, for each $S \subseteq \{1,\ldots,n\}$ with $|S| = k$, we set $S^> := \{i \in \{1,\ldots,n\} \setminus S : c_i > \min\{c_j : j \in S\}\}$ and solve the LP

$$\max\left\{cx \ : \ \sum_{i=1}^{n} w_{ij} x_i \leq W_j \ (j = 1,\ldots,m),\ 0 \leq x \leq 1,\right.$$
$$\left. x_i = 1 \ (i \in S),\ x_i = 0 \ (i \in S^>)\right\}.$$

Let x be an optimum basic solution. As x satisfies at least n constraints of the LP with equality, it has at most m fractional components. Let $S' := \{i \in \{1,\ldots,n\} : x_i = 1\}$.

Among all sets S and S' that we find, we output the best feasible solution. The running time is dominated by solving $O(n^k)$ linear programs with $O(n)$ variables and constraints.

The optimum solution either has at most $k - 1$ elements (then we find it) or corresponds to a feasible 0-1-solution z to one of these LPs, namely where S contains k of its elements with the largest profit. Then we find a solution S' (by rounding down an optimum basic solution x) which is not much worse because $cz \leq cx = \sum_{i \in S'} c_i + \sum_{i=1}^{n} c_i(x_i - \lfloor x_i \rfloor) \leq \sum_{i \in S'} c_i + m \max\{c_i : i \in \{1,\ldots,n\} \setminus (S \cup S^>)\} \leq \sum_{i \in S'} c_i + m \min\{c_i : i \in S\} \leq \sum_{i \in S'} c_i + \frac{m}{k} \sum_{i \in S} c_i \leq \sum_{i \in S'} c_i (1 + \frac{m}{k}) \leq \sum_{i \in S'} c_i (1 + \epsilon)$.   □

17.5 The Nemhauser-Ullmann Algorithm

In this section we discuss another algorithm for the KNAPSACK PROBLEM, due to Nemhauser and Ullmann [1969].

For an instance $n, c_1, \ldots, c_n, w_1, \ldots, w_n$ and W of the KNAPSACK PROBLEM and a subset $S \subseteq \{1, \ldots, n\}$ we write $w(S) := \sum_{i \in S} w_i$ and $c(S) := \sum_{i \in S} c_i$.

Definition 17.15. *Given an instance* $n, c_1, \ldots, c_n, w_1, \ldots, w_n$ *and* W *of the* KNAPSACK PROBLEM, *a subset* $S \subseteq \{1, \ldots, n\}$ *is called* **Pareto optimal** *if there is no other subset* $S' \subseteq \{1, \ldots, n\}$ *such that* $c(S) \leq c(S')$ *and* $w(S) \geq w(S')$ *and at least one of these inequalities is strict.*

In this sense, a subset is dominated if and only if it is not Pareto optimal. Pareto optimality is an important concept in multi-objective optimization. The DYNAMIC PROGRAMMING KNAPSACK ALGORITHM enumerates Pareto optimal solutions for the sub-instance consisting of the first j items (for $j = 1, \ldots, n$). One can do essentially the same in a slightly different way, exploiting the following fact:

Lemma 17.16. *Let* $j \in \{1, \ldots, n\}$, *and let* $\mathcal{P}_j$ *denote the set of all Pareto optimal sets of the instance reduced to the items* $1, \ldots, j$, *and* $\mathcal{P}_0 := \{\emptyset\}$. *Then for every* $j = 1, \ldots, n$ *and every* $S \in \mathcal{P}_j$ *we have* $S \setminus \{j\} \in \mathcal{P}_{j-1}$.

Proof: If $S \setminus \{j\}$ is dominated by some $S' \subseteq \{1, \ldots, j-1\}$, then S is dominated by $S' \cup (\{j\} \cap S)$. $\qquad\square$

The following algorithm computes these sets $\mathcal{P}_0, \ldots, \mathcal{P}_n$ and keeps them as sorted lists.

NEMHAUSER-ULLMANN ALGORITHM

Input: Nonnegative integers $n, c_1, \ldots, c_n, w_1, \ldots, w_n$ and W.

Output: A subset $S \subseteq \{1, \ldots, n\}$ such that $\sum_{j \in S} w_j \leq W$ and $\sum_{j \in S} c_j$ is maximum.

---

①   Let $\mathcal{P}_0 := \{\emptyset\}$.

②   **For** $j := 1$ **to** n **do**:
Set $\mathcal{P}_{j-1}^+ := \{S \cup \{j\} : S \in \mathcal{P}_{j-1}\}$ and $\mathcal{P}_j := \emptyset$.
Let S_1 be the first element of $\mathcal{P}_{j-1}$ and S_2 be the first element of $\mathcal{P}_{j-1}^+$.
While $S_1 \neq$ end or $S_2 \neq$ end **do**:
If $(w(S_1) \leq w(S_2))$ or $(w(S_1) = w(S_2)$ and $c(S_1) \geq c(S_2))$
(where $w(\text{end}) := \infty$)
then:
If $\mathcal{P}_j = \emptyset$ or $c(S_1) > c(S)$ **then**
set $S := S_1$ and append S to $\mathcal{P}_j$.
Let S_1 be the next element of $\mathcal{P}_{j-1}$ (or $S_1 =$ end if there is none).
else:
If $\mathcal{P}_j = \emptyset$ or $c(S_2) > c(S)$ **then**
set $S := S_2$ and append S to $\mathcal{P}_j$.
Let S_2 be the next element of $\mathcal{P}_{j-1}^+$ (or $S_2 =$ end if there is none).

③   Output $S \in \mathcal{P}_n$ with $\sum_{i \in S} w_i \leq W$ such that $\sum_{i \in S} c_i$ is maximum.

See Figure 17.1 for an illustration (unfilled circles represent elements of $\mathcal{P}_{j-1}$, filled circles elements of $\mathcal{P}_{j-1}^{+}$; the two crossed solutions are dominated and thus not added to $\mathcal{P}_j$).

Theorem 17.17. *The* NEMHAUSER-ULLMANN ALGORITHM *works correctly and can be implemented to run in* $O(\sum_{i=0}^{n-1} |\mathcal{P}_i|)$ *time.*

Proof: The lists are always stored in the order so that S precedes S' iff $w(S) < w(S')$. The while loop scans the elements of $\mathcal{P}_{j-1} \cup \mathcal{P}_{j-1}^{+}$ in this order and removes dominated ones; the rest will be $\mathcal{P}_j$. Correctness thus follows from Lemma 17.16.

The running time can be obtained by not storing the elements of $\mathcal{P}_j$ explicitly, but only $c(S)$, $w(S)$, whether or not a set S contains j, and a reference to $S \setminus \{j\}$ in $\mathcal{P}_{j-1}$. This keeps the total runtime proportional to $\sum_{i=0}^{n} |\mathcal{P}_i|$. $\square$

Of course, this is not a polynomial time algorithm (cf. Exercise 16). However, it often performs very well in practice. This can be partially explained by smoothed analysis, randomly perturbing the instance:

Theorem 17.18. (Beier, Röglin and Vöcking [2007]) *Let* $\epsilon > 0$ *and* n, $c_1, \ldots, c_n$, $w_1^{\min}, \ldots, w_n^{\min}$ *and* W *an instance of the* KNAPSACK PROBLEM *with* $w_i^{\min} \leq (1 - \epsilon)W$ *for all* $i = 1, \ldots, n$. *Consider a random instance, where for each* $i = 1, \ldots, n$ *independently we choose* $w_i \in [w_i^{\min}, w_i^{\min} + \epsilon W]$ *(uniformly distributed). Then the expected number of Pareto optimal solutions is at most* $1 + \frac{n^2}{\epsilon}$.

Proof: Let $\mathcal{P}$ be the set of Pareto optimal solutions. Then

$$\text{Exp}(|\mathcal{P}|) = 1 + \lim_{k \to \infty} \sum_{i=1}^{k} \text{Prob}\left(\exists S \in \mathcal{P} : (i-1)\frac{nW}{k} < w(S) \leq i\frac{nW}{k} \right), \quad (17.1)$$

because the empty set is Pareto optimal and has weight 0, and with increasing k it becomes extremely unlikely that the weights of two solutions differ by less than $\frac{nW}{k}$.

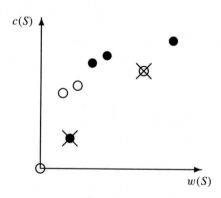

Fig. 17.1.

For a fixed number t, we let

$$c^*(t) := \max\left\{c(S^*) : S^* \subseteq \{1, \ldots, n\}, \ w(S^*) \leq t\right\} \qquad (17.2)$$

and

$$\Lambda(t) := \min\left\{w(S) - t : S \subseteq \{1, \ldots, n\}, \ c(S) > c^*(t)\right\}. \qquad (17.3)$$

Clearly, $\Lambda(t)$ is always positive. We show:

Claim: For any $t \geq 0$ and $\delta > 0$, $\mathrm{Prob}[\Lambda(t) \leq \delta] \leq \frac{n\delta}{\epsilon W}$.

Before we prove the Claim, observe that it easily implies the theorem: for any i, there exists an $S \in \mathcal{P}$ with $(i-1)\frac{nW}{k} < w(S) \leq i\frac{nW}{k}$ if and only if $\Lambda\left((i-1)\frac{nW}{k}\right) \leq \frac{nW}{k}$, and by the Claim this happens with probability at most $\frac{n}{\epsilon W} \cdot \frac{nW}{k} = \frac{n^2}{\epsilon k}$. Plugging this into (17.1) completes the proof.

To prove the Claim, define for $h = 1, \ldots, n$:

$$c^{*,h}(t) := \max\left\{c(S^*) : S^* \subseteq \{1, \ldots, n\} \setminus \{h\}, \ w(S^*) \leq t\right\}$$

and

$$\Lambda^h(t) := \min\left\{w(S) - t : h \in S \subseteq \{1, \ldots, n\}, \ c(S) > c^{*,h}(t)\right\}. \qquad (17.4)$$

Since for any sets S^* attaining (17.2) and S attaining (17.3), there is an index $h \in S \setminus S^*$, we have $\Lambda^h(t) = \Lambda(t)$ for at least one h.

For any fixed $w_1, \ldots, w_{h-1}, w_{h+1}, \ldots, w_n$ we have that $c^{*,h}(t)$ is fixed and also a set S attaining the minimum in (17.4) is independent of the random variable w_h. Therefore $\Lambda^h(t) \leq \delta$ only if $t - w(S \setminus \{h\}) < w_h \leq t + \delta - w(S \setminus \{h\})$. This happens with probability at most $\frac{\delta}{\epsilon W}$. Summing over $h = 1, \ldots, n$ completes the proof of the Claim. $\qquad \square$

One can do even better as follows. We may assume $w(\{1, \ldots, n\}) > W$. First solve the fractional knapsack problem: let $x^* \in [0, 1]^n$ be an optimum fractional solution. By Proposition 17.1 there is an item j and a number $r = \frac{c_j}{w_j}$ such that $x_i^* = 1$ for all i with $\frac{c_i}{w_i} > r$ and $x_i^* = 0$ for all i with $\frac{c_i}{w_i} < r$. For any $\gamma \geq 0$ we can consider the core problem $P(\gamma)$, which is the instance consisting of items i with $|c_i - rw_i| \leq \gamma$ and capacity $W - w(\{i : c_i > rw_i + \gamma\})$. Let S be an optimum solution of this core problem and $S^\gamma := S \cup \{i : c_i > rw_i + \gamma\}$. If γ is large enough, solving the core problem (proposed by Balas and Zemel [1980]) suffices:

Lemma 17.19.   *If $c^\top x^* - c(S^\gamma) \leq \gamma$, then S^γ is an optimum solution to the original instance.*

Proof: Let S^* be an optimum solution, and suppose that $c(S^*) > c(S^\gamma)$. Then there is an index $j \notin S^*$ with $c_j > rw_j + \gamma$ or an index $j \in S^*$ with $c_j < rw_j - \gamma$. Let $x_i := 1$ for $i \in S^*$ and $x_i := 0$ for $i \notin S^*$. In the first case, increasing x_j to 1 and decreasing some x_i to at least x_i^* (so that $w^\top x$ remains constant) will increase the profit by more than $(rw_j + \gamma) - rw_j$. In the first case, decreasing x_j to 0 and increasing some x_i to at most x_i^* (so that $w^\top x$ remains constant) will increase the profit by more than $rw_j - (rw_j - \gamma)$. In both cases we get a fractional solution x' with $c^\top x^* \geq c^\top x' \geq c^\top x + \gamma = c(S^*) + \gamma > c(S^\gamma) + \gamma$.   □

Now sort the items so that $|c_1 - rw_1| \leq \cdots \leq |c_n - rw_n|$. Run the NEMHAUSER-ULLMANN ALGORITHM with this order of items. After each iteration $i = 1, \ldots, n$, let $P_i := \{j \in \{i+1, \ldots, n\} : c_j > rw_j\}$, and let S_i be a solution attaining $\max\{c(S) : S \subseteq \{1, \ldots, i\}, w(S) \leq W - w(P_i)\}$.

By Lemma 17.19 we can stop as soon as $c^\top x^* - c(S_i \cup P_i) < |c_{i+1} - rw_{i+1}|$. This often happens quite fast (cf. Beier and Vöcking [2006]). Therefore the KNAPSACK PROBLEM, although NP-hard, can often be solved optimally in practice.

Exercises

1. Consider the fractional multi-knapsack problem defined as follows. An instance consists of nonnegative integers m and n, numbers w_j, c_{ij} and W_i ($1 \leq i \leq m$, $1 \leq j \leq n$). The task is to find numbers $x_{ij} \in [0, 1]$ with $\sum_{i=1}^{m} x_{ij} = 1$ for all j and $\sum_{j=1}^{n} x_{ij} w_j \leq W_i$ for all i such that $\sum_{i=1}^{m} \sum_{j=1}^{n} x_{ij} c_{ij}$ is minimum. Can one find a combinatorial polynomial-time algorithm for this problem (without using LINEAR PROGRAMMING)?

   Hint: Reduction to a MINIMUM COST FLOW PROBLEM.

2. Consider the following greedy algorithm for the KNAPSACK PROBLEM (similar to the one in Proposition 17.6). Sort the indices such that $\frac{c_1}{w_1} \geq \cdots \geq \frac{c_n}{w_n}$. Set $S := \emptyset$. For $i := 1$ to n do: If $\sum_{j \in S \cup \{i\}} w_j \leq W$ then set $S := S \cup \{i\}$. Show that this is not a k-factor approximation algorithm for any k.

3. Show that the algorithm in Proposition 17.6 does not have an approximation ratio better than 2.

4. Proposition 17.6 and Exercise 3 show that the integrality ratio of the LP

$$\min \left\{ c^\top x : w^\top x \leq W, 0 \leq x \leq 1 \right\}$$

   is 2. Show that, in contrast, the LP relaxation

$$\min \left\{ c^\top x : w^\top x \geq W, 0 \leq x \leq 1 \right\} \tag{17.5}$$

   has unbounded integrality ratio.

* 5. One can strengthen the LP (17.5) by knapsack cover inequalities: for every $I \subseteq \{1, \ldots, n\}$, any feasible integer solution satisfies

$$\sum_{j \notin I} \min \left\{ w_j, W - w(I) \right\} x_j \geq W - w(I),$$

   where $w(I) = \sum_{i \in I} w_i$. Show that the integrality ratio of the resulting LP is 2.

Hint: If x is an optimum LP solution, let $I := \{i : x_i \geq \frac{1}{2}\}$. Order $\{1, \ldots, n\} \setminus I = \{j_1, \ldots, j_k\}$ so that $w_{j_1} \geq \cdots \geq w_{j_k}$. Let $0 \leq \alpha < 1$. Let $S_\alpha := \{j_h : \lceil \alpha + \sum_{l=1}^{h} 2x_{j_l} \rceil > \lceil \alpha + \sum_{l=1}^{h} 2x_{j_l} \rceil\}$ Set $x_i := 1$ for $i \in I \cup S_\alpha$ and $x_i := 0$ otherwise. Show that for every α this is a feasible solution, using that the numbers $f(\alpha) := \sum_{j \in S_\alpha} \min\{w_j, W - w(I)\}$ (for $\alpha \in [0, 1)$) differ by at most $W - w(I)$, and $\int_0^1 f(\alpha)\, d\alpha = \sum_{j \notin I} \min\{w_j, W - w(I)\} 2x_j$. (Carr et al. [2000])

6. Find an exact $O(2^{n/2})$-algorithm for the KNAPSACK PROBLEM.

7. Find an exact $O(nW)$-algorithm for the KNAPSACK PROBLEM.

8. Consider the following problem: given nonnegative integers n, $c_1, \ldots, c_n$, $w_1, \ldots, w_n$ and W, find a subset $S \subseteq \{1, \ldots, n\}$ such that $\sum_{j \in S} w_j \geq W$ and $\sum_{j \in S} c_j$ is minimum. How can this problem be solved by a pseudopolynomial algorithm?

* 9. Can one solve the integral multi-knapsack problem (see Exercise 1) in pseudopolynomial time if m is fixed?

10. Let $c \in \{0, \ldots, k\}^m$ and $s \in [0, 1]^m$. How can one decide in $O(mk)$ time whether $\max\{cx : x \in \mathbb{Z}_+^m, sx \leq 1\} \leq k$?

11. Consider the two Lagrangean relaxations of Exercise 21 of Chapter 5. Show that one of them can be solved in linear time while the other one reduces to m instances of the KNAPSACK PROBLEM.

12. Consider the following variant of the PARTITION problem: Given $a_1, \ldots, a_n \in \mathbb{Z}_+$, find a set $S \subseteq \{1, \ldots, n\}$ such that $\max\left\{\sum_{i \in S} a_i, \sum_{i \in \{1, \ldots, n\} \setminus S} a_i\right\}$ is minimum.

Show that this problem has a fully polynomial approximation scheme.

13. Give a polynomial-time algorithm for the ϵ-DOMINANCE PROBLEM restricted to matroids.

* 14. Prove the if-part of Theorem 17.11.

15. Find a pseudopolynomial algorithm for the m-DIMENSIONAL KNAPSACK PROBLEM for any fixed $m \in \mathbb{N}$.

Note: This generalizes Exercise 7.

16. Show that there are instances of the KNAPSACK PROBLEM with n items where all 2^n subsets are Pareto optimal.

References

General Literature

Garey, M.R., and Johnson, D.S. [1979]: Computers and Intractability; A Guide to the Theory of *NP*-Completeness. Freeman, San Francisco 1979, Chapter 4

Kellerer, H., Pferschy, U., and Pisinger, D. [2004]: Knapsack Problems. Springer, Berlin 2004

Martello, S., and Toth, P. [1990]: Knapsack Problems; Algorithms and Computer Implementations. Wiley, Chichester 1990

Papadimitriou, C.H., and Steiglitz, K. [1982]: Combinatorial Optimization; Algorithms and Complexity. Prentice-Hall, Englewood Cliffs 1982, Sections 16.2, 17.3, and 17.4

Cited References

Balas, E., and Zemel, E. [1980]: An algorithm for large zero-one knapsack problems. Operations Research 28 (1980), 1130–1154

Beier, R., Röglin, H., and Vöcking, B. [2007]: The smoothed number of Pareto optimal solutions in bicriteria integer optimization. In: Integer Programming and Combinatorial Optimization; Proceedings of the 12th International IPCO Conference; LNCS 4513 (M. Fischetti, D.P. Williamson, eds.), Springer, Berlin 2007, pp. 53–67

Beier, R., and Vöcking, B. [2006]: An experimental study of random knapsack problems. Algorithmica 45 (2006), 121–136

Bellman, R. [1956]: Notes on the theory of dynamic programming IV – maximization over discrete sets. Naval Research Logistics Quarterly 3 (1956), 67–70

Bellman, R. [1957]: Comment on Dantzig's paper on discrete variable extremum problems. Operations Research 5 (1957), 723–724

Blum, M., Floyd, R.W., Pratt, V., Rivest, R.L., and Tarjan, R.E. [1973]: Time bounds for selection. Journal of Computer and System Sciences 7 (1973), 448–461

Carr, R.D., Fleischer, L.K., Leung, V.J., and Phillips, C.A. [2000]: Strengthening integrality gaps for capacitated network design and covering problems. Proceedings of the 11th Annual ACM-SIAM Symposium on Discrete Algorithms (2000), 106–115

Chan, T.M. [2018]: Approximation schemes for 0-1 knapsack. Proceedings of the 1st Symposium on Simplicity in Algorithms (2018), Article 5

Dantzig, G.B. [1957]: Discrete variable extremum problems. Operations Research 5 (1957), 266–277

Frieze, A.M. and Clarke, M.R.B. [1984]: Approximation algorithms for the m-dimensional 0-1 knapsack problem: worst case and probabilistic analyses. European Journal of Operations Research 15 (1984), 100–109

Garey, M.R., and Johnson, D.S. [1978]: Strong NP-completeness results: motivation, examples, and implications. Journal of the ACM 25 (1978), 499–508

Gens, G.V., and Levner, E.V. [1979]: Computational complexity of approximation algorithms for combinatorial problems. In: Mathematical Foundations of Computer Science; LNCS 74 (J. Becvar, ed.), Springer, Berlin 1979, pp. 292–300

Ibarra, O.H., and Kim, C.E. [1975]: Fast approximation algorithms for the knapsack and sum of subset problem. Journal of the ACM 22 (1975), 463–468

Karp, R.M. [1972]: Reducibility among combinatorial problems. In: Complexity of Computer Computations (R.E. Miller, J.W. Thatcher, eds.), Plenum Press, New York 1972, pp. 85–103

Kellerer, H., and Pferschy, U. [2004]: Improved dynamic programming in connection with an FPTAS for the knapsack problem. Journal on Combinatorial Optimization 8 (2004), 5–11

Korte, B., and Schrader, R. [1981]: On the existence of fast approximation schemes. In: Nonlinear Programming; Vol. 4 (O. Mangaserian, R.R. Meyer, S.M. Robinson, eds.), Academic Press, New York 1981, pp. 415–437

Lawler, E.L. [1979]: Fast approximation algorithms for knapsack problems. Mathematics of Operations Research 4 (1979), 339–356

Nemhauser, G.L., and Ullmann, Z. [1969]: Discrete dynamic programming and capital allocation. Management Science 15 (1969), 494–505

Pisinger, D. [1999]: Linear time algorithms for knapsack problems with bounded weights. Journal of Algorithms 33 (1999), 1–14

Sahni, S. [1975]: Approximate algorithms for the 0/1 knapsack problem. Journal of the ACM 22 (1975), 115–124

Vygen, J. [1997]: The two-dimensional weighted median problem. Zeitschrift für Angewandte Mathematik und Mechanik 77 (1997), Supplement, S433–S436

Woeginger, G.J. [2000]: When does a dynamic programming formulation guarantee the existence of a fully polynomial time approximation scheme (FPTAS)? INFORMS Journal on Computing 12 (2000), 57–74

18 Bin-Packing

Suppose we have n objects, each of a given size, and some bins of equal capacity. We want to assign the objects to the bins, using as few bins as possible. Of course the total size of the objects assigned to one bin should not exceed its capacity.

Without loss of generality, the capacity of the bins is 1. Then the problem can be formulated as follows:

BIN-PACKING PROBLEM

Instance: A list of nonnegative numbers $a_1, \ldots, a_n \leq 1$.

Task: Find a $k \in \mathbb{N}$ and an assignment $f : \{1, \ldots, n\} \to \{1, \ldots, k\}$ with $\sum_{i : f(i)=j} a_i \leq 1$ for all $j \in \{1, \ldots, k\}$ such that k is minimum.

There are not many combinatorial optimization problems whose practical relevance is more obvious. For example, the simplest version of the cutting stock problem is equivalent: We are given many beams of equal length (say 1 meter) and numbers $a_1, \ldots, a_n$. We want to cut as few of the beams as possible into pieces such that at the end we have beams of lengths $a_1, \ldots, a_n$.

Although an instance I is some ordered list where numbers may appear more than once, we write $x \in I$ for some element in the list I which is equal to x. By $|I|$ we mean the number of elements in the list I. We shall also use the abbreviation $\text{SUM}(a_1, \ldots, a_n) := \sum_{i=1}^{n} a_i$. This is an obvious lower bound: $\lceil \text{SUM}(I) \rceil \leq \text{OPT}(I)$ holds for any instance I.

In Section 18.1 we prove that the BIN-PACKING PROBLEM is strongly *NP*-hard and discuss some simple approximation algorithms. We shall see that no algorithm can achieve a performance ratio better than $\frac{3}{2}$ (unless $P = NP$). However, one can achieve an arbitrarily good performance ratio asymptotically: in Sections 18.2 and 18.3 we describe a fully polynomial asymptotic approximation scheme. This uses the ELLIPSOID METHOD and results of Chapter 17.

© Springer-Verlag GmbH Germany 2018
Bernhard Korte and Jens Vygen, *Combinatorial Optimization*, Algorithms and Combinatorics 21, https://doi.org/10.1007/978-3-662-56039-6_18

18.1 Greedy Heuristics

In this section we shall analyse some greedy heuristics for the BIN-PACKING PROBLEM. There is no hope for an exact polynomial-time algorithm as the problem is *NP*-hard:

Theorem 18.1. *The following problem is NP-complete: given an instance I of the* BIN-PACKING PROBLEM, *decide whether I has a solution with two bins.*

Proof: Membership in *NP* is trivial. We transform the PARTITION problem (which is *NP*-complete by Corollary 15.28) to the above decision problem. Given an instance $c_1, \ldots, c_n$ of PARTITION, consider the instance $a_1, \ldots, a_n$ of the BIN-PACKING PROBLEM, where

$$a_i = \frac{2c_i}{\sum_{j=1}^{n} c_j}.$$

Obviously two bins suffice if and only if there is a subset $S \subseteq \{1, \ldots, n\}$ such that $\sum_{j \in S} c_j = \sum_{j \notin S} c_j$. $\quad\square$

Corollary 18.2. *Unless $P = NP$, there is no ρ-factor approximation algorithm for the* BIN-PACKING PROBLEM *for any $\rho < \frac{3}{2}$.* $\quad\square$

For any fixed k, there is a pseudopolynomial algorithm which decides for a given instance I whether k bins suffice (Exercise 1). However, in general this problem is strongly *NP*-complete:

Theorem 18.3. (Garey and Johnson [1975]) *The following problem is strongly NP-complete: given an instance I of the* BIN-PACKING PROBLEM *and a number B, decide whether I can be solved with B bins.*

Proof: Transformation from 3-DIMENSIONAL MATCHING (Theorem 15.26).
    Given an instance U, V, W, T of 3DM, we construct a bin-packing instance I with $4|T|$ items. Namely, the set of items is

$$S := \bigcup_{t=(u,v,w) \in T} \{t, (u, t), (v, t), (w, t)\}.$$

Let $U = \{u_1, \ldots, u_n\}$, $V = \{v_1, \ldots, v_n\}$ and $W = \{w_1, \ldots, w_n\}$. For each $x \in U \cup V \cup W$ we choose some $t_x \in T$ such that $(x, t_x) \in S$. For each $t = (u_i, v_j, w_k) \in T$, the sizes of the items are now defined as follows:

$$
\begin{array}{lll}
t & \text{has size} & \dfrac{1}{C}(10N^4 + 8 - iN - jN^2 - kN^3) \\[2ex]
(u_i, t) & \text{has size} & \begin{cases} \dfrac{1}{C}(10N^4 + iN + 1) & \text{if } t = t_{u_i} \\[1ex] \dfrac{1}{C}(11N^4 + iN + 1) & \text{if } t \neq t_{u_i} \end{cases}
\end{array}
$$

$$(v_j, t) \quad \text{has size} \begin{cases} \frac{1}{C}(10N^4 + jN^2 + 2) & \text{if } t = t_{v_j} \\ \frac{1}{C}(11N^4 + jN^2 + 2) & \text{if } t \neq t_{v_j} \end{cases}$$

$$(w_k, t) \quad \text{has size} \begin{cases} \frac{1}{C}(10N^4 + kN^3 + 4) & \text{if } t = t_{w_k} \\ \frac{1}{C}(8N^4 + kN^3 + 4) & \text{if } t \neq t_{w_k} \end{cases}$$

where $N := 100n$ and $C := 40N^4 + 15$. This defines an instance $I = (a_1, \ldots, a_{4|T|})$ of the BIN-PACKING PROBLEM. We set $B := |T|$ and claim that I has a solution with at most B bins if and only if the initial 3DM instance is a yes-instance, i.e. there is a subset M of T with $|M| = n$ such that for distinct $(u, v, w), (u', v', w') \in M$ one has $u \neq u'$, $v \neq v'$ and $w \neq w'$.

First assume that there is such a solution M of the 3DM instance. Since the solvability of I with B bins is independent of the choice of the t_x ($x \in U \cup V \cup W$), we may redefine them such that $t_x \in M$ for all x. Now for each $t = (u, v, w) \in T$ we pack $t, (u, t), (v, t), (w, t)$ into one bin. This yields a solution with $|T|$ bins.

Conversely, let f be a solution of I with $B = |T|$ bins. Since $\text{SUM}(I) = |T|$, each bin must be completely full. Since all item sizes are strictly between $\frac{1}{5}$ and $\frac{1}{3}$, each bin must contain four items.

Consider one bin $k \in \{1, \ldots, B\}$. Since $C \sum_{i: f(i)=k} a_i = C \equiv 15 \pmod{N}$, the bin must contain one $t = (u, v, w) \in T$, one $(u', t') \in U \times T$, one $(v', t'') \in V \times T$, and one $(w', t''') \in W \times T$. Since $C \sum_{i: f(i)=k} a_i = C \equiv 15 \pmod{N^2}$, we have $u = u'$. Similarly, by considering the sum modulo N^3 and modulo N^4, we obtain $v = v'$ and $w = w'$. Furthermore, either $t' = t_u$ and $t'' = t_v$ and $t''' = t_w$ (case 1) or $t' \neq t_u$ and $t'' \neq t_v$ and $t''' \neq t_w$ (case 2).

We define M to consist of those $t \in T$ for which t is assigned to a bin where case 1 holds. Obviously M is a solution to the 3DM instance.

Note that all the numbers in the constructed bin-packing instance I are polynomially large, more precisely $O(n^4)$. Since 3DM is NP-complete (Theorem 15.26), the theorem is proved. $\qquad\square$

This proof is due to Papadimitriou [1994]. Even with the assumption $P \neq NP$ the above result does not exclude the possibility of an absolute approximation algorithm, for example one which needs at most one more bin than the optimum solution. Whether such an algorithm exists is an open question.

The simplest heuristic for bin-packing may be the following:

NEXT-FIT ALGORITHM (NF)

Input: An instance $a_1, \ldots, a_n$ of the BIN-PACKING PROBLEM.

Output: A solution (k, f).

① Set $k := 1$ and $S := 0$.

② **For** $i := 1$ **to** n **do**:
  If $S + a_i > 1$ **then** set $k := k + 1$ and $S := 0$.
  Set $f(i) := k$ and $S := S + a_i$.

Let us denote by $NF(I)$ the number k of bins this algorithm uses for instance I.

Theorem 18.4. *The* NEXT-FIT ALGORITHM *runs in* $O(n)$ *time. For any instance* $I = a_1, \ldots, a_n$ *we have*

$$NF(I) \leq 2\lceil SUM(I) \rceil - 1 \leq 2 OPT(I) - 1.$$

Proof: The time bound is obvious. Let $k := NF(I)$, and let f be the assignment found by the NEXT-FIT ALGORITHM. For $j = 1, \ldots, \lfloor \frac{k}{2} \rfloor$ we have

$$\sum_{i:f(i)\in\{2j-1,2j\}} a_i > 1.$$

Adding these inequalities we get

$$\left\lfloor \frac{k}{2} \right\rfloor < SUM(I).$$

Since the left-hand side is an integer, we conclude that

$$\frac{k-1}{2} \leq \left\lfloor \frac{k}{2} \right\rfloor \leq \lceil SUM(I) \rceil - 1.$$

This proves $k \leq 2\lceil SUM(I) \rceil - 1$. The second inequality is trivial. $\square$

The instances $2\epsilon, 1 - \epsilon, 2\epsilon, 1 - \epsilon, \ldots, 2\epsilon$ for very small $\epsilon > 0$ show that this bound is best possible. So the NEXT-FIT ALGORITHM is a 2-factor approximation algorithm. Naturally the performance ratio becomes better if the numbers involved are small:

Proposition 18.5.   *Let* $0 < \gamma < 1$. *For any instance* $I = a_1, \ldots, a_n$ *with* $a_i \leq \gamma$ *for all* $i \in \{1, \ldots, n\}$ *we have*

$$NF(I) \leq \left\lceil \frac{SUM(I)}{1 - \gamma} \right\rceil.$$

Proof: We have $\sum_{i:f(i)=j} a_i > 1 - \gamma$ for $j = 1, \ldots, NF(I) - 1$. By adding these inequalities we get $(NF(I) - 1)(1 - \gamma) < SUM(I)$ and thus

$$NF(I) - 1 \leq \left\lceil \frac{SUM(I)}{1 - \gamma} \right\rceil - 1.$$

$\square$

A second approach in designing an efficient approximation algorithm could be the following:

FIRST-FIT ALGORITHM (FF)

Input: An instance $a_1, \ldots, a_n$ of the BIN-PACKING PROBLEM.

Output: A solution (k, f).

①   **For** $i := 1$ **to** n **do:**

$$\text{Set } f(i) := \min \left\{ j \in \mathbb{N} : \sum_{h < i : f(h) = j} a_h + a_i \leq 1 \right\}.$$

②   Set $k := \max_{i \in \{1, \ldots, n\}} f(i)$.

Of course the FIRST-FIT ALGORITHM cannot be worse than NEXT-FIT. So FIRST-FIT is another 2-factor approximation algorithm. Indeed, it is better:

Theorem 18.6. (Dósa and Sgall [2013]) *For all instances I of the* BIN-PACKING PROBLEM,

$$\mathrm{FF}(I) \leq \left\lfloor \frac{17}{10} \mathrm{OPT}(I) \right\rfloor .$$

Furthermore, there exist instances I with $\mathrm{OPT}(I)$ arbitrarily large where equality holds.

This is a slight strengthening of previous results by Johnson et al. [1974] and Garey et al. [1976]. We omit the complicated proof.

Proposition 18.5 shows that the NEXT-FIT (and thus the FIRST-FIT)Ä ALGO-RITHM behaves well if the pieces are small. So it is natural to treat the large pieces first. The following modification of the FIRST-FIT ALGORITHM scans the n numbers in decreasing order:

FIRST-FIT-DECREASING ALGORITHM (FFD)

Input: An instance $a_1, \ldots, a_n$ of the BIN-PACKING PROBLEM.

Output: A solution (k, f).

①   Sort the numbers such that $a_1 \geq a_2 \geq \ldots \geq a_n$.

②   Apply the FIRST-FIT ALGORITHM.

Theorem 18.7. (Simchi-Levi [1994]) *The* FIRST-FIT-DECREASING ALGORITHM *is a $\frac{3}{2}$-factor approximation algorithm for the* BIN-PACKING PROBLEM.

Proof: Let I be an instance and $k := \mathrm{FFD}(I)$. Consider the j-th bin for $j := \lceil \frac{2}{3} k \rceil$. If it contains an item of size $> \frac{1}{2}$, then each bin with smaller index did not have space for this item, thus has been assigned an item before. As the items are considered in nonincreasing order, there are at least j items of size $> \frac{1}{2}$. Thus $\mathrm{OPT}(I) \geq j \geq \frac{2}{3}k$.

Otherwise the j-th bin, and thus each bin with greater index, contains no item of size $> \frac{1}{2}$. Hence the bins $j, j+1, \ldots, k$ contain at least $2(k-j)+1$ items, none of which fits into bins $1, \ldots, j-1$. Note that $2(k-j)+1 \geq 2(k-(\frac{2}{3}k+\frac{2}{3}))+1 = \frac{2}{3}k - \frac{1}{3} \geq j-1$. Thus $\mathrm{OPT}(I) \geq \mathrm{SUM}(I) > j-1$, i.e. $\mathrm{OPT}(I) \geq j \geq \frac{2}{3}k$.   □

By Corollary 18.2 this is best possible (for FFD, consider the instance 0.4, 0.4, 0.3, 0.3, 0.3, 0.3). However, the asymptotic performance guarantee is better: Johnson [1973] proved that $\mathrm{FFD}(I) \leq \frac{11}{9}\mathrm{OPT}(I) + 4$ for all instances I (see also Johnson [1974]). Baker [1985] gave a simpler proof showing $\mathrm{FFD}(I) \leq \frac{11}{9}\mathrm{OPT}(I) + 3$. Yue [1991] improved this to $\mathrm{FFD}(I) \leq \frac{11}{9}\mathrm{OPT}(I) + 1$. The question was finally settled by Dósa et al. [2013], who proved the following:

Theorem 18.8. (Dósa et al. [2013]) *For all instances I of the* BIN-PACKING PROBLEM,

$$\mathrm{FFD}(I) \leq \frac{11}{9}\mathrm{OPT}(I) + \frac{2}{3},$$

and this bound is tight.

The proof is too involved to be presented here. However, we present a class of instances I with $\mathrm{OPT}(I)$ arbitrarily large and $\mathrm{FFD}(I) = \frac{11}{9}\mathrm{OPT}(I)$. (This example is taken from Garey and Johnson [1979].)

Namely, let $\epsilon > 0$ be small enough and $I = \{a_1, \ldots, a_{30m}\}$ with

$$
a_i = \begin{cases}
\frac{1}{2} + \epsilon & \text{if } 1 \leq i \leq 6m, \\[1.2em]
\frac{1}{4} + 2\epsilon & \text{if } 6m < i \leq 12m, \\[1.2em]
\frac{1}{4} + \epsilon & \text{if } 12m < i \leq 18m, \\[1.2em]
\frac{1}{4} - 2\epsilon & \text{if } 18m < i \leq 30m.
\end{cases}
$$

The optimum solution consists of

$6m$ bins containing $\dfrac{1}{2} + \epsilon, \dfrac{1}{4} + \epsilon, \dfrac{1}{4} - 2\epsilon,$

$3m$ bins containing $\dfrac{1}{4} + 2\epsilon, \dfrac{1}{4} + 2\epsilon, \dfrac{1}{4} - 2\epsilon, \dfrac{1}{4} - 2\epsilon.$

The FFD-solution consists of

$6m$ bins containing $\quad \dfrac{1}{2} + \epsilon, \ \dfrac{1}{4} + 2\epsilon,$

$2m$ bins containing $\quad \dfrac{1}{4} + \epsilon, \ \dfrac{1}{4} + \epsilon, \ \dfrac{1}{4} + \epsilon,$

$3m$ bins containing $\quad \dfrac{1}{4} - 2\epsilon, \ \dfrac{1}{4} - 2\epsilon, \ \dfrac{1}{4} - 2\epsilon, \ \dfrac{1}{4} - 2\epsilon.$

So $\mathrm{OPT}(I) = 9m$ and $\mathrm{FFD}(I) = 11m$.

There are several other algorithms for the BIN-PACKING PROBLEM, some of them having a better asymptotic performance ratio than $\frac{11}{9}$. In the next section we show that an asymptotic performance ratio arbitrarily close to 1 can be achieved.

In some applications one has to pack the items in the order they arrive without knowing the subsequent items. Algorithms that do not use any information about the subsequent items are called online algorithms. For example, NEXT-FIT and FIRST-FIT are online algorithms, but the FIRST-FIT-DECREASING ALGORITHM is not an online algorithm. The best known online algorithm for the BIN-PACKING PROBLEM has an asymptotic performance ratio of 1.59 (Seiden [2002]). On the other hand, van Vliet [1992] proved that there is no online asymptotic 1.54-factor approximation algorithm for the BIN-PACKING PROBLEM. A weaker lower bound is the subject of Exercise 6.

18.2 An Asymptotic Approximation Scheme

In this section we show that for any $\epsilon > 0$ there is a linear-time algorithm which guarantees to find a solution with at most $(1 + \epsilon)\,\mathrm{OPT}(I) + \frac{1}{\epsilon^2}$ bins.

We start by considering instances with not too many different numbers. We denote the different numbers in our instance I by $s_1, \ldots, s_m$. Let I contain exactly b_i copies of s_i $(i = 1, \ldots, m)$.

Let $T_1, \ldots, T_N$ be all the possibilities of how a single bin can be packed:

$$\{T_1, \ldots, T_N\} := \left\{ (k_1, \ldots, k_m) \in \mathbb{Z}_+^m : \sum_{i=1}^{m} k_i s_i \leq 1 \right\}$$

We write $T_j = (t_{j1}, \ldots, t_{jm})$. Then our BIN-PACKING PROBLEM is equivalent to the following integer programming formulation (due to Eisemann [1957]):

$$
\begin{aligned}
\min \quad & \sum_{j=1}^{N} x_j \\
\text{s.t.} \quad & \sum_{j=1}^{N} t_{ji} x_j \ \geq \ b_i && (i = 1, \ldots, m) \\
& x_j \ \in \ \mathbb{Z}_+ && (j = 1, \ldots, N).
\end{aligned}
\tag{18.1}
$$

We actually want $\sum_{j=1}^{N} t_{ji}x_j = b_i$, but relaxing this constraint makes no difference. The LP relaxation of (18.1) is:

$$\min \quad \sum_{j=1}^{N} x_j$$

$$\text{s.t.} \quad \sum_{j=1}^{N} t_{ji}x_j \geq b_i \qquad (i = 1, \ldots, m) \tag{18.2}$$

$$x_j \geq 0 \qquad (j = 1, \ldots, N).$$

This is sometimes called the configuration LP because there is a variable for each bin configuration. The following theorem says that by rounding a solution of the LP relaxation (18.2) one obtains a solution of (18.1), i.e. of the BIN-PACKING PROBLEM, which is not much worse:

Theorem 18.9. (Fernandez de la Vega and Lueker [1981]) *Let I be an instance of the* BIN-PACKING PROBLEM *with only m different numbers. Let x be a feasible (not necessarily optimum) solution of (18.2) with at most m nonzero components. Then a solution of the* BIN-PACKING PROBLEM *with at most* $\left\lceil \sum_{j=1}^{N} x_j \right\rceil + \left\lfloor \frac{m-1}{2} \right\rfloor$ *bins can be found in $O(|I|)$ time.*

Proof: Consider $\lfloor x \rfloor$, which results from x by rounding down each component. $\lfloor x \rfloor$ does not in general pack I completely (it might pack some numbers more often than necessary, but this does not matter). The remaining pieces form an instance I'. Observe that

$$\text{SUM}(I') \leq \sum_{j=1}^{N} (x_j - \lfloor x_j \rfloor) \sum_{i=1}^{m} t_{ji}s_i \leq \sum_{j=1}^{N} x_j - \sum_{j=1}^{N} \lfloor x_j \rfloor.$$

So it is sufficient to pack I' into at most $\lceil \text{SUM}(I') \rceil + \left\lfloor \frac{m-1}{2} \right\rfloor$ bins, because then the total number of bins used is no more than

$$\sum_{j=1}^{N} \lfloor x_j \rfloor + \lceil \text{SUM}(I') \rceil + \left\lfloor \frac{m-1}{2} \right\rfloor \leq \left\lceil \sum_{j=1}^{N} x_j \right\rceil + \left\lfloor \frac{m-1}{2} \right\rfloor.$$

We consider two packing methods for I'. Firstly, the vector $\lceil x \rceil - \lfloor x \rfloor$ certainly packs at least the elements of I'. The number of bins used is at most m since x has at most m nonzero components. Secondly, we can obtain a packing of I' using at most $2\lceil \text{SUM}(I') \rceil - 1$ bins by applying the NEXT-FIT ALGORITHM (Theorem 18.4). Both packings can be obtained in linear time.

The better of these two packings uses at most $\min\{m, 2\lceil \text{SUM}(I') \rceil - 1\} \leq \lceil \text{SUM}(I') \rceil + \frac{m-1}{2}$ bins. The theorem is proved. $\square$

Corollary 18.10. (Fernandez de la Vega and Lueker [1981]) *Let m and $\gamma > 0$ be fixed constants. Let I be an instance of the* BIN-PACKING PROBLEM *with only m different numbers, none of which is less than γ. Then we can find a solution with at most* $\mathrm{OPT}(I) + \left\lfloor \frac{m-1}{2} \right\rfloor$ *bins in $O(|I|)$ time.*

Proof: By the SIMPLEX ALGORITHM (Theorem 3.14) we can find an optimum basic solution x^* of (18.2), i.e. a vertex of the polyhedron. Since any vertex satisfies N of the constraints with equality (Proposition 3.9), x^* has at most m nonzero components.

The time needed to determine x^* depends on m and N only. Observe that $N \leq (m+1)^{\frac{1}{\gamma}}$, because there can be at most $\frac{1}{\gamma}$ elements in each bin. So x^* can be found in constant time.

Since $\left\lceil \sum_{j=1}^{N} x_j^* \right\rceil \leq \mathrm{OPT}(I)$, an application of Theorem 18.9 completes the proof. $\square$

Using the ELLIPSOID METHOD (Theorem 4.18) leads to the same result. This is not best possible: one can even determine the exact optimum in polynomial time for fixed m and γ, since INTEGER PROGRAMMING with a constant number of variables can be solved in polynomial time (Lenstra [1983]). A simpler exact algorithm for fixed m is discussed in Exercise 7. However, we do not need these stronger results here. We shall apply Theorem 18.9 again in the next section and obtain the same performance guarantee in polynomial time even if m and γ are not fixed (in the proof of Theorem 18.14).

We are now able to formulate the algorithm of Fernandez de la Vega and Lueker [1981]. Roughly it proceeds as follows. First we partition the n numbers into $m + 2$ groups according to their size. We pack the group with the largest ones using one bin for each number. Then we pack the m middle groups by first rounding the size of each number to the largest number in its group and then applying Corollary 18.10. Finally we pack the group with the smallest numbers.

FERNANDEZ-DE-LA-VEGA-LUEKER ALGORITHM

Input: An instance $I = a_1, \ldots, a_n$ of the BIN-PACKING PROBLEM. A number $\epsilon > 0$.

Output: A solution (k, f) for I.

① Set $\gamma := \frac{\epsilon}{\epsilon+1}$ and $h := \lceil \epsilon\, \mathrm{SUM}(I) \rceil$.

② Let $I_1 = L, M, R$ be a rearrangement of the list I, where $M = K_0, y_1, K_1, y_2, \ldots, K_{m-1}, y_m$ and $L, K_0, K_1, \ldots, K_{m-1}$ and R are again lists, such that the following properties hold:
- (a) For all $x \in L$: $x < \gamma$.
- (b) For all $x \in K_0$: $\gamma \leq x \leq y_1$.
- (c) For all $x \in K_i$: $y_i \leq x \leq y_{i+1}$ $(i = 1, \ldots, m-1)$.
- (d) For all $x \in R$: $y_m \leq x$.
- (e) $|K_1| = \cdots = |K_{m-1}| = |R| = h - 1$ and $|K_0| \leq h - 1$.

(k, f) is now determined by the following three packing steps:

③   Find a packing S_R of R using $|R|$ bins.

④   Consider the instance Q consisting of the numbers $y_1, y_2, \ldots, y_m$, each appearing h times. Find a packing S_Q of Q using at most $\frac{m+1}{2}$ more bins than necessary (using Corollary 18.10). Transform S_Q into a packing S_M of M.

⑤   As long as a bin of S_R or S_M has room amounting to at least γ, fill it with elements of L. Finally, find a packing of the rest of L using the NEXT-FIT ALGORITHM.

In ④ we used a slightly weaker bound than the one obtained in Corollary 18.10. This does not hurt here, and we shall need the above form in Section 18.3. The above algorithm is an asymptotic approximation scheme. More precisely:

Theorem 18.11. (Fernandez de la Vega and Lueker [1981]) *For each $0 < \epsilon \le \frac{1}{2}$ and each instance I of the BIN-PACKING PROBLEM, the FERNANDEZ-DE-LA-VEGA-LUEKER ALGORITHM returns a solution using at most $(1+\epsilon)\operatorname{OPT}(I) + \frac{1}{\epsilon^2}$ bins. The running time is $O(n\frac{1}{\epsilon^2})$ plus the time needed to solve (18.2). For fixed ϵ, the running time is $O(n)$.*

Proof: In ②, we first determine L in $O(n)$ time. Then we set $m := \left\lfloor \frac{|I| - |L|}{h} \right\rfloor$. Since $\gamma(|I| - |L|) \le \operatorname{SUM}(I)$, we have

$$m \;\le\; \frac{|I| - |L|}{h} \;\le\; \frac{|I| - |L|}{\epsilon \operatorname{SUM}(I)} \;\le\; \frac{1}{\gamma\epsilon} \;=\; \frac{\epsilon + 1}{\epsilon^2}.$$

We know that y_i must be the $(|I| + 1 - (m - i + 1)h)$-th smallest element $(i = 1, \ldots, m)$. So by Corollary 17.4 we can find each y_i in $O(n)$ time. We finally determine $K_0, K_1, \ldots, K_{m-1}, R$, each in $O(n)$ time. So ② can be done in $O(mn)$ time. Note that $m = O(\frac{1}{\epsilon^2})$.

Steps ③, ④ and ⑤ – except the solution of (18.2) – can easily be implemented to run in $O(n)$ time. For fixed ϵ, (18.2) can also be solved optimally in $O(n)$ time (Corollary 18.10).

We now prove the performance guarantee. Let k be the number of bins that the algorithm uses. We write $|S_R|$ and $|S_M|$ for the number of bins used in the packing of R and M, respectively.

We have

$$|S_R| \;\le\; |R| \;=\; h - 1 \;<\; \epsilon \operatorname{SUM}(I) \;\le\; \epsilon \operatorname{OPT}(I).$$

Secondly, observe that $\operatorname{OPT}(Q) \le \operatorname{OPT}(I)$: the i-th largest element of I is greater than or equal to the i-th largest element of Q for all $i = 1, \ldots, hm$. Hence by ④ (Corollary 18.10) we have

$$|S_M| \;=\; |S_Q| \;\le\; \operatorname{OPT}(Q) + \frac{m+1}{2} \;\le\; \operatorname{OPT}(I) + \frac{m+1}{2}.$$

In ⑤ we can pack some elements of L into bins of S_R and S_M. Let L' be the list of the remaining elements in L.

Case 1: L' is nonempty. Then the total size of the elements in each bin, except possibly for the last one, exceeds $1 - \gamma$, so we have $(1 - \gamma)(k - 1) < \mathrm{SUM}(I) \le \mathrm{OPT}(I)$. We conclude that

$$k \le \frac{1}{1 - \gamma} \mathrm{OPT}(I) + 1 = (1 + \epsilon)\,\mathrm{OPT}(I) + 1.$$

Case 2: L' is empty. Then

$$
\begin{aligned}
k &\le |S_R| + |S_M| \\
&< \epsilon\,\mathrm{OPT}(I) + \mathrm{OPT}(I) + \frac{m + 1}{2} \\
&\le (1 + \epsilon)\,\mathrm{OPT}(I) + \frac{\epsilon + 1 + \epsilon^2}{2\epsilon^2} \\
&\le (1 + \epsilon)\,\mathrm{OPT}(I) + \frac{1}{\epsilon^2},
\end{aligned}
$$

because $\epsilon \le \frac{1}{2}$. $\square$

Of course the running time grows exponentially in $\frac{1}{\epsilon}$. However, Karmarkar and Karp showed how to obtain a fully polynomial asymptotic approximation scheme. This is the subject of the next section.

18.3 The Karmarkar-Karp Algorithm

The algorithm of Karmarkar and Karp [1982] works just as the algorithm in the preceding section, but instead of solving the LP relaxation (18.2) optimally as in Corollary 18.10, they solve it with a constant absolute error.

The fact that the number of variables grows exponentially in $\frac{1}{\epsilon}$ might not prevent us from solving the LP: Gilmore and Gomory [1961] developed the column generation technique and obtained a variant of the SIMPLEX ALGORITHM which solves (18.2) quite efficiently in practice. Similar ideas lead to a theoretically efficient algorithm if one uses the GRÖTSCHEL-LOVÁSZ-SCHRIJVER ALGORITHM instead.

In both above-mentioned approaches the dual LP plays a major role. The dual of (18.2) is:

$$
\begin{aligned}
\max \quad & yb \\
\text{s.t.} \quad & \sum_{i=1}^{m} t_{ji} y_i \le 1 && (j = 1, \ldots, N) \\
& y_i \ge 0 && (i = 1, \ldots, m).
\end{aligned}
\tag{18.3}
$$

It has only m variables, but an exponential number of constraints. However, the number of constraints does not matter as long as we can solve the SEPARATION PROBLEM in polynomial time. It will turn out that the SEPARATION PROBLEM is equivalent to a KNAPSACK PROBLEM. Since we can solve KNAPSACK PROBLEMS with an arbitrarily small error, we can also solve the WEAK SEPARATION PROBLEM in polynomial time. This idea enables us to prove:

Lemma 18.12. (Karmarkar and Karp [1982]) *Let I be an instance of the BIN-PACKING PROBLEM with only m different numbers, none of which is less than γ. Let $\delta > 0$. Then a feasible solution y^* of the dual LP (18.3) differing from the optimum by at most δ can be found in $O\left(m^6 \log^2 \frac{mn}{\gamma\delta} + \frac{m^5 n}{\delta} \log \frac{mn}{\gamma\delta}\right)$ time.*

Proof: We may assume that $\delta = \frac{1}{p}$ for some natural number p. We apply the GRÖTSCHEL-LOVÁSZ-SCHRIJVER ALGORITHM (Theorem 4.19). Let $\mathcal{D}$ be the polyhedron of (18.3). We have

$$B\left(x_0, \frac{\gamma}{2}\right) \subseteq [0, \gamma]^m \subseteq \mathcal{D} \subseteq [0, 1]^m \subseteq B(x_0, \sqrt{m}),$$

where x_0 is the vector all of whose components are $\frac{\gamma}{2}$.

We shall prove that we can solve the WEAK SEPARATION PROBLEM for (18.3), i.e. $\mathcal{D}$ and b, and $\frac{\delta}{2}$ in $O\left(\frac{nm}{\delta}\right)$ time, independently of the size of the input vector y. By Theorem 4.19, this implies that the WEAK OPTIMIZATION PROBLEM can be solved in $O\left(m^6 \log^2 \frac{m\|b\|}{\gamma\delta} + \frac{m^5 n}{\delta} \log \frac{m\|b\|}{\gamma\delta}\right)$ time, proving the lemma since $\|b\| \leq n$.

To show how to solve the WEAK SEPARATION PROBLEM, let $y \in \mathbb{Q}^m$ be given. We may assume $0 \leq y \leq 1$ since otherwise the task is trivial. Now observe that y is feasible if and only if

$$\max\{yx : x \in \mathbb{Z}^m_+, \, xs \leq 1\} \; \leq \; 1, \qquad (18.4)$$

where $s = (s_1, \ldots, s_m)$ is the vector of the item sizes.

(18.4) is a kind of KNAPSACK PROBLEM, so we cannot hope to solve it exactly. But this is not necessary, as the WEAK SEPARATION PROBLEM only calls for an approximate solution.

Write $y' := \lfloor \frac{2n}{\delta} y \rfloor$ (the rounding is done componentwise). The problem

$$\max\{y'x : x \in \mathbb{Z}^m_+, \, xs \leq 1\} \qquad (18.5)$$

can be solved optimally by dynamic programming, very similarly to the DYNAMIC PROGRAMMING KNAPSACK ALGORITHM in Section 17.2 (see Exercise 10 of Chapter 17): let $F(0) := 0$ and

$$F(k) \; := \; \min\{F(k - y_i') + s_i : i \in \{1, \ldots, m\}, \, y_i' \leq k\}$$

for $k = 1, \ldots, \frac{4n}{\delta}$. $F(k)$ is the minimum size of a set of items with total cost k (with respect to y').

Now the maximum in (18.5) is less than or equal to $\frac{2n}{\delta}$ if and only if $F(k) > 1$ for all $k \in \{\frac{2n}{\delta} + 1, \ldots, \frac{4n}{\delta}\}$. The total time needed to decide this is $O\left(\frac{mn}{\delta}\right)$. There are two cases:

Case 1: The maximum in (18.5) is less than or equal to $\frac{2n}{\delta}$. Then $\frac{\delta}{2n} y'$ is a feasible solution of (18.3). Furthermore, $by - b\frac{\delta}{2n} y' \leq b\frac{\delta}{2n} \mathbb{1} = \frac{\delta}{2}$. The task of the WEAK SEPARATION PROBLEM is done.

Case 2: There exists an $x \in \mathbb{Z}_+^m$ with $xs \leq 1$ and $y'x > \frac{2n}{\delta}$. Such an x can easily be computed from the numbers $F(k)$ in $O\left(\frac{mn}{\delta}\right)$ time. We have $yx \geq \frac{\delta}{2n} y'x > 1$. Thus x corresponds to a bin configuration that proves that y is infeasible. Since we have $zx \leq 1$ for all $z \in \mathcal{D}$, this is a separating hyperplane, and thus we are done. $\qquad\square$

Lemma 18.13. (Karmarkar and Karp [1982]) *Let I be an instance of the* BIN-PACKING PROBLEM *with only m different numbers, none of which is less than γ. Let $\delta > 0$. Then a feasible solution x of the primal LP (18.2) differing from the optimum by at most δ and having at most m nonzero components can be found in time polynomial in n, $\frac{1}{\delta}$ and $\frac{1}{\gamma}$.*

Proof: We first solve the dual LP (18.3) approximately, using Lemma 18.12. We obtain a vector y^* with $y^*b \geq \mathrm{OPT}(18.3) - \delta$. Now let $T_{k_1}, \ldots, T_{k_{N'}}$ be those bin configurations that appeared as a separating hyperplane in Case 2 of the previous proof, plus the unit vectors (the bin configurations containing just one element). Note that N' is bounded by the number of iterations in the GRÖTSCHEL-LOVÁSZ-SCHRIJVER ALGORITHM (Theorem 4.19), so $N' = O\left(m^2 \log \frac{mn}{\gamma\delta}\right)$.

Consider the LP

$$\max \quad yb$$

$$\text{s.t.} \quad \sum_{i=1}^{m} t_{k_j i} y_i \;\leq\; 1 \qquad (j = 1, \ldots, N') \tag{18.6}$$

$$y_i \;\geq\; 0 \qquad (i = 1, \ldots, m).$$

Observe that the above procedure for (18.3) (in the proof of Lemma 18.12) is also a valid application of the GRÖTSCHEL-LOVÁSZ-SCHRIJVER ALGORITHM for (18.6): the oracle for the WEAK SEPARATION PROBLEM can always give the same answer as above. Therefore we have $y^*b \geq \mathrm{OPT}(18.6) - \delta$. Consider

$$\min \quad \sum_{j=1}^{N'} x_{k_j}$$

$$\text{s.t.} \quad \sum_{j=1}^{N'} t_{k_j i} x_{k_j} \;\geq\; b_i \qquad (i = 1, \ldots, m) \tag{18.7}$$

$$x_{k_j} \;\geq\; 0 \qquad (j = 1, \ldots, N')$$

which is the dual of (18.6). The LP (18.7) arises from (18.2) by eliminating the variables x_j for $j \in \{1, \ldots, N\} \setminus \{k_1, \ldots, k_{N'}\}$ (forcing them to be zero). In other words, only N' of the N bin configurations can be used.

We have

$$\text{OPT}(18.7) - \delta \;=\; \text{OPT}(18.6) - \delta \;\leq\; y^* b \;\leq\; \text{OPT}(18.3) \;=\; \text{OPT}(18.2).$$

So it is sufficient to solve (18.7). But (18.7) is an LP of polynomial size: it has N' variables and m constraints; none of the entries of the matrix is larger than $\frac{1}{\gamma}$, and none of the entries of the right-hand side is larger than n. So by Khachiyan's Theorem 4.18, it can be solved in polynomial time. We obtain an optimum basic solution x (x is a vertex of the polyhedron, so x has at most m nonzero components). $\square$

Now we apply the FERNANDEZ-DE-LA-VEGA-LUEKER ALGORITHM with just one modification: we replace the exact solution of (18.2) by an application of Lemma 18.13. We summarize:

Theorem 18.14. (Karmarkar and Karp [1982]) *There is a fully polynomial asymptotic approximation scheme for the BIN-PACKING PROBLEM.*

Proof: We apply Lemma 18.13 with $\delta = 1$, obtaining an optimum solution x of (18.7) with at most m nonzero components. We have $\mathbb{1}x \leq \text{OPT}(18.2) + 1$. An application of Theorem 18.9 yields an integral solution using at most $\lceil \text{OPT}(18.2) \rceil + 1 + \frac{m-1}{2}$ bins, as required in ④ of the FERNANDEZ-DE-LA-VEGA-LUEKER ALGORITHM.

So the statement of Theorem 18.11 remains valid. Since $m \leq \frac{2}{\epsilon^2}$ and $\frac{1}{\gamma} \leq \frac{2}{\epsilon}$ (we may assume $\epsilon \leq 1$), the running time for finding x is polynomial in n and $\frac{1}{\epsilon}$. $\square$

The running time obtained in this way is worse than $O\left(\epsilon^{-40}\right)$ and completely out of the question for practical purposes. Karmarkar and Karp [1982] showed how to reduce the number of variables in (18.7) to m (while changing the optimum value only slightly) and thereby improve the running time (see Exercise 13). Plotkin, Shmoys and Tardos [1995] achieved a running time of $O(n \log \epsilon^{-1} + \epsilon^{-6} \log \epsilon^{-1})$.

Many generalizations have been considered. The two-dimensional bin packing problem, asking for packing a given set of axis-parallel rectangles into a minimum number of unit squares without rotation, does not have an asymptotic approximation scheme unless $P = NP$ (Bansal et al. [2006]). See Jansen, Prädel and Schwarz [2009], Bansal and Khan [2014], and the references therein, for related results.

Exercises

1. Let k be fixed. Describe a pseudopolynomial algorithm which – given an instance I of the BIN-PACKING PROBLEM – finds a solution for this instance using no more than k bins or decides that no such solution exists.

2. Consider the BIN-PACKING PROBLEM restricted to instances $a_1, \ldots, a_n$ with $a_i > \frac{1}{3}$ for $i = 1, \ldots, n$.
   (a) Reduce the problem to the CARDINALITY MATCHING PROBLEM.
   (b) Show how to solve the problem in $O(n \log n)$ time.

3. Consider the QUADRATIC ASSIGNMENT PROBLEM: given matrices $A, B \in \mathbb{R}_+^{n \times n}$, find a permutation π on $\{1, \ldots, n\}$ such that $\sum_{i,j=1}^{n} a_{i,j} b_{\pi(i), \pi(j)}$ is minimum. Show that this problem has no constant-factor approximation algorithm unless $P = NP$, even for the case that A is a 0-1-matrix and the entries of B define a metric.
   Hint: Use Theorem 18.3.
   (Queyranne [1986])

4. Find an instance I of the BIN-PACKING PROBLEM, where $FF(I) = 17$ while $OPT(I) = 10$.

5. Implement the FIRST-FIT ALGORITHM and the FIRST-FIT-DECREASING ALGORITHM to run in $O(n \log n)$ time.

6. Show that there is no online algorithm for the BIN-PACKING PROBLEM with performance ratio less than $\frac{4}{3}$.
   Hint: Note that we do not assume $P \neq NP$. There is no such algorithm, regardless of the running time. Consider the list consisting of n elements of size $\frac{1}{2} - \epsilon$ followed by n elements of size $\frac{1}{2} + \epsilon$.

7. Show that BIN PACKING with fixed number of different item sizes can be solved in time polynomial in the number of items.
   Hint: Compute which subsets of items can be packed into i bins, for $i = 1, 2 \ldots$, using dynamic programming.
   Note: This is not a polynomial-time algorithm since the instance can be represented by storing just the item sizes and the number (in binary encoding) of items of each size. However, Goemans and Rothvoß [2014] found a polynomial-time algorithm.

8. Show that any bin packing instance with only m different item sizes has a solution with at most 2^m different bin configurations.
   Hint: For a solution with more bin configurations, there are two bins for which the quantities of packed items have the same parities.
   (Eisenbrand and Shmonin [2006])

9. Show that ② of the FERNANDEZ-DE-LA-VEGA-LUEKER ALGORITHM can be implemented to run in $O\left(n \log \frac{1}{\epsilon}\right)$ time.

10. Consider the LP (18.3), which has a variable y_i for each $i = 1, \ldots, m$ (i.e., each item size). Assume $s_1 > \cdots > s_m$. Show that then there exists an optimum solution with $y_1 \geq \cdots \geq y_m$.
    (Caprara [2008])

* 11. Prove that for any $\epsilon > 0$ there exists a polynomial-time algorithm which for any instance $I = (a_1, \ldots, a_n)$ of the BIN-PACKING PROBLEM finds a packing using the optimum number of bins but may violate the capacity constraints by ϵ, i.e. an $f : \{1, \ldots, n\} \to \{1, \ldots, OPT(I)\}$ with $\sum_{f(i)=j} a_i \leq 1 + \epsilon$ for all $j \in \{1, \ldots, OPT(I)\}$.

Hint: Use ideas of Section 18.2.

(Hochbaum and Shmoys [1987])

12. Consider the following MULTIPROCESSOR SCHEDULING PROBLEM. Given a finite set A of tasks, a positive number $t(a)$ for each $a \in A$ (the processing time), and a number m of processors. Find a partition $A = A_1 \,\dot\cup\, A_2 \,\dot\cup\, \cdots \,\dot\cup\, A_m$ of A into m pairwise disjoint sets such that $\max_{i=1}^{m} \sum_{a \in A_i} t(a)$ is minimum.

   (a) Show that this problem is strongly *NP*-hard.

   (b) Show that a greedy algorithm which successively assigns jobs (in arbitrary order) to the currently least used machine is a 2-factor approximation algorithm.

   (c) Show that for each fixed m the problem has a fully polynomial approximation scheme.

   (Horowitz and Sahni [1976])

  * (d) Use Exercise 11 to show that the MULTIPROCESSOR SCHEDULING PROBLEM has an approximation scheme.

   (Hochbaum and Shmoys [1987])

   Note: This problem has been the subject of the first paper on approximation algorithms (Graham [1966]). Many variations of scheduling problems have been studied; see e.g. Graham et al. [1979] or Lawler et al. [1993].

 * 13. Consider the LP (18.6) in the proof of Lemma 18.13. All but m constraints can be omitted without changing its optimum value. We are not able to find these m constraints in polynomial time, but we can find m constraints such that deleting all the others does not increase the optimum value too much (e.g., by at most one). How?

   Hint: Let $D^{(0)}$ be the LP (18.6) and iteratively construct LPs $D^{(1)}, D^{(2)}, \ldots$ by deleting more and more constraints. At each iteration, a solution $y^{(i)}$ of $D^{(i)}$ is given with $by^{(i)} \geq \text{OPT}\left(D^{(i)}\right) - \delta$. The set of constraints is partitioned into $m+1$ sets of approximately equal size, and for each of the sets we test whether the set can be deleted. This test is performed by considering the LP after deletion, say $\overline{D}$, and applying the GRÖTSCHEL-LOVÁSZ-SCHRIJVER ALGORITHM. Let $\overline{y}$ be a solution of $\overline{D}$ with $b\overline{y} \geq \text{OPT}\left(\overline{D}\right) - \delta$. If $b\overline{y} \leq by^{(i)} + \delta$, the test is successful, and we set $D^{(i+1)} := \overline{D}$ and $y^{(i+1)} := \overline{y}$. Choose δ appropriately.

   (Karmarkar and Karp [1982])

14. Find an appropriate choice of ϵ as a function of SUM(I) such that the resulting modification of the KARMARKAR-KARP ALGORITHM is a polynomial-time algorithm which guarantees to find a solution with at most $\text{OPT}(I) + O((\text{OPT}(I))^{2/3})$.

   (Johnson [1982])

   Note: Hoberg and Rothvoß [2017] showed how to find a solution with at most $\text{OPT}(I) + O(\log \text{OPT}(I))$ bins.

References

General Literature

Coffman, E.G., Garey, M.R., and Johnson, D.S. [1996]: Approximation algorithms for bin-packing; a survey. In: Approximation Algorithms for *NP*-Hard Problems (D.S. Hochbaum, ed.), PWS, Boston, 1996

Cited References

Baker, B.S. [1985]: A new proof for the First-Fit Decreasing bin-packing algorithm. Journal of Algorithms 6 (1985), 49–70

Bansal, N., Correa, J.R., Kenyon, C., and Sviridenko, M. [2006]: Bin packing in multiple dimensions: inapproximability results and approximation schemes. Mathematics of Operations Research 31 (2006), 31–49

Bansal, N. and Khan, A. [2014]: Improved approximation algorithm for two-dimensional bin packing. Proceedings of the 25th Annual ACM-SIAM Symposium on Discrete Algorithms (2014), 13–25

Caprara, A. [2008]: Packing d-dimensional bins in d stages. Mathematics of Operations Research 33 (2008), 203–215

Dósa, G., Li, R., Han, X., and Tuza, Z. [2013]: Tight absolute bound for First Fit Descreasing bin-packing: $FFD(L) \leq 11/9\,OPT(L) + 6/9$. Theoretical Computer Science 510 (2013), 13–61

Dósa, G., and Sgall, J. [2013]: First fit bin packing: a tight analysis. Proceedings of the 30th International Symposium on Theoretical Aspects of Computer Science (2013), 538–549

Eisemann, K. [1957]: The trim problem. Management Science 3 (1957), 279–284

Eisenbrand, F., and Shmonin, G. [2006]: Carathéodory bounds for integer cones. Operations Research Letters 34 (2006), 564–568

Fernandez de la Vega, W., and Lueker, G.S. [1981]: Bin packing can be solved within $1 + \epsilon$ in linear time. Combinatorica 1 (1981), 349–355

Garey, M.R., Graham, R.L., Johnson, D.S., and Yao, A.C. [1976]: Resource constrained scheduling as generalized bin packing. Journal of Combinatorial Theory A 21 (1976), 257–298

Garey, M.R., and Johnson, D.S. [1975]: Complexity results for multiprocessor scheduling under resource constraints. SIAM Journal on Computing 4 (1975), 397–411

Garey, M.R., and Johnson, D.S. [1979]: Computers and Intractability; A Guide to the Theory of *NP*-Completeness. Freeman, San Francisco 1979, p. 127

Gilmore, P.C., and Gomory, R.E. [1961]: A linear programming approach to the cutting-stock problem. Operations Research 9 (1961), 849–859

Goemans, M.X., and Rothvoß, T. [2014]: Polynomiality for bin packing with a constant number of item types. Proceedings of the 25th Annual ACM-SIAM Symposium on Discrete Algorithms (2014), 830–839

Graham, R.L. [1966]: Bounds for certain multiprocessing anomalies. Bell Systems Technical Journal 45 (1966), 1563–1581

Graham, R.L., Lawler, E.L., Lenstra, J.K., and Rinnooy Kan, A.H.G. [1979]: Optimization and approximation in deterministic sequencing and scheduling: a survey. In: Discrete Optimization II; Annals of Discrete Mathematics 5 (P.L. Hammer, E.L. Johnson, B.H. Korte, eds.), North-Holland, Amsterdam 1979, pp. 287–326

Hochbaum, D.S., and Shmoys, D.B. [1987]: Using dual approximation algorithms for scheduling problems: theoretical and practical results. Journal of the ACM 34 (1987), 144–162

Hoberg, R., and Rothvoß, T. [2017]: A logarithmic additive integrality gap for bin packing. Proceedings of the 28th Annual ACM-SIAM Symposium on Discrete Algorithms (2017), 2616–2625

Horowitz, E., and Sahni, S.K. [1976]: Exact and approximate algorithms for scheduling nonidentical processors. Journal of the ACM 23 (1976), 317–327

Jansen, K., Prädel, L., and Schwarz, U.M. [2009]: Two for one: tight approximation of 2D bin packing. In: Algorithms and Data Structures – Proceedings of the 11th Algorithms and Data Structures Symposium; LNCS 5664 (F. Dehne, M. Gavrilova, J.-R. Sack, C.D. Tóth, eds.), Springer, Berlin 2009, pp. 399–410

Johnson, D.S. [1973]: Near-Optimal Bin Packing Algorithms. Doctoral Thesis, Dept. of Mathematics, MIT, Cambridge, MA, 1973

Johnson, D.S. [1974]: Fast algorithms for bin-packing. Journal of Computer and System Sciences 8 (1974), 272–314

Johnson, D.S. [1982]: The *NP*-completeness column; an ongoing guide. Journal of Algorithms 3 (1982), 288–300, Section 3

Johnson, D.S., Demers, A., Ullman, J.D., Garey, M.R., and Graham, R.L. [1974]: Worst-case performance bounds for simple one-dimensional packing algorithms. SIAM Journal on Computing 3 (1974), 299–325

Karmarkar, N., and Karp, R.M. [1982]: An efficient approximation scheme for the one-dimensional bin-packing problem. Proceedings of the 23rd Annual IEEE Symposium on Foundations of Computer Science (1982), 312–320

Lawler, E.L., Lenstra, J.K., Rinnooy Kan, A.H.G., and Shmoys, D.B. [1993]: Sequencing and scheduling: algorithms and complexity. In: Handbooks in Operations Research and Management Science; Vol. 4 (S.C. Graves, A.H.G. Rinnooy Kan, P.H. Zipkin, eds.), Elsevier, Amsterdam 1993

Lenstra, H.W. [1983]: Integer Programming with a fixed number of variables. Mathematics of Operations Research 8 (1983), 538–548

Papadimitriou, C.H. [1994]: Computational Complexity. Addison-Wesley, Reading 1994, pp. 204–205

Plotkin, S.A., Shmoys, D.B., and Tardos, É. [1995]: Fast approximation algorithms for fractional packing and covering problems. Mathematics of Operations Research 20 (1995), 257–301

Queyranne, M. [1986]: Performance ratio of polynomial heuristics for triangle inequality quadratic assignment problems. Operations Research Letters 4 (1986), 231–234

Seiden, S.S. [2002]: On the online bin packing problem. Journal of the ACM 49 (2002), 640–671

Simchi-Levi, D. [1994]: New worst-case results for the bin-packing problem. Naval Research Logistics 41 (1994), 579–585

van Vliet, A. [1992]: An improved lower bound for on-line bin packing algorithms. Information Processing Letters 43 (1992), 277–284

Yue, M. [1991]: A simple proof of the inequality $FFD(L) \leq \frac{11}{9} \text{OPT}(L) + 1, \forall L$, for the FFD bin-packing algorithm. Acta Mathematicae Applicatae Sinica 7 (1991), 321–331

19 Multicommodity Flows and Edge-Disjoint Paths

The MULTICOMMODITY FLOW PROBLEM is a generalization of the MAXIMUM FLOW PROBLEM. Given a digraph with edge capacities, we now ask for an s-t-flow for several pairs (s, t) (we speak of several commodities), such that the total flow through any edge does not exceed the capacity. We specify the pairs (s, t) by a second digraph; for technical reasons we have an edge from t to s when we ask for an s-t-flow. Formally we have:

DIRECTED MULTICOMMODITY FLOW PROBLEM

Instance: A pair (G, H) of digraphs on the same vertices.
Capacities $u : E(G) \to \mathbb{R}_+$ and demands $b : E(H) \to \mathbb{R}_+$.

Task: Find a family $(x^f)_{f \in E(H)}$, where x^f is an s-t-flow of value $b(f)$ in G for each $f = (t, s) \in E(H)$, and

$$\sum_{f \in E(H)} x^f(e) \leq u(e) \quad \text{for all } e \in E(G).$$

There is also an undirected version which we shall discuss later. The edges of G are called **supply edges**, the edges of H **demand edges** or **commodities**. Endpoints of demand edges are called **terminals**. If $u \equiv 1$, $b \equiv 1$ and x is forced to be integral, we have the EDGE-DISJOINT PATHS PROBLEM. Sometimes one also has edge weights and asks for a minimum cost multicommodity flow. But here we are only interested in feasible solutions.

Of course, the problem can be solved in polynomial time by means of LINEAR PROGRAMMING (cf. Theorem 4.18). However the LP formulations are quite large, so it is also interesting that we have a combinatorial algorithm for solving the problem approximately; see Section 19.2. This algorithm uses an LP formulation as a motivation. Moreover, LP duality yields a useful good characterization of our problem as shown in Section 19.1. However, in contrast to single-commodity flows, the dual LP does not correspond to a minimum cut problem. We will study the max-flow min-cut ratio in Sections 19.3 and 19.4.

© Springer-Verlag GmbH Germany 2018
Bernhard Korte and Jens Vygen, *Combinatorial Optimization*, Algorithms and Combinatorics 21, https://doi.org/10.1007/978-3-662-56039-6_19

In many applications one is interested in integral flows, or paths, and the EDGE-DISJOINT PATHS PROBLEM is the proper formulation. We have considered a special case of this problem in Section 8.2, where we had a necessary and sufficient condition for the existence of k edge-disjoint (or internally disjoint) paths from s to t for two given vertices s and t (Menger's Theorems 8.9 and 8.10). We shall prove that the general EDGE-DISJOINT PATHS PROBLEM is NP-hard, both in the directed and undirected case. Nevertheless there are some interesting special cases that can be solved in polynomial time, as we shall see in Sections 19.5 and 19.6.

19.1 Multicommodity Flows

We concentrate on the DIRECTED MULTICOMMODITY FLOW PROBLEM but mention that all results of this section also hold for the undirected version:

UNDIRECTED MULTICOMMODITY FLOW PROBLEM

Instance: A pair (G, H) of undirected graphs on the same vertices. Capacities $u : E(G) \to \mathbb{R}_+$ and demands $b : E(H) \to \mathbb{R}_+$.

Task: Find a family $(x^f)_{f \in E(H)}$, where x^f is an s-t-flow of value $b(f)$ in $(V(G), \{(v, w), (w, v) : \{v, w\} \in E(G)\})$ for each $f = \{t, s\} \in E(H)$, and

$$\sum_{f \in E(H)} \left(x^f((v, w)) + x^f((w, v)) \right) \leq u(e)$$

for all $e = \{v, w\} \in E(G)$.

Both versions of the MULTICOMMODITY FLOW PROBLEM have a natural formulation as an LP (cf. the LP formulation of the MAXIMUM FLOW PROBLEM in Section 8.1). Hence they can be solved in polynomial time (Theorem 4.18). Today exact polynomial-time algorithms which do not use LINEAR PROGRAMMING are known only for some special cases.

We shall now mention a different LP formulation of the MULTICOMMODITY FLOW PROBLEM which will prove useful:

Lemma 19.1. *Let (G, H, u, b) be an instance of the (DIRECTED or UNDIRECTED) MULTICOMMODITY FLOW PROBLEM. Let C be the set of circuits of $G + H$ that contain exactly one demand edge. Let M be a 0-1-matrix whose columns correspond to the elements of C and whose rows correspond to the edges of G, where $M_{e,C} = 1$ iff $e \in C$. Similarly, let N be a 0-1-matrix whose columns correspond to the elements of C and whose rows correspond to the edges of H, where $N_{f,C} = 1$ iff $f \in C$.*

Then each solution of the MULTICOMMODITY FLOW PROBLEM corresponds to at least one point in the polytope

$$\left\{ y \in \mathbb{R}^C : y \geq 0, \ My \leq u, \ Ny = b \right\}, \tag{19.1}$$

and each point in this polytope corresponds to a unique solution of the MULTI-COMMODITY FLOW PROBLEM.

Proof: To simplify our notation we consider the directed case only; the undirected case follows by substituting each undirected edge by the subgraph shown in Figure 8.2.

Let $(x^f)_{f \in E(H)}$ be a solution of the MULTICOMMODITY FLOW PROBLEM. For each $f = (t, s) \in E(H)$ the s-t-flow x^f can be decomposed into a set $\mathcal{P}$ of s-t-paths and a set $\mathcal{Q}$ of circuits (Theorem 8.8): for each demand edge f we can write

$$x^f(e) = \sum_{P \in \mathcal{P} \cup \mathcal{Q}: e \in E(P)} w(P)$$

for $e \in E(G)$, where $w : \mathcal{P} \cup \mathcal{Q} \to \mathbb{R}_+$. We set $y_{P+f} := w(P)$ for $P \in \mathcal{P}$ and $y_C := 0$ for $f \in C \in \mathcal{C}$ with $C - f \notin \mathcal{P}$. This obviously yields a vector $y \geq 0$ with $My \leq u$ and $Ny = b$.

Conversely, let $y \geq 0$ with $My \leq u$ and $Ny = b$. Setting

$$x^f(e) := \sum_{C \in \mathcal{C}: e, f \in E(C)} y_C$$

yields a solution of the MULTICOMMODITY FLOW PROBLEM. $\square$

With the help of LP duality we can now derive a necessary and sufficient condition for the solvability of the MULTICOMMODITY FLOW PROBLEM. We shall also mention the connection to the EDGE-DISJOINT PATHS PROBLEM.

Definition 19.2.   *An instance* (G, H) *of the* (DIRECTED *or* UNDIRECTED) EDGE-DISJOINT PATHS PROBLEM *satisfies the* **distance criterion** *if for each* $z : E(G) \to \mathbb{R}_+$

$$\sum_{f=(t,s) \in E(H)} \mathrm{dist}_{(G,z)}(s, t) \leq \sum_{e \in E(G)} z(e). \tag{19.2}$$

An instance (G, H, u, b) *of the* MULTICOMMODITY FLOW PROBLEM *satisfies the* **distance criterion** *if for each* $z : E(G) \to \mathbb{R}_+$

$$\sum_{f=(t,s) \in E(H)} b(f) \, \mathrm{dist}_{(G,z)}(s, t) \leq \sum_{e \in E(G)} u(e) z(e).$$

(In the undirected case, (t, s) *must be replaced by* $\{t, s\}$.)

The left-hand side of the distance criterion can be interpreted as a lower bound on the cost of a solution (with respect to edge costs z), while the right-hand side is an upper bound on the maximum possible cost.

Theorem 19.3.   *The distance criterion is necessary and sufficient for the solvability of the* MULTICOMMODITY FLOW PROBLEM *(in both the directed and the undirected case).*

Proof: We again consider only the directed case, the undirected case follows via the substitution of Figure 8.2. By Lemma 19.1, the MULTICOMMODITY FLOW PROBLEM has a solution if and only if the polyhedron $\{y \in \mathbb{R}^{\mathcal{C}}_+ : My \leq u, Ny = b\}$ is nonempty. By Corollary 3.25, this polyhedron is empty if and only if there are vectors z, w with $z \geq 0$, $zM + wN \geq 0$ and $zu + wb < 0$. (M and N are defined as above.)

The inequality $zM + wN \geq 0$ implies

$$-w_f \leq \sum_{e \in P} z_e$$

for each demand edge $f = (t, s)$ and each s-t-path P in G, so $-w_f \leq \mathrm{dist}_{(G,z)}(s, t)$. Hence there exist vectors z, w with $z \geq 0$, $zM + wN \geq 0$ and $zu + wb < 0$ if and only if there exists a vector $z \geq 0$ with

$$zu - \sum_{f=(t,s) \in E(H)} \mathrm{dist}_{(G,z)}(s, t)\, b(f) \;<\; 0.$$

This completes the proof.   □

In Section 19.2 we shall show how the LP description of Lemma 19.1 and its dual can be used to design an algorithm for the MULTICOMMODITY FLOW PROBLEM.

Theorem 19.3 implies that the distance criterion is necessary for the solvability of the EDGE-DISJOINT PATHS PROBLEM, since this can be considered as a MULTICOMMODITY FLOW PROBLEM with $b \equiv 1$, $u \equiv 1$ and with integrality constraints. Another important necessary condition is the following:

Definition 19.4.   *An instance (G, H) of the* (DIRECTED *or* UNDIRECTED) EDGE-DISJOINT PATHS PROBLEM *satisfies the* **cut criterion** *if for each $X \subseteq V(G)$*

- $|\delta^+_G(X)| \geq |\delta^-_H(X)|$ *in the directed case, or*
- $|\delta_G(X)| \geq |\delta_H(X)|$ *in the undirected case.*

An instance (G, H, u, b) of the (DIRECTED *or* UNDIRECTED) MULTICOMMODITY FLOW PROBLEM *satisfies the cut criterion if for each $X \subseteq V(G)$*

- $u(\delta^+_G(X)) \geq b(\delta^-_H(X))$ *in the directed case, or*
- $u(\delta_G(X)) \geq b(\delta_H(X))$ *in the undirected case.*

Corollary 19.5.   *For an instance (G, H) of the* (DIRECTED *or* UNDIRECTED) EDGE-DISJOINT PATHS PROBLEM, *the following implications hold: (G, H) has a solution $\Rightarrow (G, H)$ satisfies the distance criterion $\Rightarrow (G, H)$ satisfies the cut criterion.*

Proof: The first implication follows from Theorem 19.3. For the second implication observe that the cut criterion is just a special case of the distance criterion,

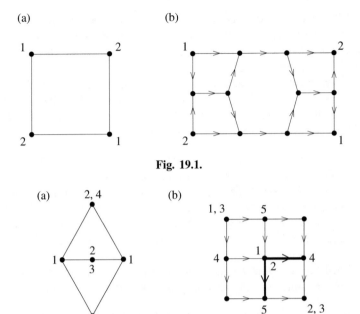

Fig. 19.1.

Fig. 19.2.

where weight functions of the type

$$z(e) := \begin{cases} 1 & \text{if } e \in \delta^+(X) \text{ (directed case) or } e \in \delta(X) \text{ (undirected case)} \\ 0 & \text{otherwise} \end{cases}$$

for $X \subseteq V(G)$ are considered.   □

None of the implications can be reversed in general. Figure 19.1 shows examples where there is no (integral) solution but there is a fractional solution, i.e. a solution of the multicommodity flow relaxation. So here the distance criterion is satisfied. In the figures of this section demand edges are indicated by equal numbers at their endpoints. In the directed case, one should orient the demand edges so that they are realizable. (A demand edge (t, s) or $\{t, s\}$ is called **realizable** if t is reachable from s in the supply graph.)

The two examples shown in Figure 19.2 satisfy the cut criterion (this is easily checked), but not the distance criterion: in the undirected example choose $z(e) = 1$ for all $e \in E(G)$, in the directed example choose $z(e) = 1$ for the bold edges and $z(e) = 0$ otherwise.

Nagamochi and Ibaraki [1989] showed that if a directed instance (G, H) has a fractional but no integral solution, then there is an instance (G', H') that has no fractional solution although the cut criterion holds, such that G' is a subgraph of G and H' is a subgraph of H.

We remark that it is *NP*-hard to check whether a given instance of the UNDI-RECTED EDGE-DISJOINT PATHS PROBLEM satisfies the cut criterion (Exercise 2). This is in contrast to the distance criterion, which can be checked in polynomial time by linear programming.

For some classes of instances of the EDGE-DISJOINT PATHS PROBLEM the cut criterion is sufficient for the existence of a solution. We will see some examples later in this chapter. A first example is Menger's Theorem. To generalize this slightly, we use the following construction.

Lemma 19.6. *Let* (G, H) *be an instance of the* (DIRECTED *or* UNDIRECTED) EDGE-DISJOINT PATHS PROBLEM *and* $v \in V(G)$.

Let (G', H') *result from* (G, H) *by adding a new vertex* x *and doing the following for each demand edge* f *incident to* v: *If* $f = \{v, w\}$, *then replace* f *by* $\{v, x\}$ *and add a supply edge* $\{w, x\}$. *If* $f = (v, w)$, *then replace* f *by* (v, x) *and add a supply edge* (x, w). *If* $f = (w, v)$, *then replace* f *by* (x, v), *and add a supply edge* (w, x).

Then (G', H') *has a solution if and only if* (G, H) *has a solution. Moreover,* (G', H') *satisfies the cut criterion if and only if* (G, H) *does.*

Proof: Any solution to (G', H') yields a solution to (G, H) by omitting the new supply edges. Conversely, any solution to (G, H) can be transformed into a solution to (G', H') by appending the new supply edges.

For the cut criterion, we observe that for any set $X \subseteq V(G') = V(G) \cup \{x\}$ we have

$$|\delta_{G'}^+(X)| - |\delta_{H'}^-(X)| \geq |\delta_G^+(X \setminus \{x\})| - |\delta_H^-(X \setminus \{x\})|$$

or

$$|\delta_{G'}(X)| - |\delta_{H'}(X)| \geq |\delta_G(X \setminus \{x\})| - |\delta_H(X \setminus \{x\})|,$$

respectively, with equality if $|X \cap \{v, x\}| = 1$. □

Theorem 19.7. *Let* (G, H) *be an instance of the* (DIRECTED *or* UNDIRECTED) EDGE-DISJOINT PATHS PROBLEM *with a vertex* v *such that*

(a) $f \in \delta^+(v)$ *for all* $f \in E(H)$, *or*
(b) $f \in \delta^-(v)$ *for all* $f \in E(H)$ *in the directed case, or*
(c) $f \in \delta(v)$ *for all* $f \in E(H)$ *in the undirected case.*

Then (G, H) *has a solution if and only if the cut criterion holds.*

Proof: By Lemma 19.6 we can reduce this to the case where H contains only one set of parallel edges. In this case the sufficiency of the cut criterion is a reformulation of Menger's Theorem 8.9. □

19.2 Algorithms for Multicommodity Flows

The definition of the MULTICOMMODITY FLOW PROBLEM directly gives rise to an LP formulation of polynomial size. Although this yields a polynomial-time algorithm it cannot be used for solving large instances: the number of variables is enormous. The LP description (19.1) given by Lemma 19.1 looks even worse since it has an exponential number of variables. Nevertheless this description proves much more useful in practice. We shall explain this now.

Since we are interested in a feasible solution only, we consider the LP

$$\max\{0y : y \geq 0, \, My \leq u, \, Ny = b\}$$

and its dual $\min\{zu + wb : z \geq 0, \, zM + wN \geq 0\}$ which we can rewrite as

$$\min\{zu + wb : z \geq 0, \, \text{dist}_{(G,z)}(s, t) \geq -w(f) \text{ for all } f = (t, s) \in E(H)\}.$$

(In the undirected case replace (t, s) by $\{t, s\}$.) This dual LP has only $|E(G)| + |E(H)|$ variables but an exponential number of constraints. However, this is not important since the SEPARATION PROBLEM can be solved by $|E(H)|$ shortest path computations; as only nonnegative vectors z have to be considered, we can use DIJKSTRA'S ALGORITHM here. If the dual LP is unbounded, then this proves infeasibility of the primal LP. Otherwise we can solve the dual LP, but this does not provide a primal solution in general.

Ford and Fulkerson [1958] suggested to use the above consideration to solve the primal LP directly, in combination with the SIMPLEX ALGORITHM. Since most variables are zero at each iteration of the SIMPLEX ALGORITHM, one only keeps track of those variables for which the nonnegativity constraint $y_C \geq 0$ does not belong to the current set J of active rows. The other variables are not stored explicitly but "generated" when they are needed (when the nonnegativity constraint becomes inactive). The decision of which variable has to be generated in each step is equivalent to the SEPARATION PROBLEM for the dual LP, so in our case it reduces to a SHORTEST PATH PROBLEM. This column generation technique can be quite efficient in practice.

Even with these techniques there are many practical instances that cannot be solved optimally. However, the above scheme also gives rise to an approximation algorithm. Let us first formulate our problem as an optimization problem:

MAXIMUM MULTICOMMODITY FLOW PROBLEM

Instance: A pair (G, H) of digraphs on the same vertices.
Capacities $u : E(G) \to \mathbb{R}_+$.

Task: Find a family $(x^f)_{f \in E(H)}$, where x^f is an s-t-flow in G for each $f = (t, s) \in E(H)$, $\sum_{f \in E(H)} x^f(e) \leq u(e)$ for all $e \in E(G)$, and the total flow value $\sum_{f \in E(H)} \text{value}(x^f)$ is maximum.

There are other interesting formulations. For example one can look for flows satisfying the greatest possible fraction of given demands (the CONCURRENT FLOW

PROBLEM), or for flows satisfying given demands but violating capacities as slightly as possible. Moreover one can consider costs on edges. We consider only the MAXIMUM MULTICOMMODITY FLOW PROBLEM; other problems can be attacked with similar techniques.

We again consider our LP formulation

$$\max \left\{ \sum_{P \in \mathcal{P}} y(P) : y \geq 0, \quad \sum_{P \in \mathcal{P}: e \in E(P)} y(P) \leq u(e) \text{ for all } e \in E(G) \right\},$$

where $\mathcal{P}$ is the family of the s-t-paths in G for all $(t, s) \in E(H)$, and its dual

$$\min \left\{ zu : z \geq 0, \quad \sum_{e \in E(P)} z(e) \geq 1 \text{ for all } P \in \mathcal{P} \right\}.$$

We shall describe a primal-dual algorithm based on these formulations which turns out to be a fully polynomial approximation scheme. This algorithm always has a primal vector $y \geq 0$ that is not necessarily a feasible primal solution since capacity constraints might be violated. Initially $y = 0$. At the end we shall multiply y by a constant in order to meet all constraints. To store y efficiently we keep track of the family $\mathcal{P}' \subseteq \mathcal{P}$ of those paths P with $y(P) > 0$; in contrast to $\mathcal{P}$ the cardinality of $\mathcal{P}'$ will be polynomially bounded.

The algorithm also has a dual vector $z \geq 0$. Initially, $z(e) = \delta$ for all $e \in E(G)$, where δ depends on n and the error parameter ϵ. In each iteration, it finds a maximally violated dual constraint (corresponding to a shortest s-t-path for $(t, s) \in E(H)$, with respect to edge lengths z) and increases z and y along this path:

MULTICOMMODITY FLOW APPROXIMATION SCHEME

Input: A pair (G, H) of digraphs on the same vertices.
           Capacities $u : E(G) \to \mathbb{R}_+ \setminus \{0\}$. A number ϵ with $0 < \epsilon \leq \frac{1}{2}$.

Output: Numbers $y : \mathcal{P} \to \mathbb{R}_+$ with $\sum_{P \in \mathcal{P}: e \in E(P)} y(P) \leq u(e)$ for all $e \in E(G)$.

---

① Set $y(P) := 0$ for all $P \in \mathcal{P}$.
   Set $\delta := (n(1 + \epsilon))^{-\lceil \frac{5}{\epsilon} \rceil}(1 + \epsilon)$ and $z(e) := \delta$ for all $e \in E(G)$.

② Let $P \in \mathcal{P}$ such that $z(E(P))$ is minimum.
   If $z(E(P)) \geq 1$, then go to ④.

③ Let $\gamma := \min_{e \in E(P)} u(e)$.
   Set $y(P) := y(P) + \gamma$.
   Set $z(e) := z(e) \left(1 + \frac{\epsilon \gamma}{u(e)}\right)$ for all $e \in E(P)$.
   Go to ②.

④   Let $\xi := \max\limits_{e \in E(G)} \dfrac{1}{u(e)} \sum\limits_{P \in \mathcal{P}: e \in E(P)} y(P)$.

Set $y(P) := \dfrac{y(P)}{\xi}$ for all $P \in \mathcal{P}$.

This algorithm is due to Young [1995] and Garg and Könemann [2007], based on earlier work of Shahrokhi and Matula [1990], Shmoys [1996], and others.

Theorem 19.8.   (Garg and Könemann [2007])   *The* MULTICOMMODITY FLOW APPROXIMATION SCHEME *produces a feasible solution with total flow value at least* $\frac{1}{1+\epsilon}$ OPT(G, H, u). *Its running time is* $O\left(\frac{1}{\epsilon^2}km(m + n\log n)\log n\right)$, *where* $k = |E(H)|, n = |V(G)|$ *and* $m = |E(G)|$, *so it is a fully polynomial approximation scheme.*

Proof: In each iteration the value $z(e)$ increases by a factor $1 + \epsilon$ for at least one edge e (the bottleneck edge). Since an edge e with $z(e) \geq 1$ is never used anymore in any path, the total number of iterations is $t \leq m\lceil\log_{1+\epsilon}(\frac{1}{\delta})\rceil$. In each iteration we have to solve k instances of the SHORTEST PATH PROBLEM with nonnegative weights to determine P. Using DIJKSTRA'S ALGORITHM (Theorem 7.4) we get an overall running time of $O(tk(m + n\log n)) = O\left(km(m + n\log n)\log_{1+\epsilon}(\frac{1}{\delta})\right)$. The stated running time now follows from observing that, for $0 < \epsilon \leq 1$,

$$\log_{1+\epsilon}\left(\frac{1}{\delta}\right) = \frac{\log(\frac{1}{\delta})}{\log(1+\epsilon)} \leq \frac{\lceil\frac{5}{\epsilon}\rceil\log(2n)}{\frac{\epsilon}{2}} = O\left(\frac{\log n}{\epsilon^2}\right);$$

here we used $\log(1 + \epsilon) \geq \frac{\epsilon}{2}$ for $0 < \epsilon \leq 1$.

We also have to check that the maximum number of bits needed to store any number occurring in the computation is bounded by a polynomial in $\log n +$ size$(u) +$ size$(\epsilon) + \frac{1}{\epsilon}$. This is clear for the y-variables. The number δ can be stored with $O(\frac{1}{\epsilon}$ size$(n(1 + \epsilon)) +$ size$(\epsilon)) = O(\frac{1}{\epsilon}(\log n +$ size$(\epsilon)))$ bits. To deal with the z-variables we assume that u is integral; otherwise we multiply all capacities by the product of the denominators in the beginning (cf. Proposition 4.1). Then the denominator of the z-variables is bounded at any time by the product of all capacities and the denominator of δ. Since the numerator is at most twice the denominator we have shown that the size of all numbers is indeed polynomial in the input size and $\frac{1}{\epsilon}$.

The feasibility of the solution is guaranteed by ④.

Note that every time we add γ units of flow on edge e we increase the weight $z(e)$ by a factor $\left(1 + \frac{\epsilon\gamma}{u(e)}\right)$. This value is at least $(1+\epsilon)^{\frac{\gamma}{u(e)}}$ because $1+\epsilon a \geq (1+\epsilon)^a$ holds for $0 \leq a \leq 1$ (both sides of this inequality are equal for $a \in \{0, 1\}$, and the left-hand side is linear in a while the right-hand side is convex). Since e is not used once $z(e) \geq 1$, we cannot add more than $u(e)(1 + \log_{1+\epsilon}(\frac{1}{\delta}))$ units of flow on edge e. Hence

$$\xi \leq 1 + \log_{1+\epsilon}\left(\frac{1}{\delta}\right) = \log_{1+\epsilon}\left(\frac{1+\epsilon}{\delta}\right). \tag{19.3}$$

Let $z^{(i)}$ denote the vector z after iteration i, and let P_i and γ_i be the path P and the number γ in iteration i. We have $z^{(i)}u = z^{(i-1)}u + \epsilon\gamma_i\sum_{e\in E(P_i)} z^{(i-1)}(e)$, so $(z^{(i)} - z^{(0)})u = \epsilon\sum_{j=1}^i \gamma_j\alpha(z^{(j-1)})$, where $\alpha(z) := \min_{P\in\mathcal{P}} z(E(P))$. Let us write $\beta := \min\left\{zu : z \in \mathbb{R}_+^{E(G)}, \alpha(z) \geq 1\right\}$. Then $(z^{(i)} - z^{(0)})u \geq \beta\alpha(z^{(i)} - z^{(0)})$ and thus $(\alpha(z^{(i)}) - \delta n)\beta \leq \alpha(z^{(i)} - z^{(0)})\beta \leq (z^{(i)} - z^{(0)})u$. We obtain

$$\alpha(z^{(i)}) \leq \delta n + \frac{\epsilon}{\beta}\sum_{j=1}^i \gamma_j\alpha(z^{(j-1)}). \tag{19.4}$$

We now prove

$$\delta n + \frac{\epsilon}{\beta}\sum_{j=1}^i \gamma_j\alpha(z^{(j-1)}) \leq \delta n e^{\left(\frac{\epsilon}{\beta}\sum_{j=1}^i \gamma_j\right)}. \tag{19.5}$$

by induction on i (here e denotes the base of the natural logarithm). The case $i = 0$ is trivial. For $i > 0$ we have

$$\delta n + \frac{\epsilon}{\beta}\sum_{j=1}^i \gamma_j\alpha(z^{(j-1)}) = \delta n + \frac{\epsilon}{\beta}\sum_{j=1}^{i-1} \gamma_j\alpha(z^{(j-1)}) + \frac{\epsilon}{\beta}\gamma_i\alpha(z^{(i-1)})$$

$$\leq \left(1 + \frac{\epsilon}{\beta}\gamma_i\right)\delta n e^{\left(\frac{\epsilon}{\beta}\sum_{j=1}^{i-1}\gamma_j\right)},$$

using (19.4) and the induction hypothesis. Using $1 + x < e^x$ for all $x > 0$ the proof of (19.5) is complete.

In particular we conclude from (19.4), (19.5) and the stopping criterion that

$$1 \leq \alpha(z^{(t)}) \leq \delta n e^{\left(\frac{\epsilon}{\beta}\sum_{j=1}^t \gamma_j\right)},$$

hence $\sum_{j=1}^t \gamma_j \geq \frac{\beta}{\epsilon}\ln\left(\frac{1}{\delta n}\right)$. Now observe that the total flow value that the algorithm computes is $\sum_{P\in\mathcal{P}} y(P) = \frac{1}{\xi}\sum_{j=1}^t \gamma_j$. By the above and (19.3) this is at least

$$\frac{\beta\ln\left(\frac{1}{\delta n}\right)}{\epsilon\log_{1+\epsilon}\left(\frac{1+\epsilon}{\delta}\right)} = \frac{\beta\ln(1+\epsilon)}{\epsilon} \cdot \frac{\ln\left(\frac{1}{\delta n}\right)}{\ln\left(\frac{1+\epsilon}{\delta}\right)}$$

$$= \frac{\beta\ln(1+\epsilon)}{\epsilon} \cdot \frac{(\lceil\frac{5}{\epsilon}\rceil - 1)\ln(n(1+\epsilon))}{\lceil\frac{5}{\epsilon}\rceil\ln(n(1+\epsilon))}$$

$$\geq \frac{\beta(1 - \frac{\epsilon}{5})\ln(1+\epsilon)}{\epsilon}$$

by the choice of δ. Recall that β is the optimum value of the dual LP, and hence, by the LP Duality Theorem 3.20, the optimum value of a primal solution. Moreover, $\ln(1 + \epsilon) \geq \epsilon - \frac{\epsilon^2}{2}$ (this inequality is trivial for $\epsilon = 0$ and the derivative of the left-hand side is greater than that of the right-hand side for every $\epsilon > 0$). Hence

$$\frac{(1 - \frac{\epsilon}{5})\ln(1 + \epsilon)}{\epsilon} \geq \left(1 - \frac{\epsilon}{5}\right)\left(1 - \frac{\epsilon}{2}\right) = \frac{1 + \frac{3}{10}\epsilon - \frac{6}{10}\epsilon^2 + \frac{1}{10}\epsilon^3}{1 + \epsilon} \geq \frac{1}{1 + \epsilon}$$

for $\epsilon \leq \frac{1}{2}$. We conclude that the algorithm finds a solution whose total flow value is at least $\frac{1}{1+\epsilon}$ OPT(G, H, u). $\square$

A different algorithm which gives the same running time (by a more complicated analysis) was published before by Grigoriadis and Khachiyan [1996]. Fleischer [2000] improved the running time of the above algorithm by a factor of k. She observed that it is sufficient to compute an approximate shortest path in ②, and used this fact to show that it is not necessary to do a shortest path computation for each $(t, s) \in E(H)$ in each iteration. See also Karakostas [2008], Müller, Radke and Vygen [2011], Bienstock and Iyengar [2006], and Chudak and Eleutério [2005].

19.3 Sparsest Cut and Max-Flow Min-Cut Ratio

Consider the following problem:

GENERALIZED SPARSEST CUT PROBLEM

Instance: An instance (G, H, u, b) of the UNDIRECTED MULTICOMMODITY FLOW PROBLEM.

Task: Find a set $X \subset V(G)$ with $b(\delta_H(X)) > 0$ and $\frac{u(\delta_G(X))}{b(\delta_H(X))}$ minimum.

The special case where H is the complete undirected graph and $b(f) = 1$ for all $f \in E(H)$ is called the SPARSEST CUT PROBLEM: here we ask for a nonempty proper subset X of the vertex set such that $\frac{u(\delta_G(X))}{|X||V(G)\setminus X|}$ is minimum. The corresponding special case of the UNDIRECTED MULTICOMMODITY FLOW PROBLEM is called the UNIFORM MULTICOMMODITY FLOW PROBLEM.

The minimum value of $\frac{u(\delta_G(X))}{b(\delta_H(X))}$ is clearly an upper bound on the optimum value of the instance (G, H, u, b) of the CONCURRENT FLOW PROBLEM, which can be written as

$$\max\left\{\lambda \ : \ y(P) \geq 0 \ (P \in \mathcal{P}), \ \sum_{P \in \mathcal{P}_f} y(P) \geq \lambda b(f) \ (f \in E(H)),\right.$$

$$\left.\sum_{P \in \mathcal{P}: e \in E(P)} y(P) \leq u(e) \ (e \in E(G))\right\}, \tag{19.6}$$

where $\mathcal{P}_f$ denotes the family of the s-t-paths in G, for $f = \{t, s\} \in E(H)$, and $\mathcal{P} = \bigcup_{f \in E(H)} \mathcal{P}_f$.

If we introduce decision variables $z(e) \in \{0, 1\}$ for $e \in E(G)$, with the intended meaning that z is the incidence vector of $\delta_G(X)$, we can write the GENERALIZED SPARSEST CUT PROBLEM as an integer nonlinear program

$$\min\left\{\frac{u^\top z}{b^\top w} \;:\; z \in \{0, 1\}^{E(G)}, \; w \in \{0, 1\}^{E(H)}, \; b^\top w > 0, \right.$$

$$\left. \sum_{e \in E(P)} z(e) \geq w(f) \;\; (P \in \mathcal{P}_f, \; f \in E(H)) \right\}. \tag{19.7}$$

Proposition 19.9. *The* GENERALIZED SPARSEST CUT PROBLEM *is equivalent to (19.7): the optimum values are the same, and from a solution to the latter we can obtain in linear time an at least as good solution of the former (and vice versa).*

Proof: For a set $X \subset V(G)$ with $b(\delta_H(X)) > 0$ let z and w be the incidence vectors of $\delta_G(X)$ and $\delta_H(X)$, respectively. Then (z, w) is a feasible solution of (19.7) and $\frac{u^\top z}{b^\top w} = \frac{u(\delta_G(X))}{b(\delta_H(X))}$.

Conversely, if (z, w) is a feasible solution of (19.7), then let $X_1, \ldots, X_p$ be the connected components of $(V(G), \{e \in E(G) : z(e) = 0\})$. We have $\min_{i=1}^{p} \frac{u(\delta_G(X_i))}{b(\delta_H(X_i))} \leq \frac{\sum_{i=1}^{p} u(\delta_G(X_i))}{\sum_{i=1}^{p} b(\delta_H(X_i))} \leq \frac{2u^\top z}{2b^\top w} = \frac{u^\top z}{b^\top w}$. $\square$

Any feasible solution to (19.7) can be scaled (by multiplying all variables with an appropriate positive constant) to a feasible solution of

$$\min\left\{u^\top z \;:\; z \in \mathbb{R}_+^{E(G)}, \; w \in \mathbb{R}_+^{E(H)}, \; b^\top w = 1, \right.$$

$$\left. \sum_{e \in E(P)} z(e) \geq w(f) \;\; (P \in \mathcal{P}_f, \; f \in E(H)) \right\}, \tag{19.8}$$

of course with the same objective function value. So (19.8) can be viewed as an LP relaxation.

Lemma 19.10. *The linear program (19.8) can be solved in polynomial time.*

Proof: If t is not reachable from s in G for some $f = \{t, s\} \in E(H)$ with $b(f) > 0$, the LP is easy to solve, and the optimum value is zero. Otherwise there is an optimum solution (z, w) with $w(f) = \text{dist}_{(G,z)}(s, t)$ for all $f = \{t, s\} \in E(H)$. We add new variables $d(\{a, b\})$ for $a, b \in V(G)$ and replace the constraints $\sum_{e \in E(P)} z(e) \geq w(f)$ for $P \in \mathcal{P}_f$ and $f \in E(H)$ equivalently by $d(e) \leq z(e)$ for $e \in E(G)$, $w(f) \leq d(f)$ for $f \in E(H)$, and $d(\{a, c\}) \leq d(\{a, b\}) + d(\{b, c\})$ for $a, b, c \in V(G)$. We obtain an LP of polynomial size and can apply Theorem 4.18. $\square$

But there is also a combinatorial fully polynomial approximation scheme similar to the one in Section 19.2, because (19.8) is the dual LP of (19.6), the CONCURRENT FLOW PROBLEM (cf. Exercise 5).

The maximum ratio of the optimum values of (19.7) and (19.8) is called the **max-flow min-cut ratio** of the multicommodity flow instance. It can be as large as $\Theta(\log n)$, where $n = |V(G)|$, even in the uniform case (i.e., when H is the complete graph and $b \equiv 1$). This is demonstrated by an expander graph with unit capacities (cf. Exercise 6). In the next section we show that it is never worse.

19.4 The Leighton-Rao Theorem

We will use a solution (z, w) to the LP (19.8) in order to compute a sparse cut. If (z, w) is an optimum solution, then $W := u^\top z$ is a lower bound on the sparsity $\min_{X \subset V(G)} \frac{u(\delta_G(X))}{|X||V(G)\setminus X|}$. We will not need w in the following. Following Leighton and Rao [1999], we first partition our graph into low diameter sets such that the capacity of the multicut is rather small.

Lemma 19.11. (Leighton and Rao [1999]) *Let G be an undirected graph with edge weights $z : E(G) \to \mathbb{R}_+$ and capacities $u : E(G) \to \mathbb{R}_+$. The diameter of a nonempty set $X \subseteq V(G)$ is defined as $\max_{v,v' \in X} \operatorname{dist}_{(G,z)}(v, v')$. Let $d > 0$.*

Then $V(G)$ can be partitioned into sets $V_1, \ldots, V_k$, each with diameter at most d, such that the total capacity of edges in the multicut is at most $\frac{8W \log n}{d}$, where $W := \sum_{e \in E(G)} u(e)z(e)$.

Moreover, such a partition can be computed in $O(m \log n)$ time, where $n := |V(G)|$ and $m := |E(G)|$.

Proof: We first deal with some trivial cases. Let $U := u(E(G))$. If $U = 0$ or $d \leq \frac{8W \log n}{U}$, the partition into singletons does the job. If $W = 0$, then we take the connected components of $(V(G), \{e \in E(G) : z(e) = 0\})$.

Otherwise we have $W > 0$ and $\epsilon := \frac{2W \log n}{Ud} < \frac{1}{4}$. Set $z'(e) := \lceil Uz(e)/W \rceil$ for $e \in E(G)$. Note that $U \leq \sum_{e \in E(G)} z'(e)u(e) < 2U$.

Then we apply the following procedure, starting with $i := 1$ and $G_1 := G$. Choose $v_i \in V(G_i)$. For $x \in \mathbb{Z}_+$ let $U_x^i := \frac{2U}{n-1} + \sum_{\xi=0}^{x-1} u(\delta_{G_i}(B_\xi^i))$, where B_ξ^i denotes the set of vertices at distance at most ξ from v_i in (G_i, z'). Choose $r_i \geq 0$ smallest such that $U_{r_i+1}^i < 2^\epsilon U_{r_i}^i$. Set $V_i := B_{r_i}^i$. Set $G_{i+1} := G_i - V_i$, increment i, and iterate until no vertices are left.

We get a partition $V(G) = V_1 \,\dot\cup\, \cdots \,\dot\cup\, V_k$. For each i we have $U_{r_i}^i \geq 2^{\epsilon r_i} U_0^i = 2^{\epsilon r_i} \frac{2U}{n-1}$ and hence $2^{\epsilon r_i} \frac{2U}{n-1} \leq U_{r_i}^i \leq \frac{2U}{n-1} + \sum_{e \in E(G)} z'(e)u(e) \leq \frac{2Un}{n-1}$. Thus $r_i \leq \frac{\log n}{\epsilon}$. For any two vertices $v, v' \in V_i$ we have $\operatorname{dist}_{(G,z)}(v, v') \leq \frac{W}{U} \operatorname{dist}_{(G,z')}(v, v') \leq \frac{W}{U}(\operatorname{dist}_{(G,z')}(v_i, v) + \operatorname{dist}_{(G,z')}(v_i, v')) \leq \frac{W}{U} 2r_i \leq \frac{2W \log n}{U\epsilon} = d$.

We have $u(\delta_{G_i}(V_i)) = U_{r_i+1}^i - U_{r_i}^i < (2^\epsilon - 1)U_{r_i}^i \leq \epsilon U_{r_i}^i$ for $i = 1, \ldots, k$. Summation yields

$$u(\delta_G(V_1, \ldots, V_k)) = \sum_{i=1}^{k} u(\delta_{G_i}(V_i))$$

$$\leq \sum_{i=1}^{k} \epsilon U_{r_i}^i$$

$$= \epsilon \left(\frac{2Uk}{n-1} + \sum_{i=1}^{k} \sum_{\xi=0}^{r_i-1} u(\delta_{G_i}(B_\xi^i)) \right)$$

$$\leq \epsilon \left(2U + \sum_{e \in E(G)} z'(e)u(e) \right)$$

$$< 4\epsilon U.$$

This procedure can be implemented similarly to DIJKSTRA'S ALGORITHM. In each iteration we only need to consider $x = 0$ and the values $x > 0$ for which $\delta_{G_i}(B_x^i) \neq \delta_{G_i}(B_{x-1}^i)$. After each of these steps we check whether $r_i = x$. Otherwise we compute the value $x' = x + 1 + \lfloor 1/(2^\epsilon - 1) - U_x^i/u(\delta_{G_i}(B_x^i)) \rfloor$. If $\delta_{G_i}(B_\xi^i)$ remains constant for $\xi = x, x+1, \ldots, x'$, then x' is the smallest integer for which $u(\delta_{G_i}(B_{x'}^i)) = u(\delta_{G_i}(B_x^i)) < (2^\epsilon - 1)(U_x^i + (x' - x)u(\delta_{G_i}(B_x^i))) = (2^\epsilon - 1)U_{x'}^i$, and thus the number r_i that we look for.

Hence we have $O(\sum_{w \in B_{r_i}^i} (1 + |\delta(w)|))$ operations in iteration i, and $O(m + n \log n)$ total time with Fibonacci heaps. $\qquad\square$

Now we can prove the main theorem of Leighton and Rao [1999]:

Theorem 19.12. *Let G be a graph with edge weights $z : E(G) \to \mathbb{R}_+$ and capacities $u : E(G) \to \mathbb{R}_+$. Suppose that $\sum_{\{v,v'\} \in \binom{V(G)}{2}} \mathrm{dist}_{(G,z)}(v, v') = 1$. Let $n := |V(G)|$. Then we can find a nonempty proper subset $X \subseteq V(G)$ with*

$$\frac{u(\delta(X))}{|X||V(G) \setminus X|} \leq 36\, W \log n$$

in polynomial time, where again $W := \sum_{e \in E(G)} u(e)z(e)$.

Proof: We first apply Lemma 19.11 with $d = \frac{1}{n^2}$. We get a partition $V(G) = V_1 \dot\cup \cdots \dot\cup V_k$.

If none of the sets $V_1, \ldots, V_k$ has more than $\frac{2n}{3}$ vertices, then we successively unite the two smallest sets until only two are left. Let X and $V(G) \setminus X$ be the resulting sets. Then $|X||V(G) \setminus X| \geq \frac{n}{3} \cdot \frac{2n}{3}$ and $u(\delta(X)) \leq u(\delta(V_1, \ldots, V_k)) \leq 8\, W \log n/d = 8\, Wn^2 \log n$, and hence $\frac{u(\delta(X))}{|X||V(G)\setminus X|} \leq 36\, W \log n$, as required.

Otherwise one of the sets, say V_1, has more than $\frac{2n}{3}$ vertices and diameter at most $\frac{1}{n^2}$.

Let B_ξ denote the set of vertices with distance at most ξ from V_1. Then

$$\int_{\xi \geq 0} |B_\xi||V(G) \setminus B_\xi|\, d\xi \geq |V_1| \int_{\xi \geq 0} |V(G) \setminus B_\xi|\, d\xi$$

$$= |V_1| \sum_{v \in V(G)} \mathrm{dist}_{(G,z)}(v, V_1)$$

$$
= \frac{|V_1|}{n-1} \sum_{\{v,v'\}\in\binom{V(G)}{2}} (\text{dist}_{(G,z)}(v, V_1) + \text{dist}_{(G,z)}(v', V_1))
$$

$$
\geq \frac{|V_1|}{n-1} \sum_{\{v,v'\}\in\binom{V(G)}{2}} (\text{dist}_{(G,z)}(v, v') - d)
$$

$$
= \frac{|V_1|}{n-1} \left(1 - \binom{n}{2}d\right)
$$

$$
= \frac{n+1}{2n(n-1)}|V_1|
$$

$$
> \frac{1}{3}.
$$

We obtain

$$
\int_{\xi\geq 0} u(\delta_G(B_\xi))\, d\xi \leq \sum_{e\in E(G)} u(e)z(e) = W < 3\, W \int_{\xi\geq 0} |B_\xi||V(G)\setminus B_\xi|\, d\xi.
$$

Hence there exists a $\xi \geq 0$ with $B_\xi \subset V(G)$ and $\frac{u(\delta_G(B_\xi))}{|B_\xi||V(G)\setminus B_\xi|} \leq 3\, W$. $\square$

We conclude:

Theorem 19.13. (Leighton and Rao [1999]) *There is an $O(\log n)$-approximation algorithm for the* SPARSEST CUT PROBLEM, *where $n = |V(G)|$. The max-flow min-cut ratio for the* UNIFORM MULTICOMMODITY FLOW PROBLEM *is $O(\log n)$.*

Proof: Given an instance (G, u), we use Lemma 19.10 to obtain an LP solution (z, w). Without loss of generality $w(f) = \text{dist}_{(G,z)}(s, t)$ for all $f = \{t, s\} \in E(H)$. Then we apply Theorem 19.12 to (G, u) and z. As W is the LP value and thus a lower bound on the optimum, this yields both statements. $\square$

This result was generalized to arbitrary instances of the UNDIRECTED MULTI-COMMODITY FLOW PROBLEM by Linial, London and Rabinovich [1995] and Aumann and Rabani [1998]; in general the max-flow min-cut ratio is $O(\log |E(H)|)$.

The approximation ratio in the first statement has been improved to $O(\sqrt{\log n})$ by Arora, Rao and Vazirani [2009] using a semidefinite programming relaxation. Faster algorithms with the same performance ratio have been proposed by Arora, Hazan and Kale [2004] and Sherman [2009].

A sparsest cut algorithm can be used for designing divide-and-conquer approximation algorithms for many optimization problems on graphs. See Exercise 8 for an example. See also Leighton and Rao [1999] and Shmoys [1996] for many further examples and references.

19.5 Directed Edge-Disjoint Paths Problem

We start by noting that the problem is *NP*-hard already in a quite restricted version:

Theorem 19.14. (Even, Itai and Shamir [1976]) *The* DIRECTED EDGE-DISJOINT PATHS PROBLEM *is NP-hard even if G is acyclic and H just consists of two sets of parallel edges.*

Proof: We polynomially transform SATISFIABILITY to our problem. Given a family $\mathcal{Z} = \{Z_1, \ldots, Z_m\}$ of clauses over $X = \{x_1, \ldots, x_n\}$, we construct an instance (G, H) of the DIRECTED EDGE-DISJOINT PATHS PROBLEM such that G is acyclic, H just consists of two sets of parallel edges and (G, H) has a solution if and only if $\mathcal{Z}$ is satisfiable.

G contains $2m$ vertices $\lambda^1, \ldots, \lambda^{2m}$ for each literal λ and additional vertices s and t, $v_1, \ldots, v_{n+1}$ and $Z_1, \ldots, Z_m$. There are edges (v_i, x_i^1), $(v_i, \overline{x_i}^1)$, (x_i^{2m}, v_{i+1}), $(\overline{x_i}^{2m}, v_{i+1})$, (x_i^j, x_i^{j+1}) and $(\overline{x_i}^j, \overline{x_i}^{j+1})$ for $i = 1, \ldots, n$ and $j = 1, \ldots, 2m - 1$. Next, there are edges (s, x_i^{2j-1}) and $(s, \overline{x_i}^{2j-1})$ for $i = 1, \ldots, n$ and $j = 1, \ldots, m$. Moreover, there are edges (Z_j, t) and (λ^{2j}, Z_j) for $j = 1, \ldots, m$ and all literals λ of the clause Z_j. See Figure 19.3 for an illustration.

Let H consist of an edge (v_{n+1}, v_1) and m parallel edges (t, s).

We show that any solution of (G, H) corresponds to a truth assignment satisfying all clauses (and vice versa). Namely, the v_1-v_{n+1}-path must pass through either

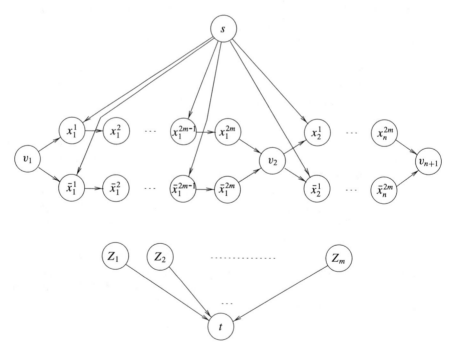

Fig. 19.3.

all x_i^j (meaning x_i is *false*) or all $\overline{x_i}^j$ (meaning x_i is *true*) for each i. One s-t-path must pass through each Z_j. This is possible if and only if the above defined truth assignment satisfies Z_j.   □

Fortune, Hopcroft and Wyllie [1980] showed that the DIRECTED EDGE-DISJOINT PATHS PROBLEM can be solved in polynomial time if G is acyclic and $|E(H)| = k$ for some fixed k. If G not acyclic, they proved that the problem is *NP*-hard already for $|E(H)| = 2$. On the positive side we have:

Theorem 19.15.   (Nash-Williams [1969])   *Let (G, H) be an instance of the DI-RECTED EDGE-DISJOINT PATHS PROBLEM where $G + H$ is Eulerian and H just consists of two sets of parallel edges. Then (G, H) has a solution if and only if the cut criterion holds.*

Proof: We first find a set of paths realizing the first set of parallel edges in H by Menger's Theorem 8.9. After deleting these paths (and the corresponding demand edges), the remaining instance satisfies the prerequisites of Proposition 8.12 and thus has a solution.   □

By Lemma 19.6, this can be extended to the case where $E(H)$ can be partitioned into two sets, each of which contains only one vertex with nonzero in-degree or only one vertex with nonzero out-degree. If $G + H$ is Eulerian and $|E(H)| = 3$, there is also a polynomial-time algorithm (Ibaraki and Poljak [1991]). On the other hand there is the following negative result:

Theorem 19.16.   (Vygen [1995])   *The DIRECTED EDGE-DISJOINT PATHS PROB-LEM is NP-hard even if G is acyclic, $G + H$ is Eulerian, and H consists just of three sets of parallel edges.*

Proof: We reduce the problem of Theorem 19.14 to this one. So let (G, H) be an instance of the DIRECTED EDGE-DISJOINT PATHS PROBLEM, where G is acyclic, and H consists just of two sets of parallel edges.

For each $v \in V(G)$ we define

$$\alpha(v) \ := \ \max(0, |\delta_{G+H}^+(v)| - |\delta_{G+H}^-(v)|) \text{ and}$$
$$\beta(v) \ := \ \max(0, |\delta_{G+H}^-(v)| - |\delta_{G+H}^+(v)|).$$

We have

$$\sum_{v \in V(G)} (\alpha(v) - \beta(v)) \ = \ \sum_{v \in V(G)} \left(|\delta_{G+H}^+(v)| - |\delta_{G+H}^-(v)| \right) \ = \ 0,$$

implying

$$\sum_{v \in V(G)} \alpha(v) \ = \ \sum_{v \in V(G)} \beta(v) \ =: \ q.$$

We now construct an instance (G', H') of the DIRECTED EDGE-DISJOINT PATHS PROBLEM. G' results from G by adding two vertices s and t as well as

$\alpha(v)$ parallel edges (s, v) and $\beta(v)$ parallel edges (v, t) for each vertex v. H' consists of all edges of H and q parallel edges (t, s).

This construction can obviously be done in polynomial time. In particular, the number of edges in $G + H$ at most quadruples. Furthermore, G' is acyclic, $G' + H'$ is Eulerian, and H' just consists of three sets of parallel edges. Thus it remains to show that (G, H) has a solution if and only if (G', H') has a solution.

Each solution of (G', H') implies a solution of (G, H) simply by omitting the s-t-paths. So let $\mathcal{P}$ be a solution of (G, H). Let G'' be the graph resulting from G' by deleting all edges used by $\mathcal{P}$. Let H'' be the subgraph of H' just consisting of the q edges from t to s. (G'', H'') satisfies the prerequisites of Proposition 8.12 and thus has a solution. Combining $\mathcal{P}$ with a solution of (G'', H'') produces a solution of (G', H'). $\qquad\square$

Since a solution to an instance of the DIRECTED EDGE-DISJOINT PATHS PROBLEM consists of edge-disjoint circuits, it is natural to ask how many edge-disjoint circuits a digraph has. At least for planar digraphs we have a good characterization. Namely, we consider the planar dual graph and ask for the maximum number of edge-disjoint directed cuts. We have the following well-known min-max theorem (which we prove very similarly to Theorem 6.14):

Theorem 19.17. (Lucchesi and Younger [1978]) *Let G be a connected but not strongly connected digraph. Then the maximum number of edge-disjoint directed cuts in G equals the minimum cardinality of an edge set that contains at least one element of each directed cut.*

Proof: Let A be the matrix whose columns are indexed by the edges and whose rows are the incidence vectors of all directed cuts. Consider the LP

$$\min\{\mathbb{1}x : Ax \geq \mathbb{1}, x \geq 0\},$$

and its dual

$$\max\{\mathbb{1}y : yA \leq \mathbb{1}, y \geq 0\}.$$

Then we have to prove that both the primal and the dual LP have integral optimum solutions. By Corollary 5.16 it suffices to show that the system $Ax \geq \mathbb{1}$, $x \geq 0$ is TDI. We use Lemma 5.24.

Let $c : E(G) \to \mathbb{Z}_+$, and let y be an optimum solution of $\max\{\mathbb{1}y : yA \leq c, y \geq 0\}$ for which

$$\sum_X y_{\delta^+(X)} |X|^2 \tag{19.9}$$

is as large as possible, where the sum is over all rows of A. We claim that the set system $(V(G), \mathcal{F})$ with $\mathcal{F} := \{X : y_{\delta^+(X)} > 0\}$ is cross-free. To see this, suppose $X, Y \in \mathcal{F}$ with $X \cap Y \neq \emptyset$, $X \setminus Y \neq \emptyset$, $Y \setminus X \neq \emptyset$ and $X \cup Y \neq V(G)$. Then $\delta^+(X \cap Y)$ and $\delta^+(X \cup Y)$ are also directed cuts (by Lemma 2.1(b)). Let $\epsilon := \min\{y_{\delta^+(X)}, y_{\delta^+(Y)}\}$. Set $y'_{\delta^+(X)} := y_{\delta^+(X)} - \epsilon$, $y'_{\delta^+(Y)} := y_{\delta^+(Y)} - \epsilon$, $y'_{\delta^+(X \cap Y)} := y_{\delta^+(X \cap Y)} + \epsilon$, $y'_{\delta^+(X \cup Y)} := y_{\delta^+(X \cup Y)} + \epsilon$, and $y'_S := y_S$ for all other directed cuts S. Since y' is a feasible dual solution, it is also optimum and contradicts the choice of y, because (19.9) is larger for y'.

Now let A' be the submatrix of A consisting of the rows corresponding to the elements of $\mathcal{F}$. A' is the two-way cut-incidence matrix of a cross-free family. So by Theorem 5.29 A' is totally unimodular, as required. $\qquad\square$

For a combinatorial proof, see Lovász [1976]. Frank [1981] gives an algorithmic proof.

Note that the sets of edges meeting all directed cuts are precisely the sets of edges whose contraction makes the digraph strongly connected. In the planar dual digraph, these sets correspond to the sets of edges meeting all directed circuits. Such sets are known as **feedback edge sets**, the minimum cardinality of a feedback edge set is the **feedback number** of the graph. The problem to determine the feedback number is *NP*-hard in general (Karp [1972]), and probably hard to approximate (Guruswami et al. [2011]), but polynomially solvable for planar graphs.

Corollary 19.18. *In a planar digraph the maximum number of edge-disjoint circuits equals the minimum number of edges meeting all circuits.*

Proof: Let G be a digraph which, without loss of generality, is connected and contains no articulation vertex. Consider the planar dual of G and Corollary 2.44, and apply the Lucchesi-Younger Theorem 19.17. $\qquad\square$

A polynomial-time algorithm for determining the feedback number for planar graphs can be composed of the planarity algorithm (Theorem 2.40), the GRÖTSCHEL-LOVÁSZ-SCHRIJVER ALGORITHM (Theorem 4.21) and an algorithm for the MAXIMUM FLOW PROBLEM to solve the SEPARATION PROBLEM (Exercise 10). An application to the EDGE-DISJOINT PATHS PROBLEM is the following:

Corollary 19.19. *Let (G, H) be an instance of the* DIRECTED EDGE-DISJOINT PATHS PROBLEM, *where G is acyclic and $G + H$ is planar. Then (G, H) has a solution if and only if deleting any $|E(H)| - 1$ edges of $G + H$ does not make $G + H$ acyclic.* $\qquad\square$

In particular, the distance criterion is necessary and sufficient in this case, and the problem can be solved in polynomial time.

19.6 Undirected Edge-Disjoint Paths Problem

The following lemma establishes a connection between directed and undirected problems.

Lemma 19.20. *Let (G, H) be an instance of the* DIRECTED EDGE-DISJOINT PATHS PROBLEM, *where G is acyclic and $G + H$ is Eulerian. Consider the instance (G', H') of the* UNDIRECTED EDGE-DISJOINT PATHS PROBLEM *which results from (G, H) by neglecting the orientations. Then each solution of (G', H') is also a solution of (G, H), and vice versa.*

Proof: It is trivial that each solution of (G, H) is also a solution of (G', H'). We prove the other direction by induction on $|E(G)|$. If G has no edges, we are done.

Now let $\mathcal{P}$ be a solution of (G', H'). Since G is acyclic, G must contain a vertex v for which $\delta_G^-(v) = \emptyset$ and $\delta_G^+(v) \neq \emptyset$. Since $G + H$ is Eulerian, we have $|\delta_H^-(v)| = |\delta_G^+(v)| + |\delta_H^+(v)|$.

For each demand edge incident to v there must be an undirected path in $\mathcal{P}$ starting at v. Thus $|\delta_G^+(v)| \geq |\delta_H^-(v)| + |\delta_H^+(v)|$. This implies $|\delta_H^+(v)| = 0$ and $|\delta_G^+(v)| = |\delta_H^-(v)|$. Therefore each edge incident to v must be used by $\mathcal{P}$ with the correct orientation.

Now let G_1 be the graph which results from G by deleting the edges incident to v. Let H_1 result from H by replacing each edge $f = (t, v)$ incident to v by (t, w), and deleting it if $t = w$, where w is the first inner vertex of the path in $\mathcal{P}$ which realizes f.

Obviously G_1 is acyclic and $G_1 + H_1$ is Eulerian. Let $\mathcal{P}_1$ arise from $\mathcal{P}$ by deleting all the edges incident to v. $\mathcal{P}_1$ is a solution of (G_1', H_1'), the undirected problem corresponding to (G_1, H_1).

By the induction hypothesis, $\mathcal{P}_1$ is a solution of (G_1, H_1) as well. So by adding the initial edges we obtain that $\mathcal{P}$ is a solution of (G, H).   □

We conclude:

Theorem 19.21. (Vygen [1995]) *The* UNDIRECTED EDGE-DISJOINT PATHS PROBLEM *is NP-hard even if $G + H$ is Eulerian and H just consists of three sets of parallel edges.*

Proof: We reduce the problem of Theorem 19.16 to the undirected case by applying Lemma 19.20.   □

Another special case in which the UNDIRECTED EDGE-DISJOINT PATHS PROBLEM is *NP*-hard is when $G+H$ is planar (Middendorf and Pfeiffer [1993]). However, if $G + H$ is known to be planar and Eulerian, then the problem becomes tractable:

Theorem 19.22. (Seymour [1981]) *Let (G, H) be an instance of the* UNDIRECTED EDGE-DISJOINT PATHS PROBLEM, *where $G+H$ is planar and Eulerian. Then (G, H) has a solution if and only if the cut criterion holds.*

Proof: We only have to prove the sufficiency of the cut criterion. We may assume that $G + H$ is connected. Let D be the planar dual of $G + H$. Let $F \subseteq E(D)$ be the set of dual edges corresponding to the demand edges. Then the cut criterion, together with Theorem 2.43, implies that $|F \cap E(C)| \leq |E(C) \setminus F|$ for each circuit C in D. So by Proposition 12.8, F is a minimum T-join, where $T := \{x \in V(D) : |F \cap \delta(x)| \text{ is odd}\}$.

Since $G+H$ is Eulerian, by Corollary 2.45 D is bipartite, so by Theorem 12.16 there are $|F|$ edge-disjoint T-cuts $C_1, \ldots, C_{|F|}$. Since by Proposition 12.15 each T-cut intersects F, each of $C_1, \ldots C_{|F|}$ must contain exactly one edge of F.

Back in $G + H$, the duals of $C_1, \ldots, C_{|F|}$ are edge-disjoint circuits, each containing exactly one demand edge. But this means that we have a solution of the EDGE-DISJOINT PATHS PROBLEM. □

This theorem also implies a polynomial-time algorithm (Exercise 15). In fact, Matsumoto, Nishizeki and Saito [1986] proved that the UNDIRECTED EDGE-DISJOINT PATHS PROBLEM with $G + H$ planar and Eulerian can be solved in $O\left(n^{\frac{5}{2}} \log n\right)$ time. Sebő [1993] showed that for any fixed integer k, the UNDI-RECTED EDGE-DISJOINT PATHS PROBLEM can be solved in polynomial time if $G + H$ is planar and H consists of only k sets of parallel edges. Another well-known result is the Okamura-Seymour Theorem:

Theorem 19.23. (Okamura and Seymour [1981]) *Let (G, H) be an instance of the* UNDIRECTED EDGE-DISJOINT PATHS PROBLEM, *where $G + H$ is Eulerian, G is planar, and all terminals lie on the outer face. Then (G, H) has a solution if and only if the cut criterion holds.*

Proof: We show the sufficiency of the cut criterion by induction on $|V(G)| + |E(G)|$. If $|V(G)| \leq 2$, this is obvious.

We may assume that G is 2-connected, for otherwise we may apply the induction hypothesis to the blocks of G (splitting up demand edges joining different blocks at articulation vertices). We fix some planar embedding of G; by Proposition 2.31 the outer face is bounded by some circuit C.

Call a set $X \subset V(G)$ **critical** if $\emptyset \neq X \cap V(C) \neq V(C)$ and $|\delta_G(X)| = |\delta_H(X)|$.

If there is no critical set, then for any edge $e \in E(C)$ the instance $(G - e, H + e)$ satisfies the cut criterion. This is because $|\delta_G(X)| - |\delta_H(X)|$ is even for all $X \subseteq V(G)$ (as $G + H$ is Eulerian). By the induction hypothesis, $(G - e, H + e)$ has a solution which immediately implies a solution for (G, H).

So suppose there is a critical set. We need the following:

Claim: Let G' be a connected subgraph of G, and let H' be any graph with all endpoints of edges in $V(C)$. Let $k := \min\{|\delta_{G'}(Y)| - |\delta_{H'}(Y)| : \emptyset \neq Y \subset V(G)\} \in \{-2, 0\}$. Then there exists a set $X \subset V(G)$ for which $C[X]$ is a path and $|\delta_{G'}(X)| - |\delta_{H'}(X)| = k$.

To prove the claim, let $\emptyset \neq X \subset V(G)$ with $|\delta_{G'}(X)| - |\delta_{H'}(X)| = k$ such that the total number of connected components in $G'[X]$ and $G'[V(G) \setminus X]$ is minimum. We first show that then $G'[X]$ and $G'[V(G) \setminus X]$ are both connected.

Suppose not, say $G'[X]$ is disconnected (the other case is symmetric), with connected components $X_1, \ldots, X_l$. Then $k = |\delta_{G'}(X)| - |\delta_{H'}(X)| \geq \sum_{i=1}^{l}(|\delta_{G'}(X_i)| - |\delta_{H'}(X_i)|)$, and hence $|\delta_G(X_i)| - |\delta_H(X_i)| = k$ for some $i \in \{1, \ldots, l\}$. But replacing X by X_i reduces the number of connected components in $G'[X]$ without increasing the number of connected components in $G'[V(G) \setminus X]$. This contradicts the choice of X.

So $G'[X]$ and $G'[V(G) \setminus X]$, and hence also $G[X]$ and $G[V(G) \setminus X]$ are connected. Also note that $\emptyset \neq X \cap V(C) \neq V(C)$ as $|\delta_{H'}(X)| = |\delta_{G'}(X)| - k \geq |\delta_{G'}(X)| > 0$. Since G is planar, $C[X]$ is a path. The claim is proved.

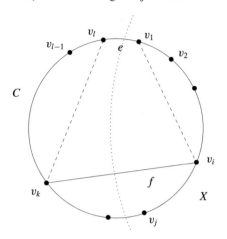

Fig. 19.4.

Applying the claim to G and H, let X be a critical set such that $C[X]$ is a path of minimum length. Let the vertices of C be numbered $v_1, \ldots, v_l$ cyclically, where $V(C) \cap X = \{v_1, \ldots, v_j\}$. Let $e := \{v_l, v_1\}$.

Choose $f = \{v_i, v_k\} \in E(H)$ such that $1 \le i \le j < k \le l$ (i.e. $v_i \in X$, $v_k \notin X$) and k is as large as possible (see Figure 19.4). Now consider $G' := G - e$ and $H' := (V(H), (E(H) \setminus \{f\}) \cup \{\{v_i, v_1\}, \{v_l, v_k\}\})$. (The cases $i = 1$ or $k = l$ are not excluded, in this case no loops should be added.)

We claim that (G', H') satisfies the cut criterion. Then by induction, (G', H') has a solution, and this can easily be transformed to a solution of (G, H).

Suppose, then, that (G', H') does not satisfy the cut criterion, i.e. $|\delta_{G'}(Y)| < |\delta_{H'}(Y)|$ for some $Y \subseteq V(G)$. By the claim we may assume that $C[Y]$ is a path. By possibly interchanging Y and $V(G) \setminus Y$ we may also assume $v_i \notin Y$. Since $|\delta_{H'}(Y)| - |\delta_{G'}(Y)| > 0 = |\delta_H(Y)| - |\delta_G(Y)|$, there are three cases:

(a) $v_1 \in Y$, $v_i, v_k, v_l \notin Y$;
(b) $v_1, v_l \in Y$, $v_i, v_k \notin Y$;
(c) $v_l \in Y$, $v_1, v_i, v_k \notin Y$.

In each case we have $Y \cap V(C) \subseteq \{v_{k+1}, \ldots, v_{i-1}\}$, so by the choice of f we have $E_H(X, Y) = \emptyset$. Furthermore, $|\delta_G(Y)| = |\delta_H(Y)|$. By applying Lemma 2.1(c) twice, we have

$$
\begin{aligned}
|\delta_H(X)| + |\delta_H(Y)| &= |\delta_G(X)| + |\delta_G(Y)| \\
&= |\delta_G(X \cap Y)| + |\delta_G(X \cup Y)| + 2|E_G(X, Y)| \\
&\ge |\delta_H(X \cap Y)| + |\delta_H(X \cup Y)| + 2|E_G(X, Y)| \\
&= |\delta_H(X)| + |\delta_H(Y)| - 2|E_H(X, Y)| + 2|E_G(X, Y)| \\
&= |\delta_H(X)| + |\delta_H(Y)| + 2|E_G(X, Y)| \\
&\ge |\delta_H(X)| + |\delta_H(Y)| .
\end{aligned}
$$

So equality must hold throughout. This implies $|\delta_G(X \cap Y)| = |\delta_H(X \cap Y)|$ and $E_G(X, Y) = \emptyset$.

So case (c) is impossible (because here $e \in E_G(X, Y)$); i.e. $v_1 \in Y$. Therefore $X \cap Y$ is nonempty and $C[X \cap Y]$ is a shorter path than $C[X]$, contradicting the choice of X.   □

This proof yields a polynomial-time algorithm (Exercise 16) for the UNDIRECTED EDGE-DISJOINT PATHS PROBLEM in this special case. It can be implemented in $O(n^2)$ time (Becker and Mehlhorn [1986]) and indeed in linear time (Wagner and Weihe [1995]). Frank [1985] obtained a polynomial-time algorithm even if the condition that $G + H$ is Eulerian is relaxed to requiring that $|\delta_{G+H}(v)|$ is even for all vertices v that do not lie on the outer face. However, the Eulerian condition cannot be dropped completely:

Theorem 19.24.   (Schwärzler [2009])   *The* UNDIRECTED EDGE-DISJOINT PATHS PROBLEM *is NP-hard even if G is planar and all terminals lie on the outer face.*

Naves [2009] strengthened this by requiring that H just consist of two sets of parallel edges. These *NP*-completeness results also hold if G is a planar acyclic digraph.

We prepare another main result of this section by a theorem concerning orientations of mixed graphs, i.e. graphs with directed and undirected edges. Given a mixed graph G, can we orient its undirected edges such that the resulting digraph is Eulerian? The following theorem answers this question:

Theorem 19.25.   (Ford and Fulkerson [1962])   *Let G be an undirected graph and H a digraph with $V(G) = V(H)$. Then G has an orientation G' such that the digraph $G' + H$ is Eulerian if and only if*

- $|\delta_H^+(v)| + |\delta_H^-(v)| + |\delta_G(v)|$ *is even for all $v \in V(G)$, and*
- $|\delta_H^+(X)| - |\delta_H^-(X)| \leq |\delta_G(X)|$ *for all $X \subseteq V(G)$.*

Proof: The necessity of the conditions is obvious. We prove the sufficiency by induction on $|E(G)|$. If $E(G) = \emptyset$, the statement is trivial.

Assuming that G and H satisfy the conditions and $E(G) \neq \emptyset$, we will orient some edge e to e' such that $G - e$ and $H + e'$ satisfy the conditions. Since every degree is even, $|\delta_G(X)| - (|\delta_H^+(X)| - |\delta_H^-(X)|)$ must be even for all $X \subseteq V(G)$. This value decreases by 2 if e' leaves X and remains the same otherwise.

Therefore we call a set X critical if $|\delta_G(X)| = |\delta_H^+(X)| - |\delta_H^-(X)| > 0$. If there is no critical set, we orient any undirected edge arbitrarily and apply induction. Otherwise let X be any critical set and $e \in \delta_G(X)$. We orient e such that the resulting arc e' enters X and claim that the conditions continue to hold.

Suppose, indirectly, that there is a $Y \subseteq V(G)$ with $|\delta_{H+e'}^+(Y)| - |\delta_{H+e'}^-(Y)| > |\delta_{G-e}(Y)|$. This means that Y was critical before orienting e, and e' now leaves Y.

Applying Lemma 2.1(a) and (b) for $|\delta_H^+|$ and $|\delta_H^-|$ and Lemma 2.1(c) for $|\delta_G|$ we have (before orienting e):

$$
\begin{aligned}
0+0 &= |\delta_G(X)| - |\delta_H^+(X)| + |\delta_H^-(X)| + |\delta_G(Y)| - |\delta_H^+(Y)| + |\delta_H^-(Y)| \\
&= |\delta_G(X \cap Y)| - |\delta_H^+(X \cap Y)| + |\delta_H^-(X \cap Y)| \\
&\quad + |\delta_G(X \cup Y)| - |\delta_H^+(X \cup Y)| + |\delta_H^-(X \cup Y)| + 2|E_G(X, Y)| \\
&\geq 0 + 0 + 2|E_G(X, Y)| \geq 0.
\end{aligned}
$$

So we have equality throughout and conclude that $E_G(X, Y) = \emptyset$, contradicting the existence of e. $\qquad\square$

Corollary 19.26. *An undirected Eulerian graph can be oriented such that a directed Eulerian graph arises.* $\qquad\square$

Of course this corollary can be proved more easily by orienting the edges according to their occurrence in an Eulerian walk.

We now return to the EDGE-DISJOINT PATHS PROBLEM.

Theorem 19.27. (Rothschild and Whinston [1966]) *Let (G, H) be an instance of the* UNDIRECTED EDGE-DISJOINT PATHS PROBLEM, *where $G + H$ is Eulerian and H is the union of two stars (i.e. two vertices meet all the demand edges). Then (G, H) has a solution if and only if the cut criterion holds.*

Proof: We show that the cut criterion is sufficient. By Lemma 19.6 we can assume that H contains only two sets of parallel edges. Note that the construction preserves the property that $G + H$ is Eulerian.

Now we orient the edges of H arbitrarily such that parallel edges have the same orientation (and call the result H'). The two graphs H' and G satisfy the prerequisites of Theorem 19.25 because the cut criterion implies $|\delta_{H'}^+(X)| - |\delta_{H'}^-(X)| \leq |\delta_G(X)|$ for all $X \subseteq V(G)$. Therefore we can orient the edges of G in order to get a digraph G' such that $G' + H'$ is Eulerian.

We regard (G', H') as an instance of the DIRECTED EDGE-DISJOINT PATHS PROBLEM. (G', H') satisfies the (directed) cut criterion as for $X \subseteq V(G)$ we have $2|\delta_{G'}^+(X)| = |\delta_G(X)| + (|\delta_{G'}^+(X)| - |\delta_{G'}^-(X)|) = |\delta_G(X)| + (|\delta_{H'}^-(X)| - |\delta_{H'}^+(X)|) \geq |\delta_H(X)| + (|\delta_{H'}^-(X)| - |\delta_{H'}^+(X)|) = 2|\delta_{H'}^-(X)|$. But now Theorem 19.15 guarantees a solution which – by neglecting the orientations – is also a solution for (G, H). $\square$

This implies a half-integral solution for any instance (G, H) in which H is the union of two stars and the cut criterion is satisfied (simply double all edges).

The same theorem holds if H (neglecting parallel edges) is K_4 or C_5 (the circuit of length 5) (Lomonosov [1979], Seymour [1980]). In the K_5 case, the distance criterion is sufficient (Karzanov [1987]). See also Hirai [2010]. However, if H is allowed to have three sets of parallel edges with pairwise distinct endpoints, the problem becomes NP-hard, as we have seen in Theorem 19.21.

On the other hand, Robertson and Seymour have found a polynomial-time algorithm for a fixed number of demand edges:

Theorem 19.28. (Robertson and Seymour [1995]) *For fixed k, there are polynomial-time algorithms for the* UNDIRECTED VERTEX-DISJOINT PATHS PROBLEM *and the* UNDIRECTED EDGE-DISJOINT PATHS PROBLEM *restricted to instances where $|E(H)| \leq k$.*

Note that the UNDIRECTED VERTEX-DISJOINT PATHS PROBLEM is also *NP*-hard; see Exercise 27. Theorem 19.28 is part of Robertson's and Seymour's important series of papers on graph minors which is far beyond the scope of this book. The theorem was proved for the vertex-disjoint case; here Robertson and Seymour proved that there either exists an irrelevant vertex (which can be deleted without affecting solvability) or the graph has a tree-decomposition of small width (in which case there is a simple polynomial-time algorithm; see Exercise 26). This is a very deep result; but Kawarabayashi and Wollan [2010] found a substantially shorter proof. The edge-disjoint case then follows easily; see Exercise 27. Although the running time is $O(n^2 m)$, the constant depending on k grows extremely fast and is beyond practical use already for $k = 3$. Recently, Kawarabayashi, Kobayashi and Reed [2012] improved the running time to $O(n^2)$.

Exercises

1. Let (G, H) be an instance of the EDGE-DISJOINT PATHS PROBLEM, directed or undirected, violating the distance criterion (19.2) for some $z : E(G) \to \mathbb{R}_+$. Prove that then there is also some $z : E(G) \to \mathbb{Z}_+$ violating (19.2). Moreover, give examples where there is no $z : E(G) \to \{0, 1\}$ violating (19.2).

2. Show that it is *NP*-hard to decide whether a given instance (G, H) of the UNDIRECTED EDGE-DISJOINT PATHS PROBLEM satisfies the cut criterion.
   Hint: Give a polynomial reduction from the following problem: Given an undirected graph G' and $X \subseteq V(G')$, find a set $Y \subseteq V(G')$ with $|\delta_{G'}(Y)| > |\delta_{G'}(X)|$ or decide that none exists. This problem is *NP*-hard by Theorem 16.6. Set $E(H) := \delta_{G'}(X)$ and $E(G) := E(G') \setminus \delta(X)$.
   Note: The problem is obviously in *coNP* but is not known to be *coNP*-complete. (Sebő [unpublished])

3. Let $k \in \mathbb{N}$ be fixed. Show that for a given instance (G, H, u, b) of the (DIRECTED or UNDIRECTED) MULTICOMMODITY FLOW PROBLEM
   (a) with at most k terminals
   (b) where H has a vertex cover of size k
   one can decide in polynomial time whether the cut criterion is satisfied.

* 4. For an instance (G, H) of the UNDIRECTED EDGE-DISJOINT PATHS PROBLEM we consider the multicommodity flow relaxation and solve

$$\min \{\lambda : \lambda \in \mathbb{R}, \ y \geq 0, \ My \leq \lambda \mathbb{1}, \ Ny = \mathbb{1}\},$$

where M and N are defined as in Lemma 19.1. Let (y^*, λ^*) be an optimum solution. Now we are looking for an integral solution, i.e. an s-t-path P_f for each demand edge $f = \{t, s\} \in E(H)$, such that the maximum load on a supply edge is minimum (by the load of an edge we mean the number of paths using it). We do this by randomized rounding: independently for each demand edge we choose a path P with probability y_P.

Let $0 < \epsilon \le 1$, and suppose that $\lambda^* \ge 3 \ln \frac{|E(G)|}{\epsilon}$. Prove that then with probability at least $1 - \epsilon$ the above randomized rounding yields an integral solution with maximum load at most $\lambda^* + \sqrt{3\lambda^* \ln \frac{|E(G)|}{\epsilon}}$.

Hint: Use the following facts from probability theory: if $B(m, N, p)$ is the probability of at least m successes in N independent Bernoulli trials, each with success probability p, then

$$B((1 + \beta)Np, N, p) < e^{-\frac{1}{3}\beta^2 Np}$$

for all $0 < \beta \le 1$. Moreover, the probability of at least m successes in N independent Bernoulli trials with success probabilities $p_1, \ldots, p_N$ is at most $B\left(m, N, \frac{1}{N}(p_1 + \cdots + p_N)\right)$.
(Raghavan and Thompson [1987])

5. Describe an approximation scheme for the CONCURRENT FLOW PROBLEM by using the MULTICOMMODITY FLOW APPROXIMATION SCHEME as subroutine.

   Hint: Use binary search to narrow bounds on the optimum value and modify the instance in order to bound the possible flow for each commodity proportionally to its demand.

   Note: One can also modify the MULTICOMMODITY FLOW APPROXIMATION SCHEME in order to solve the CONCURRENT FLOW PROBLEM directly. See the references in Section 19.2.

6. Let G be an expander graph like in Theorem 16.42, $n := |V(G)|$, and $u(e) := 1$ for all $e \in E(G)$. Show that the max-flow min-cut ratio of this instance of the UNIFORM MULTICOMMODITY FLOW PROBLEM is $\Omega(\log n)$.

   Hint: Compare left-hand side and right-hand side of the inequalities corresponding to the distance criterion and the cut criterion.

7. Let G be an undirected graph, $n := V(G)$, and $u : E(G) \to \mathbb{R}_+$. For $0 < b \le \frac{1}{2}$, a b-balanced cut is a cut $\delta_G(X)$ with $bn \le |X| \le (1 - b)n$. Let OPT_b denote the minimum capacity of a b-balanced cut. Let now $b' \le \frac{1}{3}$ and consider the following algorithm. Let $G_0 := G$ and $i := 0$. Find an approximate sparsest cut in G_i, and let X_i be the smaller of the two vertex sets. If $|V(G_i) \setminus X_i| > (1 - b')n$, set $G_{i+1} := G - X_i$, increment i, and iterate. Otherwise output $\delta(X_0 \cup \cdots \cup X_i)$.

   (a) Show that $\delta(X_0 \cup \cdots \cup X_i)$ is a b'-balanced cut.
   (b) Show that G_i has a cut $\delta(X)$ with $\frac{u(\delta(X))}{|X||V(G_i)\setminus X|} \le \frac{2\,\text{OPT}_b}{(b-b')n^2}$ for any b with $b' < b \le \frac{1}{2}$.

(c) Show that using an $O(\log n)$-approximation algorithm for the sparsest cut problem as subroutine (cf. Theorem 19.13). we get $|\delta(X_0 \cup \cdots \cup X_i)| \le O(\frac{\mathrm{OPT}_b \log n}{b - b'})$.

(Leighton and Rao [1999])

Note: This does not yield an approximation algorithm for finding a minimum capacity b-balanced cut. For the special case $b = \frac{1}{2}$ (called bisection), Räcke [2008] found an $O(\log n)$-approximation algorithm.

8. The OPTIMAL LINEAR ARRANGEMENT PROBLEM asks for numbering the n vertices of a given undirected graph G with $v_1, \ldots, v_n$ such that the total edge length $\sum_{\{v_i, v_j\} \in E(G)} |i - j|$ is minimum. We use Exercise 7 in order to give an approximation algorithm.

(a) Show that the optimum cost of a solution is at least $(1 - 2b)n\,\mathrm{OPT}_b$ for any $0 < b < \frac{1}{2}$ for which $bn \in \mathbb{N}$.

(b) Consider the following algorithm. If there is only one vertex, the problem is trivial. Otherwise find a $\frac{1}{3}$-balanced cut $\delta(X)$ as in Exercise 7, apply the algorithm recursively to the two subgraphs induced by X and $V(G) \setminus X$, and concatenate the two linear orders. Show that the cost of the resulting solution is at most $O(\log^2 n)$ times the optimum.

(Hansen [1989])

Note: The problem is *NP*-hard. The currently best approximation algorithm, with performance ratio $O(\sqrt{\log n} \log \log n)$, is due to Charikar et al. [2010] and Feige and Lee [2007].

9. Prove that there is a polynomial-time algorithm for the DIRECTED EDGE-DISJOINT PATHS PROBLEM, where $G + H$ is Eulerian and where H just consists of two sets of parallel edges.

10. Show that in a given digraph a minimum set of edges meeting all directed cuts can be found in polynomial time. Show that for planar graphs the feedback number can be determined in polynomial time.

11. Show that in a digraph a minimum set of edges whose contraction makes the graph strongly connected can be found in polynomial time.

12. Prove that the statement of Corollary 19.18 does not hold for general (non-planar) digraphs.

13. Show that the statement of Corollary 19.19 becomes false if the condition "G is acyclic" is omitted.

Note: In this case the DIRECTED EDGE-DISJOINT PATHS PROBLEM is *NP*-hard (Vygen [1995]).

14. Consider the following greedy algorithm for the DIRECTED EDGE-DISJOINT PATHS PROBLEM: Given an instance (G, H), choose an edge $f = (t, s) \in E(H)$ with $\mathrm{dist}_G(s, t)$ minimum, let P_f be a shortest s-t-path in G, remove f from $E(H)$ and $E(P_f)$ from $E(G)$. Iterate as long as there is a realizable demand edge. The result is a solution to (G, H') for a subgraph H' of H.

Let H^* be any subgraph of H such that (G, H^*) has a solution. Show that then $|E(H')| \ge \frac{|E(H^*)|}{\sqrt{m}}$, where $m := |E(G)|$.

Hint: Consider the situation when no demand edge can be realized anymore by a path of length at most $\sqrt{m}$.
(Kleinberg [1996]; see Chekuri and Khanna [2007] for an improved analysis)

15. Prove that the UNDIRECTED EDGE-DISJOINT PATHS PROBLEM can be solved in polynomial time if $G + H$ is planar and Eulerian.

16. Show that the proof of the Okamura-Seymour Theorem leads to a polynomial-time algorithm.

17. Let (G, H) be an instance of the UNDIRECTED EDGE-DISJOINT PATHS PROBLEM. Suppose that G is planar, all terminals lie on the outer face, and each vertex not on the outer face has even degree. Furthermore, assume that

$$|\delta_G(X)| > |\delta_H(X)| \quad \text{for all } \emptyset \neq X \subset V(G).$$

Prove that (G, H) has a solution.
Hint: Use the Okamura-Seymour Theorem.

18. Generalizing Robbins' Theorem (Exercise 22(c) of Chapter 2), formulate and prove a necessary and sufficient condition for the existence of an orientation of the undirected edges of a mixed graph such that the resulting digraph is strongly connected.
(Boesch and Tindell [1980])

19. Let (G, H) be an instance of the DIRECTED EDGE-DISJOINT PATHS PROBLEM where $G + H$ is Eulerian, G is planar and acyclic, and all terminals lie on the outer face. Prove that (G, H) has a solution if and only if the cut criterion holds.
Hint: Use Lemma 19.20 and the Okamura-Seymour Theorem 19.23.

20. Prove Theorem 19.25 using network flow techniques; implying a polynomial-time algorithm.

21. Show in two ways that one can decide in polynomial time whether a pair (G, H) satisfies the conditions of Theorem 19.25: once using submodular function minimization and the equivalence of separation and optimization, and once using network flows (cf. Exercise 20).

22. Prove Nash-Williams' [1969] orientation theorem, which is a generalization of Robbins' Theorem (Exercise 22(c) of Chapter 2):
An undirected graph G can be oriented to be strongly k-edge-connected (i.e. there are k edge-disjoint s-t-paths for any pair $(s, t) \in V(G) \times V(G)$) if and only if G is $2k$-edge-connected.
Hint: To prove the sufficiency, let G' be any orientation of G. Prove that the system

$$
\begin{array}{rll}
x_e & \leq & 1 \qquad (e \in E(G')), \\
x_e & \geq & 0 \qquad (e \in E(G')), \\
\displaystyle\sum_{e \in \delta^-_{G'}(X)} x_e - \sum_{e \in \delta^+_{G'}(X)} x_e & \leq & |\delta^-_{G'}(X)| - k \qquad (\emptyset \neq X \subset V(G'))
\end{array}
$$

is TDI, as in the proof of the Lucchesi-Younger Theorem 19.17.
(Frank [1980], Frank and Tardos [1984])

23. Prove Hu's Two-Commodity Flow Theorem: an instance (G, H, u, b) of the UNDIRECTED MULTICOMMODITY FLOW PROBLEM with $|E(H)| = 2$ has a solution if and only if $\sum_{e\in\delta_G(X)} u(e) \geq \sum_{f\in\delta_H(X)} b(f)$ for all $X \subseteq V(G)$, i.e. if and only if the cut criterion holds.
    Hint: Use Theorem 19.27.
    (Hu [1963])
24. Prove that there is a polynomial-time algorithm for the UNDIRECTED EDGE-DISJOINT PATHS PROBLEM, where $G + H$ is Eulerian and where H just consists of two sets of parallel edges.
* 25. In this exercise we consider instances (G, H) of the UNDIRECTED VERTEX-DISJOINT PATHS PROBLEM where G is planar and all terminals are distinct (i.e. $e \cap f = \emptyset$ for any two demand edges e and f) and lie on the outer face. Let (G, H) be such an instance, where G is 2-connected; so let C be the circuit bounding the outer face (cf. Proposition 2.31).
    Prove that (G, H) has a solution if and only if the following conditions hold:
    • $G + H$ is planar;
    • no set $X \subseteq V(G)$ separates more than $|X|$ demand edges (we say that X separates $\{v, w\}$ if $\{v, w\} \cap X \neq \emptyset$ or if w is not reachable from v in $G - X$).
    Conclude that the UNDIRECTED VERTEX-DISJOINT PATHS PROBLEM in planar graphs with distinct terminals on the outer face can be solved in polynomial time.
    Hint: To prove the sufficiency of (a) and (b), consider the following inductive step: Let $f = \{v, w\}$ be a demand edge such that at least one of the two v-w-paths on C does not contain any other terminal. Realize f by this path and delete it.
    Note: Robertson and Seymour [1986] extended this to a necessary and sufficient condition for the solvability of the UNDIRECTED VERTEX-DISJOINT PATHS PROBLEM with two demand edges.
* 26. Let $k \in \mathbb{N}$ be fixed. Prove that there is a polynomial-time algorithm for the UNDIRECTED VERTEX-DISJOINT PATHS PROBLEM restricted to graphs of tree-width at most k (cf. Exercise 28 of Chapter 2).
    Note: Scheffler [1994] proved that there is in fact a linear-time algorithm. In contrast to that, the UNDIRECTED EDGE-DISJOINT PATHS PROBLEM is NP-hard even for graphs with tree-width 2 (Nishizeki, Vygen and Zhou [2001]).
27. Prove that the DIRECTED VERTEX-DISJOINT PATHS PROBLEM and the UNDIRECTED VERTEX-DISJOINT PATHS PROBLEM are NP-hard. Prove that the vertex-disjoint part of Theorem 19.28 implies its edge-disjoint part.

References

General Literature

Frank, A. [1990]: Packing paths, circuits and cuts – a survey. In: Paths, Flows, and VLSI-Layout (B. Korte, L. Lovász, H.J. Prömel, A. Schrijver, eds.), Springer, Berlin 1990, pp. 47–100

Naves, G., and Sebő, A. [2009]: Multiflow feasibility: an annotated tableau. In: Research Trends in Combinatorial Optimization (W.J. Cook, L. Lovász, J. Vygen, eds.), Springer, Berlin 2009, pp. 261–283

Ripphausen-Lipa, H., Wagner, D., and Weihe, K. [1995]: Efficient algorithms for disjoint paths in planar graphs. In: Combinatorial Optimization; DIMACS Series in Discrete Mathematics and Theoretical Computer Science 20 (W. Cook, L. Lovász, P. Seymour, eds.), AMS, Providence 1995

Schrijver, A. [2003]: Combinatorial Optimization: Polyhedra and Efficiency. Springer, Berlin 2003, Chapters 70–76

Shmoys, D.B. [1996]: Cut problems and their application to divide-and-conquer. In: Approximation Algorithms for NP-Hard Problems (D.S. Hochbaum, ed.), PWS, Boston, 1996

Cited References

Aumann, Y. and Rabani, Y. [1998]: An $O(\log k)$ approximate min-cut max-flow theorem and approximation algorithm. SIAM Journal on Computing 27 (1998), 291–301

Arora, S., Rao, S., and Vazirani, U. [2009]: Expander flows, geometric embeddings and graph partitioning. Journal of the ACM 56 (2009), Article 5

Arora, S., Hazan, E., and Kale, S. [2004]: $O(\sqrt{\log n})$ approximation to SPARSEST CUT in $\tilde{O}(n^2)$ time. Proceedings of the 45th Annual IEEE Symposium on Foundations of Computer Science (2004), 238–247

Becker, M., and Mehlhorn, K. [1986]: Algorithms for routing in planar graphs. Acta Informatica 23 (1986), 163–176

Bienstock, D., and Iyengar, G. [2006]: Solving fractional packing problems in $O^*(\frac{1}{\epsilon})$ iterations. SIAM Journal on Computing 35 (2006), 825–854

Boesch, F., and Tindell, R. [1980]: Robbins's theorem for mixed multigraphs. American Mathematical Monthly 87 (1980), 716–719

Charikar, M., Hajiaghayi, M.T., Karloff, H. and Rao, S. [2010]: ℓ_2^2 spreading metrics for vertex ordering problems. Algorithmica 56 (2010), 577–604

Chekuri, C., and Khanna, S. [2007]: Edge-disjoint paths revisited. ACM Transactions on Algorithms 3 (2007), Article 46

Chudak, F.A., and Eleutério, V. [2005]: Improved approximation schemes for linear programming relaxations of combinatorial optimization problems. In: Integer Programming and Combinatorial Optimization; Proceedings of the 11th International IPCO Conference; LNCS 3509 (M. Jünger, V. Kaibel, eds.), Springer, Berlin 2005, pp. 81–96

Even, S., Itai, A., and Shamir, A. [1976]: On the complexity of timetable and multicommodity flow problems. SIAM Journal on Computing 5 (1976), 691–703

Feige, U., and Lee, J.R. [2007]: An improved approximation ratio for the minimum linear arrangement problem. Information Processing Letters 101 (2007), 26–29

Fleischer, L.K. [2000]: Approximating fractional multicommodity flow independent of the number of commodities. SIAM Journal on Discrete Mathematics 13 (2000), 505–520

Ford, L.R., and Fulkerson, D.R. [1958]: A suggested computation for maximal multicommodity network flows. Management Science 5 (1958), 97–101

Ford, L.R., and Fulkerson, D.R. [1962]: Flows in Networks. Princeton University Press, Princeton 1962

Fortune, S., Hopcroft, J., and Wyllie, J. [1980]: The directed subgraph homeomorphism problem. Theoretical Computer Science 10 (1980), 111–121

Frank, A. [1980]: On the orientation of graphs. Journal of Combinatorial Theory B 28 (1980), 251–261

Frank, A. [1981]: How to make a digraph strongly connected. Combinatorica 1 (1981), 145–153

Frank, A. [1985]: Edge-disjoint paths in planar graphs. Journal of Combinatorial Theory B 39 (1985), 164–178

Frank, A., and Tardos, É. [1984]: Matroids from crossing families. In: Finite and Infinite Sets; Vol. I (A. Hajnal, L. Lovász, and V.T. Sós, eds.), North-Holland, Amsterdam, 1984, pp. 295–304

Garg, N., and Könemann, J. [2007]: Faster and simpler algorithms for multicommodity flow and other fractional packing problems. SIAM Journal on Computing 37 (2007), 630–652

Grigoriadis, M.D., and Khachiyan, L.G. [1996]: Coordination complexity of parallel price-directive decomposition. Mathematics of Operations Research 21 (1996), 321–340

Guruswami, V., Håstad, J., Manokaran, R., Raghavendra, P., and Charikar, M. [2011]: Beating the random ordering is hard: every ordering CSP is approximation resistant. SIAM Journal on Computing 40 (2011), 878–914

Hansen, M.D. [1989]: Approximation algorithms for geometric embeddings in the plane with applications to parallel processing problems. Proceedings of the 30th Annual IEEE Symposium on Foundations of Computer Science (1989), 604–609

Hirai, H. [2010]: Metric packing for $K_3 + K_3$. Combinatorica 30 (2010), 295–326

Hu, T.C. [1963]: Multi-commodity network flows. Operations Research 11 (1963), 344–360

Ibaraki, T., and Poljak, S. [1991]: Weak three-linking in Eulerian digraphs. SIAM Journal on Discrete Mathematics 4 (1991), 84–98

Karakostas, G. [2008]: Faster approximation schemes for fractional multicommodity flow problems. ACM Transactions on Algorithms 4 (2008), Article 13

Karp, R.M. [1972]: Reducibility among combinatorial problems. In: Complexity of Computer Computations (R.E. Miller, J.W. Thatcher, eds.), Plenum Press, New York 1972, pp. 85–103

Karzanov, A.V. [1987]: Half-integral five-terminus flows. Discrete Applied Mathematics 18 (1987) 263–278

Kawarabayashi, K., Kobayashi, Y., and Reed, B. [2012] The disjoint paths problem in quadratic time. Journal of Combinatorial Theory B 102 (2012), 424–435

Kawarabayashi, K., and Wollan, P. [2010]: A shorter proof of the graph minor algorithm: the unique linkage theorem. Proceedings of the 42th Annual ACM Symposium on Theory of Computing (2010), 687–694

Kleinberg, J. [1996]: Approximation algorithms for disjoint paths problems. PhD thesis, MIT, Cambridge 1996

Leighton, T., and Rao, S. [1999]: Multicommodity max-flow min-cut theorems and their use in designing approximation algorithms. Journal of the ACM 46 (1999), 787–832

Linial, N., London, E., and Rabinovich, Y. [1995]: The geometry of graphs and some of its algorithmic applications. Combinatorica 15 (1995), 215–245

Lomonosov, M. [1979]: Multiflow feasibility depending on cuts. Graph Theory Newsletter 9 (1979), 4

Lovász, L. [1976]: On two minimax theorems in graph. Journal of Combinatorial Theory B 21 (1976), 96–103

Lucchesi, C.L., and Younger, D.H. [1978]: A minimax relation for directed graphs. Journal of the London Mathematical Society II 17 (1978), 369–374

Matsumoto, K., Nishizeki, T., and Saito, N. [1986]: Planar multicommodity flows, maximum matchings and negative cycles. SIAM Journal on Computing 15 (1986), 495–510

Middendorf, M., and Pfeiffer, F. [1993]: On the complexity of the disjoint path problem. Combinatorica 13 (1993), 97–107

Müller, D., Radke, K., and Vygen, J. [2011]: Faster min-max resource sharing in theory and practice. Mathematical Programming Computation 3 (2011), 1–35

Nagamochi, H., and Ibaraki, T. [1989]: On max-flow min-cut and integral flow properties for multicommodity flows in directed networks. Information Processing Letters 31 (1989), 279–285

Nash-Williams, C.S.J.A. [1969]: Well-balanced orientations of finite graphs and unobtrusive odd-vertex-pairings. In: Recent Progress in Combinatorics (W. Tutte, ed.), Academic Press, New York 1969, pp. 133–149

Naves, G. [2009]: The hardness of routing two pairs on one face. Les cahiers Leibniz, Technical Report No. 177, Grenoble 2009

Nishizeki, T., Vygen, J., and Zhou, X. [2001]: The edge-disjoint paths problem is NP-complete for series-parallel graphs. Discrete Applied Mathematics 115 (2001), 177–186

Okamura, H., and Seymour, P.D. [1981]: Multicommodity flows in planar graphs. Journal of Combinatorial Theory B 31 (1981), 75–81

Räcke, H. [2008]: Optimal hierarchical decompositions for congestion minimization in networks. Proceedings of the 40th Annual ACM Symposium on Theory of Computing (2008), 255–264

Raghavan, P., and Thompson, C.D. [1987]: Randomized rounding: a technique for provably good algorithms and algorithmic proofs. Combinatorica 7 (1987), 365–374

Robertson, N., and Seymour, P.D. [1986]: Graph minors VI; Disjoint paths across a disc. Journal of Combinatorial Theory B 41 (1986), 115–138

Robertson, N., and Seymour, P.D. [1995]: Graph minors XIII; The disjoint paths problem. Journal of Combinatorial Theory B 63 (1995), 65–110

Rothschild, B., and Whinston, A. [1966]: Feasibility of two-commodity network flows. Operations Research 14 (1966), 1121–1129

Scheffler, P. [1994]: A practical linear time algorithm for disjoint paths in graphs with bounded tree-width. Technical Report No. 396/1994, FU Berlin, Fachbereich 3 Mathematik

Schwärzler, W. [2009]: On the complexity of the planar edge-disjoint paths problem with terminals on the outer boundary. Combinatorica 29 (2009), 121–126

Sebő, A. [1993]: Integer plane multiflows with a fixed number of demands. Journal of Combinatorial Theory B 59 (1993), 163–171

Seymour, P.D. [1980]: Four-terminus flows. Networks 10 (1980), 79–86

Seymour, P.D. [1981]: On odd cuts and multicommodity flows. Proceedings of the London Mathematical Society (3) 42 (1981), 178–192

Shahrokhi, F., and Matula, D.W. [1990]: The maximum concurrent flow problem. Journal of the ACM 37 (1990), 318–334

Sherman, J. [2009]: Breaking the multicommodity flow barrier for $O(\sqrt{\log n})$-approximations to sparsest cut. Proceedings of the 50th Annual IEEE Symposium on Foundations of Computer Science (2009), 363–372

Vygen, J. [1995]: NP-completeness of some edge-disjoint paths problems. Discrete Applied Mathematics 61 (1995), 83–90

Wagner, D., and Weihe, K. [1995]: A linear-time algorithm for edge-disjoint paths in planar graphs. Combinatorica 15 (1995), 135–150

Young, N. [1995]: Randomized rounding without solving the linear program. Proceedings of the 6th Annual ACM-SIAM Symposium on Discrete Algorithms (1995), 170–178

20 Network Design Problems

Connectivity is a very important concept in combinatorial optimization. In Chapter 8 we showed how to compute the connectivity between each pair of vertices of an undirected graph. Now we are looking for subgraphs that satisfy certain connectivity requirements. The general problem is:

SURVIVABLE NETWORK DESIGN PROBLEM

Instance: An undirected graph G with weights $c : E(G) \rightarrow \mathbb{R}_+$, and a connectivity requirement $r_{xy} \in \mathbb{Z}_+$ for each (unordered) pair of vertices x, y.

Task: Find a minimum weight spanning subgraph H of G such that for each x, y there are at least r_{xy} edge-disjoint paths from x to y in H.

Practical applications arise for example in the design of telecommunication networks which can "survive" certain edge failures.

A related problem allows edges to be picked arbitrarily often (see Goemans and Bertsimas [1993], Bertsimas and Teo [1997]). However, this can be regarded as a special case since G can have many parallel edges.

We first consider the STEINER TREE PROBLEM, which is a well-known special case. Here we have a set $T \subseteq V(G)$ of so-called terminals such that $r_{xy} = 1$ if $x, y \in T$ and $r_{xy} = 0$ otherwise. We look for a shortest network connecting all terminals; such a network is called a connector, and a minimal connector is a Steiner tree:

Definition 20.1. *Let G be an undirected graph and $T \subseteq V(G)$. A* **connector** *for T is a connected graph Y with $T \subseteq V(Y)$. A* **Steiner tree** *for T in G is a tree S with $T \subseteq V(S) \subseteq V(G)$ and $E(S) \subseteq E(G)$. The elements of T are called* **terminals***, those of $V(S) \setminus T$ are the* **Steiner points** *of S.*

Sometimes it is also required that all leaves of a Steiner tree are terminals; evidently this can always be achieved by deleting edges. In Section 20.1 we discuss an exact algorithm for the STEINER TREE PROBLEM, while Sections 20.2 and 20.3 are devoted to approximation algorithms.

© Springer-Verlag GmbH Germany 2018
Bernhard Korte and Jens Vygen, *Combinatorial Optimization*, Algorithms and Combinatorics 21, https://doi.org/10.1007/978-3-662-56039-6_20

In Section 20.4 we turn to the general SURVIVABLE NETWORK DESIGN PROB-LEM, and we give two approximation algorithms in Sections 20.5 and 20.6. While the first one is faster, the second one can always guarantee a performance ratio of 2 in polynomial time.

Finally, in Section 20.7 we discuss a network design problem that can be solved exactly in polynomial time: the VPN problem.

20.1 Steiner Trees

In this section we consider the following problem:

STEINER TREE PROBLEM

Instance: An undirected graph G, weights $c : E(G) \to \mathbb{R}_+$, and a set $T \subseteq V(G)$.

Task: Find a Steiner tree S for T in G whose weight $c(E(S))$ is minimum.

We have already dealt with the special cases $T = V(G)$ (spanning tree) and $|T| = 2$ (shortest path) in Chapters 6 and 7. While we had a polynomial-time algorithm in both of these cases, the general problem is *NP*-hard.

Theorem 20.2. (Karp [1972]) *The* STEINER TREE PROBLEM *is NP-hard, even for unit weights.*

Proof: We describe a transformation from VERTEX COVER which is *NP*-complete by Corollary 15.24. Given a graph G as part of a VERTEX COVER instance, we may assume that it contains more than one edge. We consider the graph H with vertices $V(H) := V(G) \cup E(G)$ and edges $\{v, e\}$ for $v \in e \in E(G)$ and $\{v, w\}$ for $v, w \in V(G)$, $v \neq w$. See Figure 20.1 for an illustration. We set $c(e) := 1$ for all $e \in E(H)$ and $T := E(G)$.

Given a vertex cover $X \subseteq V(G)$ of G, we can connect X in H by a tree of $|X|-1$ edges and join each of the vertices in T by an edge. We obtain a Steiner tree with $|X|-1+|E(G)|$ edges. On the other hand, let $(T \cup X, F)$ be a Steiner tree for T in H. Then X is a vertex cover in G and $|F| = |T \cup X| - 1 = |X| + |E(G)| - 1$.

Hence G has a vertex cover of cardinality k if and only if H has a Steiner tree for T with $k + |E(G)| - 1$ edges.   □

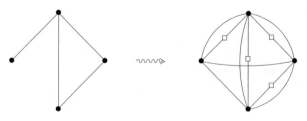

Fig. 20.1.

This transformation yields also the following stronger result:

Theorem 20.3. (Bern and Plassmann [1989])   *The* STEINER TREE PROBLEM *is MAXSNP-hard, even for unit weights.*

Proof: The transformation in the above proof is not an L-reduction in general, but we claim that it is one if G has bounded degree. By Theorem 16.46 the MINIMUM VERTEX COVER PROBLEM for graphs with maximum degree 4 is *MAXSNP*-hard.

For each Steiner tree $(T \cup X, F)$ in H and the corresponding vertex cover X in G we have

$$\begin{aligned} |X| - \text{OPT}(G) &= (|F| - |E(G)| + 1) - (\text{OPT}(H, T) - |E(G)| + 1) \\ &= |F| - \text{OPT}(H, T). \end{aligned}$$

Moreover, $\text{OPT}(H, T) \le 2|T| - 1 = 2|E(G)| - 1$ and $\text{OPT}(G) \ge \frac{|E(G)|}{4}$ if G has maximum degree 4. Hence $\text{OPT}(H, T) < 8\,\text{OPT}(G)$, and we conclude that the transformation is indeed an L-reduction. $\square$

Two other variants of the STEINER TREE PROBLEM are also *NP*-hard: the EU-CLIDEAN STEINER TREE PROBLEM (Garey, Graham and Johnson [1977]) and the MANHATTAN STEINER TREE PROBLEM (Garey and Johnson [1977]). Both ask for a network (set of straight line segments) of minimum total length which connects a given set of points in the plane. The difference between these two problems is that only horizontal and vertical line segments are allowed in the MANHATTAN STEINER TREE PROBLEM. In contrast to the *MAXSNP*-hard STEINER TREE PROB-LEM in graphs both geometric versions have an approximation scheme. A variant of this algorithm, which is due to Arora [1998], also solves the EUCLIDEAN TSP and some other geometric problems and will be presented in Section 21.2 (cf. Exercise 10 of Chapter 21). The STEINER TREE PROBLEM in planar graphs also has an approximation scheme, found by Borradaile, Klein and Mathieu [2009].

Hanan [1966] showed that the MANHATTAN STEINER TREE PROBLEM can be reduced to the STEINER TREE PROBLEM in finite grid graphs: there always exists an optimum solution where all line segments lie on the grid induced by the coordinates of the terminals. The MANHATTAN STEINER TREE PROBLEM is important in VLSI design, where electrical components must be connected with horizontal and vertical wires; see Korte, Prömel and Steger [1990], Martin [1992] and Hetzel [1995]. Here one looks for many disjoint Steiner trees. This is a generalization of the DISJOINT PATHS PROBLEM discussed in Chapter 19.

We shall now describe a dynamic programming algorithm due to Dreyfus and Wagner [1972]. This algorithm solves the STEINER TREE PROBLEM exactly but has exponential running time in general.

The DREYFUS-WAGNER ALGORITHM computes an optimum Steiner tree for each subset of T, starting with the two-element sets. It uses the following recursion formulas:

Lemma 20.4. *Let (G, c, T) be an instance of the* STEINER TREE PROBLEM. *For each $U \subseteq T$ and $x \in V(G) \setminus U$ we define*

$$p(U) \quad := \quad \min\{c(E(S)) : S \text{ is a Steiner tree for } U \text{ in } G\};$$
$$q(U \cup \{x\}, x) \quad := \quad \min\{c(E(S')) + c(E(S'')) : \emptyset \neq U' \subset U,$$
$$S' \text{ is a Steiner tree for } U' \cup \{x\} \text{ in } G,$$
$$S'' \text{ is a Steiner tree for } (U \setminus U') \cup \{x\} \text{ in } G\}.$$

Then we have for all $U \subseteq V(G)$, $|U| \geq 2$ and $x \in V(G) \setminus U$:

(a) $q(U \cup \{x\}, x) = \min_{\emptyset \neq U' \subset U} \left(p(U' \cup \{x\}) + p((U \setminus U') \cup \{x\}) \right)$,

(b) $p(U \cup \{x\}) = \min \Big\{ \min_{y \in U} \left(p(U) + \text{dist}_{(G,c)}(x, y) \right)$,

$$\min_{y \in V(G) \setminus U} \left(q(U \cup \{y\}, y) + \text{dist}_{(G,c)}(x, y) \right) \Big\}.$$

Proof: (a) is trivial, and so is the inequality "$\leq$" of (b).

To prove the inequality "$\geq$" of (b), consider an optimum Steiner tree S for $U \cup \{x\}$; we may assume that all leaves are terminals. If $|\delta_S(x)| \geq 2$, then

$$p(U \cup \{x\}) = c(E(S)) \geq q(U \cup \{x\}, x) = q(U \cup \{x\}, x) + \text{dist}_{(G,c)}(x, x),$$

where the inequality follows from letting U' be the set of terminals in one connected component of $S - x$.

If $|\delta_S(x)| = 1$, then let y be the nearest vertex from x in S that belongs to U or has $|\delta_S(y)| \geq 3$. We distinguish two cases: if $y \in U$, then

$$p(U \cup \{x\}) = c(E(S)) \geq p(U) + \text{dist}_{(G,c)}(x, y),$$

otherwise

$$p(U \cup \{x\}) = c(E(S)) \geq \min_{y \in V(G) \setminus U} \left(q(U \cup \{y\}, y) + \text{dist}_{(G,c)}(x, y) \right).$$

In (b), the minimum over these three formulas is computed. $\square$

These recursion formulas immediately suggest the following dynamic programming algorithm:

DREYFUS-WAGNER ALGORITHM

Input: An undirected graph G, weights $c : E(G) \rightarrow \mathbb{R}_+$, and a set $T \subseteq V(G)$.

Output: The length $p(T)$ of an optimum Steiner tree for T in G.

---

① **If** $|T| \leq 1$ **then** set $p(T) := 0$ and **stop**.
Compute $\text{dist}_{(G,c)}(x, y)$ for all $x, y \in V(G)$.
Set $p(\{x, y\}) := \text{dist}_{(G,c)}(x, y)$ for all $x, y \in V(G)$.

②   **For** $k := 2$ **to** $|T| - 1$ **do:**
       For all $U \subseteq T$ with $|U| = k$ and all $x \in V(G) \setminus U$ **do:**
          Set $q(U \cup \{x\}, x) := \min\limits_{\emptyset \neq U' \subset U} \left(p(U' \cup \{x\}) + p((U \setminus U') \cup \{x\}) \right)$.
       For all $U \subseteq T$ with $|U| = k$ and all $x \in V(G) \setminus U$ **do:**
          Set $p(U \cup \{x\}) := \min\Big\{ \min\limits_{y \in U} \left(p(U) + \mathrm{dist}_{(G,c)}(x, y) \right),$

$$\min\limits_{y \in V(G) \setminus U} \left(q(U \cup \{y\}, y) + \mathrm{dist}_{(G,c)}(x, y) \right) \Big\}.$$

Theorem 20.5. (Dreyfus and Wagner [1972]) *The* DREYFUS-WAGNER ALGO-RITHM *correctly determines the length of an optimum Steiner tree in* $O\left(3^t n + 2^t n^2 + mn + n^2 \log n\right)$ *time, where* $n = |V(G)|$, $m = |E(G)|$, *and* $t = |T|$.

Proof: The correctness follows from Lemma 20.4. ① consists of solving an ALL PAIRS SHORTEST PATHS PROBLEM, which can be done in $O(mn + n^2 \log n)$ time by Theorem 7.8.

The first recursion in ② requires $O\left(3^t n\right)$ time since there are 3^t possibilities to partition T into $U', U \setminus U'$, and $T \setminus U$. The second recursion in ② obviously requires $O\left(2^t n^2\right)$ time. $\qquad\qquad\qquad\qquad\qquad \square$

In the present form the DREYFUS-WAGNER ALGORITHM computes the length of an optimum Steiner tree, but not the Steiner tree itself. However, this can easily be achieved by storing some additional information and backtracking. We have already discussed this in detail with respect to DIJKSTRA'S ALGORITHM (Section 7.1). Hougardy, Silvanus and Vygen [2017] proposed a variant that has the same worst-case running time but is much faster in practice.

Note that the algorithm in general requires exponential time and exponential space. For a bounded number of terminals it is an $O(n^3)$-algorithm. There is another interesting special case where it runs in polynomial time (and space): if G is a planar graph and all terminals lie on the outer face, then the DREYFUS-WAGNER ALGORITHM can be modified to run in $O\left(n^3 t^2\right)$ time (Exercise 4). For the general case (with many terminals), the running time has been improved by Fuchs et al. [2007] and Vygen [2011]. If all edge weights are small integers, the algorithm of Björklund et al. [2007] is better.

Since we cannot hope for an exact polynomial-time algorithm for the general STEINER TREE PROBLEM, approximation algorithms are valuable. One idea underlying some of these algorithms is to approximate the optimum Steiner tree for T in G by a minimum weight spanning tree in the subgraph of the metric closure of G induced by T.

Theorem 20.6. *Let* G *be a connected graph with weights* $c : E(G) \to \mathbb{R}_+$, *and let* $(\bar{G}, \bar{c})$ *be the metric closure. Moreover, let* $T \subseteq V(G)$. *If* S *is an optimum Steiner tree for* T *in* G, *and* M *is a minimum weight spanning tree in* $\bar{G}[T]$ *(with respect to* $\bar{c}$), *then* $\bar{c}(E(M)) \leq 2c(E(S))$.

Proof: Consider the graph H containing two copies of each edge of S. H is Eulerian, so by Theorem 2.24 there exists an Eulerian walk W in H. The first appearance of the elements of T in W defines an order of T, and thus a tour W' in $\bar{G}[T]$. Since $\bar{c}$ satisfies the triangle inequality $(\bar{c}(\{x, z\}) \leq \bar{c}(\{x, y\}) + \bar{c}(\{y, z\}) \leq c(\{x, y\}) + c(\{y, z\})$ for all $x, y, z)$,

$$\bar{c}(W') \leq c(W) = c(E(H)) = 2c(E(S)).$$

Since W' contains a spanning tree of $\bar{G}[T]$ (just delete one edge) the theorem is proved.   □

This theorem was published by Gilbert and Pollak [1968] (referring to E.F. Moore), Choukhmane [1978], Kou, Markowsky and Berman [1981], and Takahashi and Matsuyama [1980]. It immediately suggests the following 2-factor approximation algorithm:

KOU-MARKOWSKY-BERMAN ALGORITHM

Input: A connected undirected graph G, weights $c : E(G) \to \mathbb{R}_+$, and a set $T \subseteq V(G)$.

Output: A Steiner tree for T in G.

① Compute the metric closure $(\bar{G}, \bar{c})$ and a shortest path P_{st} for all $s, t \in T$.

② Find a minimum weight spanning tree M in $\bar{G}[T]$ (with respect to $\bar{c}$).

Set $E(S) := \bigcup_{\{x,y\} \in E(M)} E(P_{xy})$ and $V(S) := \bigcup_{\{x,y\} \in E(M)} V(P_{xy})$.

③ Output a minimal connected spanning subgraph of S.

Theorem 20.7.   (Kou, Markowsky and Berman [1981])   *The* KOU-MARKOWSKY-BERMAN ALGORITHM *is a 2-factor approximation algorithm for the* STEINER TREE PROBLEM *and runs in* $O(k(m + n \log n))$ *time, where* $k = |T|$, $n = |V(G)|$, *and* $m = |E(G)|$.

Proof: The correctness and the performance guarantee follow directly from Theorem 20.6. ① can be implemented by running DIJKSTRA'S ALGORITHM with t as source, for each $t \in T$. ② can be done in $O(k^2)$ time using PRIM'S ALGORITHM (Theorem 6.6). ③ can be implemented with BFS with $O(kn)$ time.   □

Mehlhorn [1988] and Kou [1990] proposed an $O(m + n \log n)$-implementation of this algorithm. The idea is to compute, instead of $\bar{G}[T]$, a similar graph whose minimum weight spanning trees are also minimum weight spanning trees in $\bar{G}[T]$.

The minimum weight spanning tree itself yields a 2-factor approximation for any metric instance of the STEINER TREE PROBLEM. An algorithm with a better performance ratio was not known until Zelikovsky [1993] came up with an $\frac{11}{6}$-factor approximation algorithm for the STEINER TREE PROBLEM. The performance ratio has subsequently been improved to 1.75 by Berman and Ramaiyer [1994], to 1.65 by Karpinski and Zelikovsky [1997] , to 1.60 by Hougardy and Prömel [1999], to 1.55 by Robins and Zelikovsky [2005] (see Section 20.2), and finally to 1.39 by Byrka et al. [2013] (see Section 20.3).

On the other hand, by Theorem 20.3 and Corollary 16.40, an approximation scheme cannot exist unless $P = NP$. Indeed, Chlebík and Chlebíková [2008] showed that, unless $P = NP$, there is no 1.01-factor approximation algorithm for the STEINER TREE PROBLEM.

An algorithm which computes optimum Steiner trees and is quite efficient, especially for the MANHATTAN STEINER TREE PROBLEM, was developed by Warme, Winter and Zachariasen [2000].

20.2 The Robins-Zelikovsky Algorithm

Definition 20.8. *A **full Steiner tree** for a terminal set T in a graph G is a tree Y in G where T is the set of leaves of Y. Every minimal Steiner tree for T can be decomposed into full Steiner trees for subsets of T, its **full components**. Unions of full components each of which has at most k terminals are called **k-restricted** (with respect to the given terminal set). More precisely, a graph Y is called k-restricted (in G with respect to T) if there are full Steiner trees Y_i for $T \cap V(Y_i)$ in G with $|T \cap V(Y_i)| \le k$ $(i = 1, \dots, t)$, such that the graph arising from $(\{(i, v) : i = 1, \dots, t, \ v \in V(Y_i)\}, \{\{(i, v), (i, w)\} : i = 1, \dots, t, \ \{v, w\} \in E(Y_i)\})$ by contracting the set $\{i : v \in V(Y_i)\} \times \{v\}$ for each $v \in T$ is connected, $V(Y) = \bigcup_{i=1}^{t} V(Y_i)$, and $E(Y)$ is the disjoint union of the sets $E(Y_i)$. Note that parallel edges may arise.*

*We define the **k-Steiner ratio** as*

$$\rho_k := \sup_{(G,c,T)} \left\{ \frac{\min\{c(E(Y)) : Y \ k\text{-restricted connector for } T\}}{\min\{c(E(Y)) : Y \ \text{Steiner tree for } T\}} \right\},$$

where the supremum is taken over all instances of the STEINER TREE PROBLEM.

For example, 2-restricted connectors are composed of paths between terminals. So optimum 2-restricted connectors for T in (G, c) correspond to minimum weight spanning trees in $(\bar{G}[T], \bar{c})$; thus $\rho_2 \le 2$ by Theorem 20.6. The stars with unit weights show that in fact $\rho_2 = 2$ (and in general $\rho_k \ge \frac{k}{k-1}$). If we restrict the supremum to instances of the MANHATTAN STEINER TREE PROBLEM, the ratio is better, e.g. $\frac{3}{2}$ for $k = 2$ (Hwang [1976]).

Theorem 20.9. (Du, Zhang and Feng [1991]) $\rho_{2^s} \le \frac{s+1}{s}$.

Proof: Let (G, c, T) be an instance and Y an optimum Steiner tree in the metric closure. W.l.o.g. Y is a full Steiner tree (otherwise handle full components separately). Moreover, by duplicating vertices and adding edges of length zero, we may assume Y to be a full binary tree whose leaves are the terminals. One vertex, the root, has degree 2, and all other Steiner points have degree 3. We say that a vertex $v \in V(Y)$ is at level i if its distance from the root is i. All terminals are at the same level h (the height of the binary tree).

We define s 2^s-restricted connectors for T, which have total length at most $(s + 1)c(E(Y))$. For $v \in V(Y)$, let $P(v)$ be a path in Y from v to some leaf, such that all these paths are pairwise edge-disjoint (for example, descend once to the left, and then always to the right).

For $i = 1, \ldots, s$, let Y_i be the union of the following full components:

- the subtree of Y induced by the vertices up to level i, plus $P(v)$ for every vertex v on level i;
- for each vertex u on level $ks + i$: the subtree induced by the successors of u up to level $(k + 1)s + i$, plus $P(v)$ for every vertex v in the subtree at level $(k + 1)s + i$ $(k = 0, \ldots, \lfloor \frac{h-1-i}{s} \rfloor - 1)$;
- for each vertex u on level $\lfloor \frac{h-1-i}{s} \rfloor s + i$: the subtree induced by all successors of u.

Clearly, each of these trees is 2^s-restricted, and the union of the trees in Y_i is Y, i.e. is a connector for T. Moreover, each edge of Y is contained once in each set Y_i, not counting the appearance in a path $P(v)$. Moreover, each $P(v)$ is used in only one Y_i. Thus each edge appears at most $s + 1$ times.   □

In particular, $\rho_k \to 1$ as $k \to \infty$. Thus we cannot expect to find the optimum k-restricted connector in polynomial time for fixed k. Indeed, this problem is NP-hard for every fixed $k \geq 4$ (Garey and Johnson [1977]). The bound of Theorem 20.9 is tight: Borchers and Du [1997] proved that $\rho_k = \frac{(s+1)2^s+t}{s2^s+t}$ for all $k \geq 2$, where $k = 2^s + t$ and $0 \leq t < 2^s$.

We will present an algorithm that starts with a minimum spanning tree in the subgraph of the metric closure induced by T, and tries to improve it by using k-restricted full Steiner trees. However, the algorithm only decides to include at most half of such a Steiner tree, its so-called loss. For each Steiner tree Y we define a **loss** of Y to be a minimum cost edge set F connecting each Steiner point of degree at least 3 to a terminal. See Figure 20.2 for an example of a full Steiner tree with a loss (bold edges), assuming that the cost of an edge is proportional to its length.

Proposition 20.10. *Let Y be a full Steiner tree for T, $c : E(Y) \to \mathbb{R}_+$, and let F be a loss of Y. Then $c(F) \leq \frac{1}{2}c(E(Y))$.*

Proof: Let r be an arbitrary vertex of Y. Let U be the set of Steiner points $v \in V(Y) \setminus T$ of degree at least 3. For each $v \in U$, let $P(v)$ be a minimum cost one among the (at least two) paths from v to a vertex $w \in U \cup T$ such that w has larger distance from r than v. The union of the edge sets of these paths connects each element of U to a terminal and has at most half the total cost.   □

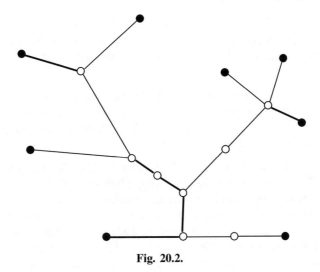

Fig. 20.2.

Instead of explicitly contracting losses of k-restricted full Steiner trees, we adjust the costs as follows:

Proposition 20.11. *Let G be a complete graph, $T \subseteq V(G)$, $c : E(G) \rightarrow \mathbb{R}_+$ and $k \geq 2$. Let $S \subseteq T$ with $|S| \leq k$, let Y be a Steiner tree for S in G and L a loss of Y. Let $c'(e) := 0$ for $e \in L$ and $c'(e) := c(e)$ otherwise. We define $c/(Y, L) : E(G) \rightarrow \mathbb{R}_+$ by $c/(Y, L)(\{v, w\}) := \min\{c(\{v, w\}), \text{dist}_{(Y,c')}(v, w)\}$ for $v, w \in S$, $v \neq w$, and $c/(Y, L)(e) := c(e)$ for all other edges. Then:*

(a) *There exists a spanning tree M in $G[S]$ with $c/(Y, L)(E(M)) + c(L) \leq c(E(Y))$.*
(b) *For each k-restricted connector H' of T in G there is a k-restricted connector H of T in G with $c(E(H)) \leq c/(Y, L)(E(H')) + c(L)$.*

Proof: (a): We may assume that L is minimal, so every connected component of $(V(Y), L)$ contains at most one element of S. Partition $E(Y) \setminus L$ into edge sets of maximal paths in Y all whose internal vertices have degree 2 in Y. For every such path P there are two terminals $s, t \in S$ such that the s-t-path Q in Y contains only edges of P and L. The disjoint union of these paths (one for each such P) contains each edge of $E(Y) \setminus L$ exactly once and each edge of L at least once. So it is connected, and it corresponds to a spanning tree M in $G[S]$ with $\sum_{\{s,t\} \in E(M)} \text{dist}_{(Y,c')}(s, t) \leq c(E(Y) \setminus L)$, implying (a).

(b): For the second statement, let H' be a k-restricted connector of T. Replace each edge $e = \{v, w\} \in E(H')$ with $c/(Y, L)(e) < c(e)$ by a shortest v-w-path in (Y, c'), and eliminate parallel edges. Then the resulting graph H is a k-restricted connector of T and satisfies $c(E(H)) = c'(E(H)) + c(E(H) \cap L) \leq c/(Y, L)(E(H')) + c(L)$. $\square$

We will repeatedly modify the cost function by adding edges corresponding to full components. The following observation says that the reduction of the cost of a minimum spanning tree does not increase when other edges have been inserted before:

Lemma 20.12. (Zelikovsky [1993], Berman and Ramaiyer [1994]) *Let G be a graph, $c : E(G) \to \mathbb{R}_+$, $T \subseteq V(G)$, (T, U) another graph, $c' : U \to \mathbb{R}_+$. Let $m : 2^U \to \mathbb{R}_+$, where $m(X)$ is the cost of a minimum spanning tree in $(T, E(G[T]) \cup X)$. Then m is supermodular.*

Proof: Let $A \subseteq U$ and $f \in U$. We run KRUSKAL'S ALGORITHM in parallel on $G[T]$ and on $G[T] + f$, examining edges of $G[T]$ in the same order (with nondecreasing cost). Both versions run exactly the same way, except that the first one does not choose f, while the second one does not choose the first edge that closes a circuit in $G + f$ containing f. Thus the minimum costs of the spanning trees in the two graphs differ by $\min\{\gamma : G[T] + f$ contains a circuit containing f whose edges have cost at most $\gamma\} - c'(f)$. Clearly, this difference can only decrease if we consider $G[T] + A$ and $(G[T] + A) + f$ instead of $G[T]$ and $G[T] + f$. Hence

$$m(A) - m(A \cup \{f\}) \leq m(\emptyset) - m(\{f\}).$$

Now let $X, Y \subseteq U$, $Y \setminus X = \{y_1, \ldots, y_k\}$, and write $m_i(Z) := m((X \cap Y) \cup \{y_1, \ldots, y_{i-1}\} \cup Z)$ for $i = 1, \ldots, k$. By applying the above to m_i we get

$$
\begin{aligned}
m(X) - m(X \cup Y) &= \sum_{i=1}^{k}(m_i(X \setminus Y) - m_i((X \setminus Y) \cup \{y_i\})) \\
&\leq \sum_{i=1}^{k}(m_i(\emptyset) - m_i(\{y_i\})) \\
&= m(X \cap Y) - m(Y),
\end{aligned}
$$

i.e. supermodularity.   □

We now describe the algorithm. Denote by $mst(c)$ the minimum cost of a spanning tree in the subgraph of $(\bar{G}, c)$ induced by T.

ROBINS-ZELIKOVSKY ALGORITHM

Input: An undirected graph G, weights $c : E(G) \to \mathbb{R}_+$, and a set $T \subseteq V(G)$ of terminals. A number $k \geq 2$.

Output: A Steiner tree for T in G.

①   Compute the metric closure $(\bar{G}, \bar{c})$ of (G, c).

②   Choose a subset S of at most k terminals and a pair $K = (Y, L)$, where Y is an optimum Steiner tree for S and L is a loss of Y, such that $\frac{mst(\bar{c}) - mst(\bar{c}/K)}{\bar{c}(L)}$ is maximum and at least 1.
   If such a choice is impossible, **then go to** ④.

③   Set $\bar{c} := \bar{c}/K$.
   Go to ②.

④   Compute a minimum spanning tree in the subgraph of $(\bar{G}, \bar{c})$ induced by
   T.
   Replace all edges by shortest paths in (G, c'), where $c'(e) := 0$ if $e \in L$
   for some L chosen in the algorithm and $c'(e) = c(e)$ otherwise.
   Finally compute a minimal connected subgraph spanning T.

Suppose that the algorithm stops in iteration $t + 1$, and let $K_i := (Y_i, L_i)$ be
the Steiner tree and its loss chosen in the i-th iteration $(i = 1, \ldots, t)$. Let c_0 be
the cost function $\bar{c}$ after ①, and let $c_i := c_{i-1}/K_i$ be the cost function $\bar{c}$ after i
iterations $(i = 1, \ldots, t)$.

Lemma 20.13.   *The algorithm computes a Steiner tree for T of weight at most*
$mst(c_t) + \sum_{i=1}^{t} c(L_i)$.

Proof: This follows from Proposition 20.11(b).   □

Let Y^* be an optimum k-restricted connector for T, let $Y_1^*, \ldots, Y_{t^*}^*$ be k-
restricted full Steiner trees whose union is Y^*, let L_j^* be a loss of Y_j^* and $K_j^* = (Y_j^*, L_j^*)$ $(j = 1, \ldots, t^*)$, and let $L^* := L_1^* \cup \cdots \cup L_{t^*}^*$. We write c/K^* instead of
$c/K_1^*/\cdots/K_{t^*}^*$.

Lemma 20.14. $mst(c/K^*) + c(L^*) = c(E(Y^*))$.

Proof: The inequality "≤" follows from Proposition 20.11(a); the other one from
Proposition 20.11(b).   □

Lemma 20.15. $mst(c_t) \le c(E(Y^*)) \le mst(c_0)$.

Proof: $c(E(Y^*)) \le mst(c_0)$ is trivial. When the algorithm terminates, $mst(c_t) - mst(c_t/K_j^*) \le c(L_j^*)$ for $j = 1, \ldots, t^*$. Hence, using Lemma 20.12,

$$
\begin{aligned}
mst(c_t) - mst(c/K^*) &\le mst(c_t) - mst(c_t/K^*) \\
&= \sum_{j=1}^{t^*} (mst(c_t/K_1^*/\cdots/K_{j-1}^*) - mst(c_t/K_1^*/\cdots/K_j^*)) \\
&\le \sum_{j=1}^{t^*} (mst(c_t) - mst(c_t/K_j^*)) \\
&\le \sum_{j=1}^{t^*} c(L_j^*),
\end{aligned}
$$

implying $mst(c_t) \le mst(c/K^*) + c(L^*)$. Now use Lemma 20.14.   □

Lemma 20.16. $mst(c_t) + \sum_{i=1}^{t} c(L_i) \le c(E(Y^*))(1 + \frac{\ln 3}{2})$.

Proof: Let $i \in \{1, \ldots, t\}$. By the choice of L_i in iteration i of the algorithm,

$$
\frac{mst(c_{i-1}) - mst(c_i)}{c(L_i)} \geq \max_{j=1,\ldots,t^*} \frac{mst(c_{i-1}) - mst(c_{i-1}/K_j^*)}{c(L_j^*)}
$$

$$
\geq \frac{\sum_{j=1}^{t^*} (mst(c_{i-1}) - mst(c_{i-1}/K_j^*))}{\sum_{j=1}^{t^*} c(L_j^*)}
$$

$$
\geq \frac{\sum_{j=1}^{t^*} (mst(c_{i-1}/K_1^*/\cdots/K_{j-1}^*) - mst(c_{i-1}/K_1^*/\cdots/K_j^*))}{c(L^*)}
$$

$$
= \frac{mst(c_{i-1}) - mst(c_{i-1}/K^*)}{c(L^*)}
$$

$$
\geq \frac{mst(c_{i-1}) - mst(c/K^*)}{c(L^*)}
$$

(we used Lemma 20.12 in the third and monotonicity in the last inequality). Moreover, the left-hand side is at least 1. Hence

$$
\sum_{i=1}^{t} c(L_i) \leq \sum_{i=1}^{t} (mst(c_{i-1}) - mst(c_i)) \frac{c(L^*)}{\max\{c(L^*), mst(c_{i-1}) - mst(c/K^*)\}}
$$

$$
\leq \int_{mst(c_t)}^{mst(c_0)} \frac{c(L^*)}{\max\{c(L^*), x - mst(c/K^*)\}} \, dx.
$$

As $c(E(Y^*)) = mst(c/K^*) + c(L^*)$ by Lemma 20.14 and $mst(c_t) \leq c(E(Y^*)) \leq mst(c_0)$ by Lemma 20.15, we compute

$$
\sum_{i=1}^{t} c(L_i) \leq \int_{mst(c_t)}^{mst(c/K^*)+c(L^*)} 1 \, dx + \int_{c(L^*)}^{mst(c_0)-mst(c/K^*)} \frac{c(L^*)}{x} \, dx
$$

$$
= c(E(Y^*)) - mst(c_t) + c(L^*) \ln \frac{mst(c_0) - mst(c/K^*)}{c(L^*)}.
$$

Since $mst(c_0) \leq 2\,\mathrm{OPT}(G, c, T) \leq c(E(Y^*)) + mst(c/K^*) + c(L^*)$ (by Lemma 20.14), we obtain

$$
mst(c_t) + \sum_{i=1}^{t} c(L_i) \leq c(E(Y^*)) \left(1 + \frac{c(L^*)}{c(E(Y^*))} \ln \left(1 + \frac{c(E(Y^*))}{c(L^*)} \right) \right).
$$

As $0 \leq c(L^*) \leq \frac{1}{2} c(E(Y^*))$ (by Proposition 20.10) and $\max\{x \ln(1 + \frac{1}{x}) : 0 < x \leq \frac{1}{2}\}$ is attained for $x = \frac{1}{2}$ (for the derivative $\ln(1 + \frac{1}{x}) - \frac{1}{x+1}$ is always positive), we conclude that $mst(c_t) + \sum_{i=1}^{t} c(L_i) \leq c(E(Y^*))(1 + \frac{\ln 3}{2})$.   □

This proof is essentially due to Gröpl et al. [2001]. We conclude:

Theorem 20.17. (Robins and Zelikovsky [2005]) *The Robins-Zelikovsky Algorithm has a performance guarantee of $\rho_k(1 + \frac{\ln 3}{2})$ and runs in polynomial time for each fixed k. For k sufficiently large, the performance guarantee is less than 1.55.*

Proof: By Lemma 20.13, the algorithm outputs a Steiner tree of cost at most $mst(c_t) + \sum_{i=1}^{t} c(L_i)$. By Lemma 20.16, this is at most $\rho_k(1 + \frac{\ln 3}{2})$. Choosing $k = \min\{|V(G)|, 2^{2^{233}}\}$ and applying Theorem 20.9, we get a performance ratio of $\rho_k(1 + \frac{\ln 3}{2}) \leq \frac{2234}{2233}(1 + \frac{\ln 3}{2}) < 1.55$.

There are at most n^k possible subsets S, and for each one there are at most n^{k-2} choices for the (at most $k - 2$) Steiner points of degree at least 3 in an optimum Steiner tree Y for S. Then, given Y, there are at most $(2k - 3)^{k-2}$ choices for a loss (up to inclusion of edges with zero cost). Hence each iteration takes $O(n^{2k})$ time (for fixed k), and there are at most n^{2k-2} iterations.   □

20.3 Rounding the Directed Component LP

In this section we present the currently best known approximation algorithm for the STEINER TREE PROBLEM, due to Byrka et al. [2013], with approximation ratio arbitrarily close to $\ln 4$.

Let (G, c, T) be an instance of the STEINER TREE PROBLEM. Again we restrict ourselves to k-restricted full components for some large enough k. But now we orient them: Let C^k be the set of pairs (t, R) for all $t \in R \subseteq T$ with $|R| \leq k$; they are called the (directed) components. For $C = (t, R) \in C^k$ let $c(C)$ denote the minimum cost of a Steiner tree for R in (G, c). Note that, by Theorem 20.5, $c(C)$ for $C \in C^k$ can be computed in polynomial time for each fixed k.

Fix an arbitrary terminal $r \in T$ as root. For $r \in U \subset T$ we denote by $\delta^+_{C^k}(U)$ the set of all $(t, R) \in C^k$ with $t \in U$ and $R \setminus U \neq \emptyset$. The so-called "directed component LP" then reads as follows.

$$\min \quad \sum_{C \in C^k} c(C) x_C$$

$$\text{s.t.} \quad \sum_{C \in \delta^+_{C^k}(U)} x_C \geq 1 \quad (r \in U \subset T) \tag{20.1}$$

$$x_C \geq 0 \quad (C \in C^k)$$

This LP is a relaxation of the problem of finding a minimum cost k-restricted Steiner tree, because any k-restricted Steiner tree can be oriented as an arborescence rooted at r, and we can set $x_{(t,R)} := 1$ whenever R is a full component with root t.

The LP (20.1) can be solved in polynomial time for each fixed k, because the number of variables is $O(k|T|^k)$, and the cut constraints can be replaced by requiring an r-s-flow of value 1 for all $s \in T \setminus \{r\}$ in an auxiliary digraph with vertices $T \cup \{v_C : C \in C^k\}$ and edges (t, v_C) and (v_C, s) for $s \in R \setminus \{t\}$ with capacity x_C for each $C = (t, R) \in C^k$ (Exercise 9).

The algorithm of Byrka et al. [2013] solves the LP (20.1), selects a component $C = (t, R)$, where $C \in C^k$ is chosen with probability $\frac{x_C}{\sum_{C' \in C^k} x_{C'}}$, contracts R, and iterates until only one terminal is left. For $C' \subseteq C^k$ we denote by $(G, c, T)/C'$ the instance which results from contracting R for each $(t, R) \in C'$ (and vertices that result from contraction are terminals).

ITERATIVE RANDOMIZED ROUNDING ALGORITHM

Input: An undirected graph G, weights $c : E(G) \to \mathbb{R}_+$, and a set $T \subseteq V(G)$ of terminals. A number $k \geq 2$.

Output: A Steiner tree for T in G.

① $C' := \emptyset$.

②   Compute an optimum solution x to the LP (20.1) for $(G, c, T)/C'$.

③   Randomly choose a component $C \in C^k$, where each C is chosen with probability $\frac{x_C}{\sum_{C' \in C^k} x_{C'}}$.

④   Set $C' := C' \cup \{C\}$.

⑤   **If** more than one terminal is left, **then go to** ②.

⑥   Let $V(Y) = \bigcup_{C \in C'} V(Y_C)$ and $E(Y) = \bigcup_{C \in C'} E(Y_C)$, where, for $C = (t, R) \in C'$, Y_C is an optimum Steiner tree for R in (G, c).
Return a minimal connected subgraph of Y containing all terminals.

The algorithm can be derandomized, but the above randomized version is easier to analyze. Let C'_t be the set C' after t iterations.

We will use a tree (T, W) to measure progress; this tree is chosen randomly as follows.

Let Y_k^* be an optimum k-restricted Steiner tree for T. Transform Y_k^* into a full Steiner tree Y^* of the same cost, in which all Steiner nodes have degree 3 except one Steiner node r^* that has degree 2 (as in the proof of Theorem 20.9; see Figure 20.3, where contracting the dotted edges in Y^* on the right-hand side yields Y_k^* on the left-hand side; terminals are shown as squares). We can regard Y^* as a tree rooted at r^* whose leaves are the terminals. For each vertex $v \in V(Y^*) \setminus T$ independently, we choose one of the two edges to the children of v as "losing edge", each with probability $\frac{1}{2}$. Let L be the set of losing edges. We can regard (Y^*, L) as a knock-out tournament, where the terminals are the players, the Steiner

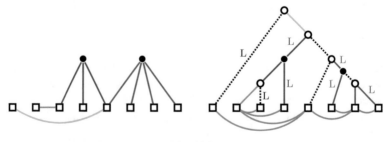

Fig. 20.3.

nodes are the matches, and a player plays along its path to r^* but drops out once it passes a losing edge. Now let W be the set of terminal pairs that play each other during this tournament; formally,

$$W := \left\{\{u, v\} \in \binom{T}{2} : \text{the } u\text{-}v\text{-path in } Y^* \text{ contains only one losing edge}\right\}.$$

(T, W) is a tree; it is called the *witness tree* (the bent edges at the bottom in Figure 20.3, right-hand side). Now for $e \in E(Y^*)$ we let $W(e)$ be the set of edges $\{u, v\} \in W$ for which the u-v-path in Y^* contains e. The set $W(e)$ corresponds to the matches that the winner of the match v will play after v, where v is the endpoint of e that is farther away from r^*. Thus, for a losing edge e we have $|W(e)| = 1$, and the probability that $|W(e)|$ has more than q elements is at most 2^{-q} (for all $q \in \mathbb{N}$).

We measure progress by marking edges of the witness tree so that the unmarked edges of W connect the terminals (after contracting the components chosen in previous iterations). The set of possibilities for edges of the witness tree to be marked when component $C = (t, R)$ is chosen is denoted by $\mathcal{B}_W(R)$, the family of *bridge sets* of R:

$$\mathcal{B}_W(R) := \{B \subseteq W : (W \setminus B)/R \text{ is a tree}\}.$$

We want to show that for connecting the terminals (after contracting the components chosen in previous iterations) we do not need edges $e \in E(Y^*)$ for which all edges of $W(e)$ are marked:

Lemma 20.18. *Let (T, W) be a tree, $t \in \mathbb{N}$, and $C_1, \ldots, C_t \in \mathcal{C}^k$. Let $W_0 := W$, $B_i \in \mathcal{B}_{(T, W_{i-1})/C_1/\cdots/C_{i-1}}(R)$ (where $C_i = (t, R)$) and $W_i := W_{i-1} \setminus B_i$ for $i = 1, \ldots, t$. Then all terminals are in the same connected component of $(V(Y^*), \{e \in E(Y^*) : W(e) \cap W_t \neq \emptyset\})/C_1/\cdots/C_t$.*

Proof: First we get by induction on t that $(T, W_t)/C_1/\cdots/C_t$ is a tree, using the definition of the bridge sets.

Now suppose $(V(Y^*), \{e \in E(Y^*) : W(e) \cap W_t \neq \emptyset\})/C_1/\cdots/C_t$ has terminals in different connected components. Then there is a subset $U \subset V(Y^*)$ with $U \cap T \neq \emptyset$ and $T \setminus U \neq \emptyset$ such that $R \subseteq U$ or $R \cap U = \emptyset$ for all $(t, R) \in \{C_1, \ldots, C_t\}$ and $W(e) \cap W_t = \emptyset$ for all $e \in \delta_{Y^*}(U)$. Since $(T, W_t)/C_1/\cdots/C_t$ is connected, there is an edge $\{u, v\} \in W_t$ with $u \in U$ and $v \notin U$. Then the u-v-path in Y^* must contain an edge $e \in \delta_{Y^*}(U)$. But then $\{u, v\} \in W(e)$, a contradiction. $\square$

In order to show that we can mark edges fast, we need the following Bridge Lemma:

Lemma 20.19. *Let (T, W) be a tree, $d : W \to \mathbb{R}_+$, and x a feasible solution to the directed component LP (20.1). Then*

$$\sum_{e \in W} d(e) \leq \sum_{C=(t,R) \in \mathcal{C}^k} x_C \max\left\{\sum_{e \in B} d(e) : B \in \mathcal{B}_W(R)\right\}.$$

Proof: For $e \in \binom{T}{2}$ let $\bar{d}(e) := \max\{d(f) : f \in E(P_e)\}$ where P_e is the path in (T, W) connecting the endpoints of e. Since the tree (T, W) satisfies condition (b) of Theorem 6.3 with respect to the complete graph on T with weights $\bar{d}$, it is a minimum cost spanning tree with respect to $\bar{d}$.

Let $C = (t, R) \in C^k$ and $B_C \in \mathcal{B}_W(R)$ maximizing $\sum_{e \in B_C} d(e)$. For $u \in T$ let $r(u)$ be the element of R that is in the same connected component as u in $(T, W \setminus B_C)$. We construct a spanning tree (R, S_C) by setting $S_C := \{\{r(u), r(v)\} : \{u, v\} \in B_C\}$.

Note that $\bar{d}(\{r(u), r(v)\}) = d(\{u, v\})$ for all $\{u, v\} \in B_C$ as B_C maximizes $\sum_{e \in B_C} d(e)$. Hence $\sum_{e \in S_C} \bar{d}(e) = \sum_{e \in B_C} d(e)$. We can orient (R, S_C) as an arborescence rooted at t and denote the oriented edge set by $\vec{S_C}$.

Now, for $e = (v, w) \in T \times T$ with $v \neq w$, let $y_e := \sum_{C \in C^k : e \in \vec{S_C}} x_C$. Then $\sum_{e \in U \times (T \setminus U)} y_e \geq \sum_{C \in \delta_{C^k}^+} x_C \geq 1$ for all $r \in U \subset T$, so y is a feasible solution to the LP in Corollary 6.15, applied to the complete digraph on vertex set T. Since (T, W) is a minimum cost spanning tree with respect to $\bar{d}$ (and can be oriented as an arborescence rooted at r),

$$\sum_{e \in W} \bar{d}(e) \leq \sum_{e = (v,w) \in T \times T : v \neq w} \bar{d}(\{v, w\}) y_e = \sum_{e \in \binom{T}{2}} \bar{d}(e) \sum_{C \in C^k : e \in S_C} x_C$$

$$= \sum_{C \in C^k} x_C \sum_{e \in S_C} \bar{d}(e) = \sum_{C \in C^k} x_C \sum_{e \in B_C} d(e).$$

Observing $d(e) = \bar{d}(e)$ for $e \in W$ concludes the proof. $\qquad\square$

We want to show that there is a policy of marking edges (choosing the bridge set) such that every edge of W (that is not marked yet) is marked with probability at least $\frac{1}{M}$ in each iteration:

Lemma 20.20. *Let (T, W) be a tree and let x be a feasible solution to the LP (20.1). Let $\mathcal{P} := \{(C, B) : C = (t, R) \in C^k, B \in \mathcal{B}_W(R)\}$. Then there is a number $M > 0$ and a probability distribution $p : \mathcal{P} \to [0, 1]$ such that $\sum_{B \in \mathcal{B}_W(R)} p(C, B) = \frac{x_C}{M}$ for $C = (t, R) \in C^k$ and $\sum_{(C,B) \in \mathcal{P} : e \in B} p(C, B) \geq \frac{1}{M}$ for $e \in W$.*

Proof: Setting $M := \sum_{C \in C^k} x_C$ and substituting $p(C, B) = \frac{x_C}{M} q(C, B)$ for $(C, B) \in \mathcal{P}$, the assertion is equivalent to saying that the following linear inequality system has a solution:

$$\sum_{B \in \mathcal{B}_W(R)} q(C, B) \leq 1 \quad (C = (t, R) \in C^k)$$

$$\sum_{(C,B) \in \mathcal{P} : e \in B} x_C q(C, B) \geq 1 \quad (e \in W)$$

$$q(C, B) \geq 0 \quad ((C, B) \in \mathcal{P})$$

Suppose not. Then, by Corollary 3.25, there are $z \in \mathbb{R}_+^{\mathcal{C}^k}$ and $d \in \mathbb{R}_+^W$ with

$$\sum_{C \in \mathcal{C}^k} z_C \;<\; \sum_{e \in W} d_e$$

$$z_C \;\geq\; x_C \sum_{e \in B} d_e \quad ((C, B) \in \mathcal{P}).$$

This implies

$$\sum_{C \in \mathcal{C}^k} x_C \max \left\{ \sum_{e \in B} d_e : B \in \mathcal{B}_W(R) \right\} \;\leq\; \sum_{C \in \mathcal{C}^k} z_C \;<\; \sum_{e \in W} d_e,$$

contradicting the Bridge Lemma 20.19. $\square$

Note that we apply this lemma in iteration i to the tree $(T, W_{i-1})/C_1/\cdots/C_{i-1}$ (cf. Lemma 20.18).

The number M can differ in each iteration. For the purpose of the analysis, let M be a global upper bound, and modify the LP solutions by increasing the value of the variable $x_{(r,\{r\})}$ (this dummy component has cost zero), so that $\sum_{C \in \mathcal{C}^k} x_C = M$ in every iteration. Now many iterations would choose the dummy component (and do nothing), which does not change what happens afterwards. This little trick allows us to work with a single value of M.

Lemma 20.21. *Let $X \subseteq W$, and suppose each element is marked with probability at least $\frac{1}{M}$ in each iteration. Then the expected number of iterations until all elements of X are marked is at most $M H(|X|)$, where $H(i) = 1 + \frac{1}{2} + \cdots + \frac{1}{i}$ is the i-th Harmonic number.*

Proof: Let $\tau(q)$ be (an upper bound on) the expected number of iterations until all elements of a q-element set X are marked. By Lemma 20.20 we have $\tau(1) \leq 1 + (1 - \frac{1}{M}) + (1 - \frac{1}{M})^2 + \cdots = M$. We use induction on q, so let now $|X| = q > 1$. For $i = 0, \ldots, q+1$ let λ_i be the probability that at least i elements of X are marked in the first iteration. We have $\lambda_0 = 1$, $\lambda_{q+1} = 0$, and (by Lemma 20.20) $\sum_{i=1}^q \lambda_i \geq \frac{q}{M}$. Hence

$$\tau(q) \;\leq\; 1 + \sum_{i=0}^q (\lambda_i - \lambda_{i+1}) \tau(q - i)$$

$$\leq\; 1 + (1 - \lambda_1)\tau(q) + \sum_{i=1}^q (\lambda_i - \lambda_{i+1}) M H(q - i)$$

$$=\; 1 + (1 - \lambda_1)\tau(q) - \sum_{i=1}^q \lambda_i M \frac{1}{q - i + 1} + \lambda_1 M H(q)$$

$$\leq \ 1+(1-\lambda_1)\tau(q)-\sum_{i=1}^{q}\lambda_i M\frac{1}{q}+\lambda_1 MH(q)$$

$$\leq \ (1-\lambda_1)\tau(q)+\lambda_1 MH(q),$$

and since $\lambda_1 > 0$ this implies $\tau(q) \leq MH(q)$.   □

The special case when each iteration marks only one element is known as the coupon collector problem. Now we have everything we need:

Theorem 20.22. (Byrka et al. [2013]) *The* ITERATIVE RANDOMIZED ROUNDING ALGORITHM *computes a Steiner tree Y for T with expected cost at most* (ln 4) *times the cost of an optimum k-restricted Steiner tree. By choosing k large enough we get an approximation ratio of 1.39.*

Proof: For an edge $e \in E(Y^*)$ of the optimum Steiner tree Y^* let $\tau(e)$ denote the expected number of iterations until all elements of $W(e)$ are marked. As the probability that $|W(e)|$ has more than q elements is at most 2^{-q} (for all $q \in \mathbb{N}$), we have, using Lemma 20.21,

$$\tau(e) \leq \sum_{q\geq 1}\mathrm{Prob}(|W(e)|=q)MH(q) \leq \sum_{q\geq 1}2^{-q}MH(q) = M\sum_{q\geq 1}\frac{2^{1-q}}{q} = M\ln 4.$$

Let f_t be the LP value after t iterations, and Y_t^* the forest that results from Y^* by removing the edges e for which all elements of $W(e)$ are marked after t iterations. By Lemma 20.18 we have $f_t \leq \rho_k \sum_{e\in E(Y_t^*)} c(e)$, where ρ_k is again the k-Steiner ratio (cf. Definition 20.8). Therefore the expected cost of the output of the algorithm can be bounded by

$$\sum_{t\geq 1}\frac{1}{M}\mathrm{Exp}(f_{t-1}) \ \leq \ \sum_{t\geq 1}\frac{\rho_k}{M}\mathrm{Exp}\left(\sum_{e\in E(Y_{t-1}^*)}c(e)\right)$$

$$= \ \frac{\rho_k}{M}\sum_{e\in E(Y^*)}\tau(e)c(e)$$

$$\leq \ \rho_k(\ln 4)\sum_{e\in E(Y^*)}c(e).$$

By choosing k large enough, so that $\rho_k \leq \frac{1.39}{\ln 4}$ (by Theorem 20.9, $k = 2^{375}$ suffices), the theorem follows.   □

Goemans et al. [2012] proved that one can avoid resolving the LP in each iteration and thereby showed that the LP (20.1) has integrality ratio at most ln 4.

20.4 Survivable Network Design

Before turning to the general SURVIVABLE NETWORK DESIGN PROBLEM we mention two more special cases. If all connectivity requirements r_{xy} are 0 or 1, the problem is called the STEINER FOREST PROBLEM (of course the STEINER TREE PROBLEM is a special case). The first approximation algorithm for the STEINER FOREST PROBLEM was found by Agrawal, Klein and Ravi [1995].

Another interesting special case is the problem of finding a minimum weight k-edge-connected subgraph (here $r_{xy} = k$ for all x, y). See Exercise 11 for a combinatorial 2-factor approximation algorithm for this special case and for references related to this problem.

When considering the general SURVIVABLE NETWORK DESIGN PROBLEM, given connectivity requirements r_{xy} for all $x, y \in V(G)$, it is useful to define a function $f : 2^{V(G)} \to \mathbb{Z}_+$ by $f(\emptyset) := f(V(G)) := 0$ and $f(S) := \max_{x \in S,\, y \in V(G) \setminus S} r_{xy}$ for $\emptyset \neq S \subset V(G)$. Then our problem can be formulated as the following integer linear program:

$$
\begin{array}{lll}
\min & \displaystyle\sum_{e \in E(G)} c(e) x_e & \\[2mm]
\text{s.t.} & \displaystyle\sum_{e \in \delta(S)} x_e \ \geq\ f(S) & (S \subseteq V(G)) \\[2mm]
& x_e \ \in\ \{0, 1\} & (e \in E(G)).
\end{array}
\tag{20.2}
$$

We shall not deal with this integer program in the general form but rather make use of an important property of f:

Definition 20.23. *A function $f : 2^U \to \mathbb{Z}_+$ is called* **proper** *if it satisfies the following three conditions:*

- $f(S) = f(U \setminus S)$ *for all $S \subseteq U$;*
- $f(A \cup B) \leq \max\{f(A), f(B)\}$ *for all $A, B \subseteq U$ with $A \cap B = \emptyset$;*
- $f(\emptyset) = 0$.

It is obvious that f as constructed above is proper. Proper functions were introduced by Goemans and Williamson [1995] who gave a 2-factor approximation algorithm for proper functions f with $f(S) \in \{0, 1\}$ for all S. For proper functions f with $f(S) \in \{0, 2\}$ for all S, Klein and Ravi [1993] gave a 3-factor approximation algorithm.

The following property of proper functions is essential:

Proposition 20.24. *A proper function $f : 2^U \to \mathbb{Z}_+$ is* **weakly supermodular**, *i.e. at least one of the following conditions hold for all $A, B \subseteq U$:*

- $f(A) + f(B) \ \leq\ f(A \cup B) + f(A \cap B)$.
- $f(A) + f(B) \ \leq\ f(A \setminus B) + f(B \setminus A)$.

Proof: By definition we have

$$f(A) \leq \max\{f(A \setminus B), f(A \cap B)\}; \qquad\qquad (20.3)$$

$$f(B) \leq \max\{f(B \setminus A), f(A \cap B)\}; \qquad\qquad (20.4)$$

$$f(A) = f(U \setminus A) \leq \max\{f(B \setminus A), f(U \setminus (A \cup B))\} \qquad (20.5)$$
$$= \max\{f(B \setminus A), f(A \cup B)\};$$

$$f(B) = f(U \setminus B) \leq \max\{f(A \setminus B), f(U \setminus (A \cup B))\} \qquad (20.6)$$
$$= \max\{f(A \setminus B), f(A \cup B)\}.$$

Now we distinguish four cases, depending on which of the four numbers $f(A \setminus B)$, $f(B \setminus A)$, $f(A \cap B)$, $f(A \cup B)$ is the smallest. If $f(A \setminus B)$ is the smallest, we add (20.3) and (20.6). If $f(B \setminus A)$ is the smallest, we add (20.4) and (20.5). If $f(A \cap B)$ is the smallest, we add (20.3) and (20.4). If $f(A \cup B)$ is the smallest, we add (20.5) and (20.6). $\square$

In the rest of this section we show how to solve the LP relaxation of (20.2):

$$\min \quad \sum_{e \in E(G)} c(e)x_e$$

$$\text{s.t.} \quad \sum_{e \in \delta(S)} x_e \geq f(S) \qquad (S \subseteq V(G)) \qquad (20.7)$$

$$x_e \geq 0 \qquad\qquad (e \in E(G))$$
$$x_e \leq 1 \qquad\qquad (e \in E(G)).$$

We do not know how to solve this LP in polynomial time for arbitrary functions f, not even for arbitrary weakly supermodular functions. Therefore we restrict ourselves to the case when f is proper. By Theorem 4.21 it suffices to solve the SEPARATION PROBLEM. We use a Gomory-Hu tree:

Lemma 20.25.   *Let G be an undirected graph with capacities $u \in \mathbb{R}_+^{E(G)}$, and let $f : 2^{V(G)} \to \mathbb{Z}_+$ be a proper function. Let H be a Gomory-Hu tree for (G, u). Then for each $\emptyset \neq S \subset V(G)$ we have:*

(a) $\sum_{e' \in \delta_G(S)} u(e') \geq \max_{e \in \delta_H(S)} \sum_{e' \in \delta_G(C_e)} u(e')$;
(b) $f(S) \leq \max_{e \in \delta_H(S)} f(C_e)$;

where C_e and $V(H) \setminus C_e$ are the two connected components of $H - e$.

Proof: (a): By definition of the Gomory-Hu tree, $\delta_G(C_e)$ is a minimum capacity x-y-cut for each $e = \{x, y\} \in E(H)$, and for $\{x, y\} \in \delta_H(S)$ the left-hand side of (a) is the capacity of some x-y-cut.

To prove (b), let $X_1, \ldots, X_l$ be the connected components of $H - S$. Since $H[X_i]$ is connected and H is a tree, we have for each $i \in \{1, \ldots, l\}$:

$$V(H) \setminus X_i = \bigcup_{e \in \delta_H(X_i)} C_e$$

(if necessary, replace C_e by $V(H) \setminus C_e$). Since f is proper we have

$$f(X_i) = f(V(G) \setminus X_i) = f(V(H) \setminus X_i) = f\left(\bigcup_{e \in \delta_H(X_i)} C_e\right) \leq \max_{e \in \delta_H(X_i)} f(C_e).$$

Since $\delta_H(X_i) \subseteq \delta_H(S)$, we conclude that

$$f(S) = f(V(G) \setminus S) = f\left(\bigcup_{i=1}^{l} X_i\right) \leq \max_{i \in \{1,\dots,l\}} f(X_i) \leq \max_{e \in \delta_H(S)} f(C_e).$$
$\square$

Now we can show how to solve the SEPARATION PROBLEM for (20.7) by considering the fundamental cuts of a Gomory-Hu tree. Note that storing the proper function f explicitly would require exponential space, so we assume that f is given by an oracle.

Theorem 20.26. *Let G be an undirected graph, $x \in \mathbb{R}_+^{E(G)}$, and let $f : 2^{V(G)} \to \mathbb{Z}_+$ be a proper function (given by an oracle). One can find a set $S \subseteq V(G)$ with $\sum_{e \in \delta_G(S)} x_e < f(S)$ or decide that none exists in $O\left(n^4 + n\theta\right)$ time. Here $n = |V(G)|$ and θ is the time required by the oracle for f.*

Proof: We first compute a Gomory-Hu tree H for G, where the capacities are given by x. H can be computed in $O(n^4)$ time by Theorem 8.38.

By Lemma 20.25(b) we have that for each $\emptyset \neq S \subset V(G)$ there exists an $e \in \delta_H(S)$ with $f(S) \leq f(C_e)$. From Lemma 20.25(a) we get $f(S) - \sum_{e' \in \delta_G(S)} x_{e'} \leq f(C_e) - \sum_{e' \in \delta_G(C_e)} x_{e'}$. We conclude

$$\max_{\emptyset \neq S \subset V(G)} \left(f(S) - \sum_{e' \in \delta_G(S)} x_{e'}\right) = \max_{e \in E(H)} \left(f(C_e) - \sum_{e' \in \delta_G(C_e)} x_{e'}\right). \quad (20.8)$$

Hence the SEPARATION PROBLEM for (20.7) can be solved by checking only $n-1$ cuts. $\square$

It is worthwhile to compare (20.8) with Theorem 12.19.

In contrast to the LP relaxation (20.7) we cannot hope to find an optimum integral solution in polynomial time: by Theorem 20.2 this would imply $P = NP$. So we consider approximation algorithms for (20.2).

In the following section we describe a primal-dual approximation algorithm which consecutively adds edges in most violated cuts. This combinatorial algorithm performs well if the maximum connectivity requirement $k := \max_{S \subseteq V(G)} f(S)$ is not too large. In particular it is a 2-factor approximation algorithm for the case $k = 1$, which includes the STEINER FOREST PROBLEM. In Section 20.6 we describe a 2-factor approximation algorithm for the general case. However, this algorithm has the drawback that it uses the above solution of the LP relaxation (20.7), which has a polynomial running time but is too inefficient for practical purposes.

20.5 A Primal-Dual Approximation Algorithm

The algorithm which we present in this section was developed in the papers of Williamson et al. [1995], Gabow, Goemans and Williamson [1998], and Goemans et al. [1994], in this order.

Suppose an undirected graph G with weights $c : E(G) \to \mathbb{R}_+$ and a proper function f are given. We are looking for an edge set F whose incidence vector satisfies (20.2).

The algorithm proceeds in $k := \max_{S \subseteq V(G)} f(S)$ phases. Since f is proper we have $k = \max_{v \in V(G)} f(\{v\})$, so k can be computed easily. In phase p $(1 \leq p \leq k)$ the proper function f_p is considered, where $f_p(S) := \max\{f(S) + p - k, 0\}$. It will be guaranteed that after phase p the current edge set F (or, more precisely, its incidence vector) satisfies (20.2) with respect to f_p. Let us start with some definitions.

Definition 20.27. *Given some proper function g, some $F \subseteq E(G)$ and $X \subseteq V(G)$, we say that X is **violated** with respect to (g, F) if $|\delta_F(X)| < g(X)$. The minimal violated sets with respect to (g, F) are the **active** sets with respect to (g, F). $F \subseteq E(G)$ **satisfies** g if no set is violated with respect to (g, F). We say that F **almost satisfies** g if $|\delta_F(X)| \geq g(X) - 1$ for all $X \subseteq V(G)$.*

Throughout the algorithm, the current function f_p will be almost satisfied by the current set F. The active sets will play a central role. A key observation is the following:

Lemma 20.28. *Given some proper function g, some $F \subseteq E(G)$ almost satisfying g, and two violated sets A and B. Then either $A \setminus B$ and $B \setminus A$ are both violated or $A \cap B$ and $A \cup B$ are both violated. In particular, the active sets with respect to (g, F) are pairwise disjoint.*

Proof: Directly from Proposition 20.24 and Lemma 2.1, parts (c) and (d). $\square$

This lemma shows in particular that there can be at most $n = |V(G)|$ active sets. We now show how to compute the active sets; similarly to the proof of Theorem 20.26 we use a Gomory-Hu tree.

Theorem 20.29. (Gabow, Goemans and Williamson [1998]) *Given a proper function g (by an oracle) and a set $F \subseteq E(G)$ almost satisfying g. Then the active sets with respect to (g, F) can be computed in $O\left(n^4 + n^2\theta\right)$ time. Here $n = |V(G)|$ and θ is the time required by the oracle for g.*

Proof: We first compute a Gomory-Hu tree H for $(V(G), F)$ (and unit capacities). H can be computed in $O(n^4)$ time by Theorem 8.38. By Lemma 20.25 we have for each $\emptyset \neq S \subset V(G)$:

$$|\delta_F(S)| \geq \max_{e \in \delta_H(S)} |\delta_F(C_e)| \qquad (20.9)$$

and

$$g(S) \leq \max_{e \in \delta_H(S)} g(C_e), \tag{20.10}$$

where C_e and $V(H) \setminus C_e$ are the two connected components of $H - e$.

Let A be an active set. By (20.10), there exists an edge $e = \{s, t\} \in \delta_H(A)$ with $g(A) \leq g(C_e)$. By (20.9), $|\delta_F(A)| \geq |\delta_F(C_e)|$. So we have

$$1 = g(A) - |\delta_F(A)| \leq g(C_e) - |\delta_F(C_e)| \leq 1,$$

because F almost satisfies g. We must have equality throughout, in particular $|\delta_F(A)| = |\delta_F(C_e)|$. So $\delta_F(A)$ is a minimum s-t-cut in $(V(G), F)$. Let us assume w.l.o.g. that A contains t but not s.

Let G' be the digraph $(V(G), \{(v, w), (w, v) : \{v, w\} \in F\})$. Consider a maximum s-t-flow f in G' and the residual graph G'_f. Form an acyclic digraph G'' from G'_f by contracting the set S of vertices reachable from s to a vertex v_S, contracting the set T of vertices from which t is reachable to a vertex v_T, and contracting each strongly connected component X of $G'_f - (S \cup T)$ to a vertex v_X. There is a one-to-one correspondence between the minimum s-t-cuts in G' and the directed v_T-v_S-cuts in G'' (cf. Exercise 5 of Chapter 8; this follows easily from the Max-Flow-Min-Cut Theorem 8.6 and Lemma 8.3). In particular, A is the union of sets X with $v_X \in V(G'')$. Since $g(A) > |\delta_F(A)| = |\delta_{G'}^-(A)| = \text{value}(f)$ and g is proper, there exists a vertex $v_X \in V(G'')$ with $X \subseteq A$ and $g(X) > \text{value}(f)$.

We now show how to find A. If $g(T) > \text{value}(f)$, then set $Z := T$, else let v_Z be any vertex of G'' with $g(Z) > \text{value}(f)$ and $g(Y) \leq \text{value}(f)$ for all vertices $v_Y \in V(G'') \setminus \{v_Z\}$ from which v_Z is reachable. Let

$$B := T \cup \bigcup \{Y : v_Z \text{ is reachable from } v_Y \text{ in } G''\}.$$

Since

$$\text{value}(f) < g(Z) = g(V(G) \setminus Z) \leq \max\{g(V(G) \setminus B), g(B \setminus Z)\}$$
$$= \max\{g(B), g(B \setminus Z)\}$$

and

$$g(B \setminus Z) \leq \max\{g(Y) : v_Y \in V(G'') \setminus \{v_Z\}, Y \subseteq B\} \leq \text{value}(f)$$

we have $g(B) > \text{value}(f) = |\delta_{G'}^-(B)| = |\delta_F(B)|$, so B is violated with respect to (g, F). Since B is not a proper subset of A (as A is active) and both A and B contain T, we conclude from Lemma 20.28 that $A \subseteq B$. But then $Z = X$, as v_Z is the only vertex with $Z \subseteq B$ and $g(Z) > \text{value}(f)$, and A contains all sets Y for which v_Z is reachable from v_Y (as $\delta_{G'_f}^-(A) = \emptyset$). Hence $A = B$.

For a given pair (s, t) a set B as above (if existent) can be found in linear time by constructing G'' (using the STRONGLY CONNECTED COMPONENT ALGORITHM) and then finding a topological order of G'' (cf. Theorem 2.20), starting with v_T. We repeat the above procedure for all ordered pairs (s, t) such that $\{s, t\} \in E(H)$.

In this way we obtain a list of at most $2n - 2$ candidates for active sets. The running time is evidently dominated by finding $O(n)$ times a maximum flow in G' plus asking $O(n^2)$ times the oracle for g. Finally we can eliminate those violated sets among the candidates that are not minimal in $O(n^2)$ time. □

The running time can be improved if $\max_{S \subseteq V(G)} g(S)$ is small (see Exercise 13). We now turn to the description of the algorithm.

PRIMAL-DUAL ALGORITHM FOR NETWORK DESIGN

Input: An undirected graph G, weights $c : E(G) \to \mathbb{R}_+$, and an oracle for a proper function $f : 2^{V(G)} \to \mathbb{Z}_+$.

Output: A set $F \subseteq E(G)$ satisfying f.

---

① **If** $E(G)$ does not satisfy f, **then stop** (the problem is infeasible).

② Set $F := \emptyset$, $k := \max\limits_{v \in V(G)} f(\{v\})$, and $p := 1$.

③ Set $i := 0$.
Set $\pi(v) := 0$ for all $v \in V(G)$.
Let $\mathcal{A}$ be the family of active sets with respect to (F, f_p), where f_p is defined by $f_p(S) := \max\{f(S) + p - k, 0\}$ for all $S \subseteq V(G)$.

④ **While** $\mathcal{A} \neq \emptyset$ **do**:
Set $i := i + 1$.
$$\text{Set } \epsilon := \min \left\{ \frac{c(e) - \pi(v) - \pi(w)}{|\{A \in \mathcal{A} : e \in \delta_G(A)\}|} : e = \{v, w\} \in \bigcup_{A \in \mathcal{A}} \delta_G(A) \setminus F \right\},$$
and let e_i be some edge attaining this minimum.
Increase $\pi(v)$ by ϵ for all $v \in \bigcup\limits_{A \in \mathcal{A}} A$.
Set $F := F \cup \{e_i\}$.
Update $\mathcal{A}$.

⑤ **For** $j := i$ **down to** 1 **do**:
If $F \setminus \{e_j\}$ satisfies f_p **then** set $F := F \setminus \{e_j\}$.

⑥ **If** $p = k$ **then stop, else** set $p := p + 1$ and **go to** ③.

The feasibility check in ① can be done in $O\left(n^4 + n\theta\right)$ time by Theorem 20.26. Before we discuss how to implement ③ and ④, let us show that the output F is indeed feasible with respect to f. Let us denote by F_p the set F at the end of phase p (and $F_0 := \emptyset$).

Lemma 20.30. *At each stage of phase p the set F almost satisfies f_p and $F \setminus F_{p-1}$ is a forest. At the end of phase p, F_p satisfies f_p.*

Proof: Since $f_1(S) = \max\{0, f(S) + 1 - k\} \leq \max\{0, \max_{v \in S} f(\{v\}) + 1 - k\} \leq 1$ (as f is proper), the empty set almost satisfies f_1.

After ④ there are no active sets, so F satisfies f_p. In ⑤, this property is explicitly maintained. Hence each F_p satisfies f_p and thus almost satisfies f_{p+1} ($p = 0, \dots, k-1$). To see that $F \setminus F_{p-1}$ is a forest, observe that each edge added to F belongs to $\delta(A)$ for some active set A and must be the first edge of $\delta(A)$ added to F in this phase (as $|\delta_{F_{p-1}}(A)| = f_{p-1}(A)$). Hence no edge creates a circuit in $F \setminus F_{p-1}$. $\qquad\square$

So Theorem 20.29 can be applied to determine $\mathcal{A}$. The number of iterations within each phase is at most $n - 1$. The only remaining implementation issue we have to discuss is how to determine ϵ and e_i in ④.

Lemma 20.31.   *Determining ϵ and e_i in ④ of the algorithm can be done in $O(mn)$ time per phase.*

Proof: At each iteration of a phase we do the following. First we assign a number to each vertex according to which active set it belongs to (or zero if none). This can be done in $O(n)$ time (note that the active sets are pairwise disjoint by Lemma 20.28). For each edge e the number of active sets containing exactly one endpoint of e can now be determined in $O(1)$ time. So ϵ and e_i can be determined in $O(m)$ time. There are at most $n - 1$ iterations per phase, so the time bound is proved. $\qquad\square$

We remark that Gabow, Goemans and Williamson [1998] improved this bound to $O\left(n^2\sqrt{\log\log n}\right)$ with a sophisticated implementation.

Theorem 20.32.   (Goemans et al. [1994])   *The* PRIMAL-DUAL ALGORITHM FOR NETWORK DESIGN *returns a set F satisfying f in $O\left(kn^5 + kn^3\theta\right)$ time, where $k = \max_{S\subseteq V(G)} f(S)$, $n = |V(G)|$ and θ is the time required by the oracle for f.*

Proof: The feasibility of F is guaranteed by Lemma 20.30 since $f_k = f$.

An oracle for each f_p of course uses the oracle for f and thus takes time $\theta + O(1)$. Computing the active sets takes $O\left(n^4 + n^2\theta\right)$ time (Theorem 20.29), and this is done $O(nk)$ times. Determining ϵ and e_i can be done in $O(n^3)$ time per phase (Lemma 20.31). Everything else can easily be done in $O(kn^2)$ time. $\quad\square$

Exercise 13 shows how to improve the running time to $O\left(k^3n^3 + kn^3\theta\right)$. It can be improved to $O\left(k^2n^3 + kn^2\theta\right)$ by using a different clean-up step (⑤ of the algorithm) and a more sophisticated implementation (Gabow, Goemans and Williamson [1998]). For fixed k and $\theta = O(n)$ this means that we have an $O\left(n^3\right)$-algorithm. For the special case of the SURVIVABLE NETWORK DESIGN PROBLEM (f is determined by connectivity requirements r_{xy}) the running time can be improved to $O\left(k^2n^2\sqrt{\log\log n}\right)$.

Now we analyze the performance guarantee of the algorithm and justify that we have called it a primal-dual algorithm. The dual of (20.7) is:

$$\max \quad \sum_{S \subseteq V(G)} f(S)\, y_S - \sum_{e \in E(G)} z_e$$

$$\text{s.t.} \quad \sum_{S:e \in \delta(S)} y_S \;\le\; c(e) + z_e \qquad (e \in E(G)) \qquad (20.11)$$

$$y_S \;\ge\; 0 \qquad (S \subseteq V(G))$$

$$z_e \;\ge\; 0 \qquad (e \in E(G)).$$

This dual LP is essential for the analysis of the algorithm.

We show how the algorithm in each phase p implicitly constructs a feasible dual solution $y^{(p)}$. Starting with $y^{(p)} = 0$, in each iteration of this phase $y_A^{(p)}$ is increased by ϵ for each $A \in \mathcal{A}$. Moreover we set

$$z_e^{(p)} := \begin{cases} \displaystyle\sum_{S:\, e \in \delta(S)} y_S^{(p)} & \text{if } e \in F_{p-1} \\[2em] 0 & \text{otherwise} \end{cases}.$$

There is no point in constructing this dual solution explicitly in the algorithm. The variables $\pi(v) = \sum_{S:v \in S} y_S$ $(v \in V(G))$ contain all information that is needed.

Lemma 20.33. (Williamson et al. [1995]) *For each p, $(y^{(p)}, z^{(p)})$ as defined above is a feasible solution of (20.11).*

Proof: The nonnegativity constraints are obviously satisfied. The constraints for $e \in F_{p-1}$ are satisfied by definition of $z_e^{(p)}$.

Moreover, by ④ of the algorithm we have

$$\sum_{S:e \in \delta(S)} y_S^{(p)} \;\le\; c(e) \qquad \text{for each } e \in E(G) \setminus F_{p-1},$$

since e is added to F when equality is reached and after that sets S with $e \in \delta(S)$ are no longer violated with respect to (F, f_p) (recall that $F \setminus \{e\} \supseteq F_{p-1}$ almost satisfies f_p by Lemma 20.30). $\qquad \square$

Let us denote by $\mathrm{OPT}(G, c, f)$ the optimum value of the integer linear program (20.2). Next we show:

Lemma 20.34. (Goemans et al. [1994]) *For each $p \in \{1, \ldots, k\}$ we have*

$$\sum_{S \subseteq V(G)} y_S^{(p)} \;\le\; \frac{1}{k - p + 1}\, \mathrm{OPT}(G, c, f).$$

Proof: $\text{OPT}(G, c, f)$ is greater than or equal to the optimum value of the LP relaxation (20.7), and this is bounded from below by the objective value of any feasible dual solution (by the Duality Theorem 3.20). Since $(y^{(p)}, z^{(p)})$ is feasible for the dual LP (20.11) by Lemma 20.33 we conclude that

$$\text{OPT}(G, c, f) \geq \sum_{S \subseteq V(G)} f(S)\, y_S^{(p)} - \sum_{e \in E(G)} z_e^{(p)}.$$

Now observe that, for each $S \subseteq V(G)$, y_S can only become positive if S is violated with respect to (f_p, F_{p-1}). So we may conclude that

$$y_S^{(p)} > 0 \;\Rightarrow\; |\delta_{F_{p-1}}(S)| \leq f(S) + p - k - 1.$$

We thus obtain:

$$
\begin{aligned}
\text{OPT}(G, c, f) \;\geq\;& \sum_{S \subseteq V(G)} f(S)\, y_S^{(p)} - \sum_{e \in E(G)} z_e^{(p)} \\
=\;& \sum_{S \subseteq V(G)} f(S)\, y_S^{(p)} - \sum_{e \in F_{p-1}} \left(\sum_{S : e \in \delta(S)} y_S^{(p)} \right) \\
=\;& \sum_{S \subseteq V(G)} f(S)\, y_S^{(p)} - \sum_{S \subseteq V(G)} |\delta_{F_{p-1}}(S)|\, y_S^{(p)} \\
=\;& \sum_{S \subseteq V(G)} (f(S) - |\delta_{F_{p-1}}(S)|)\, y_S^{(p)} \\
\geq\;& \sum_{S \subseteq V(G)} (k - p + 1)\, y_S^{(p)}. \qquad \square
\end{aligned}
$$

Lemma 20.35.   (Williamson et al. [1995])   *At each iteration of any phase p we have*

$$\sum_{A \in \mathcal{A}} |\delta_{F_p \setminus F_{p-1}}(A)| \;\leq\; 2\,|\mathcal{A}|.$$

Proof: We consider some particular iteration of phase p, which we call the current iteration. Let $\mathcal{A}$ denote the family of active sets at the beginning of this iteration, and let

$$H \;:=\; (F_p \setminus F_{p-1}) \cap \bigcup_{A \in \mathcal{A}} \delta(A).$$

Note that all the edges of H must have been added during or after the current iteration.

Let $e \in H$. $F_p \setminus \{e\}$ does not satisfy f_p, because otherwise e would have been deleted in the clean-up step ⑤ of phase p. So let X_e be a violated set with respect to $(f_p, F_p \setminus \{e\})$ with minimum cardinality. Since $F_p \setminus \{e\} \supseteq F_{p-1}$ almost satisfies f_p we have $\delta_{F_p \setminus F_{p-1}}(X_e) = \{e\}$.

We claim that the family $\mathcal{X} := \{X_e : e \in H\}$ is laminar. For suppose that there are two edges $e, e' \in H$ (say e was added before e') for which $X_e \setminus X_{e'}$, $X_{e'} \setminus X_e$, and $X_e \cap X_{e'}$ are all nonempty. Since X_e and $X_{e'}$ are violated at the beginning of the current iteration, either $X_e \cup X_{e'}$ and $X_e \cap X_{e'}$ are both violated or $X_e \setminus X_{e'}$ and $X_{e'} \setminus X_e$ are both violated (by Lemma 20.28). In the first case we have

$$
\begin{aligned}
1 + 1 \ &\leq\ |\delta_{F_p \setminus F_{p-1}}(X_e \cup X_{e'})| + |\delta_{F_p \setminus F_{p-1}}(X_e \cap X_{e'})| \\
&\leq\ |\delta_{F_p \setminus F_{p-1}}(X_e)| + |\delta_{F_p \setminus F_{p-1}}(X_{e'})|\ =\ 1 + 1
\end{aligned}
$$

by the submodularity of $|\delta_{F_p \setminus F_{p-1}}|$ (Lemma 2.1(c)). We conclude $|\delta_{F_p \setminus F_{p-1}}(X_e \cup X_{e'})| = |\delta_{F_p \setminus F_{p-1}}(X_e \cap X_{e'})| = 1$, contradicting the minimal choice of X_e or of $X_{e'}$ because $X_e \cap X_{e'}$ could have been chosen instead. In the second case, by Lemma 2.1(d), $|\delta_{F_p \setminus F_{p-1}}(X_e \setminus X_{e'})| = |\delta_{F_p \setminus F_{p-1}}(X_{e'} \setminus X_e)| = 1$, and the smaller one of $X_e \setminus X_{e'}$ and $X_{e'} \setminus X_e$ contradicts the minimal choice of X_e or $X_{e'}$.

Now consider a tree-representation (T, φ) of $\mathcal{X}$, where T is an arborescence (cf. Proposition 2.14). For each $e \in H$, X_e is violated at the beginning of the current iteration because e has not been added yet at that time. So by Lemma 20.28 we have $A \subseteq X_e$ or $A \cap X_e = \emptyset$ for all $A \in \mathcal{A}$. Hence $\{\varphi(a) : a \in A\}$ contains only one element, denoted by $\varphi(A)$, for each $A \in \mathcal{A}$. We call a vertex $v \in V(T)$ **occupied** if $v = \varphi(A)$ for some $A \in \mathcal{A}$.

We assert that all vertices of T with out-degree 0 are occupied. Namely, for such a vertex v, $\varphi^{-1}(v)$ is a minimal element of $\mathcal{X}$. A minimal element of $\mathcal{X}$ is violated at the beginning of the current iteration, so it contains an active set and must thus be occupied. Hence the average out-degree of the occupied vertices is less than one.

Observe that there is a one-to-one correspondence between H, $\mathcal{X}$, and $E(T)$ (see Figure 20.4; (a) shows H, the elements of $\mathcal{A}$ (squares) and those of $\mathcal{X}$ (circles); (b) shows T). We conclude that for each $v \in V(T)$

$$
|\delta_T(v)| \ =\ |\delta_H(\{x \in V(G) : \varphi(x) = v\})| \ \geq\ \sum_{A \in \mathcal{A} : \varphi(A) = v} |\delta_{F_p \setminus F_{p-1}}(A)|.
$$

(a)   (b)

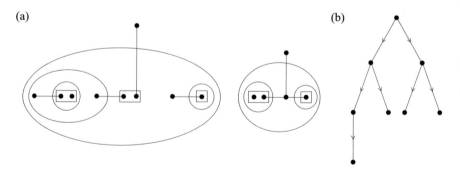

Fig. 20.4.

By summing over all occupied vertices we obtain:

$$\sum_{A \in \mathcal{A}} |\delta_{F_p \backslash F_{p-1}}(A)| \leq \sum_{\substack{v \in V(T) \text{ occupied}}} |\delta_T(v)|$$
$$< 2 |\{v \in V(T) : v \text{ occupied}\}|$$
$$\leq 2 |\mathcal{A}|.$$

$\square$

The proof of the next lemma shows the role of the complementary slackness conditions:

Lemma 20.36. (Williamson et al. [1995]) *For each $p \in \{1, \ldots, k\}$ we have*

$$\sum_{e \in F_p \backslash F_{p-1}} c(e) \leq 2 \sum_{S \subseteq V(G)} y_S^{(p)}.$$

Proof: In each phase p the algorithm maintains the primal complementary slackness conditions

$$e \in F \backslash F_{p-1} \Rightarrow \sum_{S:e \in \delta(S)} y_S^{(p)} = c(e).$$

So we have

$$\sum_{e \in F_p \backslash F_{p-1}} c(e) = \sum_{e \in F_p \backslash F_{p-1}} \left(\sum_{S:e \in \delta(S)} y_S^{(p)} \right) = \sum_{S \subseteq V(G)} y_S^{(p)} |\delta_{F_p \backslash F_{p-1}}(S)|.$$

Thus it remains to show that

$$\sum_{S \subseteq V(G)} y_S^{(p)} |\delta_{F_p \backslash F_{p-1}}(S)| \leq 2 \sum_{S \subseteq V(G)} y_S^{(p)}. \tag{20.12}$$

At the beginning of phase p we have $y^{(p)} = 0$, so (20.12) holds. In each iteration, the left-hand side increases by $\sum_{A \in \mathcal{A}} \epsilon |\delta_{F_p \backslash F_{p-1}}(A)|$, while the right-hand side increases by $2\epsilon |\mathcal{A}|$. So Lemma 20.35 shows that (20.12) is not violated.

$\square$

In (20.12) the dual complementary slackness conditions

$$y_S^{(p)} > 0 \Rightarrow |\delta_{F_p}(S)| = f_p(S)$$

appear. $|\delta_{F_p}(S)| \geq f_p(S)$ holds throughout, while (20.12) roughly means that $|\delta_{F_p}(S)| \leq 2f_p(S)$ is satisfied on the average. As we shall see, this implies a performance ratio of 2 in the case $k = 1$.

Theorem 20.37. (Goemans et al. [1994]) *The PRIMAL-DUAL ALGORITHM FOR NETWORK DESIGN returns a set F which satisfies f and whose weight is at most $2H(k) \text{OPT}(G, c, f)$ in $O\left(kn^5 + kn^3\theta\right)$ time, where $n = |V(G)|$, $k = \max_{S \subseteq V(G)} f(S)$, $H(k) = 1 + \frac{1}{2} + \cdots + \frac{1}{k}$, and θ is the time required by the oracle for f.*

Proof: The correctness and the running time have been proved in Theorem 20.32. The weight of F is

$$
\sum_{e \in F} c(e) = \sum_{p=1}^{k} \left(\sum_{e \in F_p \setminus F_{p-1}} c(e) \right)
$$

$$
\leq \sum_{p=1}^{k} \left(2 \sum_{S \subseteq V(G)} y_S^{(p)} \right)
$$

$$
\leq 2 \sum_{p=1}^{k} \frac{1}{k-p+1} \operatorname{OPT}(G, c, f)
$$

$$
= 2H(k) \operatorname{OPT}(G, c, f)
$$

due to Lemma 20.36 and Lemma 20.34.   □

The primal-dual approximation algorithm presented in this section has been put into a more general framework by Bertsimas and Teo [1995]. A related, but apparently more difficult problem arises by considering vertex-connectivity instead of edge-connectivity (one looks for a subgraph containing at least a specified number r_{ij} of internally disjoint i-j-paths for each i and j). See the remarks at the end of the next section.

20.6 Jain's Algorithm

In this section we present Jain's [2001] 2-factor approximation algorithm for the SURVIVABLE NETWORK DESIGN PROBLEM. Although it has a much better performance guarantee than the PRIMAL-DUAL ALGORITHM FOR NETWORK DESIGN it is of less practical value since it is based on the equivalence of optimization and separation (cf. Section 4.6).

The algorithm starts by solving the LP relaxation (20.7). In fact, it causes no difficulty to have integral capacities $u : E(G) \to \mathbb{N}$ on the edges, i.e. we are allowed to pick some edges more than once:

$$
\begin{aligned}
\min \quad & \sum_{e \in E(G)} c(e) x_e \\
\text{s.t.} \quad & \sum_{e \in \delta(S)} x_e \geq f(S) && (S \subseteq V(G)) \\
& x_e \geq 0 && (e \in E(G)) \\
& x_e \leq u(e) && (e \in E(G)).
\end{aligned}
\tag{20.13}
$$

Of course we are eventually looking for an integral solution. By solving the LP relaxation of an integer program and rounding up one gets a 2-factor approximation

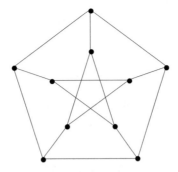

Fig. 20.5.

algorithm if the LP relaxation always has a half-integral optimum solution (see Exercise 6 of Chapter 16 for an example).

However, (20.13) does not have this property. To see this, consider the Petersen graph (Figure 20.5) with $u(e) = c(e) = 1$ for all edges e and $f(S) = 1$ for all $\emptyset \neq S \subset V(G)$. Here the optimum value of the LP (20.13) is 5 ($x_e = \frac{1}{3}$ for all e is an optimum solution), and every solution of value 5 satisfies $\sum_{e \in \delta(v)} x_e = 1$ for all $v \in V(G)$. Thus an optimum half-integral solution must have $x_e = \frac{1}{2}$ for the edges e of a Hamiltonian circuit and $x_e = 0$ otherwise. However, the Petersen graph is not Hamiltonian.

Nevertheless the solution of the LP relaxation (20.13) gives rise to a 2-factor approximation algorithm. The key observation is that for every optimum basic solution x there is an edge e with $x_e \geq \frac{1}{2}$ (Theorem 20.39). The algorithm will then round up and fix only these components and consider the remaining problem which has at least one edge less.

We need some preparation. For a set $S \subseteq V(G)$ we denote by χ^S the incidence vector of $\delta_G(S)$ with respect to $E(G)$. For any feasible solution x of (20.13) we call a set $S \subseteq V(G)$ **tight** if $\chi^S x = f(S)$.

Lemma 20.38. (Jain [2001]) *Let G be a graph, $m := |E(G)|$, and $f : 2^{V(G)} \to \mathbb{Z}_+$ a weakly supermodular function. Let x be a basic solution of the LP (20.13), and suppose that $0 < x_e < 1$ for each $e \in E(G)$. Then there exists a laminar family $\mathcal{B}$ of m tight subsets of $V(G)$ such that the vectors χ^B, $B \in \mathcal{B}$, are linearly independent in $\mathbb{R}^{E(G)}$.*

Proof: Let $\mathcal{B}$ be a laminar family of tight subsets of $V(G)$ such that the vectors χ^B, $B \in \mathcal{B}$, are linearly independent. Suppose that $|\mathcal{B}| < m$; we show how to extend $\mathcal{B}$.

Since x is a basic solution of (20.13), i.e. a vertex of the polytope, there are m linearly independent inequality constraints satisfied with equality (Proposition 3.9). Since $0 < x_e < 1$ for each $e \in E(G)$, these constraints correspond to a family $\mathcal{S}$ (not necessarily laminar) of m tight subsets of $V(G)$ such that the vectors χ^S ($S \in \mathcal{S}$) are linearly independent. Since $|\mathcal{B}| < m$, there exists a tight set $S \subseteq V(G)$

such that the vectors χ^B, $B \in \mathcal{B} \cup \{S\}$ are linearly independent. Choose S such that

$$\gamma(S) := |\{B \in \mathcal{B} : B \text{ crosses } S\}|$$

is minimal, where we say that B crosses S if $B \cap S \neq \emptyset$ and $B \setminus S \neq \emptyset$ and $S \setminus B \neq \emptyset$.

If $\gamma(S) = 0$, then we can add S to $\mathcal{B}$ and we are done. So assume that $\gamma(S) > 0$, and let $B \in \mathcal{B}$ be a set crossing S. Since f is weakly supermodular we have

$$
\begin{aligned}
f(S \setminus B) + f(B \setminus S) \;\; &\geq \;\; f(S) + f(B) \\
&= \;\; \sum_{e \in \delta_G(S)} x_e + \sum_{e \in \delta_G(B)} x_e \\
&= \;\; \sum_{e \in \delta_G(S \setminus B)} x_e + \sum_{e \in \delta_G(B \setminus S)} x_e + 2 \sum_{e \in E_G(S \cap B, V(G) \setminus (S \cup B))} x_e
\end{aligned}
$$

or

$$
\begin{aligned}
f(S \cap B) + f(S \cup B) \;\; &\geq \;\; f(S) + f(B) \\
&= \;\; \sum_{e \in \delta_G(S)} x_e + \sum_{e \in \delta_G(B)} x_e \\
&= \;\; \sum_{e \in \delta_G(S \cap B)} x_e + \sum_{e \in \delta_G(S \cup B)} x_e + 2 \sum_{e \in E_G(S \setminus B, B \setminus S)} x_e.
\end{aligned}
$$

In the first case, $S \setminus B$ and $B \setminus S$ are both tight and $E_G(S \cap B, V(G) \setminus (S \cup B)) = \emptyset$, implying $\chi^{S \setminus B} + \chi^{B \setminus S} = \chi^S + \chi^B$. In the second case, $S \cap B$ and $S \cup B$ are both tight and $E_G(S \setminus B, B \setminus S) = \emptyset$, implying $\chi^{S \cap B} + \chi^{S \cup B} = \chi^S + \chi^B$.

Hence there is at least one set T among $S \setminus B$, $B \setminus S$, $S \cap B$ and $S \cup B$ that is tight and has the property that the vectors χ^B, $B \in \mathcal{B} \cup \{T\}$, are linearly independent. We finally show that $\gamma(T) < \gamma(S)$; this yields a contradiction to the choice of S.

Since B crosses S but not T, it suffices to show that there is no $C \in \mathcal{B}$ which crosses T but not S. Indeed, since T is one of the sets $S \setminus B$, $B \setminus S$, $S \cap B$ and $S \cup B$, any set C crossing T but not S must cross B. Since $\mathcal{B}$ is laminar and $B \in \mathcal{B}$ this implies $C \notin \mathcal{B}$. □

Now we can prove the crucial theorem for JAIN'S ALGORITHM:

Theorem 20.39. (Jain [2001]) *Let G be a graph and $f : 2^{V(G)} \to \mathbb{Z}_+$ a weakly supermodular function that is not identically zero. Let x be a basic solution of the LP (20.13). Then there exists an edge $e \in E(G)$ with $x_e \geq \frac{1}{2}$.*

Proof: We may assume $x_e > 0$ for each edge e, since otherwise we can delete e. In fact we assume $0 < x_e < \frac{1}{2}$ for all $e \in E(G)$ and will deduce a contradiction.

By Lemma 20.38 there exists a laminar family $\mathcal{B}$ of $m := |E(G)|$ tight subsets of $V(G)$ such that the vectors χ^B, $B \in \mathcal{B}$, are linearly independent. The linear independence implies in particular that none of the χ^B is the zero vector, hence

$0 < \chi^B x = f(B)$ and thus $f(B) \geq 1$ for all $B \in \mathcal{B}$. Moreover, $\bigcup_{B \in \mathcal{B}} \delta_G(B) = E(G)$. By the assumption that $x_e < \frac{1}{2}$ for each $e \in E(G)$ we have $|\delta_G(B)| \geq 2f(B) + 1 \geq 3$ for all $B \in \mathcal{B}$.

Let (T, φ) be a tree-representation of $\mathcal{B}$. For each vertex t of the arborescence T we denote by T_t the maximal subgraph of T which is an arborescence rooted at t (T_t contains t and all its successors). Moreover, let $B_t := \{v \in V(G) : \varphi(v) \in V(T_t)\}$. By definition of the tree-representation we have $B_r = V(G)$ for the root r of T and $\mathcal{B} = \{B_t : t \in V(T) \setminus \{r\}\}$.

Claim: For each $t \in V(T)$ we have $\sum_{v \in B_t} |\delta_G(v)| \geq 2|V(T_t)| + 1$, with equality only if $|\delta_G(B_t)| = 2f(B_t) + 1$.

We prove the claim by induction on $|V(T_t)|$. If $\delta_T^+(t) = \emptyset$ (i.e. $V(T_t) = \{t\}$), then B_t is a minimal element of $\mathcal{B}$ and thus $\sum_{v \in B_t} |\delta_G(v)| = |\delta_G(B_t)| \geq 3 = 2|V(T_t)| + 1$, with equality only if $|\delta_G(B_t)| = 3$ (implying $f(B_t) = 1$).

For the induction step let $t \in V(T)$ with $\delta_T^+(t) \neq \emptyset$, say $\delta_T^+(t) = \{(t, s_1), \ldots, (t, s_k)\}$, where k is the number of children of t. Let $E_1 := \bigcup_{i=1}^k \delta_G(B_{s_i}) \setminus \delta_G(B_t)$ and $E_2 := \delta_G \left(B_t \setminus \bigcup_{i=1}^k B_{s_i} \right)$ (see Figure 20.6 for an illustration).

Note that $E_1 \cup E_2 \neq \emptyset$, since otherwise $\chi^{B_t} = \sum_{i=1}^k \chi^{B_{s_i}}$, contradicting the assumption that the vectors χ^B, $B \in \mathcal{B}$, are linearly independent (note that either $B_t \in \mathcal{B}$ or $t = r$ and then $\chi^{B_t} = 0$). Moreover, we have

$$|\delta_G(B_t)| + 2|E_1| = \sum_{i=1}^k |\delta_G(B_{s_i})| + |E_2| \tag{20.14}$$

and, since $B_{s_1}, \ldots, B_{s_k}$ and B_t are tight,

$$f(B_t) + 2 \sum_{e \in E_1} x_e = \sum_{i=1}^k f(B_{s_i}) + \sum_{e \in E_2} x_e. \tag{20.15}$$

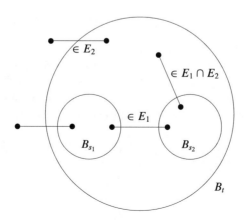

Fig. 20.6.

Furthermore, by the induction hypothesis,

$$
\begin{aligned}
\sum_{v \in B_t} |\delta_G(v)| &\geq \sum_{i=1}^{k} \sum_{v \in B_{s_i}} |\delta_G(v)| + |E_2| \\
&\geq \sum_{i=1}^{k} (2|V(T_{s_i})| + 1) + |E_2| \tag{20.16} \\
&= 2|V(T_t)| - 2 + k + |E_2|.
\end{aligned}
$$

Now we distinguish three cases.

Case 1: $k + |E_2| \geq 3$. Then by (20.16)

$$
\sum_{v \in B_t} |\delta_G(v)| \geq 2|V(T_t)| + 1,
$$

with equality only if $k + |E_2| = 3$ and $|\delta_G(B_{s_i})| = 2f(B_{s_i}) + 1$ for $i = 1, \ldots, k$. We have to show that then $|\delta_G(B_t)| = 2f(B_t) + 1$.

By (20.14) we have

$$
\begin{aligned}
|\delta_G(B_t)| + 2|E_1| &= \sum_{i=1}^{k} |\delta_G(B_{s_i})| + |E_2| = 2\sum_{i=1}^{k} f(B_{s_i}) + k + |E_2| \\
&= 2\sum_{i=1}^{k} f(B_{s_i}) + 3,
\end{aligned}
$$

hence $|\delta_G(B_t)|$ is odd. Moreover with (20.15) we conclude that

$$
\begin{aligned}
|\delta_G(B_t)| + 2|E_1| &= 2\sum_{i=1}^{k} f(B_{s_i}) + 3 = 2f(B_t) + 4\sum_{e \in E_1} x_e - 2\sum_{e \in E_2} x_e + 3 \\
&< 2f(B_t) + 2|E_1| + 3,
\end{aligned}
$$

because $E_1 \cup E_2 \neq \emptyset$. We have $|\delta_G(B_t)| = 2f(B_t) + 1$, as required.

Case 2: $k = 2$ and $E_2 = \emptyset$. Then $E_1 \neq \emptyset$, and by (20.15) $2\sum_{e \in E_1} x_e$ is an integer, hence $2\sum_{e \in E_1} x_e \leq |E_1| - 1$. Note that $E_1 \neq \delta_G(B_{s_1})$ since otherwise $\chi^{B_{s_2}} = \chi^{B_{s_1}} + \chi^{B_t}$, contradicting the assumption that the vectors χ^B, $B \in \mathcal{B}$, are linearly independent. Analogously, $E_1 \neq \delta_G(B_{s_2})$. For $i = 1, 2$ we get

$$
2f(B_{s_i}) = 2\sum_{e \in \delta(B_{s_i}) \setminus E_1} x_e + 2\sum_{e \in E_1} x_e < |\delta_G(B_{s_i}) \setminus E_1| + |E_1| - 1 = |\delta_G(B_{s_i})| - 1.
$$

By the induction hypothesis this implies $\sum_{v \in B_{s_i}} |\delta_G(v)| > 2|V(T_{s_i})| + 1$, and as in (20.16) we get

$$\sum_{v \in B_t} |\delta_G(v)| \geq \sum_{i=1}^{2} \sum_{v \in B_{s_i}} |\delta_G(v)| \geq \sum_{i=1}^{2} (2|V(T_{s_i})| + 2)$$

$$= 2|V(T_t)| + 2.$$

Case 3: $k = 1$ and $|E_2| \leq 1$. Note that $k = 1$ implies $E_1 \subseteq E_2$, hence $|E_2| = 1$. By (20.15) we have

$$\sum_{e \in E_2 \setminus E_1} x_e - \sum_{e \in E_1} x_e = \sum_{e \in E_2} x_e - 2 \sum_{e \in E_1} x_e = f(B_t) - f(B_{s_1}).$$

This is a contradiction since the right-hand side is an integer but the left-hand side is not; hence Case 3 cannot occur.

The claim is proved. For $t = r$ we get $\sum_{v \in V(G)} |\delta_G(v)| \geq 2|V(T)| + 1$, i.e. $2|E(G)| > 2|V(T)|$. Since on the other hand $|V(T)| - 1 = |E(T)| = |\mathcal{B}| = |E(G)|$ we have a contradiction. $\square$

This was generalized by Nagarajan, Ravi and Singh [2010]. With this theorem the following algorithm is natural:

JAIN'S ALGORITHM

Input: An undirected graph G, weights $c : E(G) \rightarrow \mathbb{R}_+$, capacities $u : E(G) \rightarrow \mathbb{N}$ and a proper function $f : 2^{V(G)} \rightarrow \mathbb{Z}_+$ (given by an oracle).

Output: A vector $x \in \mathbb{Z}_+^{E(G)}$ with $\sum_{e \in \delta_G(S)} x_e \geq f(S)$ for all $S \subseteq V(G)$ and $x_e \leq u(e)$ for all $e \in E(G)$.

①   Set $x_e := 0$ if $c(e) > 0$ and $x_e := u(e)$ if $c(e) = 0$ for all $e \in E(G)$.

②   Find an optimum basic solution y to the LP (20.13) with respect to c, u' and f', where $u'(e) := u(e) - x_e$ for all $e \in E(G)$ and
$$f'(S) := f(S) - \sum_{e \in \delta_G(S)} x_e \quad \text{for all } S \subseteq V(G).$$
If $y_e = 0$ for all $e \in E(G)$, **then stop.**

③   Set $x_e := x_e + \lceil y_e \rceil$ for all $e \in E(G)$ with $y_e \geq \frac{1}{2}$.
Go to ②.

Theorem 20.40.   (Jain [2001]) JAIN'S ALGORITHM *finds an integral solution to the LP (20.13) whose cost is at most twice the optimal value of the LP. It can be implemented to run in polynomial time. Hence it is a 2-factor approximation algorithm for the* SURVIVABLE NETWORK DESIGN PROBLEM.

Proof: First note that f' is weakly supermodular at any stage: f is proper, hence weakly supermodular by Proposition 20.24. Now for $A, B \subseteq V(G)$ we have

$\sum_{e\in\delta_G(A)} x_e + \sum_{e\in\delta_G(B)} x_e \geq \sum_{e\in\delta_G(A\cap B)} x_e + \sum_{e\in\delta_G(A\cup B)} x_e$ and $\sum_{e\in\delta_G(A)} x_e + \sum_{e\in\delta_G(B)} x_e \geq \sum_{e\in\delta_G(A\setminus B)} x_e + \sum_{e\in\delta_G(B\setminus A)} x_e$. Hence f' is weakly supermodular.

After the first iteration we have $f'(S) \leq \sum_{e\in\delta_G(S)} \frac{1}{2} \leq \frac{|E(G)|}{2}$ for all $S \subseteq V(G)$. By Theorem 20.39 each subsequent iteration increases at least one x_e by at least 1. Since each x_e is increased by at most $\frac{|E(G)|}{2}$ after the first iteration, the total number of iterations is bounded by $\frac{|E(G)|^2}{2}$.

The only implementation problem is ②. By Theorem 4.21 it suffices to solve the SEPARATION PROBLEM. For a given vector $y \in \mathbb{R}_+^{E(G)}$ we have to decide whether $\sum_{e\in\delta_G(S)} y_e \geq f'(S) = f(S) - \sum_{e\in\delta_G(S)} x_e$ for all $S \subseteq V(G)$, and if not, find a violated cut. Since f is proper this can be done in $O(n^4 + n\theta)$ time by Theorem 20.26, where $n = |V(G)|$ and θ is the time bound of the oracle for f.

We finally prove the performance ratio of 2, by induction on the number of iterations. If the algorithm terminates within the first iteration, then the solution has zero cost and is thus optimal.

Otherwise let $x^{(1)}$ and $y^{(1)}$ be the vectors x and y after the first iteration, and let $x^{(t)}$ be the vector x at termination. Let $z_e := y_e^{(1)}$ if $y_e^{(1)} < \frac{1}{2}$ and $z_e = 0$ otherwise. We have $cx^{(1)} \leq 2c\left(y^{(1)} - z\right)$. Let $f^{(1)}$ be the residual function defined by $f^{(1)}(S) := f(S) - \sum_{e\in\delta_G(S)} x_e^{(1)}$. Since z is a feasible solution for $f^{(1)}$, we know from the induction hypothesis that $c\left(x^{(t)} - x^{(1)}\right) \leq 2cz$. We conclude:

$$cx^{(t)} \leq cx^{(1)} + c\left(x^{(t)} - x^{(1)}\right) \leq 2c\left(y^{(1)} - z\right) + 2cz = 2cy^{(1)}.$$

Since $cy^{(1)}$ is a lower bound on the cost of an optimum solution we are done. $\square$

No better approximation ratio is known, even for the STEINER FOREST PROB-LEM. Melkonian and Tardos [2004] extended Jain's technique to a directed network design problem. Fleischer, Jain and Williamson [2006], Cheriyan and Vetta [2007], and Chuzhoy and Khanna [2009] showed how some vertex-connectivity constraints can be taken into account. However, results by Kortsarz, Krauthgamer and Lee [2004] and Chakraborty, Chuzhoy and Khanna [2008] indicate that the vertex-connectivity version of the SURVIVABLE NETWORK DESIGN PROBLEM is hard to approximate.

20.7 The VPN Problem

In many applications the demands are not exactly known in advance, but we need to design a network that meets all demands in a certain set. There are many versions of this problem, often called robust network design. Here we consider the case where each pair of terminals should be connected by a fixed path, and the capacity allocated on every edge must be sufficient for any fractional matching on the terminal set. More precisely, let

$$D(W) := \left\{ (d_f)_{f \in \binom{W}{2}} : d_f \geq 0 \ (f \in \binom{W}{2}), \ \sum_{w \in W \setminus \{v\}} d_{\{v,w\}} \leq 1 \ (v \in W) \right\}$$

denote the set of fractional matchings in the complete graph with vertex set W. Then we define:

VPN PROBLEM

Instance: An undirected graph G with weights $c : E(G) \to \mathbb{R}_+$, and a set $W \subseteq V(G)$ of terminals.

Task: Find a v-w-path P_f for each unordered pair $f = \{v, w\}$ of terminals such that $\sum_{e \in E(G)} c(e)u(e)$ is minimum, where $u(e) = \max_{d \in D(W)} \sum_{f \in \binom{W}{2}:e \in E(P_f)} d_f$ is the capacity to be installed at edge $e \in E(G)$.

One might think of a virtual private network that a company rents for communication between its locations; hence the name. See Exercise 23 for a slight generalization. This is sometimes called oblivious routing because the paths are chosen independently of the concrete demands $d \in D(W)$. The interesting feature of the VPN PROBLEM is that it can be solved exactly in polynomial time. This follows directly from the following theorem:

Theorem 20.41. (Goyal, Olver and Shepherd [2013]) *For any instance of the* VPN PROBLEM *there exists an optimum solution in which, for some $s \in V(G)$ (the hub) and s-w-paths P_w ($w \in W$), we install capacity $u(e) := |\{w \in W : e \in E(P_w)\}|$ at each edge $e \in E(G)$, and $E(P_f) \subseteq E(P_v) \cup E(P_w)$ for all $f = \{v, w\} \in \binom{W}{2}$.*

Such a solution may be called a hub solution. We first prove the following lower bound. Let $\tau(T \triangle \{w\})$ denote the minimum weight of a $(T \triangle \{w\})$-join in (G, c).

Lemma 20.42. *For any instance of the* VPN PROBLEM *there is a set $T \subseteq V(G)$ of odd cardinality such that any solution has cost at least $\sum_{w \in W} \tau(T \triangle \{w\})$.*

Proof: Let $(P_f)_{f \in \binom{W}{2}}$ be a solution to the given instance, and let $u(e) := \max_{d \in D(W)} \sum_{f \in \binom{W}{2}:e \in E(P_f)} d_f$ be the capacity that this solution needs to allocate for edge $e \in E(G)$. We show that there is a set $T \subseteq V(G)$ of odd cardinality such that $\sum_{e \in E(G)} c(e)u(e) \geq \sum_{w \in W} \tau(T \triangle \{w\})$.

For every $t \in W$, let $\mathcal{P}_t := \{E(P_{\{w,t\}}) : w \in W\}$ denote the set of edge sets of paths to t, where we set $P_{\{t,t\}} := (\{t\}, \emptyset)$. For $e \in E(G)$ we write $l(e, \mathcal{P}_t) := |\{P \in \mathcal{P}_t : e \in P\}|$ and $y(e, \mathcal{P}_t) := \min\{l(e, \mathcal{P}_t), |W| - l(e, \mathcal{P}_t)\} = |\{P \in \mathcal{P}_t : e \in E(P) \triangle H_t\}|$, where

$$H_t := \left\{ e \in E(G) : l(e, \mathcal{P}_t) \geq \frac{|W|}{2} \right\}$$

is the set of ("heavy") edges that at least half of the paths to t use. Finally, let $T_t := \text{odd}(H_t) \triangle \{t\}$, where $\text{odd}(H_t)$ is the set of odd-degree vertices in $(V(G), H_t)$.

Let $e \in E(G)$. Then we claim that

$$d_f^e := \begin{cases} \frac{1}{|W|}\left(\frac{y(e, \mathcal{P}_v)}{l(e, \mathcal{P}_v)} + \frac{y(e, \mathcal{P}_w)}{l(e, \mathcal{P}_w)}\right) & \text{if } e \in E(P_f) \\ 0 & \text{otherwise} \end{cases}$$

for $f = \{v, w\} \in \binom{W}{2}$ defines a fractional matching. Indeed, for any $v \in W$,

$$\begin{aligned}
\sum_{w \in W \setminus \{v\}} d_{\{v,w\}}^e &= \sum_{w \in W \setminus \{v\} : e \in E(P_{\{v,w\}})} \frac{1}{|W|}\left(\frac{y(e, \mathcal{P}_v)}{l(e, \mathcal{P}_v)} + \frac{y(e, \mathcal{P}_w)}{l(e, \mathcal{P}_w)}\right) \\
&\leq \sum_{w \in W \setminus \{v\} : e \in E(P_{\{v,w\}})} \frac{1}{|W|}\left(\frac{|W| - l(e, \mathcal{P}_v)}{l(e, \mathcal{P}_v)} + \frac{l(e, \mathcal{P}_w)}{l(e, \mathcal{P}_w)}\right) \\
&= \sum_{w \in W \setminus \{v\} : e \in E(P_{\{v,w\}})} \frac{1}{|W|} \cdot \frac{|W|}{l(e, \mathcal{P}_v)} \\
&= 1.
\end{aligned}$$

So $d^e \in D(W)$, and hence

$$\begin{aligned}
u(e) &\geq \sum_{f \in \binom{W}{2} : e \in E(P_f)} d_f^e \\
&= \frac{1}{|W|} \sum_{f = \{v,w\} \in \binom{W}{2} : e \in E(P_f)} \left(\frac{y(e, \mathcal{P}_v)}{l(e, \mathcal{P}_v)} + \frac{y(e, \mathcal{P}_w)}{l(e, \mathcal{P}_w)}\right) \\
&= \frac{1}{|W|} \sum_{v \in W} \sum_{w \in W \setminus \{v\} : e \in E(P_{\{v,w\}})} \frac{y(e, \mathcal{P}_v)}{l(e, \mathcal{P}_v)} \\
&= \frac{1}{|W|} \sum_{v \in W} y(e, \mathcal{P}_v).
\end{aligned}$$

We conclude

$$\begin{aligned}
\sum_{e \in E(G)} c(e)u(e) &\geq \frac{1}{|W|} \sum_{v \in W} \sum_{e \in E(G)} c(e)y(e, \mathcal{P}_v) \\
&= \frac{1}{|W|} \sum_{v \in W} \sum_{w \in W} c(E(P_{\{v,w\}}) \triangle H_v) \\
&\geq \frac{1}{|W|} \sum_{v \in W} \sum_{w \in W} \tau(T_v \triangle \{w\}) \\
&\geq \min_{v \in W} \sum_{w \in W} \tau(T_v \triangle \{w\}).
\end{aligned}$$

$\square$

This proof by Goyal, Olver and Shepherd [2013] is based on earlier partial results by Gupta et al. [2001] and Grandoni et al. [2008]. Next we show the following lemma.

Lemma 20.43. *Let H be an undirected graph and $T \subseteq V(H)$ such that $|T|$ is even and $E(H)$ contains k edge-disjoint T-joins. Then, for every $t \in T$ there exists an $s \in V(H) \setminus \{t\}$ such that H contains k edge-disjoint s-t-paths.*

Proof: Consider a Gomory-Hu tree G of H. The vertex set of at least one connected component of $G - t$, say C, has odd intersection with T (because $|T|$ is even and $t \in T$). Let $f = \{s, t\} \in E(G)$ with $s \in C$. Then $\delta_H(C)$ is a minimum cardinality s-t-cut in H (by definition of the Gomory-Hu tree). Since $|C \cap T|$ is odd, every T-join intersects $\delta(C)$, and so $|\delta_H(C)| \geq k$. Hence, by Menger's Theorem 8.9, H contains k edge-disjoint s-t-paths.   □

This proof is due to A. Sebő (cf. Goyal, Olver and Shepherd [2013]). Now we can prove the main theorem of this section:

Proof of Theorem 20.41:   Fix an instance (G, c, W) of the VPN PROBLEM. Let $T \subseteq V(G)$ be as in Lemma 20.42, i.e., $|T|$ is odd and any solution to the instance has cost at least $\sum_{w \in W} \tau(T \triangle \{w\})$. Let J_w be a minimum cost $(T \triangle \{w\})$-join, and let H be the graph with $V(H) = V(G) \dot\cup \{t\}$ whose edge set is the disjoint union of the J_w ($w \in W$) plus an edge $\{t, w\}$ for each $w \in W$.

In other words, $E(H)$ is the disjoint union of the $(T \triangle \{t\})$-joins $J_w \cup \{\{w, t\}\}$ ($w \in W$). By Lemma 20.43 there exists a vertex $s \in V(G)$ such that H contains $|W|$ edge-disjoint s-t-paths. They correspond to edge-disjoint paths P_w ($w \in W$) in G, where P_w connects s and w. So s serves as hub.   □

Corollary 20.44. *The VPN PROBLEM can be solved in $O(|W|(m + n \log n))$ time, where $n = |V(G)|$ and $m = |E(G)|$.*

Proof: Compute shortest paths from each $w \in W$ to all $s \in V(G)$ by DIJKSTRA'S ALGORITHM (Theorem 7.4). Take a vertex s which minimizes $\sum_{w \in W} \text{dist}_{(G, c)}(s, w)$. This vertex s serves as hub in a hub solution as in Theorem 20.41.   □

Exercises

1. Let (G, c, T) be an instance of the STEINER TREE PROBLEM where G is a complete graph and $c : E(G) \to \mathbb{R}_+$ satisfies the triangle inequality. Prove that there exists an optimum Steiner tree for T with at most $|T| - 2$ Steiner points.

2. In the DIRECTED STEINER TREE PROBLEM, we are given a digraph G with weights $c : E(G) \to \mathbb{R}_+$, a root $r \in V(G)$, and a set $T \subseteq V(G)$ of terminals; the task is to find a minimum weight arborescence in G that is rooted at r and

contains all terminals. Show that a k-factor approximation algorithm for this problem would imply a k-factor approximation algorithm for the MINIMUM WEIGHT SET COVER PROBLEM.

3. Prove that the STEINER TREE PROBLEM is *MAXSNP*-hard even for complete graphs with edge weights 1 and 2 only.
   Hint: Modify the proof of Theorem 20.3. What if G is disconnected?
   (Bern, Plassmann [1989])

4. Formulate an $O(n^3 t^2)$ algorithm for the STEINER TREE PROBLEM in planar graphs with all terminals lying on the outer face, and prove its correctness.
   Hint: Show that in the DREYFUS-WAGNER ALGORITHM it suffices to consider sets $U \subseteq T$ that are consecutive, i.e. there is a path P whose vertices all lie on the outer face such that $V(P) \cap T = U$ (we assume w.l.o.g. that G is 2-connected).
   (Erickson, Monma and Veinott [1987])

5. Describe an algorithm for the STEINER TREE PROBLEM which runs in $O(n^3)$ time for instances (G, c, T) with $|V(G) \setminus T| \le k$, where k is some constant.

6. Prove the following strengthening of Theorem 20.6: if (G, c, T) is an instance of the STEINER TREE PROBLEM with $|T| \ge 2$, $(\bar{G}, \bar{c})$ the metric closure, S an optimum Steiner tree for T in G, and M a minimum weight spanning tree in $\bar{G}[T]$ with respect to $\bar{c}$, then

$$\bar{c}(M) \le 2\left(1 - \frac{1}{b}\right)c(S),$$

where b is the number of leaves (vertices of degree 1) of S. Show that this bound is tight.

7. Prove that the 4-Steiner ratio ρ_4 is $\frac{3}{2}$.

8. Sharpen the inequality of Proposition 20.10 for the cases $|T| = 3$ and $|T| = 4$.

9. Show that (20.1) can be solved in polynomial time for each fixed k.

* 10. Show that the family of bridge sets, $\mathcal{B}_W(R)$, is the set of bases of a matroid for any tree (T, W) and any $R \subseteq T$.
    (Goemans et al. [2012])

11. Find a combinatorial 2-factor approximation algorithm for the SURVIVABLE NETWORK DESIGN PROBLEM with $r_{ij} = k$ for all i, j (i.e. the MINIMUM WEIGHT k-EDGE-CONNECTED SUBGRAPH PROBLEM).
    Hint: Replace each edge by a pair of oppositely directed edges (with the same weight) and apply either Exercise 26 of Chapter 13 or Theorem 6.18 directly.
    (Khuller and Vishkin [1994])
    Note: For more results for similar problems, see Khuller and Raghavachari [1996], Gabow [2005], Jothi, Raghavachari and Varadarajan [2003], and Gabow et al. [2009].

12. Show that in the special case of the SURVIVABLE NETWORK DESIGN PROBLEM the LP relaxation (20.7) can be reformulated as a linear program of polynomial size.

13. Prove the following strengthening of Theorem 20.29. Given a proper function g (by an oracle) and a set $F \subseteq E(G)$ almost satisfying g, the active sets with

respect to (g, F) can be computed in $O\left(k^2n^2 + n^2\theta\right)$ time, where $n = |V(G)|$, $k = \max_{S \subseteq V(G)} g(S)$, and θ is the time required by the oracle for g.

Hint: The idea is to stop the flow computations whenever the value of the maximum flow is at least k, because cuts with k or more edges are not relevant here.

The GOMORY-HU ALGORITHM (see Section 8.6) is modified as follows. At each step, each vertex of the tree T is a forest (rather than a subset of vertices). The edges of the forests correspond to maximum flow problems for which the value of a maximum flow is at least k. At each iteration of the modified GOMORY-HU ALGORITHM, we pick two vertices s and t of different connected components of the forest corresponding to one vertex of T. If the value of the maximum flow is at least k, we insert an edge $\{s, t\}$ into the forest. Otherwise we split the vertex as in the original Gomory-Hu procedure. We stop when all vertices of T are trees. We finally replace each vertex in T by its tree.

It is clear that the modified Gomory-Hu tree also satisfies the properties (20.9) and (20.10). If the flow computations are done by the FORD-FULKERSON ALGORITHM, stopping after the k-th augmenting path, then the $O(k^2n^2)$ bound can be achieved.

Note: This leads to an overall running time of $O\left(k^3n^3 + kn^3\theta\right)$ of the PRIMAL-DUAL ALGORITHM FOR NETWORK DESIGN.

(Gabow et al. [1998])

* 14. Consider the SURVIVABLE NETWORK DESIGN PROBLEM, which we have seen to be a special case of (20.2).

    (a) Consider a maximum spanning tree T in the complete graph having cost r_{ij} on edge $\{i, j\}$. Show that if a set of edges satisfies the connectivity requirements of the edges of T, then it satisfies all connectivity requirements.

    (b) When determining the active sets at the beginning of phase p, we only need to look for one augmenting i-j-path for each $\{i, j\} \in E(T)$ (we can use the i-j-flow of the preceding phase). If there is no augmenting i-j-path, then there are at most two candidates for active sets. Among those $O(n)$ candidates we can find the active sets in $O(n^2)$ time.

    (c) Show that updating these data structures can be done in $O(kn^2)$ total time per phase.

    (d) Conclude that the active sets can be computed in an overall running time of $O(k^2n^2)$.

(Gabow, Goemans and Williamson [1998])

15. Show that the clean-up step ⑤ of the PRIMAL-DUAL ALGORITHM FOR NETWORK DESIGN is crucial: without ⑤, the algorithm does not even achieve any finite performance ratio for $k = 1$.

16. No algorithm for the MINIMUM WEIGHT T-JOIN PROBLEM with a better worst-case complexity than $O(n^3)$ for dense graphs (cf. Corollary 12.12) is known. Let G be an undirected graph, $c : E(G) \rightarrow \mathbb{R}_+$ and $T \subseteq V(G)$ with

$|T|$ even. Consider the integer linear program (20.2), where we set $f(S) := 1$ if $|S \cap T|$ is odd and $f(S) := 0$ otherwise.

(a) Prove that our primal-dual algorithm applied to (20.2) returns a forest in which each connected component contains an even number of elements of T.

(b) Prove that any optimum solution to (20.2) is a minimum weight T-join plus possibly some zero weight edges.

(c) The primal-dual algorithm can be implemented in $O(n^2 \log n)$ time if $f(S) \in \{0, 1\}$ for all S. Show that this implies a 2-factor approximation algorithm for the MINIMUM WEIGHT T-JOIN PROBLEM with nonnegative weights, with the same running time.

Hint: By (a), the algorithm returns a forest F. For each connected component C of F consider $\bar{G}[V(C) \cap T]$ and find a tour whose weight is at most twice the weight of C (cf. the proof of Theorem 20.6). Now take every second edge of the tour. (A similar idea is the basis of CHRISTOFIDES' ALGORITHM, see Section 21.1.)

(Goemans and Williamson [1995])

17. Find a 2-factor approximation algorithm for the POINT-TO-POINT CONNEC-TION PROBLEM: Given an undirected graph G with edge weights $c : E(G) \to \mathbb{R}_+$ and sets $S, T \subseteq V(G)$ with $S \cap T = \emptyset$ and $|S| = |T| \geq 1$, find a set $F \subseteq E(G)$ of minimum cost such that there is a bijection $\pi : S \to T$ and paths from s to $\pi(s)$ for all $s \in S$ in $(V(G), F)$.

Hint: Show that $f(X) = 1$ if $|X \cap S| \neq |X \cap T|$ and $f(X) = 0$ otherwise defines a proper function.

(Goemans and Williamson [1995])

18. An instance of the price-collecting Steiner forest problem consists of an instance of the STEINER FOREST PROBLEM plus a penalty $\pi_{\{v,w\}} \in \mathbb{R}_+$ for each terminal pair. The goal is to find a spanning forest H which minimizes $c(E(H)) + \pi(H)$, where $\pi(H)$ is the sum of penalties of terminal pairs that are not in the same connected component of H. The natural LP relaxation is:

$$\min \left\{ c^\top x + \pi^\top z : \sum_{e \in \delta(U)} x_e + z_{\{v,w\}} \geq 1 \ (v \in U \subset V(G) \setminus \{w\}), \ x, z \geq 0 \right\}.$$

Consider the following threshold rounding approach: for some $0 \leq \alpha < 1$, set $x'_e := \frac{1}{1-\alpha}$ for all e; then x' is a feasible solution for the Steiner forest problem with the terminal pairs whose variable has value at most α. Apply a 2-approximation algorithm to x'.

(a) Show that for $\alpha = \frac{1}{3}$ this yields a 3-factor approximation algorithm.

(b) Show that by choosing the best α, one can obtain a $(1/(1 - e^{-1/2}))$-factor approximation algorithm.

(Goemans [unpublished]; see also Könemann et al. [2017])

19. Find an optimum basic solution x for (20.13), where G is the Petersen graph (Figure 20.5) and $f(S) = 1$ for all $\emptyset \neq S \subset V(G)$. Find a maximal laminar

family $\mathcal{B}$ of tight sets with respect to x such that the vectors χ^B, $B \in \mathcal{B}$, are linearly independent (cf. Lemma 20.38).

20. Prove that the optimum value of (20.13) can be arbitrarily close to half the value of an optimum integral solution.

    Note: By JAIN'S ALGORITHM (cf. the proof of Theorem 20.40) it cannot be less than half.

21. Show that in a slight variant of JAIN'S ALGORITHM the number of iterations in which we have to solve an LP can be bounded by
    (a) $2|V(G)|^2$;
    (b) $2|V(G)|$.
    Here we set $x_e := x_e + \lfloor y_e \rfloor$ for all e if some $y_e \geq 1$, otherwise we update x as before.

    Hint: Conclude from Lemma 20.38 that in the second case all but $2|V(G)| - 2$ edges can be deleted. For (b), delete one more edge in each iteration.

22. Let $T(m, n)$ be an upper bound on the time needed for solving the LP in ② of JAIN'S ALGORITHM, but not necessarily producing an optimum basic solution. Here $m = |E(G)|$ and $n = |V(G)|$. Describe a 2-factor approximation algorithm for the SURVIVABLE NETWORK DESIGN PROBLEM that runs in $O(m^2 T(m, n))$ time.

23. For a finite set W and $b \in \mathbb{R}_+^W$ let

$$D(b) := \left\{ (d_f)_{f \in \binom{W}{2}} : d_f \geq 0 \ (f \in \binom{W}{2}), \ \sum_{w \in W \setminus \{v\}} d_{\{v, w\}} \leq b_v \ (v \in W) \right\}$$

be the set of fractional b-matchings in the complete graph with vertex set W. Then a generalization of the VPN PROBLEM gets such a vector b as additional input and sets $u(e) = \max_{d \in D(b)} \sum_{f \in \binom{W}{2}: e \in E(P_f)} d_f$. Show that Theorem 20.41 generalizes to this setting, and that the problem can still be solved in polynomial time.

Hint: Prove the theorem by reducing this generalization to the case where $b_w = 1$ for all $w \in W$.

References

General Literature

Cheng, X., and Du, D.-Z. [2001]: Steiner Trees in Industry. Kluwer, Dordrecht 2001

Du, D.-Z., Smith, J.M., and Rubinstein, J.H. [2000]: Advances in Steiner Trees. Kluwer, Boston 2000

Goemans, M.X., and Williamson, D.P. [1996]: The primal-dual method for approximation algorithms and its application to network design problems. In: Approximation Algorithms for *NP*-Hard Problems. (D.S. Hochbaum, ed.), PWS, Boston, 1996

Grötschel, M., Monma, C.L., and Stoer, M. [1995]: Design of survivable networks. In: Handbooks in Operations Research and Management Science; Volume 7; Network Models (M.O. Ball, T.L. Magnanti, C.L. Monma, G.L. Nemhauser, eds.), Elsevier, Amsterdam 1995

Gupta, A., and Könemann, J. [2011]: Approximation algorithms for network design: a survey. Surveys in Operations Research and Management Science 16 (2011) 3–20

Hwang, F.K., Richards, D.S., and Winter, P. [1992]: The Steiner Tree Problem; Annals of Discrete Mathematics 53. North-Holland, Amsterdam 1992

Kerivin, H., and Mahjoub, A.R. [2005]: Design of survivable networks: a survey. Networks 46 (2005), 1–21

Lau, L.C., Ravi, R., and Singh, M. [2011]: Iterative Methods in Combinatorial Optimization. Cambridge University Press 2011, Chapter 10

Prömel, H.J., and Steger, A. [2002]: The Steiner Tree Problem. Vieweg, Braunschweig 2002

Stoer, M. [1992]: Design of Survivable Networks. Springer, Berlin 1992

Vazirani, V.V. [2001]: Approximation Algorithms. Springer, Berlin 2001, Chapters 22 and 23

Cited References

Agrawal, A., Klein, P.N., and Ravi, R. [1995]: When trees collide: an approximation algorithm for the generalized Steiner tree problem in networks. SIAM Journal on Computing 24 (1995), 440–456

Arora, S. [1998]: Polynomial time approximation schemes for Euclidean traveling salesman and other geometric problems. Journal of the ACM 45 (1998), 753–782

Berman, P., and Ramaiyer, V. [1994]: Improved approximations for the Steiner tree problem. Journal of Algorithms 17 (1994), 381–408

Bern, M., and Plassmann, P. [1989]: The Steiner problem with edge lengths 1 and 2. Information Processing Letters 32 (1989), 171–176

Bertsimas, D., and Teo, C. [1995]: From valid inequalities to heuristics: a unified view of primal-dual approximation algorithms in covering problems. Operations Research 46 (1998), 503–514

Bertsimas, D., and Teo, C. [1997]: The parsimonious property of cut covering problems and its applications. Operations Research Letters 21 (1997), 123–132

Björklund, A., Husfeldt, T., Kaski, P., and Koivisto, M. [2007]: Fourier meets Möbius: fast subset convolution. Proceedings of the 39th Annual ACM Symposium on Theory of Computing (2007), 67–74

Borchers, A., and Du, D.-Z. [1997]: The k-Steiner ratio in graphs. SIAM Journal on Computing 26 (1997), 857–869

Borradaile, G., Klein, P., and Mathieu, C., [2009]: An $O(n \log n)$ approximation scheme for Steiner tree in planar graphs. ACM Transactions on Algorithms 5 (2009), Article 31

Byrka, J., Grandoni, F., Rothvoß, T., and Sanità, L. [2013]: Steiner tree approximation via iterative randomized rounding. Journal of the ACM 60 (2013), Article 6

Chakraborty, T., Chuzhoy, J., and Khanna, S. [2008]: Network design for vertex connectivity. Proceedings of the 40th Annual ACM Symposium on Theory of Computing (2008), 167–176

Cheriyan, J., and Vetta, A. [2007]: Approximation algorithms for network design with metric costs. SIAM Journal on Discrete Mathematics 21 (2007), 612–636

Chlebík, M., and Chlebíková, J. [2008]: The Steiner tree problem on graphs: Inapproximability results. Theoretical Computer Science 406 (2008), 207–214

Choukhmane, E. [1978]: Une heuristique pour le problème de l'arbre de Steiner. RAIRO Recherche Opérationnelle 12 (1978), 207–212 [in French]

Chuzhoy, J., and Khanna, S. [2009]: An $O(k^3 \log n)$-approximation algorithm for vertex-connectivity survivable network design. Proceedings of the 50th Annual IEEE Symposium on Foundations of Computer Science (2009), 437–441

Dreyfus, S.E., and Wagner, R.A. [1972]: The Steiner problem in graphs. Networks 1 (1972), 195–207

Du, D.-Z., Zhang, Y., and Feng, Q. [1991]: On better heuristic for Euclidean Steiner minimum trees. Proceedings of the 32nd Annual IEEE Symposium on Foundations of Computer Science (1991), 431–439. See also Mathematical Programming 57 (1992) 193–202

Erickson, R.E., Monma, C.L., and Veinott, A.F., Jr. [1987]: Send-and-split method for minimum concave-cost network flows. Mathematics of Operations Research 12 (1987), 634–664

Fleischer, L., Jain, K., and Williamson, D.P. [2006]: Iterative rounding 2-approximation algorithms for minimum-cost vertex connectivity problems. Journal of Computer and System Sciences 72 (2006), 838–867

Fuchs, B., Kern, W., Mölle, D., Richter, S., Rossmanith, P., and Wang, X. [2007]: Dynamic programming for minimum Steiner trees. Theory of Computing Systems 41 (2007), 493–500

Gabow, H.N. [2005]: An improved analysis for approximating the smallest k-edge connected spanning subgraph of a multigraph. SIAM Journal on Discrete Mathematics 19 (2005), 1–18

Gabow, H.N., Goemans, M.X., and Williamson, D.P. [1998]: An efficient approximation algorithm for the survivable network design problem. Mathematical Programming B 82 (1998), 13–40

Gabow, H.N., Goemans, M.X., Tardos, É., and Williamson, D.P. [2009]: Approximating the smallest k-edge connected spanning subgraph by LP-rounding. Networks 53 (2009), 345–357

Garey, M.R., Graham, R.L., and Johnson, D.S. [1977]: The complexity of computing Steiner minimal trees. SIAM Journal of Applied Mathematics 32 (1977), 835–859

Garey, M.R., and Johnson, D.S. [1977]: The rectilinear Steiner tree problem is NP-complete. SIAM Journal on Applied Mathematics 32 (1977), 826–834

Gilbert, E.N., and Pollak, H.O. [1968]: Steiner minimal trees. SIAM Journal on Applied Mathematics 16 (1968), 1–29

Goemans, M.X., and Bertsimas, D.J. [1993]: Survivable networks, linear programming and the parsimonious property, Mathematical Programming 60 (1993), 145–166

Goemans, M.X., Goldberg, A.V., Plotkin, S., Shmoys, D.B., Tardos, É., and Williamson, D.P. [1994]: Improved approximation algorithms for network design problems. Proceedings of the 5th Annual ACM-SIAM Symposium on Discrete Algorithms (1994), 223–232

Goemans, M.X., Olver, N., Rothvoß, T., and Zenklusen, R. [2012]: Matroids and integrality gaps for hypergraphic Steiner tree relaxations. Proceedings of the 44th Annual ACM Symposium on Theory of Computing (2012), 1161–1176

Goemans, M.X., and Williamson, D.P. [1995]: A general approximation technique for constrained forest problems. SIAM Journal on Computing 24 (1995), 296–317

Goyal, N., Olver, N., and Shepherd, F.B. [2013]: The VPN conjecture is true. Journal of the ACM 60 (2013), Article 17

Grandoni, F., Kaibel, V., Oriolo, G., and Skutella, M. [2008]: A short proof of the VPN tree routing conjecture on ring networks. Operations Research Letters 36 (2008), 361–365

Gröpl, C., Hougardy, S., Nierhoff, T., and Prömel, H.J. [2001]: Approximation algorithms for the Steiner tree problem in graphs. In: Cheng and Du [2001], pp. 235–279

Gupta, A., Kleinberg, J., Kumar, A., Rastogi, R., and Yener, B. [2001]: Provisioning a virtual private network: a network design problem for multicommodity flow. Proceedings of the 33th Annual ACM Symposium on Theory of Computing (2001), 389–398

Hanan, M. [1966]: On Steiner's problem with rectilinear distance. SIAM Journal on Applied Mathematics 14 (1966), 255–265

Hetzel, A. [1995]: Verdrahtung im VLSI-Design: Spezielle Teilprobleme und ein sequentielles Lösungsverfahren. Ph.D. thesis, University of Bonn, 1995 [in German]

Hougardy, S., and Prömel, H.J. [1999]: A 1.598 approximation algorithm for the Steiner tree problem in graphs. Proceedings of the 10th Annual ACM-SIAM Symposium on Discrete Algorithms (1999), 448–453

Hougardy, S., Silvanus, J., and Vygen, J. [2017]: Dijkstra meets Steiner: a fast exact goal-oriented Steiner tree algorithm. Mathematical Programming Computation 9 (2017), 135–202

Hwang, F.K. [1976]: On Steiner minimal trees with rectilinear distance. SIAM Journal on Applied Mathematics 30 (1976), 104–114

Jain, K. [2001]: A factor 2 approximation algorithm for the generalized Steiner network problem. Combinatorica 21 (2001), 39–60

Jothi, R., Raghavachari, B., and Varadarajan, S. [2003]: A 5/4-approximation algorithm for minimum 2-edge-connectivity. Proceedings of the 14th Annual ACM-SIAM Symposium on Discrete Algorithms (2003), 725–734

Karp, R.M. [1972]: Reducibility among combinatorial problems. In: Complexity of Computer Computations (R.E. Miller, J.W. Thatcher, eds.), Plenum Press, New York 1972, pp. 85–103

Karpinski, M., and Zelikovsky, A. [1997]: New approximation algorithms for Steiner tree problems. Journal of Combinatorial Optimization 1 (1997), 47–65

Khuller, S., and Raghavachari, B. [1996]: Improved approximation algorithms for uniform connectivity problems. Journal of Algorithms 21 (1996), 434–450

Khuller, S., and Vishkin, U. [1994]: Biconnectivity augmentations and graph carvings. Journal of the ACM 41 (1994), 214–235

Klein, P.N., and Ravi, R. [1993]: When cycles collapse: a general approximation technique for constrained two-connectivity problems. Proceedings of the 3rd Integer Programming and Combinatorial Optimization Conference (1993), 39–55

Könemann, J., Olver, N., Pashkovish, K., Ravi, R., Swamy, C., Vygen, J. [2017]: On the integrality gap of the prize-collecting Steiner forest LP. Proceedings of APPROX 2017, Article 17

Korte, B., Prömel, H.J., and Steger, A. [1990]: Steiner trees in VLSI-layout. In: Paths, Flows, and VLSI-Layout (B. Korte, L. Lovász, H.J. Prömel, A. Schrijver, eds.), Springer, Berlin 1990, pp. 185–214

Kortsarz, G., Krauthgamer, R., and Lee, J.R. [2004]: Hardness of approximation for vertex-connectivity network design problems. SIAM Journal on Computing 33 (2004), 704–720

Kou, L.T. [1990]: On efficient implementation of an approximation algorithm for the Steiner tree problem. Acta Informatica 27 (1990), 369–380

Kou, L., Markowsky, G., and Berman, L. [1981]: A fast algorithm for Steiner trees. Acta Informatica 15 (1981), 141–145

Martin, A. [1992]: Packen von Steinerbäumen: Polyedrische Studien und Anwendung. Ph.D. thesis, Technical University of Berlin 1992 [in German]

Mehlhorn, K. [1988]: A faster approximation algorithm for the Steiner problem in graphs. Information Processing Letters 27 (1988), 125–128

Melkonian, V., and Tardos, É. [2004]: Algorithms for a network design problem with crossing supermodular demands. Networks 43 (2004), 256–265

Nagarajan, V., Ravi, R., and Singh, M. [2010]: Simpler analysis of LP extreme points for traveling salesman and survivable network design problems. Operations Research Letters 38 (2010), 156–160

Robins, G., and Zelikovsky, A. [2005]: Tighter bounds for graph Steiner tree approximation. SIAM Journal on Discrete Mathematics 19 (2005), 122–134

Takahashi, M., and Matsuyama, A. [1980]: An approximate solution for the Steiner problem in graphs. Mathematica Japonica 24 (1980), 573–577

Vygen, J. [2011]: Faster algorithm for optimum Steiner trees. Information Processing Letters 111 (2011), 1075–1079

Warme, D.M., Winter, P., and Zachariasen, M. [2000]: Exact algorithms for plane Steiner tree problems: a computational study. In: Advances in Steiner trees (D.-Z. Du, J.M. Smith, J.H. Rubinstein, eds.), Kluwer Academic Publishers, Boston, 2000, pp. 81–116

Williamson, D.P., Goemans, M.X., Mihail, M., and Vazirani, V.V. [1995]: A primal-dual approximation algorithm for generalized Steiner network problems. Combinatorica 15 (1995), 435–454

Zelikovsky, A.Z. [1993]: An 11/6-approximation algorithm for the network Steiner problem. Algorithmica 9 (1993), 463–470

21 The Traveling Salesman Problem

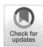

In Chapter 15 we introduced the TRAVELING SALESMAN PROBLEM (TSP) and showed that it is *NP*-hard (Theorem 15.43). The TSP is perhaps the best-studied *NP*-hard combinatorial optimization problem, and there are many techniques which have been applied. We start by discussing approximation algorithms in Sections 21.1 and 21.2. In practice, so-called local search algorithms (discussed in Section 21.3) find better solutions for large instances although they do not have a finite performance ratio.

We study the traveling salesman polytope (the convex hull of the incidence vectors of all tours in K_n) in Section 21.4. Using a cutting plane approach (cf. Section 5.5) combined with a branch-and-bound scheme one can solve TSP instances with several thousand cities optimally. We shall discuss this in Section 21.6 after we have shown how to obtain good lower bounds in Section 21.5. Note that all these ideas and techniques can also be applied to other combinatorial optimization problems. We present them with the TSP since this is a problem where these techniques have proved to be most effective.

We consider the symmetric TSP only, although the ASYMMETRIC TSP (where the distance from i to j can be different to the distance from j to i) is also interesting (cf. Exercise 4).

21.1 Approximation Algorithms for the TSP

In this and the next section we investigate the approximability of the TSP. We start with the following negative result:

Theorem 21.1. (Sahni and Gonzalez [1976]) *Unless $P = NP$ there is no k-factor approximation algorithm for the* TSP *for any $k \geq 1$.*

Proof: Suppose there is a k-factor approximation algorithm A for the TSP. Then we prove that there is a polynomial-time algorithm for the HAMILTONIAN CIRCUIT problem. Since the latter is *NP*-complete by Theorem 15.25, this implies $P = NP$.

© Springer-Verlag GmbH Germany 2018
Bernhard Korte and Jens Vygen, *Combinatorial Optimization*, Algorithms and Combinatorics 21, https://doi.org/10.1007/978-3-662-56039-6_21

Given a graph G, we construct an instance of the TSP with $n = |V(G)|$ cities: the distances are defined as $c : E(K_n) \to \mathbb{Z}_+$,

$$
c(\{i, j\}) := \begin{cases} 1 & \text{if } \{i, j\} \in E(G) \\ 2 + (k - 1)n & \text{if } \{i, j\} \notin E(G). \end{cases}
$$

Now we apply A to this instance. If the returned tour has length n, then this tour is a Hamiltonian circuit in G. Otherwise the returned tour has length at least $n + 1 + (k - 1)n = kn + 1$. If $\text{OPT}(K_n, c)$ is the length of an optimum tour, then $\frac{kn+1}{\text{OPT}(K_n,c)} \leq k$ since A is a k-factor approximation algorithm. Hence $\text{OPT}(K_n, c) > n$, showing that G has no Hamiltonian circuit. □

In most practical applications the distances of the TSP instances satisfy the triangle inequality:

METRIC TSP

Instance: A complete graph K_n with weights $c : E(K_n) \to \mathbb{R}_+$ such that $c(\{x, y\}) + c(\{y, z\}) \geq c(\{x, z\})$ for all $x, y, z \in V(K_n)$.

Task: Find a Hamiltonian circuit in K_n of minimum weight.

In other words, (K_n, c) is its own metric closure.

Theorem 21.2.   *The* METRIC TSP *is strongly NP-hard.*

Proof: The proof of Theorem 15.43 shows that the TSP is *NP*-hard even if all distances are 1 or 2. □

One can immediately think of several heuristics to generate reasonably good solutions. One of the simplest is the so-called nearest neighbour heuristic: given an instance (K_n, c) of the TSP, choose $v_1 \in V(K_n)$ arbitrarily. Then for $i = 2, \ldots, n$ choose v_i among $V(K_n) \setminus \{v_1, \ldots, v_{i-1}\}$ such that $c(\{v_{i-1}, v_i\})$ is minimum. In other words, at each step the nearest unvisited city is chosen.

The nearest neighbour heuristic is not a constant-factor approximation algorithm for the METRIC TSP. For infinitely many n there are instances (K_n, c) for which the nearest neighbour heuristic returns a tour of length $\frac{1}{3} \text{OPT}(K_n, c) \log n$ (Rosenkrantz, Stearns and Lewis [1977]). See also Hurkens and Woeginger [2004].

The rest of this section is devoted to approximation algorithms for the METRIC TSP. These algorithms first construct a closed walk containing all vertices (but some vertices may be repeated). As the following lemma shows, this is sufficient if the triangle inequality holds.

Lemma 21.3.   *Given an instance (K_n, c) of the* METRIC TSP *and a connected Eulerian graph G spanning $V(K_n)$, possibly with parallel edges. Then we can construct a tour of weight at most $c(E(G))$ in linear time.*

Proof: By Theorem 2.25 we can find an Eulerian walk in G in linear time. The order in which the vertices appear in this walk (we ignore all but the first occurrence of a vertex) defines a tour. The triangle inequality immediately implies that this tour is no longer than $c(E(G))$.   □

We have already encountered this idea when approximating the STEINER TREE PROBLEM (Theorem 20.6).

DOUBLE-TREE ALGORITHM

Input: An instance (K_n, c) of the METRIC TSP.

Output: A tour.

①   Find a minimum weight spanning tree T in K_n with respect to c.

②   Let G be the graph containing two copies of each edge of T. G satisfies the prerequisites of Lemma 21.3.
      Construct a tour as in the proof of Lemma 21.3.

Theorem 21.4. *The* DOUBLE-TREE ALGORITHM *is a 2-factor approximation algorithm for the* METRIC TSP. *Its running time is* $O(n^2)$.

Proof: The running time follows from Theorem 6.6. We have $c(E(T)) \leq$ OPT(K_n, c) since by deleting one edge of any tour we get a spanning tree. Therefore $c(E(G)) = 2c(E(T)) \leq 2\,\text{OPT}(K_n, c)$. The theorem now follows from Lemma 21.3.   □

For Euclidean instances (cf. Section 21.2) one can find an optimum tour in the metric closure of (T, c) in ② in $O(n^3)$ time instead of applying Lemma 21.3 (Burkard, Deĭneko and Woeginger [1998]). The performance guarantee of the DOUBLE-TREE ALGORITHM is tight (Exercise 6). The best known approximation algorithm for the METRIC TSP is due to Christofides [1976]:

CHRISTOFIDES' ALGORITHM

Input: An instance (K_n, c) of the METRIC TSP.

Output: A tour.

①   Find a minimum weight spanning tree T in K_n with respect to c.

②   Let W be the set of vertices having odd degree in T.
      Find a minimum weight W-join J in K_n with respect to c.

③   Let $G := (V(K_n), E(T) \,\dot\cup\, J)$. G satisfies the prerequisites of Lemma 21.3.
      Construct a tour as in the proof of Lemma 21.3.

Because of the triangle inequality a minimum weight perfect matching in $K_n[W]$ can be taken as J in ②.

Theorem 21.5. (Christofides [1976]) CHRISTOFIDES' ALGORITHM *is a $\frac{3}{2}$-factor approximation algorithm for the* METRIC TSP. *Its running time is* $O(n^3)$.

Proof: The time bound is a consequence of Theorem 12.10. As in the proof of Theorem 21.4 we have $c(E(T)) \leq \text{OPT}(K_n, c)$. Since each tour is the union of two W-joins we also have $c(J) \leq \frac{1}{2}\text{OPT}(K_n, c)$. We conclude $c(E(G)) = c(E(T)) + c(J) \leq \frac{3}{2}\text{OPT}(K_n, c)$, and the result follows from Lemma 21.3.   □

It is not known whether there is an approximation algorithm with a better performance ratio. On the other hand, there is the following negative result:

Theorem 21.6. (Papadimitriou and Yannakakis [1993]) *The* METRIC TSP *is MAXSNP-hard.*

Proof: We describe an L-reduction from the MINIMUM VERTEX COVER PROBLEM for graphs with maximum degree 4 (which is *MAXSNP*-hard by Theorem 16.46) to the METRIC TSP.

Given an undirected graph G with maximum degree 4, we construct an instance (H, c) of the METRIC TSP as follows:

For each $e = \{v, w\} \in E(G)$ we introduce a subgraph H_e of twelve vertices and 14 edges as shown in Figure 21.1. Four vertices of H_e denoted by $(e, v, 1)$, $(e, v, 2)$, $(e, w, 1)$ and $(e, w, 2)$ have a special meaning. The graph H_e has the property that it has a Hamiltonian path from $(e, v, 1)$ to $(e, v, 2)$ and another one from $(e, w, 1)$ to $(e, w, 2)$, but it has no Hamiltonian path from (e, v, i) to (e, w, j) for any $i, j \in \{1, 2\}$.

Now let H be the complete graph on the vertex set $V(H) := \bigcup_{e \in E(G)} V(H_e)$. For $\{x, y\} \in E(H)$ we set

$$c(\{x, y\}) := \begin{cases} 1 & \text{if } \{x, y\} \in E(H_e) \text{ for some } e \in E(G); \\ \text{dist}_{H_e}(x, y) & \text{if } x, y \in V(H_e) \text{ for some } e \in E(G) \\ & \text{but } \{x, y\} \notin E(H_e); \\ 4 & \text{if } x = (e, v, i) \text{ and } y = (f, v, j) \text{ with } e \neq f; \\ 5 & \text{otherwise.} \end{cases}$$

Fig. 21.1.

Fig. 21.2.

This construction is illustrated by Figure 21.2 (only edges of length 1 or 4 are shown).

(H, c) is an instance of the METRIC TSP. We claim that it has the following properties:

(a) For each vertex cover X of G there is a tour of length $15|E(G)| + |X|$.
(b) Given any tour T, one can construct another tour T' in polynomial time which is at most as long and contains a Hamiltonian path of each subgraph H_e $(e \in E(G))$.
(c) Given a tour of length $15|E(G)| + k$, we can construct a vertex cover of cardinality k in G in polynomial time.

(a) and (c) imply that we have an L-reduction, because the optimum tour length is $15|E(G)| + \tau(G) \le 15(4\tau(G)) + \tau(G)$ as G has maximum degree 4.

To prove (a), let X be a vertex cover of G, and let $(E_x)_{x \in X}$ be a partition of $E(G)$ with $E_x \subseteq \delta(x)$ $(x \in X)$. Then for each $x \in X$ the subgraph induced by $\bigcup_{e \in E_x} V(H_e)$ obviously contains a Hamiltonian path with $11|E_x|$ edges of length 1 and $|E_x| - 1$ edges of length 4. Adding $|X|$ edges to the union of these Hamiltonian paths yields a tour with only $|X|$ edges of length 5, $|E(G)| - |X|$ edges of length 4, and $11|E(G)|$ edges of length 1.

To prove (b), let T be any tour and $e \in E(G)$ such that T does not contain a Hamiltonian path in H_e. Let $\{x, y\} \in E(T)$, $x \notin V(H_e)$, $y \in V(H_e)$, and let z be the first vertex outside $V(H_e)$ when traversing T from y without passing x. Then we delete the piece of the tour between x and z and replace it by $\{x, (e, v, 1)\}$, a Hamiltonian path in H_e from $(e, v, 1)$ to $(e, v, 2)$, and the edge $\{(e, v, 2), z\}$ (where $v \in e$ is chosen arbitrarily). At all other places where T contains vertices of H_e we shortcut T. We claim that the resulting tour T' is not longer than T.

First suppose that $k := |\delta_T(V(H_e))| \ge 4$. Then the total weight of the edges incident to $V(H_e)$ in T is at least $4k + (12 - \frac{k}{2})$. In T' the total weight of the edges incident to $V(H_e)$ is at most $5 + 5 + 11$, and we have added another $\frac{k}{2} - 1$ edges by shortcutting. Since $5 + 5 + 11 + 5(\frac{k}{2} - 1) \le 4k + (12 - \frac{k}{2})$, the tour has not become longer.

Now suppose that $|\delta_T(V(H_e))| = 2$ but T contains an edge $\{x, y\}$ with $x, y \in V(H_e)$ but $\{x, y\} \notin E(H_e)$. Then the total length of the edges of T incident to

$V(H_e)$ is at least 21, as can be checked easily. Since in T' the total length of the edges incident to $V(H_e)$ is at most $5 + 5 + 11 = 21$, the tour has not become longer.

We finally prove (c). Let T be a tour of length $15|E(G)| + k$, for some k. By (b) we may assume that T contains a Hamiltonian path of each H_e ($e \in E(G)$), say from $(e, v, 1)$ to $(e, v, 2)$; here we set $v(e) := v$. Then $X := \{v(e) : e \in E(G)\}$ is a vertex cover of G. Since T contains exactly $11|E(G)|$ edges of length 1, $|E(G)|$ edges of length 4 or 5, and at least $|X|$ edges of length 5, we conclude that $|X| \leq k$. □

So by Corollary 16.40 an approximation scheme cannot exist unless $P = NP$. Karpinski, Lampis and Schmied [2013] showed that even the existence of a $\frac{124}{123}$-factor approximation algorithm would imply $P = NP$.

Papadimitriou and Yannakakis [1993] proved that the problem remains MAXSNP-hard even if all weights are 1 or 2. For this special case Berman and Karpinski [2006] found an $\frac{8}{7}$-factor approximation algorithm. See Vygen [2012] for a survey on further approximation algorithms for special cases.

21.2 Euclidean TSP

In this section we consider the TSP for Euclidean instances.

EUCLIDEAN TSP

Instance: A finite set $V \subseteq \mathbb{R}^2$, $|V| \geq 3$.

Task: Find a Hamiltonian circuit T in the complete graph on V such that the total length $\sum_{\{v,w\} \in E(T)} ||v - w||_2$ is minimum.

Here $||v - w||_2$ denotes the Euclidean distance between v and w. We often identify an edge with the straight line segment joining its endpoints. Every optimum tour can thus be regarded as a polygon (it cannot cross itself).

The EUCLIDEAN TSP is evidently a special case of the METRIC TSP, and it is also strongly NP-hard (Garey, Graham and Johnson [1976], Papadimitriou [1977]). However, one can make use of the geometric nature as follows:

Suppose we have a set of n points in the unit square, partition it by a regular grid such that each region contains few points, find an optimum tour within each region and then patch the subtours together. This method has been proposed by Karp [1977] who showed that it leads to $(1 + \epsilon)$-approximate solutions on almost all randomly generated instances in the unit square. Arora [1998] developed this further and found an approximation scheme for the EUCLIDEAN TSP, which we present in this section. A similar approximation scheme has been proposed by Mitchell [1999].

Let ϵ with $0 < \epsilon < 1$ be fixed throughout this section. We show how to find a tour in polynomial time such that its length exceeds the length of an optimal tour by a factor of at most $1 + \epsilon$. We begin by rounding the coordinates:

Definition 21.7. *Let $V \subseteq \mathbb{R}^2$ be an instance of the* EUCLIDEAN *TSP, $n := |V|$. Let $L := \max_{v,w \in V} ||v - w||_2$ and $v_x^{\min} := \min\{v_x : (v_x, v_y) \in V\}$ and $v_y^{\min} := \min\{v_y : (v_x, v_y) \in V\}$. Then*

$$V' := \left\{\left(\left(1 + 2\left\lfloor\frac{5n}{\epsilon L}(v_x - v_x^{\min})\right\rfloor\right), 1 + 2\left\lfloor\frac{5n}{\epsilon L}(v_y - v_y^{\min})\right\rfloor\right) : (v_x, v_y) \in V\right\}$$

is called the **well-rounded** *instance corresponding to V.*

Note that V' may contain fewer elements than V. If $|V'| = 2$, we say that the circuit of length two containing the edge V' twice is the only tour for V'. The following result says that it is sufficient to deal with well-rounded instances:

Proposition 21.8. *Let $V \subseteq \mathbb{R}^2$ be an instance of the* EUCLIDEAN *TSP, $n = |V|$, and V' its well-rounded instance. Then:*

(a) *for all $(v_x, v_y) \in V'$, v_x and v_y are odd integers between 1 and $1 + \frac{10n}{\epsilon}$.*
(b) *from any $(1 + \frac{\epsilon}{2})$-optimal tour for V' we can construct a $(1 + \epsilon)$-optimal tour for V in $O(|V|)$ time.*

Proof: (a) is obvious. Given a tour for V' whose length l' is at most $(1 + \frac{\epsilon}{2})$ OPT(V'), we construct a tour for the original instance V in a straightforward way. The length l of this tour is no more than $\left(\frac{l'}{2} + 2n\right)\frac{\epsilon L}{5n}$. Furthermore,

$$\text{OPT}(V') \leq 2\left(\frac{5n}{\epsilon L}\text{OPT}(V) + 2n\right).$$

Altogether we have

$$l \leq \frac{\epsilon L}{5n}\left(2n + \left(1 + \frac{\epsilon}{2}\right)\left(\frac{5n}{\epsilon L}\text{OPT}(V) + 2n\right)\right) = \left(1 + \frac{\epsilon}{2}\right)\text{OPT}(V) + \frac{4\epsilon L}{5} + \frac{\epsilon^2 L}{5}.$$

Since OPT$(V) \geq 2L$ and $\epsilon \leq 1$, we conclude that $l \leq (1 + \epsilon)$ OPT(V). $\square$

So from now on we shall deal with well-rounded instances only. All coordinates are within the square $[0, 2^N] \times [0, 2^N]$, where $N := \lceil \log(2 + \frac{10n}{\epsilon}) \rceil$. Now we partition this square successively by a regular grid: for $i = 1, \ldots, N - 1$ let $G_i := X_i \cup Y_i$, where

$$X_i := \left\{\left[\left(0, k2^{N-i}\right), \left(2^N, k2^{N-i}\right)\right] : k = 0, \ldots, 2^i - 1\right\},$$
$$Y_i := \left\{\left[\left(j2^{N-i}, 0\right), \left(j2^{N-i}, 2^N\right)\right] : j = 0, \ldots, 2^i - 1\right\}.$$

(The notation $[(x, y), (x', y')]$ denotes the line segment between (x, y) and (x', y').)

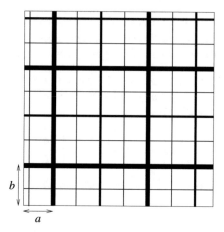

Fig. 21.3.

More precisely, we consider **shifted grids**: Let $a, b \in \{0, 2, \ldots, 2^N - 2\}$ be even integers. For $i = 1, \ldots, N - 1$ let $G_i^{(a,b)} := X_i^{(b)} \cup Y_i^{(a)}$, where

$$X_i^{(b)} := \left\{ \left[\left(0, (b + k2^{N-i}) \bmod 2^N \right), \left(2^N, (b + k2^{N-i}) \bmod 2^N \right) \right] : \right.$$
$$\left. k = 0, \ldots, 2^i - 1 \right\},$$

$$Y_i^{(a)} := \left\{ \left[\left((a + j2^{N-i}) \bmod 2^N, 0 \right), \left((a + j2^{N-i}) \bmod 2^N, 2^N \right) \right] : \right.$$
$$\left. j = 0, \ldots, 2^i - 1 \right\}.$$

($x \bmod y$ denotes the unique number z with $0 \leq z < y$ and $\frac{x-z}{y} \in \mathbb{Z}$.) Note that $G_{N-1} = G_{N-1}^{(a,b)}$ does not depend on a or b.

A line l is said to be at **level** 1 if $l \in G_1^{(a,b)}$, and at level i if $l \in G_i^{(a,b)} \setminus G_{i-1}^{(a,b)}$ ($i = 2, \ldots, N - 1$). See Figure 21.3, where thicker lines are at smaller levels. The **regions** of the grid $G_i^{(a,b)}$ are the sets

$$\left\{ (x, y) \in [0, 2^N) \times [0, 2^N) \ : \ (x - a - j2^{N-i}) \bmod 2^N < 2^{N-i}, \right.$$
$$\left. (y - b - k2^{N-i}) \bmod 2^N < 2^{N-i} \right\}$$

for $j, k \in \{0, \ldots, 2^i - 1\}$. For $i < N - 1$, some of the regions may be disconnected and consist of two or four rectangles. Since all lines are defined by even coordinates, no line contains a point of our well-rounded EUCLIDEAN TSP instance. Furthermore, each region of G_{N-1} contains at most one of these points, for any a, b.

For a polygon T and a line l of G_{N-1} we denote by $cr(T, l)$ the number of times T crosses l. The following proposition will prove useful:

Proposition 21.9. *For an optimum tour T of a well-rounded instance V of the* EUCLIDEAN *TSP,* $\sum_{l \in G_{N-1}} cr(T, l) \leq \mathrm{OPT}(V)$.

Proof: Consider an edge of T of length s. Let x and y be the length of the horizontal and vertical projection, respectively (thus $s^2 = x^2 + y^2$). The edge crosses lines of G_{N-1} exactly $\frac{x+y}{2} \leq s$ times. Summing over all edges of T yields the stated inequality. □

Set $C := 7 + \left\lceil \frac{36}{\epsilon} \right\rceil$ and $P := N \left\lceil \frac{6}{\epsilon} \right\rceil$. For each line we now define **portals**: if $l = \left[\left(0, (b + k2^{N-i}) \bmod 2^N \right), \left(2^N, (b + k2^{N-i}) \bmod 2^N \right) \right]$ is a horizontal line at level i, we define the set of its portals to be

$$\left\{ \left(\left(a + \frac{h}{P} 2^{N-i} \right) \bmod 2^N, (b + k2^{N-i}) \bmod 2^N \right) : h = 0, \ldots, P2^i \right\}.$$

Portals of vertical lines are defined analogously.

Definition 21.10. *Let $V \subseteq [0, 2^N] \times [0, 2^N]$ be a well-rounded instance of the* EUCLIDEAN *TSP. Let $a, b \in \{0, 2, \ldots, 2^N - 2\}$ be given, and let the shifted grids, C, P, and the portals be defined as above. A **Steiner tour** is a closed walk of straight line segments containing V such that its intersection with each line of the grids is a subset of portals. A Steiner tour is **light** if for each i and for each region of $G_i^{(a,b)}$, the tour crosses each edge of the region at most C times.*

Note that Steiner tours are not necessarily polygons; they may cross themselves. To make a Steiner tour light we shall make frequent use of the following Patching Lemma:

Lemma 21.11. *Let $V \subset \mathbb{R}^2$ be an* EUCLIDEAN *TSP instance and T a tour for V. Let l be a segment of length s of a line not containing any point in V. Then there exists a tour for V whose length exceeds the length of T by at most $6s$ and which crosses l at most twice.*

Proof: For clearer exposition, assume l to be a vertical line segment. Suppose T crosses l exactly k times, say with edges $e_1, \ldots, e_k$. Let $k \geq 3$; otherwise the assertion is trivial. We subdivide each of $e_1, \ldots, e_k$ by two new vertices without increasing the tour length. In other words, we replace e_i by a path of length 3 with two new inner vertices $p_i, q_i \in \mathbb{R}^2$ very close to l, where p_i is to the left of l and q_i is to the right of l $(i = 1, \ldots, k)$. Let the resulting tour be T'.

Let $t := \lfloor \frac{k-1}{2} \rfloor$ (then $k - 2 \leq 2t \leq k - 1$), and let T'' result from T' by deleting the edges $\{p_1, q_1\}, \ldots, \{p_{2t}, q_{2t}\}$.

Let P consist of a shortest tour through $p_1, \ldots, p_k$ plus a minimum cost perfect matching of $p_1, \ldots, p_{2t}$. Analogously, let Q consist of a shortest tour through $q_1, \ldots, q_k$ plus a minimum cost perfect matching of $q_1, \ldots, q_{2t}$. The total length of the edges is at most $3s$ in each of P and Q.

Then $T'' + P + Q$ crosses l at most $k - 2t \leq 2$ times, and is connected and Eulerian, and all vertices except $p_1, \ldots, p_k, q_1, \ldots, q_k$ have degree 2. We now proceed as in Lemma 21.3. By Euler's Theorem 2.24 there exists an Eulerian walk in $T'' + P + Q$. By shortcutting paths this can be converted to a tour for V, without increasing the length or the number of crossings with l, and without changing the position of any crossing. $\qquad\square$

The following theorem is the key idea of the algorithm:

Theorem 21.12. (Arora [1998]) *Let $V \subseteq [0, 2^N] \times [0, 2^N]$ be a well-rounded instance of the* EUCLIDEAN *TSP. If a and b are randomly chosen out of $\{0, 2, \ldots, 2^N - 2\}$, then with probability at least $\frac{1}{2}$ a light Steiner tour exists whose length is at most $(1 + \epsilon)\, \mathrm{OPT}(V)$.*

Proof: Let T be an optimum tour for V. We introduce Steiner points whenever the tour crosses a line.

Now all the Steiner points are moved to portals. The nearest portal from a Steiner point on a line on level i can be as far away as $\frac{2^{N-i-1}}{P}$. Since a line l is at level i with probability $p(l, i) := \begin{cases} 2^{i-N} & \text{if } i > 1 \\ 2^{2-N} & \text{if } i = 1 \end{cases}$, the expected total tour length increase by moving all Steiner points on l to portals is at most

$$\sum_{i=1}^{N-1} p(l, i) \cdot cr(T, l) \cdot 2 \cdot \frac{2^{N-i-1}}{P} = N \cdot \frac{cr(T, l)}{P}.$$

Now we modify the Steiner tour so that it becomes light. By a segment of a horizontal or vertical line of $G_i^{(a,b)}$ we mean a segment between the point $\big((a + j2^{N-i}), (b + k2^{N-i})\big)$ and the point $\big((a + (j+1)2^{N-i}), (b + k2^{N-i})\big)$ or $\big((a + j2^{N-i}), (b + (k+1)2^{N-i})\big)$, respectively (all coordinates must be taken mod 2^N), for $j, k \in \{0, \ldots, 2^i - 1\}$. Note that such a segment can consist of two separate line segments. Consider the following procedure:

> **For $i := N - 1$ down to 1 do:**
>   Apply the Patching Lemma 21.11 to each segment of a horizontal line of $G_i^{(a,b)}$ which is crossed more than $C - 4$ times.
>   Apply the Patching Lemma 21.11 to each segment of a vertical line of $G_i^{(a,b)}$ which is crossed more than $C - 4$ times.

Two remarks must be made. For segments of a horizontal or vertical line that consist of two separate parts the Patching Lemma is applied to each part, so the total number of crossings afterwards may be 4.

Furthermore, observe that the application of the Patching Lemma to a vertical line segment l in iteration i may introduce new crossings (Steiner points) on a horizontal line segment which has one endpoint on l. These new crossings are at portals and will not be considered anymore in subsequent iterations of the above procedure, because they are on lines of higher level.

For each line l, the number of applications of the Patching Lemma to l is at most $\frac{cr(T,l)}{C-7}$, since each time the number of crossings decreases by at least $C-7$ (at least $C-3$ crossings are replaced by at most 4). For a line l, let $c(l, i, a, b)$ be the total number of times the Patching Lemma is applied to l at iteration i of the above procedure. Note that $c(l, i, a, b)$ is independent of the level of l as long as it is at most i.

Then the total increase in tour length due to applications of the Patching Lemma to line l is $\sum_{i \geq level(l)} c(l, i, a, b) \cdot 6 \cdot 2^{N-i}$. Furthermore, we note that

$$\sum_{i \geq level(l)} c(l, i, a, b) \leq \frac{cr(T, l)}{C - 7}.$$

Since l is at level j with probability $p(l, j)$, the expected total increase in tour length by the above procedure is at most

$$\sum_{j=1}^{N-1} p(l, j) \sum_{i \geq j} c(l, i, a, b) \cdot 6 \cdot 2^{N-i} = 6 \sum_{i=1}^{N-1} c(l, i, a, b) \cdot 2^{N-i} \sum_{j=1}^{i} p(l, j)$$

$$\leq \frac{12\, cr(T, l)}{C - 7}.$$

After the above procedure, each line segment (and therefore each edge of a region) is crossed by the tour at most $C-4$ times, not counting the new crossings introduced by the procedure (see the remark above). These additional crossings are all at one of the endpoints of the line segments. But if a tour crosses through the same point three or more times, two of the crossings can be removed without increasing the tour length or introducing any additional crossings. (Removing two out of three parallel edges of a connected Eulerian graph results in a connected Eulerian graph.) So we have at most four additional crossings for each edge of each region (at most two for each endpoint), and the tour is indeed light.

So – using Proposition 21.9 – the expectation of the total tour length increase is at most

$$\sum_{l \in G_{N-1}} N \frac{cr(T, l)}{P} + \sum_{l \in G_{N-1}} \frac{12\, cr(T, l)}{C - 7} \leq \mathrm{OPT}(V) \left(\frac{N}{P} + \frac{12}{C - 7} \right) \leq \mathrm{OPT}(V) \frac{\epsilon}{2}.$$

Hence the probability that the tour length increase is at most $\mathrm{OPT}(V)\epsilon$ must be at least $\frac{1}{2}$.   □

With this theorem we can finally describe ARORA'S ALGORITHM. The idea is to enumerate all light Steiner tours, using dynamic programming. A **subproblem** consists of a region r of a grid $G_i^{(a,b)}$ with $1 \leq i \leq N-1$, a set A of even cardinality, each element of which is assigned to a portal on one of the edges of r (such that no more than C elements are assigned to each edge), and a perfect matching M of the complete graph on A. So for each region, we have less than $(P + 2)^{4C}(4C)!$ subproblems (up to renaming the elements of A). A solution to such a subproblem is a set of $|M|$ paths $\{P_e : e \in M\}$ which form the intersection of some light Steiner tour for V with r, such that $P_{\{v,w\}}$ has the endpoints v and w, and each point of $V \cap r$ belongs to exactly one of the paths. A solution is optimum if the total length of the paths is shortest possible.

ARORA'S ALGORITHM

Input: A well-rounded instance $V \subseteq [0, 2^N] \times [0, 2^N]$ of the EUCLIDEAN
            TSP. A number $0 < \epsilon < 1$.

Output: A tour which is optimal up to a factor of $(1 + \epsilon)$.

①   Choose a and b randomly out of $\{0, 2, \ldots, 2^N - 2\}$.
     Set $R_0 := \{([0, 2^N] \times [0, 2^N], V)\}$.

②   **For** $i := 1$ **to** $N - 1$ **do**:
        Construct $G_i^{(a,b)}$. Set $R_i := \emptyset$.
        For each $(r, V_r) \in R_{i-1}$ for which $|V_r| \geq 2$ **do**:
            Construct the four regions r_1, r_2, r_3, r_4 of $G_i^{(a,b)}$ with
               $r_1 \cup r_2 \cup r_3 \cup r_4 = r$ and add $(r_j, V_r \cap r_j)$ to R_i ($j = 1, 2, 3, 4$).

③   **For** $i := N - 1$ **down to** 1 **do**:
        For each region $r \in R_i$ **do**: Solve all its subproblems optimally.
        If $|V_r| \leq 1$ **then** this is done directly,
                   else the already computed optimum solutions of the
                subproblems for the four subregions are used.

④   Compute an optimum light Steiner tour for V using the optimum solutions
     of the subproblems for the four subregions.
     Remove the Steiner points to obtain a tour which is no longer.

Theorem 21.13. ARORA'S ALGORITHM *finds a tour which is, with probability at least $\frac{1}{2}$, no longer than $(1 + \epsilon)\,\mathrm{OPT}(V)$. The running time is $O(n(\log n)^c)$ for a constant c (depending linearly on $\frac{1}{\epsilon}$).*

Proof: The algorithm chooses a and b randomly and then computes an optimum light Steiner tour. By Theorem 21.12, this is no longer than $(1 + \epsilon)\,\mathrm{OPT}(V)$ with probability at least $\frac{1}{2}$. The final removal of the Steiner points can only make the tour shorter.

To estimate the running time, consider the following arborescence A: the root is the region in R_0, and each region $r \in R_i$ has 0 or 4 children (its subregions in R_{i+1}). Let S be the set of vertices in A that have 4 children which are all leaves. Since the interiors of these regions are pairwise disjoint and each contains at least two points of V, we have $|S| \leq \frac{n}{2}$. Since each vertex of A is either a leaf or an ancestor of at least one vertex in S, we have at most $N\frac{n}{2}$ vertices that are not leaves and thus at most $\frac{5}{2}Nn$ vertices altogether.

For each region, at most $(P + 2)^{4C}(4C)!$ subproblems arise. Subproblems corresponding to regions with at most one point can be solved directly in $O(C)$ time. For other subproblems, all possible multisets of portals on the four edges between the subregions and all possible orders in which the portals can be traversed, are considered. All these at most $(P+2)^{4C}(8C)!$ possibilities can then be evaluated in constant time using the stored solutions of the subproblems.

So for all subproblems of one region, the running time is $O((P + 2)^{8C}(4C)! (8C)!)$. Observe that this also holds for disconnected regions: since the tour may not go from one connected component of a region to another, the problem can only become easier.

Since at most $\frac{5}{2}Nn$ regions are considered, $N = O(\log \frac{n}{\epsilon})$ (the instance is well-rounded), $C = O(\frac{1}{\epsilon})$ and $P = O(\frac{N}{\epsilon})$, we obtain an overall running time of

$$O\left(n \log \frac{n}{\epsilon}(P + 2)^{8C}(8C)^{12C}\right) = O\left(n(\log n)^{O\left(\frac{1}{\epsilon}\right)} O\left(\frac{1}{\epsilon}\right)^{O\left(\frac{1}{\epsilon}\right)}\right).$$

$\square$

Of course, ARORA's ALGORITHM can easily be derandomized by trying all possible values of a and b. This adds a factor of $O\left(\frac{n^2}{\epsilon^2}\right)$ to the running time. We conclude:

Corollary 21.14. *There is an approximation scheme for the* EUCLIDEAN *TSP. For each fixed $\epsilon > 0$, a $(1 + \epsilon)$-approximate solution can be determined in $O\left(n^3(\log n)^c\right)$ time for some constant c.* $\square$

Rao and Smith [1998] improved the running time to $O(n \log n)$ for each fixed $\epsilon > 0$. However, the constants involved are still quite large for reasonable values of ϵ, and thus the practical value seems to be limited. Klein [2008] found an approximation scheme with linear running time for each fixed $\epsilon > 0$ for instances that are the metric closure of a planar graph with nonnegative edge weights. The techniques presented in this section also apply to other geometric problems; see for instance Exercise 10.

21.3 Local Search

In general, the most successful technique for obtaining good solutions for TSP instances in practice is local search. The idea is as follows. We start with any tour found by some heuristic. Then we try to improve our solution by certain "local" modifications. For example, we could cut our tour into two pieces by deleting two edges and then join the pieces to a different tour.

Local search is an algorithmic principle rather than an algorithm. In particular, two decisions must be made in advance:

- Which are the modifications allowed, i.e. how is the neighbourhood of a solution defined?
- When do we actually modify our solution? (One possibility here is to allow improvements only.)

To give a concrete example, we describe the well-known k-OPT ALGORITHM for the TSP. Let $k \geq 2$ be a fixed integer.

---

k-OPT ALGORITHM

Input: An instance (K_n, c) of the TSP.

Output: A tour T.

①   Let T be any tour.

②   Let S be the family of k-element subsets of $E(T)$.

③   **For all** $S \in S$ and all tours T' with $E(T') \supseteq E(T) \setminus S$ **do:**
       If $c(E(T')) < c(E(T))$ **then** set $T := T'$ and **go to** ②.

A faster implementation of this algorithm was proposed by de Berg et al. [2016].

A tour is called **k-opt** if it cannot be improved by the k-OPT ALGORITHM. For any fixed k there are TSP instances and k-opt tours that are not $(k + 1)$-opt. For example, the tour shown in Figure 21.4 is 2-opt but not 3-opt (with respect to Euclidean distances). It can be improved by exchanging three edges (the tour (a, b, e, c, d, a) is optimum).

The tour shown on the right-hand side of Figure 21.5 is 3-opt with respect to the weights shown on the left-hand side. Edges not shown have weight 4. However, a 4-exchange immediately produces the optimum solution. Note that the triangle inequality holds.

Indeed, the situation is much worse: a tour produced by the k-OPT ALGORITHM for an n-city instance can be longer than the optimum tour by a factor of $\frac{1}{4}n^{\frac{1}{2k}}$ (for all k and infinitely many n). On the other hand a 2-opt tour is never worse

Fig. 21.4.

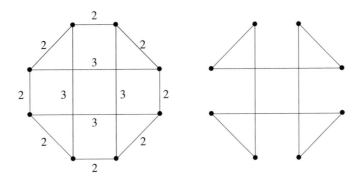

Fig. 21.5.

than $4\sqrt{n}$ times the optimum. However, the worst-case running time of the k-OPT ALGORITHM is exponential for all k, and this holds even for 2-OPT on Euclidean instances. These results are due to Chandra, Karloff and Tovey [1999], and Englert, Röglin and Vöcking [2014].

Another question is how to choose k in advance. Of course, instances (K_n, c) are solved optimally by the k-OPT ALGORITHM with $k = n$, but the running time grows exponentially in k. Often $k = 3$ is a good choice. Lin and Kernighan [1973] proposed an effective heuristic where k is not fixed but rather determined by the algorithm. Their idea is based on the following concept:

Definition 21.15.   *Given an instance* (K_n, c) *of the* TSP *and a tour* T. *An* **alternating walk** *is a sequence of vertices (cities)* $P = (x_0, x_1, \ldots, x_{2m})$ *such that* $\{x_i, x_{i+1}\} \neq \{x_j, x_{j+1}\}$ *for all* $0 \leq i < j < 2m$, *and for* $i = 0, \ldots, 2m - 1$ *we have* $\{x_i, x_{i+1}\} \in E(T)$ *if and only if* i *is even.* P *is* **closed** *if in addition* $x_0 = x_{2m}$.

The **gain** *of* P *is defined by*

$$g(P) := \sum_{i=0}^{m-1} (c(\{x_{2i}, x_{2i+1}\}) - c(\{x_{2i+1}, x_{2i+2}\})).$$

P *is called* **proper** *if* $g((x_0, \ldots, x_{2i})) > 0$ *for all* $i \in \{1, \ldots, m\}$. *We use the abbreviation* $E(P) = \{\{x_i, x_{i+1}\} : i = 0, \ldots, 2m - 1\}$.

Note that vertices may occur more than once in an alternating walk. In the example shown in Figure 21.4, (a, e, b, c, e, d, a) is a proper closed alternating walk. Given a tour T, we are of course interested in those closed alternating walks P for which $E(T) \triangle E(P)$ again defines a tour.

Lemma 21.16.   (Lin and Kernighan [1973])   *If there is a closed alternating walk* P *with* $g(P) > 0$, *then*

(a) $c(E(T) \triangle E(P)) = c(E(T)) - g(P)$;
(b) *there is a proper closed alternating walk* Q *with* $E(Q) = E(P)$.

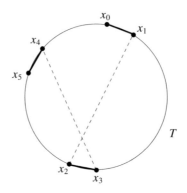

Fig. 21.6.

Proof: Part (a) follows from the definition. To see (b), let $P = (x_0, x_1, \ldots, x_{2m})$, and let k be the largest index for which $g((x_0, \ldots, x_{2k}))$ is minimum. We claim that $Q := (x_{2k}, x_{2k+1}, \ldots, x_{2m-1}, x_0, x_1, \ldots, x_{2k})$ is proper. For $i = k+1, \ldots, m$ we have

$$g((x_{2k}, x_{2k+1}, \ldots, x_{2i})) = g((x_0, x_1, \ldots, x_{2i})) - g((x_0, x_1, \ldots, x_{2k})) > 0$$

by definition of k. For $i = 1, \ldots, k$ we have

$$\begin{aligned}
&g((x_{2k}, x_{2k+1}, \ldots, x_{2m-1}, x_0, x_1, \ldots, x_{2i})) \\
= \;& g((x_{2k}, x_{2k+1}, \ldots, x_{2m})) + g((x_0, x_1, \ldots, x_{2i})) \\
\geq \;& g((x_{2k}, x_{2k+1}, \ldots, x_{2m})) + g((x_0, x_1, \ldots, x_{2k})) \\
= \;& g(P) > 0,
\end{aligned}$$

again by definition of k. So Q is indeed proper. $\square$

We now go ahead with the description of the algorithm. Given any tour T, it looks for a proper closed alternating walk P, replaces T by $(V(T), E(T) \triangle E(P))$, and iterates. At each iteration it exhaustively checks all possibilities until some proper closed alternating walk is found, or until one of the two parameters p_1 and p_2 prevent it from doing so. See also Figure 21.6 for an illustration.

LIN-KERNIGHAN ALGORITHM

Input: An instance (K_n, c) of the TSP. A spanning subgraph G of K_n. Two parameters $p_1 \in \mathbb{N}$ (backtracking depth) and $p_2 \in \mathbb{N}$ (infeasibility depth).

Output: A tour T.

①   Let T be any tour.

②   Set $X_0 := V(K_n)$, $i := 0$ and $g^* := 0$.

③   **If** $X_i = \emptyset$ **and** $g^* > 0$ **then:**
   Set $T := (V(T), E(T) \triangle E(P^*))$ and **go to** ②.
   If $X_i = \emptyset$ **and** $g^* = 0$ **then:**
   Set $i := \min\{i - 1, p_1\}$. **If** $i < 0$ **then stop, else go to** ③.

④   Choose $x_i \in X_i$ and set $X_i := X_i \setminus \{x_i\}$.
   If i **is odd,** $i \geq 3$, $(V(T), E(T) \triangle E((x_0, x_1, \ldots, x_{i-1}, x_i, x_0)))$ **is a tour**
   and $g((x_0, x_1, \ldots, x_{i-1}, x_i, x_0)) > g^*$ **then:**
   Set $P^* := (x_0, x_1, \ldots, x_{i-1}, x_i, x_0)$ and $g^* := g(P^*)$.

⑤   **If** i **is odd then:**
   Set $X_{i+1} := \{x \in \Gamma_G(x_i) \setminus \{x_0\} :$
   $\{x_i, x\} \notin E(T) \cup E((x_0, x_1, \ldots, x_i)),$
   $g((x_0, x_1, \ldots, x_{i-1}, x_i, x)) > g^*\}.$
   If i **is even and** $i \leq p_2$ **then:**
   Set $X_{i+1} := \{x \in V(K_n) : \{x_i, x\} \in E(T) \setminus E((x_0, x_1, \ldots, x_i))\}.$
   If i **is even and** $i > p_2$ **then:**
   Set $X_{i+1} := \{x \in V(K_n) : \{x_i, x\} \in E(T) \setminus E((x_0, x_1, \ldots, x_i)),$
   $(V(T), E(T) \triangle E((x_0, x_1, \ldots, x_i, x, x_0)))$ is a tour$\}.$
   Set $i := i + 1$. **Go to** ③.

Note that in ④ if $(V(T), E(T) \triangle E((x_0, x_1, \ldots, x_{i-1}, x_i, x_0)))$ is a tour, then $(x_0, x_1, \ldots, x_{i-1}, x_i, x_0)$ is a closed alternating walk.

Lin and Kernighan have proposed the parameters $p_1 = 5$ and $p_2 = 2$. These are the smallest values which guarantee that the algorithm finds a favourable 3-exchange:

Theorem 21.17. (Lin and Kernighan [1973]) *The* LIN-KERNIGHAN ALGO-RITHM

(a) *for* $G = K_n$, $p_1 = \infty$ *and* $p_2 = \infty$ *finds a proper closed alternating walk* P *such that* $(V(T), E(T) \triangle E(P))$ *is a tour, if one exists.*
(b) *for* $G = K_n$, $p_1 = 5$ *and* $p_2 = 2$ *returns a tour which is 3-opt.*

Proof: Let T be the tour the algorithm ends with. Then g^* must have been zero all the time since the last tour change. This implies that, in the case $p_1 = p_2 = \infty$, the algorithm has completely enumerated all proper alternating walks. In particular, (a) holds.

In the case $p_1 = 5$ and $p_2 = 2$, the algorithm has at least enumerated all proper closed alternating walks of length 4 or 6. Suppose there exists a favourable 2-exchange or 3-exchange resulting in a tour T'. Then the edges $E(T) \triangle E(T')$ form a closed alternating walk P with at most six edges and $g(P) > 0$. By Lemma 21.16, P is w.l.o.g. proper and the algorithm would have found it. This proves (b). □

We should remark that this procedure has no chance of finding a "non-sequential" exchange such as the 4-exchange shown in Figure 21.5. In this example the tour cannot be improved by the LIN-KERNIGHAN ALGORITHM, but a (non-sequential) 4-exchange would provide the optimum solution.

So a refinement of the LIN-KERNIGHAN ALGORITHM could be as follows. If the algorithm stops, we try (by some heuristics) to find a favourable non-sequential 4-exchange. If we are successful, we continue with the new tour, otherwise we give up.

The graph G should be sparse (otherwise computing X_{i+1} may take too long) but also contain a good tour. For example, the Delaunay triangulation often works well for Euclidean instances.

The LIN-KERNIGHAN ALGORITHM is far more effective than e.g. the 3-OPT ALGORITHM. While being at least as good (and usually much better), the expected running time (with $p_1 = 5$ and $p_2 = 2$) also compares favourably: Lin and Kernighan report an empirical running time of about $O(n^{2.2})$. However, it seems unlikely that the worst-case running time is polynomial; for a precise formulation of this statement (and a proof), see Exercise 13 (Papadimitriou [1992]).

Almost all local search heuristics used for the TSP in practice are based on this algorithm. Although the worst-case behaviour is worse than CHRISTOFIDES' ALGORITHM, the LIN-KERNIGHAN ALGORITHM typically produces much better solutions, usually within a few percent of the optimum. For very efficient variants, see Applegate, Cook and Rohe [2003] and Helsgaun [2009].

By Exercise 13 of Chapter 9 there is no local search algorithm for the TSP which has polynomial-time complexity per iteration and always finds an optimum solution, unless $P = NP$ (here one iteration is taken to be the time between two changes of the current tour). We now show that one cannot even decide whether a given tour is optimum. To do this we first consider the following restriction of the HAMILTONIAN CIRCUIT problem:

RESTRICTED HAMILTONIAN CIRCUIT

Instance: An undirected graph G and some Hamiltonian path in G.

Question: Does G contain a Hamiltonian circuit?

Lemma 21.18. RESTRICTED HAMILTONIAN CIRCUIT *is NP-complete.*

Proof: Given an instance G of the HAMILTONIAN CIRCUIT PROBLEM (which is NP-complete, see Theorem 15.25), we construct an equivalent instance of RESTRICTED HAMILTONIAN CIRCUIT.

Assume $V(G) = \{1, \ldots, n\}$. We take n copies of the "diamond graph" shown in Figure 21.7 and join them vertically with edges $\{S_i, N_{i+1}\}$ $(i = 1, \ldots, n-1)$.

It is clear that the resulting graph contains a Hamiltonian path from N_1 to S_n. We now add edges $\{W_i, E_j\}$ and $\{W_j, E_i\}$ for any edge $\{i, j\} \in E(G)$. Let us call the resulting graph H. It is obvious that any Hamiltonian circuit in G induces a Hamiltonian circuit in H.

Furthermore, a Hamiltonian circuit in H must traverse all the diamond subgraphs in the same way: either all from E_i to W_i or all from S_i to N_i. But the latter is impossible, so H is Hamiltonian if and only if G is.   □

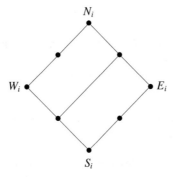

Fig. 21.7.

Theorem 21.19. (Papadimitriou and Steiglitz [1977]) *The problem of deciding whether a given tour is optimum for a given* METRIC TSP *instance is coNP-complete.*

Proof: Membership in *coNP* is clear, since an optimum tour serves as a certificate for suboptimality.

We shall now transform RESTRICTED HAMILTONIAN CIRCUIT to the complement of our problem. Namely, given a graph G and a Hamiltonian path P in G, we first check whether the ends of P are connected by an edge. If so, we are done. Otherwise we define

$$c_{ij} := \begin{cases} 1 & \text{if } \{i, j\} \in E(G) \\ 2 & \text{if } \{i, j\} \notin E(G). \end{cases}$$

The triangle inequality is of course satisfied. Moreover, P defines a tour of cost $n + 1$, which is optimum if and only if there is no Hamiltonian circuit in G. $\square$

Corollary 21.20. *Unless $P = NP$, no local search algorithm for the* TSP *having polynomial-time complexity per iteration can be exact.*

Proof: An exact local search algorithm includes the decision whether the initial tour is optimum. $\square$

Local search of course also applies to many other combinatorial optimization problems. The SIMPLEX ALGORITHM can also be regarded as a local search algorithm. Although local search algorithms have proved to be very successful in practice, almost no theoretical evidence for their efficiency is known, except in some special cases (see, e.g., Exercise 10 in Chapter 16 and Sections 22.6 and 22.8). For many *NP*-hard problems and interesting neighbourhoods (including those discussed in this section) it is not even known whether a local optimum can be computed in polynomial time; see Exercise 13. The book edited by Aarts and Lenstra [1997] contains more examples of local search heuristics. Michiels, Aarts and Korst [2007] describe more theoretical results on local search.

21.4 The Traveling Salesman Polytope

Dantzig, Fulkerson and Johnson [1954] were the first to solve a TSP instance of non-trivial size optimally. They started by solving an LP relaxation of a suitable integer linear programming formulation, and then successively added cutting planes. This was the starting point of the analysis of the traveling salesman polytope:

Definition 21.21. *For $n \geq 3$ we denote by $Q(n)$ the* **traveling salesman polytope**, *i.e. the convex hull of the incidence vectors of tours in the complete graph K_n.*

Although no complete description of the traveling salesman polytope is known, there are several interesting results, some of which are also relevant for practical computations. Since $\sum_{e \in \delta(v)} x_e = 2$ for all $v \in V(K_n)$ and all $x \in Q(n)$, the dimension of $Q(n)$ is at most $|E(K_n)| - |V(K_n)| = \binom{n}{2} - n = \frac{n(n-3)}{2}$. In order to prove that in fact $\dim(Q(n)) = \frac{n(n-3)}{2}$, we need the following graph-theoretical lemma:

Lemma 21.22. *For any $k \geq 1$:*

(a) *The edge set of K_{2k+1} can be partitioned into k tours.*
(b) *The edge set of K_{2k} can be partitioned into $k-1$ tours and one perfect matching.*

Proof: (a): Suppose the vertices are numbered $0, \ldots, 2k - 1, x$. Consider the tours

$$T_i \;=\; (x, i, i + 1, i - 1, i + 2, i - 2, i + 3, \ldots,$$
$$i - k + 2, i + k - 1, i - k + 1, i + k, x)$$

for $i = 0, \ldots, k - 1$ (everything is meant modulo $2k$). See Figure 21.8 for an illustration. Since $|E(K_{2k+1})| = k(2k + 1)$ it suffices to show that these tours are pairwise edge-disjoint. This is clear with respect to the edges incident to x. Moreover, for $\{a, b\} \in E(T_i)$ with $a, b \neq x$ we have $a + b \in \{2i, 2i + 1\}$, as is easily seen.

(b): Suppose the vertices are numbered $0, \ldots, 2k - 2, x$. Consider the tours

$$T_i \;=\; (x, i, i + 1, i - 1, i + 2, i - 2, i + 3, \ldots,$$
$$i + k - 2, i - k + 2, i + k - 1, i - k + 1, x)$$

for $i = 0, \ldots, k - 2$ (everything is meant modulo $2k - 1$). The same argument as above shows that these tours are pairwise edge-disjoint. After deleting them, the remaining graph is 1-regular and thus provides a perfect matching. $\square$

Theorem 21.23. (Grötschel and Padberg [1979])

$$\dim(Q(n)) \;=\; \frac{n(n-3)}{2}.$$

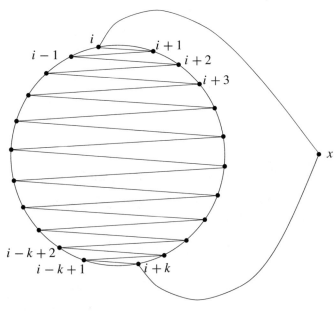

Fig. 21.8.

Proof: For $n = 3$ the statement is trivial. Let $n \geq 4$, and let $v \in V(K_n)$ be an arbitrary vertex.

Case 1: n is even, say $n = 2k + 2$ for some integer $k \geq 1$. By Lemma 21.22(a) $K_n - v$ is the union of k edge-disjoint tours $T_0, \ldots, T_{k-1}$. Now let T_{ij} arise from T_i by replacing the j-th edge $\{a, b\}$ by two edges $\{a, v\}, \{v, b\}$ ($i = 0, \ldots, k - 1$; $j = 1, \ldots, n - 1$). Consider the matrix whose rows are the incidence vectors of these $k(n - 1)$ tours. Then the columns corresponding to edges not incident with v form a square matrix

$$\begin{pmatrix} A & 0 & 0 & \cdots & 0 \\ 0 & A & 0 & \cdots & 0 \\ 0 & 0 & A & \cdots & 0 \\ \cdots & \cdots & \cdots & \cdots & \cdots \\ 0 & 0 & 0 & \cdots & A \end{pmatrix}, \quad \text{where} \quad A = \begin{pmatrix} 0 & 1 & 1 & \cdots & 1 \\ 1 & 0 & 1 & \cdots & 1 \\ 1 & 1 & 0 & \cdots & 1 \\ \cdots & \cdots & \cdots & \cdots & \cdots \\ 1 & 1 & 1 & \cdots & 0 \end{pmatrix}.$$

Since this matrix is nonsingular, the incidence vectors of the $k(n - 1)$ tours are linearly independent, implying $\dim(Q(n)) \geq k(n - 1) - 1 = \frac{n(n-3)}{2}$.

Case 2: n is odd, so $n = 2k + 3$ with $k \geq 1$ integer. By Lemma 21.22(b) $K_n - v$ is the union of k tours and one perfect matching M. From the tours, we construct $k(n - 1)$ tours in K_n as in Case 1. Now we complete the perfect matching M arbitrarily to a tour T in K_{n-1}. For each edge $e = \{a, b\}$ of M, we replace e in T by the two edges $\{a, v\}$ and $\{v, b\}$. In this way we obtain another $k + 1$ tours. Similarly as above, the incidence vectors of all the $k(n - 1) + k + 1 = kn + 1$ tours are linearly independent, proving $\dim(Q(n)) \geq kn + 1 - 1 = \frac{n(n-3)}{2}$. □

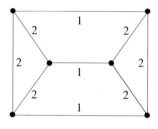

Fig. 21.9.

The integral points of $Q(n)$, i.e. the tours, can be described nicely:

Proposition 21.24. *The incidence vectors of the tours in K_n are exactly the integral vectors x satisfying*

$$0 \leq x_e \leq 1 \qquad\qquad (e \in E(K_n)); \qquad\qquad (21.1)$$

$$\sum_{e \in \delta(v)} x_e = 2 \qquad\qquad (v \in V(K_n)); \qquad\qquad (21.2)$$

$$\sum_{e \in E(K_n[X])} x_e \leq |X| - 1 \qquad\qquad (\emptyset \neq X \subset V(K_n)). \qquad\qquad (21.3)$$

Proof: Obviously the incidence vector of any tour satisfies these constraints. Any integral vector satisfying (21.1) and (21.2) is the incidence vector of a perfect simple 2-matching, i.e. the union of vertex-disjoint circuits covering all the vertices. The constraints (21.3) prevent circuits with less than n edges. $\qquad\square$

The constraints (21.3) are usually called **subtour inequalities**, and the polytope defined by (21.1), (21.2), (21.3) is called the **subtour polytope**. In general the subtour polytope is not integral, as the instance in Figure 21.9 shows (edges not shown have weight 3): the shortest tour has length 10, but the optimum fractional solution ($x_e = 1$ if e has weight 1, and $x_e = \frac{1}{2}$ if e has weight 2) has total weight 9.

The following equivalent descriptions of the subtour polytope will be useful:

Proposition 21.25. *Let $V(K_n) = \{1, \ldots, n\}$. Let $x \in [0, 1]^{E(K_n)}$ with $\sum_{e \in \delta(v)} x_e = 2$ for all $v \in V(K_n)$. Then the following statements are equivalent:*

$$\sum_{e \in E(K_n[X])} x_e \leq |X| - 1 \qquad\qquad (\emptyset \neq X \subset V(K_n)); \qquad\qquad (21.3)$$

$$\sum_{e \in E(K_n[X])} x_e \leq |X| - 1 \qquad\qquad (\emptyset \neq X \subseteq V(K_n) \setminus \{1\}); \qquad (21.4)$$

$$\sum_{e \in \delta(X)} x_e \geq 2 \qquad\qquad (\emptyset \neq X \subset V(K_n)). \qquad\qquad (21.5)$$

Proof: For any $\emptyset \neq X \subset V(K_n)$ we have

$$\sum_{e\in\delta(V(K_n)\setminus X)} x_e \;=\; \sum_{e\in\delta(X)} x_e \;=\; \sum_{v\in X}\sum_{e\in\delta(v)} x_e \;-\; 2\sum_{e\in E(K_n[X])} x_e$$

$$= \; 2|X| \;-\; 2\sum_{e\in E(K_n[X])} x_e,$$

which implies the equivalence of (21.3), (21.4) and (21.5). $\square$

Corollary 21.26.   *The* SEPARATION PROBLEM *for subtour inequalities can be solved in polynomial time.*

Proof: Using (21.5) and regarding x as edge capacities we have to decide whether there is a cut in (K_n, x) with capacity less than 2. Therefore the SEPARATION PROBLEM reduces to the problem of finding a minimum cut in an undirected graph with nonnegative capacities. By Theorem 8.42 this problem can be solved in $O(n^3)$ time. $\square$

Since any tour is a perfect simple 2-matching, the convex hull of all perfect simple 2-matchings contains the traveling salesman polytope. So by Theorem 12.3 we have:

Proposition 21.27.   *Any $x \in Q(n)$ satisfies the inequalities*

$$\sum_{e\in E(K_n[X])\cup F} x_e \;\leq\; |X| + \frac{|F|-1}{2} \qquad \text{for } X \subseteq V(K_n),\ F \subseteq \delta(X) \text{ with } |F| \text{ odd.}$$

(21.6)

The constraints (21.6) are called **2-matching inequalities**. It is sufficient to consider inequalities (21.6) for the case when F is a matching; the other 2-matching inequalities are implied by these (Exercise 17). For the 2-matching inequalities, the SEPARATION PROBLEM can be solved in polynomial time by Theorem 12.21. So by the ELLIPSOID METHOD (Theorem 4.21) we can optimize a linear function over the polytope defined by (21.1), (21.2), (21.3), and (21.6) in polynomial time (Exercise 16). The 2-matching inequalities are generalized by the so-called **comb inequalities**, illustrated in Figure 21.10:

Proposition 21.28.   (Chvátal [1973], Grötschel and Padberg [1979]) *Let $T_1, \ldots, T_s \subseteq V(K_n)$ be s pairwise disjoint sets, $s \geq 3$ odd, and $H \subseteq V(K_n)$ with $T_i \cap H \neq \emptyset$ and $T_i \setminus H \neq \emptyset$ for $i = 1, \ldots, s$. Then any $x \in Q(n)$ satisfies*

$$\sum_{e\in\delta(H)} x_e \;+\; \sum_{i=1}^{s}\sum_{e\in\delta(T_i)} x_e \;\geq\; 3s+1.$$

(21.7)

Fig. 21.10.

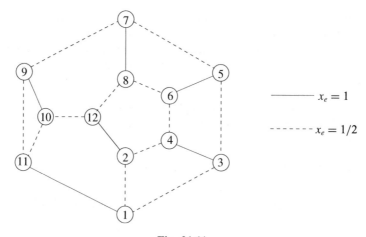

$x_e = 1$

$x_e = 1/2$

Fig. 21.11.

Proof: Let x be the incidence vector of any tour. For any $i \in \{1, \dots, s\}$ we have

$$\sum_{e \in \delta(T_i)} x_e + \sum_{e \in \delta(H) \cap E(K_n[T_i])} x_e \geq 3,$$

since the tour must enter and leave $T_i \setminus H$ as well as $T_i \cap H$. Summing these s inequalities we get

$$\sum_{e \in \delta(H)} x_e + \sum_{i=1}^{s} \sum_{e \in \delta(T_i)} x_e \geq 3s.$$

Since the left-hand side is an even integer, the theorem follows.   □

The fractional solution x shown in Figure 21.11 (edges e with $x_e = 0$ are omitted) is an example of a violated comb inequality in K_{12}: consider $H = \{1, 2, 3, 4, 5, 6\}$, $T_1 = \{1, 11\}$, $T_2 = \{2, 12\}$ and $T_3 = \{5, 6, 7, 8\}$. It is easy to check that the corresponding comb inequality is violated. Note that the in-equalities (21.1), (21.2), (21.3), (21.6) are satisfied, and x is optimum with respect to weights $c(e) := 1 - x_e$ (total weight 3), while the optimum tours have weight $\frac{7}{2}$.

Let us mention just one further class, the **clique tree inequalities**:

Theorem 21.29. (Grötschel and Pulleyblank [1986]) *Let $H_1, \ldots, H_r$ be pairwise disjoint subsets of $V(G)$ (the handles), and let $T_1, \ldots, T_s$ ($s \geq 1$) be pairwise disjoint nonempty proper subsets of $V(G)$ (the teeth) such that*

- *for each handle, the number of teeth it intersects is odd and at least three;*
- *each tooth T contains at least one vertex not belonging to any handle;*
- *$G := K_n[H_1 \cup \cdots \cup H_r \cup T_1 \cup \cdots \cup T_s]$ is connected, but $G - (T_i \cap H_j)$ is disconnected whenever $T_i \cap H_j \neq \emptyset$.*

Let t_j denote the number of handles intersecting T_j ($j = 1, \ldots, s$). Then any $x \in Q(n)$ satisfies

$$\sum_{i=1}^{r} \sum_{e \in E(K_n[H_i])} x_e + \sum_{j=1}^{s} \sum_{e \in E(K_n[T_j])} x_e \leq \sum_{i=1}^{r} |H_i| + \sum_{j=1}^{s} (|T_j| - t_j) - \frac{s+1}{2}. \quad (21.8)$$

We omit the proof which is rather technical. Clique tree inequalities include (21.3) and (21.6) (Exercise 18). They have been generalized further, e.g. to bipartition inequalities by Boyd and Cunningham [1991]. There is a polynomial-time algorithm for the SEPARATION PROBLEM for the clique tree inequalities (21.8) with a fixed number of handles and teeth (Carr [1997]), but none is known for general clique tree inequalities. Even for the SEPARATION PROBLEM for comb inequalities no polynomial-time algorithm is known.

All the inequalities (21.1), (21.4) (restricted to the case where $3 \leq |X| \leq n - 3$) and (21.6) (restricted to the case where F is a matching) define distinct facets of $Q(n)$ (for $n \geq 6$). The proof that the trivial inequalities (21.1) define facets consists of finding $\dim(Q(n))$ linearly independent tours with $x_e = 1$ (and the same for $x_e = 0$) for some fixed edge e. The proof is similar to that of Theorem 21.23, we refer to Grötschel and Padberg [1979]. Even all the inequalities (21.8) define facets of $Q(n)$ ($n \geq 6$). The proof is quite involved, see Grötschel and Padberg [1979], or Grötschel and Pulleyblank [1986].

The number of facets of $Q(n)$ grows fast: already $Q(10)$ has more than 50 billion facets. Any polyhedron that projects to $Q(n)$ (i.e., any extended formulation) has $2^{\Omega(n^{1/2})}$ facets (Fiorini et al. [2015]). This bound was improved to $2^{\Omega(n)}$ by Rothvoß [2017]. Any semidefinite program that projects to $Q(n)$ has $2^{\Omega(n^{1/13})}$ variables (Lee, Raghavendra and Steurer [2015]).

No complete description of $Q(n)$ is known, and it appears very unlikely that one exists. Consider the following problem:

TSP FACETS

Instance: An integer n and an integral inequality $ax \leq \beta$.

Question: Does the given inequality define a facet of $Q(n)$?

The following result shows that a complete description of the traveling salesman polytope is unlikely:

Theorem 21.30. (Karp and Papadimitriou [1982]) *If TSP* FACETS *is in NP, then NP = coNP.*

Moreover, it is *NP*-complete to decide whether two given vertices of $Q(n)$ are adjacent, i.e. belong to a common face of dimension one (Papadimitriou [1978]).

21.5 Lower Bounds

Suppose we have found some tour heuristically, e.g. by the LIN-KERNIGHAN ALGORITHM. We do not know in advance whether this tour is optimum or at least close to the optimum. Is there any way to guarantee that our tour is no more than x percent away from the optimum? In other words, is there a lower bound for the optimum?

Lower bounds can be found by considering any LP relaxation of an integer programming formulation of the TSP, e.g. by taking the inequalities (21.1), (21.2), (21.3), (21.6). However, this LP is not easy to solve (though there is a polynomial-time algorithm via the ELLIPSOID METHOD). A more reasonable lower bound is obtained by taking just (21.1), (21.2), (21.6), i.e. finding a minimum weight perfect simple 2-matching (cf. Exercise 1 of Chapter 12).

However, the most efficient method known is the use of Lagrangean relaxation (cf. Section 5.6). Lagrangean relaxation was first applied to the TSP by Held and Karp [1970, 1971]. Their method is based on the following notion:

Definition 21.31. *Given a complete graph K_n with $V(K_n) = \{1, \ldots, n\}$, a* **1-tree** *is a graph consisting of a spanning tree on the vertices $\{2, \ldots, n\}$ and two edges incident to vertex 1.*

The tours are exactly the 1-trees T with $|\delta_T(i)| = 2$ for $i = 1, \ldots, n$. We know spanning trees well, and 1-trees are not much different. For example we have:

Proposition 21.32. *The convex hull of the incidence vectors of all 1-trees is the set of vectors $x \in [0, 1]^{E(K_n)}$ with*

$$\sum_{e \in E(K_n)} x_e = n, \quad \sum_{e \in \delta(1)} x_e = 2, \quad \sum_{e \in E(K_n[X])} x_e \leq |X| - 1 \quad (\emptyset \neq X \subseteq \{2, \ldots, n\}).$$

Proof: This follows directly from Theorem 6.13. □

Observe that any linear objective function can easily be optimized over the set of 1-trees: just find a minimum weight spanning tree on $\{2, \ldots, n\}$ (cf. Section 6.1) and add the two cheapest edges incident to the vertex 1. Now Lagrangean relaxation yields the following lower bound:

Proposition 21.33. (Held and Karp [1970]) *Given an instance (K_n, c) of the TSP with $V(K_n) = \{1, \ldots, n\}$, and $\lambda = (\lambda_2, \ldots, \lambda_n) \in \mathbb{R}^{n-1}$. Then*

$$LR(K_n, c, \lambda) \; := \; \min \left\{ c(E(T)) + \sum_{i=2}^{n} (|\delta_T(i)| - 2)\, \lambda_i : T \text{ is a 1-tree} \right\}$$

is a lower bound for the length of an optimum tour, which can be computed in the time needed to solve a MINIMUM SPANNING TREE PROBLEM *on $n - 1$ vertices.*

Proof: An optimum tour T is a 1-tree with $|\delta_T(i)| = 2$ for all i, proving that $LR(K_n, c, \lambda)$ is a lower bound. Given $\lambda = (\lambda_2, \ldots, \lambda_n)$, we choose λ_1 arbitrarily and replace the weights c by $c'(\{i, j\}) := c(\{i, j\}) + \lambda_i + \lambda_j$ $(1 \leq i < j \leq n)$. Then all we have to do is to find a minimum weight 1-tree with respect to c'. $\square$

Note that the Lagrange multipliers λ_i $(i = 2, \ldots, n)$ are not restricted to the nonnegative numbers because the additional constraints $|\delta_T(i)| = 2$ are equalities. The λ_i can be determined by some subgradient optimization procedure; cf. Section 5.6. The maximum possible value

$$HK(K_n, c) \; := \; \max\{LR(K_n, c, \lambda) : \lambda \in \mathbb{R}^{n-1}\}$$

(the Lagrangean dual) is called the **Held-Karp bound** for (K_n, c). We have:

Theorem 21.34. (Held and Karp [1970]) *For any instance (K_n, c) of the TSP with $V(K_n) = \{1, \ldots, n\}$,*

$$HK(K_n, c) \; = \; \min \Big\{ cx \; : \; 0 \leq x_e \leq 1 \qquad (e \in E(K_n)),$$

$$\sum_{e \in \delta(v)} x_e = 2 \qquad (v \in V(K_n)),$$

$$\sum_{e \in E(K_n[I])} x_e \leq |I| - 1 \qquad (\emptyset \neq I \subseteq \{2, \ldots, n\}) \Big\}.$$

Proof: This follows directly from Theorem 5.37 and Proposition 21.32. $\square$

In other words, the Held-Karp bound equals the optimum LP value over the subtour polytope (cf. Proposition 21.25). This helps us to estimate the quality of the Held-Karp bound for the METRIC TSP. We also use the idea of CHRISTOFIDES' ALGORITHM again:

Theorem 21.35. (Wolsey [1980]) *For any* METRIC TSP *instance, the Held-Karp bound is at least $\frac{2}{3}$ of the length of an optimum tour.*

Proof: Let (K_n, c) be a METRIC TSP instance, and let T be a minimum weight 1-tree in (K_n, c). We have

$$c(E(T)) \;=\; LR(K_n, c, 0) \;\le\; HK(K_n, c).$$

Let $W \subseteq V(K_n)$ consist of the vertices having odd degree in T. Since each vector x in the subtour polytope of (K_n, c) satisfies $\sum_{e \in \delta(X)} x_e \ge 2$ for all $\emptyset \ne X \subset V(K_n)$, the polyhedron

$$\left\{ x : x_e \ge 0 \text{ for all } e \in E(K_n), \sum_{e \in \delta(X)} x_e \ge 2 \text{ for all } X \text{ with } |X \cap W| \text{ odd} \right\}$$

contains the subtour polytope. Therefore, by Theorem 21.34,

$$\min \left\{ cx : x_e \ge 0 \text{ for all } e \in E(K_n), \sum_{e \in \delta(X)} x_e \ge 1 \text{ for all } X \text{ with } |X \cap W| \text{ odd} \right\}$$

$$\le \frac{1}{2} HK(K_n, c).$$

But now observe that by Theorem 12.18, the left-hand side is the minimum weight of a W-join J in (K_n, c). So $c(E(T)) + c(J) \le \frac{3}{2} HK(K_n, c)$. Since the graph $G := (V(K_n), E(T) \cup J)$ is connected and Eulerian, this is an upper bound on the length of an optimum tour (by Lemma 21.3).   □

A different proof is due to Shmoys and Williamson [1990]. It is not known whether this bound is tight. The instance of Figure 21.9 on page 612 (edges not shown have weight 3) is an example where the Held-Karp bound (9) is strictly less than the length of an optimum tour (which is 10). There are instances of the METRIC TSP where $\frac{HK(K_n, c)}{OPT(K_n, c)}$ is arbitrarily close to $\frac{3}{4}$ (Exercise 19). In other words, the integrality ratio of the subtour polytope for the METRIC TSP is between $\frac{4}{3}$ and $\frac{3}{2}$.

In practice, the Held-Karp bound is usually much better; see e.g. Johnson, McGeoch and Rothberg [1996] or Applegate et al. [2006].

21.6 Branch-and-Bound

Branch-and-bound is a technique for simulating a complete enumeration of all possible solutions without having to consider them one by one. For many NP-hard combinatorial optimization problems it is the best known framework for obtaining an optimum solution. It has been proposed by Land and Doig [1960] and applied to the TSP first by Little et al. [1963].

To apply the BRANCH-AND-BOUND method to a combinatorial optimization (say minimization) problem, we need two steps:

- "branch": a given subset of the possible solutions (tours for the TSP) can be partitioned into at least two nonempty subsets.
- "bound": for a subset obtained by branching iteratively, a lower bound on the cost of any solution within this subset can be computed.

The general procedure then is as follows:

BRANCH-AND-BOUND

Input: An instance of a minimization problem.

Output: An optimum solution S^*.

① Set the initial tree $T := (\{S\}, \emptyset)$, where S is the set of all feasible solutions. Mark S active.
  Set the upper bound $U := \infty$ (or apply a heuristic in order to get a better upper bound).

② Choose an active vertex X of the tree T (if there is none, **stop**).
  Mark X non-active.
  ("branch") Find a partition $X = X_1 \dot{\cup} \ldots \dot{\cup} X_t$.

③ **For each** $i = 1, \ldots, t$ **do:**
    ("bound") Find a lower bound L on the cost of any solution in X_i.
    If $|X_i| = 1$ (say $X_i = \{S\}$) and $\mathrm{cost}(S) < U$ **then:**
      Set $U := \mathrm{cost}(S)$ and $S^* := S$.
    If $|X_i| > 1$ and $L < U$ **then:**
      Set $T := (V(T) \cup \{X_i\}, E(T) \cup \{\{X, X_i\}\})$ and mark X_i active.

④ **Go to** ②.

It should be clear that the above method always finds an optimum solution. The implementation (and the efficiency) of course depends very much on the actual problem. We shall discuss a possible implementation for the TSP.

The easiest way to perform the branching is to choose an edge e and write $X = X_e \cup (X \setminus X_e)$, where X_e consists of those solutions in X that contain edge e. Then we can write any vertex X of the tree as

$$S_{A,B} = \{S \in \mathcal{S} : A \subseteq S, B \cap S = \emptyset\} \qquad \text{for some } A, B \subseteq E(G).$$

For these $X = S_{A,B}$, the TSP with the additional constraint that all edges of A, but none of B, belong to the tour, can be written as an unconstrained TSP by modifying the weights c accordingly: namely we set

$$c'_e := \begin{cases} c_e & \text{if } e \in A \\ c_e + C & \text{if } e \notin A \cup B \\ c_e + 2C & \text{if } e \in B \end{cases}$$

with $C := \sum_{e \in E(G)} c_e + 1$. Then the tours in $\mathcal{S}_{A,B}$ are exactly the tours whose modified weight is less than $(n + 1 - |A|)C$. Furthermore, the original and the modified weight of any tour in $\mathcal{S}_{A,B}$ differ by exactly $(n - |A|)C$.

Then the Held-Karp bound (cf. Section 21.5) can be used to implement the "bound"-step.

The above BRANCH-AND-BOUND method for the TSP has been used to solve fairly large instances of the TSP (up to about 100 cities).

BRANCH-AND-BOUND is also often used to solve integer programs, especially when the variables are binary (restricted to be 0 or 1). Here the most natural branching step consists of taking one variable and trying both possible values for it. A lower bound can easily be determined by solving the corresponding LP relaxation.

In the worst case, BRANCH-AND-BOUND is no better than the complete enumeration of all possible solutions. In practice, the efficiency depends not only on how the "branch" and "bound" steps are implemented. It is also important to have a good strategy for choosing the active vertex X in ② of the algorithm. Furthermore, a good heuristic at the beginning (and thus a good upper bound to start with) can help to keep the branch-and-bound tree T small.

BRANCH-AND-BOUND is often combined with a cutting plane method (see Section 5.5), based on the results of Section 21.4. One proceeds as follows. Since we have an exponential number of constraints (which do not even describe $Q(n)$ completely), we start by solving the LP

$$\min \left\{ cx : 0 \le x_e \le 1 \ (e \in E(K_n)), \ \sum_{e \in \delta(v)} x_e = 2 \ (v \in V(K_n)) \right\},$$

i.e. with constraints (21.1) and (21.2). This polyhedron contains the perfect simple 2-matchings as integral vectors. Suppose we have a solution x^* of the above LP. There are three cases:

(a) x^* is the incidence vector of a tour;
(b) We find some violated subtour inequality (21.3), 2-matching inequality (21.6), comb inequality (21.7), or clique tree inequality (21.8).
(c) No violated inequality can be found (in particular no subtour inequality is violated), but x^* is not integral.

If x^* is integral but not the incidence vector of a tour, some subtour inequality must be violated by Proposition 21.24.

In case (a) we are done. In case (b) we simply add the violated inequality (or possibly several violated inequalities) to our LP and solve the new LP. In case (c), all we have is a (usually very good) lower bound for the length of a tour. Using this bound (and the fractional solution), we may start a BRANCH-AND-BOUND procedure. Because of the tight lower bound we hopefully can fix many variables in advance and thus considerably reduce the branching steps necessary to obtain an optimum solution. Moreover, at each node of the branch-and-bound tree, we can again look for violated inequalities.

This method – called **branch-and-cut** – has been used to solve TSP instances with more than 10000 cities up to optimality. Of course, many sophisticated ideas not described here are necessary to obtain an efficient implementation. In particular, good heuristics to detect violated inequalities are essential. See Applegate et al. [2003, 2006] and Jünger and Naddef [2001] for more information and further references.

These successes in solving large instances optimally are in contrast to poor worst-case running times. Woeginger [2002] gives a survey on subexponential exact algorithms for *NP*-hard problems; see also Exercise 1.

Exercises

1. Describe an exact algorithm for the TSP by means of dynamic programming. If the vertices (cities) are numbered $1, \ldots, n$, we denote by $\gamma(A, x)$ the minimum cost of a 1-x-path P with $V(P) = A \cup \{1\}$, for all $A \subseteq \{2, 3, \ldots, n\}$ and $x \in A$. The idea is now to compute all these numbers $\gamma(A, x)$. Compare the running time of this algorithm with the naive enumeration of all tours.
   (Bellman [1962], Held and Karp [1962])
   Note: This is the exact TSP algorithm with the best known worst-case running time. For the EUCLIDEAN TSP, Hwang, Chang and Lee [1993] described an exact algorithm using planar separators with a subexponential running time $O(c^{\sqrt{n}\log n})$.

2. Suppose the n cities of a TSP instance are partitioned into m clusters such that the distance between two cities is zero if and only if they belong to the same cluster.
   (a) Prove that there exists an optimum tour with at most $m(m - 1)$ edges of positive weight.
   (b) Prove that such a TSP can be solved in polynomial time if m is fixed.
   (Triesch, Nolles and Vygen [1994])

3. Consider the following problem. A truck starting at some depot d_1 must visit certain customers $c_1, \ldots, c_n$ and finally return to d_1. Between visiting two customers it must visit one of the depots $d_1, \ldots, d_k$. Given nonnegative symmetric distances between the customers and depots, we look for the shortest possible tour.
   (a) Show that this problem is *NP*-complete.
   (b) Show that it can be solved in polynomial time if k is fixed. (*Hint:* Use Exercise 2.)
   (Triesch, Nolles and Vygen [1994])

4. Consider the ASYMMETRIC TSP with triangle inequality: given a number $n \in \mathbb{N}$ and distances $c((i, j)) \in \mathbb{R}_+$ for $i, j \in \{1, \ldots, n\}, i \neq j$, satisfying the triangle inequality $c((i, j)) + c((j, k)) \geq c((i, k))$ for all distinct $i, j, k \in \{1, \ldots, n\}$, find a permutation $\pi : \{1, \ldots, n\} \to \{1, \ldots, n\}$ such that $\sum_{i=1}^{n-1} c((\pi(i), \pi(i + 1))) + c((\pi(n), \pi(1)))$ is minimum.

Describe an algorithm that always finds a solution whose cost is at most $\log n$ times the optimum.

Hint: First find a digraph H with $V(H) = \{1, \ldots, n\}$, $|\delta_H^-(v)| = |\delta_H^+(v)| = 1$ for all $v \in V(H)$ and minimum cost $c(E(H))$. Contract the circuits of H and iterate.

(Frieze, Galbiati and Maffioli [1982])

Note: Svensson, Tarnawski and Végh [2017] devised a constant-factor approximation algorithm.

5. Describe a polynomial-time algorithm which optimally solves any TSP instance that is the metric closure of a weighted tree.

* 6. Find instances of the EUCLIDEAN TSP for which the DOUBLE-TREE ALGORITHM returns a tour whose length is arbitrarily close to twice the optimum.

7. Let G be a complete bipartite graph with bipartition $V(G) = A \,\dot\cup\, B$, where $|A| = |B|$. Let $c : E(G) \to \mathbb{R}_+$ be a cost function with $c(\{a, b\}) + c(\{b, a'\}) + c(\{a', b'\}) \geq c(\{a, b'\})$ for all $a, a' \in A$ and $b, b' \in B$. Now the task is to find a Hamiltonian circuit in G of minimum cost. This problem is called the METRIC BIPARTITE TSP.

   (a) Prove that, for any k, if there is a k-factor approximation algorithm for the METRIC BIPARTITE TSP then there is also a k-factor approximation algorithm for the METRIC TSP.

   (b) Find a 2-factor approximation algorithm for the METRIC BIPARTITE TSP. (*Hint:* Combine Exercise 27 of Chapter 13 with the idea of the DOUBLE-TREE ALGORITHM.)

   (Frank et al. [1998], Chalasani, Motwani and Rao [1996])

* 8. Find instances of the METRIC TSP for which CHRISTOFIDES' ALGORITHM returns a tour whose length is arbitrarily close to $\frac{3}{2}$ times the optimum.

9. Consider the (metric) s-t-path TSP: Given an instance of METRIC TSP and two vertices s and t, we look for a Hamiltonian s-t-path of minimum weight.

   (a) Describe a 2-factor approximation algorithm, generalizing the DOUBLE-TREE ALGORITHM.

   * (b) Describe a $\frac{5}{3}$-factor approximation algorithm, generalizing CHRISTOFIDES' ALGORITHM.

   (Hoogeveen [1991])

   Note: Traub and Vygen [2018] found a $(\frac{3}{2} + \epsilon)$-factor approximation algorithm for any $\epsilon > 0$.

10. Show that the results of Section 21.2 extend to the EUCLIDEAN STEINER TREE PROBLEM. Describe an approximation scheme for this problem.

11. Prove that in the LIN-KERNIGHAN ALGORITHM a set X_i never contains more than one element for any odd i with $i > p_2 + 3$.

12. Consider the following decision problem:

ANOTHER HAMILTONIAN CIRCUIT

Instance: A graph G and a Hamiltonian circuit in G.

Question: Is there another Hamiltonian circuit in G?

(a) Show that this problem is *NP*-complete. (*Hint:* Recall the proof of Lemma 21.18.)

* (b) Prove that for a 3-regular graph G and $e \in E(G)$, the number of Hamiltonian circuits containing e is even.

(c) Show that ANOTHER HAMILTONIAN CIRCUIT for 3-regular graphs is in P. (Nevertheless no polynomial-time algorithm is known for finding another Hamiltonian circuit, given a 3-regular graph G and a Hamiltonian circuit in G.)

13. Let $(X, (S_x)_{x \in X}, c, \text{goal})$ be a discrete optimization problem with neighbourhood $N_x(y) \subseteq S_x$ for $y \in S_x$ and $x \in X$. Suppose that we can do the following in polynomial time: for each $x \in X$ find an element of S_x, and for each $y \in S_x$ find an element $y' \in N_x(y)$ with better cost or decide that none exists. Then the problem together with this neighbourhood is said to belong to the class *PLS* (for polynomial local search). Prove that if a problem in *PLS* exists for which it is *NP*-hard to compute a local optimum for a given instance, then $NP = coNP$.

Hint: Design a nondeterministic algorithm for any *coNP*-complete problem. (Johnson, Papadimitriou and Yannakakis [1988])

Note: The TSP is *PLS*-complete with the k-opt and also with the Lin-Kernighan neighbourhood (Krentel [1989], Papadimitriou [1992]), i.e. if one can find a local optimum in polynomial time, one can do so for every problem and neighbourhood in *PLS* (and this would imply another proof of Theorem 4.18 due to the correctness of the SIMPLEX ALGORITHM).

14. Show that $Q(n) = \{x \in \{0, 1\}^{E(K_n)} : x_e = y_{ij} + y_{ji} \ (e = \{i, j\} \in E(K_n)),$ $\sum_{j \neq i} y_{ij} = \sum_{j \neq i} y_{ji} = 1 \ (i = 1, \ldots, n), \ u_j - u_i \geq 1 - n(1 - y_{ij}) \ (1 \leq i \leq n, 2 \leq j \leq n, i \neq j), \ y_{ij} \in \{0, 1\} \ (1 \leq i, j \leq n, i \neq j)\}$.

(Miller, Tucker and Zemlin [1960])

Note: Kaibel and Weltge [2015] showed that every description of $Q(n)$ has additional variables (as the above and in Exercise 15) or exponentially many constraints (as in Proposition 21.24).

15. Describe a polytope P with a polynomial number of variables and constraints (a so-called extended formulation) such that the projection of P to some of the variables is the subtour polytope.

16. Show that one can optimize any linear function over the polytope defined by (21.1), (21.2), (21.3), (21.6).

Hint: Use Theorem 21.23 to reduce the dimension in order to obtain a full-dimensional polytope. Find a point in the interior and apply Theorem 4.21.

17. Consider the 2-matching inequalities (21.6) in Proposition 21.27. Show that it is irrelevant whether one requires additionally that F is a matching: if a vector in the subtour polytope satisfies the inequalities (21.6) when F is a matching, it satisfies all inequalities (21.6).

18. Show that the subtour inequalities (21.3), the 2-matching inequalities (21.6) and the comb inequalities (21.7) are special cases of the clique tree inequalities (21.8).

19. Prove that there are instances (K_n, c) of the METRIC TSP where $\frac{HK(K_n,c)}{OPT(K_n,c)}$ is arbitrarily close to $\frac{3}{4}$.

    Hint: Replace the edges of weight 1 in Figure 21.9 by long paths and consider the metric closure.

    Note: Hougardy [2014] showed that this even holds for the EUCLIDEAN TSP.

20. Consider the TSP on n cities. For any weight function $w : E(K_n) \to \mathbb{R}_+$ let c_w^* be the length of an optimum tour with respect to w. Prove: if $L_1 \leq c_{w_1}^*$ and $L_2 \leq c_{w_2}^*$ for two weight functions w_1 and w_2, then also $L_1 + L_2 \leq c_{w_1+w_2}^*$, where the sum of the two weight functions is taken componentwise.

21. Let T be an optimum tour for an instance (K_n, c) of the METRIC TSP, and let T' be a shortest tour different from T. Show that

$$\frac{c(E(T')) - c(E(T))}{c(E(T))} \leq \frac{2}{n}.$$

    (Papadimitriou and Steiglitz [1978])

22. Let $x \in [0, 1]^{E(K_n)}$ with $\sum_{e \in \delta(v)} x_e = 2$ for all $v \in V(K_n)$. Prove that if there exists a violated subtour constraint, i.e. a set $S \subset V(K_n)$ with $\sum_{e \in \delta(S)} x_e < 2$, then there exists one with $x_e < 1$ for all $e \in \delta(S)$.

    (Crowder and Padberg [1980])

23. For a family $\mathcal{F}$ of (not necessarily distinct) subsets of $\{1, \ldots, n\}$ and a vector $x \in \mathbb{R}^{E(K_n)}$ we write $\mathcal{F}(x) := \sum_{X \in \mathcal{F}} \sum_{e \in \delta(X)} x_e$ and $\mu_{\mathcal{F}}$ for the minimum of $\mathcal{F}(x)$ taken over all incidence vectors of tours in K_n. An inequality of the form $\mathcal{F}(x) \geq \mu_{\mathcal{F}}$ is called a *hypergraph inequality*. (21.5) and (21.7) are examples. Show that the TSP polytope can be described by degree constraints and hypergraph inequalities, i.e. there are families $\mathcal{F}_1, \ldots, \mathcal{F}_k$ such that $Q(n) =$

$$\left\{ x \in \mathbb{R}^{E(K_n)} : \sum_{e \in \delta(v)} x_e = 2 \ (v \in V(K_n)), \ \mathcal{F}_i(x) \leq \mu_{\mathcal{F}_i} \ (i = 1, \ldots, k) \right\}.$$

    Hint: Rewrite any facet-defining inequality using the fact that $\sum_{e \in \delta(\{v,w\})} x_e = 4 - 2x_{\{v,w\}}$ for each x satisfying the degree constraints.

    (Applegate et al. [2006])

References

General Literature

Applegate, D.L., Bixby, R., Chvátal, V., and Cook, W.J. [2006]: The Traveling Salesman Problem: A Computational Study. Princeton University Press 2006

Cook, W.J., Cunningham, W.H., Pulleyblank, W.R., and Schrijver, A. [1998]: Combinatorial Optimization. Wiley, New York 1998, Chapter 7

Gutin, G., and Punnen, A.P. [2002]: The Traveling Salesman Problem and Its Variations. Kluwer, Dordrecht 2002

Jünger, M., Reinelt, G., and Rinaldi, G. [1995]: The traveling salesman problem. In: Handbooks in Operations Research and Management Science; Volume 7; Network Models (M.O. Ball, T.L. Magnanti, C.L. Monma, G.L. Nemhauser, eds.), Elsevier, Amsterdam 1995

Jungnickel, D. [2013]: Graphs, Networks and Algorithms. Fourth Edition. Springer, Berlin 2013, Chapter 15

Lawler, E.L., Lenstra J.K., Rinnooy Kan, A.H.G., and Shmoys, D.B. [1985]: The Traveling Salesman Problem. Wiley, Chichester 1985

Papadimitriou, C.H., and Steiglitz, K. [1982]: Combinatorial Optimization; Algorithms and Complexity. Prentice-Hall, Englewood Cliffs 1982, Section 17.2, Chapters 18 and 19

Reinelt, G. [1994]: The Traveling Salesman; Computational Solutions for TSP Applications. Springer, Berlin 1994

Vygen, J. [2012]: New approximation algorithms for the TSP. OPTIMA 90 (2012), 1–12

Cited References

Aarts, E., and Lenstra, J.K. [1997]: Local Search in Combinatorial Optimization. Wiley, Chichester 1997

Applegate, D., Bixby, R., Chvátal, V., and Cook, W. [2003]: Implementing the Dantzig-Fulkerson-Johnson algorithm for large traveling salesman problems. Mathematical Programming B 97 (2003), 91–153

Applegate, D., Cook, W., and Rohe, A. [2003]: Chained Lin-Kernighan for large traveling salesman problems. INFORMS Journal on Computing 15 (2003), 82–92

Arora, S. [1998]: Polynomial time approximation schemes for Euclidean traveling salesman and other geometric problems. Journal of the ACM 45 (1998), 753–782

Bellman, R. [1962]: Dynamic programming treatment of the travelling salesman problem. Journal of the ACM 9 (1962), 61–63

de Berg, M., Buchin, K., Jansen, B.M.P., and Woeginger, G. [2016]: Fine-grained complexity analysis of two classic TSP variants. Proceedings of ICALP 2016, Article 5

Berman, P., and Karpinski, M. [2006]: 8/7-approximation algorithm for (1,2)-TSP. Proceedings of the 17th Annual ACM-SIAM Symposium on Discrete Algorithms (2006), 641–648

Boyd, S.C., and Cunningham, W.H. [1991]: Small traveling salesman polytopes. Mathematics of Operations Research 16 (1991), 259–271

Burkard, R.E., Deĭneko, V.G., and Woeginger, G.J. [1998]: The travelling salesman and the PQ-tree. Mathematics of Operations Research 23 (1998), 613–623

Carr, R. [1997]: Separating clique trees and bipartition inequalities having a fixed number of handles and teeth in polynomial time. Mathematics of Operations Research 22 (1997), 257–265

Chalasani, P., Motwani, R., and Rao, A. [1996]: Algorithms for robot grasp and delivery. Proceedings of the 2nd International Workshop on Algorithmic Foundations of Robotics (1996), 347–362

Chandra, B., Karloff, H., and Tovey, C. [1999]: New results on the old k-opt algorithm for the traveling salesman problem. SIAM Journal on Computing 28 (1999), 1998–2029

Christofides, N. [1976]: Worst-case analysis of a new heuristic for the traveling salesman problem. Technical Report 388, Graduate School of Industrial Administration, Carnegie-Mellon University, Pittsburgh 1976

Chvátal, V. [1973]: Edmonds' polytopes and weakly hamiltonian graphs. Mathematical Programming 5 (1973), 29–40

Crowder, H., and Padberg, M.W. [1980]: Solving large-scale symmetric travelling salesman problems to optimality. Management Science 26 (1980), 495–509

Dantzig, G., Fulkerson, R., and Johnson, S. [1954]: Solution of a large-scale traveling-salesman problem. Operations Research 2 (1954), 393–410

Englert, M., Röglin, H., and Vöcking, B. [2014]: Worst case and probabilistic analysis of the 2-opt algorithm for the TSP. Algorithmica 68 (2014), 190–264

Fiorini, S., Massar, S., Pokutta, S., Tiwary, H.R., and de Wolf, R. [2015]: Exponential lower bounds for polytopes in combinatorial optimization. Journal of the ACM 62 (2015), Article 17

Frank, A., Triesch, E., Korte, B., and Vygen, J. [1998]: On the bipartite travelling salesman problem. Report No. 98866, Research Institute for Discrete Mathematics, University of Bonn, 1998

Frieze, A.M., Galbiati, G., and Maffioli, F. [1982]: On the worst-case performance of some algorithms for the asymmetric traveling salesman problem. Networks 12 (1982), 23–39

Garey, M.R., Graham, R.L., and Johnson, D.S. [1976]: Some NP-complete geometric problems. Proceedings of the 8th Annual ACM Symposium on the Theory of Computing (1976), 10–22

Grötschel, M., and Padberg, M.W. [1979]: On the symmetric travelling salesman problem. Mathematical Programming 16 (1979), 265–302

Grötschel, M., and Pulleyblank, W.R. [1986]: Clique tree inequalities and the symmetric travelling salesman problem. Mathematics of Operations Research 11 (1986), 537–569

Held, M., and Karp, R.M. [1962]: A dynamic programming approach to sequencing problems. Journal of SIAM 10 (1962), 196–210

Held M., and Karp, R.M. [1970]: The traveling-salesman problem and minimum spanning trees. Operations Research 18 (1970), 1138–1162

Held, M., and Karp, R.M. [1971]: The traveling-salesman problem and minimum spanning trees; part II. Mathematical Programming 1 (1971), 6–25

Helsgaun, K. [2009]: General k-opt submoves for the Lin-Kernighan TSP heuristic. Mathematical Programming Computation 1 (2009), 119–163

Hoogeveen, J.A. [1991]: Analysis of Christofides' heuristic: some paths are more difficult than cycles. Operations Research Letters 10 (1991), 291–295

Hougardy, S. [2014]: On the integrality ratio of the subtour LP for Euclidean TSP. Operations Research Letters 42 (2014), 495–499

Hurkens, C.A.J., and Woeginger, G.J. [2004]: On the nearest neighbour rule for the traveling salesman problem. Operations Research Letters 32 (2004), 1–4

Hwang, R.Z., Chang, R.C., and Lee, R.C.T. [1993]: The searching over separators strategy to solve some *NP*-hard problems in subexponential time. Algorithmica 9 (1993), 398–423

Johnson, D.S., McGeoch, L.A., and Rothberg, E.E. [1996]: Asymptotic experimental analysis for the Held-Karp traveling salesman bound. Proceedings of the 7th Annual ACM-SIAM Symposium on Discrete Algorithms (1996), 341–350

Johnson, D.S., Papadimitriou, C.H., and Yannakakis, M. [1988]: How easy is local search? Journal of Computer and System Sciences 37 (1988), 79–100

Jünger, M., and Naddef, D. [2001]: Computational Combinatorial Optimization. Springer, Berlin 2001

Kaibel, V., and Weltge, S. [2015]: Lower bounds on the sizes of integer programs without additional variables. Mathematical Programming B 154 (2015), 407–425

Karp, R.M. [1977]: Probabilistic analysis of partitioning algorithms for the TSP in the plane. Mathematics of Operations Research 2 (1977), 209–224

Karp, R.M., and Papadimitriou, C.H. [1982]: On linear characterizations of combinatorial optimization problems. SIAM Journal on Computing 11 (1982), 620–632

Karpinski, M., Lampis, M., and Schmied, R. [2013]: New inapproximability bounds for TSP. Algorithms and Computation; Proceedings of ISAAC 2013; LNCS 8283 (L. Cai, S.-W. Chen, T.-W. Lam, eds.), Springer, Berlin 2013, pp. 568–578

Klein, P.N. [2008]: A linear-time approximation scheme for TSP in undirected planar graphs with edge-weights. SIAM Journal on Computing 37 (2008), 1926–1952

Krentel, M.W. [1989]: Structure in locally optimal solutions. Proceedings of the 30th Annual IEEE Symposium on Foundations of Computer Science (1989), 216–221

Land, A.H., and Doig, A.G. [1960]: An automatic method of solving discrete programming problems. Econometrica 28 (1960), 497–520

Lee, J.R., Raghavendra, P., and Steurer, D. [2015]: Lower bounds on the size of semidefinite programming relaxations. Proceedings of the 47th Annual ACM Symposium on Theory of Computing (2015), 567–576

Lin, S., and Kernighan, B.W. [1973]: An effective heuristic algorithm for the traveling-salesman problem. Operations Research 21 (1973), 498–516

Little, J.D.C., Murty, K.G., Sweeny, D.W., and Karel, C. [1963]: An algorithm for the traveling salesman problem. Operations Research 11 (1963), 972–989

Michiels, W., Aarts, E., and Korst, J. [2007]: Theoretical Aspects of Local Search. Springer, Berlin 2007

Miller, C.E., Tucker, A.W., and Zemlin, R.A. [1960]: Integer programming formulations of traveling salesman problems. Journal of the ACM 7 (1960), 326–329

Mitchell, J. [1999]: Guillotine subdivisions approximate polygonal subdivisions: a simple polynomial-time approximation scheme for geometric TSP, *k*-MST, and related problems. SIAM Journal on Computing 28 (1999), 1298–1309

Papadimitriou, C.H. [1977]: The Euclidean traveling salesman problem is *NP*-complete. Theoretical Computer Science 4 (1977), 237–244

Papadimitriou, C.H. [1978]: The adjacency relation on the travelling salesman polytope is *NP*-complete. Mathematical Programming 14 (1978), 312–324

Papadimitriou, C.H. [1992]: The complexity of the Lin-Kernighan heuristic for the traveling salesman problem. SIAM Journal on Computing 21 (1992), 450–465

Papadimitriou, C.H., and Steiglitz, K. [1977]: On the complexity of local search for the traveling salesman problem. SIAM Journal on Computing 6 (1), 1977, 76–83

Papadimitriou, C.H., and Steiglitz, K. [1978]: Some examples of difficult traveling salesman problems. Operations Research 26 (1978), 434–443

Papadimitriou, C.H., and Yannakakis, M. [1993]: The traveling salesman problem with distances one and two. Mathematics of Operations Research 18 (1993), 1–12

Rao, S.B., and Smith, W.D. [1998]: Approximating geometric graphs via "spanners" and "banyans". Proceedings of the 30th Annual ACM Symposium on Theory of Computing (1998), 540–550

Rosenkrantz, D.J. Stearns, R.E., and Lewis, P.M. [1977]: An analysis of several heuristics for the traveling salesman problem. SIAM Journal on Computing 6 (1977), 563–581

Rothvoß, T. [2017]: The matching polytope has exponential extension complexity. Journal of the ACM 64 (2017), Article 41

Sahni, S., and Gonzalez, T. [1976]: *P*-complete approximation problems. Journal of the ACM 23 (1976), 555–565

Shmoys, D.B., and Williamson, D.P. [1990]: Analyzing the Held-Karp TSP bound: a monotonicity property with application. Information Processing Letters 35 (1990), 281–285

Svensson, O., Tarnawski, J., and Végh, L. [2017]: A constant-factor approximation algorithm for the asymmetric traveling salesman problem. arXiv:1708.04215

Traub, V., and Vygen, J. [2017]: Approaching $\frac{3}{2}$ for the s-t-path TSP. Proceedings of the 29th Annual ACM-SIAM Symposium on Discrete Algorithms (2018), 1854–1864

Triesch, E., Nolles, W., and Vygen, J. [1994]: Die Einsatzplanung von Zementmischern und ein Traveling Salesman Problem In: Operations Research; Reflexionen aus Theorie und Praxis (B. Werners, R. Gabriel, eds.), Springer, Berlin 1994 [in German]

Woeginger, G.J. [2002]: Exact algorithms for *NP*-hard problems. OPTIMA 68 (2002), 2–8

Wolsey, L.A. [1980]: Heuristic analysis, linear programming and branch and bound. Mathematical Programming Study 13 (1980), 121–134

22 Facility Location

Many economic decisions involve selecting and/or placing certain facilities to serve given demands efficiently. Examples include manufacturing plants, storage facilities, depots, warehouses, libraries, fire stations, hospitals, base stations for wireless services (like TV broadcasting or mobile phone service), etc. The problems have in common that a set of facilities, each with a certain position, has to be chosen, and the objective is to meet the demand (of customers, users etc.) best. Facility location problems, which occur also in less obvious contexts, indeed have numerous applications.

The most widely studied model in discrete facility location is the so-called UNCAPACITATED FACILITY LOCATION PROBLEM, also known as plant location problem or warehouse location problem. It is introduced in Section 22.1. Although it has been intensively studied since the 1960s (see, e.g., Stollsteimer [1963], Balinski and Wolfe [1963], Kuehn and Hamburger [1963], Manne [1964]), no approximation algorithm was known for this problem until 1997. Since then several quite different approaches have been used to prove an upper bound on the approximation ratio. We will present them in this chapter, and also consider extensions to more general problems, such as capacitated variants, the k-MEDIAN PROBLEM, and the UNIVERSAL FACILITY LOCATION PROBLEM.

22.1 The Uncapacitated Facility Location Problem

The most basic problem, for which we shall present many results, is the UNCAPACITATED FACILITY LOCATION PROBLEM. It is defined as follows.

© Springer-Verlag GmbH Germany 2018
Bernhard Korte and Jens Vygen, *Combinatorial Optimization*, Algorithms and Combinatorics 21, https://doi.org/10.1007/978-3-662-56039-6_22

UNCAPACITATED FACILITY LOCATION PROBLEM

Instance: A finite set $\mathcal{D}$ of customers (clients), a finite set $\mathcal{F}$ of (potential) facilities, a fixed cost $f_i \in \mathbb{R}_+$ for opening each facility $i \in \mathcal{F}$, and a service cost $c_{ij} \in \mathbb{R}_+$ for each $i \in \mathcal{F}$ and $j \in \mathcal{D}$.

Task: Find a subset X of facilities (called **open**) and an assignment $\sigma :$ $\mathcal{D} \to X$ of the customers to open facilities, such that the sum of facility costs and service costs

$$\sum_{i \in X} f_i + \sum_{j \in \mathcal{D}} c_{\sigma(j)j}$$

is minimum.

In many practical applications, service costs come from a metric c on $\mathcal{D} \cup \mathcal{F}$ (e.g., when they are proportional to geometric distances or travel times). In this case we have

$$c_{ij} + c_{i'j} + c_{i'j'} \geq c_{ij'} \quad \text{for all } i, i' \in \mathcal{F} \text{ and } j, j' \in \mathcal{D}. \tag{22.1}$$

Conversely, if this condition holds, we can define $c_{ii} := 0$ and $c_{ii'} := \min_{j \in \mathcal{D}}(c_{ij} + c_{i'j})$ for $i, i' \in \mathcal{F}$, $c_{jj} := 0$ and $c_{jj'} := \min_{i \in \mathcal{F}}(c_{ij} + c_{ij'})$ for $j, j' \in \mathcal{D}$, and $c_{ji} := c_{ij}$ for $j \in \mathcal{D}$ and $i \in \mathcal{F}$, and obtain a (semi)metric c on $\mathcal{D} \cup \mathcal{F}$. Therefore we speak of *metric* service costs if (22.1) is satisfied. The above problem restricted to instances with metric service costs is called the METRIC UNCAPACITATED FACILITY LOCATION PROBLEM.

Proposition 22.1. *The* METRIC UNCAPACITATED FACILITY LOCATION PROBLEM *is strongly NP-hard.*

Proof: We consider the MINIMUM WEIGHT SET COVER PROBLEM with unit weights (which is strongly *NP*-hard as a consequence of Corollary 15.24). Any instance $(U, \mathcal{S})$ can be transformed to an instance of the METRIC UNCAPACITATED FACILITY LOCATION PROBLEM as follows: let $\mathcal{D} := U$, $\mathcal{F} := \mathcal{S}$, $f_i := 1$ for $i \in \mathcal{S}$, $c_{ij} := 1$ for $j \in i \in \mathcal{S}$ and $c_{ij} := 3$ for $j \in U \setminus \{i\}$, $i \in \mathcal{S}$. Then, for $k \leq |\mathcal{S}|$, the resulting instance has a solution of cost $|\mathcal{D}| + k$ if and only if $(U, \mathcal{S})$ has a set cover of cardinality k. $\square$

The number 3 in the above proof can be replaced by any number greater than 1 but not greater than 3 (otherwise (22.1) would be violated). Indeed, a similar construction shows that metric service costs are necessary to obtain approximation algorithms: if we set $c_{ij} := \infty$ for $j \in U \setminus \{i\}$ and $i \in \mathcal{S}$ in the above proof, we see that any approximation algorithm for the UNCAPACITATED FACILITY LOCATION PROBLEM would imply an approximation algorithm for set covering with the same performance ratio (and there is no constant-factor approximation for set covering unless $P = NP$; see Section 16.1). Guha and Khuller [1999] and Sviridenko [unpublished] extended the above construction to show that a 1.463-factor approximation algorithm for the METRIC UNCAPACITATED FACILITY LOCATION PROBLEM (even with service costs 1 and 3 only) would imply $P = NP$ (see Vygen [2005] for details).

Conversely, let an instance of the UNCAPACITATED FACILITY LOCATION PROBLEM be given. Setting $U := \mathcal{D}$, $\mathcal{S} = 2^{\mathcal{D}}$, and $c(D) := \min_{i \in \mathcal{F}}(f_i + \sum_{j \in D} c_{ij})$ for $D \subseteq \mathcal{D}$ yields an equivalent instance of the MINIMUM WEIGHT SET COVER PROBLEM. Although this instance has exponential size, we can run the GREEDY ALGORITHM FOR SET COVER and obtain a solution of cost at most $(1 + \frac{1}{2} + \cdots + \frac{1}{|\mathcal{D}|})$ times the optimum in polynomial time (cf. Theorem 16.3), as proposed by Hochbaum [1982]:

Namely, in each step, we have to find a pair $(D, i) \in 2^{\mathcal{D}} \times \mathcal{F}$ with minimum $\frac{f_i + \sum_{j \in D} c_{ij}}{|D|}$, open i, assign all customers in D to i and disregard them henceforth. Although there are exponentially many choices, it is easy to find a best one as it suffices to consider pairs (D_k^i, i) for $i \in \mathcal{F}$ and $k \in \{1, \ldots, |\mathcal{D}|\}$, where D_k^i is the set of the first k customers in a linear order with nondecreasing c_{ij}. Clearly, other pairs cannot be more effective.

Jain et al. [2003] showed that the performance guarantee of this greedy algorithm is $\Omega(\log n / \log \log n)$ even for metric instances, where $n = |\mathcal{D}|$. Indeed, before the paper of Shmoys, Tardos and Aardal [1997] no constant-factor approximation algorithm was known even for metric service costs. Since then, this has changed dramatically. The following sections show different techniques for obtaining constant-factor approximations for the METRIC UNCAPACITATED FACILITY LOCATION PROBLEM.

An even more restricted problem is given in the special case when facilities and customers are points in the plane and service costs are geometric distances. Here Arora, Raghavan and Rao [1998] showed that the problem has an approximation scheme, i.e. a k-factor approximation algorithm for any $k > 1$, similarly to the algorithm in Section 21.2. This result was improved by Kolliopoulos and Rao [2007], but the algorithm seems to be still too slow for practical purposes.

In the rest of this chapter we assume general metric service costs. For a given instance $\mathcal{D}$, $\mathcal{F}$, f_i, c_{ij} and a given nonempty subset X of facilities, a best assignment $\sigma : \mathcal{D} \to X$ satisfying $c_{\sigma(j)j} = \min_{i \in X} c_{ij}$ can be computed easily. Therefore we will often call a nonempty set $X \subseteq \mathcal{F}$ a feasible solution, with facility cost $c_F(X) := \sum_{i \in X} f_i$ and service cost $c_S(X) := \sum_{j \in \mathcal{D}} \min_{i \in X} c_{ij}$. The task is to find a nonempty subset $X \subseteq \mathcal{F}$ such that $c_F(X) + c_S(X)$ is minimum.

22.2 Rounding Linear Programming Solutions

The UNCAPACITATED FACILITY LOCATION PROBLEM can be formulated as an integer linear program as follows:

$$\min \quad \sum_{i \in \mathcal{F}} f_i y_i + \sum_{i \in \mathcal{F}} \sum_{j \in \mathcal{D}} c_{ij} x_{ij}$$

$$
\begin{array}{rll}
\text{s.t.} & x_{ij} \leq y_i & (i \in \mathcal{F}, j \in \mathcal{D}) \\[2mm]
& \sum_{i \in \mathcal{F}} x_{ij} = 1 & (j \in \mathcal{D}) \\[2mm]
& x_{ij} \in \{0, 1\} & (i \in \mathcal{F}, j \in \mathcal{D}) \\[1mm]
& y_i \in \{0, 1\} & (i \in \mathcal{F})
\end{array}
$$

By relaxing the integrality constraints we get the linear program:

$$\min \quad \sum_{i \in \mathcal{F}} f_i y_i + \sum_{i \in \mathcal{F}} \sum_{j \in \mathcal{D}} c_{ij} x_{ij}$$

$$
\begin{array}{rll}
\text{s.t.} & x_{ij} \leq y_i & (i \in \mathcal{F}, j \in \mathcal{D}) \\[2mm]
& \sum_{i \in \mathcal{F}} x_{ij} = 1 & (j \in \mathcal{D}) \\[2mm]
& x_{ij} \geq 0 & (i \in \mathcal{F}, j \in \mathcal{D}) \\[1mm]
& y_i \geq 0 & (i \in \mathcal{F})
\end{array}
\tag{22.2}
$$

This was first formulated by Balinski [1965]. The dual of this LP is:

$$\max \quad \sum_{j \in \mathcal{D}} v_j$$

$$
\begin{array}{rll}
\text{s.t.} & v_j - w_{ij} \leq c_{ij} & (i \in \mathcal{F}, j \in \mathcal{D}) \\[2mm]
& \sum_{j \in \mathcal{D}} w_{ij} \leq f_i & (i \in \mathcal{F}) \\[2mm]
& w_{ij} \geq 0 & (i \in \mathcal{F}, j \in \mathcal{D})
\end{array}
\tag{22.3}
$$

LP rounding algorithms solve these linear programs (cf. Theorem 4.18) and round the fractional solution of the primal LP suitably. Shmoys, Tardos and Aardal [1997] obtained the first constant-factor approximation by this technique:

SHMOYS-TARDOS-AARDAL ALGORITHM

Input: An instance $(\mathcal{D}, \mathcal{F}, (f_i)_{i \in \mathcal{F}}, (c_{ij})_{i \in \mathcal{F}, j \in \mathcal{D}})$ of the UNCAPACITATED FACILITY LOCATION PROBLEM.

Output: A solution $X \subseteq \mathcal{F}$ and $\sigma : \mathcal{D} \to X$.

---

① Compute an optimum solution (x^*, y^*) to (22.2) and an optimum solution (v^*, w^*) to (22.3).

② Let $k := 1$, $X := \emptyset$, and $U := \mathcal{D}$.

③   Let $j_k \in U$ such that $v_{j_k}^*$ is minimum.
    Let $i_k \in \mathcal{F}$ with $x_{i_k j_k}^* > 0$ and f_{i_k} minimum. Set $X := X \cup \{i_k\}$.
    Let $N_k := \{j \in U : \exists i \in \mathcal{F} : x_{ij_k}^* > 0, x_{ij}^* > 0\}$.
    Set $\sigma(j) := i_k$ for all $j \in N_k$.
    Set $U := U \setminus N_k$.

④   Set $k := k + 1$.
    If $U \neq \emptyset$ **then go to** ③.

Theorem 22.2.   (Shmoys, Tardos and Aardal [1997]) *The above is a 4-factor approximation algorithm for the* METRIC UNCAPACITATED FACILITY LOCATION PROBLEM.

Proof: By complementary slackness (Corollary 3.23), $x_{ij}^* > 0$ implies $v_j^* - w_{ij}^* = c_{ij}$, and thus $c_{ij} \leq v_j^*$. Hence the service cost for customer $j \in N_k$ is at most

$$c_{i_k j} \leq c_{ij} + c_{ij_k} + c_{i_k j_k} \leq v_j^* + 2v_{j_k}^* \leq 3v_j^*,$$

where i is a facility with $x_{ij}^* > 0$ and $x_{ij_k}^* > 0$.

The facility cost f_{i_k} can be bounded by

$$f_{i_k} \leq \sum_{i \in \mathcal{F}} x_{ij_k}^* f_i = \sum_{i \in \mathcal{F}: x_{ij_k}^* > 0} x_{ij_k}^* f_i \leq \sum_{i \in \mathcal{F}: x_{ij_k}^* > 0} y_i^* f_i.$$

As $x_{ij_k}^* > 0$ implies $x_{ij_{k'}}^* = 0$ for $k \neq k'$, the total facility cost is at most $\sum_{i \in \mathcal{F}} y_i^* f_i$.

Summing up, the total cost is $3 \sum_{j \in \mathcal{D}} v_j^* + \sum_{i \in \mathcal{F}} y_i^* f_i$, which is at most four times the LP value, and hence at most four times the optimum.   □

The performance ratio was improved to 1.736 by Chudak and Shmoys [2003] and to 1.582 by Sviridenko [2002].

22.3 Primal-Dual Algorithms

Jain and Vazirani [2001] proposed a different approximation algorithm. It is a primal-dual algorithm in the classical sense: it computes feasible primal and dual solutions (to the LPs presented in Section 22.2) simultaneously. The primal solution is integral, and the performance guarantee will follow from approximate complementary slackness conditions.

One can view the algorithm as continuously raising all dual variables (starting with zero) and freezing v_j when $j \in \mathcal{D}$ is tentatively connected. At any stage, let $w_{ij} := \max\{0, v_j - c_{ij}\}$. Initially all facilities are closed. We tentatively open facilities and connect customers when the following two events occur:

- $v_j = c_{ij}$ for some tentatively open facility i and unconnected customer j. Then set $\sigma(j) := i$ and freeze v_j.

- $\sum_{j \in \mathcal{D}} w_{ij} = f_i$ for some facility i which is not (yet) tentatively open.
  Then tentatively open i. For all unconnected customers $j \in \mathcal{D}$ with $v_j \geq c_{ij}$:
  set $\sigma(j) := i$ and freeze v_j.

Several events can occur at the same time and are then processed in arbitrary order. This continues until all customers are connected.

Now let V be the set of facilities that are tentatively open, and let E be the set of pairs $\{i, i'\}$ of distinct tentatively open facilities such that there is a customer j with $w_{ij} > 0$ and $w_{i'j} > 0$. Choose a maximal stable set X in the graph (V, E). Open the facilities in X and connect each customer to a closest open facility.

Actually, X can be chosen greedily while tentatively opening facilities. Then the algorithm can be described more formally as follows. Here Y is the set of facilities that are not (yet) tentatively open.

JAIN-VAZIRANI ALGORITHM

Input: An instance $(\mathcal{D}, \mathcal{F}, (f_i)_{i \in \mathcal{F}}, (c_{ij})_{i \in \mathcal{F}, j \in \mathcal{D}})$ of the UNCAPACITATED FACILITY LOCATION PROBLEM.

Output: A solution $X \subseteq \mathcal{F}$ and $\sigma : \mathcal{D} \to X$.

①   Set $X := \emptyset$, $Y := \mathcal{F}$ and $U := \mathcal{D}$.

②   Set $t_1 := \min\{c_{ij} : i \in \mathcal{F} \setminus Y, j \in U\}$.
     Set $t_2 := \min\{\tau : \exists i \in Y : \omega(i, \tau, U) = f_i\}$, where
$$\omega(i, \tau, U) := \sum_{j \in U} \max\{0, \tau - c_{ij}\} + \sum_{j \in \mathcal{D} \setminus U} \max\{0, v_j - c_{ij}\}.$$
     Set $t := \min\{t_1, t_2\}$.

③   **For** $i \in \mathcal{F} \setminus Y$ and $j \in U$ with $c_{ij} = t$ **do**:
     Set $v_j := t$ and $U := U \setminus \{j\}$.

④   **For** $i \in Y$ with $\omega(i, t, U) = f_i$ **do:**
     Set $Y := Y \setminus \{i\}$.
     If there are no $i' \in X$ and $j \in \mathcal{D} \setminus U$ with $v_j > c_{ij}$ and $v_j > c_{i'j}$
       then set $X := X \cup \{i\}$.
     For $j \in U$ with $c_{ij} \leq t$ **do**: Set $v_j := t$ and $U := U \setminus \{j\}$.

⑤   **If** $U \neq \emptyset$ **then go to** ②.

⑥   **For** $j \in \mathcal{D}$ **do**: Let $i \in X$ with $c_{ij} = \min\{c_{i'j} : i' \in X\}$. Set $\sigma(j) := i$.

Theorem 22.3. (Jain and Vazirani [2001]) *For metric instances I, the* JAIN-VAZIRANI ALGORITHM *opens a set X of facilities with $3c_F(X) + c_S(X) \leq 3\,\text{OPT}(I)$. In particular, it is a 3-factor approximation algorithm for the* METRIC UNCAPACITATED FACILITY LOCATION PROBLEM. *It can be implemented to run in $O(m \log m)$ time, where $m = |\mathcal{F}||\mathcal{D}|$.*

Proof: First observe that t is non-decreasing in the course of the algorithm.

The algorithm computes a primal solution X and σ, and numbers v_j, $j \in \mathcal{D}$, which together with $w_{ij} := \max\{0, v_j - c_{ij}\}$, $i \in \mathcal{F}$, $j \in \mathcal{D}$, constitute a feasible solution to the dual LP (22.3). Hence $\sum_{j \in \mathcal{D}} v_j \le \mathrm{OPT}(I)$. For each open facility i, all customers j with $w_{ij} > 0$ are connected to i, and $f_i = \sum_{j \in \mathcal{D}} w_{ij}$. Moreover, we claim that the service cost for each customer j is at most $3(v_j - w_{\sigma(j)j})$.

We distinguish two cases. If $c_{\sigma(j)j} = v_j - w_{\sigma(j)j}$, this is clear. Otherwise $c_{\sigma(j)j} > v_j$ and $w_{\sigma(j)j} = 0$. Hence $c_{ij} > v_j$ for all $i \in X$. This means that when j is removed from U in ③ or ④, there is an $i \in \mathcal{F} \setminus (Y \cup X)$ with $c_{ij} \le v_j$ and there are $j' \in \mathcal{D} \setminus U$ and $i' \in X$ with $w_{ij'} > 0$ and $w_{i'j'} > 0$. Hence $c_{ij'} = v_{j'} - w_{ij'} < v_{j'}$ and $c_{i'j'} = v_{j'} - w_{i'j'} < v_{j'}$. Note that $v_{j'} \le v_j$, because j' is removed from U before j. We conclude that $c_{\sigma(j)j} \le c_{i'j} \le c_{i'j'} + c_{ij'} + c_{ij} < v_{j'} + v_{j'} + v_j \le 3v_j = 3(v_j - w_{\sigma(j)j})$.

For the running time we observe that the number of iterations is at most $|\mathcal{D}| + 1$ as at least one customer is removed from U in each iteration, maybe except the first one if $f_i = 0$ for some $i \in \mathcal{F}$. The total time for computing t_1 in ②, and for ③, is $O(m \log m)$ if we sort all c_{ij} once in advance. Next, note that

$$t_2 = \min\left\{ \frac{t_2^i}{|U_i|} : i \in Y \right\}, \text{ where}$$

$$t_2^i = f_i + \sum_{j \in \mathcal{D} \setminus U : v_j > c_{ij}} (c_{ij} - v_j) + \sum_{j \in U_i} c_{ij}$$

and U_i is the set of unconnected customers whose service cost to i is at most the new value of t. As this number is in fact what we want to compute, we proceed as follows.

We maintain t_2, t_2^i and $|U_i|$ ($i \in Y$) throughout; initially $t_2 = \infty$, $t_2^i = f_i$ and $|U_i| = 0$ for all i. When a new customer j is connected and $v_j > c_{ij}$ for some $i \in Y$, then t_2^i is reduced by v_j and $|U_i|$ is reduced by one, which may also imply a change of t_2. However, we also have to increase $|U_i|$ by one and increase t_2^i by c_{ij} (and possibly change t_2) when t reaches c_{ij} for some $i \in Y$ and $j \in U$. This can be done by changing the definition of t_1 in ② to $t_1 := \min\{c_{ij} : i \in \mathcal{F}, j \in U\}$ and performing these updates before ⑤ for all $i \in Y$ and $j \in U$ with $c_{ij} = t$. Note that there are $O(m)$ such updates overall, each of which takes constant time.

The if-statement in ④ can be implemented in $O(|\mathcal{D}|)$ time if we store for each $j \in \mathcal{D} \setminus U$ the facility $i' \in X$ with $v_j > c_{i'j}$ (if any; there can be at most one). □

A better primal-dual algorithm has been proposed by Jain et al. [2003]. One idea is to relax the feasibility of the dual variables. We interpret the dual variables as the customers' budgets, which they use to pay their service costs and contribute to facility opening costs. We open a facility when the offered contributions suffice to pay the opening cost. Connected customers do not increase their budget anymore, but they can still offer a certain amount to other facilities if these are closer and re-connecting would save service cost. The algorithm proceeds as follows.

DUAL FITTING ALGORITHM

Input: An instance $(\mathcal{D}, \mathcal{F}, (f_i)_{i \in \mathcal{F}}, (c_{ij})_{i \in \mathcal{F}, j \in \mathcal{D}})$ of the UNCAPACITATED FACILITY LOCATION PROBLEM.

Output: A solution $X \subseteq \mathcal{F}$ and $\sigma : \mathcal{D} \to X$.

①   Let $X := \emptyset$ and $U := \mathcal{D}$.

②   Set $t_1 := \min\{c_{ij} : i \in X, j \in U\}$.
    Set $t_2 := \min\{\tau : \exists i \in \mathcal{F} \setminus X : \omega(i, \tau, U) = f_i\}$, where
$$\omega(i, \tau, U) := \sum_{j \in U} \max\{0, \tau - c_{ij}\} + \sum_{j \in \mathcal{D} \setminus U} \max\{0, c_{\sigma(j)j} - c_{ij}\}.$$
    Set $t := \min\{t_1, t_2\}$.

③   **For** $i \in X$ and $j \in U$ with $c_{ij} = t$ **do**:
    Set $\sigma(j) := i$, $v_j := t$ and $U := U \setminus \{j\}$.

④   **For** $i \in \mathcal{F} \setminus X$ with $\omega(i, t, U) = f_i$ **do**:
    Set $X := X \cup \{i\}$.
    For $j \in \mathcal{D} \setminus U$ with $c_{ij} < c_{\sigma(j)j}$ **do**: Set $\sigma(j) := i$.
    For $j \in U$ with $c_{ij} \leq t$ **do**: Set $\sigma(j) := i$, $v_j := t$ and $U := U \setminus \{j\}$.

⑤   **If** $U \neq \emptyset$ **then go to** ②.

Theorem 22.4. *The above algorithm computes numbers v_j, $j \in \mathcal{D}$, and a feasible solution X, σ of cost $\sum_{j \in \mathcal{D}} v_j$. It can be implemented to run in $O(|\mathcal{F}|^2 |\mathcal{D}|)$ time.*

Proof: The first statement is evident. The running time can be obtained as for the JAIN-VAZIRANI ALGORITHM. However, we have to update all t_2^i whenever a customer is re-connected, i.e. whenever a new facility is opened. $\qquad \square$

We will find a number γ such that $\sum_{j \in D} v_j \leq \gamma(f_i + \sum_{j \in D} c_{ij})$ for all pairs $(i, D) \in \mathcal{F} \times 2^{\mathcal{D}}$ (i.e. $(\frac{v_j}{\gamma})_{j \in \mathcal{D}}$ is a feasible solution to the dual LP in Exercise 3). This will imply the performance ratio γ. Of course, we have to assume service costs to be metric.

Consider $i \in \mathcal{F}$ and $D \subseteq \mathcal{D}$ with $|D| = d$. Number the customers in D in the order in which they are removed from U in the algorithm; w.l.o.g. $D = \{1, \ldots, d\}$. We have $v_1 \leq v_2 \leq \cdots \leq v_d$.

Let $k \in D$. Note that k is connected at time $t = v_k$ in the algorithm, and consider the situation when t is set to v_k in ② for the first time. For $j = 1, \ldots, k - 1$ let

$$r_{j,k} := \begin{cases} c_{i(j,k)j} & \text{if } j \text{ is connected to } i(j, k) \in \mathcal{F} \text{ at this time} \\ v_k & \text{otherwise, i.e. if } v_j = v_k \end{cases}.$$

We now write down valid inequalities for these variables. First, for $j = 1, \ldots, d - 2$,

$$r_{j,j+1} \geq r_{j,j+2} \geq \cdots \geq r_{j,d} \tag{22.4}$$

because the service cost decreases if customers are re-connected. Next, for $k = 1, \ldots, d$,

$$\sum_{j=1}^{k-1} \max\{0, r_{j,k} - c_{ij}\} + \sum_{l=k}^{d} \max\{0, v_k - c_{il}\} \leq f_i. \tag{22.5}$$

To see this, we consider two cases. If $i \in \mathcal{F} \setminus X$ at the considered time, (22.5) holds by the choice of t in ②. Otherwise i was inserted into X before, and at that time $\sum_{j \in U} \max\{0, v_j - c_{ij}\} + \sum_{j \in \mathcal{D} \setminus U} \max\{0, c_{\sigma(j)j} - c_{ij}\} = f_i$. Later the left-hand side can only become smaller.

Finally, for $1 \leq j < k \leq d$,

$$v_k \leq r_{j,k} + c_{ij} + c_{ik}, \tag{22.6}$$

which is trivial if $r_{j,k} = v_k$, and otherwise follows from the choice of t_1 in ② by observing that the right-hand side of (22.6) is at least $c_{i(j,k)k}$ due to metric service costs, and that facility $i(j, k)$ is open at the considered time.

To prove a performance ratio, we consider the following optimization problem for $\gamma_F \geq 1$ and $d \in \mathbb{N}$. As we want to make a statement for all instances, we consider f_i, c_{ij} and v_j ($j = 1, \ldots, d$) and $r_{j,k}$ ($1 \leq j < k \leq d$) as variables:

$$
\begin{array}{lrcll}
\max & \dfrac{\sum_{j=1}^{d} v_j - \gamma_F f_i}{\sum_{j=1}^{d} c_{ij}} & & & \\[2ex]
\text{s.t.} & v_j & \leq & v_{j+1} & (1 \leq j < d) \\
& r_{j,k} & \geq & r_{j,k+1} & (1 \leq j < k < d) \\
& r_{j,k} + c_{ij} + c_{ik} & \geq & v_k & (1 \leq j < k \leq d) \\
& \sum_{j=1}^{k-1} \max\{0, r_{j,k} - c_{ij}\} + & & & \\
& \sum_{l=k}^{d} \max\{0, v_k - c_{il}\} & \leq & f_i & (1 \leq k \leq d) \\
& \sum_{j=1}^{d} c_{ij} & > & 0 & \\
& f_i & \geq & 0 & \\
& v_j, c_{ij} & \geq & 0 & (1 \leq j \leq d) \\
& r_{j,k} & \geq & 0 & (1 \leq j < k \leq d)
\end{array}
\tag{22.7}
$$

Note that this optimization problem can be easily re-formulated as a linear program (Exercise 6); it is often referred to as the *factor-revealing LP*. Its optimum values imply performance guarantees for the DUAL FITTING ALGORITHM:

Theorem 22.5. *Let $\gamma_F \geq 1$, and let γ_S be the supremum of the optimum values of the factor-revealing LP (22.7) over all $d \in \mathbb{N}$. Let an instance of the METRIC UNCAPACITATED FACILITY LOCATION PROBLEM be given, and let $X^* \subseteq \mathcal{F}$ be any solution. Then the cost of the solution produced by the DUAL FITTING ALGORITHM on this instance is at most $\gamma_F c_F(X^*) + \gamma_S c_S(X^*)$.*

Proof: The algorithm produces numbers v_j and, implicitly, $r_{j,k}$ for all $j, k \in \mathcal{D}$ with $v_j \leq v_k$ and $j \neq k$. For each pair $(i, D) \in \mathcal{F} \times 2^{\mathcal{D}}$, the numbers $f_i, c_{ij}, v_j, r_{j,k}$ satisfy conditions (22.4), (22.5) and (22.6) and thus constitute a feasible solution to (22.7) unless $\sum_{j=1}^{d} c_{ij} = 0$. Hence $\sum_{j=1}^{d} v_j - \gamma_F f_i \leq \gamma_S \sum_{j=1}^{d} c_{ij}$ (this follows directly from (22.5) and (22.6) if $c_{ij} = 0$ for all $j \in D$). Choosing $\sigma^* : \mathcal{D} \to X^*$ such that $c_{\sigma^*(j)j} = \min_{i \in X^*} c_{ij}$, and summing over all pairs $(i, \{j \in \mathcal{D} : \sigma^*(j) = i\})$ $(i \in X^*)$, we get

$$\sum_{j \in \mathcal{D}} v_j \leq \gamma_F \sum_{i \in X^*} f_i + \gamma_S \sum_{j \in \mathcal{D}} c_{\sigma^*(j)j} = \gamma_F c_F(X^*) + \gamma_S c_S(X^*).$$

As the solution computed by the algorithm has total cost at most $\sum_{j \in \mathcal{D}} v_j$, this proves the theorem. $\qquad\square$

To apply this, we observe:

Lemma 22.6.   *Consider the factor-revealing LP (22.7) for some $d \in \mathbb{N}$.*

(a) *For $\gamma_F = 1$, the optimum is at most 2.*
(b) (Jain et al. [2003]) *For $\gamma_F = 1.61$, the optimum is at most 1.61.*
(c) (Mahdian, Ye and Zhang [2006]) *For $\gamma_F = 1.11$, the optimum is at most 1.78.*

Proof: Here we only prove (a). For a feasible solution we have

$$d \left(f_i + \sum_{j=1}^{d} c_{ij} \right) \geq \sum_{k=1}^{d} \left(\sum_{j=1}^{k-1} r_{j,k} + \sum_{l=k}^{d} v_k \right)$$
$$\geq \sum_{k=1}^{d} d v_k - (d-1) \sum_{j=1}^{d} c_{ij}, \qquad (22.8)$$

implying that $d \sum_{j=1}^{d} v_j \leq d f_i + (2d-1) \sum_{j=1}^{d} c_{ij}$, i.e. $\sum_{j=1}^{d} v_j \leq f_i + 2 \sum_{j=1}^{d} c_{ij}$. $\qquad\square$

The proofs of (b) and (c) are quite long and technical. (a) directly implies that $(\frac{v_j}{2})_{j \in \mathcal{D}}$ is a feasible dual solution, and the DUAL FITTING ALGORITHM is a 2-factor approximation algorithm. (b) implies a performance ratio of 1.61. Even better results can be obtained by combining the DUAL FITTING ALGORITHM with scaling and greedy augmentation, techniques presented in the next section. For later use we summarize what follows from Theorem 22.5 and Lemma 22.6:

Corollary 22.7.   *Let $(\gamma_F, \gamma_S) \in \{(1, 2), (1.61, 1.61), (1.11, 1.78)\}$. Let an instance of the* METRIC UNCAPACITATED FACILITY LOCATION PROBLEM *be given, and let $\emptyset \neq X^* \subseteq \mathcal{F}$ be any solution. Then the cost of the solution produced by the* DUAL FITTING ALGORITHM *on this instance is at most $\gamma_F c_F(X^*) + \gamma_S c_S(X^*)$.* $\qquad\square$

22.4 Scaling and Greedy Augmentation

Many approximation results are asymmetric in terms of facility cost and service cost. Often the service cost can be reduced by opening additional facilities. Indeed, this can be exploited to improve several performance guarantees.

Proposition 22.8. *Let* $\emptyset \neq X, X^* \subseteq \mathcal{F}$. *Then* $\sum_{i \in X^*}(c_S(X) - c_S(X \cup \{i\})) \geq c_S(X) - c_S(X^*)$.

In particular, there exists an $i \in X^*$ *with* $\frac{c_S(X) - c_S(X \cup \{i\})}{f_i} \geq \frac{c_S(X) - c_S(X^*)}{c_F(X^*)}$.

Proof: For $j \in \mathcal{D}$ let $\sigma(j) \in X$ such that $c_{\sigma(j)j} = \min_{i \in X} c_{ij}$, and let $\sigma^*(j) \in X^*$ such that $c_{\sigma^*(j)j} = \min_{i \in X^*} c_{ij}$. Then $c_S(X) - c_S(X \cup \{i\}) \geq \sum_{j \in \mathcal{D}: \sigma^*(j)=i}(c_{\sigma(j)j} - c_{ij})$ for all $i \in X^*$. Summation yields the lemma. □

By **greedy augmentation** of a set X we mean iteratively picking an element $i \in \mathcal{F}$ maximizing $\frac{c_S(X) - c_S(X \cup \{i\})}{f_i}$ and adding it to X until $c_S(X) - c_S(X \cup \{i\}) \leq f_i$ for all $i \in \mathcal{F}$. We need the following lemma:

Lemma 22.9. (Charikar and Guha [2005]) *Let* $\emptyset \neq X, X^* \subseteq \mathcal{F}$. *Apply greedy augmentation to* X, *obtaining a set* $Y \supseteq X$. *Then*

$$c_F(Y) + c_S(Y) \leq$$
$$c_F(X) + c_F(X^*) \ln\left(\max\left\{1, \frac{c_S(X) - c_S(X^*)}{c_F(X^*)}\right\}\right) + c_F(X^*) + c_S(X^*).$$

Proof: If $c_S(X) \leq c_F(X^*) + c_S(X^*)$, the above inequality evidently holds even with X in place of Y. Greedy augmentation never increases the cost.

Otherwise, let $X = X_0, X_1, \ldots, X_k$ be the sequence of augmented sets, such that k is the first index for which $c_S(X_k) \leq c_F(X^*) + c_S(X^*)$. By renumbering facilities we may assume $X_i \setminus X_{i-1} = \{i\}$ $(i = 1, \ldots, k)$. By Proposition 22.8,

$$\frac{c_S(X_{i-1}) - c_S(X_i)}{f_i} \geq \frac{c_S(X_{i-1}) - c_S(X^*)}{c_F(X^*)}$$

for $i = 1, \ldots, k$. Hence $f_i \leq c_F(X^*)\frac{c_S(X_{i-1}) - c_S(X_i)}{c_S(X_{i-1}) - c_S(X^*)}$ (note that $c_S(X_{i-1}) > c_S(X^*)$), and

$$c_F(X_k) + c_S(X_k) \leq c_F(X) + c_F(X^*) \sum_{i=1}^{k} \frac{c_S(X_{i-1}) - c_S(X_i)}{c_S(X_{i-1}) - c_S(X^*)} + c_S(X_k).$$

As the right-hand side increases with increasing $c_S(X_k)$ (the derivative is $1 - \frac{c_F(X^*)}{c_S(X_{k-1}) - c_S(X^*)} > 0$), we do not make the right-hand side smaller if we replace

$c_S(X_k)$ by $c_F(X^*) + c_S(X^*)$. Using $x - 1 \geq \ln x$ for $x > 0$, we get

$$
\begin{aligned}
c_F(X_k) + c_S(X_k) \ &\leq\ c_F(X) + c_F(X^*) \sum_{i=1}^{k} \left(1 - \frac{c_S(X_i) - c_S(X^*)}{c_S(X_{i-1}) - c_S(X^*)}\right) + c_S(X_k) \\
&\leq\ c_F(X) - c_F(X^*) \sum_{i=1}^{k} \ln \frac{c_S(X_i) - c_S(X^*)}{c_S(X_{i-1}) - c_S(X^*)} + c_S(X_k) \\
&=\ c_F(X) - c_F(X^*) \ln \frac{c_S(X_k) - c_S(X^*)}{c_S(X) - c_S(X^*)} + c_S(X_k) \\
&=\ c_F(X) + c_F(X^*) \ln \frac{c_S(X) - c_S(X^*)}{c_F(X^*)} + c_F(X^*) + c_S(X^*).
\end{aligned}
$$

$\square$

    This can be used to improve several of the previous performance guarantees. Sometimes it is good to combine greedy augmentation with scaling. We get the following general result:

Theorem 22.10. *Suppose there are positive constants β, γ_S, γ_F and an algorithm A which, for every instance, computes a solution X such that $\beta c_F(X) + c_S(X) \leq \gamma_F c_F(X^*) + \gamma_S c_S(X^*)$ for each $\emptyset \neq X^* \subseteq \mathcal{F}$. Let $\delta \geq \frac{1}{\beta}$.*
    Then scaling facility costs by δ, applying A to the modified instance, and applying greedy augmentation to the result with respect to the original instance yields a solution of cost at most $\max\{\frac{\gamma_F}{\beta} + \ln(\beta\delta), 1 + \frac{\gamma_S - 1}{\beta\delta}\}$ times the optimum.

Proof: Let X^* be the set of open facilities of an optimum solution to the original instance. We have $\beta\delta c_F(X) + c_S(X) \leq \gamma_F \delta c_F(X^*) + \gamma_S c_S(X^*)$. If $c_S(X) \leq c_S(X^*) + c_F(X^*)$, then we have $\beta\delta(c_F(X) + c_S(X)) \leq \gamma_F \delta c_F(X^*) + \gamma_S c_S(X^*) + (\beta\delta - 1)(c_S(X^*) + c_F(X^*))$, so X is a solution that costs at most $\max\{1 + \frac{\gamma_F \delta - 1}{\beta\delta}, 1 + \frac{\gamma_S - 1}{\beta\delta}\}$ times the optimum. Note that $1 + \frac{\gamma_F \delta - 1}{\beta\delta} \leq \frac{\gamma_F}{\beta} + \ln(\beta\delta)$ as $1 - \frac{1}{x} \leq \ln x$ for all $x > 0$.

    Otherwise we apply greedy augmentation to X and get a solution of cost at most

$$
\begin{aligned}
&c_F(X) + c_F(X^*) \ln \frac{c_S(X) - c_S(X^*)}{c_F(X^*)} + c_F(X^*) + c_S(X^*) \\
&\leq\ c_F(X) + c_F(X^*) \ln \frac{(\gamma_S - 1)c_S(X^*) + \gamma_F \delta c_F(X^*) - \beta\delta c_F(X)}{c_F(X^*)} \\
&\quad + c_F(X^*) + c_S(X^*).
\end{aligned}
$$

The derivative of this expression with respect to $c_F(X)$ is

$$
1 - \frac{\beta\delta c_F(X^*)}{(\gamma_S - 1)c_S(X^*) + \gamma_F \delta c_F(X^*) - \beta\delta c_F(X)},
$$

which is zero for $c_F(X) = \frac{\gamma_F - \beta}{\beta} c_F(X^*) + \frac{\gamma_S - 1}{\beta\delta} c_S(X^*)$. Hence we get a solution of cost at most

$$\left(\frac{\gamma_F}{\beta} + \ln(\beta\delta)\right) c_F(X^*) + \left(1 + \frac{\gamma_S - 1}{\beta\delta}\right) c_S(X^*).$$

$\square$

With Corollary 22.7 we can apply this result to the DUAL FITTING ALGORITHM with $\beta = \gamma_F = 1$ and $\gamma_S = 2$: by setting $\delta = 1.76$ we obtain a performance guarantee of 1.57. With $\beta = 1$, $\gamma_F = 1.11$ and $\gamma_S = 1.78$ (cf. Corollary 22.7) we can do even better:

Corollary 22.11. (Mahdian, Ye and Zhang [2006]) *Multiply all facility costs by $\delta = 1.504$, apply the* DUAL FITTING ALGORITHM, *scale back the facility costs, and apply greedy augmentation. Then this algorithm has a performance guarantee of 1.52.* $\square$

Byrka and Aardal [2007] showed that the performance ratio of this algorithm is not better than 1.494. They also found a 1.500-factor approximation algorithm by combining the above with an LP rounding algorithm (Byrka and Aardal [2007]). This was further improved by Li [2013] to 1.488, which is the currently best known performance ratio.

For the special case when all service costs are between 1 and 3, greedy augmentation yields an even better performance ratio. Let α be the solution of the equation $\alpha + 1 = \ln\frac{2}{\alpha}$; we have $0.463 \leq \alpha \leq 0.4631$. A simple calculation shows that $\alpha = \frac{\alpha}{\alpha+1}\ln\frac{2}{\alpha} = \max\{\frac{\xi}{\xi+1}\ln\frac{2}{\xi} : \xi > 0\}$.

Theorem 22.12. (Guha and Khuller [1999]) *Consider the* UNCAPACITATED FACILITY LOCATION PROBLEM *restricted to instances where all service costs are within the interval $[1, 3]$. This problem has a $(1 + \alpha + \epsilon)$-factor approximation algorithm for every $\epsilon > 0$.*

Proof: Let $\epsilon > 0$, and let $k := \lceil\frac{1}{\epsilon}\rceil$. Enumerate all solutions $X \subseteq \mathcal{F}$ with $|X| \leq k$.

We compute another solution as follows. We first open one facility i with minimum opening cost f_i, and then apply greedy augmentation to obtain a solution Y. We claim that the best solution costs at most $1 + \alpha + \epsilon$ times the optimum.

Let X^* be an optimum solution and $\xi = \frac{c_F(X^*)}{c_S(X^*)}$. We may assume that $|X^*| > k$, as otherwise we have found X^* above. Then $c_F(\{i\}) \leq \frac{1}{k}c_F(X^*)$. Moreover, as the service costs are between 1 and 3, $c_S(\{i\}) \leq 3|\mathcal{D}| \leq 3c_S(X^*)$.

By Lemma 22.9, the cost of Y is at most

$$\frac{1}{k}c_F(X^*) + c_F(X^*)\ln\left(\max\left\{1, \frac{2c_S(X^*)}{c_F(X^*)}\right\}\right) + c_F(X^*) + c_S(X^*)$$
$$= c_S(X^*)\left(\frac{\xi}{k} + \xi\ln\left(\max\left\{1, \frac{2}{\xi}\right\}\right) + \xi + 1\right)$$

$$\leq\ c_S(X^*)(1+\xi)\left(1+\epsilon+\frac{\xi}{\xi+1}\ln\left(\max\left\{1,\frac{2}{\xi}\right\}\right)\right)$$
$$\leq\ (1+\alpha+\epsilon)(1+\xi)c_S(X^*)$$
$$=\ (1+\alpha+\epsilon)(c_F(X^*)+c_S(X^*)).\qquad\qquad\qquad\square$$

This performance guarantee seems to be best possible in view of the following:

Theorem 22.13. *If there is an $\epsilon > 0$ and a $(1+\alpha-\epsilon)$-factor approximation algorithm for the* UNCAPACITATED FACILITY LOCATION PROBLEM *restricted to instances with service costs 1 and 3 only, then $P = NP$.*

This has been shown by Sviridenko [unpublished] (based on results by Feige [1998] and Guha and Khuller [1999]) and can be found in the survey by Vygen [2005].

22.5 Bounding the Number of Facilities

The k-FACILITY LOCATION PROBLEM is the UNCAPACITATED FACILITY LOCA-TION PROBLEM with the additional constraint that no more than k facilities may be opened, where k is a natural number that is part of the instance. A special case, where facility opening costs are zero, is the well-known k-MEDIAN PROBLEM. In this section we describe an approximation algorithm for the METRIC k-FACILITY LOCATION PROBLEM.

For problems which become much easier if a certain type of constraints is omitted, Lagrangean relaxation (cf. Section 5.6) is a common technique. Here we relax the bound on the number of open facilities and add a constant λ to each facility opening cost.

Theorem 22.14. (Jain and Vazirani [2001]) *If there is a constant γ_S and a polynomial-time algorithm A, such that for every instance of the* METRIC UNCA-PACITATED FACILITY LOCATION PROBLEM *A computes a solution X such that $c_F(X) + c_S(X) \leq c_F(X^*) + \gamma_S c_S(X^*)$ for every $\emptyset \neq X^* \subseteq \mathcal{F}$, then there is a $(2\gamma_S)$-factor approximation algorithm for the* METRIC k-FACILITY LOCATION PROBLEM *with integral data.*

Proof: Let an instance of the METRIC k-FACILITY LOCATION PROBLEM be given. We assume that service costs are integers within $\{0, 1, \ldots, c_{max}\}$ and facility opening costs are integers within $\{0, 1, \ldots, f_{max}\}$.

First we check if there is a solution with zero cost, and find one if it exists. This is easy; see the proof of Lemma 22.15. Hence we assume that the cost of any solution is at least 1. Let X^* be an optimum solution (we will use it for analysis only).

Let $A(\lambda) \subseteq \mathcal{F}$ be the solution computed by A for the instance where all facility opening costs are increased by λ but the constraint on the number of facilities is omitted. We have $c_F(A(\lambda)) + |A(\lambda)|\lambda + c_S(A(\lambda)) \leq c_F(X^*) + |X^*|\lambda + \gamma_S c_S(X^*)$, and hence

$$c_F(A(\lambda)) + c_S(A(\lambda)) \leq c_F(X^*) + \gamma_S c_S(X^*) + (k - |A(\lambda)|)\lambda \qquad (22.9)$$

for all $\lambda \geq 0$. If $|A(0)| \leq k$, then $A(0)$ is a feasible solution costing at most γ_S times the optimum, and we are done.

Otherwise $|A(0)| > k$, and note that $|A(f_{max} + \gamma_S|\mathcal{D}|c_{max} + 1)| = 1 \leq k$. Set $\lambda' := 0$ and $\lambda'' := f_{max} + \gamma_S|\mathcal{D}|c_{max} + 1$, and apply binary search, maintaining $|A(\lambda'')| \leq k < |A(\lambda')|$. After $O(\log|\mathcal{D}| + \log f_{max} + \log c_{max})$ iterations, in each of which we set one of λ', λ'' to their arithmetic mean depending on whether $|A(\frac{\lambda'+\lambda''}{2})| \leq k$ or not, we have $\lambda'' - \lambda' \leq \frac{1}{|\mathcal{D}|^2}$. (Note that this binary search works although $\lambda \mapsto |A(\lambda)|$ is not monotonic in general.)

If $|A(\lambda'')| = k$, then (22.9) implies that $A(\lambda'')$ is a feasible solution costing at most γ_S times the optimum, and we are done. However, we will not always encounter such a λ'', because $\lambda \mapsto |A(\lambda)|$ is not always monotonic and can jump by more than 1 (Archer, Rajagopalan and Shmoys [2003] showed how to fix this by perturbing costs, but were unable to do it in polynomial time).

Thus we consider $X := A(\lambda')$ and $Y := A(\lambda'')$ and assume henceforth $|X| > k > |Y|$. Let $\alpha := \frac{k-|Y|}{|X|-|Y|}$ and $\beta := \frac{|X|-k}{|X|-|Y|}$.

Choose a subset X' of X with $|X'| = |Y|$ such that $\min_{i \in X'} c_{ii'} = \min_{i \in X} c_{ii'}$ for each $i' \in Y$, where we write $c_{ii'} := \min_{j \in \mathcal{D}}(c_{ij} + c_{i'j})$.

We open either all elements of X' (with probability α) or all elements of Y (with probability $\beta = 1 - \alpha$). In addition, we choose a set of $k - |Y|$ facilities of $X \setminus X'$, uniformly at random. We open each of these unless it is already open (note that X and Y are not necessarily disjoint). Then the total expected facility cost is at most $\alpha c_F(X) + \beta c_F(Y)$.

Let $j \in \mathcal{D}$, and let i' be a closest facility in X, and let i'' be a closest facility in Y. Connect j to i' if it is open, else to i'' if it is open. If neither i' nor i'' is open, connect j to a facility $i''' \in X'$ minimizing $c_{i''i'''}$.

This yields an expected service cost $\alpha c_{i'j} + \beta c_{i''j}$ if $i' \in X'$ and at most

$$\begin{aligned}
&\alpha c_{i'j} + (1-\alpha)\beta c_{i''j} + (1-\alpha)(1-\beta)c_{i'''j} \\
\leq\; &\alpha c_{i'j} + \beta^2 c_{i''j} + \alpha\beta\left(c_{i''j} + \min_{j' \in \mathcal{D}}(c_{i''j'} + c_{i'''j'})\right) \\
\leq\; &\alpha c_{i'j} + \beta^2 c_{i''j} + \alpha\beta(c_{i''j} + c_{i''j} + c_{i'j}) \\
=\; &\alpha(1+\beta)c_{i'j} + \beta(1+\alpha)c_{i''j}
\end{aligned}$$

if $i' \in X \setminus X'$.

Thus the total expected service cost is at most

$$(1 + \max\{\alpha, \beta\})(\alpha c_S(X) + \beta c_S(Y)) \leq \left(2 - \frac{1}{|\mathcal{D}|}\right)(\alpha c_S(X) + \beta c_S(Y)).$$

Overall, using (22.9), we get an expected cost of at most

$$
\left(2 - \frac{1}{|\mathcal{D}|}\right)\left(\alpha(c_F(X) + c_S(X)) + \beta(c_F(Y) + c_S(Y))\right)
$$
$$
\leq \left(2 - \frac{1}{|\mathcal{D}|}\right)\left(c_F(X^*) + \gamma_S c_S(X^*) + (\lambda'' - \lambda')\frac{(|X| - k)(k - |Y|)}{|X| - |Y|}\right)
$$
$$
\leq \left(2 - \frac{1}{|\mathcal{D}|}\right)\left(c_F(X^*) + \gamma_S c_S(X^*) + (\lambda'' - \lambda')\frac{|X| - |Y|}{4}\right)
$$
$$
\leq \left(2 - \frac{1}{|\mathcal{D}|}\right)\left(c_F(X^*) + \gamma_S c_S(X^*) + \frac{1}{4|\mathcal{D}|}\right)
$$
$$
\leq \left(2 - \frac{1}{|\mathcal{D}|}\right)\left(1 + \frac{1}{4|\mathcal{D}|}\right)\left(c_F(X^*) + \gamma_S c_S(X^*)\right)
$$
$$
\leq \left(2 - \frac{1}{2|\mathcal{D}|}\right)\left(c_F(X^*) + \gamma_S c_S(X^*)\right)
$$

and thus at most $2\gamma_S(c_F(X^*) + c_S(X^*))$.

Note that the expected cost is easy to compute even under the condition that a subset Z is opened with probability 1 and $k - |Z|$ randomly chosen facilities of some other set are opened. Hence one can derandomize this algorithm by the method of conditional probabilities: first open X' or Y depending on where the bound on the expected cost is at most $\left(2 - \frac{1}{|\mathcal{D}|}\right)(\alpha(c_F(X) + c_S(X)) + \beta(c_F(Y) + c_S(Y)))$, and then successively open facilities of $X \setminus X'$ such that this bound continues to hold.   □

In particular, by the DUAL FITTING ALGORITHM (Corollary 22.7), we obtain a 4-factor approximation algorithm for the METRIC k-FACILITY LOCATION PROBLEM with integral data. The first constant-factor approximation algorithm for the METRIC k-FACILITY LOCATION PROBLEM was due to Charikar et al. [2002].

The running time of the binary search is weakly polynomial and works for integral data only. However we can make it strongly polynomial by discretizing the input data:

Lemma 22.15. *For any instance I of the* METRIC k-FACILITY LOCATION PROBLEM, $\gamma_{\max} \geq 1$ *and* $0 < \epsilon \leq 1$, *we can decide whether* OPT$(I) = 0$ *and otherwise generate another instance I' in* $O(|\mathcal{F}||\mathcal{D}| \log(|\mathcal{F}||\mathcal{D}|))$ *time, such that all service and facility costs are integers in* $\left\{0, 1, \ldots, \left\lceil \frac{2\gamma_{\max}(k+|\mathcal{D}|)^3}{\epsilon} \right\rceil\right\}$, *and for each* $1 \leq \gamma \leq \gamma_{\max}$, *each solution to I' with cost at most γ OPT(I') is a solution to I with cost at most $\gamma(1 + \epsilon)$ OPT(I).*

Proof: Let $n := k + |\mathcal{D}|$. Given an instance I, we first compute an upper bound and a lower bound on OPT(I) differing by a factor of at most $2n^2 - 1$ as follows. For each $B \in \{f_i : i \in \mathcal{F}\} \cup \{c_{ij} : i \in \mathcal{F}, j \in \mathcal{D}\}$ we consider the bipartite graph $G_B := (\mathcal{D} \cup \mathcal{F}, \{\{i, j\} : i \in \mathcal{F}, j \in \mathcal{D}, f_i \leq B, c_{ij} \leq B\})$.

The smallest B for which the elements of $\mathcal{D}$ belong to at most k different connected components of G_B, each of which contains at least one facility, is a lower bound on OPT(I). This number B can be found in $O(|\mathcal{F}||\mathcal{D}| \log(|\mathcal{F}||\mathcal{D}|))$ time by a straightforward variant of KRUSKAL'S ALGORITHM for minimum spanning trees.

Moreover, for this B we can choose an arbitrary facility in each connected component of G_B that contains an element of $\mathcal{D}$, and connect each customer with service cost at most $(2|\mathcal{D}| - 1)B$ (using the assumption that service costs are metric). Thus OPT(I) $\leq kB + (2|\mathcal{D}| - 1)|\mathcal{D}|B < (2n^2 - 1)B$ unless $B = 0$, in which case we are done.

Thus we can ignore facility and service costs exceeding $B' := 2\gamma_{\max} n^2 B$. We obtain I' from I by setting each c_{ij} to $\lceil \frac{\min\{B', c_{ij}\}}{\delta} \rceil$ and each f_i to $\lceil \frac{\min\{B', f_i\}}{\delta} \rceil$, where $\delta = \frac{\epsilon B}{n}$. Now all input numbers are integers in $\{0, 1, \ldots, \lceil \frac{2\gamma_{\max} n^3}{\epsilon} \rceil\}$.

We have

$$\text{OPT}(I') \leq \frac{\text{OPT}(I)}{\delta} + n = \frac{\text{OPT}(I) + \epsilon B}{\delta} < \frac{(2n^2 - 1)B + \epsilon B}{\delta} \leq \frac{2n^2 B}{\delta} = \frac{B'}{\gamma_{\max}\delta},$$

and thus a solution to I' of cost at most γ OPT(I') contains no element of cost $\lceil \frac{B'}{\delta} \rceil$, and hence is a solution to I of cost at most

$$\delta\gamma \, \text{OPT}(I') \leq \gamma(\text{OPT}(I) + \epsilon B) \leq \gamma(1 + \epsilon) \, \text{OPT}(I).$$

$\square$

Corollary 22.16. *There is a strongly polynomial 4-factor approximation algorithm for the* METRIC k-FACILITY LOCATION PROBLEM.

Proof: Apply Lemma 22.15 with $\gamma_{\max} = 4$ and $\epsilon = \frac{1}{4|\mathcal{D}|}$, and apply Theorem 22.14 with the DUAL FITTING ALGORITHM to the resulting instance. We have $\gamma_S = 2$ by Corollary 22.7 and get a solution of total cost at most

$$\left(2 - \frac{1}{2|\mathcal{D}|}\right)\left(1 + \frac{1}{4|\mathcal{D}|}\right)\left(c_F(X^*) + \gamma_S c_S(X^*)\right) \leq 4\left(c_F(X^*) + c_S(X^*)\right)$$

for any $\emptyset \neq X^* \subseteq \mathcal{F}$. $\square$

Zhang [2007] found a 3.733-factor approximation algorithm for the METRIC k-FACILITY LOCATION PROBLEM. This algorithm uses local search techniques similar to those presented in the next section.

22.6 Local Search

As discussed in Section 21.3, local search is a technique that is often applied successfully in practice, although usually no good performance guarantees can be shown. It was therefore a surprise to learn that facility location problems can be approximated well by local search. This was first explored by Korupolu, Plaxton and Rajaraman [2000] and led to several strong results subsequently. We shall present some of them in this and the next section.

For the METRIC k-MEDIAN PROBLEM, local search yields the best known performance ratio. Before presenting this result, we consider the simplest possible local search algorithm: we start with an arbitrary feasible solution (i.e., a set of k facilities) and improve it by single swaps. Note that we have to consider only the service cost, as facility costs are zero in the k-MEDIAN PROBLEM. Moreover, it causes no loss of generality to assume that a solution must contain exactly k facilities.

Theorem 22.17. (Arya et al. [2004]) *Consider an instance of the* METRIC k-MEDIAN PROBLEM. *Let X be a feasible solution and X^* an optimum solution. If $c_S((X \setminus \{x\}) \cup \{y\}) \geq c_S(X)$ for all $x \in X$ and $y \in X^*$, then $c_S(X) \leq 5c_S(X^*)$.*

Proof: Let us consider optimum assignments σ and σ^* of the customers to the k facilities in X and X^*, respectively. We say that $x \in X$ *captures* $y \in X^*$ if $|\{j \in \mathcal{D} : \sigma(j) = x, \sigma^*(j) = y\}| > \frac{1}{2}|\{j \in \mathcal{D} : \sigma^*(j) = y\}|$. Each $y \in X^*$ is captured by at most one $x \in X$.

Let $\pi : \mathcal{D} \to \mathcal{D}$ be a bijection such that for all $j \in \mathcal{D}$:

- $\sigma^*(\pi(j)) = \sigma^*(j)$; and
- if $\sigma(\pi(j)) = \sigma(j)$ then $\sigma(j)$ captures $\sigma^*(j)$.

Such a mapping π can be obtained easily by ordering, for each $y \in X^*$, the elements of $\{j \in \mathcal{D} : \sigma^*(j) = y\} = \{j_0, \ldots, j_{t-1}\}$ such that customers j with identical $\sigma(j)$ are consecutive, and setting $\pi(j_k) := j_{k'}$, where $k' = (k + \lfloor \frac{t}{2} \rfloor) \bmod t$.

Let us define a *swap* to be an element of $X \times X^*$. For a swap (x, y) we call x the source and y the target. We will define k swaps such that each $y \in X^*$ is the target of exactly one of them.

If an $x \in X$ captures only one facility $y \in X^*$, we consider a swap (x, y). If there are l such swaps, then there are $k - l$ elements left in X and in X^*. Some of the remaining elements of X (at most $\frac{k-l}{2}$) may capture at least two facilities of X^*; we will not consider these. For each remaining facility $y \in X^*$ we choose an $x \in X$ such that x does not capture any facility, and such that each $x \in X$ is source of at most two such swaps.

We now analyze the swaps one by one. Consider the swap (x, y), and let $X' := (X \setminus \{x\}) \cup \{y\}$. Then $c_S(X') \geq c_S(X)$. Transform $\sigma : \mathcal{D} \to X$ to a new assignment $\sigma' : \mathcal{D} \to X'$ by reassigning customers as follows:

Customers $j \in \mathcal{D}$ with $\sigma^*(j) = y$ are assigned to y. Customers $j \in \mathcal{D}$ with $\sigma(j) = x$ and $\sigma^*(j) = y' \in X^* \setminus \{y\}$ are assigned to $\sigma(\pi(j))$; note that $\sigma(\pi(j)) \neq x$ as x does not capture y'. For all other customers, the assignment does not change.

We have

$$
\begin{aligned}
0 \;\leq\; & c_S(X') - c_S(X) \\
\leq\; & \sum_{j \in \mathcal{D}:\sigma^*(j)=y} (c_{\sigma^*(j)j} - c_{\sigma(j)j}) + \sum_{j \in \mathcal{D}:\sigma(j)=x,\sigma^*(j)\neq y} (c_{\sigma(\pi(j))j} - c_{\sigma(j)j})
\end{aligned}
$$

$$\leq \sum_{j\in\mathcal{D}:\sigma^*(j)=y}(c_{\sigma^*(j)j}-c_{\sigma(j)j}) + \sum_{j\in\mathcal{D}:\sigma(j)=x}(c_{\sigma(\pi(j))j}-c_{\sigma(j)j})$$

as $c_{\sigma(\pi(j))j} \geq \min_{i\in X}c_{ij} = c_{\sigma(j)j}$ by definition of σ.

We now sum over all swaps. Note that each facility of X^* is the target of exactly one swap, thus the sum of the first terms is $c_S(X^*) - c_S(X)$. Moreover, each $x \in X$ is the source of at most two swaps. Hence

$$\begin{aligned}
0 &\leq \sum_{j\in\mathcal{D}}(c_{\sigma^*(j)j}-c_{\sigma(j)j}) + 2\sum_{j\in\mathcal{D}}(c_{\sigma(\pi(j))j}-c_{\sigma(j)j})\\
&\leq c_S(X^*) - c_S(X) + 2\sum_{j\in\mathcal{D}}(c_{\sigma^*(j)j}+c_{\sigma^*(j)\pi(j)}+c_{\sigma(\pi(j))\pi(j)}-c_{\sigma(j)j})\\
&= c_S(X^*) - c_S(X) + 2\sum_{j\in\mathcal{D}}(c_{\sigma^*(j)j}+c_{\sigma^*(\pi(j))\pi(j)})\\
&= c_S(X^*) - c_S(X) + 4c_S(X^*),
\end{aligned}$$

because π is a bijection. $\qquad\square$

Thus a local optimum is a 5-approximation. However, this does not make any statement about the running time to achieve a local optimum; conceivably, the number of steps to reach a local optimum could be exponential. However, by discretizing costs we obtain a strongly polynomial running time:

Corollary 22.18. Let $0 < \epsilon \leq 1$. Then the following is a strongly polynomial $(5+\epsilon)$-factor approximation algorithm for the METRIC k-MEDIAN PROBLEM: transform the instance according to Lemma 22.15 with $\gamma_{\max}=5$ and $\frac{\epsilon}{5}$ in place of ϵ, start with any set of k facilities, and apply swaps decreasing the service cost as long as possible.

Proof: As each service cost of the new instance is an integer in the set $\{0, 1, \ldots, \lceil\frac{50(k+|\mathcal{D}|)^3}{\epsilon}\rceil\}$, we can apply at most $|\mathcal{D}|\lceil\frac{50(k+|\mathcal{D}|)^3}{\epsilon}\rceil$ successive swaps each of which reduces the total service cost. $\qquad\square$

Using multiswaps the performance ratio can be improved significantly:

Theorem 22.19. (Arya et al. [2004]) Consider an instance of the METRIC k-MEDIAN PROBLEM, and let $p \in \mathbb{N}$. Let X be a feasible solution and X^* an optimum solution. If $c_S((X\setminus A)\cup B) \geq c_S(X)$ for all $A \subseteq X$ and $B \subseteq X^*$ with $|A|=|B|\leq p$, then $c_S(X) \leq (3+\frac{2}{p})c_S(X^*)$.

Proof: Let σ and σ^* again be optimum assignments of the customers to the k facilities in X and X^*, respectively. For each $A \subseteq X$, let $C(A)$ be the set of facilities in X^* that are captured by A, i.e.

$$C(A) := \left\{y\in X^* : |\{j\in\mathcal{D}:\sigma(j)\in A, \sigma^*(j)=y\}| > \frac{1}{2}|\{j\in\mathcal{D}:\sigma^*(j)=y\}|\right\}.$$

We partition $X = A_1 \dot\cup \cdots \dot\cup A_r$ and $X^* = B_1 \dot\cup \cdots \dot\cup B_r$ as follows:

> Let $\{x \in X : C(\{x\}) \neq \emptyset\} =: \{x_1, \ldots, x_s\} =: \bar{X}$.
> Set $r := \max\{s, 1\}$.
> **For** $i = 1$ **to** $r - 1$ **do:**
> > Set $A_i := \{x_i\}$.
> > **While** $|A_i| < |C(A_i)|$ **do:**
> > > Add an element $x \in X \setminus (A_1 \cup \cdots \cup A_i \cup \bar{X})$ to A_i.
> > Set $B_i := C(A_i)$.
> Set $A_r := X \setminus (A_1 \cup \cdots \cup A_{r-1})$ and $B_r := X^* \setminus (B_1 \cup \cdots \cup B_{r-1})$.

It is clear that this algorithm guarantees $|A_i| = |B_i| \geq 1$ for $i = 1, \ldots, r$, and that the sets $A_1, \ldots, A_r$ are pairwise disjoint and $B_1, \ldots, B_r$ are pairwise disjoint. Note that adding an element is always possible if $|A_i| < |C(A_i)|$, because then

$$
\begin{aligned}
& |X \setminus (A_1 \cup \cdots \cup A_i \cup \bar{X})| \\
= {} & |X| - |A_1| - \cdots - |A_i| - |\{x_{i+1}, \ldots, x_r\}| \\
> {} & |X^*| - |C(A_1)| - \cdots - |C(A_i)| - |C(\{x_{i+1}\})| - \cdots - |C(\{x_r\})| \\
= {} & |X^* \setminus (C(A_1) \cup \cdots \cup C(A_i) \cup C(\{x_{i+1}\}) \cup \cdots \cup C(\{x_r\}))| \\
\geq {} & 0.
\end{aligned}
$$

Let $\pi : \mathcal{D} \to \mathcal{D}$ be a bijection such that for all $j \in \mathcal{D}$:

- $\sigma^*(\pi(j)) = \sigma^*(j)$;
- if $\sigma(\pi(j)) = \sigma(j)$ then $\sigma(j)$ captures $\sigma^*(j)$; and
- if $\sigma(j) \in A_i$ and $\sigma(\pi(j)) \in A_i$ for some $i \in \{1, \ldots, r\}$, then A_i captures $\sigma^*(j)$.

Such a mapping π can be obtained almost identically as in the proof of Theorem 22.17.

We now define a set of swaps (A, B) with $|A| = |B| \leq p$, $A \subseteq X$ and $B \subseteq X^*$. Each swap will be associated with a positive weight. The swap (A, B) means that X is replaced by $X' := (X \setminus A) \cup B$; we say that A is the source set and B is the target set.

For each $i \in \{1, \ldots, r\}$ with $|A_i| \leq p$, we consider the swap (A_i, B_i) with weight 1. For each $i \in \{1, \ldots, r\}$ with $|A_i| = q > p$, we consider the swap $(\{x\}, \{y\})$ with weight $\frac{1}{q-1}$ for each $x \in A_i \setminus \{x_i\}$ and $y \in B_i$. Each $y \in X^*$ appears in the target set of swaps of total weight 1, and each $x \in X$ appears in the source set of swaps of total weight at most $\frac{p+1}{p}$.

We reassign customers as in the case of single swaps. More precisely, for a swap (A, B) we reassign all $j \in \mathcal{D}$ with $\sigma^*(j) \in B$ to $\sigma^*(j)$ and all $j \in \mathcal{D}$ with $\sigma^*(j) \notin B$ and $\sigma(j) \in A$ to $\sigma(\pi(j))$. Note that we have $B \supseteq C(A)$ for each of the considered swaps (A, B). Thus, for all $j \in \mathcal{D}$ with $\sigma(j) \in A$ and $\sigma^*(j) \notin B$ we have $\sigma(\pi(j)) \notin A$. Therefore we can bound the increase in cost due to the swap as follows:

$$
\begin{aligned}
0 \;\le\; & c_S(X') - c_S(X) \\
\le\; & \sum_{j\in\mathcal{D}:\sigma^*(j)\in B}(c_{\sigma^*(j)j} - c_{\sigma(j)j}) + \sum_{j\in\mathcal{D}:\sigma(j)\in A,\sigma^*(j)\notin B}(c_{\sigma(\pi(j))j} - c_{\sigma(j)j}) \\
\le\; & \sum_{j\in\mathcal{D}:\sigma^*(j)\in B}(c_{\sigma^*(j)j} - c_{\sigma(j)j}) + \sum_{j\in\mathcal{D}:\sigma(j)\in A}(c_{\sigma(\pi(j))j} - c_{\sigma(j)j})
\end{aligned}
$$

as $c_{\sigma(\pi(j))j} \ge c_{\sigma(j)j}$ by definition of σ. Hence taking the weighted sum over all swaps yields

$$
\begin{aligned}
0 \;\le\; & \sum_{j\in\mathcal{D}}(c_{\sigma^*(j)j} - c_{\sigma(j)j}) + \frac{p+1}{p}\sum_{j\in\mathcal{D}}(c_{\sigma(\pi(j))j} - c_{\sigma(j)j}) \\
\le\; & c_S(X^*) - c_S(X) + \frac{p+1}{p}\sum_{j\in\mathcal{D}}(c_{\sigma^*(j)j} + c_{\sigma^*(j)\pi(j)} + c_{\sigma(\pi(j))\pi(j)} - c_{\sigma(j)j}) \\
=\; & c_S(X^*) - c_S(X) + \frac{p+1}{p}\sum_{j\in\mathcal{D}}(c_{\sigma^*(j)j} + c_{\sigma^*(\pi(j))\pi(j)}) \\
=\; & c_S(X^*) - c_S(X) + 2\,\frac{p+1}{p}c_S(X^*),
\end{aligned}
$$

because π is a bijection.   □

Arya et al. [2004] also showed that this performance guarantee is tight. Like Corollary 22.18, Lemma 22.15 and Theorem 22.19 imply a $(3+\epsilon)$-factor approximation algorithm for any $\epsilon > 0$. This is the currently best known performance ratio for the METRIC k-MEDIAN PROBLEM.

We can apply similar techniques to the METRIC UNCAPACITATED FACILITY LOCATION PROBLEM and obtain a simple approximation algorithm based on local search:

Theorem 22.20.   (Arya et al. [2004])   *Consider an instance of the* METRIC UN-CAPACITATED FACILITY LOCATION PROBLEM. *Let X and X^* be any feasible solutions. If neither $X\setminus\{x\}$ nor $X\cup\{y\}$ nor $(X\setminus\{x\})\cup\{y\}$ is better than X for any $x \in X$ and $y \in \mathcal{F}\setminus X$, then $c_S(X) \le c_F(X^*) + c_S(X^*)$ and $c_F(X) \le c_F(X^*) + 2c_S(X^*)$.*

Proof: We use the same notation as in the previous proofs. In particular, let σ and σ^* be optimum assignments of the customers to X and X^*, respectively.

The first inequality is easily proved by considering, for each $y \in X^*$, the operation of adding y to X, which increases the cost by at most $f_y + \sum_{j\in\mathcal{D}:\sigma^*(j)=y}(c_{\sigma^*(j)j} - c_{\sigma(j)j})$. Summing these values up yields that $c_F(X^*) + c_S(X^*) - c_S(X)$ is nonnegative.

Let again $\pi : \mathcal{D} \to \mathcal{D}$ be a bijection such that for all $j \in \mathcal{D}$:

- $\sigma^*(\pi(j)) = \sigma^*(j)$;
- if $\sigma(\pi(j)) = \sigma(j)$ then $\sigma(j)$ captures $\sigma^*(j)$ and $\pi(j) = j$.

Such a mapping π can be obtained as in the proof of Theorem 22.17 after fixing $\pi(j) := j$ for $|\{j \in \mathcal{D} : \sigma^*(j) = y, \sigma(j) = x\}| - |\{j \in \mathcal{D} : \sigma^*(j) = y, \sigma(j) \neq x\}|$ elements $j \in \mathcal{D}$ with $\sigma^*(j) = y$ and $\sigma(j) = x$ for any pair $x \in X$, $y \in X^*$ where x captures y.

To bound the facility cost of X, let $x \in X$, and let $\mathcal{D}_x := \{j \in \mathcal{D} : \sigma(j) = x\}$. If x does not capture any $y \in X^*$, we consider dropping x and reassigning each $j \in \mathcal{D}_x$ to $\sigma(\pi(j)) \in X \setminus \{x\}$. Hence

$$0 \le -f_x + \sum_{j \in \mathcal{D}_x} (c_{\sigma(\pi(j))j} - c_{xj}). \tag{22.10}$$

If the set $C(\{x\})$ of facilities captured by x is nonempty, let $y \in C(\{x\})$ be a nearest facility in $C(\{x\})$ (i.e. $\min_{j \in \mathcal{D}}(c_{xj} + c_{yj})$ is minimum). We consider the addition of each facility $y' \in C(\{x\}) \setminus \{y\}$, which increases the cost by at least zero and at most

$$f_{y'} + \sum_{j \in \mathcal{D}_x : \sigma^*(j)=y', \pi(j)=j} (c_{\sigma^*(j)j} - c_{xj}). \tag{22.11}$$

Moreover, we consider the swap $(\{x\}, \{y\})$. For $j \in \mathcal{D}_x$ we reassign j to $\sigma(\pi(j))$ if $\pi(j) \neq j$, and to y otherwise.

The new service cost for $j \in \mathcal{D}_x$ is at most $c_{\sigma(\pi(j))j}$ in the first case, $c_{\sigma^*(j)j}$ if $\pi(j) = j$ and $\sigma^*(j) = y$, and

$$c_{yj} \le c_{xj} + \min_{k \in \mathcal{D}}(c_{xk} + c_{yk}) \le c_{xj} + \min_{k \in \mathcal{D}}(c_{xk} + c_{\sigma^*(j)k}) \le 2c_{xj} + c_{\sigma^*(j)j}$$

otherwise, where the second inequality holds because x captures $\sigma^*(j)$ if $\pi(j) = j$.

Altogether, the swap from x to y increases the cost by at least zero and at most

$$f_y - f_x - \sum_{j \in \mathcal{D}_x} c_{xj} + \sum_{j \in \mathcal{D}_x : \pi(j) \neq j} c_{\sigma(\pi(j))j}$$
$$+ \sum_{j \in \mathcal{D}_x : \pi(j)=j, \sigma^*(j)=y} c_{\sigma^*(j)j} + \sum_{j \in \mathcal{D}_x : \pi(j)=j, \sigma^*(j) \neq y} (2c_{xj} + c_{\sigma^*(j)j}). \tag{22.12}$$

Adding (22.11) and (22.12), both of which are nonnegative, yields

$$
\begin{aligned}
0 \le{}& \sum_{y' \in C(\{x\})} f_{y'} - f_x + \sum_{j \in \mathcal{D}_x : \pi(j) \neq j} (c_{\sigma(\pi(j))j} - c_{xj}) \\
&+ \sum_{j \in \mathcal{D}_x : \pi(j)=j, \sigma^*(j)=y} (c_{\sigma^*(j)j} - c_{xj}) + \sum_{j \in \mathcal{D}_x : \pi(j)=j, \sigma^*(j) \neq y} 2c_{\sigma^*(j)j} \\
\le{}& \sum_{y' \in C(\{x\})} f_{y'} - f_x + \sum_{j \in \mathcal{D}_x : \pi(j) \neq j} (c_{\sigma(\pi(j))j} - c_{xj}) + 2 \sum_{j \in \mathcal{D}_x : \pi(j)=j} c_{\sigma^*(j)j}.
\end{aligned}
$$
$$\tag{22.13}$$

Summing (22.10) and (22.13), respectively, over all $x \in X$ yields

$$
\begin{aligned}
0 \; \leq \; & \sum_{x \in X} \sum_{y' \in C(\{x\})} f_{y'} - c_F(X) + \sum_{j \in \mathcal{D}: \pi(j) \neq j} (c_{\sigma(\pi(j))j} - c_{\sigma(j)j}) \\
& + 2 \sum_{j \in \mathcal{D}: \pi(j) = j} c_{\sigma^*(j)j} \\
\leq \; & c_F(X^*) - c_F(X) + \sum_{j \in \mathcal{D}: \pi(j) \neq j} (c_{\sigma^*(j)j} + c_{\sigma^*(j)\pi(j)} + c_{\sigma(\pi(j))\pi(j)} - c_{\sigma(j)j}) \\
& + 2 \sum_{j \in \mathcal{D}: \pi(j) = j} c_{\sigma^*(j)j} \\
= \; & c_F(X^*) - c_F(X) + 2c_S(X^*).
\end{aligned}
$$

$\square$

With Lemma 22.15 this implies a $(3+\epsilon)$-factor approximation algorithm for any $\epsilon > 0$. Combining this with Theorem 22.10, we get a 2.375-factor approximation algorithm (Exercise 12). Charikar and Guha [2005] proved the same performance guarantee for a very similar local search algorithm.

22.7 Capacitated Facility Location Problems

A main advantage of local search algorithms is their flexibility; they can be applied to arbitrary cost functions and even in the presence of complicated additional constraints. For most facility location problems with hard capacities, local search is the only technique that is currently known to yield a performance guarantee.

Several capacitated facility location problems have been studied. Mahdian and Pál [2003] defined the following general problem, which contains several important special cases:

UNIVERSAL FACILITY LOCATION PROBLEM

Instance: Finite sets $\mathcal{D}$ of customers and $\mathcal{F}$ of potential facilities; a metric c on $V := \mathcal{D} \cup \mathcal{F}$, i.e. distances $c_{ij} \geq 0$ $(i, j \in V)$ with $c_{ii} = 0$, $c_{ij} = c_{ji}$ and $c_{ij} + c_{jk} \geq c_{ik}$ for all $i, j, k \in V$; a demand $d_j \geq 0$ for each $j \in \mathcal{D}$; and for each $i \in \mathcal{F}$ a cost function $f_i : \mathbb{R}_+ \to \mathbb{R}_+ \cup \{\infty\}$ which is left-continuous and non-decreasing.

Task: Find $x_{ij} \in \mathbb{R}_+$ for $i \in \mathcal{F}$ and $j \in \mathcal{D}$, with $\sum_{i \in \mathcal{F}} x_{ij} = d_j$ for all $j \in \mathcal{D}$, such that $c(x) := c_F(x) + c_S(x)$ is minimum, where

$$
c_F(x) := \sum_{i \in \mathcal{F}} f_i \left(\sum_{j \in \mathcal{D}} x_{ij} \right) \qquad \text{and} \qquad c_S(x) := \sum_{i \in \mathcal{F}} \sum_{j \in \mathcal{D}} c_{ij} x_{ij}.
$$

$f_i(z)$ can be interpreted as the cost of installing capacity z at facility i. We have to specify how the functions f_i are given. We assume an oracle that, for each $i \in \mathcal{F}$, $u, c \in \mathbb{R}_+$ and $t \in \mathbb{R}$, computes $f_i(u)$ and $\max\{\delta \in \mathbb{R} : u + \delta \geq 0, f_i(u + \delta) - f_i(u) + c|\delta| \leq t\}$. This is a natural assumption as this oracle can be implemented trivially for the most important special cases of the UNIVERSAL FACILITY LOCATION PROBLEM. These are:

- the METRIC UNCAPACITATED FACILITY LOCATION PROBLEM. Here $d_j = 1$ ($j \in \mathcal{D}$), and $f_i(0) = 0$ and $f_i(z) = t_i$ for some $t_i \in \mathbb{R}_+$ and all $z > 0$ ($i \in \mathcal{F}$).
- the METRIC CAPACITATED FACILITY LOCATION PROBLEM. Here $f_i(0) = 0$, $f_i(z) = t_i$ for $0 < z \leq u_i$ and $f_i(z) = \infty$ for $z > u_i$, where $u_i, t_i \in \mathbb{R}_+$ ($i \in \mathcal{F}$).
- the METRIC SOFT-CAPACITATED FACILITY LOCATION PROBLEM. Here $d_j = 1$ ($j \in \mathcal{D}$), and $f_i(z) = \lceil \frac{z}{u_i} \rceil t_i$ for some $u_i \in \mathbb{N}$, $t_i \in \mathbb{R}_+$ and all $z \geq 0$ ($i \in \mathcal{F}$).

Note that in the first and third case there is always an optimum integral solution. While this is trivial in the first case, it follows easily in the third case by taking an arbitrary optimum solution y and applying the following observation to $d_j = 1$ for $j \in \mathcal{D}$ and $z_i = \max\{z : f_i(z) \leq f_i(\sum_{j \in \mathcal{D}} y_{ij})\} \in \mathbb{Z}_+$ for $i \in \mathcal{F}$:

Proposition 22.21. *Let $\mathcal{D}$ and $\mathcal{F}$ be finite sets, $d_j \geq 0$ ($j \in \mathcal{D}$), $z_i \geq 0$ ($i \in \mathcal{F}$) and $c_{ij} \geq 0$ ($i \in \mathcal{F}, j \in \mathcal{D}$) such that $\sum_{j \in \mathcal{D}} d_j \leq \sum_{i \in \mathcal{F}} z_i$. Then an optimum solution to*

$$\min \left\{ \sum_{i \in \mathcal{F}, j \in \mathcal{D}} c_{ij} x_{ij} : x \geq 0, \sum_{i \in \mathcal{F}} x_{ij} = d_j \ (j \in \mathcal{D}), \sum_{j \in \mathcal{D}} x_{ij} \leq z_i \ (i \in \mathcal{F}) \right\} \tag{22.14}$$

can be found in $O(n^3 \log n)$ time, where $n = |\mathcal{D}| + |\mathcal{F}|$. If all d_j and z_i are integral, then there exists an integral optimum solution.

Proof: (22.14) is equivalent to the instance (G, b, c) of the HITCHCOCK PROBLEM, defined by $G := (A \,\dot\cup\, B, A \times B)$, $A := \{v_j : j \in \mathcal{D}\} \,\dot\cup\, \{0\}$, $B := \{w_i : i \in \mathcal{F}\}$, $b(v_j) := d_j$ for $j \in \mathcal{D}$, $b(w_i) = -z_i$ for $i \in \mathcal{F}$, $b(0) := \sum_{i \in \mathcal{F}} z_i - \sum_{j \in \mathcal{D}} d_j$, $c(v_j, w_i) := c_{ij}$ and $c(0, w_i) := 0$ for $i \in \mathcal{F}$ and $j \in \mathcal{D}$. Thus (22.14) can be solved in $O(n^3 \log n)$ time by Theorem 9.19. If b is integral, the MINIMUM MEAN CYCLE-CANCELLING ALGORITHM and the SUCCESSIVE SHORTEST PATH ALGORITHM compute integral optimum solutions. $\square$

The soft-capacitated version can be reduced quite easily to the uncapacitated one, by a technique that has been proposed originally by Jain and Vazirani [2001]:

Theorem 22.22.   (Mahdian, Ye and Zhang [2006]) *Let γ_F and γ_S be constants and A a polynomial-time algorithm such that, for every instance of the METRIC UNCAPACITATED FACILITY LOCATION PROBLEM, A computes a solution X with $c_F(X) + c_S(X) \leq \gamma_F c_F(X^*) + \gamma_S c_S(X^*)$ for each $\emptyset \neq X^* \subseteq \mathcal{F}$. Then there is a*

($\gamma_F + \gamma_S$)-*factor approximation algorithm for the* METRIC SOFT-CAPACITATED FACILITY LOCATION PROBLEM.

Proof: Consider an instance $I = (\mathcal{F}, \mathcal{D}, (c_{ij})_{i \in \mathcal{F}, j \in \mathcal{D}}, (f_i)_{i \in \mathcal{F}})$ of the METRIC SOFT-CAPACITATED FACILITY LOCATION PROBLEM, where $f_i(z) = \lceil \frac{z}{u_i} \rceil t_i$ for $i \in \mathcal{F}$ and $z \in \mathbb{R}_+$. We transform it to the instance $I' = (\mathcal{F}, \mathcal{D}, (f_i')_{i \in \mathcal{F}}, (c_{ij}')_{i \in \mathcal{F}, j \in \mathcal{D}})$ of the METRIC UNCAPACITATED FACILITY LOCATION PROBLEM by setting $f_i' := t_i$ and $c_{ij}' := c_{ij} + \frac{t_i}{u_i}$ for $i \in \mathcal{F}$ and $j \in \mathcal{D}$. (Note that c' is metric whenever c is metric.)

We apply A to I' and find a solution $X \subseteq \mathcal{F}$ and an assignment $\sigma : \mathcal{D} \to X$. Set $x_{ij} := 1$ if $\sigma(j) = i$ and $x_{ij} := 0$ otherwise. If $\sigma^* : \mathcal{D} \to \mathcal{F}$ is an optimum solution to I and $X^* := \{i \in \mathcal{F} : \exists j \in \mathcal{D} : \sigma^*(j) = i\}$ is the set of facilities opened at least once,

$$
\begin{aligned}
c_F(x) + c_S(x) &= \sum_{i \in X} \left\lceil \frac{|\{j \in \mathcal{D} : \sigma(j) = i\}|}{u_i} \right\rceil t_i + \sum_{j \in \mathcal{D}} c_{\sigma(j)j} \\
&\leq \sum_{i \in X} t_i + \sum_{j \in \mathcal{D}} c'_{\sigma(j)j} \\
&\leq \gamma_F \sum_{i \in X^*} t_i + \gamma_S \sum_{j \in \mathcal{D}} c'_{\sigma^*(j)j} \\
&\leq (\gamma_F + \gamma_S) \sum_{i \in X^*} \left\lceil \frac{|\{j \in \mathcal{D} : \sigma^*(j) = i\}|}{u_i} \right\rceil t_i + \gamma_S \sum_{j \in \mathcal{D}} c_{\sigma^*(j)j}.
\end{aligned}
$$

$\square$

Corollary 22.23. *The* METRIC SOFT-CAPACITATED FACILITY LOCATION PROBLEM *has a 2.89-factor approximation algorithm.*

Proof: Apply Theorem 22.22 to the DUAL FITTING ALGORITHM (Corollary 22.7(c)); here $\gamma_F = 1.11$ and $\gamma_S = 1.78$. $\square$

See Exercise 11 for a better performance ratio.

When dealing with hard capacities, we have to allow the demand of customers to be split, i.e. assigned to multiple open facilities: if we do not allow splitting, we cannot expect any result as even deciding whether a feasible solution exists at all is *NP*-complete (this contains the PARTITION problem; cf. Corollary 15.28).

The first approximation algorithm for the METRIC CAPACITATED FACILITY LOCATION PROBLEM was due to Pál, Tardos and Wexler [2001], extending an earlier result for a special case by Korupolu, Plaxton and Rajaraman [2000]. The performance guarantee was then improved to 5.83 by Zhang, Chen and Ye [2005]. For the special case of uniform facility opening costs, Levi, Shmoys and Swamy [2012] obtained a 5-factor approximation algorithm by rounding an LP relaxation. For the general METRIC CAPACITATED FACILITY LOCATION PROBLEM, An, Singh and Svensson [2017] obtained the first constant-factor approximation algorithm that is based on rounding an LP relaxation.

The work by Pál, Tardos and Wexler [2001] has been generalized to the UNIVERSAL FACILITY LOCATION PROBLEM by Mahdian and Pál [2003]. They obtained a 7.88-factor approximation algorithm. In the next section we present a local search algorithm that yields a performance guarantee of 6.702 for the UNIVERSAL FACILITY LOCATION PROBLEM. But let us first note the following.

Lemma 22.24. (Mahdian and Pál [2003]) *Every instance of the* UNIVERSAL FACILITY LOCATION PROBLEM *has an optimum solution.*

Proof: If there is no solution with finite cost, any solution is optimum. Otherwise let $(x^i)_{i \in \mathbb{N}}$ be a sequence of solutions whose costs approach the infimum $c^* \in \mathbb{R}_+$ of the set of costs of feasible solutions. As this sequence is bounded, there is a subsequence $(x^{i_j})_{j \in \mathbb{N}}$ converging to some x^*. x^* is feasible. As all f_i are left-continuous and non-decreasing, we have $c(x^*) = c(\lim_{j \to \infty} x^{i_j}) \le \lim_{j \to \infty} c(x^{i_j}) = c^*$, i.e. x^* is optimum.   □

22.8 Universal Facility Location

In this section, based on Vygen [2007], we present a local search algorithm for the UNIVERSAL FACILITY LOCATION PROBLEM. It uses two operations. First, for $t \in \mathcal{F}$ and $\delta \in \mathbb{R}_+$ we consider the operation ADD(t, δ), which consists in replacing the current feasible solution x by an optimum solution y to the following problem:

$$\min\left\{ c_S(y) \; : \; y_{ij} \ge 0 \,(i \in \mathcal{F}, \; j \in \mathcal{D}), \; \sum_{i \in \mathcal{F}} y_{ij} = d_j \,(j \in \mathcal{D}), \right.$$
$$\left. \sum_{j \in \mathcal{D}} y_{ij} \le \sum_{j \in \mathcal{D}} x_{ij} \,(i \in \mathcal{F} \setminus \{t\}), \; \sum_{j \in \mathcal{D}} y_{tj} \le \sum_{j \in \mathcal{D}} x_{tj} + \delta \right\}. \tag{22.15}$$

We denote by $c^x(t, \delta) := c_S(y) - c_S(x) + f_t(\sum_{j \in \mathcal{D}} x_{tj} + \delta) - f_t(\sum_{j \in \mathcal{D}} x_{tj})$ the estimated cost of this operation; this is an upper bound on $c(y) - c(x)$.

Lemma 22.25. (Mahdian and Pál [2003]) *Let $\epsilon > 0$. Let x be a feasible solution to a given instance, and let $t \in \mathcal{F}$. Then there is an algorithm with running time $O(|V|^3 \log |V| \epsilon^{-1})$ that finds a $\delta \in \mathbb{R}_+$ with $c^x(t, \delta) \le -\epsilon c(x)$ or decides that no $\delta \in \mathbb{R}_+$ exists for which $c^x(t, \delta) \le -2\epsilon c(x)$.*

Proof: We may assume that $c(x) > 0$. Let $C := \{v \epsilon c(x) : v \in \mathbb{Z}_+, v \le \lceil \frac{1}{\epsilon} \rceil\}$. For each $\gamma \in C$ let δ_γ be the maximum $\delta \in \mathbb{R}_+$ for which $f_t(\sum_{j \in \mathcal{D}} x_{tj} + \delta) - f_t(\sum_{j \in \mathcal{D}} x_{tj}) \le \gamma$. We compute $c^x(t, \delta_\gamma)$ for all $\gamma \in C$.

Suppose there is a $\delta \in \mathbb{R}_+$ with $c^x(t, \delta) \leq -2\epsilon c(x)$. Then consider

$$\gamma := \epsilon c(x) \left\lceil \frac{1}{\epsilon c(x)} \left(f_t \left(\sum_{j \in \mathcal{D}} x_{tj} + \delta \right) - f_t \left(\sum_{j \in \mathcal{D}} x_{tj} \right) \right) \right\rceil \in C.$$

Note that $\delta_\gamma \geq \delta$ and hence $c^x(t, \delta_\gamma) < c^x(t, \delta) + \epsilon c(x) \leq -\epsilon c(x)$.

The running time is dominated by solving $|C|$ problems of the type (22.15). Hence the running time follows from Proposition 22.21. $\qquad\square$

If there is no sufficiently profitable ADD operation, the service cost can be bounded. The following result is essentially due to Pál, Tardos and Wexler [2001]:

Lemma 22.26. *Let $\epsilon > 0$, and let x, x^* be feasible solutions to a given instance, and let $c^x(t, \delta) \geq -\frac{\epsilon}{|\mathcal{F}|} c(x)$ for all $t \in \mathcal{F}$ and $\delta \in \mathbb{R}_+$. Then $c_S(x) \leq c_F(x^*) + c_S(x^*) + \epsilon c(x)$.*

Proof: Consider the (complete bipartite) digraph $G = (\mathcal{D} \dot\cup \mathcal{F}, (\mathcal{D} \times \mathcal{F}) \cup (\mathcal{F} \times \mathcal{D}))$ with edge weights $c((j, i)) := c_{ij}$ and $c((i, j)) := -c_{ij}$ for $i \in \mathcal{F}$ and $j \in \mathcal{D}$. Let $b(i) := \sum_{j \in \mathcal{D}} (x_{ij} - x_{ij}^*)$ for $i \in \mathcal{F}$, $S := \{i \in \mathcal{F} : b(i) > 0\}$ and $T := \{i \in \mathcal{F} : b(i) < 0\}$.

Define a b-flow $g : E(G) \to \mathbb{R}_+$ by $g(i, j) := \max\{0, x_{ij} - x_{ij}^*\}$ and $g(j, i) := \max\{0, x_{ij}^* - x_{ij}\}$ for $i \in \mathcal{F}$, $j \in \mathcal{D}$.

Write g as the sum of b_t-flows g_t for $t \in T$, where $b_t(t) = b(t)$, $b_t(v) = 0$ for $v \in T \setminus \{t\}$ and $0 \leq b_t(v) \leq b(v)$ for $v \in V(G) \setminus T$. (This can be done by standard flow decomposition techniques.)

For each $t \in T$, g_t defines a feasible way to reassign customers to t, i.e. a new solution x^t defined by $x_{ij}^t := x_{ij} + g_t(j, i) - g_t(i, j)$ for $i \in \mathcal{F}$, $j \in \mathcal{D}$. We have $c_S(x^t) = c_S(x) + \sum_{e \in E(G)} c(e) g_t(e)$ and hence

$$c^x(t, -b(t)) \leq \sum_{e \in E(G)} c(e) g_t(e) + f_t \left(\sum_{j \in \mathcal{D}} x_{tj}^* \right) - f_t \left(\sum_{j \in \mathcal{D}} x_{tj} \right).$$

If the left-hand side is at least $-\frac{\epsilon}{|\mathcal{F}|} c(x)$ for each $t \in T$, summation yields

$$\begin{aligned}
-\epsilon c(x) &\leq \sum_{e \in E(G)} c(e) g(e) + \sum_{t \in T} f_t \left(\sum_{j \in \mathcal{D}} x_{tj}^* \right) \\
&\leq \sum_{e \in E(G)} c(e) g(e) + c_F(x^*) \\
&= c_S(x^*) - c_S(x) + c_F(x^*).
\end{aligned}$$

$\qquad\square$

We will now describe the second type of operation. Let x be a feasible solution for a given instance of the UNIVERSAL FACILITY LOCATION PROBLEM. Let A

be an arborescence with $V(A) \subseteq \mathcal{F}$ and $\delta \in \Delta_A^x := \{\delta \in \mathbb{R}^{V(A)} : \sum_{j \in \mathcal{D}} x_{ij} + \delta_i \geq 0$ for all $i \in V(A)$, $\sum_{i \in V(A)} \delta_i = 0\}$.

Then we consider the operation PIVOT(A, δ), which consists in replacing x by a solution x' with $\sum_{j \in \mathcal{D}} x'_{ij} = \sum_{j \in \mathcal{D}} x_{ij} + \delta_i$ for $i \in V(A)$, $\sum_{j \in \mathcal{D}} x'_{ij} = \sum_{j \in \mathcal{D}} x_{ij}$ for $i \in \mathcal{F} \setminus V(A)$ and $c(x') \leq c(x) + c^x(A, \delta)$, where $c^x(A, \delta) := \sum_{i \in V(A)} c_{A,i}^x(\delta)$ and

$$c_{A,i}^x(\delta) := f_i\left(\sum_{j \in \mathcal{D}} x_{ij} + \delta_i\right) - f_i\left(\sum_{j \in \mathcal{D}} x_{ij}\right) + \left|\sum_{l \in A_i^+} \delta_l\right| c_{ip(i)}$$

for $i \in V(A)$. Here A_i^+ denotes the set of vertices reachable from i in A, and $p(i)$ is the predecessor of i in A (and arbitrary if i is the root). Such an x' can be constructed easily by moving demand along the edges in A in reverse topological order. Note that the orientation of A is irrelevant for $c^x(A, \delta)$ and used only to simplify notation.

The operation will be performed if its *estimated cost* $c^x(A, \delta)$ is sufficiently negative. This guarantees that the resulting local search algorithm stops after a polynomial number of improvement steps. We call $\sum_{i \in V(A)} \left|\sum_{l \in A_i^+} \delta_l\right| c_{ip(i)}$ the *estimated routing cost* of PIVOT(A, δ).

We now show how to find an improving PIVOT operation unless we are at an approximate local optimum:

Lemma 22.27. (Vygen [2007]) *Let $\epsilon > 0$ and A an arborescence with $V(A) \subseteq \mathcal{F}$. Let x be a feasible solution. Then there is an algorithm with running time $O(|\mathcal{F}|^4 \epsilon^{-3})$ that finds a $\delta \in \Delta_A^x$ with $c^x(A, \delta) \leq -\epsilon c(x)$ or decides that no $\delta \in \Delta_A^x$ exists for which $c^x(A, \delta) \leq -2\epsilon c(x)$.*

Proof: Number $V(A) = \{1, \ldots, n\}$ in reverse topological order, i.e. for all $(i, j) \in E(A)$ we have $i > j$. For $k \in V(A)$ with $(p(k), k) \in E(A)$ let $B(k) := \{i < k : (p(k), i) \in E(A)\}$ be the set of smaller siblings of k, and let $B(k) := \emptyset$ if k is the root of A. Let $I_k := \bigcup_{l \in B(k) \cup \{k\}} A_l^+$, $b(k) := \max(\{0\} \cup B(k))$ and $s(k) := \max(\{0\} \cup (A_k^+ \setminus \{k\}))$.

Let $C := \{v \frac{\epsilon}{n} c(x) : v \in \mathbb{Z}, -\lceil \frac{n}{\epsilon} \rceil - n \leq v \leq \lceil \frac{n}{\epsilon} \rceil + n\}$. We compute the table $(T_A^x(k, \gamma))_{k \in \{0, \ldots, n\}, \gamma \in C}$, defined as follows. Let $T_A^x(0, 0) := 0$, $T_A^x(0, \gamma) := \emptyset$ for all $\gamma \in C \setminus \{0\}$, and for $k = 1, \ldots, n$ let $T_A^x(k, \gamma)$ be an optimum solution $\delta \in \mathbb{R}^{I_k}$ of

$$\max\left\{\sum_{i \in I_k} \delta_i : \gamma' \in C, \ T_A^x(b(k), \gamma') \neq \emptyset, \ \delta_i = (T_A^x(b(k), \gamma'))_i \text{ for } i \in \bigcup_{l \in B(k)} A_l^+, \right.$$

$$\gamma'' \in C, \ T_A^x(s(k), \gamma'') \neq \emptyset, \ \delta_i = (T_A^x(s(k), \gamma''))_i \text{ for } i \in A_k^+ \setminus \{k\},$$

$$\left. \sum_{j \in \mathcal{D}} x_{kj} + \delta_k \geq 0, \ \gamma' + \gamma'' + c_{A,k}^x(\delta) \leq \gamma \right\}$$

if the set over which the maximum is taken is nonempty, and $T_A^x(k, \gamma) := \emptyset$ otherwise.

Roughly, $-\sum_{i \in I_k}(T_A^x(k, \gamma))_i$ is the minimum excess we get at the predecessor $p(k)$ of k when moving demand from each vertex in I_k to its respective predecessor or vice versa, at a total rounded estimated cost of at most γ.

Note that $T_A^x(k, 0) \neq \emptyset$ for $k = 0, \ldots, n$. Thus we can choose the minimum $\gamma \in C$ such that $T_A^x(n, \gamma) \neq \emptyset$ and $\sum_{i=1}^n (T_A^x(n, \gamma))_i \geq 0$. Then we choose $\delta \in \Delta_A^x$ such that $\delta_i = (T_A^x(n, \gamma))_i$ or $0 \leq \delta_i \leq (T_A^x(n, \gamma))_i$ for all $i = 1, \ldots, n$ and $|\sum_{l \in A_i^+} \delta_l| \leq |\sum_{l \in A_i^+}(T_A^x(n, \gamma))_l|$ for all $i = 1, \ldots, n$. This can be done by setting $\delta := T_A^x(n, \gamma)$ and repeatedly decreasing δ_i for the maximum i for which $\delta_i > 0$ and $\sum_{l \in A_k^+} \delta_l > 0$ for all vertices k on the path from n to i in A. Note that the property $c^x(A, \delta) \leq \gamma$ is maintained. It remains to show that γ is small enough.

Suppose there exists an operation $\text{PIVOT}(A, \delta)$ with $c^x(A, \delta) \leq -2\epsilon c(x)$. As $c_{A,i}^x(\delta) \geq -f_i(\sum_{j \in D} x_{ij}) \geq -c(x)$ for all $i \in V(A)$, this also implies $c_{A,i}^x(\delta) < c_F(x) \leq c(x)$. Hence $\gamma_i := \left\lceil c_{A,i}^x(\delta) \frac{n}{\epsilon c(x)} \right\rceil \frac{\epsilon c(x)}{n} \in C$ for $i = 1, \ldots, n$, and $\sum_{i \in I} \gamma_i \in C$ for all $I \subseteq \{1, \ldots, n\}$. Then an easy induction shows $\sum_{i \in I_k}(T_A^x(k, \sum_{l \in I_k} \gamma_l))_i \geq \sum_{i \in I_k} \delta_i$ for $k = 1, \ldots, n$. Hence we find a pivot operation with estimated cost at most $\sum_{i=1}^n \gamma_i < c^x(A, \delta) + \epsilon c(x) \leq -\epsilon c(x)$.

The running time can be estimated as follows. We have to compute $n|C|$ table entries, and for each entry $T_A^x(k, \gamma)$ we try all values of $\gamma', \gamma'' \in C$. This yields values δ_i for $i \in I_k \setminus \{k\}$, and the main step is to compute the maximum δ_k for which $\gamma' + \gamma'' + c_{A,k}^x(\delta) \leq \gamma$. This can be done directly with the oracle that we assumed for the functions f_i, $i \in \mathcal{F}$. The final computation of δ from $T_A^x(n, \gamma)$, $\gamma \in C$, is easily done in linear time. Hence the overall running time is $O(n|C|^3) = O(|\mathcal{F}|^4 \epsilon^{-3})$. $\square$

We consider $\text{PIVOT}(A, \delta)$ for special arborescences: stars and comets. A is called the **star centered at** v if $A = (\mathcal{F}, \{(v, w) : w \in \mathcal{F} \setminus \{v\}\})$ and the **comet with center** v **and tail** (t, s) if $A = (\mathcal{F}, \{(t, s)\} \cup \{(v, w) : w \in \mathcal{F} \setminus \{v, s\}\})$ and v, t, s are distinct elements of $\mathcal{F}$. Note that there are less than $|\mathcal{F}|^3$ stars and comets.

We will now show that an (approximate) local optimum has low facility cost.

Lemma 22.28. *Let* x, x^* *be feasible solutions to a given instance, and let* $c^x(A, \delta) \geq -\frac{\epsilon}{|\mathcal{F}|} c(x)$ *for all stars and comets* A *and* $\delta \in \Delta_A^x$. *Then* $c_F(x) \leq 4c_F(x^*) + 2c_S(x^*) + 2c_S(x) + \epsilon c(x)$.

Proof: We use the notation of Lemma 22.26 and consider the following instance of the HITCHCOCK PROBLEM:

$$\min \quad \sum_{s \in S, t \in T} c_{st} y(s, t)$$

$$\text{s.t.} \quad \sum_{t \in T} y(s, t) = b(s) \quad (s \in S) \tag{22.16}$$

$$\sum_{s \in S} y(s, t) = -b(t) \quad (t \in T)$$

$$y(s, t) \geq 0 \quad (s \in S, t \in T)$$

By Proposition 9.22 there exists an optimum solution $y : S \times T \to \mathbb{R}_+$ of (22.16) such that $F := (S \cup T, \{\{s, t\} : y(s, t) > 0\})$ is a forest.

As $(b_t(s))_{s \in S, t \in T}$ is a feasible solution of (22.16), we have

$$
\begin{aligned}
\sum_{s \in S, t \in T} c_{st} y(s, t) &\leq \sum_{s \in S, t \in T} c_{st} b_t(s) \\
&= \sum_{s \in S, t \in T} c_{st} (g_t(\delta^+(s)) - g_t(\delta^-(s))) \\
&\leq \sum_{e \in E(G)} |c(e)| g(e) \\
&\leq c_S(x^*) + c_S(x).
\end{aligned}
\tag{22.17}
$$

We will now define at most $|\mathcal{F}|$ PIVOT operations. We say that an operation PIVOT(A, δ) *closes* $s \in S$ (with respect to x and x^*) if $\sum_{j \in \mathcal{D}} x_{sj} > \sum_{j \in \mathcal{D}} x_{sj}^* + \delta_s = \sum_{j \in \mathcal{D}} x_{sj}^*$. We say that it *opens* $t \in T$ if $\sum_{j \in \mathcal{D}} x_{tj} < \sum_{j \in \mathcal{D}} x_{tj} + \delta_t \leq \sum_{j \in \mathcal{D}} x_{tj}^*$. Over all operations that we are going to define, each $s \in S$ will be closed once, and each $t \in T$ will be opened at most four times. Moreover, the total estimated routing cost will be at most $2 \sum_{s \in S, t \in T} c_{st} y(s, t)$. Thus the total estimated cost of the operations will be at most $4c_F(x^*) + 2c_S(x^*) + 2c_S(x) - c_F(x)$. This will prove the lemma.

To define the operations, orient F as a branching B each of whose components is rooted at an element of T. Write $y(e) := y(s, t)$ if $e \in E(B)$ has endpoints $s \in S$ and $t \in T$. A vertex $v \in V(B)$ is called *weak* if $y(\delta_B^+(v)) > y(\delta_B^-(v))$ and *strong* otherwise. We denote by $\Gamma_s^+(v)$, $\Gamma_w^+(v)$ and $\Gamma^+(v)$ the set of strong, weak, and all children of $v \in V(B)$ in B, respectively.

Let $t \in T$, and let $\Gamma_w^+(t) = \{w_1, \ldots, w_k\}$ be the weak children of t ordered such that $r(w_1) \leq \cdots \leq r(w_k)$, where $r(w_i) := \max\left\{0, y(w_i, t) - \sum_{t' \in \Gamma_w^+(w_i)} y(w_i, t')\right\}$. Moreover, order $\Gamma_s^+(t) = \{s_1, \ldots, s_l\}$ such that $y(s_1, t) \geq \cdots \geq y(s_l, t)$.

Let us first assume $k > 0$. For $i = 1, \ldots, k - 1$ consider a PIVOT operation with the star centered at w_i, routing

- at most $2y(w_i, t')$ units of demand from w_i to each weak child t' of w_i,
- $y(w_i, t')$ units from w_i to each strong child t' of w_i, and
- $r(w_i)$ units from w_i to $\Gamma_s^+(w_{i+1})$,

closing w_i and opening a subset of $\Gamma^+(w_i) \cup \Gamma_s^+(w_{i+1})$. The estimated routing cost is at most

$$
\sum_{t' \in \Gamma_w^+(w_i)} c_{w_i t'} 2y(w_i, t') + \sum_{t' \in \Gamma_s^+(w_i)} c_{w_i t'} y(w_i, t') + c_{t w_i} r(w_i)
$$

$$
+ c_{t w_{i+1}} r(w_{i+1}) + \sum_{t' \in \Gamma_s^+(w_{i+1})} c_{w_{i+1} t'} y(w_{i+1}, t'),
$$

as $r(w_i) \leq r(w_{i+1}) \leq \sum_{t' \in \Gamma_s^+(w_{i+1})} y(w_{i+1}, t')$.

To define more PIVOT operations related to t, we distinguish three cases.

Case 1: t is strong or $l = 0$. Then consider:

(1) a PIVOT operation with the star centered at w_k, routing
- $y(w_k, t')$ units of demand from w_k to each child t' of w_k, and
- $y(w_k, t)$ units from w_k to t,

closing w_k and opening t and the children of w_k, and
(2) a PIVOT operation with the star centered at t, routing
- at most $2y(s, t)$ units from each strong child s of t to t,

closing the strong children of t and opening t.

(In the case $l = 0$ the second PIVOT can be omitted.)

Case 2: t is weak, $l \geq 1$, and $y(w_k, t) + y(s_1, t) \geq \sum_{i=2}^{l} y(s_i, t)$. Then consider:

(1) a PIVOT operation with the star centered at w_k, routing
- $y(w_k, t')$ units of demand from w_k to each child t' of w_k, and
- $y(w_k, t)$ units from w_k to t,

closing w_k, opening the children of w_k, and opening t,
(2) a PIVOT operation with the star centered at s_1, routing
- $y(s_1, t')$ units from s_1 to each child t' of s_1, and
- $y(s_1, t)$ units from s_1 to t,

closing s_1, opening the children of s_1, and opening t, and
(3) a PIVOT operation with the star centered at t, routing
- at most $2y(s_i, t)$ units from s_i to t for $i = 2, \ldots, l$,

closing $s_2, \ldots, s_l$ and opening t.

Case 3: t is weak, $l \geq 1$, and $y(w_k, t) + y(s_1, t) < \sum_{i=2}^{l} y(s_i, t)$. Then consider:

(1) a PIVOT operation with the comet with center w_k and tail (t, s_1), routing
- $y(w_k, t')$ units of demand from w_k to each child t' of w_k,
- $y(w_k, t)$ units from w_k to t, and
- at most $2y(s_1, t)$ units from s_1 to t,

closing w_k and s_1 and opening t and the children of w_k,
(2) a PIVOT operation with the star centered at t, routing
- at most $2y(s_i, t)$ units from s_i to t for each odd element i of $\{2, \ldots, l\}$,

closing the odd elements of $\{s_2, \ldots, s_l\}$ and opening t, and
(3) a PIVOT operation with the star centered at t, routing
- at most $2y(s_i, t)$ units from s_i to t for each even element i of $\{2, \ldots, l\}$,

closing the even elements of $\{s_2, \ldots, s_l\}$ and opening t.

In the case $k = 0$ we consider the same PIVOT operations, except that (1) is omitted in Case 1 and 2 (where $y(w_0, t) := 0$) and replaced by a PIVOT with the star centered at t in Case 3, routing at most $2y(s_1, t)$ units from s_1 to t, closing s_1 and opening t.

We collect all these PIVOT operations for all $t \in T$. Then, altogether, we have closed each $s \in S$ once and opened each $t \in T$ at most four times, with a total estimated routing cost of at most $2 \sum_{\{s,t\} \in E(F)} c_{st} y(s,t)$, which is at most $2c_S(x^*) + 2c_S(x)$ by (22.17). If none of the operations has an estimated cost of less than $-\frac{\epsilon}{|\mathcal{F}|} c(x)$, we have $-\epsilon c(x) \leq -c_F(x) + 4c_F(x^*) + 2c_S(x^*) + 2c_S(x)$, as required. $\qquad\square$

From the previous results we can conclude:

Theorem 22.29. *Let* $0 < \epsilon \leq 1$*, and let* x, x^* *be feasible solutions to a given instance, and let* $c^x(t, \delta) > -\frac{\epsilon}{8|\mathcal{F}|} c(x)$ *for* $t \in \mathcal{F}$ *and* $\delta \in \mathbb{R}_+$ *and* $c^x(A, \delta) > -\frac{\epsilon}{8|\mathcal{F}|} c(x)$ *for all stars and comets* A *and* $\delta \in \Delta_A^x$. *Then* $c(x) \leq (1+\epsilon)(7c_F(x^*) + 5c_S(x^*))$.

Proof: By Lemma 22.26 we have $c_S(x) \leq c_F(x^*) + c_S(x^*) + \frac{\epsilon}{8}c(x)$, and by Lemma 22.28 we have $c_F(x) \leq 4c_F(x^*) + 2c_S(x^*) + 2c_S(x) + \frac{\epsilon}{8}c(x)$. Hence $c(x) = c_F(x) + c_S(x) \leq 7c_F(x^*) + 5c_S(x^*) + \frac{\epsilon}{2}c(x)$, implying $c(x) \leq (1 + \epsilon)(7c_F(x^*) + 5c_S(x^*))$. $\qquad\square$

We finally apply a standard scaling technique and obtain the main result of this section:

Theorem 22.30. (Vygen [2007]) *For every* $\epsilon > 0$ *there is a polynomial-time* $\left(\frac{\sqrt{41}+7}{2} + \epsilon\right)$*-factor approximation algorithm for the* UNIVERSAL FACILITY LOCATION PROBLEM.

Proof: We may assume $\epsilon \leq \frac{1}{3}$. Let $\beta := \frac{\sqrt{41}-5}{2} \approx 0.7016$. Set $f_i'(z) := \beta f_i(z)$ for all $z \in \mathbb{R}_+$ and $i \in \mathcal{F}$, and consider the modified instance.

Let x be any initial feasible solution. Apply the algorithms of Lemma 22.25 and Lemma 22.27 with $\frac{\epsilon}{16|\mathcal{F}|}$ in place of ϵ. They either find an ADD or PIVOT operation that reduces the cost of the current solution x by at least $\frac{\epsilon}{16|\mathcal{F}|} c(x)$, or they conclude that the prerequisites of Theorem 22.29 are fulfilled.

If x is the resulting solution, c_F' and c_F denote the facility cost of the modified and original instance, respectively, and x^* is any feasible solution, then $c_F(x) + c_S(x) = \frac{1}{\beta}c_F'(x) + c_S(x) \leq \frac{1}{\beta}(6c_F'(x^*) + 4c_S(x^*) + \frac{3\epsilon}{8}c(x)) + c_F'(x^*) + c_S(x^*) + \frac{\epsilon}{8}c(x) \leq (6 + \beta)c_F(x^*) + (1 + \frac{4}{\beta})c_S(x^*) + \frac{3\epsilon}{4}c(x) = (6 + \beta)(c_F(x^*) + c_S(x^*)) + \frac{3\epsilon}{4}c(x)$. Hence $c(x) \leq (1 + \epsilon)(6 + \beta)c(x^*)$.

Each iteration reduces the cost by a factor of at least $\frac{1}{1 - \frac{\epsilon}{16|\mathcal{F}|}}$, hence after $\frac{1}{-\log(1 - \frac{\epsilon}{16|\mathcal{F}|})} < \frac{16|\mathcal{F}|}{\epsilon}$ iterations the cost reduces at least by a factor of 2 (note that $\log x < x - 1$ for $0 < x < 1$). This implies a weakly polynomial running time. $\qquad\square$

In particular, as $\frac{\sqrt{41}+7}{2} < 6.702$, we have a 6.702-factor approximation algorithm. This is the best performance guarantee known today.

Exercises

1. Show that the k-MEDIAN PROBLEM (without requiring metric service costs) has no constant-factor approximation algorithm unless $P = NP$.
2. Consider an instance of the UNCAPACITATED FACILITY LOCATION PROBLEM. Prove that $c_S : 2^{\mathcal{F}} \to \mathbb{R}_+ \cup \{\infty\}$ is supermodular, where $c_S(X) := \sum_{j \in D} \min_{i \in X} c_{ij}$.
3. Consider a different integer programming formulation of the UNCAPACITATED FACILITY LOCATION PROBLEM with a 0/1-variable z_S for each pair $S \in \mathcal{F} \times 2^{\mathcal{D}}$:

$$\min \sum_{S=(i,D)\in\mathcal{F}\times 2^{\mathcal{D}}} \left(f_i + \sum_{j \in D} c_{ij} \right) z_S$$

$$\text{s.t.} \sum_{S=(i,D)\in\mathcal{F}\times 2^{\mathcal{D}}:j\in D} z_S \geq 1 \quad (j \in \mathcal{D})$$

$$z_S \in \{0, 1\} \quad (S \in \mathcal{F} \times 2^{\mathcal{D}})$$

Consider the natural LP relaxation and its dual. Show how to solve them in polynomial time (despite their exponential size). Show that the optimum LP value is the same as that of (22.2) and (22.3).
4. Consider the LP relaxation of a simple special case of the METRIC CAPACITATED FACILITY LOCATION PROBLEM, in which each facility can serve up to u customers ($u \in \mathbb{N}$): This LP is obtained by extending (22.2) by the constraints $y_i \leq 1$ and $\sum_{j \in \mathcal{D}} x_{ij} \leq u y_i$ for $i \in \mathcal{F}$.
   Show that this class of LPs has an unbounded integrality ratio, i.e. the ratio of the cost of an optimum integral solution over the optimum LP value can be arbitrarily large.
   (Shmoys, Tardos and Aardal [1997])
5. Consider the UNCAPACITATED FACILITY LOCATION PROBLEM with the property that each customer $j \in \mathcal{D}$ is associated with a demand $d_j > 0$ and service costs per unit demand are metric, i.e. $\frac{c_{ij}}{d_j} + \frac{c_{i'j}}{d_j} + \frac{c_{i'j'}}{d_{j'}} \geq \frac{c_{ij'}}{d_{j'}}$ for $i, i' \in \mathcal{F}$ and $j, j' \in \mathcal{D}$. Modify the approximation algorithms for the case of unit demands and show that the same performance guarantees can be obtained in this more general case.
6. Show that (22.7) can indeed be reformulated equivalently as a linear program.
7. Consider the factor-revealing LP (22.7) for $\gamma_F = 1$. Show that the supremum of the optima for all $d \in \mathbb{N}$ is 2.
   (Jain et al. [2003])
8. Consider an instance of the METRIC UNCAPACITATED FACILITY LOCATION PROBLEM. Now the task is to find a set $X \subseteq \mathcal{F}$ such that $\sum_{i \in X} f_i + \sum_{j \in \mathcal{D}} \min_{i \in X} c_{ij}^2$ is minimum. Find a constant-factor approximation algorithm for this problem. Try to achieve a performance ratio less than 3.

9. Combine Theorem 22.3 and Theorem 22.10 to show that the JAIN-VAZIRANI ALGORITHM combined with scaling and greedy augmentation has a performance ratio of 1.853.

* 10. The MAX-k-COVER PROBLEM is defined as follows. Given a set system $(U, \mathcal{F})$ and a natural number k, find a subset $\mathcal{S} \subseteq \mathcal{F}$ with $|\mathcal{S}| = k$ and $|\bigcup \mathcal{S}|$ maximum. Prove that the natural greedy algorithm (iteratively picking a set covering as many new elements as possible) is an $(\frac{e}{e-1})$-factor approximation algorithm for the MAX-k-COVER PROBLEM.

11. Show that there is a 2-factor approximation algorithm for the METRIC SOFT-CAPACITATED FACILITY LOCATION PROBLEM.
   Hint: Combine the proof of Theorem 22.22 with the analysis of the DUAL FITTING ALGORITHM; here (22.6) can be strengthened.
   (Mahdian, Ye and Zhang [2006])

12. Combine local search (Theorem 22.20) with discretizing costs (Lemma 22.15) and scaling and greedy augmentation (Theorem 22.10) to obtain a 2.375-factor approximation algorithm for the METRIC UNCAPACITATED FACILITY LOCATION PROBLEM.

13. Consider the special case of the UNIVERSAL FACILITY LOCATION PROBLEM where the cost functions f_i are linear for all $i \in \mathcal{F}$. Describe a 3-factor approximation algorithm for this case.

14. Let $\alpha_0, \alpha_1, \ldots, \alpha_r \in \mathbb{R}_+$ with $\alpha_1 = \max_{i=1}^{r} \alpha_i$ and $S := \sum_{i=0}^{r} \alpha_i$. Show that there exists a partition $\{2, \ldots, r\} = I_0 \dot\cup I_1$ with $\alpha_k + \sum_{i \in I_k} 2\alpha_i \leq S$ for $k = 0, 1$.
   Hint: Sort the list and take every second element.

* 15. Consider a local search algorithm for the METRIC CAPACITATED FACILITY LOCATION PROBLEM which, in addition to the algorithm in Section 22.8, has an additional operation, namely a PIVOT on forests that are the disjoint union of two stars. It can be proved that this operation can be implemented in polynomial time in this special case. Show that with this additional operation one can obtain a performance ratio of 5.83.
   Hint: Modify the proof of Lemma 22.28 using this new operation. Use Exercise 14.
   (Zhang, Chen and Ye [2005])

16. Consider the special case of the SOFT-CAPACITATED FACILITY LOCATION PROBLEM in which the capacity u_i of each facility i is 1 or 2. Show that this problem can be solved optimally in polynomial time.

References

General Literature

Cornuéjols, G., Nemhauser, G.L., and Wolsey, L.A. [1990]: The uncapacitated facility location problem. In: Discrete Location Theory (P.B. Mirchandani, R.L. Francis, eds.), Wiley, New York 1990, pp. 119–171

Shmoys, D.B. [2000]: Approximation algorithms for facility location problems. Proceedings of the 3rd International Workshop on Approximation Algorithms for Combinatorial Optimization; LNCS 1913 (K. Jansen, S. Khuller, eds.), Springer, Berlin 2000, pp. 27–33

Vygen, J. [2005]: Approximation algorithms for facility location problems (lecture notes). Report No. 05950-OR, Research Institute for Discrete Mathematics, University of Bonn, 2005

Cited References

An, H.-C., Singh, M., and Svensson, O. [2017]: LP-based algorithms for capacitated facility location. SIAM Journal on Computing 46 (2017), 272–306

Archer, A., Rajagopalan, R., and Shmoys, D.B. [2003]: Lagrangian relaxation for the k-median problem: new insights and continuity properties. In: Algorithms – Proceedings of the 11th European Symposium on Algorithms (ESA); LNCS 2832 (G. di Battista, U. Zwick, eds.), Springer, Berlin 2003, pp. 31–42

Arora, S., Raghavan, P., and Rao, S. [1998]: Approximation schemes for Euclidean k-medians and related problems. Proceedings of the 30th Annual ACM Symposium on Theory of Computing (1998), 106–113

Arya, V., Garg, N., Khandekar, R., Meyerson, A., Munagala, K., and Pandit, V. [2004]: Local search heuristics for k-median and facility location problems. SIAM Journal on Computing 33 (2004), 544–562

Balinski, M.L. [1965]: Integer programming: methods, uses, computation. Management Science 12 (1965), 253–313

Balinski, M.L., and Wolfe, P. [1963]: On Benders decomposition and a plant location problem. Working paper ARO-27. Mathematica, Princeton 1963

Byrka, J., and Aardal, K. [2007]: The approximation gap for the metric facility location problem is not yet closed. Operations Research Letters 35 (2007), 379–384

Byrka, J., and Aardal, K. [2007]: The approximation gap for the metric facility location problem is not yet closed. Operations Research Letters 35 (2007), 379–384

Charikar, M., and Guha, S. [2005]: Improved combinatorial algorithms for the facility location and k-median problems. SIAM Journal on Computing 34 (2005), 803–824

Charikar, M., Guha, S., Tardos, É., and Shmoys, D.B. [2002]: A constant-factor approximation algorithm for the k-median problem. Journal of Computer and System Sciences 65 (2002), 129–149

Chudak, F.A., and Shmoys, D.B. [2003]: Improved approximation algorithms for the uncapacitated facility location problem. SIAM Journal on Computing 33 (2003), 1–25

Feige, U. [1998]: A threshold of $\ln n$ for the approximating set cover. Journal of the ACM 45 (1998), 634–652

Guha, S., and Khuller, S. [1999]: Greedy strikes back: improved facility location algorithms. Journal of Algorithms 31 (1999), 228–248

Hochbaum, D.S. [1982]: Heuristics for the fixed cost median problem. Mathematical Programming 22 (1982), 148–162

Jain, K., Mahdian, M., Markakis, E., Saberi, A., and Vazirani, V.V. [2003]: Greedy facility location algorithms analyzed using dual fitting with factor-revealing LP. Journal of the ACM 50 (2003), 795–824

Jain, K., and Vazirani, V.V. [2001]: Approximation algorithms for metric facility location and k-median problems using the primal-dual schema and Lagrangian relaxation. Journal of the ACM 48 (2001), 274–296

Kolliopoulos, S.G., and Rao, S. [2007]: A nearly linear-time approximation scheme for the Euclidean k-median problem.

Korupolu, M., Plaxton, C., and Rajaraman, R. [2000]: Analysis of a local search heuristic for facility location problems. Journal of Algorithms 37 (2000), 146–188

Kuehn, A.A., and Hamburger, M.J. [1963]: A heuristic program for locating warehouses. Management Science 9 (1963), 643–666

Levi, R., Shmoys, D.B., and Swamy, C. [2012]: LP-based approximation algorithms for capacitated facility location. Mathematical Programming A 131 (2012), 365–379

Li, S. [2013]: A 1.488-approximation algorithm for the uncapacitated facility location problem. Information and Computation 222 (2013), 45–58

Mahdian, M., and Pál, M. [2003]: Universal facility location. In: Algorithms – Proceedings of the 11th European Symposium on Algorithms (ESA); LNCS 2832 (G. di Battista, U. Zwick, eds.), Springer, Berlin 2003, pp. 409–421

Mahdian, M., Ye, Y., and Zhang, J. [2006]: Approximation algorithms for metric facility location problems. SIAM Journal on Computing 36 (2006), 411–432

Manne, A.S. [1964]: Plant location under economies-of-scale-decentralization and computation. Management Science 11 (1964), 213–235

Pál, M., Tardos, É., and Wexler, T. [2001]: Facility location with nonuniform hard capacities. Proceedings of the 42nd Annual IEEE Symposium on the Foundations of Computer Science (2001), 329–338

Shmoys, D.B., Tardos, É., and Aardal, K. [1997]: Approximation algorithms for facility location problems. Proceedings of the 29th Annual ACM Symposium on Theory of Computing (1997), 265–274

Stollsteimer, J.F. [1963]: A working model for plant numbers and locations. Journal of Farm Economics 45 (1963), 631–645

Sviridenko, M. [2002]: An improved approximation algorithm for the metric uncapacitated facility location problem. In: Integer Programming and Combinatorial Optimization; Proceedings of the 9th International IPCO Conference; LNCS 2337 (W. Cook, A. Schulz, eds.), Springer, Berlin 2002, pp. 240–257

Vygen, J. [2007]: From stars to comets: improved local search for universal facility location. Operations Research Letters 35 (2007), 427–433

Zhang, J., Chen, B., and Ye, Y. [2005]: A multi-exchange local search algorithm for the capacitated facility location problem. Mathematics of Operations Research 30 (2005), 389–403

Zhang, P. [2007]: A new approximation algorithm for the k-facility location problem. Theoretical Computer Science 384 (2007), 126–135

Notation Index

$\mathbb{N}$	set of natural numbers $\{1, 2, 3, \ldots\}$	
$\mathbb{Z}$ $(\mathbb{Z}_+)$	set of (nonnegative) integers	
$\mathbb{Q}$ $(\mathbb{Q}_+)$	set of (nonnegative) rationals	
$\mathbb{R}$ $(\mathbb{R}_+)$	set of (nonnegative) real numbers	
$\subset$	proper subset	
$\subseteq$	subset or equal	
$\dot{\cup}$	disjoint union	
$X \triangle Y$	symmetric difference of sets X and Y	
$\lVert x \rVert_2$	Euclidean norm of a vector x	
$\lVert x \rVert_\infty$	infinity-norm of a vector x	
$x \bmod y$	the unique number z with $0 \le z < y$ and $\frac{x-z}{y} \in \mathbb{Z}$	
$x^\top, A^\top$	transpose of vector x and matrix A	
$\lceil x \rceil$	smallest integer greater than or equal to x	
$\lfloor x \rfloor$	greatest integer less than or equal to x	
$f = O(g)$	O-notation	4
$f = \Omega(g)$	Ω-notation	4
$f = \Theta(g)$	Θ-notation	4
size(x)	encoding length of x; length of the binary string x	6, 76, 386
$\log x$	logarithm of x with basis 2	6
$V(G)$	vertex set of graph G	15
$E(G)$	edge set of graph G	15
$G[X]$	subgraph of G induced by $X \subseteq V(G)$	16
$G - v$	subgraph of G induced by $V(G) \setminus \{v\}$	16
$G - e$	graph obtained by deleting edge e from G	16
$G + e$	graph obtained by adding edge e to G	16
$G + H$	sum of graphs G and H	16
G/X	the graph resulting from G by contracting the subset X of vertices	16
$E(X, Y)$	set of edges between $X \setminus Y$ and $Y \setminus X$	16
$E^+(X, Y)$	set of directed edges from $X \setminus Y$ to $Y \setminus X$	16
$\delta(X), \delta(v)$	$E(X, V(G) \setminus X), E(\{v\}, V(G) \setminus \{v\})$	16
$\Gamma(X), \Gamma(v)$	set of neighbours of vertex set X, of vertex v	16
$\delta^+(X), \delta^+(v)$	set of edges leaving vertex set X, vertex v	16
$\delta^-(X), \delta^-(v)$	set of edges entering vertex set X, vertex v	16
2^S	power set of S	17

© Springer-Verlag GmbH Germany 2018
Bernhard Korte and Jens Vygen, *Combinatorial Optimization*, Algorithms and Combinatorics 21, https://doi.org/10.1007/978-3-662-56039-6

Author Index

© Springer-Verlag GmbH Germany 2018

Bernhard Korte and Jens Vygen, *Combinatorial Optimization*, Algorithms and Combinatorics 21, https://doi.org/10.1007/978-3-662-56039-6

Subject Index

© Springer-Verlag GmbH Germany 2018
Bernhard Korte and Jens Vygen, *Combinatorial Optimization*, Algorithms
and Combinatorics 21, https://doi.org/10.1007/978-3-662-56039-6

Printed in the United States
By Bookmasters